McGraw-Hill Series in Electrical and Computer Engineering

Senior Consulting Editor

Stephen W. Director, *Carnegie-Mellon University*

Circuits and Systems
Communications and Signal Processing
Computer Engineering
Control Theory
Electromagnetics
Electronics and VLSI Circuits
Introductory
Power and Energy
Radar and Antennas

Previous Consulting Editors

Ronald N. Bracewell, Colin Cherry, James F. Gibbons, Willis W. Harman, Hubert Heffner, Edward W. Herold, John G. Linvill, Simon Ramo, Ronald A. Rohrer, Anthony E. Siegman, Charles Susskind, Frederick E. Terman, John G. Truxal, Ernst Weber, and John R. Whinnery

Power and Energy

Senior Consulting Editor

Stephen W. Director, *Carnegie-Mellon University*

Chapman: *Electric Machinery Fundamentals*
Elgerd: *Electric Energy Systems Theory*
Fitzgerald, Kingsley, and Umans: *Electric Machinery*
Gönen: *Electric Power Distribution System Engineering*
Grainger and Stevenson: *Power System Analysis*
Krause and Wasynczuk: *Electromechanical Motion Devices*
Stevenson: *Elements of Power System Analysis*

POWER SYSTEM ANALYSIS

John J. Grainger

Professor, Department of Electrical and Computer Engineering
North Carolina State University

William D. Stevenson, Jr.

Late Professor of Electrical Engineering
North Carolina State University

McGraw Hill Education (India) Private Limited
CHENNAI

McGraw Hill Education Offices
Chennai New York St Louis San Francisco Auckland Bogotá Caracas
Kuala Lumpur Lisbon London Madrid Mexico City Milan Montreal
San Juan Santiago Singapore Sydney Tokyo Toronto

McGraw Hill Education (India) Private Limited

POWER SYSTEM ANALYSIS

34th reprint 2017
RCAACRLMRYDQD

McGraw Hill Education (India) Edition 2003

Reprinted in India by arrangement with McGraw-Hill Global Education Holdings, LLC, New York

Sales territories: India, Pakistan, Nepal, Bangladesh, Sri Lanka and Bhutan

Library of Congress Cataloging-in-Publication Data

Grainger, John J.
Power system analysis / John J. Grainger, William D. Stevenson.
p. cm.
Includes index.
ISBN 0-07-061293-5
1. Electric power distribution. 2. Electric power systems.
Power system analysis. III. Title.
TK3001.G73 1994
620.319—dc20 93-39219

ISBN-13: 978-0-07-058515-7
ISBN-10: 0-07-058515-6

Published by McGraw Hill Education (India) Private Limited,
444/1, Sri Ekambara Naicker Industrial Estate, Alapakkam, Porur, Chennai 600 116,
Tamil Nadu, India, and printed at Adarsh Printers, New Delhi 110 032

Visit us at: www.mheducation.co.in

ABOUT THE AUTHORS

John J. Grainger is Professor of Electrical and Computer Engineering at North Carolina State Unversity. He is a graduate of the National University of Ireland and received his M.S.E.E. and Ph.D. degrees at the University of Wisconsin-Madison.

Dr. Grainger is the founding Director of the Electric Power Research Center at North Carolina State University, a joint university/industry cooperative research center in electric power systems engineering. He leads the Center's major research programs in transmission and distribution systems planning, design, automation, and control areas, as well as power system dynamics.

Professor Grainger has also taught at the University of Wisconsin-Madison, The Illinois Institute of Technology, Marquette University, and North Carolina State University. His industrial experience has been with the Electricity Supply Board of Ireland; Commonwealth Edison Company, Chicago; Wisconsin Electric Power Company, Milwaukee; and Carolina Power & Light Company, Raleigh. Dr. Grainger is an active consultant with the Pacific Gas and Electric Company, San Francisco; Southern California Edison Company, Rosemead; and many other power industry organizations. His educational and technical involvements include the IEEE Power Engineering Society, The American Society of Engineering Education, the American Power Conference, CIRED, and CIGRE.

Dr. Grainger is the author of numerous papers in the IEEE Power Engineering Society's *Transactions* and was recognized by the IEEE Transmission and Distribution Committee for the 1985 Prize Paper Award.

In 1984, Professor Grainger was chosen by the Edison Electric Institute for the EEI Power Engineering Educator Award.

William D. Stevenson, Jr. (deceased) was a professor and the Associate Head of the Electrical Engineering Department of North Carolina State University A Fellow of the Institute of Electrical and Electronics Engineers, he worked in private industry and taught at both Clemson University and Princeton University. Dr. Stevenson also served as a consulting editor in electrical power engineering for the *McGraw-Hill Encyclopedia of Science and Technology*. He was the recipient of several teaching and professional awards.

To the Memory of
William D. Stevenson, Jr.
1912–1988
True friend and colleague

CONTENTS

PREFACE

This book embodies the principles and objectives of *Elements of Power System Analysis*, the long-standing McGraw-Hill textbook by Professor William D. Stevenson, Jr., who was for many years my friend and colleague emeritus at North Carolina State University. Sadly, Professor Stevenson passed away on May 1, 1988, shortly after planning this joint venture. In my writing I have made great efforts to continue the *student-oriented* style and format of his own famous textbook that has guided the education of numerous power system engineering students for a considerable number of years.

The aim here is to instill confidence and understanding of those concepts of power system analysis that are likely to to be encountered in the study and practice of electric power engineering. The presentation is tutorial with emphasis on a thorough understanding of fundamentals and underlying principles. The approach and level of treatment are directed toward the senior undergraduate and first-year graduate student of electrical engineering at technical colleges and universities. The coverage, however, is quite comprehensive and spans a wide range of topics commonly encountered in electric power system engineering practice. In this regard, electric utility and other industry-based engineers will find this textbook of much benefit in their everyday work.

Modern power systems have grown larger and more geographically expansive with many interconnections between neighboring systems. Proper planning, operation, and control of such large-scale systems require advanced computer-based techniques, many of which are explained in a tutorial manner by means of numerical examples throughout this book. The senior undergraduate engineering student about to embark on a career in the electric power industry will most certainly benefit from the exposure to these techniques, which are presented here in the detail appropriate to an introductory level. Likewise, electric utility engineers, even those with a previous course in power system analysis, may find that the explanations of these commonly used analytic techniques more adequately prepare them to move beyond routine work.

Power System Analysis can serve as a basis for two semesters of undergraduate study or for first-semester graduate study. The wide range of topics facilitates versatile selection of chapters and sections for completion in the semester or quarter time frame. Familiarity with the basic principles of electric circuits, phasor algebra, and the rudiments of differential equations is assumed. The reader should also have some understanding of matrix operations and notation as they are used throughout the text. The coverage includes newer topics such as *state estimation* and *unit commitment*, as well as more detailed presentations and newer approaches to traditional subjects such as transformers, synchronous machines, and network faults. Where appropriate, summary tables allow quick reference of important ideas. Basic concepts of computer-based algorithms are presented so that students can implement their own computer programs.

Chapters 2 and 3 are devoted to the transformer and synchronous machine, respectively, and should complement material covered in other electric circuits and machines courses. Transmission-line parameters and calculations are studied in Chapters 4 through 6. Network models based on the admittance and impedance representations are developed in Chapters 7 and 8, which also introduce gaussian elimination, Kron reduction, triangular factorization, and the $\mathbf{Z}_{\text{bus}}$ building algorithm. The power-flow problem, symmetrical components, and unsymmetrical faults are presented in Chapters 9 through 12; whereas Chapter 13 provides a self-contained development of economic dispatch and the basics of unit commitment. Contingency analysis and external equivalents are the subjects of Chapter 14. Power system state estimation is covered in Chapter 15, while power system stability is introduced in Chapter 16. Homework problems and exercises are provided at the end of each chapter.

I am most pleased to acknowledge the assistance given to me by a number of people with whom I have been associated within the Department of Electrical and Computer Engineering at North Carolina State University. Dr. Stan S. H. Lee, my colleague and friend for many years, has always willingly given his time and effort when I needed help, advice, or suggestions at the various stages of development of this textbook. A number of the homework problems and solutions were contributed by him and by Dr. Gamini Wickramasekara, one of my former graduate students at North Carolina State University. Dr. Michael J. Gorman, another of my recent graduate students, gave unstintingly of himself in developing the computer-based figures and solutions for many of the numerical examples throughout the various chapters of the text. Mr. W. Adrian Buie, a recent graduate of the Department of Electrical and Computer Engineering, undertook the challenge of committing the entire textbook to the computer and produced a truly professional manuscript; in this regard, Mr. Barry W. Tyndall was also most helpful in the early stages of the writing. My loyal secretary, Mrs. Paulette Cannady-Kea, has always enthusiastically assisted in the overall project. I am greatly indebted and extremely grateful to each and all of these individuals for their generous efforts.

Also within the Department of Electrical and Computer Engineering at North Carolina State University, the successive leadership of Dr. Larry K. Monteith (now Chancellor of the University), Dr. Nino A. Masnari (now Director of the Engineering Research Center for Advanced Electronic Materials Processing), and Dr. Ralph K. Cavin III (presently Head of the Department), along with my faculty colleagues, particularly Dr. Alfred J. Goetze, provided an environment of support that I am very pleased to record.

The members of my family, especially my wife, Barbara, have been a great source of patient understanding and encouragement during the preparation of this book. I ask each of them, and my friend Anne Stevenson, to accept my sincere thanks.

McGraw-Hill and I would like to thank the following reviewers for their many helpful comments and suggestions: Vernon D. Albertson, University of Minnesota; David R. Brown, University of Texas at Austin; Mehdi Etezadi-Amoli, University of Nevada, Reno; W. Mack Grady, University of Texas at Austin; Clifford Grigg, Rose-Hulman Institute of Technology; William H. Kersting, New Mexico State University; Kenneth Kruempel, Iowa State University; Mangalore A. Pai, University of Illinois, Urbana-Champaign; Arun G. Phadke, Virginia Polytechnic Institute and State University; B. Don Russell, Texas A & M University; Peter W. Sauer, University of Illinois, Urbana-Champaign; and Ernie L. Stagliano, Jr., Drexel University.

John J. Grainger

POWER SYSTEM ANALYSIS

CHAPTER 1

BASIC CONCEPTS

Normal and abnormal conditions of operation of the system are the concern of the power system engineer who must be very familiar with steady-state ac circuits, particularly three-phase circuits. The purpose of this chapter is to review a few of the fundamental ideas of such circuits; to establish the notation used throughout the book; and to introduce the expression of values of voltage, current, impedance, and power in per unit. Modern power system analysis relies almost exclusively on nodal network representation which is introduced in the form of the bus admittance and the bus impedance matrices.

1.1 INTRODUCTION

The waveform of voltage at the buses of a power system can be assumed to be purely sinusoidal and of constant frequency. In developing most of the theory in this book, we are concerned with the phasor representations of sinusoidal voltages and currents and use the capital letters V and I to indicate these phasors (with appropriate subscripts where necessary). Vertical bars enclosing V and I, that is, $|V|$ and $|I|$, designate the magnitudes of the phasors. Magnitudes of complex numbers such as impedance Z and admittance Y are also indicated by vertical bars. Lowercase letters generally indicate instantaneous values. Where a generated voltage [electromotive force (emf)] is specified, the letter E rather than V is often used for voltage to emphasize the fact that an emf rather than a general potential difference between two points is being considered.

If a voltage and a current are expressed as functions of time, such as

$$v = 141.4\cos(\omega t + 30°)$$

and

$$i = 7.07\cos\omega t$$

their maximum values are obviously $V_{max} = 141.4$ V and $I_{max} = 7.07$ A, respectively. Vertical bars are not needed when the subscript max with V and I is used to indicate *maximum* value. The term *magnitude* refers to root-mean-square (or rms) values, which equal the maximum values divided by $\sqrt{2}$. Thus, for the above expressions for v and i

$$|V| = 100\text{ V} \qquad \text{and} \qquad |I| = 5\text{ A}$$

These are the values read by the ordinary types of voltmeters and ammeters. Another name for the rms value is the *effective value*. The average power expended in a resistor by a current of magnitude $|I|$ is $|I|^2R$.

To express these quantities as phasors, we employ Euler's identity $\varepsilon^{j\theta} = \cos\theta + j\sin\theta$, which gives

$$\cos\theta = \text{Re}\{\varepsilon^{j\theta}\} = \text{Re}\{\cos\theta + j\sin\theta\} \tag{1.1}$$

where Re means *the real part of*. We now write

$$v = \text{Re}\{\sqrt{2}\,100\varepsilon^{j(\omega t + 30°)}\} = \text{Re}\{100\varepsilon^{j30°}\sqrt{2}\,\varepsilon^{j\omega t}\}$$

$$i = \text{Re}\{\sqrt{2}\,5\varepsilon^{j(\omega t + 0°)}\} = \text{Re}\{5\varepsilon^{j0°}\sqrt{2}\,\varepsilon^{j\omega t}\}$$

If the current is the reference phasor, we have

$$I = 5\varepsilon^{j0°} = 5\underline{/0°} = 5 + j0\text{ A}$$

and the voltage which leads the reference phasor by 30° is

$$V = 100\varepsilon^{j30°} = 100\underline{/30°} = 86.6 + j50\text{ V}$$

Of course, we might not choose as the reference phasor either the voltage or the current whose instantaneous expressions are v and i, respectively, in which case their phasor expressions would involve other angles.

In circuit diagrams it is often most convenient to use polarity marks in the form of plus and minus signs to indicate the terminal assumed positive when specifying voltage. An arrow on the diagram specifies the direction assumed positive for the flow of current. In the single-phase equivalent of a three-phase circuit single-subscript notation is usually sufficient, but double-subscript notation is usually simpler when dealing with all three phases.

1.2 SINGLE-SUBSCRIPT NOTATION

Figure 1.1 shows an ac circuit with an emf represented by a circle. The emf is E_g, and the voltage between nodes a and o is identified as V_t. The current in the circuit is I_L and the voltage across Z_L is V_L. To specify these voltages as phasors, however, the + and − markings, called *polarity marks*, on the diagram and an arrow for current direction are necessary.

In an ac circuit the terminal marked + is positive with respect to the terminal marked − for half a cycle of voltage and is negative with respect to the other terminal during the next half cycle. We mark the terminals to enable us to say that the voltage between the terminals is positive at any instant when the terminal marked plus is actually at a higher potential than the terminal marked minus. For instance, in Fig. 1.1 the instantaneous voltage v_t is positive when the terminal marked plus is actually at a higher potential than the terminal marked with a negative sign. During the next half cycle the positively marked terminal is actually negative, and v_t is negative. Some authors use an arrow but must specify whether the arrow points toward the terminal which would be labeled plus or toward the terminal which would be labeled minus in the convention described above.

The current arrow performs a similar function. The subscript, in this case L, is not necessary unless other currents are present. Obviously, the actual direction of current flow in an ac circuit reverses each half cycle. The arrow points in the direction which is to be called positive for current. When the current is actually flowing in the direction opposite to that of the arrow, the current is negative. The phasor current is

$$I_L = \frac{V_t - V_L}{Z_A} \tag{1.2}$$

and

$$V_t = E_g - I_L Z_g \tag{1.3}$$

Since certain nodes in the circuit have been assigned letters, the voltages may be designated by the single-letter subscripts identifying the node whose voltages are expressed with respect to a reference node. In Fig. 1.1 the instantaneous voltage v_a and the phasor voltage V_a express the voltage of node a with respect to the reference node o, and v_a is positive when a is at a higher

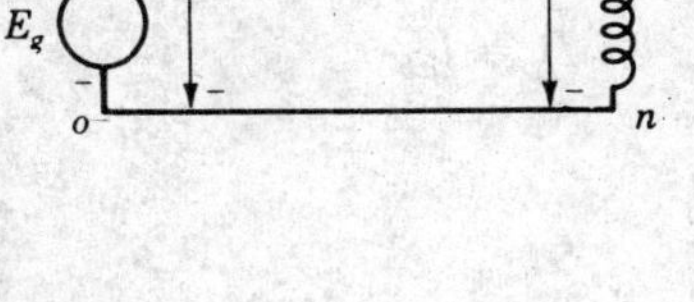

FIGURE 1.1
An ac circuit with emf E_g and load impedance Z_L.

potential than o. Thus,

$$v_a = v_t \qquad v_b = v_L$$

$$V_a = V_t \qquad V_b = V_L$$

1.3 DOUBLE-SUBSCRIPT NOTATION

The use of polarity marks for voltages and direction arrows for currents can be avoided by double-subscript notation. The understanding of three-phase circuits is considerably clarified by adopting a system of double subscripts. The convention to be followed is quite simple.

In denoting a current the order of the subscripts assigned to the symbol for current defines the direction of the flow of current when the current is considered to be positive. In Fig. 1.1 the arrow pointing from a to b defines the positive direction for the current I_L associated with the arrow. The instantaneous current i_L is positive when the current is actually in the direction from a to b, and in double-subscript notation this current is i_{ab}. The current i_{ab} is equal to $-i_{ba}$.

In double-subscript notation the letter subscripts on a voltage indicate the nodes of the circuit between which the voltage exists. We shall follow the convention which says that the first subscript denotes the voltage of that node with respect to the node identified by the second subscript. This means that the instantaneous voltage v_{ab} across Z_A of the circuit of Fig. 1.1 is the voltage of node a with respect to node b and that v_{ab} is positive during that half cycle when a is at a higher potential than b. The corresponding phasor voltage is V_{ab}, which is related to the current I_{ab} flowing from node a to node b by

$$V_{ab} = I_{ab}Z_A \qquad \text{and} \qquad I_{ab} = Y_A V_{ab} \tag{1.4}$$

where Z_A is the complex impedance (also called Z_{ab}) and $Y_A = 1/Z_A$ is the complex admittance (also called Y_{ab}).

Reversing the order of the subscripts of either a current or a voltage gives a current or a voltage 180° out of phase with the original; that is,

$$V_{ba} = V_{ab}\varepsilon^{j180^\circ} = V_{ab}\underline{/\ 180^\circ} = -V_{ab}$$

The relation of single- and double-subscript notation for the circuit of Fig. 1.1 is summarized as follows:

$$V_t = V_a = V_{ao} \qquad V_L = V_b = V_{bo} \qquad I_L = I_{ab}$$

In writing Kirchhoff's voltage law, the order of the subscripts is the order of tracing a closed path around the circuit. For Fig. 1.1

$$V_{oa} + V_{ab} + V_{bn} = 0 \tag{1.5}$$

Nodes n and o are the same in this circuit, and n has been introduced to identify the path more precisely. Replacing V_{oa} by $-V_{ao}$ and noting that $V_{ab} = I_{ab}Z_A$ yield

$$-V_{ao} + I_{ab}Z_A + V_{bn} = 0 \tag{1.6}$$

and so

$$I_{ab} = \frac{V_{ao} - V_{bn}}{Z_A} = (V_{ao} - V_{bn})Y_A \tag{1.7}$$

1.4 POWER IN SINGLE-PHASE AC CIRCUITS

Although the fundamental theory of the transmission of energy describes the travel of energy in terms of the interaction of electric and magnetic fields, the power system engineer is usually more concerned with describing the rate of change of energy with respect to time (which is the definition of *power*) in terms of voltage and current. The unit of power is a *watt*. The power in watts being absorbed by a load at any instant is the product of the instantaneous voltage drop across the load in volts and the instantaneous current into the load in amperes. If the terminals of the load are designated a and n, and if the voltage and current are expressed by

$$v_{an} = V_{\max} \cos \omega t \qquad \text{and} \qquad i_{an} = I_{\max} \cos(\omega t - \theta)$$

the instantaneous power is

$$p = v_{an} i_{an} = V_{\max} I_{\max} \cos \omega t \cos(\omega t - \theta) \tag{1.8}$$

The angle θ in these equations is positive for current *lagging* the voltage and negative for *leading* current. A positive value of p expresses the rate at which energy is being *absorbed* by the part of the system between the points a and n. The instantaneous power is obviously positive when both v_{an} and i_{an} are positive and becomes negative when v_{an} and i_{an} are opposite in sign. Figure 1.2 illustrates this point. Positive power calculated as $v_{an}i_{an}$ results when current is flowing in the direction of a voltage drop and is the rate of transfer of energy to the load. Conversely, negative power calculated as $v_{an}i_{an}$ results when current is flowing in the direction of a voltage rise and means energy is being transferred from the load into the system to which the load is connected. If v_{an} and i_{an} are in phase, as they are in a purely resistive load, the instantaneous power will never become negative. If the current and voltage are out of phase by 90°, as in a purely inductive or purely capacitive ideal circuit element, the instantaneous

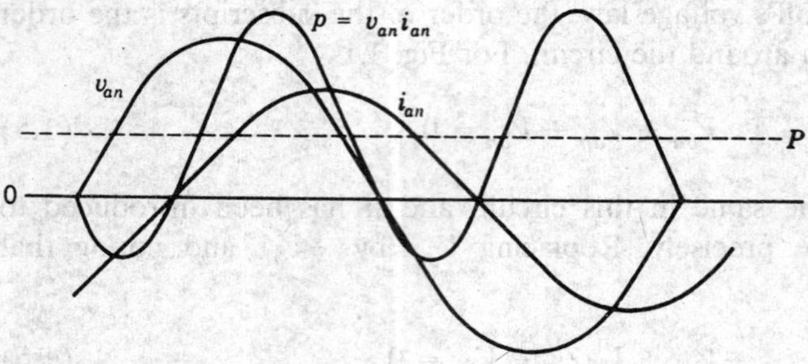

FIGURE 1.2
Current, voltage, and power plotted versus time.

power will have equal positive and negative half cycles and its average value will always be zero.

By using trigonometric identities the expression of Eq. (1.8) is reduced to

$$p = \frac{V_{max} I_{max}}{2} \cos\theta(1 + \cos 2\omega t) + \frac{V_{max} I_{max}}{2} \sin\theta \sin 2\omega t \qquad (1.9)$$

where $V_{max} I_{max}/2$ may be replaced by the product of the rms voltage and current, that is, by $|V_{an}|\,|I_{an}|$ or $|V|\,|I|$.

Another way of looking at the expression for instantaneous power is to consider the component of the current in phase with v_{an} and the component 90° out of phase with v_{an}. Figure 1.3(*a*) shows a parallel circuit for which Fig. 1.3(*b*) is the phasor diagram. The component of i_{an} in phase with v_{an} is i_R, and from Fig. 1.3(*b*), $|I_R| = |I_{an}|\cos\theta$. If the maximum value of i_{an} is I_{max}, the maximum value of i_R is $I_{max}\cos\theta$. The instantaneous current i_R must be in phase with v_{an}. For $v_{an} = V_{max}\cos\omega t$

$$i_R = \underbrace{I_{max}\cos\theta}_{\max i_R}\cos\omega t \qquad (1.10)$$

Similarly, the component of i_{an} lagging v_{an} by 90° is i_X with maximum value

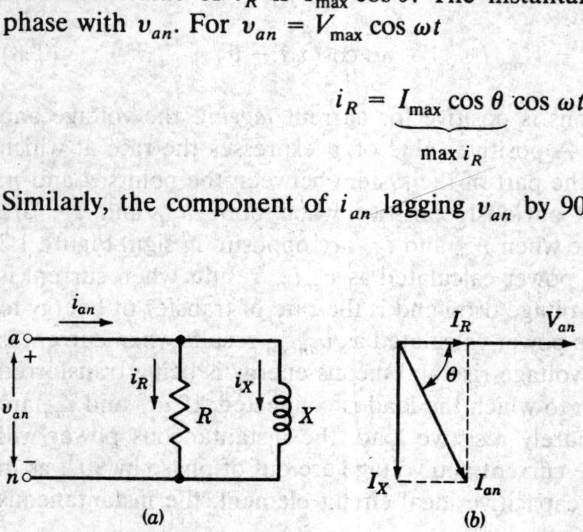

FIGURE 1.3
Parallel *RL* circuit and the corresponding phasor diagram.

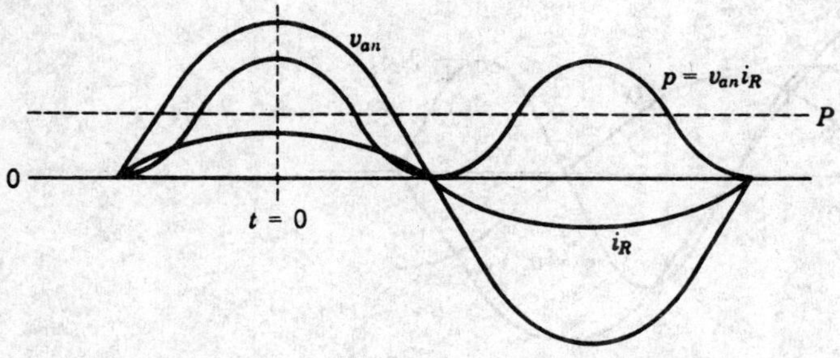

FIGURE 1.4
Voltage, current in phase with the voltage, and the resulting power plotted versus time.

$I_{max} \sin \theta$. Since i_X must lag v_{an} by 90°,

$$i_X = \underbrace{I_{max} \sin \theta}_{\max i_X} \sin \omega t \tag{1.11}$$

Then,

$$v_{an}i_R = V_{max} I_{max} \cos \theta \cos^2 \omega t$$

$$= \frac{V_{max} I_{max}}{2} \cos \theta (1 + \cos 2\omega t) \tag{1.12}$$

which is the instantaneous power in the resistance and the first term in Eq. (1.9). Figure 1.4 shows $v_{an}i_R$ plotted versus t.

Similarly,

$$v_{an}i_X = V_{max} I_{max} \sin \theta \sin \omega t \cos \omega t$$

$$= \frac{V_{max} I_{max}}{2} \sin \theta \sin 2\omega t \tag{1.13}$$

which is the instantaneous power in the inductance and the second term in Eq. (1.9). Figure 1.5 shows v_{an}, i_X and their product plotted versus t.

Examination of Eq. (1.9) shows that the term containing $\cos \theta$ is always positive and has an average value of

$$P = \frac{V_{max} I_{max}}{2} \cos \theta \tag{1.14}$$

or when rms values of voltage and current are substituted,

$$P = |V|\,|I| \cos \theta \tag{1.15}$$

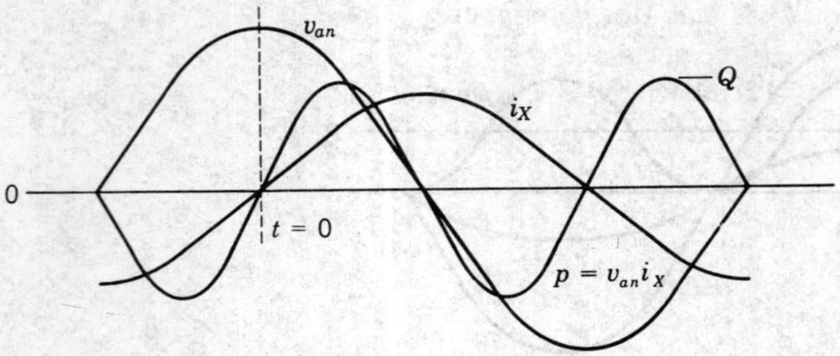

FIGURE 1.5
Voltage, current lagging the voltage by 90°, and the resulting power plotted versus time.

P is the quantity to which the word *power* refers when not modified by an adjective identifying it otherwise. P, the *average* power, is also called the *real* or *active* power. The fundamental unit for both instantaneous and average power is the watt, but a watt is such a small unit in relation to power system quantities that P is usually measured in kilowatts or megawatts.

The cosine of the phase angle θ between the voltage and the current is called the *power factor*. An inductive circuit is said to have a lagging power factor, and a capacitive circuit is said to have a leading power factor. In other words, the terms *lagging power factor* and *leading power factor* indicate, respectively, whether the current is lagging or leading the applied voltage.

The second term of Eq. (1.9), the term containing $\sin\theta$, is alternately positive and negative and has an average value of zero. This component of the instantaneous power p is called the *instantaneous reactive power* and expresses the flow of energy alternately toward the load and away from the load. The maximum value of this pulsating power, designated Q, is called *reactive power* or *reactive voltamperes* and is very useful in describing the operation of a power system, as becomes increasingly evident in further discussion. The reactive power is

$$Q = \frac{V_{\max} I_{\max}}{2} \sin\theta \tag{1.16}$$

or

$$Q = |V|\,|I| \sin\theta \tag{1.17}$$

The square root of the sum of the squares of P and Q is equal to the product of $|V|$ and $|I|$, for

$$\sqrt{P^2 + Q^2} = \sqrt{(|V|\,|I| \cos\theta)^2 + (|V|\,|I| \sin\theta)^2} = |V|\,|I| \tag{1.18}$$

Of course, P and Q have the same dimensional units, but it is usual to designate the units for Q as *vars* (for voltamperes reactive). The more practical units for Q are kilovars or megavars.

In a simpie series circuit where Z is equal to $R + jX$ we can substitute $|I|\,|Z|$ for $|V|$ in Eqs. (1.15) and (1.17) to obtain

$$P = |I|^2|Z|\cos\theta \tag{1.19}$$

and

$$Q = |I|^2|Z|\sin\theta \tag{1.20}$$

Recognizing that $R = |Z|\cos\theta$ and $X = |Z|\sin\theta$, we then find

$$P = |I|^2R \quad \text{and} \quad Q = |I|^2X \tag{1.21}$$

Equations (1.15) and (1.17) provide another method of computing the power factor since we see that $Q/P = \tan\theta$. The power factor is therefore

$$\cos\theta = \cos\left(\tan^{-1}\frac{Q}{P}\right)$$

or from Eqs. (1.15) and (1.18)

$$\cos\theta = \frac{P}{\sqrt{P^2 + Q^2}}$$

If the instantaneous power expressed by Eq. (1.9) is the power in a predominantly capacitive circuit with the same impressed voltage, θ becomes negative, making $\sin\theta$ and Q negative. If capacitive and inductive circuits are in parallel, the instantaneous reactive power for the RL circuit is 180° out of phase with the instantaneous reactive power of the RC circuit. The net reactive power is the difference between Q for the RL circuit and Q for the RC circuit. A positive value is assigned to Q drawn by an inductive load and a negative sign to Q drawn by a capacitive load.

Power system engineers usually think of a capacitor as a generator of positive reactive power rather than a load requiring negative reactive power. This concept is very logical, for a capacitor drawing *negative* Q in parallel with an inductive load reduces the Q which would otherwise have to be supplied by the system to the inductive load. In other words, the capacitor *supplies* the Q required by the inductive load. This is the same as considering a capacitor as a device that delivers a lagging current rather than as a device which draws a leading current, as shown in Fig. 1.6. An adjustable capacitor in parallel with an inductive load, for instance, can be adjusted so that the leading current to the capacitor is exactly equal in magnitude to the component of current in the inductive load which is lagging the voltage by 90°. Thus, the resultant current is in phase with the voltage. The inductive circuit still requires positive reactive

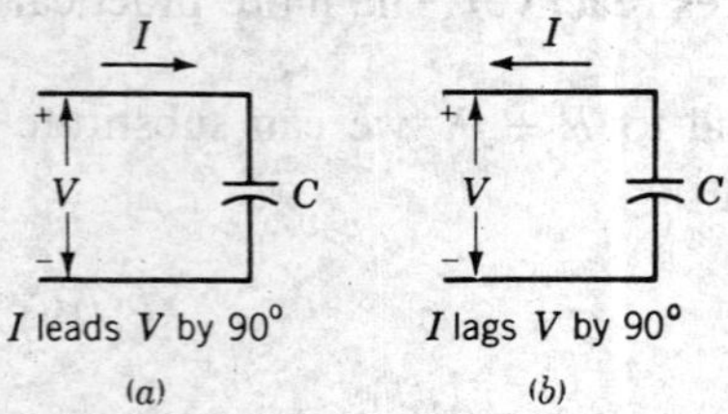

FIGURE 1.6
Capacitor considered as: (*a*) a passive circuit element drawing leading current; (*b*) a generator supplying lagging current.

power, but the net reactive power is zero. It is for this reason that the power system engineer finds it convenient to consider the capacitor to be supplying reactive power to the inductive load. When the words *positive* and *negative* are not used, positive reactive power is assumed.

1.5 COMPLEX POWER

If the phasor expressions for voltage and current are known, the calculation of real and reactive power is accomplished conveniently in complex form. If the voltage across and the current into a certain load or part of a circuit are expressed by $V = |V|\underline{/\alpha}$ and $I = |I|\underline{/\beta}$, respectively, the product of voltage times the conjugate of current in polar form is

$$VI^* = |V|\varepsilon^{j\alpha} \times |I|\varepsilon^{-j\beta} = |V|\,|I|\varepsilon^{j(\alpha-\beta)} = |V|\,|I|\underline{/\alpha - \beta} \qquad (1.22)$$

This quantity, called the *complex power*, is usually designated by S. In rectangular form

$$S = VI^* = |V|\,|I|\cos(\alpha - \beta) + j|V|\,|I|\sin(\alpha - \beta) \qquad (1.23)$$

Since $\alpha - \beta$, the phase angle between voltage and current, is θ in the previous equations,

$$S = P + jQ \qquad (1.24)$$

Reactive power Q will be positive when the phase angle $\alpha - \beta$ between voltage and current is positive, that is, when $\alpha > \beta$, which means that current is lagging the voltage. Conversely, Q will be negative for $\beta > \alpha$, which indicates that current is leading the voltage. This agrees with the selection of a positive sign for the reactive power of an inductive circuit and a negative sign for the reactive power of a capacitive circuit. To obtain the proper sign for Q, it is necessary to calculate S as VI^* rather than V^*I, which would reverse the sign for Q.

1.6 THE POWER TRIANGLE

Equation (1.24) suggests a graphical method of obtaining the overall P, Q, and phase angle for several loads in parallel since $\cos\theta$ is $P/|S|$. A power triangle can be drawn for an inductive load, as shown in Fig. 1.7. For several loads in

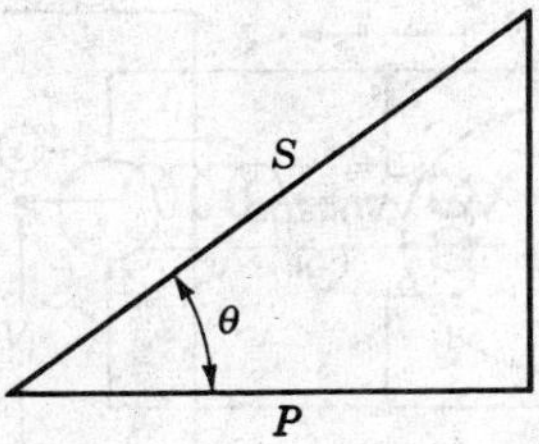

FIGURE 1.7
Power triangle for an inductive load.

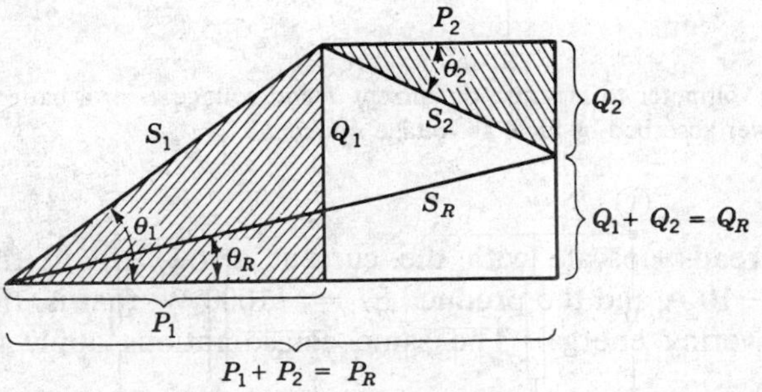

FIGURE 1.8
Power triangle for combined loads. Note that Q_2 is negative.

parallel the total P will be the sum of the average powers of the individual loads, which should be plotted along the horizontal axis for a graphical analysis. For an inductive load Q will be drawn vertically upward since it is positive. A capacitive load will have negative reactive power, and Q will be vertically downward. Figure 1.8 illustrates the power triangle composed of P_1, Q_1, and S_1 for a lagging power-factor load having a phase angle θ_1 combined with the power triangle composed of P_2, Q_2, and S_2, which is for a capacitive load with a negative θ_2. These two loads in parallel result in the triangle having sides $P_1 + P_2$, $Q_1 + Q_2$, and hypotenuse S_R. In general, $|S_R|$ is *not* equal to $|S_1| + |S_2|$. The phase angle between voltage and current supplied to the combined loads is θ_R.

1.7 DIRECTION OF POWER FLOW

The relation among P, Q, and bus voltage V, or generated voltage E, with respect to the signs of P and Q is important when the flow of power in a system is considered. The question involves the direction of flow of power, that is, whether power is being *generated* or *absorbed* when a voltage and a current are specified.

The question of delivering power to a circuit or absorbing power from a circuit is rather obvious for a dc system. Consider the current and voltage of Fig. 1.9(a) where dc current I is flowing through a battery. If the voltmeter V_m and the ammeter A_m both read upscale to show $E = 100$ V and $I = 10$ A, the battery is being charged (absorbing energy) at the rate given by the product $EI = 1000$ W. On the other hand, if the ammeter connections have to be

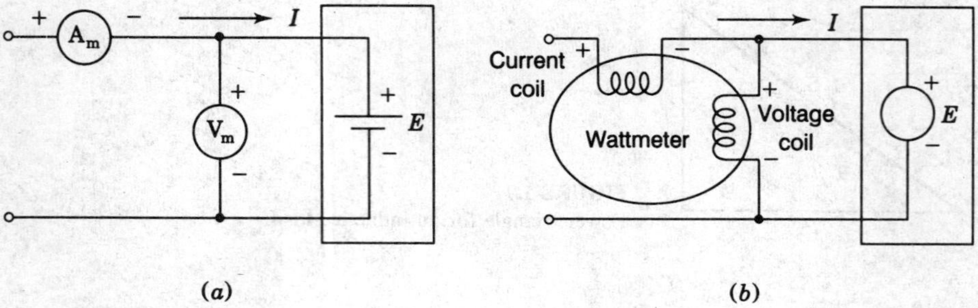

FIGURE 1.9
Connections of: (*a*) ammeter and voltmeter to measure dc current I and voltage E of a battery; (*b*) wattmeter to measure real power absorbed by ideal ac voltage source E.

reversed in order that it reads upscale with the current arrow still in the direction shown, then $I = -10$ A and the product $EI = -1000$ W; that is, the battery is discharging (delivering energy). The same considerations apply to the ac circuit relationships.

For an ac system Fig. 1.9(*b*) shows within the box an ideal voltage source E (constant magnitude, constant frequency, zero impedance) with polarity marks to indicate, as usual, the terminal which is positive during the half cycle of positive instantaneous voltage. Similarly, the arrow indicates the direction of current I into the box during the positive half cycle of current. The *wattmeter* of Fig. 1.9(*b*) has a current coil and a voltage coil corresponding, respectively, to the ammeter A_m and the voltmeter V_m of Fig. 1.9(*a*). To measure active power the coils must be correctly connected so as to obtain an upscale reading. By definition we know that the power absorbed inside the box is

$$S = VI^* = P + jQ = |V|\,|I|\cos\theta + j|V|\,|I|\sin\theta \qquad (1.25)$$

where θ is the phase angle by which I lags V. Hence, if the wattmeter reads upscale for the connections shown in Fig. 1.9(*b*), $P = |V|\,|I|\cos\theta$ is positive and real power is being *absorbed* by E. If the wattmeter tries to deflect downscale, then $P = |V|\,|I|\cos\theta$ is negative and reversing the connections of the current coil or the voltage coil, but not both, causes the meter to read upscale, indicating that *positive* power is being *supplied* by E inside the box. This is equivalent to saying that *negative* power is being *absorbed* by E. If the wattmeter is replaced by a *varmeter*, similar considerations apply to the sign of the reactive power Q absorbed or supplied by E. In general, we can determine the P and Q absorbed or supplied by any ac circuit simply by regarding the circuit as enclosed in a box with entering current I and voltage V having the polarity shown in Table 1.1. Then, the numerical values of the real and imaginary parts of the product $S = VI^*$ determine the P and Q absorbed or supplied by the enclosed circuit or network. When current I lags voltage V by

TABLE 1.1
Direction of P and Q flow where $S = VI^* = P + jQ$

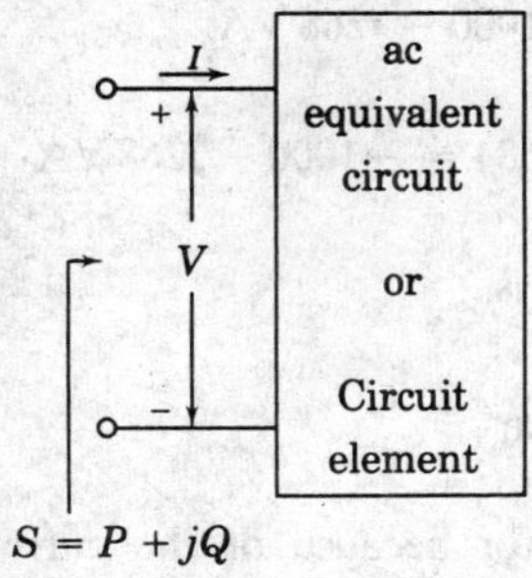

	If $P > 0$, circuit absorbs real power
	If $P < 0$, circuit supplies real power
	If $Q > 0$, circuit absorbs relative power (I lags V)
	If $Q < 0$, circuit supplies reactive power (I leads V)

an angle θ between 0° and 90°, we find that $P = |V|\,|I|\cos\theta$ and $Q = |V|\,|I|\sin\theta$ are both positive, indicating watts and vars are being absorbed by the inductive circuit inside the box. When I leads V by an angle between 0° and 90°, P is still positive but θ and $Q = |V|\,|I|\sin\theta$ are both negative, indicating that negative vars are being absorbed or positive vars are being supplied by the capacitive circuit inside the box.

Example 1.1. Two ideal voltage sources designated as machines 1 and 2 are connected, as shown in Fig. 1.10. If $E_1 = 100\angle 0°$ V, $E_2 = 100\angle 30°$ V, and $Z = 0 + j5\ \Omega$, determine (*a*) whether each machine is generating or consuming real power and the amount, (*b*) whether each machine is receiving or supplying reactive power and the amount, and (*c*) the P and Q absorbed by the impedance.

Solution

$$I = \frac{E_1 - E_2}{Z} = \frac{100 + j0 - (86.6 + j50)}{j5}$$

$$= \frac{13.4 - j50}{j5} = -10 - j2.68 = 10.35\angle 195°\ \text{A}$$

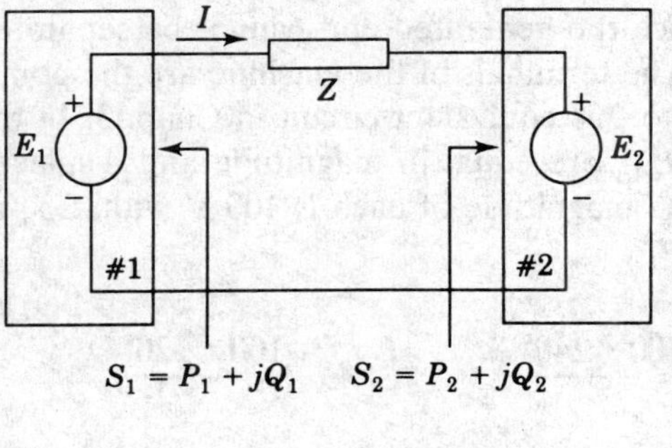

FIGURE 1.10
Ideal voltage sources connected through impedance Z.

The current entering box 1 is $-I$ and that entering box 2 is I so that

$$S_1 = E_1(-I)^* = P_1 + jQ_1 = 100(10 + j2.68)^* = 1000 - j268\text{ VA}$$

$$S_2 = E_2 I^* = P_2 + jQ_2 = (86.6 + j50)(-10 + j2.68) = -1000 - j268\text{ VA}$$

The reactive power absorbed in the series impedance is

$$|I|^2 X = 10.35^2 \times 5 = 536\text{ var}$$

Machine 1 may be expected to be a generator because of the current direction and polarity markings. However, since P_1 is positive and Q_1 is negative, the machine consumes energy at the rate of 1000 W and supplies reactive power of 268 var. The machine is actually a motor.

Machine 2, expected to be a motor, has negative P_2 and negative Q_2. Therefore, this machine generates energy at the rate of 1000 W and supplies reactive power of 268 var. The machine is actually a generator.

Note that the supplied reactive power of 268 + 268 is equal to 536 var, which is required by the inductive reactance of 5 Ω. Since the impedance is purely reactive, no P is consumed by the impedance, and all the watts generated by machine 2 are transferred to machine 1.

1.8 VOLTAGE AND CURRENT IN BALANCED THREE-PHASE CIRCUITS

Electric power systems are supplied by three-phase generators. Ideally, the generators are supplying balanced three-phase loads, which means loads with identical impedances in all three phases. Lighting loads and small motors are, of course, single-phase, but distribution systems are designed so that overall the phases are essentially balanced. Figure 1.11 shows a Y-connected generator with neutral marked o supplying a balanced-Y load with neutral marked n. In discussing this circuit, we assume that the impedances of the connections between the terminals of the generator and the load, as well as the impedance of the direct connection between o and n, are negligible.

The equivalent circuit of the three-phase generator consists of an emf in each of the three phases, as indicated by circles on the diagram. Each emf is in series with a resistance and inductive reactance composing the impedance Z_d. Points a', b', and c' are fictitious since the generated emf cannot be separated from the impedance of each phase. The terminals of the machine are the points a, b, and c. Some attention is given to this equivalent circuit in Chap. 3. In the generator the emfs $E_{a'o}$, $E_{b'o}$, and $E_{c'o}$ are equal in magnitude and displaced from each other 120° in phase. If the magnitude of each is 100 V with $E_{a'o}$ as reference,

$$E_{a'o} = 100\angle 0^\circ \text{ V} \qquad E_{b'o} = 100\angle 240^\circ \text{ V} \qquad E_{c'o} = 100\angle 120^\circ \text{ V}$$

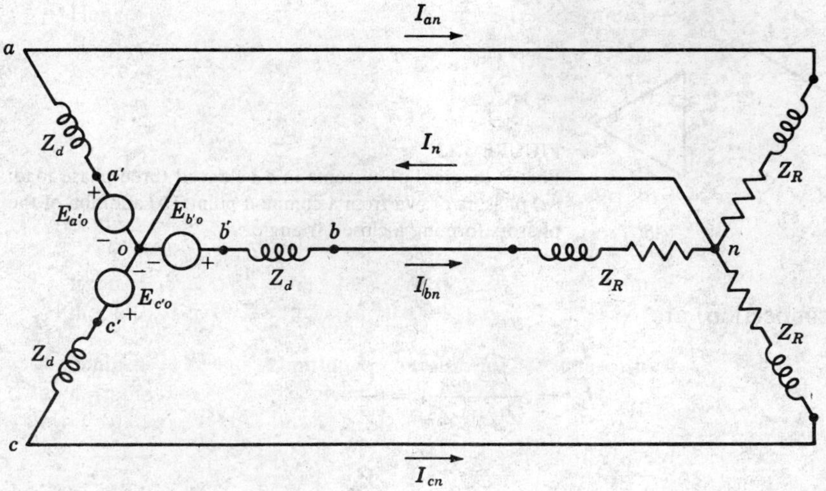

FIGURE 1.11
Circuit diagram of a Y-connected generator connected to a balanced-Y load.

provided the phase sequence is *abc*, which means that $E_{a'o}$ leads $E_{b'o}$ by 120° and $E_{b'o}$ in turn leads $E_{c'o}$ by 120°. The circuit diagram gives no indication of phase sequence, but Fig. 1.12 shows these emfs with phase sequence *abc*.

At the generator terminals (and at the load in this case) the terminal voltages to neutral are

$$
\begin{aligned}
V_{ao} &= E_{a'o} - I_{an}Z_d \\
V_{bo} &= E_{b'o} - I_{bn}Z_d \qquad (1.26) \\
V_{co} &= E_{c'o} - I_{cn}Z_d
\end{aligned}
$$

Since *o* and *n* are at the same potential, V_{ao}, V_{bo}, and V_{co} are equal to V_{an}, V_{bn}, and V_{cn}, respectively, and the line currents (which are also the phase currents

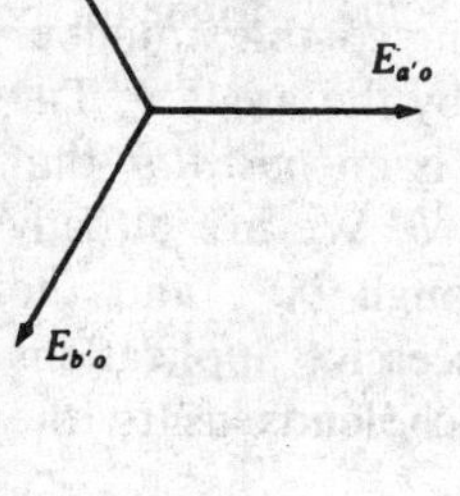

FIGURE 1.12
Phasor diagram of the emfs of the circuit shown in Fig. 1.11.

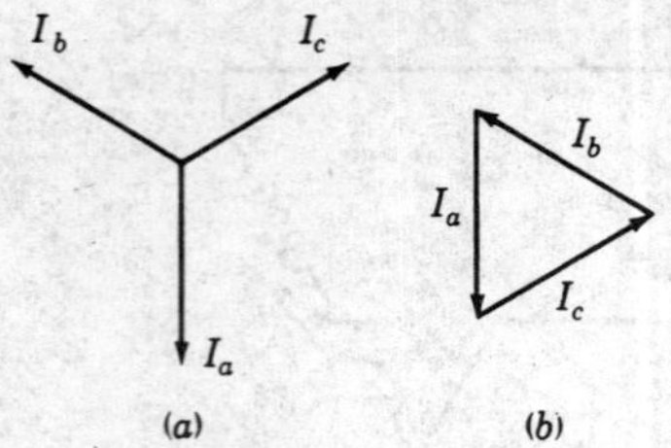

FIGURE 1.13
Phasor diagram of currents in a balanced three-phase load: (*a*) phasors drawn from a common point; (*b*) addition of the phasors forming a closed triangle.

for a Y connection) are

$$I_{an} = \frac{E_{a'o}}{Z_d + Z_R} = \frac{V_{an}}{Z_R}$$
$$I_{bn} = \frac{E_{b'o}}{Z_d + Z_R} = \frac{V_{bn}}{Z_R} \tag{1.27}$$
$$I_{cn} = \frac{E_{c'o}}{Z_d + Z_R} = \frac{V_{cn}}{Z_R}$$

Since $E_{a'o}$, $E_{b'o}$, and $E_{c'o}$ are equal in magnitude and 120° apart in phase, and since the impedances seen by these emfs are identical, the currents will also be equal in magnitude and displaced 120° from each other in phase. The same must also be true of V_{an}, V_{bn}, and V_{cn}. In this case we describe the voltages and currents as *balanced*. Figure 1.13(*a*) shows three line currents of a balanced system. In Fig. 1.13(*b*) these currents form a closed triangle and it is obvious that their sum is zero. Therefore, I_n must be zero in the connection shown in Fig. 1.11 between the neutrals of the generator and load. Then, the connection between n and o may have any impedance, or even be open, and n and o will remain at the same potential. If the load is not balanced, the sum of the currents will not be zero and a current will flow between o and n. For the unbalanced condition o and n will not be at the same potential unless they are connected by zero impedance.

Because of the phase displacement of the voltages and currents in a balanced three-phase system, it is convenient to have a shorthand method of indicating the rotation of a phasor through 120°. The result of the multiplication of two complex numbers is the product of their magnitudes and the sum of their angles. If the complex number expressing a phasor is multiplied by a complex number of unit magnitude and angle θ, the resulting complex number represents a phasor equal to the original phasor displaced by the angle θ. The complex number of unit magnitude and associated angle θ is an *operator* that rotates the phasor on which it operates through the angle θ. We are already familiar with the operator j, which causes rotation through 90°, and the operator -1, which causes rotation through 180°. Two successive applications of the operator j cause rotation through 90° + 90°, which leads us to the conclusion that $j \times j$ causes rotation through 180°, and thus we recognize that j^2 is equal to -1. Other powers of the operator j are found by similar analysis.

The letter a is commonly used to designate the operator that causes a rotation of 120° in the counterclockwise direction. Such an operator is a complex number of unit magnitude with an angle of 120° and is defined by

$$a = 1\angle 120° = 1\varepsilon^{j2\pi/3} = -0.5 + j0.866$$

If the operator a is applied to a phasor twice in succession, the phasor is rotated through 240°. Three successive applications of a rotate the phasor through 360°. Thus,

$$a^2 = 1\angle 240° = 1\varepsilon^{j4\pi/3} = -0.5 - j0.866$$

$$a^3 = 1\angle 360° = 1\varepsilon^{j2\pi} = 1\angle 0° = 1$$

It is evident that $1 + a + a^2 = 0$. Figure 1.14 shows phasors representing various powers and functions of a.

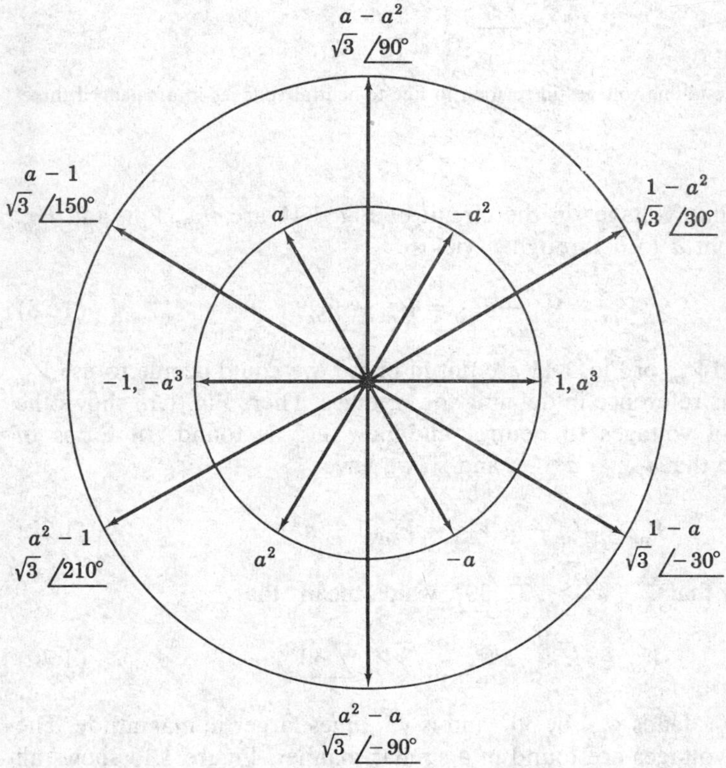

FIGURE 1.14
Phasor diagram of various powers and functions of the operator a.

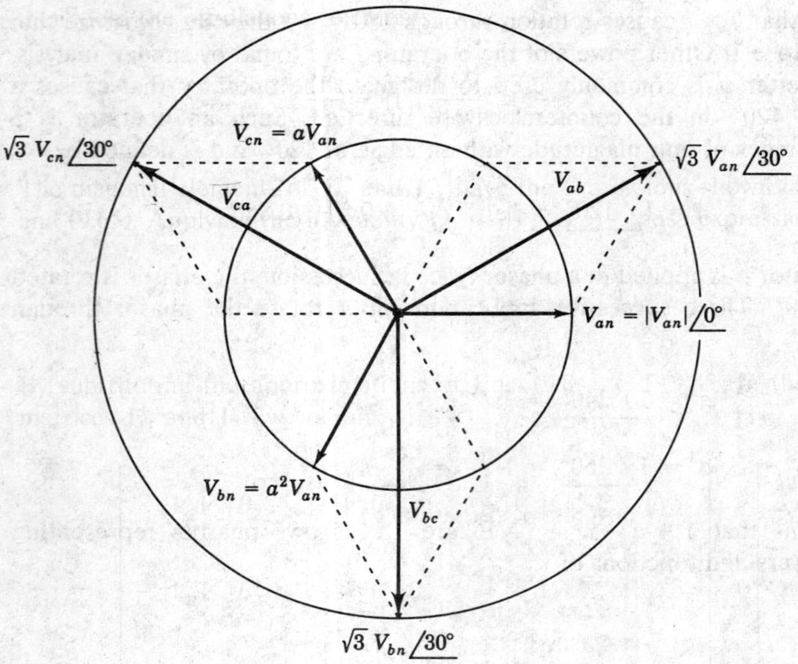

FIGURE 1.15
Phasor diagram of line-to-line voltages in relation to line-to-neutral voltages in a balanced three-phase circuit.

The line-to-line voltages in the circuit of Fig. 1.11 are V_{ab}, V_{bc}, and V_{ca}. Tracing a path from a to b through n yields

$$V_{ab} = V_{an} + V_{nb} = V_{an} - V_{bn} \tag{1.28}$$

Although $E_{a'o}$ and V_{an} of Fig. 1.11 are not in phase, we could decide to use V_{an} rather than $E_{a'o}$ as reference in defining the voltages. Then, Fig. 1.15 shows the phasor diagram of voltages to neutral and how V_{ab} is found. In terms of operator a we see that $V_{bn} = a^2V_{an}$, and so we have

$$V_{ab} = V_{an} - a^2V_{bn} = V_{an}(1 - a^2) \tag{1.29}$$

Figure 1.14 shows that $1 - a^2 = \sqrt{3}\,\underline{/30°}$, which means that

$$V_{ab} = \sqrt{3}\,V_{an}\varepsilon^{j30°} = \sqrt{3}\,V_{an}\,\underline{/\ 30°} \tag{1.30}$$

So, as a phasor, V_{ab} leads V_{an} by 30° and is $\sqrt{3}$ times larger in magnitude. The other line-to-line voltages are found in a similar manner. Figure 1.15 shows all the line-to-line voltages in relation to the line-to-neutral voltages. The fact that

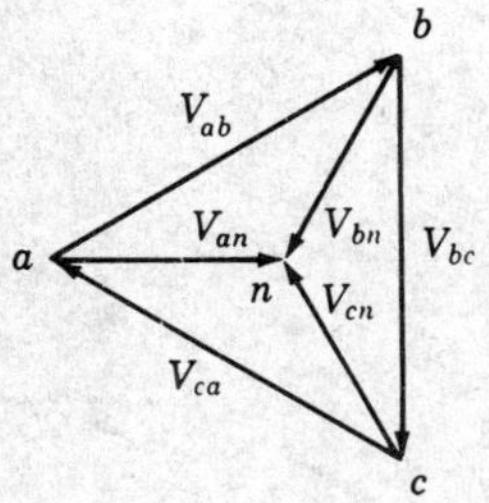

FIGURE 1.16
Alternative method of drawing the phasors of Fig. 1.15.

the magnitude of *balanced* line-to-line voltages of a three-phase circuit is always equal to $\sqrt{3}$ times the magnitude of the line-to-neutral voltages is very important.

Figure 1.16 shows another way of displaying the line-to-line and line-to-neutral voltages. The line-to-line voltage phasors are drawn to form a closed triangle oriented to agree with the chosen reference, in this case V_{an}. The vertices of the triangle are labeled so that each phasor begins and ends at the vertices corresponding to the order of the subscripts of that phasor voltage. Line-to-neutral voltage phasors are drawn to the center of the triangle. Once this phasor diagram is understood, it will be found to be the simplest way to determine the various voltages.

The order in which the vertices $a, b,$ and c of the triangle follow each other when the triangle is rotated counterclockwise about n indicates the phase sequence. The importance of phase sequence becomes clear when we discuss transformers and when *symmetrical components* are used to analyze unbalanced faults on power systems.

A separate current diagram can be drawn to relate each current properly with respect to its phase voltage.

Example 1.2. In a balanced three-phase circuit the voltage V_{ab} is $173.2\underline{/0^\circ}$ V. Determine all the voltages and the currents in a Y-connected load having $Z_L = 10\underline{/20^\circ}\ \Omega$. Assume that the phase sequence is *abc*.

Solution. With V_{ab} as reference, the phasor diagram of voltages is drawn as shown in Fig. 1.17, from which it is determined that

$$V_{ab} = 173.2\underline{/0^\circ}\ \text{V} \qquad V_{an} = 100\underline{/-30^\circ}\ \text{V}$$

$$V_{bc} = 173.2\underline{/240^\circ}\ \text{V} \qquad V_{bn} = 100\underline{/210^\circ}\ \text{V}$$

$$V_{ca} = 173.2\underline{/120^\circ}\ \text{V} \qquad V_{cn} = 100\underline{/90^\circ}\ \text{V}$$

Each current lags the voltage across its load impedance by 20° and each current magnitude is 10 A. Figure 1.18 is the phasor diagram of the currents

$$I_{an} = 10\underline{/-50^\circ}\ \text{A} \qquad I_{bn} = 10\underline{/190^\circ}\ \text{A} \qquad I_{cn} = 10\underline{/70^\circ}\ \text{A}$$

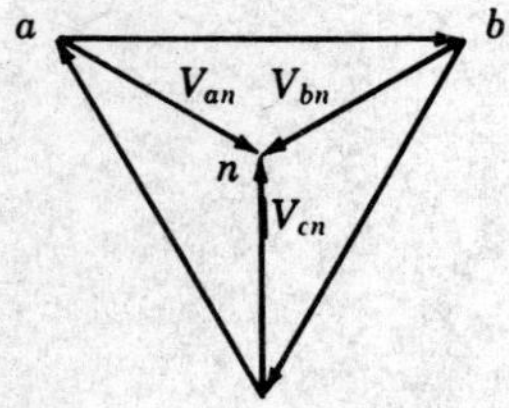

FIGURE 1.17
Phasor diagram of voltages for Example 1.2.

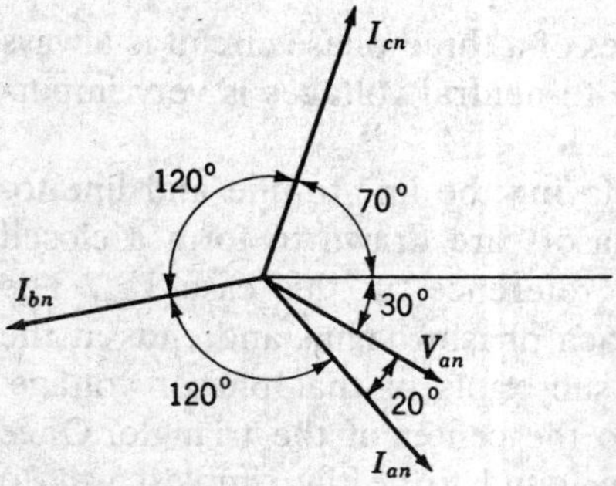

FIGURE 1.18
Phasor diagram of currents for Example 1.2.

Balanced loads are often connected in Δ, as shown in Fig. 1.19. Here it is left to the reader using the properties of the operator *a* to show that the magnitude of a line current such as I_a is equal to $\sqrt{3}$ times the magnitude of *a*-phase current I_{ab} and that I_a lags I_{ab} by 30° when the phase sequence is *abc*. Figure 1.20 shows the current relationships when I_{ab} is chosen as reference.

When solving balanced three-phase circuits, it is not necessary to work with the entire three-phase circuit diagram of Fig. 1.11. To solve the circuit a neutral connection of zero impedance is assumed to be present and to carry the sum of the three phase currents, which is zero for balanced conditions. The circuit is solved by applying Kirchhoff's voltage law around a closed path which includes one phase and neutral. Such a closed path is shown in Fig. 1.21. This

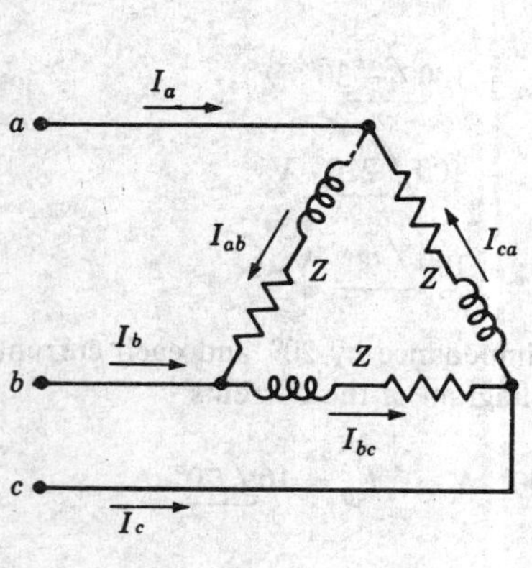

FIGURE 1.19
Circuit diagram of Δ-connected three-phase load.

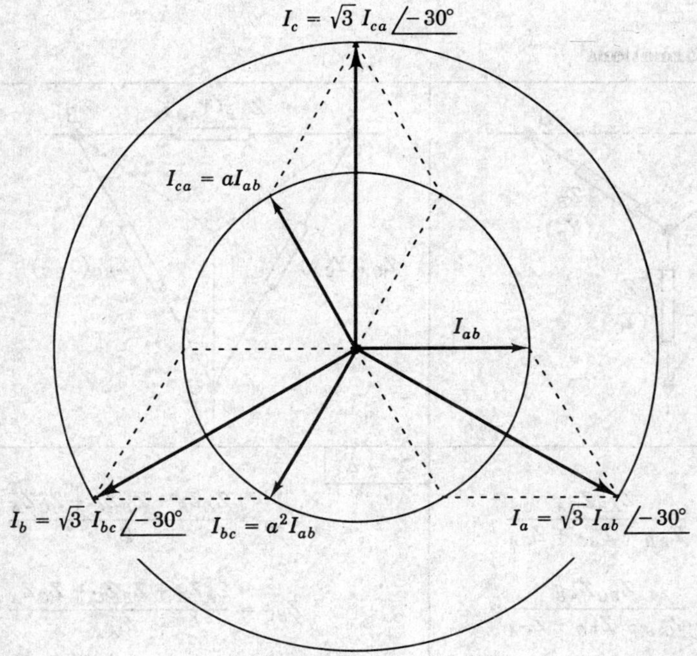

FIGURE 1.20
Phasor diagram of the line currents in relation to the phase currents in a balanced Δ-connected three-phase load.

circuit is the *single-phase* or *per-phase equivalent* of the circuit of Fig. 1.11. Calculations made for this path are extended to the whole three-phase circuit by recalling that the currents in the other two phases are equal in magnitude to the current of the phase calculated and are displaced 120° and 240° in phase. It is immaterial whether the balanced load (specified by its line-to-line voltage, total power, and power factor) is Δ- or Y-connected since the Δ can always be replaced for purposes of calculation by its equivalent Y, as shown in Table 1.2. It is apparent from the table that the general expression for a wye impedance

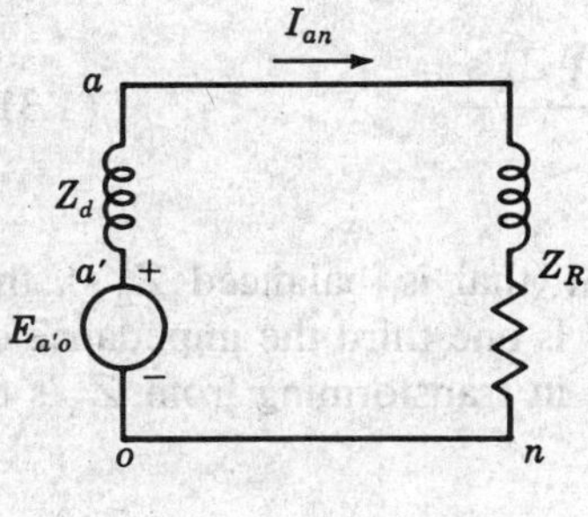

FIGURE 1.21
One phase of the circuit of Fig. 1.11.

TABLE 1.2
Y-Δ and Δ-Y transformations†

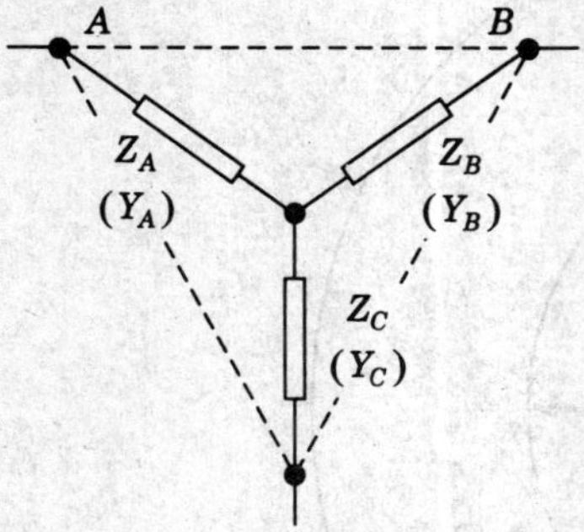

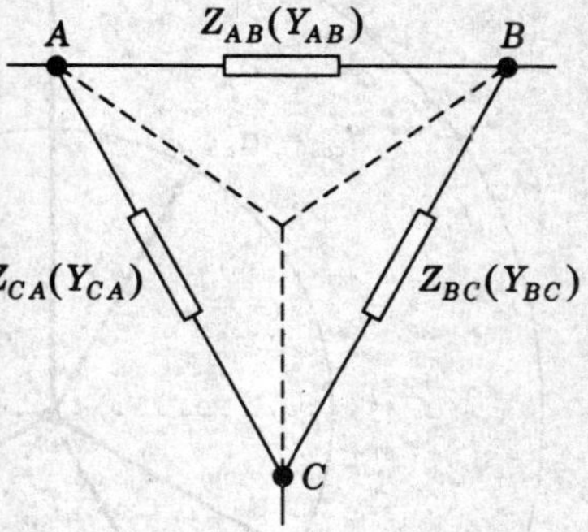

Δ → Y	Y → Δ
$Z_A = \dfrac{Z_{AB}Z_{CA}}{Z_{AB}+Z_{BC}+Z_{CA}}$	$Z_{AB} = \dfrac{Z_AZ_B+Z_BZ_C+Z_CZ_A}{Z_C}$
$Z_B = \dfrac{Z_{BC}Z_{AB}}{Z_{AB}+Z_{BC}+Z_{CA}}$	$Z_{BC} = \dfrac{Z_AZ_B+Z_BZ_C+Z_CZ_A}{Z_A}$
$Z_C = \dfrac{Z_{CA}Z_{BC}}{Z_{AB}+Z_{BC}+Z_{CA}}$	$Z_{CA} = \dfrac{Z_AZ_B+Z_BZ_C+Z_CZ_A}{Z_B}$
Δ → Y	**Y → Δ**
$Y_A = \dfrac{Y_{AB}Y_{CA}+Y_{BC}Y_{AB}+Y_{CA}Y_{BC}}{Y_{BC}}$	$Y_{AB} = \dfrac{Y_AY_B}{Y_A+Y_B+Y_C}$
$Y_B = \dfrac{Y_{AB}Y_{CA}+Y_{BC}Y_{AB}+Y_{CA}Y_{BC}}{Y_{CA}}$	$Y_{BC} = \dfrac{Y_BY_C}{Y_A+Y_B+Y_C}$
$Y_C = \dfrac{Y_{AB}Y_{CA}+Y_{BC}Y_{AB}+Y_{CA}Y_{BC}}{Y_{AB}}$	$Y_{CA} = \dfrac{Y_CY_A}{Y_A+Y_B+Y_C}$

† Admittances and impedances with the same subscripts are reciprocals of one another.

Z_Y in terms of the delta impedances Z_Δ's is

$$Z_Y = \frac{\text{product of adjacent } Z_\Delta\text{'s}}{\text{sum of } Z_\Delta\text{'s}} \tag{1.31}$$

So, when all the impedances in the Δ are equal (that is, balanced Z_Δ's), the impedance Z_Y of each phase of the equivalent Y is one-third the impedance of each phase of the Δ which it replaces. Likewise, in transforming from Z_Y's to

Z_Δ's, Table 1.2 shows that

$$Z_\Delta = \frac{\text{sum of pairwise products of } Z_Y\text{'s}}{\text{the opposite } Z_Y} \tag{1.32}$$

Similar statements apply to the admittance transformations.

Example 1.3. The terminal voltage of a Y-connected load consisting of three equal impedances of $20\angle 30°\ \Omega$ is 4.4 kV line to line. The impedance of each of the three lines connecting the load to a bus at a substation is $Z_L = 1.4\angle 75°\ \Omega$. Find the line-to-line voltage at the substation bus.

Solution. The magnitude of the voltage to neutral at the load is $4400/\sqrt{3} = 2540$ V. If V_{an}, the voltage across the load, is chosen as reference,

$$V_{an} = 2540\angle 0°\ \text{V} \quad \text{and} \quad I_{an} = \frac{2540\angle 0°}{20\angle 30°} = 127.0\angle -30°\ \text{A}$$

The line-to-neutral voltage at the substation is

$$\begin{aligned} V_{an} + I_{an}Z_L &= 2540\angle 0° + 127\angle -30° \times 1.4\angle 75° \\ &= 2540\angle 0° + 177.8\angle 45° \\ &= 2666 + j125.7 = 2670\angle 2.70°\ \text{V} \end{aligned}$$

and the magnitude of the voltage at the substation bus is

$$\sqrt{3} \times 2.67 = 4.62\ \text{kV}$$

Figure 1.22 shows the per-phase equivalent circuit and quantities involved.

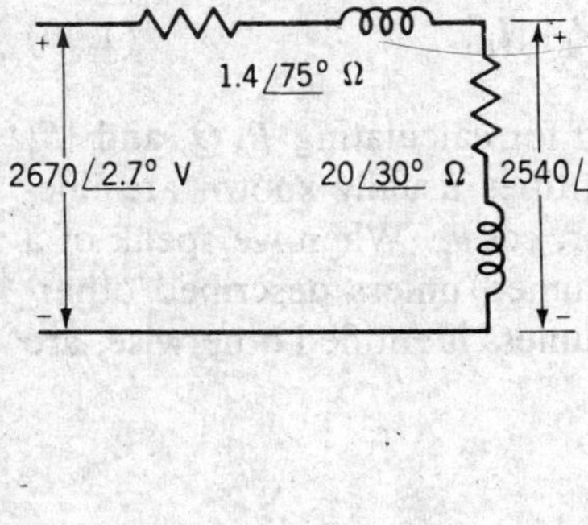

FIGURE 1.22
Per-phase equivalent circuit for Example 1.3.

1.9 POWER IN BALANCED THREE-PHASE CIRCUITS

The total power delivered by a three-phase generator or absorbed by a three-phase load is found simply by adding the power in each of the three phases. In a balanced circuit this is the same as multiplying the power in any one phase by 3 since the power is the same in all phases.

If the magnitude of the voltages to neutral V_p for a Y-connected load is

$$|V_p| = |V_{an}| = |V_{bn}| = |V_{cn}| \tag{1.33}$$

and if the magnitude of the phase current I_p for a Y-connected load is

$$|I_p| = |I_{an}| = |I_{bn}| = |I_{cn}| \tag{1.34}$$

the total three-phase power is

$$P = 3|V_p|\,|I_p|\cos\theta_p \tag{1.35}$$

where θ_p is the angle by which *phase current* I_p lags the *phase voltage* V_p, that is, the angle of the impedance in each phase. If $|V_L|$ and $|I_L|$ are the magnitudes of line-to-line voltage V_L and line current I_L, respectively,

$$|V_p| = \frac{|V_L|}{\sqrt{3}} \qquad \text{and} \qquad |I_p| = |I_L| \tag{1.36}$$

and substituting in Eq. (1.35) yields

$$P = \sqrt{3}\,|V_L|\,|I_L|\cos\theta_p \tag{1.37}$$

The total vars are

$$Q = 3|V_p|\,|I_p|\sin\theta_p \tag{1.38}$$

$$Q = \sqrt{3}\,|V_L|\,|I_L|\sin\theta_p \tag{1.39}$$

and the voltamperes of the load are

$$|S| = \sqrt{P^2 + Q^2} = \sqrt{3}\,|V_L|\,|I_L| \tag{1.40}$$

Equations (1.37), (1.39), and (1.40) are used for calculating P, Q, and $|S|$ in balanced three-phase networks since the quantities usually known are line-to-line voltage, line current, and the power factor, $\cos\theta_p$. When we speak of a three-phase system, balanced conditions are assumed unless described otherwise; and the terms *voltage*, *current*, and *power*, unless identified otherwise, are understood to mean *line-to-line voltage*, *line current*, and total *three-phase power*, respectively.

If the load is connected Δ, the voltage across each impedance is the line-to-line voltage and the magnitude of the current through each impedance is the magnitude of the line current divided by $\sqrt{3}$, or

$$|V_p| = |V_L| \quad \text{and} \quad |I_p| = \frac{|I_L|}{\sqrt{3}} \tag{1.41}$$

The total three-phase power is

$$P = 3|V_p|\,|I_p|\cos\theta_p \tag{1.42}$$

and substituting in this equation the values of $|V_p|$ and $|I_p|$ from Eq. (1.41) gives

$$P = \sqrt{3}\,|V_L|\,|I_L|\cos\theta_p \tag{1.43}$$

which is identical to Eq. (1.37). It follows that Eqs. (1.39) and (1.40) are also valid regardless of whether a particular load is connected Δ or Y.

1.10 PER-UNIT QUANTITIES

Power transmission lines are operated at voltage levels where the kilovolt (kV) is the most convenient unit to express voltage. Because of the large amount of power transmitted, kilowatts or megawatts and kilovoltamperes or megavoltamperes are the common terms. However, these quantities as well as amperes and ohms are often expressed as a percent or per unit of a *base* or *reference value* specified for each. For instance, if a base voltage of 120 kV is chosen, voltages of 108, 120, and 126 kV become 0.90, 1.00, and 1.05 per unit, or 90, 100, and 105%, respectively. The *per-unit value* of any quantity is defined as the ratio of the quantity to its base expressed as a decimal. The ratio in percent is 100 times the value in per unit. Both the percent and per-unit methods of calculation are simpler and often more informative than the use of actual amperes, ohms, and volts. The per-unit method has an advantage over the percent method because the product of two quantities expressed in per unit is expressed in per unit itself, but the product of two quantities expressed in percent must be divided by 100 to obtain the result in percent.

Voltage, current, kilovoltamperes, and impedance are so related that selection of base values for any two of them determines the base values of the remaining two. If we specify the base values of current and voltage, base impedance and base kilovoltamperes can be determined. The base impedance is that impedance which will have a voltage drop across it equal to the base voltage when the current flowing in the impedance is equal to the base value of the current. The base kilovoltamperes in single-phase systems is the product of base voltage in kilovolts and base current in amperes. Usually, base megavoltamperes and base voltage in kilovolts are the quantities selected to specify the base. For single-phase systems, or three-phase systems where the term current refers to line current, where the term voltage refers to voltage to neutral and where the term kilovoltamperes refers to kilovoltamperes per phase, the following formulas relate the various quantities:

$$\text{Base current, A} = \frac{\text{base kVA}_{1\phi}}{\text{base voltage, kV}_{LN}} \tag{1.44}$$

$$\text{Base impedance, }\Omega = \frac{\text{base voltage, V}_{LN}}{\text{base current, A}} \tag{1.45}$$

$$\text{Base impedance, }\Omega = \frac{(\text{base voltage, kV}_{LN})^2 \times 1000}{\text{base kVA}_{1\phi}} \tag{1.46}$$

$$\text{Base impedance, }\Omega = \frac{(\text{base voltage, kV}_{LN})^2}{\text{MVA}_{1\phi}} \tag{1.47}$$

$$\text{Base power, kW}_{1\phi} = \text{base kVA}_{1\phi} \tag{1.48}$$

$$\text{Base power, MW}_{1\phi} = \text{base MVA}_{1\phi} \tag{1.49}$$

$$\text{Per-unit impedance of an element} = \frac{\text{actual impedance, }\Omega}{\text{base impedance, }\Omega} \tag{1.50}$$

In these equations the subscripts $_{1\phi}$ and $_{LN}$ denote "per phase" and "line to neutral," respectively, where the equations apply to three-phase circuits. If the equations are used for a single-phase circuit, kV_{LN} means the voltage across the single-phase line, or line-to-ground voltage if one side is grounded.

Since balanced three-phase circuits are solved as a single line with a neutral return, the bases for quantities in the impedance diagram are kilovoltamperes per phase and kilovolts from line to neutral. Data are usually given as total three-phase kilovoltamperes or megavoltamperes and line-to-line kilovolts. Because of this custom of specifying line-to-line voltage and total kilovoltamperes or megavoltamperes, confusion may arise regarding the relation between the per-unit value of line voltage and the per-unit value of phase voltage. Although a line voltage may be specified as base, the voltage in the single-phase circuit required for the solution is still the voltage to neutral. The base voltage to neutral is the base voltage from line to line divided by $\sqrt{3}$. Since this is also the ratio between line-to-line and line-to-neutral voltages of a balanced three-phase system, *the per-unit value of a line-to-neutral voltage on the line-to-neutral voltage base is equal to the per-unit value of the line-to-line voltage at the same point on the line-to-line voltage base if the system is balanced.* Similarly, the three-phase kilovoltamperes is three times the kilovoltamperes per phase, and the three-phase kilovoltamperes base is three times the base kilovoltamperes per phase. Therefore, *the per-unit value of the three-phase kilovoltamperes on the three-phase kilovoltampere base is identical to the per-unit value of the kilovoltamperes per phase on the kilovoltampere-per-phase base.*

A numerical example clarifies the relationships. For instance, if

$$\text{Base kVA}_{3\phi} = 30{,}000 \text{ kVA}$$

and

$$\text{Base kV}_{LL} = 120 \text{ kV}$$

where subscripts $_{3\phi}$ and $_{LL}$ mean "three-phase" and "line to line," respectively,

$$\text{Base kVA}_{1\phi} = \frac{30{,}000}{3} = 10{,}000 \text{ kVA}$$

and

$$\text{Base kV}_{LN} = \frac{120}{\sqrt{3}} = 69.2 \text{ kV}$$

For an actual line-to-line voltage of 108 kV in a balanced three-phase set the line-to-neutral voltage is $108/\sqrt{3} = 62.3$ kV, and

$$\text{Per-unit voltage} = \frac{108}{120} = \frac{62.3}{69.2} = 0.90$$

For total three-phase power of 18,000 kW the power per phase is 6000 kW, and

$$\text{Per-unit power} = \frac{18{,}000}{30{,}000} = \frac{6{,}000}{10{,}000} = 0.6$$

Of course, megawatt and megavoltampere values may be substituted for kilowatt and kilovoltampere values throughout the above discussion. Unless otherwise specified, a given value of base voltage in a three-phase system is a line-to-line voltage, and a given value of base kilovoltamperes or base megavoltamperes is the total three-phase base.

Base impedance and base current can be computed directly from three-phase values of base kilovolts and base kilovoltamperes. If we interpret *base kilovoltamperes and base voltage in kilovolts* to mean *base kilovoltamperes for the total of the three phases* and *base voltage from line to line*, we find

$$\text{Base current, A} = \frac{\text{base kVA}_{3\phi}}{\sqrt{3} \times \text{base voltage, kV}_{LL}} \tag{1.51}$$

and from Eq. (1.46)

$$\text{Base impedance} = \frac{(\text{base voltage, kV}_{LL}/\sqrt{3})^2 \times 1000}{\text{base kVA}_{3\phi}/3} \tag{1.52}$$

$$\text{Base impedance} = \frac{(\text{base voltage, kV}_{LL})^2 \times 1000}{\text{base kVA}_{3\phi}} \tag{1.53}$$

$$\text{Base impedance} = \frac{(\text{base voltage, kV}_{LL})^2}{\text{base MVA}_{3\phi}} \tag{1.54}$$

Except for the subscripts, Eqs. (1.46) and (1.47) are identical to Eqs. (1.53) and (1.54), respectively. Subscripts have been used in expressing these relations in order to emphasize the distinction between working with three-phase quantities and quantities per phase. We use these equations without the subscripts, but we must

- Use line-to-line kilovolts with three-phase kilovoltamperes or megavoltamperes, and
- Use line-to-neutral kilovolts with kilovoltamperes or megavoltamperes per phase.

Equation (1.44) determines the base current for single-phase systems or for three-phase systems where the bases are specified in total kilovoltamperes per phase and kilvolts to neutral. Equation (1.51) determines the base current for three-phase systems where the bases are specified in total kilovoltamperes for the three phases and in kilovolts from line to line.

Example 1.4. Find the solution of Example 1.3 by working in per unit on a base of 4.4 kV, 127 A so that both voltage and current magnitudes will be 1.0 per unit. Current rather than kilovoltamperes is specified here since the latter quantity does not enter the problem.

Solution. Base impedance is

$$\frac{4400/\sqrt{3}}{127} = 20.0\ \Omega$$

and therefore the magnitude of the load impedance is also 1.0 per unit. The line impedance is

$$Z = \frac{1.4\angle 75^\circ}{20} = 0.07\angle 75^\circ \text{ per unit}$$

$$V_{an} = 1.0\angle 0^\circ + 1.0\angle -30^\circ \times 0.07\angle 75^\circ$$

$$= 1.0\angle 0^\circ + 0.07\angle 45^\circ$$

$$= 1.0495 + j0.0495 = 1.051\angle 2.70^\circ \text{ per unit}$$

$$V_{LN} = 1.051 \times \frac{4400}{\sqrt{3}} = 2670 \text{ V, or } 2.67 \text{ kV}$$

$$V_{LL} = 1.051 \times 4.4 = 4.62 \text{ kV}$$

When the problems to be solved are more complex, and particularly when transformers are involved, the advantages of calculations in per unit are more apparent. Impedance values and other parameters of a component, when given in per unit without specified bases, are generally understood to be based on the megavoltampere and kilovolt *ratings* of the component.

1.11 CHANGING THE BASE OF PER-UNIT QUANTITIES

Sometimes the per-unit impedance of a component of a system is expressed on a base other than the one selected as base for the part of the system in which the component is located. Since all impedances in any one part of a system must be expressed on the same impedance base when making computations, it is necessary to have a means of converting per-unit impedances from one base to another. Substituting the expression for base impedance given by Eq. (1.46) or (1.53) for base impedance in Eq. (1.50) gives for any circuit element

$$\text{Per-unit impedance} = \frac{(\text{actual impedance, } \Omega) \times (\text{base kVA})}{(\text{base voltage, kV})^2 \times 1000} \tag{1.55}$$

which shows that per-unit impedance is directly proportional to base kilovoltamperes and inversely proportional to the square of the base voltage. Therefore, to change from per-unit impedance on a given base to per-unit impedance on a new base, the following equation applies:

$$\text{Per-unit } Z_{\text{new}} = \text{per-unit } Z_{\text{given}}\left(\frac{\text{base kV}_{\text{given}}}{\text{base kV}_{\text{new}}}\right)^2\left(\frac{\text{base kVA}_{\text{new}}}{\text{base kVA}_{\text{given}}}\right) \tag{1.56}$$

The reader should note that this equation has nothing to do with transferring the ohmic value of impedance from one side of a transformer to another. The application of the equation is in changing the value of the per-unit impedance of any component given on a particular base to a new base.

Rather than using Eq. (1.56) directly, the change in base may also be accomplished by first converting the per-unit value on the given base to ohms and then dividing by the new base impedance.

Example 1.5. The reactance of a generator designated X'' is given as 0.25 per unit based on the generator's nameplate rating of 18 kV, 500 MVA. The base for calculations is 20 kV, 100 MVA. Find X'' on the new base.

Solution. By Eq. (1.56)

$$X'' = 0.25\left(\frac{18}{20}\right)^2\left(\frac{100}{500}\right) = 0.0405 \text{ per unit}$$

or by converting the given value to ohms and dividing by the new base impedance,

$$X'' = \frac{0.25(18^2/500)}{20^2/100} = 0.0405 \text{ per unit}$$

Resistance and reactance of a device in percent or per unit are usually available from the manufacturer. The impedance base is understood to be derived from the rated kilovoltamperes and kilovolts of the device. Tables A.1 and A.2 in the Appendix list some representative values of reactance for transformers and generators. Per-unit quantities are further discussed in Chap. 2 associated with the study of transformers.

1.12 NODE EQUATIONS

The junctions formed when two or more circuit elements (R, L, or C, or an ideal source of voltage or current) are connected to each other at their terminals are called *nodes*. Systematic formulation of equations determined at nodes of a circuit by applying Kirchhoff's current law is the basis of some excellent computer solutions of power system problems.

In order to examine some features of node equations, we begin with the simple circuit diagram of Fig. 1.23, which shows node numbers within circles. Current sources are connected at nodes ③ and ④ and all other elements are represented as admittances. Single-subscript notation is used to designate the voltage of each node with respect to the reference node ⓪. Applying Kirchhoff's current law at node ① with current away from the node equated to current into the node from the source gives

$$(V_1 - V_3)Y_c + (V_1 - V_2)Y_d + (V_1 - V_4)Y_f = 0 \tag{1.57}$$

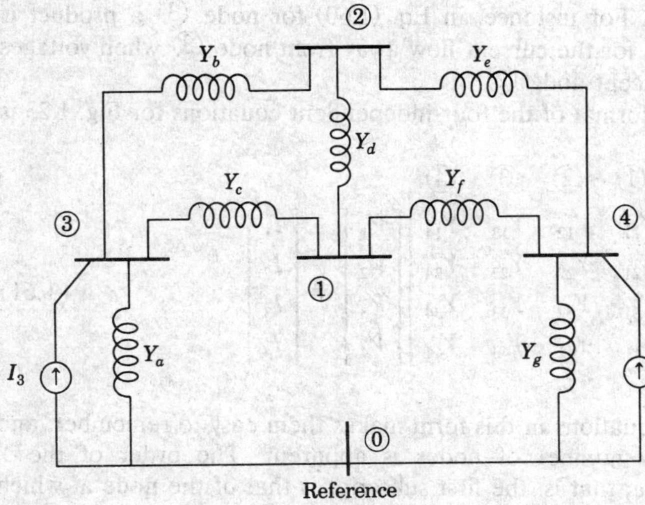

FIGURE 1.23
A circuit diagram showing current sources at nodes ③ and ④; all other elements are admittances.

and for node ③

$$V_3Y_a + (V_3 - V_2)Y_b + (V_3 - V_1)Y_c = I_3 \tag{1.58}$$

Rearranging these equations yields

At node ①: $$V_1(Y_c + Y_d + Y_f) - V_2Y_d - V_3Y_c - V_4Y_f = 0 \tag{1.59}$$

At node ③: $$-V_1Y_c - V_2Y_b + V_3(Y_a + Y_b + Y_c) = I_3 \tag{1.60}$$

Similar equations can be formed for nodes ② and ④, and the four equations can be solved simultaneously for the voltages V_1, V_2, V_3, and V_4. All branch currents can be found when these voltages are known, and a node equation formed for the reference node would yield no further information. Hence, the required number of *independent* node equations is one less than the number of nodes.

We have not written the equations for nodes ② and ④ because we can already see how to formulate node equations in standard notation. In both Eqs. (1.59) and (1.60) it is apparent that the current flowing into the network from current sources connected to a node is equated to the sum of several products. At any node one product is the voltage of that node times the sum of all the admittances which terminate on the node. This product accounts for the current that flows away from the node if the voltage is zero at each other node. Each of the other products equals the negative of the voltage at another node times the admittance connected directly between that node and the node at which the equation is formulated. For instance, in Eq. (1.60) for node ③ a product is $-V_2Y_b$, which accounts for the current flow away from node ③ when voltages are zero at all nodes except node ②.

The usual matrix format of the four-independent equations for Fig. 1.23 is

$$\begin{array}{c} \\ ① \\ ② \\ ③ \\ ④ \end{array}\begin{array}{c} \begin{array}{cccc} ① & ② & ③ & ④ \end{array} \\ \begin{bmatrix} Y_{11} & Y_{12} & Y_{13} & Y_{14} \\ Y_{21} & Y_{22} & Y_{23} & Y_{24} \\ Y_{31} & Y_{32} & Y_{33} & Y_{34} \\ Y_{41} & Y_{42} & Y_{43} & Y_{44} \end{bmatrix} \end{array} \begin{bmatrix} V_1 \\ V_2 \\ V_3 \\ V_4 \end{bmatrix} = \begin{bmatrix} I_1 \\ I_2 \\ I_3 \\ I_4 \end{bmatrix} \tag{1.61}$$

The symmetry of the equations in this form makes them easy to remember, and their extension to any number of nodes is apparent. The order of the Y subscripts is *effect-cause*; that is, the first subscript is that of the node at which the current is being expressed, and the second subscript is the same as that of the voltage causing this component of current. The Y matrix is designated $\mathbf{Y}_{\text{bus}}$ and called the *bus admittance matrix*. The *usual* rules when forming the typical elements of $\mathbf{Y}_{\text{bus}}$ are:

- The diagonal element Y_{jj} equals the sum of the admittances *directly* connected to node ⓙ.

- The off-diagonal element Y_{ij} equals the *negative* of the net admittance connected between nodes ⓘ and ⓙ.

The diagonal admittances are called the *self-admittances* at the nodes, and the off-diagonal admittances are the *mutual admittances* of the nodes. Some authors call the self- and mutual admittances of the nodes the *driving-point* and *transfer admittances* of the nodes. From the above rules $\mathbf{Y}_{\text{bus}}$ for the circuit of Fig. 1.23 becomes

$$\mathbf{Y}_{\text{bus}} = \begin{array}{c} \\ ① \\ ② \\ ③ \\ ④ \end{array}\begin{array}{c} \begin{array}{cccc} ① & ② & ③ & ④ \end{array} \\ \begin{bmatrix} (Y_c + Y_d + Y_f) & -Y_d & -Y_c & -Y_f \\ -Y_d & (Y_b + Y_d + Y_e) & -Y_b & -Y_e \\ -Y_c & -Y_b & (Y_a + Y_b + Y_c) & 0 \\ -Y_f & -Y_e & 0 & (Y_e + Y_f + Y_g) \end{bmatrix} \end{array} \tag{1.62}$$

where the encircled numbers are node numbers which almost always correspond to the subscripts of the elements Y_{ij} of $\mathbf{Y}_{\text{bus}}$. Separating out the entries for any one of the admittances, say, Y_c, we obtain

$$\mathbf{Y}_{\text{bus}} = \begin{matrix} ① \\ ② \\ ③ \\ ④ \end{matrix} \overset{\begin{matrix} ① & & ② & & ③ & & ④ \end{matrix}}{\begin{bmatrix} (Y_d + Y_f) & -Y_d & 0 & -Y_f \\ -Y_d & (Y_b + Y_d + Y_e) & -Y_b & -Y_e \\ 0 & -Y_b & (Y_a + Y_b) & 0 \\ -Y_f & -Y_e & 0 & (Y_e + Y_f + Y_g) \end{bmatrix}}$$

$$+ \begin{matrix} ① \\ ② \\ ③ \\ ④ \end{matrix} \overset{\begin{matrix} ① & ② & ③ & ④ \end{matrix}}{\begin{bmatrix} Y_c & 0 & -Y_c & 0 \\ 0 & 0 & 0 & 0 \\ -Y_c & 0 & Y_c & 0 \\ 0 & 0 & 0 & 0 \end{bmatrix}} \tag{1.63}$$

The matrix for Y_c can be written as shown in Eq. (1.63) or more compactly as follows:

$$\begin{matrix} ① \\ ② \\ ③ \\ ④ \end{matrix} \overset{\begin{matrix} ① & ② & ③ & ④ \end{matrix}}{\begin{bmatrix} Y_c & \cdot & -Y_c & \cdot \\ \cdot & \cdot & \cdot & \cdot \\ -Y_c & \cdot & Y_c & \cdot \\ \cdot & \cdot & \cdot & \cdot \end{bmatrix}} \Leftrightarrow \begin{matrix} ① \\ ③ \end{matrix} \overset{\begin{matrix} ① & ③ \end{matrix}}{\begin{bmatrix} 1 & -1 \\ -1 & 1 \end{bmatrix}} Y_c \tag{1.64}$$

While the left-hand side shows the actual matrix contributed by Y_c to $\mathbf{Y}_{\text{bus}}$, we can interpret the smaller matrix on the right as a compact storage matrix for the same contribution. The encircled numbers ① and ③ point to the rows and columns of $\mathbf{Y}_{\text{bus}}$ to which the entries Y_c and $-Y_c$ belong. The 2×2 matrix multiplying Y_c is an important *building block* in forming $\mathbf{Y}_{\text{bus}}$ for more general networks, which we consider in Chap. 7.

Inverting $\mathbf{Y}_{\text{bus}}$ yields an important matrix called the *bus impedance matrix* $\mathbf{Z}_{\text{bus}}$, which has the standard form

$$\mathbf{Z}_{\text{bus}} = \mathbf{Y}_{\text{bus}}^{-1} = \begin{matrix} ① \\ ② \\ ③ \\ ④ \end{matrix} \overset{\begin{matrix} ① & ② & ③ & ④ \end{matrix}}{\begin{bmatrix} Z_{11} & Z_{12} & Z_{13} & Z_{14} \\ Z_{21} & Z_{22} & Z_{23} & Z_{24} \\ Z_{31} & Z_{32} & Z_{33} & Z_{34} \\ Z_{41} & Z_{42} & Z_{43} & Z_{44} \end{bmatrix}} \tag{1.65}$$

The construction and properties of $\mathbf{Z}_{\text{bus}}$ are considered in Chap. 8.

1.13 THE SINGLE-LINE OR ONE-LINE DIAGRAM

In Chaps. 2 through 6 we develop the circuit models for transformers, synchronous machines, and transmission lines. Our present interest is in how to portray the assemblage of these components to model a complete system. Since a balanced three-phase system is always solved as a single-phase or per-phase equivalent circuit composed of one of the three lines and a neutral return, it is seldom necessary to show more than one phase and the neutral return when drawing a diagram of the circuit. Often the diagram is simplified further by omitting the completed circuit through the neutral and by indicating the component parts by standard symbols rather than by their equivalent circuits. Circuit parameters are not shown, and a transmission line is represented by a single line between its two ends. Such a simplified diagram of an electric system is called a *single-line* or *one-line diagram*. It indicates by a single line and standard symbols how the transmission lines and associated apparatus of an electric system are connected together.

The purpose of the one-line diagram is to supply in concise form the significant information about the system. The importance of different features of a system varies with the problem under consideration, and the amount of information included on the diagram depends on the purpose for which the diagram is intended. For instance, the location of circuit breakers and relays is unimportant in making a load study. Breakers and relays are not shown if the primary function of the diagram is to provide information for such a study. On the other hand, determination of the stability of a system under transient conditions resulting from a fault depends on the speed with which relays and circuit breakers operate to isolate the faulted part of the system. Therefore, information about the circuit breakers may be of extreme importance. Sometimes one-line diagrams include information about the current and potential transformers which connect the relays to the system or which are installed for metering. The information found on a one-line diagram must be expected to vary according to the problem at hand and according to the practice of the particular company preparing the diagram.

The American National Standards Institute (ANSI) and the Institute of Electrical and Electronics Engineers (IEEE) have published a set of standard symbols for electrical diagrams.[1] Not all authors follow these symbols consistently, especially in indicating transformers. Figure 1.24 shows a few symbols which are commonly used. The basic symbol for a machine or rotating armature is a circle, but so many adaptations of the basic symbol are listed that every piece of rotating electric machinery in common use can be indicated. For anyone who is not working constantly with one-line diagrams, it is clearer to

---

[1]See Graphic Symbols for Electrical and Electronics Diagrams, IEEE Std 315-1975.

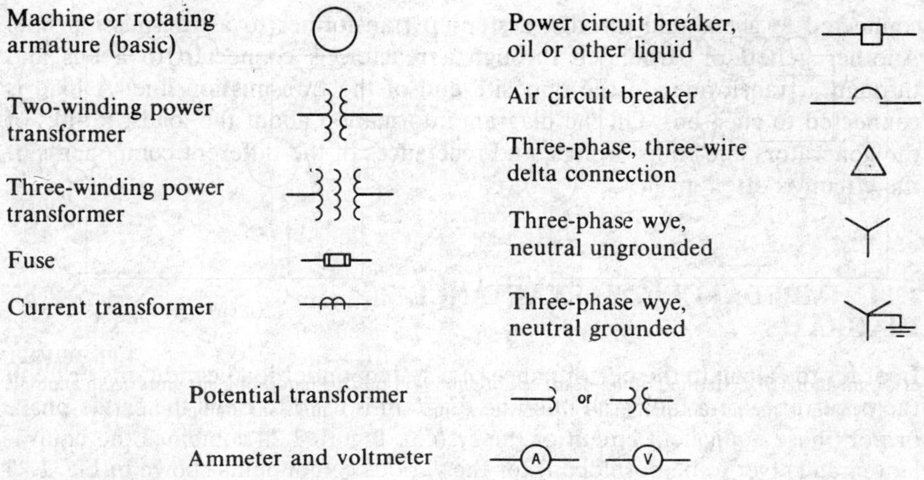

FIGURE 1.24
Apparatus symbols.

indicate a particular machine by the basic symbol followed by information on its type and rating.

It is important to know the location of points where a system is connected to ground in order to calculate the amount of current flowing when an unsymmetrical fault involving ground occurs. The standard symbol to designate a three-phase Y with the neutral solidly grounded is shown in Fig. 1.24. If a resistor or reactor is inserted between the neutral of the Y and ground to limit the flow of current to ground during a fault, the appropriate symbol for resistance or inductance may be added to the standard symbol for the grounded Y. Most transformer neutrals in transmission systems are solidly grounded. Generator neutrals are usually grounded through fairly high resistances and sometimes through inductance coils.

Figure 1.25 is the single-line diagram of a simple power system. Two generators, one grounded through a reactor and one through a resistor, are

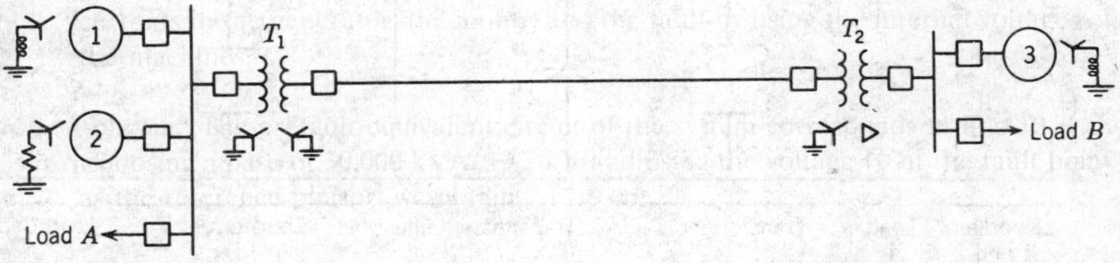

FIGURE 1.25
Single-line diagram of an electrical power system.

connected to a bus and through a step-up transformer to a transmission line. Another generator, grounded through a reactor, is connected to a bus and through a transformer to the opposite end of the transmission line. A load is connected to each bus. On the diagram information about the loads, ratings of the generators and transformers, and reactances of the different components of the circuit is often given.

1.14 IMPEDANCE AND REACTANCE DIAGRAMS

In order to calculate the performance of a system under load conditions or upon the occurrence of a fault, the one-line diagram is used to draw the single-phase or per-phase equivalent circuit of the system. Figure 1.26 combines the equivalent circuits (yet to be developed) for the various components shown in Fig. 1.25 to form the *per-phase impedance diagram* of the system. If a load study is to be made, the lagging loads A and B are represented by resistance and inductive reactance in series. The impedance diagram does not include the current-limiting impedances shown in the one-line diagram between the neutrals of the generators and ground because no current flows in the ground under balanced conditions and the neutrals of the generators are at the potential of the neutral of the system. Since the shunt current of a transformer is usually insignificant compared with the full-load current, the shunt admittance is usually omitted in the equivalent circuit of the transformer.

Resistance is often omitted when making fault calculations, even in computer programs. Of course, omission of resistance introduces some error, but the results may be satisfactory since the inductive reactance of a system is much larger than its resistance. Resistance and inductive reactance do not add directly, and impedance is not far different from the inductive reactance if the resistance is small. Loads which do not involve rotating machinery have little effect on the total line current during a fault and are usually omitted. Synchronous motor loads, however, are always included in making fault calculations

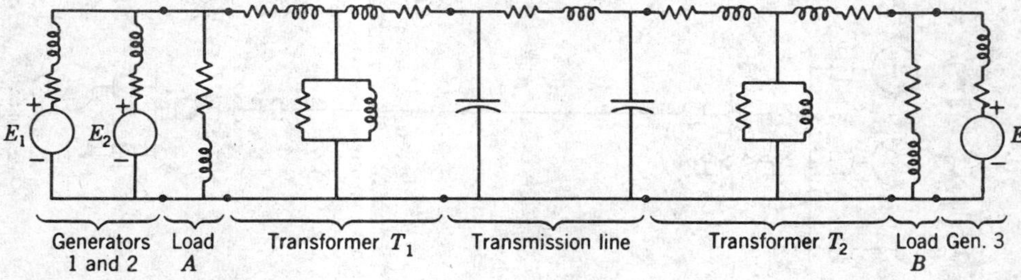

FIGURE 1.26
The per-phase impedance diagram corresponding to the single-line diagram of Fig. 1.25.

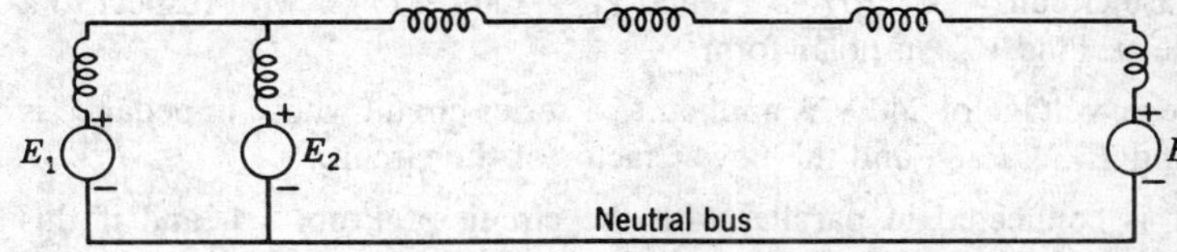

FIGURE 1.27
Per-phase reactance diagram adapted from Fig. 1.26 by omitting all loads, resistances, and shunt admittances.

since their generated emfs contribute to the short-circuit current. The diagram should take induction motors into account by a generated emf in series with an inductive reactance if the diagram is to be used to determine the current immediately after the occurrence of a fault. Induction motors are ignored in computing the current a few cycles after the fault occurs because the current contributed by an induction motor dies out very quickly after the induction motor is short-circuited.

If we decide to simplify our calculation of fault current by omitting all static loads, all resistances, the shunt admittance of each transformer, and the capacitance of the transmission line, the impedance diagram reduces to the per-phase reactance diagram of Fig. 1.27. These simplifications apply to fault calculations only as discussed in Chap. 10 and not to power-flow studies, which are the subject of Chap. 9. If a computer is available, such simplification is not necessary.

The per-phase impedance and reactance diagrams discussed here are sometimes called the *per-phase positive-sequence diagrams* since they show impedances to balanced currents in one phase of a symmetrical three-phase system. The significance of this designation is apparent in Chap. 11.

1.15 SUMMARY

This chapter reviews fundamentals of single-phase and balanced three-phase circuits and explains some of the notation to be used throughout the text. Per-unit calculations are introduced and the single-line diagram, along with its associated impedance diagram, is described. Formulation of node equations for circuits without mutual coupling is also demonstrated.

PROBLEMS

1.1. If $v = 141.4 \sin(\omega t + 30^\circ)$ V and $i = 11.31 \cos(\omega t - 30^\circ)$ A, find for each (*a*) the maximum value, (*b*) the rms value, and (*c*) the phasor expression in polar and rectangular form if voltage is the reference. Is the circuit inductive or capacitive?

1.2. If the circuit of Prob. 1.1 consists of a purely resistive and a purely reactive element, find R and X (*a*) if the elements are in series and (*b*) if the elements are in parallel.

1.3. In a single-phase circuit $V_a = 120\angle 45°$ V and $V_b = 100\angle -15°$ V with respect to a reference node o. Find V_{ba} in polar form.

1.4. A single-phase ac voltage of 240 V is applied to a series circuit whose impedance is $10\angle 60°\ \Omega$. Find R, X, P, Q, and the power factor of the circuit.

1.5. If a capacitor is connected in parallel with the circuit of Prob. 1.4, and if this capacitor supplies 1250 var, find the P and Q supplied by the 240-V source, and find the resultant power factor.

1.6. A single-phase inductive load draws 10 MW at 0.6 power-factor lagging. Draw the power triangle and determine the reactive power of a capacitor to be connected in parallel with the load to raise the power factor to 0.85.

1.7. A single-phase induction motor is operating at a very light load during a large part of every day and draws 10 A from the supply. A device is proposed to "increase the efficiency" of the motor. During a demonstration the device is placed in parallel with the unloaded motor and the current drawn from the supply drops to 8 A. When two of the devices are placed in parallel, the current drops to 6 A. What simple device will cause this drop in current? Discuss the advantages of the device. Is the efficiency of the motor increased by the device? (Recall that an induction motor draws lagging current.)

1.8. If the impedance between machines 1 and 2 of Example 1.1 is $Z = 0 - j5\ \Omega$, determine (*a*) whether each machine is generating or consuming power, (*b*) whether each machine is receiving or supplying positive reactor power and the amount, and (*c*) the value of P and Q absorbed by the impedance.

1.9. Repeat Problem 1.8 if $Z = 5 + j0\ \Omega$.

1.10. A voltage source $E_{an} = -120\angle 210°$ V and the current through the source is given by $I_{na} = 10\angle 60°$ A. Find the values of P and Q and state whether the source is delivering or receiving each.

1.11. Solve Example 1.1 if $E_1 = 100\angle 0°$ V and $E_2 = 120\angle 30°$ V. Compare the results with Example 1.1 and form some conclusions about the effect of variation of the magnitude of E_2 in this circuit.

1.12. Evaluate the following expressions in polar form:

(*a*) $a - 1$
(*b*) $1 - a^2 + a$
(*c*) $a^2 + a + j$
(*d*) $ja + a^2$

1.13. Three identical impedances of $10\angle -15°\ \Omega$ are Y-connected to balanced three-phase line voltages of 208 V. Specify all the line and phase voltages and the currents as phasors in polar form with V_{ca} as reference for a phase sequence of *abc*.

1.14. In a balanced three-phase system the Y-connected impedances are $10\angle 30°\ \Omega$. If $V_{bc} = 416\angle 90°$ V, specify I_{cn} in polar form.

1.15. The terminals of a three-phase supply are labeled a, b, and c. Between any pair a voltmeter measures 115 V. A resistor of 100 Ω and a capacitor of 100 Ω at the frequency of the supply are connected in series from a to b with the resistor connected to a. The point of connection of the elements to each other is labeled n.

Determine graphically the voltmeter reading between c and n if phase sequence is abc and if phase sequence is acb.

1.16. Determine the current drawn from a three-phase 440-V line by a three-phase 15-hp motor operating at full load, 90% efficiency, and 80% power-factor lagging. Find the values of P and Q drawn from the line.

1.17. If the impedance of each of the three lines connecting the motor of Prob. 1.16 to a bus is $0.3 + j1.0\ \Omega$, find the line-to-line voltage at the bus which supplies 440 V at the motor.

1.18. A balanced-Δ load consisting of pure resistances of 15 Ω per phase is in parallel with a balanced-Y load having phase impedances of $8 + j6\ \Omega$. Identical impedances of $2 + j5\ \Omega$ are in each of the three lines connecting the combined loads to a 110-V three-phase supply. Find the current drawn from the supply and line voltage at the combined loads.

1.19. A three-phase load draws 250 kW at a power factor of 0.707 lagging from a 440-V line. In parallel with this load is a three-phase capacitor bank which draws 60 kVA. Find the total current and resultant power factor.

1.20. A three-phase motor draws 20 kVA at 0.707 power-factor lagging from a 220-V source. Determine the kilovoltampere rating of capacitors to make the combined power factor 0.90 lagging, and determine the line current before and after the capacitors are added.

1.21. A coal mining "drag-line" machine in an open-pit mine consumes 0.92 MVA at 0.8 power-factor lagging when it digs coal, and it generates (delivers to the electric system) 0.10 MVA at 0.5 power-factor leading when the loaded shovel swings away from the pit wall. At the end of the "dig" period the change in supply current magnitude can cause tripping of a protective relay, which is constructed of solid-state circuitry. Therefore, it is desired to minimize the change in current magnitude. Consider the placement of capacitors at the machine terminals and find the amount of capacitive correction (in kvar) to eliminate the change in steady-state current magnitude. The machine is energized from a 36.5 kV, three-phase supply. Start the solution by letting Q be the total three-phase megavars of the capacitors connected across the machine terminals, and write an expression for the magnitude of the line current *drawn by* the machine in terms of Q for both the digging and generating operations.

1.22. A generator (which may be represented by an emf in series with an inductive reactance) is rated 500 MVA, 22 kV. Its Y-connected windings have a reactance of 1.1 per unit. Find the ohmic value of the reactance of the windings.

1.23. The generator of Prob. 1.22 is in a circuit for which the bases are specified as 100 MVA, 20 kV. Starting with the per-unit value given in Prob. 1.22, find the per-unit value of reactance of the generator windings on the specified base.

1.24. Draw the single-phase equivalent circuit for the motor (an emf in series with inductive reactance labeled Z_m) and its connection to the voltage supply described in Probs. 1.16 and 1.17. Show on the diagram the per-unit values of the line impedance and the voltage at the motor terminals on a base of 20 kVA, 440 V. Then using per-unit values, find the supply voltage in per unit and convert the per-unit value of the supply voltage to volts.

1.25. Write the two nodal admittance equations, similar to Eqs. (1.57) and (1.58), for the voltages at nodes ② and ④ of the circuit of Fig. 1.23. Then, arrange the nodal admittance equations for all four independent nodes of Fig. 1.23 into the $\mathbf{Y}_{\text{bus}}$ form of Eq. (1.61).

1.26. The values for the parameters of Fig. 1.23 are given in per unit as follows:

$$Y_a = -j0.8 \qquad Y_b = -j4.0 \qquad Y_c = -j4.0 \qquad Y_d = -j8.0 \qquad Y_e = -j5.0$$

$$Y_f = -j2.5 \qquad Y_g = -j0.8 \qquad I_3 = 1.0\angle -90^\circ \quad I_4 = 0.68\angle -135^\circ$$

Substituting these values in the equations determined in Prob. 1.25, compute the voltages at the nodes of Fig. 1.23. Numerically determine the corresponding $\mathbf{Z}_{\text{bus}}$ matrix.

CHAPTER 2

TRANSFORMERS

Transformers are the link between the generators of the power system and the transmission lines, and between lines of different voltage levels. Transmission lines operate at nominal voltages up to 765 kV line to line. Generators are usually built in the range of 18–24 kV with some at slightly higher rated voltages. Transformers also lower the voltages to distribution levels and finally for residential use at 240/120 V. They are highly (nearly 100%) efficient and very reliable.

In this chapter we discuss the modeling of transformers and see the great advantages of per-unit calculations. We also consider transformers that regulate voltage magnitude and phase shifting, and in this and a later chapter we shall see how these regulating transformers are used to control the flow of real and reactive power.

Figure 2.1 is the photograph of a three-phase transformer which raises the voltage of a generator to the transmission-line voltage. The transformer is rated 750 MVA, 525/22.8 kV.

2.1 THE IDEAL TRANSFORMER

Transformers consist of two or more coils placed so that they are linked by the same magnetic flux. In a power transformer the coils are placed on an iron core

FIGURE 2.1
Photograph of a three-phase transformer rated 750 MVA, 525/22.8 kV. (*Courtesy Duke Power Company.*)

in order to confine the flux so that almost all of the flux linking any one coil links all the others. Several coils may be connected in series or parallel to form one winding, the coils of which may be stacked on the core alternately with those of the other winding or windings.

Figure 2.2 shows how two windings may be placed on an iron core to form a single-phase transformer of the so-called *shell* type. The number of turns in a winding may range from several hundreds up to several thousands.

We begin our analysis by assuming that the flux varies sinusoidally in the core and that the transformer is *ideal*, which means that (1) the permeability μ of the core is infinite, (2) all of the flux is confined to the core and therefore links all of the turns of both windings, and (3) core losses and winding resistances are zero. Thus, the voltages e_1 and e_2 induced by the changing flux must equal the terminal voltages v_1 and v_2, respectively.

We can see from the relationship of the windings shown in Fig. 2.2 that instantaneous voltages e_1 and e_2 induced by the changing flux are in phase

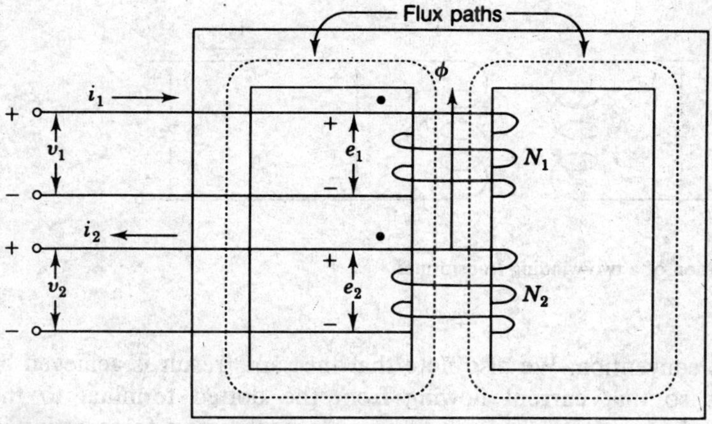

FIGURE 2.2
Two-winding transformer.

when defined by the + and − polarity marks indicated. Then, by Faraday's law

$$v_1 = e_1 = N_1 \frac{d\phi}{dt} \tag{2.1}$$

and

$$v_2 = e_2 = N_2 \frac{d\phi}{dt} \tag{2.2}$$

where ϕ is the instantaneous value of the flux and N_1 and N_2 are the number of turns on windings 1 and 2, as shown in Fig. 2.2. The flux ϕ is taken in the positive direction of coil 1 according to the *right-hand rule*, which states that if a coil is grasped in the right hand with fingers curled in the direction of current flow, the thumb extends in the direction of the flux. Since we have assumed sinusoidal variation of the flux, we can convert the voltages to phasor form after dividing Eq. (2.1) by Eq. (2.2) to yield

$$\frac{V_1}{V_2} = \frac{E_1}{E_2} = \frac{N_1}{N_2} \tag{2.3}$$

Usually, we do not know the direction in which the coils of a transformer are wound. One device to provide winding information is to place a dot at the end of each winding such that all dotted ends of windings are positive at the same time; that is, *voltage drops from dotted to unmarked terminals of all windings are in phase*. Dots are shown on the two-winding transformer in Fig. 2.2

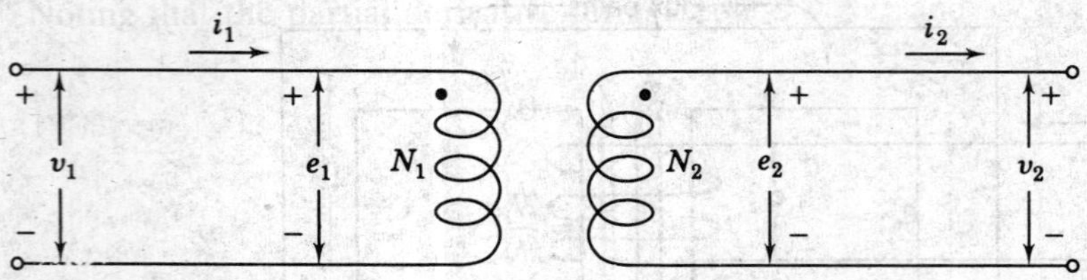

FIGURE 2.3
Schematic representation of a two-winding transformer.

according to this convention. We also note that the same result is achieved by placing the dots so that current flowing from the dotted terminal to the unmarked terminal of each winding produces a magnetomotive force acting in the same direction in the magnetic circuit. Figure 2.3 is a schematic representation of a transformer and provides the same information about the transformer as that in Fig. 2.2.

To find the relation between the currents i_1 and i_2 in the windings, we apply Ampere's law, which states that the magnetomotive force (mmf) around a closed path is given by the *line integral*

$$\oint H \cdot ds = i \tag{2.4}$$

where i = *net* current that passes through the area bounded by the closed path

H = magnetic field intensity

$H \cdot ds$ = product of the tangential component of H and the incremental distance ds along the path

In applying the law around each of the closed paths of flux shown by dotted lines in Fig. 2.2, i_1 is enclosed N_1 times and the current i_2 is enclosed N_2 times. However, $N_1 i_1$ and $N_2 i_2$ produce mmfs in opposite directions, and so

$$\oint H \cdot ds = N_1 i_1 - N_2 i_2 \tag{2.5}$$

The minus sign would change to plus if we had chosen the opposite direction for the current i_2. The integral of the field intensity H around the closed path is zero when permeability is infinite. If this were not true, flux density (being equal to μH) would be infinite. Flux density must have a finite value so that a finite e is induced in each winding by the varying flux. So, upon converting the currents to phasor form, we have

$$N_1 I_1 - N_2 I_2 = 0 \tag{2.6}$$

$$\frac{I_1}{I_2} = \frac{N_2}{N_1} \tag{2.7}$$

and I_1 and I_2 are therefore in phase. *Note then that I_1 and I_2 are in phase if we choose the current to be positive when entering the dotted terminal of one winding and leaving the dotted terminal of the other*. If the direction chosen for either current is reversed, they are 180° out of phase.

From Eq. (2.7)

$$I_1 = \frac{N_2}{N_1} I_2 \tag{2.8}$$

and in the ideal transformer I_1 must be zero if I_2 is zero.

The winding across which an impedance or other load may be connected is called the *secondary* winding, and any circuit elements connected to this winding are said to be on the secondary side of the transformer. Similarly, the winding which is toward the source of energy is called the *primary* winding on the primary side. In the power system energy often will flow in either direction through a transformer and the designation of primary and secondary loses its meaning. These terms are in general use, however, and we shall use them wherever they do not cause confusion.

If an impedance Z_2 is connected across winding 2 of Figs. 2.2 or 2.3,

$$Z_2 = \frac{V_2}{I_2} \tag{2.9}$$

and substituting for V_2 and I_2 the values found from Eqs. (2.3) and (2.7) gives

$$Z_2 = \frac{(N_2/N_1)V_1}{(N_1/N_2)I_1} \tag{2.10}$$

The impedance as measured across the primary winding is then

$$Z_2' = \frac{V_1}{I_1} = \left(\frac{N_1}{N_2}\right)^2 Z_2 \tag{2.11}$$

Thus, the impedance connected to the secondary side is *referred* to the primary side by multiplying the impedance on the secondary side of the transformer by the square of the ratio of primary to secondary voltage.

We should note also that $V_1 I_1^*$ and $V_2 I_2^*$ are equal, as shown by the following equation, which again makes use of Eqs. (2.3) and (2.7):

$$V_1 I_1^* = \frac{N_1}{N_2} V_2 \times \frac{N_2}{N_1} I_2^* = V_2 I_2^* \tag{2.12}$$

So,

$$S_1 = S_2 \tag{2.13}$$

which means that the complex power input to the primary winding equals the complex power output from the secondary winding since we are considering an ideal transformer.

Example 2.1. If $N_1 = 2000$ and $N_2 = 500$ in the circuit of Fig. 2.3, and if $V_1 = 1200\angle 0°$ V and $I_1 = 5\angle -30°$ A with an impedance Z_2 connected across winding 2, find V_2, I_2, Z_2, and the impedance Z_2', which is defined as the value of Z_2 referred to the primary side of the transformer.

Solution

$$V_2 = \frac{N_2}{N_1} V_1 = \frac{500}{2000}(1200\angle 0°) = 300\angle 0° \text{ V}$$

$$I_2 = \frac{N_1}{N_2} I_1 = \frac{2000}{500}(5\angle -30°) = 20\angle -30° \text{ A}$$

$$Z_2 = \frac{V_2}{I_2} = \frac{300\angle 0°}{20\angle -30°} = 15\angle 30° \ \Omega$$

$$Z_2' = Z_2\left(\frac{N_1}{N_2}\right)^2 = (15\angle 30°)\left(\frac{2000}{500}\right)^2 = 240\angle 30° \ \Omega$$

Alternatively,

$$Z_2' = \frac{V_1}{I_1} = \frac{1200\angle 0°}{5\angle -30°} = 240\angle 30° \ \Omega$$

2.2 MAGNETICALLY COUPLED COILS

The ideal transformer is a first step in studying a practical transformer, where (1) permeability is not infinite and inductances are therefore finite, (2) not all the flux linking any one of the windings links the other windings, (3) winding resistance is present, and (4) losses occur in the iron core due to the cyclic changing of direction of the flux. As a second step, let us consider the two coils of Fig. 2.4 which represent the windings of a transformer of the *core* type of

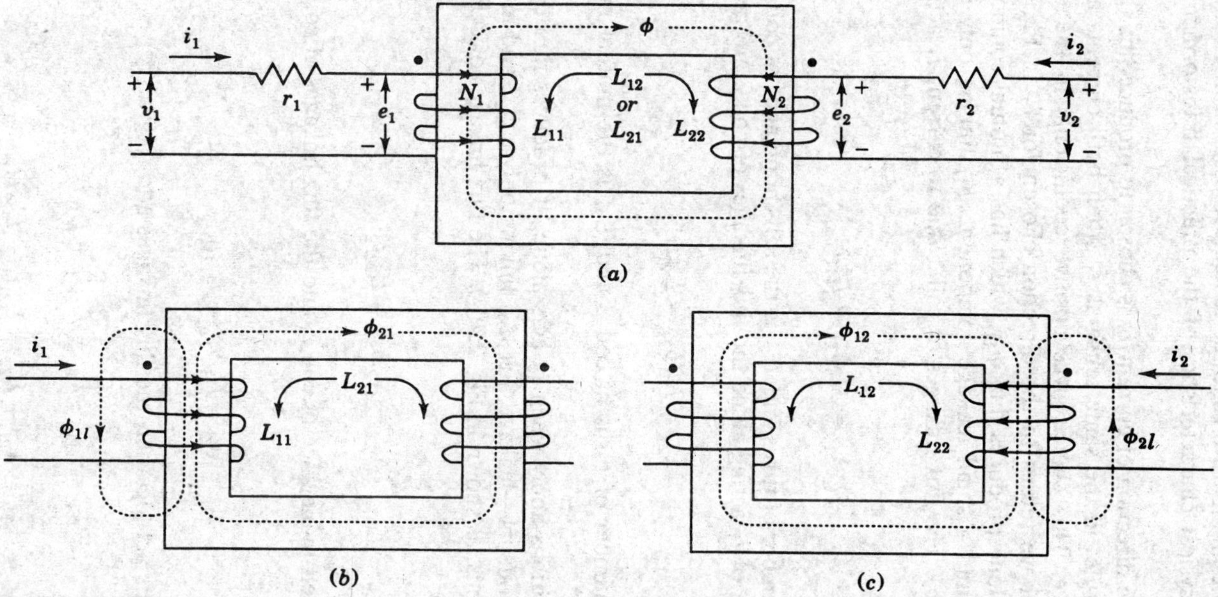

Figure 2.4
Mutually coupled coils with: (*a*) mutual flux due to currents i_1 and i_2; (*b*) leakage flux ϕ_{1l} and mutual flux ϕ_{21} due to i_1 alone; (*c*) leakage flux ϕ_{2l} and mutual flux ϕ_{12} due to i_2 alone.

construction. For the moment we continue to neglect losses in the iron core, but the other three physical characteristics of the practical transformer are now considered.

In Fig. 2.4 the direction of current i_2 is chosen to produce flux (according to the right-hand rule) in the same sense as i_1 when both currents are either positive or negative. This choice gives positive coefficients in the equations which follow. Later we return to the direction chosen for i_2 in Fig. 2.2. The current i_1 acting alone produces flux ϕ_{11}, which has a mutual component ϕ_{21} linking both coils and a small leakage component ϕ_{1l} linking only coil 1, as shown in Fig. 2.4(*b*). The flux linkages of coil 1 due to current i_1 acting alone are given by

$$\lambda_{11} = N_1\phi_{11} = L_{11}i_1 \tag{2.14}$$

where N_1 is the number of turns and L_{11} is the self-inductance of coil 1. Under the same condition of i_1 acting alone the flux linkages of coil 2 are given by

$$\lambda_{21} = N_2\phi_{21} = L_{21}i_1 \tag{2.15}$$

where N_2 is the number of turns of coil 2 and L_{21} is the mutual inductance between the coils.

Similar definitions apply when i_2 acts alone. It produces flux ϕ_{22}, which also has two components—leakage flux ϕ_{2l} linking only coil 2 and mutual flux ϕ_{12} linking both coils, as shown in Fig. 2.4(*c*). The flux linkages of coil 2 due to i_2 acting alone are

$$\lambda_{22} = N_2\phi_{22} = L_{22}i_2 \tag{2.16}$$

where L_{22} is the self-inductance of coil 2, and the flux linkages of coil 1 due to i_2 alone are

$$\lambda_{12} = N_1\phi_{12} = L_{12}i_2 \tag{2.17}$$

When both currents act together, the flux linkages add to give

$$\begin{aligned} \lambda_1 &= \lambda_{11} + \lambda_{12} = L_{11}i_1 + L_{12}i_2 \\ \lambda_2 &= \lambda_{21} + \lambda_{22} = L_{21}i_1 + L_{22}i_2 \end{aligned} \tag{2.18}$$

The order ot the subscripts of L_{12} and L_{21} is not important since mutual inductance is a single reciprocal property of the coils, and so $L_{12} = L_{21}$. The direction of the currents and the orientation of the coils determine the sign of mutual inductance, which is positive in Fig. 2.4 because i_1 and i_2 are taken to magnetize in the same sense.

When the flux linkages change with time, the voltage drops across the coils in the direction of their circulating currents are

$$v_1 = r_1 i_1 + \frac{d\lambda_1}{dt} = r_1 i_1 + L_{11}\frac{di_1}{dt} + L_{12}\frac{di_2}{dt} \tag{2.19}$$

$$v_2 = r_2 i_2 + \frac{d\lambda_2}{dt} = r_2 i_2 + L_{21}\frac{di_1}{dt} + L_{22}\frac{di_2}{dt} \tag{2.20}$$

The positive signs of Eqs. (2.19) and (2.20) are usually associated with a coil that is absorbing power from a source as if the coil were a *load*. For instance, in Fig. 2.4 if both v_2 and i_2 have positive values simultaneously, then instantaneous power is being *absorbed* by coil 2. If the voltage drop across coil 2 is now reversed so that $v_2' = -v_2$, we have

$$v_2' = -v_2 = -r_2 i_2 - \frac{d\lambda_2}{dt} = -r_2 i_2 - L_{21}\frac{di_1}{dt} - L_{22}\frac{di_2}{dt} \tag{2.21}$$

For positive instantaneous values of v_2' and i_2 power is being *supplied* by coil 2. Thus, the negative signs of Eq. (2.21) are characteristic of a coil acting as a *generator* delivering power (and energy over time) to an external load.

In the steady state, with ac voltages and currents in the coils, Eqs. (2.19) and (2.20) assume the phasor form

$$V_1 = \underbrace{(r_1 + j\omega L_{11})}_{z_{11}} I_1 + \underbrace{(j\omega L_{12})}_{z_{12}} I_2 \tag{2.22}$$

$$V_2 = \underbrace{(j\omega L_{21})}_{z_{21}} I_1 + \underbrace{(r_2 + j\omega L_{22})}_{z_{22}} I_2 \tag{2.23}$$

Here we use lowercase z_{ij} to distinguish the coil impedances from node impedances Z_{ij}. In vector-matrix form Eqs. (2.22) and (2.23) become

$$\begin{bmatrix} V_1 \\ V_2 \end{bmatrix} = \begin{bmatrix} z_{11} & z_{12} \\ z_{21} & z_{22} \end{bmatrix} \begin{bmatrix} I_1 \\ I_2 \end{bmatrix} \tag{2.24}$$

We should also note that the V's are the voltage drops across the terminals of the coils and the I's are the circulating currents in the coils. The inverse of the coefficient matrix is the matrix of admittances denoted by

$$\begin{bmatrix} y_{11} & y_{12} \\ y_{21} & y_{22} \end{bmatrix} = \begin{bmatrix} z_{11} & z_{12} \\ z_{21} & z_{22} \end{bmatrix}^{-1} = \frac{1}{(z_{11}z_{22} - z_{12}^2)} \begin{bmatrix} z_{22} & -z_{12} \\ -z_{21} & z_{11} \end{bmatrix} \tag{2.25}$$

Multiplying Eq. (2.24) by the admittance matrix gives

$$\begin{bmatrix} I_1 \\ I_2 \end{bmatrix} = \begin{bmatrix} y_{11} & y_{12} \\ y_{21} & y_{22} \end{bmatrix} \begin{bmatrix} V_1 \\ V_2 \end{bmatrix} \tag{2.26}$$

Of course, the y and z parameters with the same subscripts are not simple reciprocals of each other. If the terminals of coil 2 are open, then setting $I_2 = 0$ in Eq. (2.24) shows that the *open-circuit* input impedance to coil 1 is

$$\left.\frac{V_1}{I_1}\right|_{I_2=0} = z_{11} \tag{2.27}$$

If the terminals of coil 2 are closed, then $V_2 = 0$ and Eq. (2.26) shows that the *short-circuit* input impedance to coil 1 is

$$\left.\frac{V_1}{I_1}\right|_{V_2=0} = y_{11}^{-1} = z_{11} - \frac{z_{12}^2}{z_{22}} \tag{2.28}$$

By substituting the expressions defining the z_{ij} from Eqs. (2.22) and (2.23) into Eq. (2.28), the reader can show that the *apparent* reactance of coil 1 is reduced by the presence of closed coil 2. In Chap. 3 a similar result is found for the synchronous machine under short-circuit conditions.

An important equivalent circuit for the mutually coupled coils is shown in Fig. 2.5. The current on the coil 2 side appears as I_2/a and the terminal voltage as aV_2, where a is a positive constant. On the coil 1 side V_1 and I_1 are the same as before. By writing Kirchhoff's voltage equation around the path of each of the currents I_1 and I_2/a in Fig. 2.5, the reader should find that Eqs. (2.22) and (2.23) are satisfied exactly. The inductances in brackets in Fig. 2.5 are the *leakage inductances* L_{1l} and L_{2l} of the coils if we let $a = N_1/N_2$. This is shown

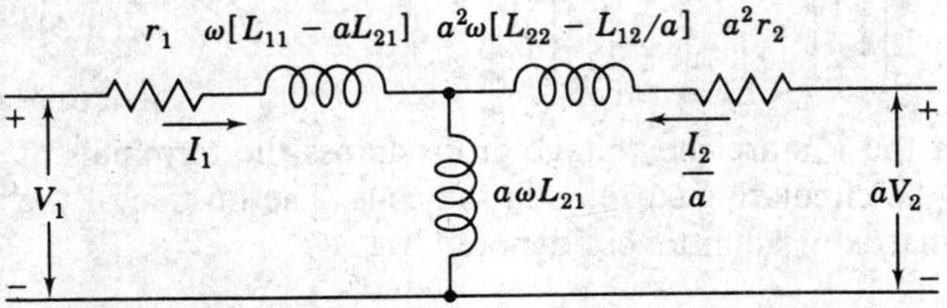

FIGURE 2.5
An ac equivalent circuit for Fig. 2.4 with secondary current and voltage redefined and $a = N_1/N_2$.

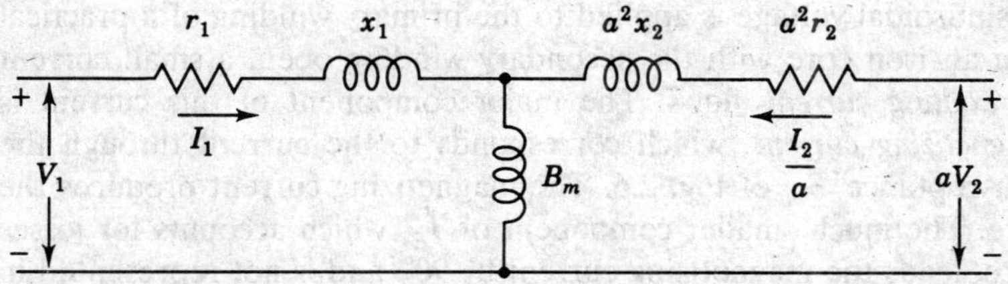

FIGURE 2.6
The equivalent circuit of Fig. 2.5 with inductance parameters renamed.

from Eqs. (2.14) through (2.17) as follows:

$$L_{1l} \triangleq L_{11} - aL_{21} = \frac{N_1\phi_{11}}{i_1} - \frac{N_1}{N_2}\frac{N_2\phi_{21}}{i_1} = \frac{N_1}{i_1}\underbrace{(\phi_{11} - \phi_{21})}_{\phi_{1l}} \tag{2.29}$$

$$L_{2l} \triangleq L_{22} - L_{12}/a = \frac{N_2\phi_{22}}{i_2} - \frac{N_2}{N_1}\frac{N_1\phi_{12}}{i_2} = \frac{N_2}{i_2}\underbrace{(\phi_{22} - \phi_{12})}_{\phi_{2l}} \tag{2.30}$$

where ϕ_{1l} and ϕ_{2l} are the leakage fluxes of the coils. Likewise, with $a = N_1/N_2$, the shunt inductance aL_{21} is a *magnetizing inductance* associated with the mutual flux ϕ_{21} linking the coils due to i_1 since

$$aL_{21} = \frac{N_1}{N_2}\frac{N_2\phi_{21}}{i_1} = \frac{N_1}{i_1}\phi_{21} \tag{2.31}$$

Defining the series *leakage reactances* $x_1 = \omega L_{1l}$ and $x_2 = \omega L_{2l}$, and the shunt *magnetizing susceptance* $B_m = (\omega a L_{21})^{-1}$, leads to the equivalent circuit of Fig. 2.6, which is the basis of the equivalent circuit of the practical transformer in Sec. 2.3.

2.3 THE EQUIVALENT CIRCUIT OF A SINGLE-PHASE TRANSFORMER

The equivalent circuit of Fig. 2.6 comes close to matching the physical characteristics of the practical transformer. However, it has three deficiencies: (1) It does not reflect any current or voltage transformation, (2) it does not provide for electrical isolation of the primary from the secondary, and (3) it does not account for the core losses.

When a sinusoidal voltage is applied to the primary winding of a practical transformer on an iron core with the secondary winding open, a small current I_E called the *exciting current* flows. The major component of this current is called the *magnetizing current*, which corresponds to the current through the magnetizing susceptance B_m of Fig. 2.6. The magnetizing current produces the flux in the core. The much smaller component of I_E, which accounts for losses in the iron core, leads the magnetizing current by 90° and is not represented in Fig. 2.6. The core losses occur due, first, to the fact that the cyclic changes of the direction of the flux in the iron require energy which is dissipated as heat and is called *hysteresis loss*. The second loss is due to the fact that circulating currents are induced in the iron due to the changing flux, and these currents produce an $|I|^2R$ loss in the iron called *eddy-current loss*. Hysteresis loss is reduced by the use of certain high grades of alloy steel for the core. Eddy-current loss is reduced by building up the core with laminated sheets of steel. With the secondary open, the transformer primary circuit is simply one of very high inductance due to the iron core. In the equivalent circuit I_E is taken fully into account by a conductance G_c in parallel with the magnetizing susceptance B_m, as shown in Fig. 2.7.

In a well-designed transformer the maximum flux density in the core occurs at the knee of the *B-H* or *saturation* curve of the transformer. So, flux density is not linear with respect to field intensity. The magnetizing current cannot be sinusoidal if it is to produce sinusoidally varying flux required for inducing sinusoidal voltages e_1 and e_2 when the applied voltage is sinusoidal. The exciting current I_E will have a third harmonic content as high as 40% and lesser amounts of higher harmonics. Since I_E is small compared to rated current, it is treated as sinusoidal for convenience, and so use of G_c and B_m is acceptable in the equivalent circuit.

Voltage and current transformation and electrical isolation of the primary from the secondary can be obtained by adding to Fig. 2.6 an ideal transformer

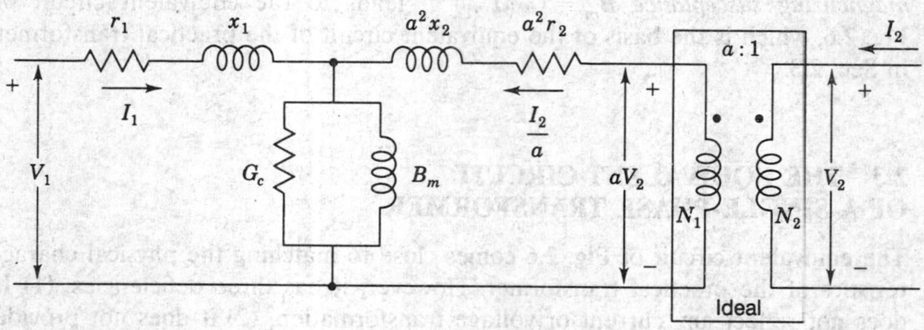

FIGURE 2.7
Equivalent circuit for a single-phase transformer with an ideal transformer of turns ratio $a = N_1/N_2$.

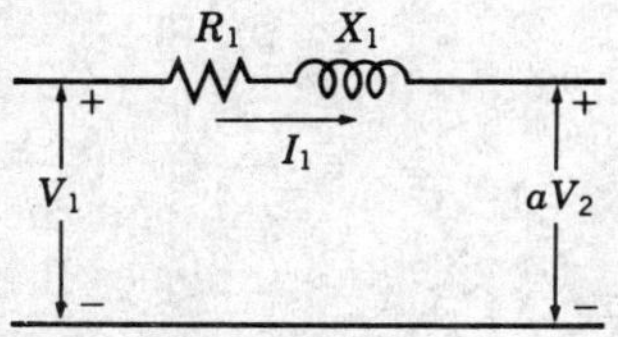

FIGURE 2.8
Transformer equivalent circuit with magnetizing current neglected.

with turns ratio $a = N_1/N_2$, as shown in Fig. 2.7. The location of the ideal transformer is not fixed. For instance, it may be moved to the left past the series elements a^2r_2 and a^2x_2, which then become the winding resistance r_2 and the leakage reactance x_2 of the secondary winding. This is in keeping with the rule established for the ideal transformer in Sec. 2.1 that whenever a branch impedance is referred from a given side to the opposite side of an ideal transformer, its impedance value is multiplied by the square of the ratio of the turns on the opposite side to the turns on the given side.

The ideal transformer may be omitted in the equivalent circuit if we refer all quantities to either the high- or the low-voltage side of the transformer. For instance, in Fig. 2.6 we say that all voltages, currents, and impedances are referred to the primary circuit of the transformer. Without the ideal transformer, we have to be careful not to create unnecessary short circuits when developing equivalents for multiwinding transformers.

Often we neglect exciting current because it is so small compared to the usual load currents and to simplify the circuit further, we let

$$R_1 = r_1 + a^2r_2 \qquad X_1 = x_1 + a^2x_2 \tag{2.32}$$

to obtain the equivalent circuit of Fig. 2.8. All impedances and voltages in the part of the circuit connected to the secondary terminals must now be referred to the primary side.

Voltage regulation is defined as the difference between the voltage magnitude at the load terminals of the transformer at full load and at no load in percent of full-load voltage with input voltage held constant. In the form of an equation

$$\text{Percent regulation} = \frac{|V_{2,\text{NL}}| - |V_{2,\text{FL}}|}{|V_{2,\text{FL}}|} \times 100 \tag{2.33}$$

where $|V_{2,\text{NL}}|$ is the magnitude of load voltage V_2 at no load and $|V_{2,\text{FL}}|$ is the magnitude of V_2 at full load with $|V_1|$ constant.

Example 2.2. A single-phase transformer has 2000 turns on the primary winding and 500 turns on the secondary. Winding resistances are $r_1 = 2.0\ \Omega$ and $r_2 = 0.125\ \Omega$. Leakage reactances are $x_1 = 8.0\ \Omega$ and $x_2 = 0.50\ \Omega$. The resistance load

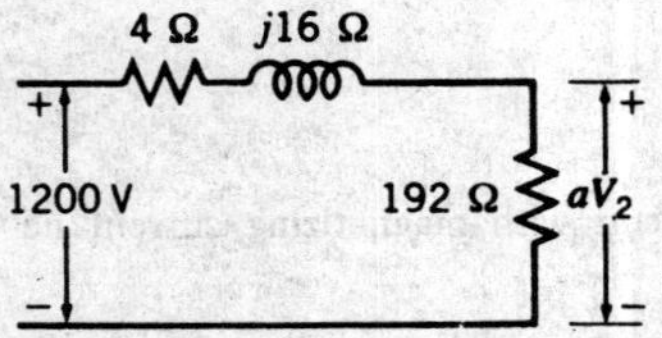

FIGURE 2.9
Circuit for Example 2.2.

Z_2 is 12 Ω. If applied voltage at the terminals of the primary winding is 1200 V, find V_2 and the voltage regulation. Neglect magnetizing current.

Solution

$$a = \frac{N_1}{N_2} = \frac{2000}{500} = 4$$

$$R_1 = 2 + 0.125(4)^2 = 4.0\ \Omega$$

$$X_1 = 8 + 0.5(4)^2 = 16\ \Omega$$

$$Z_2' = 12 \times (4)^2 = 192\ \Omega$$

The equivalent circuit is shown in Fig. 2.9, and we can calculate

$$I_1 = \frac{1200\angle 0°}{192 + 4 + j16} = 6.10\angle -4.67°\ \text{A}$$

$$aV_2 = 6.10\angle -4.67° \times 192 = 1171.6\angle -4.67°\ \text{V}$$

$$V_2 = \frac{1171.6\angle -4.67°}{4} = 292.9\angle -4.67°\ \text{V}$$

Since $V_{2,\text{NL}} = V_1/a$,

$$\text{Voltage regulation} = \frac{1200/4 - 292.9}{292.9} = 0.0242 \text{ or } 2.42\%$$

The parameters R and X of the two-winding transformer are determined by the *short-circuit test*, where impedance is measured across the terminals of one winding when the other winding is short-circuited. Usually, the low-voltage side is short-circuited and just enough voltage is applied to the high-voltage terminals to circulate rated current. This is because the current rating of the source supplying the high-voltage side can be smaller. Voltage, current, and power input are determined. Since only a small voltage is required, the exciting current is insignificant, and the calculated impedance is essentially equal to $R + jX$.

Example 2.3. A single-phase transformer is rated 15 MVA, 11.5/69 kV. If the 11.5 kV winding (designated winding 2) is short-circuited, the rated current flows when the voltage applied to winding 1 is 5.50 kV. The power input is 105.8 kW. Find R_1 and X_1 in ohms referred to the high-voltage winding.

Solution. Rated current for the 69-kV winding has the magnitude

$$\frac{|S_1|}{|V_1|} = |I_1| = \frac{15{,}000}{69} = 217.4 \text{ A}$$

Then,

$$|I_1|^2 R_1 = (217.4)^2 R_1 = 105{,}800$$

$$R_1 = 2.24\ \Omega$$

$$|Z_1| = \frac{5500}{217.4} = 25.30\ \Omega$$

$$X_1 = \sqrt{|Z_1|^2 - R_1^2} = \sqrt{(25.30)^2 - (2.24)^2} = 25.20\ \Omega$$

The example illustrates the fact that the winding resistance may often be omitted in the transformer equivalent circuit. Typically, R is less than 1%. Although exciting current may be neglected (as in Example 2.2) for most power system calculations, $G_c - jB_m$ can be calculated for the equivalent circuit by an *open-circuit test*. Rated voltage is applied to the low-voltage terminals, and the power input and currents are measured. This is because the voltage rating of the source supplying the low-voltage side can be smaller. The measured impedance includes the resistance and leakage reactance of the winding, but these values are insignificant when compared to $1/(G_c - jB_m)$.

Example 2.4. For the transformer of Example 2.3 the open-circuit test with 11.5 kV applied results in a power input of 66.7 kW and a current of 30.4 A. Find the values of G_c and B_m referred to the high-voltage winding 1. What is the efficiency of the transformer for a load of 12 MW at 0.8 power-factor lagging at rated voltage?

Solution. The turns ratio is $a = N_1/N_2 = 6$. Measurements are made on the low-voltage side. To transfer shunt admittance $Y = G_c - jB_m$ from high-voltage side 1 to low-voltage side 2, multiply by a^2 since we would divide by a^2 to transfer impedance from side 1 to side 2. Under open-circuit test conditions

$$|V_2|^2 a^2 G_c = (11.5 \times 10^3)^2 \times 36 \times G_c = 66.7 \times 10^3 \text{ W}$$

$$G_c = 14.0 \times 10^{-6} \text{ S}$$

$$|Y| = \frac{|I_2|}{|V_2|} \times \frac{1}{a^2} = \frac{30.4}{11{,}500} \times \frac{1}{36} = 73.4 \times 10^{-6} \text{ S}$$

$$B_m = \sqrt{|Y|^2 - G_c^2} = 10^{-6}\sqrt{73.4^2 - 14.0^2} = 72.05 \times 10^{-6} \text{ S}$$

Under *rated* conditions the total loss is approximately the sum of short-circuit and open-circuit test losses, and since efficiency is the ratio of the output to the input kilowatts, we have

$$\text{Efficiency} = \frac{12{,}000}{12{,}000 + (105.8 + 66.7)} \times 100 = 98.6\%$$

This example illustrates the fact that G_c is so much smaller than B_m that it may be omitted. B_m is also very small so that I_E is often neglected entirely.

2.4 PER-UNIT IMPEDANCES IN SINGLE-PHASE TRANSFORMER CIRCUITS

The ohmic values of resistance and leakage reactance of a transformer depend on whether they are measured on the high- or low-voltage side of the transformer. If they are expressed in per unit, the base kilovoltamperes is understood to be the kilovoltampere rating of the transformer. The base voltage is understood to be the voltage rating of the low-voltage winding if the ohmic values of resistance and leakage reactance are referred to the low-voltage side of the transformer. Likewise, the base voltage is taken to be the voltage rating of the high-voltage winding if the ohmic values are referred to the high-voltage side of the transformer. The per-unit impedance of a transformer is the same regardless of whether it is determined from ohmic values referred to the high-voltage or low-voltage sides of the transformers, as shown by the following example.

Example 2.5. A single-phase transformer is rated 110/440 V, 2.5 kVA. Leakage reactance measured from the low-voltage side is 0.06 Ω. Determine leakage reactance in per unit.

Solution. From Eq. (1.46) we have

$$\text{Low-voltage base impedance} = \frac{0.110^2 \times 1000}{2.5} = 4.84\ \Omega$$

In per unit

$$X = \frac{0.06}{4.84} = 0.0124 \text{ per unit}$$

If leakage reactance had been measured on the high-voltage side, the value would be

$$X = 0.06\left(\frac{440}{110}\right)^2 = 0.96\ \Omega$$

$$\text{High-voltage base impedance} = \frac{0.440^2 \times 1000}{2.5} = 77.5\ \Omega$$

In per unit

$$X = \frac{0.96}{77.5} = 0.0124 \text{ per unit}$$

A great advantage in making per-unit computations is realized by the proper selection of different bases for circuits connected to each other through a transformer. To achieve the advantage in a single-phase system, *the voltage bases for the circuits connected through the transformer must have the same ratio as the turns ratio of the transformer windings.* With such a selection of voltage bases and the same kilovoltampere base, the per-unit value of an impedance will be the same when it is expressed on the base selected for its own side of the transformer as when it is referred to the other side of the transformer and expressed on the base of that side.

So, the transformer is represented completely by its impedance ($R + jX$) in per unit when magnetizing current is neglected. No per-unit voltage transformation occurs when this system is used, and the current will also have the same per-unit value on both sides of the transformer if magnetizing current is neglected.

Example 2.6. Three parts of a single-phase electric system are designated A, B, and C and are connected to each other through transformers, as shown in Fig. 2.10. The transformers are rated as follows:

A-B	10,000 kVA, 13.8/138 kV, leakage reactance 10%
B-C	10,000 kVA, 138/69 kV, leakage reactance 8%

If the base in circuit B is chosen as 10,000 kVA, 138 kV, find the per-unit impedance of the 300-Ω resistive load in circuit C referred to circuits C, B, and A. Draw the impedance diagram neglecting magnetizing current, transformer resistances, and line impedances.

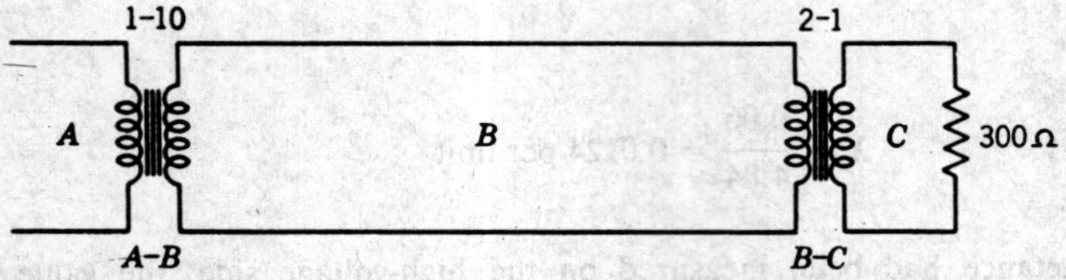

FIGURE 2.10
Circuit for Example 2.6.

Solution

$$\text{Base voltage for circuit } A: \quad 0.1 \times 138 = 13.8 \text{ kV}$$

$$\text{Base voltage for circuit } C: \quad 0.5 \times 138 = 69 \text{ kV}$$

$$\text{Base impedance of circuit } C: \quad \frac{69^2 \times 1000}{10{,}000} = 476\ \Omega$$

$$\text{Per-unit impedance of load in circuit } C: \quad \frac{300}{476} = 0.63 \text{ per unit}$$

Because the selection of base in various parts of the system is determined by the turns ratio of the transformers, and because the base kilovoltamperes is the same in all parts of the system, the per-unit impedance of the load referred to any part of the system will be the same. This is verified as follows:

$$\text{Base impedance of circuit } B: \quad \frac{138^2 \times 1000}{10{,}000} = 1900\ \Omega$$

$$\text{Impedance of load referred to circuit } B: \quad 300 \times 2^2 = 1200\ \Omega$$

$$\text{Per-unit impedance of load referred to } B: \quad \frac{1200}{1900} = 0.63 \text{ per unit}$$

$$\text{Base impedance of circuit } A: \quad \frac{13.8^2 \times 1000}{10{,}000} = 19\ \Omega$$

$$\text{Impedance of load referred to circuit } A: \quad 300 \times 2^2 \times 0.1^2 = 12\ \Omega$$

$$\text{Per-unit impedance of load referred to } A: \quad \frac{12}{19} = 0.63 \text{ per unit}$$

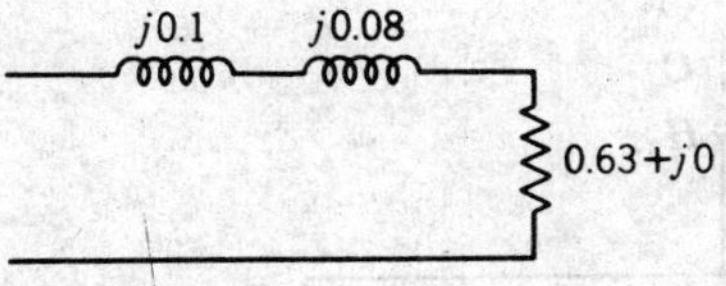

FIGURE 2.11
Impedance diagram for Example 2.6. Impedances are marked in per unit.

Since the chosen bases for kilovolts and kilovoltamperes agree with the transformer ratings, the transformer reactances in per unit are 0.08 and 0.1, respectively. Figure 2.11 is the required impedance diagram with impedances marked in per unit.

Because of the advantage previously pointed out, the principle demonstrated in the preceding example for selecting bases in various parts of the single-phase system is always followed in making computations in per unit. That is, *the kilovoltampere base should be the same in all parts of the system, and the selection of the base kilovolts in one part of the system determines the base kilovolts to be assigned, according to the turns ratios of the transformers, to the other parts of the system*. This principle allows us to combine on one impedance diagram the per-unit impedances of the entire system.

2.5 THREE-PHASE TRANSFORMERS

Three identical single-phase transformers may be connected so that the three windings of one voltage rating are Δ-connected and the three windings of the other voltage rating are Y-connected to form a three-phase transformer. Such a transformer is said to be connected Y-Δ or Δ-Y. The other possible connections are Y-Y and Δ-Δ. If each of the three single-phase transformers has three windings (a primary, secondary, and tertiary), two sets might be connected in Y and one in Δ, or two could be Δ-connected with one Y-connected. Instead of using three identical single-phase transformers, a more usual unit is a three-phase transformer where all three phases are on the same iron structure. The theory is the same for a three-phase transformer as for a three-phase bank of single-phase transformers. The three-phase unit has the advantage of requiring less iron to form the core, and is therefore more economical than three single-phase units and occupies less space. Three single-phase units have the advantage of replacement of only one unit of the three-phase bank in case of a failure rather than losing the whole three-phase bank. If a failure occurs in a Δ-Δ bank composed of three separate units, one of the single-phase transformers can be removed and the remaining two will still operate as a three-phase transformer at a reduced kilovoltampere. Such an operation is called *open delta*.

For a single-phase transformer we can continue to place a dot on one end of each winding, or alternatively, the dotted ends may be marked H_1 for the high-voltage winding and X_1 for the low-voltage winding. The opposite ends are then labeled H_2 and X_2, respectively.

Figure 2.12 shows how three single-phase transformers are connected to form a Y-Y three-phase transformer bank. In this text we shall use capital

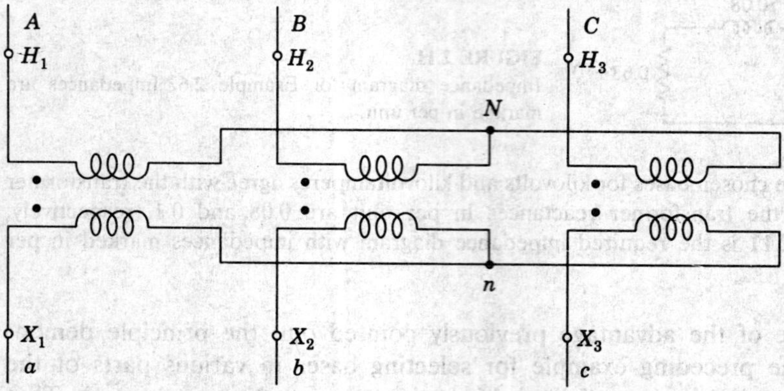

(*a*) Y-Y connection diagram

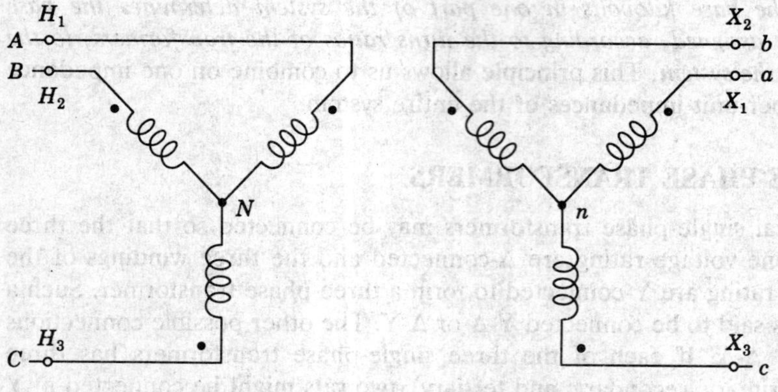

(*b*) Alternate form of connection diagram

FIGURE 2.12
Wiring diagrams for Y-Y transformer.

letters A, B, and C to identify the phases of the high-voltage windings and lowercase letters a, b, and c for the low-voltage windings. The high-voltage terminals of three-phase transformers are marked H_1, H_2, and H_3, and the low-voltage terminals are marked X_1, X_2, and X_3. In Y-Y or Δ-Δ transformers the markings are such that voltages to neutral from terminals H_1, H_2, and H_3 are *in phase* with the voltages to neutral from terminals X_1, X_2, and X_3, respectively. Of course, the Δ windings have no neutral, but the part of the system to which the Δ winding is connected will have a connection to ground. Thus, the ground can serve as the effective neutral under balanced conditions and voltages to neutral from the terminals of the Δ do exist.

To conform with the American standard, the terminals of Y-Δ and Δ-Y transformers are labeled so that the voltages from H_1, H_2, and H_3 to neutral *lead* the voltages to neutral from X_1, X_2, and X_3, respectively, by 30°. We consider this phase shift more fully in the next section.

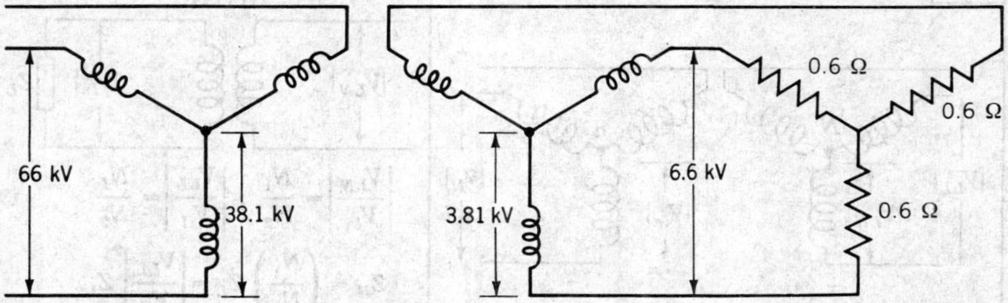

FIGURE 2.13
Y-Y transformer rated 66/6.6 kV.

Figure 2.12(*b*) provides the same information as Fig. 2.12(*a*). Windings of the primary and secondary, which are drawn in parallel directions in Fig. 2.12(*b*), are for the same single-phase transformer or on the same leg of a three-phase transformer. For instance, the winding from A to N is linked by the same flux as the winding from a to n, and V_{AN} is in phase with V_{an}. The diagrams of Fig. 2.12(*b*) are *wiring diagrams* only. They are *not* phasor diagrams.

Figure 2.13 is a schematic method of indicating winding connections of a three-phase transformer. Voltages are shown for a 66/6.6-kV, Y-Y transformer supplying 0.6-Ω resistors or impedances. Figure 2.13 shows a *balanced* system in which each phase can be treated separately, whether or not the neutral points are connected. Then, impedances would transfer from the low-voltage to the high-voltage side by the square of the ratio of line-to-neutral voltages, which is the same as the square of the ratio of line-to-line voltages; that is,

$$0.6\left(\frac{38.1}{3.81}\right)^2 = 0.6\left(\frac{66}{6.6}\right)^2 = 60\ \Omega$$

If we had used a Y-Δ transformer to obtain 6.6 kV across the resistors with the same 66-kV primary, the Δ windings would be rated 6.6 kV rather than 3.81 kV. So far as the voltage *magnitude* at the low-voltage terminals is concerned, the Y-Δ transformer could then be replaced by a Y-Y transformer bank having an effective phase-to-neutral turns ratio of $38.1 : 6.6/\sqrt{3}$, or $N_1 : N_2/\sqrt{3}$, as shown in Table 2.1, so that the same 60-Ω resistance per phase would be seen by the primary. So, we see that the criterion for the selection of base voltage involves the square of the ratio of line-to-line voltages and not the square of the turns ratio of the individual windings of the Y-Δ transformer.

This discussion leads to the conclusion that to transfer the ohmic value of impedance from the voltage level on one side of a three-phase transformer to the voltage level on the other, the multiplying factor is the square of the ratio of line-to-line voltages regardless of whether the transformer connection is Y-Y or

TABLE 2.1

Transferring ohmic values of per-phase impedances from one side of a three-phase transformer to another†

Y-Y

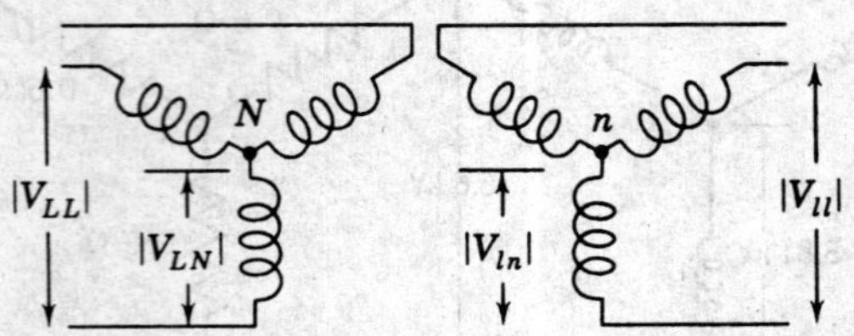

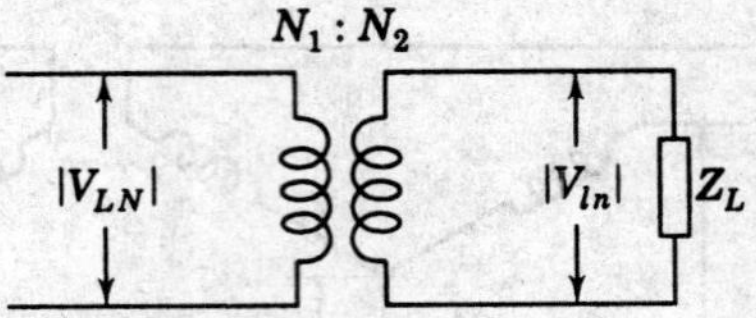

$$\left|\frac{V_{LN}}{V_{ln}}\right| = \frac{N_1}{N_2}\ ;\ \left|\frac{V_{LL}}{V_{ll}}\right| = \frac{N_1}{N_2}$$

$$Z_H = \left(\frac{N_1}{N_2}\right)^2 Z_L = \left|\frac{V_{LL}}{V_{ll}}\right|^2 Z_L$$

Y-Δ

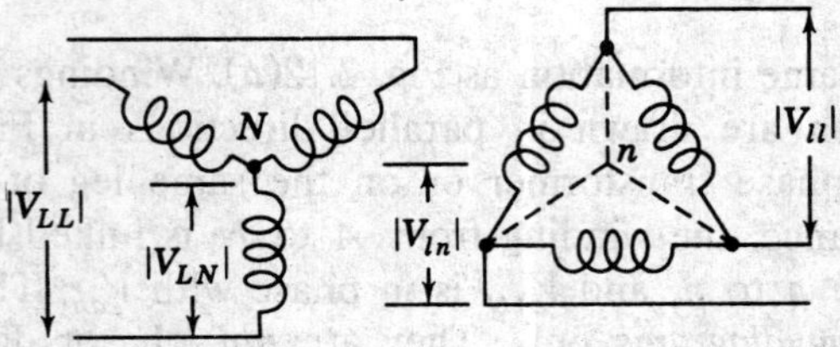

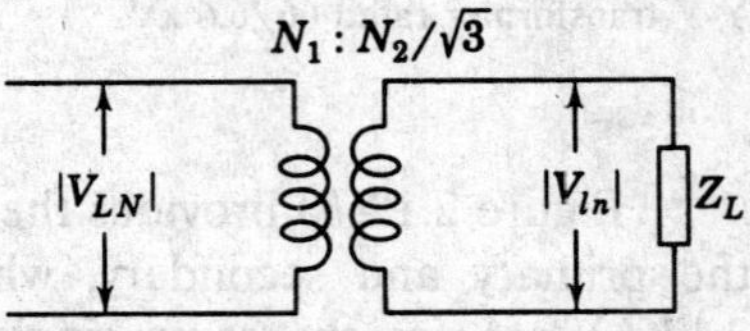

$$\left|\frac{V_{LN}}{V_{ll}}\right| = \frac{N_1}{N_2}\ ;\ \left|\frac{V_{LL}}{V_{ll}}\right| = \sqrt{3}\,\frac{N_1}{N_2}$$

$$Z_H = \left(\frac{N_1}{N_2/\sqrt{3}}\right)^2 Z_L = \left|\frac{V_{LL}}{V_{ll}}\right|^2 Z_L$$

Δ-Y

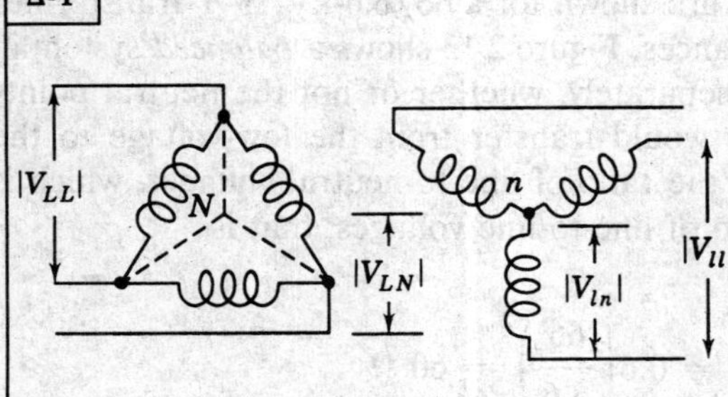

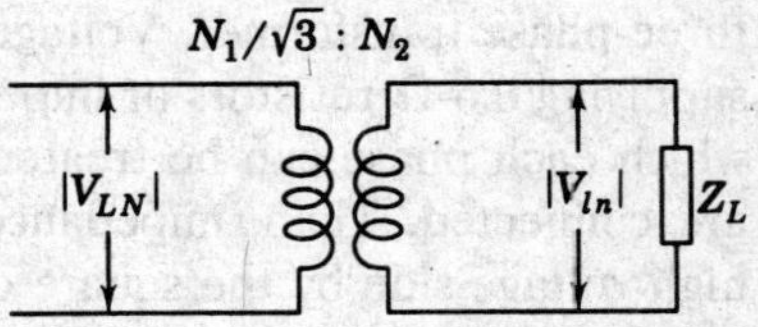

$$\left|\frac{V_{LL}}{V_{ln}}\right| = \frac{N_1}{N_2}\ ;\ \left|\frac{V_{LL}}{V_{ll}}\right| = \frac{1}{\sqrt{3}}\,\frac{N_1}{N_2}$$

$$Z_H = \left(\frac{N_1/\sqrt{3}}{N_2}\right)^2 Z_L = \left|\frac{V_{LL}}{V_{ll}}\right|^2 Z_L$$

Δ-Δ

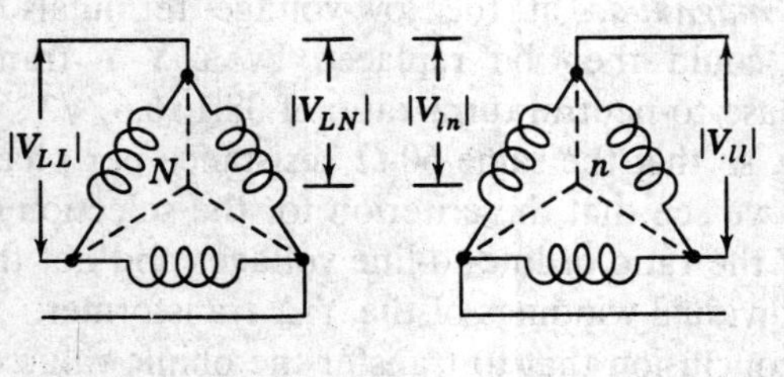

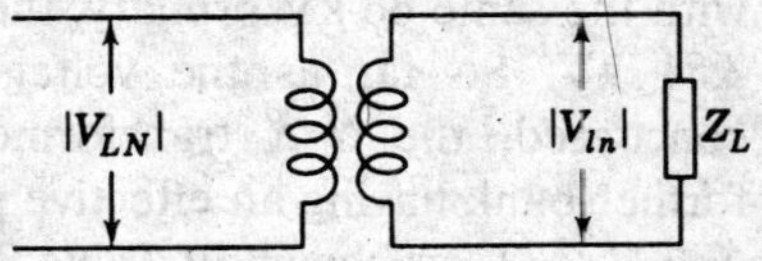

$$\left|\frac{V_{LN}}{V_{ln}}\right| = \frac{N_1/\sqrt{3}}{N_2/\sqrt{3}}\ ;\ \left|\frac{V_{LL}}{V_{ll}}\right| = \frac{N_1}{N_2}$$

$$Z_H = \left(\frac{N_1/\sqrt{3}}{N_2/\sqrt{3}}\right)^2 Z_L = \left|\frac{V_{LL}}{V_{ll}}\right|^2 Z_L$$

†Secondary load consists of balanced Y-connected impedances Z_L.

Y-Δ. This is shown in Table 2.1, which summarizes the relations for the effective turns ratio of the different types of transformer connections. Therefore, in per-unit calculations involving transformers in three-phase circuits we require *the base voltages on the two sides of the transformer to have the same ratio as the rated line-to-line voltages on the two sides of the transformer. The kilovoltampere base is the same on each side.*

Example 2.7. Three transformers, each rated 25 MVA, 38.1/3.81 kV, are connected Y-Δ with a balanced load of three 0.6-Ω, Y-connected resistors. Choose a base of 75 MVA, 66 kV for the high-voltage side of the transformer and specify the base for the low-voltage side. Determine the per-unit resistance of the load on the base for the low-voltage side. Then, determine the load resistance R_L in ohms referred to the high-voltage side and the per-unit value of this resistance on the chosen base.

Solution. Since $\sqrt{3} \times 38.1$ kV equals 66 kV, the rating of the transformer as a three-phase bank is 75 MVA, 66Y/3.81Δ kV. So, base for the low-voltage side is 75 MVA, 3.81 kV.

By Eq. (1.54) base impedance on the low-voltage side is

$$\frac{(\text{base kV}_{LL})^2}{\text{base MVA}_{3\phi}} = \frac{(3.81)^2}{75} = 0.1935\ \Omega$$

and on the low-voltage side

$$R_L = \frac{0.6}{0.1935} = 3.10 \text{ per unit}$$

Base impedance on the high-voltage side is

$$\frac{(66)^2}{75} = 58.1\ \Omega$$

The resistance referred to the high-voltage side is

$$0.6\left(\frac{66}{3.81}\right)^2 = 180\ \Omega$$

$$R_L = \frac{180}{58.1} = 3.10 \text{ per unit}$$

The resistance R and leakage reactance X of a three-phase transformer are measured by the short-circuit test as discussed for single-phase transformers. In a three-phase equivalent circuit R and X are connected in each *line* to an ideal three-phase transformer. Since R and X will have the same per-unit value whether on the low-voltage or the high-voltage side of the transformer, the per-phase equivalent circuit will account for the transformer by the per-unit impedance $R + jX$ *without* the ideal transformer, if phase-shift is not important in the calculations and all quantities in the circuit are in per unit with the proper selection of base.

Table A.1 in the Appendix lists typical values of transformer impedances, which are essentially equal to the leakage reactance since the resistance is usually less than 0.01 per unit.

Example 2.8. A three-phase transformer is rated 400 MVA, 220 Y/22Δ kV. The Y-equivalent short-circuit impedance measured on the low-voltage side of the transformer is 0.121 Ω, and because of the low resistance, this value may be considered equal to the leakage reactance. Determine the per-unit reactance of the transformer and the value to be used to represent this transformer in a system whose base on the high-voltage side of the transformer is 100 MVA, 230 kV.

Solution. On its own base the transformer reactance is

$$\frac{0.121}{(22)^2/400} = 0.10 \text{ per unit}$$

On the chosen base the reactance becomes

$$0.1\left(\frac{220}{230}\right)^2 \frac{100}{400} = 0.0228 \text{ per unit}$$

2.6 THREE-PHASE TRANSFORMERS: PHASE SHIFT AND EQUIVALENT CIRCUITS

As mentioned in Sec. 2.5, a phase shift occurs in Y-Δ transformers. We now examine phase shift in more detail, and the importance of phase sequence becomes apparent. Later in studying faults we have to deal with both positive- or *ABC*-sequence quantities and negative- or *ACB*-sequence quantities. So, we need to examine phase shift for both positive and negative sequences. Positive-sequence voltages and currents are identified by the superscript 1 and negative-sequence voltages and currents by the superscript 2. To avoid too many subscripts, we sometimes write $V_A^{(1)}$ instead of $V_{AN}^{(1)}$ for the voltage drop from terminal A to N and similarly identify other voltages and currents to neutral. In a positive-sequence set of line-to-neutral voltages $V_B^{(1)}$ lags $V_A^{(1)}$ by 120°, whereas $V_C^{(1)}$ lags $V_A^{(1)}$ by 240°; in a negative-sequence set of line-to-neutral voltages $V_B^{(2)}$ leads $V_A^{(2)}$ by 120°, whereas $V_C^{(2)}$ leads $V_A^{(2)}$ by 240°. Later on when we discuss unbalanced currents and voltages (in Chaps. 11 and 12), we must be careful to distinguish between voltages to neutral and voltages to ground since they can differ under unbalanced conditions.

Figure 2.14(a) is the schematic wiring diagram of a Y-Δ transformer, where the Y side is the high-voltage side. We recall that capital letters apply to

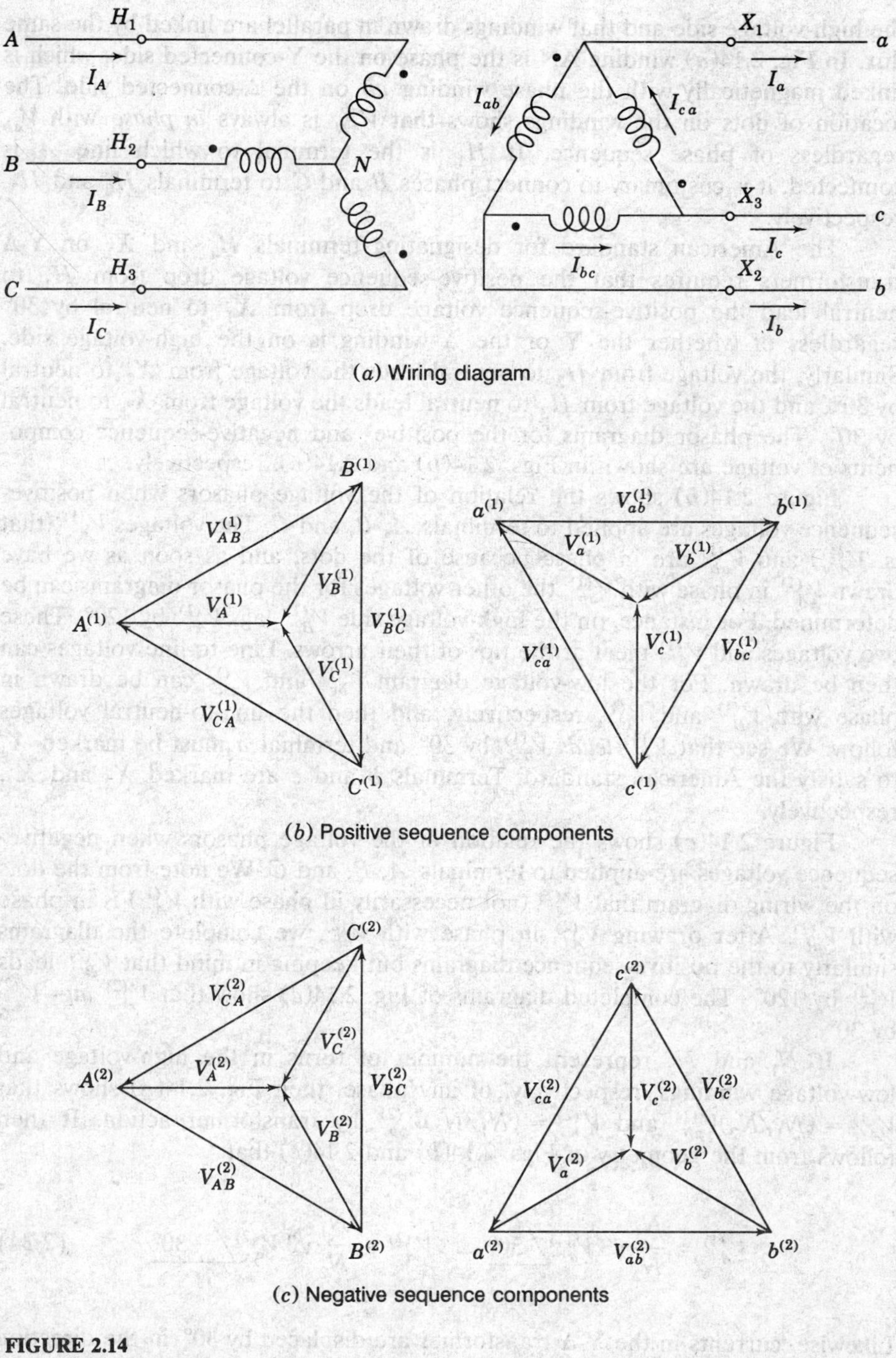

FIGURE 2.14
Wiring diagram and voltage phasors for a three-phase transformer connected Y-Δ where the Y side is the high-voltage side.

the high-voltage side and that windings drawn in parallel are linked by the same flux. In Fig. 2.14(*a*) winding AN is the phase on the Y-connected side, which is linked magnetically with the phase winding *ab* on the Δ-connected side. The location of dots on the windings shows that V_{AN} is always *in phase* with V_{ab} regardless of phase sequence. If H_1 is the terminal to which line A is connected, it is customary to connect phases B and C to terminals H_2 and H_3, respectively.

The American standard for designating terminals H_1 and X_1 on Y-Δ transformers requires that the positive-sequence voltage drop from H_1 to neutral lead the positive-sequence voltage drop from X_1 to neutral by 30° regardless of whether the Y or the Δ winding is on the high-voltage side. Similarly, the voltage from H_2 to neutral leads the voltage from X_2 to neutral by 30°, and the voltage from H_3 to neutral leads the voltage from X_3 to neutral by 30°. The phasor diagrams for the positive- and negative-sequence components of voltage are shown in Figs. 2.14(*b*) and 2.14(*c*), respectively.

Figure 2.14(*b*) shows the relation of the voltage phasors when positive-sequence voltages are applied to terminals A, B, and C. The voltages $V_A^{(1)}$ (that is, $V_{AN}^{(1)}$) and $V_{ab}^{(1)}$ are in phase because of the dots, and as soon as we have drawn $V_A^{(1)}$ in phase with $V_{ab}^{(1)}$, the other voltages for the phasor diagrams can be determined. For instance, on the high-voltage side $V_B^{(1)}$ lags $V_A^{(1)}$ by 120°. These two voltages and $V_C^{(1)}$ meet at the tips of their arrows. Line-to-line voltages can then be drawn. For the low-voltage diagram $V_{bc}^{(1)}$ and $V_{ca}^{(1)}$ can be drawn in phase with $V_B^{(1)}$ and $V_C^{(1)}$, respectively, and then the line-to-neutral voltages follow. We see that $V_A^{(1)}$ *leads* $V_a^{(1)}$ by 30° and terminal a must be marked X_1 to satisfy the American standard. Terminals b and c are marked X_2 and X_3, respectively.

Figure 2.14(*c*) shows the relation of the voltage phasors when negative-sequence voltages are applied to terminals A, B, and C. We note from the dots on the wiring diagram that $V_A^{(2)}$ (not necessarily in phase with $V_A^{(1)}$) is in phase with $V_{ab}^{(2)}$. After drawing $V_A^{(2)}$ in phase with $V_{ab}^{(2)}$, we complete the diagrams similarly to the positive-sequence diagrams but keeping in mind that $V_B^{(2)}$ leads $V_A^{(2)}$ by 120°. The completed diagrams of Fig. 2.14(*c*) show that $V_A^{(2)}$ *lags* $V_a^{(2)}$ by 30°.

If N_1 and N_2 represent the number of turns in the high-voltage and low-voltage windings, respectively, of any phase, then Fig. 2.14(*a*) shows that $V_A^{(1)} = (N_1/N_2)V_{ab}^{(1)}$ and $V_A^{(2)} = (N_1/N_2)V_{ab}^{(2)}$ by transformer action. It then follows from the geometry of Figs. 2.14(*b*) and 2.14(*c*) that

$$V_A^{(1)} = \frac{N_1}{N_2}\sqrt{3}\,V_a^{(1)}\angle 30^\circ \qquad V_A^{(2)} = \frac{N_1}{N_2}\sqrt{3}\,V_a^{(2)}\angle -30^\circ \tag{2.34}$$

Likewise, currents in the Y-Δ transformer are displaced by 30° in the direction of the voltages since the phase angles of the currents with respect to their associated voltages are determined by the load impedance. The ratio of the rated line-to-line voltage of the Y winding to the rated line-to-line voltage of the Δ winding equals $\sqrt{3}\,N_1/N_2$, so that in choosing the line-to-line voltage bases on the two sides of the transformer in the same ratio, we obtain in per unit

$$\begin{aligned} V_A^{(1)} &= V_a^{(1)} \times 1\angle 30^\circ \qquad I_A^{(1)} = I_a^{(1)} \times 1\angle 30^\circ \\ V_A^{(2)} &= V_a^{(2)} \times 1\angle -30^\circ \qquad I_A^{(2)} = I_a^{(2)} \times 1\angle -30^\circ \end{aligned} \tag{2.35}$$

Transformer impedance and magnetizing currents are handled separately from the phase shift, which can be represented by an ideal transformer. This explains why, according to Eq. (2.35), the *per-unit* magnitudes of voltage and current are exactly the same on both sides of the transformer (for instance, $|V_a^{(1)}| = |V_A^{(1)}|$).

Usually, the high-voltage winding in a Y-Δ transformer is Y-connected. Insulation costs for a given step up in voltage are thereby reduced since this connection takes advantage of the fact that the voltage transformation from the low-voltage side to the high-voltage side of the transformer is then $\sqrt{3}\,(N_1/N_2)$, where N_1 and N_2 are the same as in Eq. (2.34).

If the high-voltage windings are Δ-connected, the transformation ratio of line voltages is reduced rather than increased. Figure 2.15 is the schematic diagram for the Δ-Y transformer where the Δ side is the high-voltage side. The reader should verify that the voltage phasors are exactly the same as in Figs. 2.14(*b*) and 2.14(*c*), and Eqs. (2.34) and (2.35) are therefore still valid. These equations still hold if we reverse the directions of *all* currents on the wiring diagram.

Under normal operating conditions only positive-sequence quantities are involved and then the general rule for any Y-Δ or Δ-Y transformer is that voltage is advanced 30° when it is stepped up. As already discussed, we can indicate this phase shift in voltage by an ideal transformer of complex turns ratio $1:\varepsilon^{j\pi/6}$. Since $V_A^{(1)}/I_A^{(1)} = V_a^{(1)}/I_a^{(1)}$ in Eq. (2.35), per-unit impedance

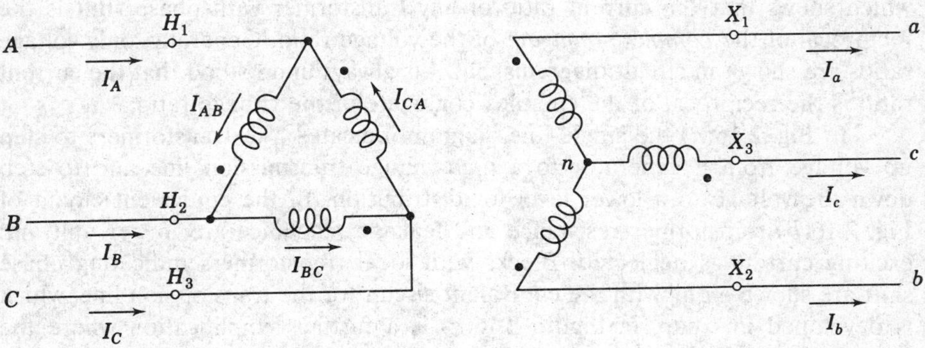

FIGURE 2.15
Wiring diagram for a three-phase transformer connected Δ-Y where the Δ side is the high-voltage side.

values are the same when moved from one side of the ideal transformer to the other. Real and reactive power flow is also not affected by the phase shift because the current phase shift compensates exactly for the voltage phase shift as far as power values are concerned. This is easily seen by writing the per-unit complex power for each side of the Y-Δ (or Δ-Y) transformer from Eqs. (2.35) as follows:

$$V_A^{(1)} I_A^{(1)*} = V_a^{(1)} \angle 30^\circ \times I_a^{(1)*} \angle -30^\circ = V_a^{(1)} I_a^{(1)*} \tag{2.36}$$

Hence, if only P and Q quantities are required, it is not necessary to include ideal transformers for the phase shift of Y-Δ and Δ-Y transformers in the impedance diagram. The only case in which the ideal transformer cannot be ignored is in any closed-loop portion of a system in which the product of all the actual transformer voltage ratios is not unity around the loop. We encounter one such case in Sec. 2.9 when parallel connections of regulating transformers are considered. In most other situations we can eliminate the ideal transformers from the per-unit impedance diagram, and then the calculated currents and voltages are proportional to the actual currents and voltages. Phase angles of the actual currents and voltages can be found if needed by noting from the one-line diagram the positions of the Y-Δ and Δ-Y transformers and by applying the rules of Eq. (2.35); namely,

> When stepping up from the low-voltage to the high-voltage side of a Δ-Y or Y-Δ transformer, *advance* positive-sequence voltages and currents by 30° and *retard* negative-sequence voltages and currents by 30°.

It is important to note from Eq. (2.36) that

$$\frac{I_A^{(1)}}{I_a^{(1)}} = \left(\frac{V_A^{(1)*}}{V_a^{(1)*}} \right)^{-1} \tag{2.37}$$

which shows that the current ratio of any transformer with phase shift is the reciprocal of the *complex conjugate* of the voltage ratio. Generally, only voltage ratios are shown in circuit diagrams, but it is always understood that the current ratio is the reciprocal of the complex conjugate of the voltage ratio.

In Fig. 2.16(a) the single-line diagram indicates Y-Δ transformers to step up voltage from a generator to a high-voltage transmission line and to step down the voltage to a lower level for distribution. In the equivalent circuit of Fig. 2.16(b) transformer resistance and leakage reactance are in per unit and exciting current is neglected. Blocks with ideal transformers indicating phase shift are shown along with the equivalent circuit for the transmission line, which is developed in Chap. 6. Figure 2.16(c) is a further simplification where the resistances, shunt capacitors, and ideal transformers are neglected. Here we rely upon the single-line diagram to remind us to account for phase shift due to the Y-Δ transformers. We must remember that positive-sequence voltages and

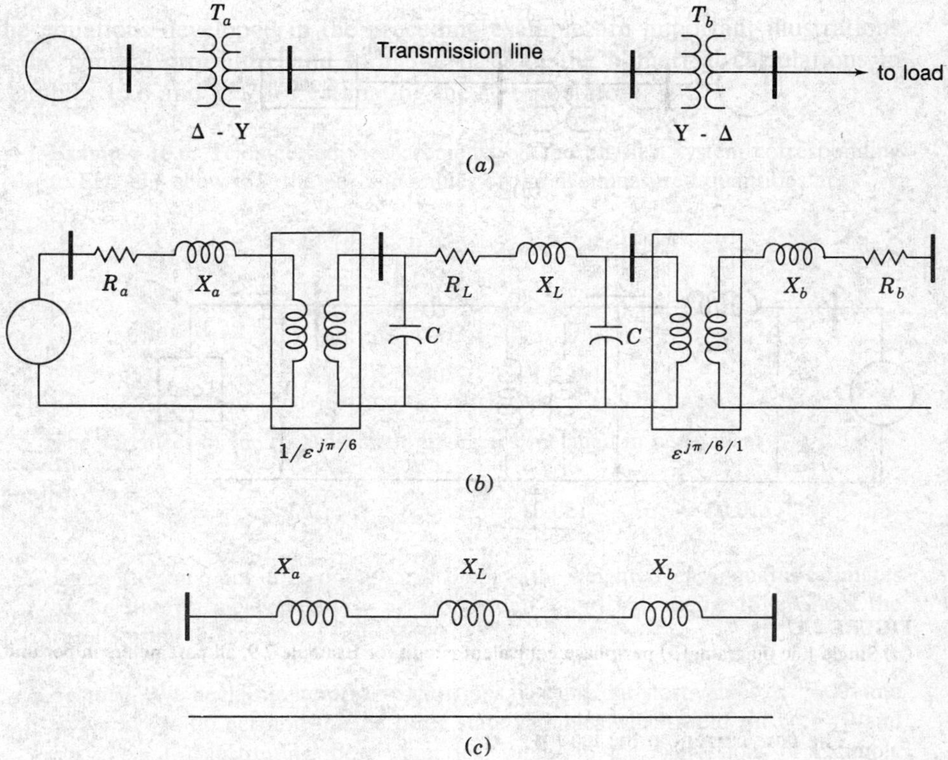

FIGURE 2.16
(*a*) Single-line diagram; (*b*) per-phase equivalent circuits with parameters in per unit; (*c*) per-phase equivalent circuit with resistance, capacitance, and ideal transformers neglected. The per-phase equivalent circuit for the transmission line is developed in Chap. 6.

currents in the higher-voltage transmission line lead the corresponding quantities in the lower-voltage generator and distribution circuits by 30°.

Example 2.9. Figure 2.17 shows a three-phase generator rated 300 MVA, 23 kV supplying a system load of 240 MVA, 0.9 power-factor lagging at 230 kV through a 330-MVA 23Δ/230Y-kV step-up transformer of leakage reactance 11%. Neglecting magnetizing current and choosing base values at the load of 100 MVA and 230 kV, find I_A, I_B, and I_C supplied to the load in per unit with V_A as reference. Specifying the proper base for the generator circuit, determine I_a, I_b, and I_c from the generator and its terminal voltage.

Solution. The current supplied to the load is

$$\frac{240{,}000}{\sqrt{3} \times 230} = 602.45 \text{ A}$$

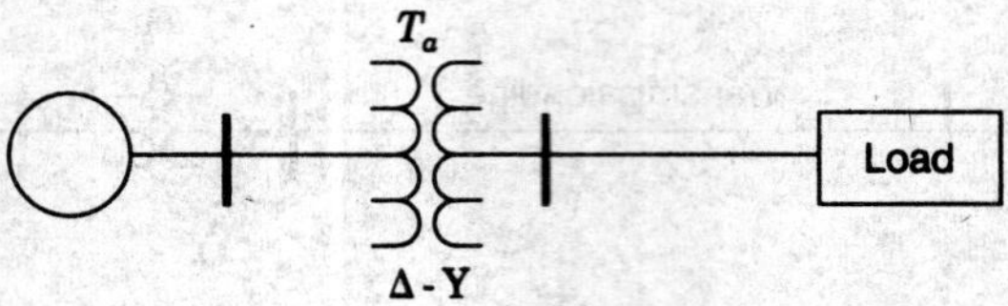

(a)

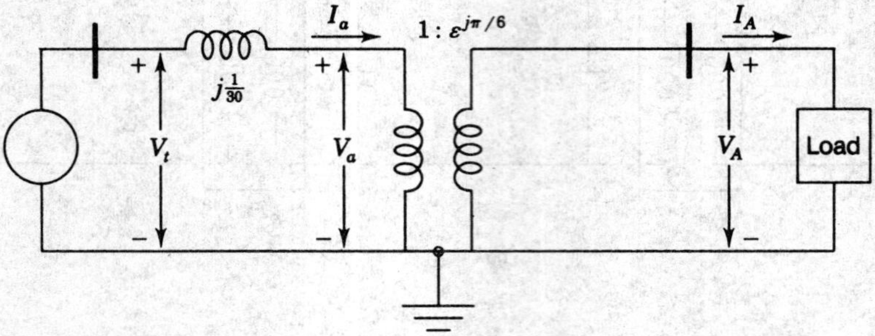

(b)

FIGURE 2.17
(a) Single-line diagram; (b) per-phase equivalent circuit for Example 2.9, all parameters in per unit.

The *base* current at the load is

$$\frac{100{,}000}{\sqrt{3} \times 230} = 251.02 \text{ A}$$

The power-factor angle of the load current is

$$\theta = \cos^{-1} 0.9 = 25.84^\circ \text{ lag}$$

Hence, with $V_A = 1.0\angle 0^\circ$ as reference in Fig. 2.17(b), the line currents into the load are

$$I_A = \frac{602.45}{251.02}\angle -25.84^\circ = 2.40\angle -25.84^\circ \text{ per unit}$$

$$I_B = 2.40\angle -25.84^\circ - 120^\circ = 2.40\angle -145.84^\circ \text{ per unit}$$

$$I_C = 2.40\angle -25.84^\circ + 120^\circ = 2.40\angle 94.16^\circ \text{ per unit}$$

Low-voltage side currents further lag by 30°, and so in per unit

$$I_a = 2.40\angle -55.84^\circ \qquad I_b = 2.40\angle 175.84^\circ \qquad I_c = 2.40\angle 64.16^\circ$$

The transformer reactance modified for the chosen base is

$$0.11 \times \frac{100}{330} = \frac{1}{30} \text{ per unit}$$

and so from Fig. 2.17(*b*) the terminal voltage of the generator is

$$V_t = V_A \angle -30^\circ + jXI_a$$

$$= 1.0\angle -30^\circ + \frac{j}{30} \times 2.40\angle -55.84^\circ$$

$$= 0.9322 - j0.4551 = 1.0374\angle -26.02^\circ \text{ per unit}$$

The base generator voltage is 23 kV, which means that the terminal voltage of the generator is 23 × 1.0374 = 23.86 kV. The real power supplied by the generator is

$$\text{Re}\{V_t I_a^*\} = 1.0374 \times 2.4 \cos(-26.02^\circ + 55.84^\circ) = 2.160 \text{ per unit}$$

which corresponds to 216 MW absorbed by the load since there are no I^2R losses. The interested reader will find the same value for $|V_t|$ by omitting the phase shift of the transformer altogether or by recalculating V_t with the reactance $j/30$ per unit on the high-voltage side of Fig. 2.17(*b*).

2.7 THE AUTOTRANSFORMER

An autotransformer differs from the ordinary transformer in that the windings of the autotransformer are electrically connected as well as coupled by a mutual flux. We examine the autotransformer by electrically connecting the windings of an ideal transformer. Figure 2.18(*a*) is a schematic diagram of an ideal transformer, and Fig. 2.18(*b*) shows how the windings are connected electrically to form an autotransformer. Here the windings are shown so that their voltages are additive although they could have been connected to oppose each other. The great disadvantage of the autotransformer is that electrical isolation is lost, but the following example demonstrates the increase in power rating obtained.

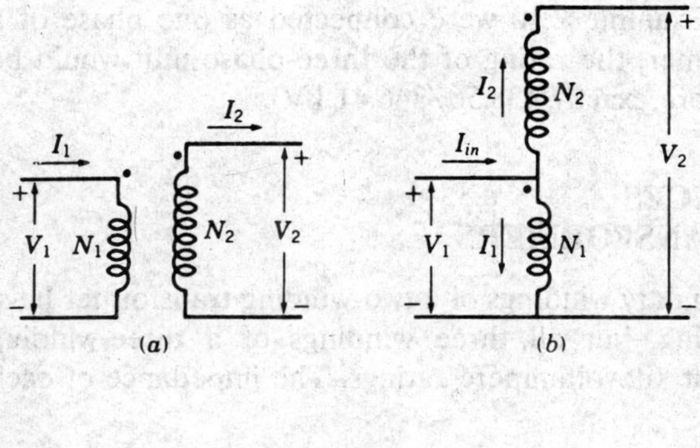

FIGURE 2.18
Schematic diagram of an ideal transformer connected: (*a*) in the usual manner; (*b*) as an autotransformer.

Example 2.10. A 90-MVA single-phase transformer rated 80/120 kV is connected as an autotransformer, as shown in Fig. 2.18(*b*). Rated voltage $|V_1| = 80$ kV is applied to the low-voltage winding of the transformer. Consider the transformer to be ideal and the load to be such that currents of rated magnitudes $|I_1|$ and $|I_2|$ flow in the windings. Determine $|V_2|$ and the kilovoltampere rating of the autotransformer.

Solution

$$|I_1| = \frac{90{,}000}{80} = 1125 \text{ A}$$

$$|I_2| = \frac{90{,}000}{120} = 750 \text{ A}$$

$$|V_2| = 80 + 120 = 200 \text{ kV}$$

The directions chosen for I_1 and I_2 in relation to the dotted terminals show that these currents are in phase. So, the input current is

$$|I_{in}| = 1125 + 750 = 1875 \text{ A}$$

Input kilovoltamperes are

$$|I_{in}| \times |V_1| = 1875 \times 80 = 150{,}000 \text{ kVA}$$

Output kilovoltamperes are

$$|I_2| \times |V_2| = 750 \times 200 = 150{,}000 \text{ kVA}$$

The increase in the kilovoltampere rating from 90,000 to 150,000 kVA and in the output voltage from 120 to 200 kV demonstrates the advantage of the autotransformer. The autotransformer provides a higher rating for the same cost, and its efficiency is greater since the losses are the same as in the ordinary connection of the same transformer.

Single-phase autotransformers can be connected for Y-Y three-phase operation or a three-phase unit can be built. Three-phase autotransformers are often used to connect two transmission lines operating at different voltage levels. If the transformer of Example 2.10 were connected as one phase of a three-phase Y-Y autotransformer, the rating of the three-phase unit would be 450 MVA, 138/345 kV (or more exactly 138.56/346.41 kV).

2.8 PER-UNIT IMPEDANCES OF THREE-WINDING TRANSFORMERS

Both the primary and the secondary windings of a two-winding transformer have the same kilovoltampere rating, but all three windings of a three-winding transformer may have different kilovoltampere ratings. The impedance of each

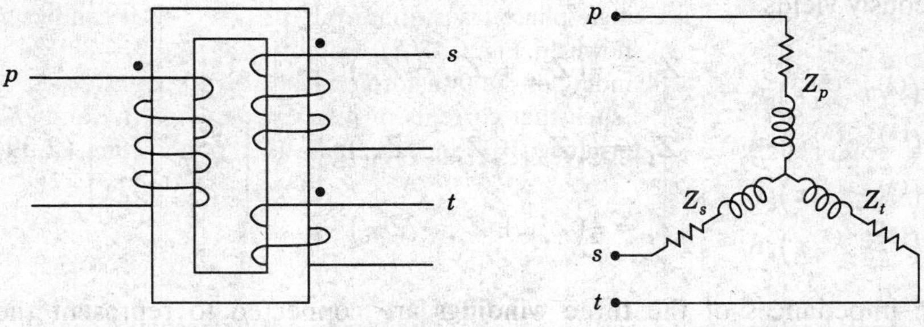

FIGURE 2.19
The (*a*) schematic diagram and (*b*) equivalent circuit of a three-winding transformer. Points p, s, and t link the circuit of the transformer to the appropriate equivalent circuits representing parts of the system connected to the primary, secondary, and tertiary windings.

winding of a three-winding transformer may be given in percent or per unit based on the rating of its own winding, or tests may be made to determine the impedances. In any case, all the per-unit impedances must be expressed on the same kilovoltampere base.

A single-phase three-winding transformer is shown schematically in Fig. 2.19(*a*), where we designate the three windings as *primary*, *secondary*, and *tertiary*. Three impedances may be measured by the standard short-circuit test, as follows:

Z_{ps} leakage impedance measured in primary with secondary short-circuited and tertiary open

Z_{pt} leakage impedance measured in primary with tertiary short-circuited and secondary open

Z_{st} leakage impedance measured in secondary with tertiary short-circuited and primary open

If the three impedances measured in ohms are referred to the voltage of one of the windings, the impedances of each separate winding referred to that same winding are related to the measured impedances so referred as follows:

$$Z_{ps} = Z_p + Z_s$$

$$Z_{pt} = Z_p + Z_t \qquad (2.38)$$

$$Z_{st} = Z_s + Z_t$$

Here Z_p, Z_s, and Z_t are the impedances of the primary, secondary, and tertiary windings, respectively, referred to the primary circuit if Z_{ps}, Z_{pt}, and Z_{st} are the measured impedances referred to the primary circuit. Solving Eqs. (2.38) simultaneously yields

$$Z_p = \tfrac{1}{2}(Z_{ps} + Z_{pt} - Z_{st})$$

$$Z_s = \tfrac{1}{2}(Z_{ps} + Z_{st} - Z_{pt}) \tag{2.39}$$

$$Z_t = \tfrac{1}{2}(Z_{pt} + Z_{st} - Z_{ps})$$

The impedances of the three windings are connected to represent the equivalent circuit of the single-phase three-winding transformer with magnetizing current neglected, as shown in Fig. 2.19(*b*). The common point is fictitious and unrelated to the neutral of the system. The points *p*, *s*, and *t* are connected to the parts of the impedance diagrams representing the parts of the system connected to the primary, secondary, and tertiary windings, respectively, of the transformer. As in two-winding transformers, conversion to per-unit impedance requires the same kilovoltampere base for all three circuits and requires voltage bases in the three circuits that are in the same ratio as the rated line-to-line voltages of the three circuits of the transformer.

When three such transformers are connected for three-phase operation, the primary and secondary windings are usually Y-connected and the tertiary windings are connected in Δ to provide a path for the third harmonic of the exciting current.

Example 2.11. The three-phase ratings of a three-winding transformer are:

Primary Y-connected, 66 kV, 15 MVA

Secondary Y-connected, 13.2 kV, 10 MVA

Tertiary Δ-connected, 2.3 kV, 5 MVA

Neglecting resistance, the leakage impedances are

$$Z_{ps} = 7\% \text{ on 15 MVA, 66 kV base}$$

$$Z_{pt} = 9\% \text{ on 15 MVA, 66 kV base}$$

$$Z_{st} = 8\% \text{ on 10 MVA, 13.2 kV base}$$

Find the per-unit impedances of the per-phase equivalent circuit for a base of 15 MVA, 66 kV in the primary circuit.

Solution. With a base of 15 MVA, 66 kV in the primary circuit, the proper bases for the per-unit impedances of the equivalent circuit are 15 MVA, 66 kV for primary-circuit quantities, 15 MVA, 13.2 kV for secondary-circuit quantities, and 15 MVA, 2.3 kV for tertiary-circuit quantities.

Since Z_{ps} and Z_{pt} are measured in the primary circuit, they are already expressed on the proper base for the equivalent circuit. No change of voltage base is required for Z_{st}. The required change in base megavoltamperes for Z_{st} is made as follows:

$$Z_{st} = 8\% \times \frac{15}{10} = 12\%$$

In per unit on the specified base

$$Z_p = \tfrac{1}{2}(j0.07 + j0.09 - j0.12) = j0.02 \text{ per unit}$$

$$Z_s = \tfrac{1}{2}(j0.07 + j0.12 - j0.09) = j0.05 \text{ per unit}$$

$$Z_t = \tfrac{1}{2}(j0.09 + j0.12 - j0.07) = j0.07 \text{ per unit}$$

Example 2.12. A constant-voltage source (infinite bus) supplies a purely resistive 5 MW, 2.3 kV three-phase load and a 7.5 MVA, 13.2 kV synchronous motor having a subtransient reactance of $X'' = 20\%$. The source is connected to the primary of the three-winding transformer described in Example 2.11. The motor and resistive load are connected to the secondary and tertiary of the transformer. Draw the impedance diagram of the system and mark the per-unit impedances for a base of 66 kV, 15 MVA in the primary. Neglect exciting current and all resistance except that of the resistive load.

Solution. The constant-voltage source can be represented by a generator having no internal impedance.

The resistance of the load is 1.0 per unit on a base of 5 MVA, 2.3 kV in the tertiary. Expressed on a 15 MVA, 2.3 kV base, the load resistance is

$$R = 1.0 \times \frac{15}{5} = 3.0 \text{ per unit}$$

The reactance of the motor on a base of 15 MVA, 13.2 kV is

$$X'' = 0.20\frac{15}{7.5} = 0.40 \text{ per unit}$$

Figure 2.20 is the required diagram. We must remember, however, the phase shift which occurs between the Y-connected primary and the Δ-connected tertiary.

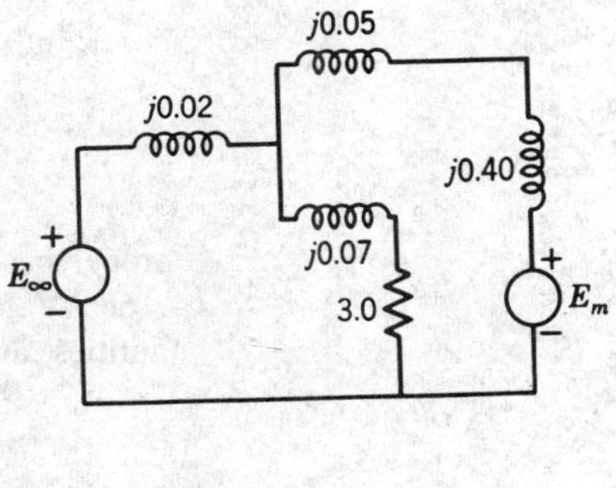

FIGURE 2.20
Impedance diagram for Example 2.11.

2.9 TAP-CHANGING AND REGULATING TRANSFORMERS

Transformers which provide a small adjustment of voltage magnitude, usually in the range of $\pm 10\%$, and others which shift the phase angle of the line voltages are important components of a power system. Some transformers regulate both the magnitude and phase angle.

Almost all transformers provide taps on windings to adjust the ratio of transformation by changing taps when the transformer is deenergized. A change in tap can be made while the transformer is energized, and such transformers are called *load-tap-changing* (LTC) *transformers or tap-changing-under-load* (TCUL) *transformers*. The tap changing is automatic and operated by motors which respond to relays set to hold the voltage at the prescribed level. Special circuits allow the change to be made without interrupting the current.

A type of transformer designed for small adjustments of voltage rather than large changes in voltage levels is called a *regulating transformer*. Figure 2.21 shows a regulating transformer for control of voltage magnitude, and Fig. 2.22 shows a regulating transformer for phase-angle control. The phasor diagram of Fig. 2.23 helps to explain the shift in phase angle. Each of the three windings to which taps are made is on the same magnetic core as the phase winding whose voltage is 90° out of phase with the voltage from neutral to the point connected to the center of the tapped winding. For instance, the voltage to neutral V_{an} is increased by a component ΔV_{an}, which is in phase or 180° out of phase with V_{bc}. Figure 2.23 shows how the three line voltages are shifted in phase angle with very little change in magnitude.

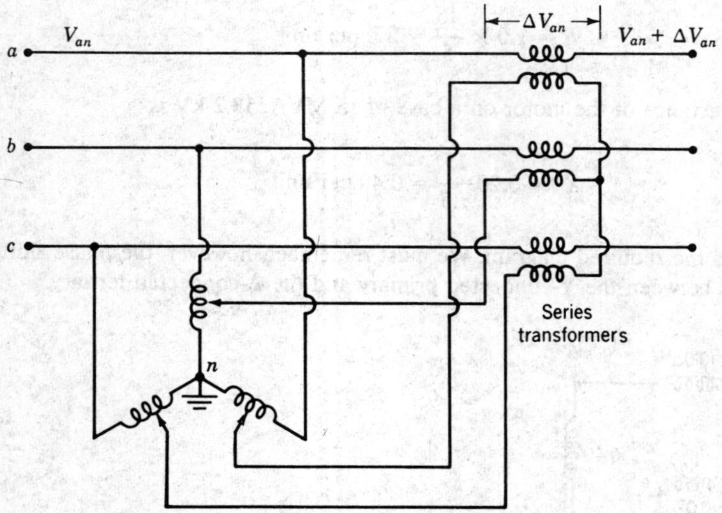

FIGURE 2.21
Regulating transformer for control of voltage magnitude.

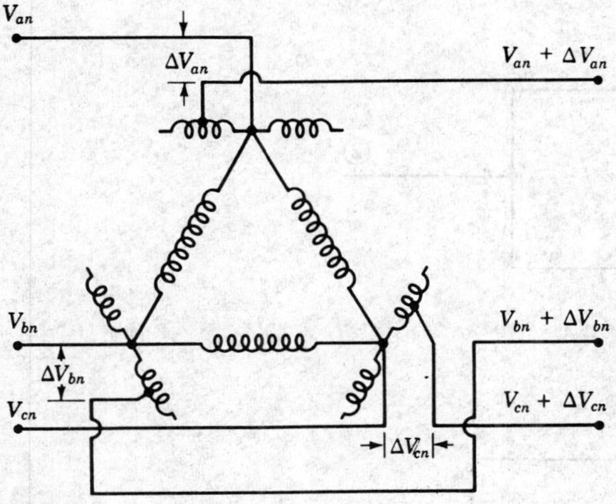

FIGURE 2.22
Regulating transformer for control of phase angle. Windings drawn parallel to each other are on the same iron core.

The procedure to determine the bus admittance matrix $\mathbf{Y}_{\text{bus}}$ in per unit for a network containing a regulating transformer is the same as the procedure to account for any transformer whose turns ratio is other than the ratio used to select the ratio of base voltages on the two sides of the transformer. We defer consideration of the procedure until Chap. 9. We can, however, investigate the usefulness of tap-changing and regulating transformers by a simple example.

If we have two buses connected by a transformer, and if the ratio of the line-to-line voltages of the transformer is the same as the ratio of the base voltages of the two buses, the per-phase equivalent circuit (with the magnetizing current neglected) is simply the transformer impedance in per unit on the chosen base connected between the buses. Figure 2.24(a) is a one-line diagram of two transformers in parallel. Let us assume that one of them has the voltage ratio $1/n$, which is also the ratio of base voltages on the two sides of the transformer, and that the voltage ratio of the other is $1/n'$. The equivalent circuit is then that of Fig. 2.24(b). We need the ideal (no impedance) transformer with the ratio $1/t$ in the per-unit reactance diagram to take care of the off-nominal turns ratio of the second transformer because base voltages are

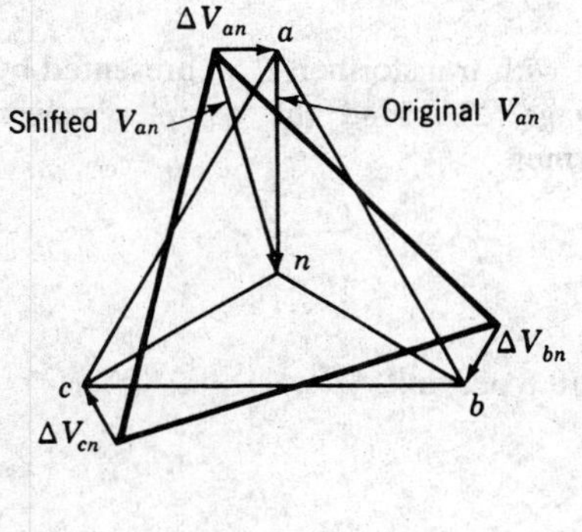

FIGURE 2.23
Phasor diagram for the regulating transformer shown in Fig. 2.22.

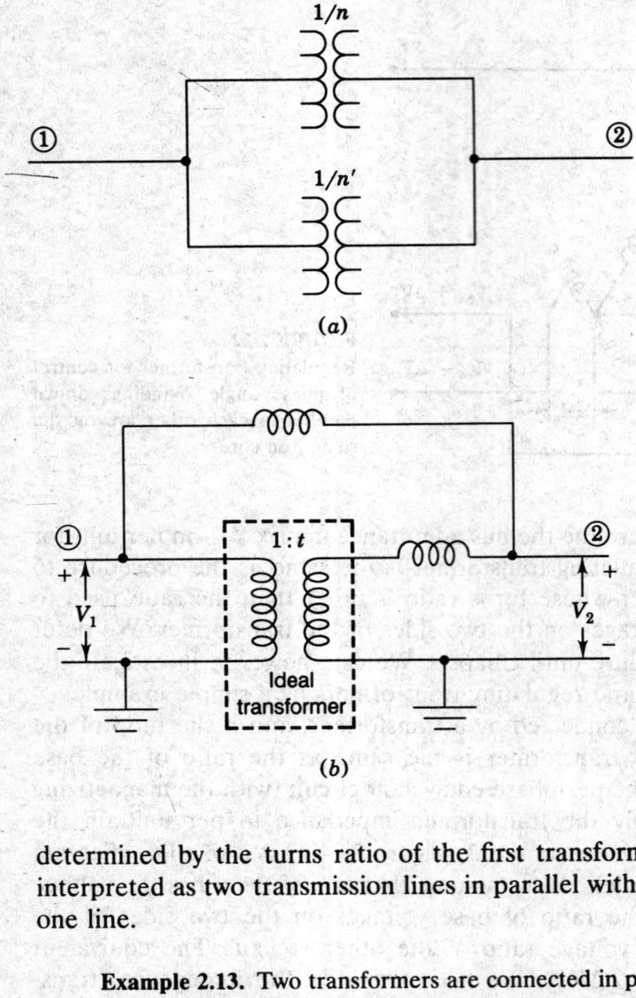

FIGURE 2.24
Transformers with differing turns ratio connected in parallel: (*a*) the single-line diagram; (*b*) the per-phase reactance diagram in per unit. The turns ratio $1/t$ is equal to n/n'.

determined by the turns ratio of the first transformer. Figure 2.24(*b*) may be interpreted as two transmission lines in parallel with a regulating transformer in one line.

Example 2.13. Two transformers are connected in parallel to supply an impedance to neutral per phase of $0.8 + j0.6$ per unit at a voltage of $V_2 = 1.0\angle 0^\circ$ per unit. Transformer T_a has a voltage ratio equal to the ratio of the base voltages on the two sides of the transformer. This transformer has an impedance of $j0.1$ per unit on the appropriate base. The second transformer T_b also has an impedance of $j0.1$ per unit on the same base but has a step-up toward the load of 1.05 times that of T_a (secondary windings on 1.05 tap).

Figure 2.25 shows the equivalent circuit with transformer T_b represented by its impedance and the insertion of a voltage ΔV. Find the complex power transmitted to the load through each transformer.

Solution. Load current is

$$\frac{1.0}{0.8 + j0.6} = 0.8 - j0.6 \text{ per unit}$$

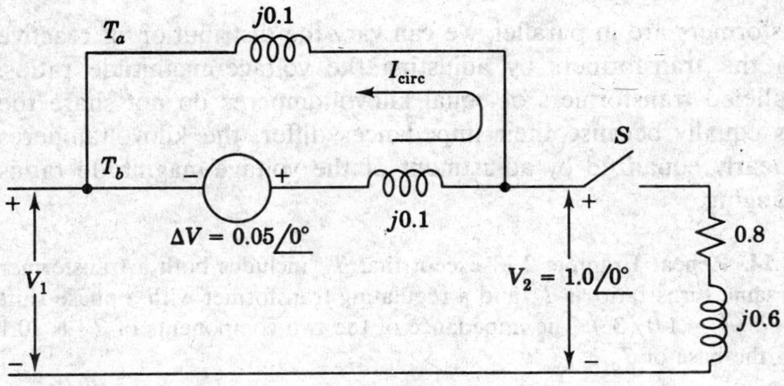

FIGURE 2.25
An equivalent circuit for Example 2.13.

An approximate solution to this problem is found by recognizing that Fig. 2.25 with switch S closed is an equivalent circuit for the problem if the voltage ΔV, which is in the branch of the circuit equivalent to transformer T_b, is equal to $t - 1$ in per unit. In other words, if T_a is providing a voltage ratio 5% higher than T_b, t equals 1.05 and ΔV equals 0.05 per unit. If we consider that the current set up by ΔV circulates around the loop indicated by I_{circ} with switch S open, and that with S closed only a very small fraction of that current goes through the load impedance (because it is much larger than the transformer impedance), then we can apply the superposition principle to ΔV and the source voltage. From ΔV alone we obtain

$$I_{circ} = \frac{0.05}{j0.2} = -j0.25 \text{ per unit}$$

and with ΔV short-circuited, the current in each path is half the load current, or $0.4 - j0.3$. Then, superimposing the circulating current gives

$$I_{T_a} = 0.4 - j0.3 - (-j0.25) = 0.4 - j0.05 \text{ per unit}$$

$$I_{T_b} = 0.4 - j0.3 + (-j0.25) = 0.4 - j0.55 \text{ per unit}$$

so that

$$S_{T_a} = 0.40 + j0.05 \text{ per unit}$$

$$S_{T_b} = 0.40 + j0.55 \text{ per unit}$$

This example shows that the transformer with the higher tap setting is supplying most of the reactive power to the load. The real power is dividing equally between the transformers. Since both transformers have the same impedance, they would share both the real and reactive power equally if they had the same turns ratio. In that case each would be represented by the same per-unit reactance of $j0.1$ between the two buses and would carry equal current.

When two transformers are in parallel, we can vary the distribution of reactive power between the transformers by adjusting the voltage-magnitude ratios. When two paralleled transformers of equal kilovoltamperes do not share the kilovoltamperes equally because their impedances differ, the kilovoltamperes may be more nearly equalized by adjustment of the voltage-magnitude ratios through tap changing.

Example 2.14. Repeat Example 2.13 except that T_b includes both a transformer having the same turns ratio as T_a and a regulating transformer with a phase shift of 3° ($t = \varepsilon^{j\pi/60} = 1.0\angle 3°$). The impedance of the two components of T_b is $j0.1$ per unit on the base of T_a.

Solution. As in Example 2.13, we can obtain an approximate solution of the problem by inserting a voltage source ΔV in series with the impedance of transformer T_b. The proper per-unit voltage is

$$t - 1 = 1.0\angle 3° - 1.0\angle 0° = (2\sin 1.5°)\angle 91.5° = 0.0524\angle 91.5°$$

$$I_{\text{circ}} = \frac{0.0524\angle 91.5°}{0.2\angle 90°} = 0.262 + j0.0069 \text{ per unit}$$

$$I_{T_a} = 0.4 - j0.3 - (0.262 + j0.007) = 0.138 - j0.307 \text{ per unit}$$

$$I_{T_b} = 0.4 - j0.3 + (0.262 + j0.007) = 0.662 - j0.293 \text{ per unit}$$

So,

$$S_{T_a} = 0.138 + j0.307 \text{ per unit}$$

$$S_{T_b} = 0.662 + j0.293 \text{ per unit}$$

The example shows that the phase-shifting transformer is useful to control the amount of real power flow but has less effect on the flow of reactive power. Both Examples 2.13 and 2.14 are illustrative of two transmission lines in parallel with a regulating transformer in one of the lines.

2.10 THE ADVANTAGES OF PER-UNIT COMPUTATIONS

When bases are specified properly for the various parts of a circuit connected by a transformer, the per-unit values of impedances determined in their own part of the system are the same when viewed from another part. Therefore, it is necessary only to compute each impedance on the base of its own part of the circuit. The great advantage of using per-unit values is that no computations are required to refer an impedance from one side of a transformer to the other.

The following points should be kept in mind:

1. A base kilovolts and base kilovoltamperes is selected in one part of the system. The base values for a three-phase system are understood to be line-to-line kilovolts and three-phase kilovoltamperes or megavoltamperes.
2. For other parts of the system, that is, on other sides of transformers, the base kilovolts for each part is determined according to the line-to-line voltage ratios of the transformers. The base kilovoltamperes will be the same in all parts of the system. It will be helpful to mark the base kilovolts of each part of the system on the one-line diagram.
3. Impedance information available for three-phase transformers will usually be in per unit or percent on the base determined by their own ratings.
4. For three single-phase transformers connected as a three-phase unit the three-phase ratings are determined from the single-phase rating of each individual transformer. Impedance in percent for the three-phase unit is the same as that for each individual transformer.
5. Per-unit impedance given on a base other than that determined for the part of the system in which the element is located must be changed to the proper base by Eq. (1.56).

Making computations for electric systems in terms of per-unit values simplifies the work greatly. A real appreciation of the value of the per-unit method comes through experience. Some of the advantages of the method are summarized briefly below:

1. Manufacturers usually specify the impedance of a piece of apparatus in percent or per unit on the base of the nameplate rating.
2. The per-unit impedances of machines of the same type and widely different rating usually lie within a narrow range although the ohmic values differ materially for machines of different ratings. For this reason when the impedance is not known definitely, it is generally possible to select from tabulated average values a per-unit impedance which will be reasonably correct. Experience in working with per-unit values brings familiarity with the proper values of per-unit impedance for different types of apparatus.
3. When impedance in ohms is specified in an equivalent circuit, each impedance must be referred to the same circuit by multiplying it by the square of the ratio of the rated voltages of the two sides of the transformer connecting the reference circuit and the circuit containing the impedance. The per-unit impedance, once expressed on the proper base, is the same referred to either side of any transformer.
4. The way in which transformers are connected in three-phase circuits does not affect the per-unit impedances of the equivalent circuit although the transformer connection does determine the relation between the voltage bases on the two sides of the transformer.

2.11 SUMMARY

The introduction in this chapter of the simplified equivalent circuit for the transformer is of great importance. Per-unit calculations are used almost continuously throughout the chapters to follow. We have seen how the transformer is eliminated in the equivalent circuit by the use of per-unit calculations. It is important to remember that $\sqrt{3}$ does not enter the detailed per-unit computations because of the specification of a base line-to-line voltage and base line-to-neutral voltage related by $\sqrt{3}$.

The concept of proper selection of base in the various parts of a circuit linked by transformers and the calculation of parameters in per unit on the base specified for the part of the circuit in which the parameters exist is fundamental in building an equivalent circuit from a single-line diagram.

PROBLEMS

2.1. A single-phase transformer rated 7.2 kVA, 1.2 kV/120 V has a primary winding of 800 turns. Determine (*a*) the turns ratio and the number of turns in the secondary winding and (*b*) the currents carried by the two windings when the transformer delivers its rated kVA at rated voltages. Hence, verify Eq. (2.7).

2.2. The transformer of Prob. 2.1 is delivering 6 kVA at its rated voltages and 0.8 power-factor lagging. (*a*) Determine the impedance Z_2 connected across its secondary terminals. (*b*) What is the value of this impedance referred to the primary side (i.e. Z_2')? (*c*) Using the value of Z_2' obtained in part (*b*), determine the magnitude of the primary current and the kVA supplied by the source.

2.3. With reference to Fig 2.2, consider that the flux density inside the center-leg of the transformer core, as a function of time t, is $B(t) = B_m \sin(2\pi ft)$, where B_m is the peak value of the sinusoidal flux density and f is the operating frequency in Hz. If the flux density is uniformly distributed over the cross-sectional area A m^2 of the center-leg, determine

(*a*) The instantaneous flux $\phi(t)$ in terms of B_m, f, A, and t.

(*b*) The instantaneous induced-voltage $e_1(t)$, according to Eq. (2.1).

(*c*) Hence, show that the rms magnitude of the induced voltage of the primary is given by $|E_1| = \sqrt{2}\,\pi f N_1 B_m A$.

(*d*) If $A = 100$ cm^2, $f = 60$ Hz, $B_m = 1.5$ T, and $N_1 = 1000$ turns, compute $|E_1|$.

2.4. For the pair of mutually coupled coils shown in Fig. 2.4, consider that $L_{11} = 1.9$ H, $L_{12} = L_{21} = 0.9$ H, $L_{22} = 0.5$ H, and $r_1 = r_2 = 0$ Ω. The system is operated at 60 Hz.

(*a*) Write the impedance form [Eq. (2.24)] of the system equations.

(*b*) Write the admittance form [Eq. (2.26)] of the system equations.

(*c*) Determine the primary voltage V_1 and the primary current I_1 when the secondary is

(i) open-circuited and has the induced voltage $V_2 = 100\angle 0°$ V.

(ii) short-circuited and carries the current $I_2 = 2\angle 90°$ A.

2.5. For the pair of mutually coupled coils shown in Fig. 2.4, develop an equivalent-T network in the form of Fig. 2.5. Use the parameter values given in Prob. 2.4 and assume that the turns ratio a equals 2. What are the values of the leakage reactances of the windings and the magnetizing susceptance of the coupled coils?

2.6. A single-phase transformer rated 1.2 kV/120 V, 7.2 kVA has the following winding parameters: $r_1 = 0.8\ \Omega$, $x_1 = 1.2\ \Omega$, $r_2 = 0.01\ \Omega$, and $x_2 = 0.01\ \Omega$. Determine
(*a*) The combined winding resistance and leakage reactance referred to the primary side, as shown in Fig. 2.8.
(*b*) The values of the combined parameters referred to the secondary side.
(*c*) The voltage regulation of the transformer when it is delivering 7.5 kVA to a load at 120 V and 0.8 power-factor lagging.

2.7. A single-phase transformer is rated 440/220 V, 5.0 kVA. When the low-voltage side is short-circuited and 35 V is applied to the high-voltage side, rated current flows in the windings and the power input is 100 W. Find the resistance and reactance of the high- and low-voltage windings if the power loss and ratio of reactance to resistance is the same in both windings.

2.8. A single-phase transformer rated 1.2 kV/120 V, 7.2 kVA yields the following test results:
Open-circuit test (primary-open)

Voltage $V_2 = 120$ V; current $I_2 = 1.2$ A; power $W_2 = 40$ W

Short-circuit test (secondary-shorted)

Voltage $V_1 = 20$ V; current $I_1 = 6.0$ A; power $W_1 = 36$ W

Determine
(*a*) The parameters $R_1 = r_1 + a^2r_2$, $X_1 = x_1 + a^2x_2$, G_c, and B_m referred to the primary side, Fig. 2.7.
(*b*) The values of the above parameters referred to the secondary side.
(*c*) The efficiency of the transformer when it delivers 6 kVA at 120 V and 0.9 power factor.

2.9. A single-phase transformer rated 1.2 kV/120 V, 7.2 kVA has primary-referred parameters $R_1 = r_1 + a^2r_2 = 1.0\ \Omega$ and $X_1 = x_1 + a^2x_2 = 4.0\ \Omega$. At rated voltage its core loss may be assumed to be 40 W for all values of the load current.
(*a*) Determine the efficiency and regulation of the transformer when it delivers 7.2 kVA at $V_2 = 120$ V and power factor of (*i*) 0.8 lagging and (*ii*) 0.8 leading.
(*b*) For a given load voltage and power factor it can be shown that the efficiency of a transformer attains its maximum value at the kVA load level which makes the I^2R winding losses equal to the core loss. Using this result, determine the maximum efficiency of the above transformer at rated voltage and 0.8 power factor, and the kVA load level at which it occurs.

2.10. A single-phase system similar to that shown in Fig. 2.10 has two transformers *A-B* and *B-C* connected by a line *B* feeding a load at the receiving end *C*. The ratings and parameter values of the components are:

Transformer *A-B*:	500 V/1.5 kV, 9.6 kVA, leakage reactance = 5%
Transformer *B-C*:	1.2 kV/120 V, 7.2 kVA, leakage reactance = 4%
Line *B*:	series impedance = $(0.5 + j3.0)\ \Omega$
Load *C*:	120 V, 6 kVA at 0.8 power-factor lagging

(*a*) Determine the value of the load impedance in ohms and the actual ohmic impedances of the two transformers referred to both their primary and secondary sides.
(*b*) Choosing 1.2 kV as the voltage base for circuit B and 10 kVA as the systemwide kVA base, express all system impedances in per unit.
(*c*) What value of sending-end voltage corresponds to the given loading conditions?

2.11. A balanced Δ-connected resistive load of 8000 kW is connected to the low-voltage, Δ-connected side of a Y-Δ transformer rated 10,000 kVA, 138/13.8 kV. Find the load resistance in ohms in each phase as measured from line to neutral on the high-voltage side of the transformer. Neglect transformer impedance and assume rated voltage is applied to the transformer primary.

2.12. Solve Prob. 2.11 if the same resistances are reconnected in Y.

2.13. Three transformers, each rated 5 kVA, 220 V on the secondary side, are connected Δ-Δ and have been supplying a balanced 15-kW purely resistive load at 220 V. A change is made which reduces the load to 10 kW, still purely resistive and balanced. Someone suggests that with two-thirds of the load, one transformer can be removed and the system can be operated open Δ. Balanced three-phase voltages will still be supplied to the load since two of the line voltages (and thus also the third) will be unchanged.

To investigate the suggestion further,
(*a*) Find each of the line currents (magnitude and angle) with the 10-kW load and the transformer between a and c removed. (Assume $V_{ab} = 220\angle 0^\circ$ V, sequence *abc*.)
(*b*) Find the kilovoltamperes supplied by each of the remaining transformers.
(*c*) What restriction must be placed on the load for open-Δ operation with these transformers?
(*d*) Think about why the individual transformer kilovoltampere values include a Q component when the load is purely resistive.

2.14. A transformer rated 200 MVA, 345Y/20.5Δ kV connects a balanced load rated 180 MVA, 22.5 kV, 0.8 power-factor lag to a transmission line. Determine
(*a*) The rating of each of three single-phase transformers which when properly connected will be equivalent to the above three-phase transformer.
(*b*) The complex impedance of the load in per unit in the impedance diagram if the base in the transmission line is 100 MVA, 345 kV.

2.15. A three-phase transformer rated 5 MVA, 115/13.2 kV has per-phase series impedance of $(0.007 + j0.075)$ per unit. The transformer is connected to a short distribution line which can be represented by a series impedance per phase of $(0.02 + j0.10)$ per unit on a base of 10 MVA, 13.2 kV. The line supplies a balanced three-phase load rated 4 MVA, 13.2 kV, with lagging power factor 0.85.
(*a*) Draw an equivalent circuit of the system indicating all impedances in per unit. Choose 10 MVA, 13.2 kVA as the base at the load.
(*b*) With the voltage at the primary side of the transformer held constant at 115 kV, the load at the receiving end of the line is disconnected. Find the voltage regulation at the load.

2.16. Three identical single-phase transformers, each rated 1.2 kV/120 V, 7.2 kVA and having a leakage reactance of 0.05 per unit, are connected together to form a three-phase bank. A balanced Y-connected load of 5 Ω per phase is connected across the secondary of the bank. Determine the Y-equivalent per-phase impedance (in ohms and in per unit) seen from the primary side when the transformer bank is connected (*a*) Y-Y, (*b*) Y-Δ, (*c*) Δ-Y, and (*d*) Δ-Δ. Use Table 2.1.

2.17. Figure 2.17(*a*) shows a three-phase generator supplying a load through a three-phase transformer rated 12 kVA/600 V Y, 600 kVA. The transformer has per-phase leakage reactance of 10%. The line-to-line voltage and the line current at the generator terminals are 11.9 kV and 20 A, respectively. The power factor seen by the generator is 0.8 lagging and the phase sequence of supply is *ABC*.

(*a*) Determine the line current and the line-to-line voltage at the load, and the per-phase (equivalent-Y) impedance of the load.

(*b*) Using the line-to-neutral voltage V_A at the transformer primary as reference, draw complete per-phase phasor diagrams of all voltages and currents. Show the correct phase relations between primary and secondary quantities.

(*c*) Compute the real and reactive power supplied by the generator and consumed by the load.

2.18. Solve Prob. 2.17 with phase sequence *ACB*.

2.19. A single-phase transformer rated 30 kVA, 1200/120 V is connected as an autotransformer to supply 1320 V from a 1200 V bus.

(*a*) Draw a diagram of the transformer connections showing the polarity marks on the windings and directions chosen as positive for current in each winding so that the currents will be in phase.

(*b*) Mark on the diagram the values of rated current in the windings and at the input and output.

(*c*) Determine the rated kilovoltamperes of the unit as an autotransformer.

(*d*) If the efficiency of the transformer connected for 1200/120 V operation at rated load unity power factor is 97%, determine its efficiency as an autotransformer with rated current in the windings and operating at rated voltage to supply a load at unity power factor.

2.20. Solve Prob. 2.19 if the transformer is to supply 1080 V from a 1200 V bus.

2.21. Two buses *a* and *b* are connected to each other through impedances $X_1 = 0.1$ and $X_2 = 0.2$ per unit in parallel. Bus *b* is a load bus supplying a current $I = 1.0\angle -30°$ per unit. The per-unit bus voltage V_b is $1.0\angle 0°$. Find *P* and *Q* into bus *b* through each of the parallel branches (*a*) in the circuit described, (*b*) if a regulating transformer is connected at bus *b* in the line of higher reactance to give a boost of 3% in voltage magnitude toward the load ($a = 1.03$), and (*c*) if the regulating transformer advances the phase 2° ($a = \varepsilon^{j\pi/90}$). Use the circulating-current method for parts (*b*) and (*c*), and assume that V_a is adjusted for each part of the problem so that V_b remains constant. Figure 2.26 is the single-line diagram showing buses *a* and *b* of the system with the regulating transformer in place. Neglect the impedance of the transformer.

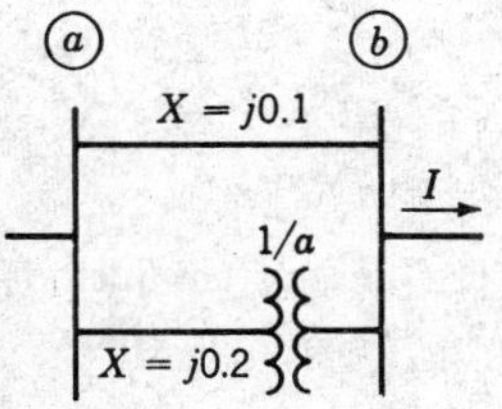

FIGURE 2.26
Circuit for Prob. 2.21.

2.22. Two reactances $X_1 = 0.08$ and $X_2 = 0.12$ per unit are in parallel between two buses a and b in a power system. If $V_a = 1.05\angle 10^\circ$ and $V_b = 1.0\angle 0^\circ$ per unit, what should be the turns ratio of the regulating transformer to be inserted in series with X_2 at bus b so that no vars flow into bus b from the branch whose reactance is X_1? Use the circulating-current method, and neglect the reactance of the regulating transformer. P and Q of the load and V_b remain constant.

2.23. Two transformers, each rated 115Y/13.2Δ kV, operate in parallel to supply a load of 35 MVA, 13.2 kV at 0.8 power-factor lagging. Transformer 1 is rated 20 MVA with $X = 0.09$ per unit, and transformer 2 is rated 15 MVA with $X = 0.07$ per unit. Find the magnitude of the current in per unit through each transformer, the megavoltampere output of each transformer, and the megavoltamperes to which the total load must be limited so that neither transformer is overloaded. If the taps on transformer 1 are set at 111 kV to give a 3.6% boost in voltage toward the low-voltage side of that transformer compared to transformer 2, which remains on the 115-kV tap, find the megavoltampere output of each transformer for the original 35-MVA total load and the maximum megavoltamperes of the total load which will not overload the transformers. Use a base of 35 MVA, 13.2 kV on the low-voltage side. The circulating-current method is satisfactory for this problem.

CHAPTER 3

THE SYNCHRONOUS MACHINE

The synchronous machine as an ac generator, driven by a turbine to convert mechanical energy into electrical energy, is the major electric power generating source throughout the world. As a motor the machine converts electrical energy to mechanical energy. We are chiefly concerned with the synchronous generator, but we shall give some consideration to the synchronous motor. We cannot treat the synchronous machine fully, but there are many books on the subject of ac machinery which provide quite adequate analysis of generators and motors.[1] Our interest is in the application and operation of the synchronous machine within a large interconnected power system. Emphasis is on principles and external behavior under both steady-state and transient conditions.

The windings of the polyphase synchronous machine constitute a group of inductively coupled electric circuits, some of which rotate relative to others so that mutual inductances are variable. The general equations developed for the flux linkages of the various windings are applicable to both steady-state and transient analysis. Only linear magnetic circuits are considered, with saturation neglected. This allows us, whenever convenient, to refer separately to the flux

[1]For a much more detailed discussion of synchronous machines, consult any of the texts on electric machinery such as A. E. Fitzgerald, C. Kingsley, Jr., and S. D. Umans, *Electric Machinery*, 4th ed., McGraw-Hill, Inc., New York, 1983.

and flux linkages produced by a component magnetomotive force (mmf)—even though in any electric machine there exists only *net physical flux* due to the resultant mmf of all the magnetizing sources. Simplified equivalent circuits are developed through which important physical relationships within the machine can be visualized. Therefore, our treatment of the synchronous machine should provide confidence in the equivalent circuits sufficient for understanding the role of the generator in our further studies of power system analysis.

3.1 DESCRIPTION OF THE SYNCHRONOUS MACHINE

The two principal parts of a synchronous machine are ferromagnetic structures. The stationary part which is essentially a hollow cylinder, called the *stator* or *armature*, has longitudinal slots in which there are coils of the armature windings. These windings carry the current supplied to an electrical load by a generator or the current received from an ac supply by a motor. The *rotor* is the part of the machine which is mounted on the shaft and rotates inside the hollow stator. The winding on the rotor, called the *field winding*, is supplied with dc current. The very high mmf produced by this current in the field winding combines with the mmf produced by currents in the armature windings. The resultant flux across the air gap between the stator and rotor generates voltages in the coils of the armature windings and provides the electromagnetic torque between the stator and rotor. Figure 3.1 shows the threading of a four-pole cylindrical rotor into the stator of a 1525-MVA generator.

The dc current is supplied to the field winding by an *exciter*, which may be a generator mounted on the same shaft or a separate dc source connected to the field winding through brushes bearing on slip rings. Large ac generators usually have exciters consisting of an ac source with solid-state rectifiers.

If the machine is a generator, the shaft is driven by a *prime mover*, which is usually a steam or hydraulic turbine. The electromagnetic torque developed in the generator when it delivers power opposes the torque of the prime mover. The difference between these two torques is due to losses in the iron core and friction. In a motor the electromagnetic torque developed in the machine (except for core and friction losses) is converted to the shaft torque which drives the mechanical load.

Figure 3.2 shows a very elementary three-phase generator. The field winding, indicated by the f-coil, gives rise to two poles N and S as marked. The axis of the field poles is called the *direct axis* or simply the *d-axis*, while the centerline of the interpolar space is called the *quadrature axis* or simply the *q-axis*. The positive direction along the d-axis leads the positive direction along the q-axis by 90° as shown. The generator in Fig. 3.2 is called a *nonsalient* or *round-rotor machine* because it has a cylindrical rotor like that of Fig. 3.1. In the actual machine the winding has a large number of turns distributed in slots around the circumference of the rotor. The strong magnetic field produced links

FIGURE 3.1
Photograph showing the threading of a four-pole cylindrical rotor into the stator of a 1525-MVA generator. (*Courtesy Utility Power Corporation, Wisconsin.*)

the stator coils to induce voltage in the armature windings as the shaft is turned by the prime mover.

The stator is shown in cross section in Fig. 3.2. Opposite sides of a coil, which is almost rectangular, are in slots a and a' 180° apart. Similar coils are in slots b and b' and slots c and c'. Coil sides in slots a, b, and c are 120° apart. The conductors shown in the slots indicate a coil of only one turn, but such a coil may have many turns and is usually in series with identical coils in adjacent slots to form a winding having ends designated a and a'. Windings with ends designated $b - b'$ and $c - c'$ are the same as the $a - a'$ winding except for their symmetrical location at angles of 120° and 240°, respectively, around the armature.

Figure 3.3 shows a *salient-pole machine* which has *four poles*. Opposite sides of an armature coil are 90° apart. So, there are two coils for each phase. Coil sides a, b, and c of adjacent coils are 60° apart. The two coils of each phase may be connected in series or in parallel.

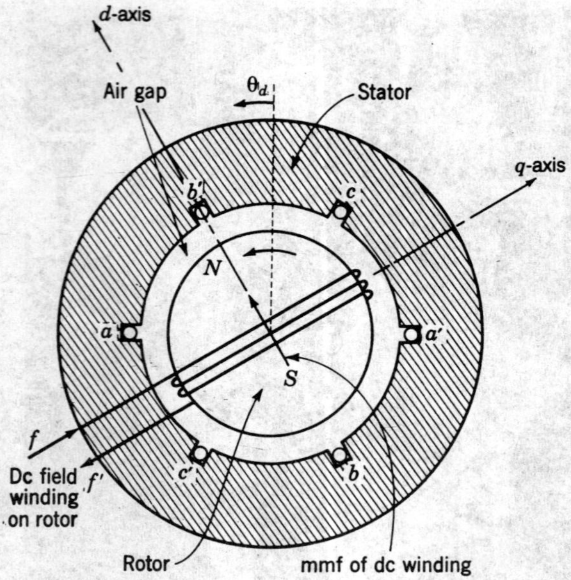

FIGURE 3.2
Elementary three-phase ac generator showing end view of the two-pole cylindrical rotor and cross section of the stator.

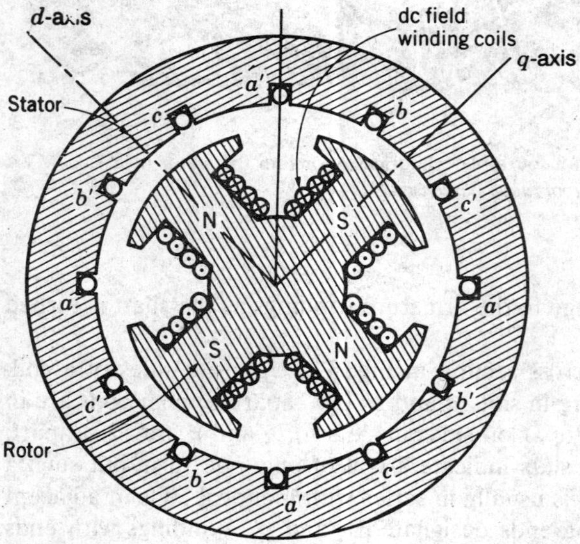

FIGURE 3.3
Cross section of an elementary stator and salient-pole rotor.

Although not shown in Fig. 3.3, salient-pole machines usually have *damper windings*, which consist of short-circuited copper bars through the pole face similar to part of a "squirrel cage" winding of an induction motor. The purpose of the damper winding is to reduce the mechanical oscillations of the rotor about *synchronous speed*, which is determined by the number of poles of the machine and the frequency of the system to which the machine is connected.

In the *two*-pole machine *one* cycle of voltage is generated for each revolution of the two-pole rotor. In the *four*-pole machine *two* cycles are generated in each coil per revolution. Since the number of cycles per revolution equals the number of pairs of poles, the frequency of the generated voltage is

$$f = \frac{P}{2}\frac{N}{60} = \frac{P}{2}f_m \text{ Hz} \tag{3.1}$$

where f = electrical frequency in Hz

P = number of poles

N = rotor speed in revolutions per minute (rpm)

$f_m = N/60$, the mechanical frequency in revolutions per second (rps).

Equation (3.1) tells us that a two-pole, 60-Hz machine operates at 3600 rpm, whereas a four-pole machine operates at 1800 rpm. Usually, fossil-fired steam turbogenerators are two-pole machines, whereas hydrogenerating units are slower machines with many pole pairs.

Since one cycle of voltage (360° of the voltage wave) is generated every time a pair of poles passes a coil, we must distinguish between *electrical* degrees used to express voltage and current and *mechanical* degrees used to express the position of the rotor. In a two-pole machine electrical and mechanical degrees are equal. In any other machine the number of electrical degrees or radians equals $P/2$ times the number of mechanical degrees or radians, as can be seen from Eq. (3.1) by multiplying both sides by 2π. In a four-pole machine, therefore, two cycles or 720 electrical degrees are produced per revolution of 360 mechanical degrees.

In this chapter all angular measurements are expressed in electrical degrees unless otherwise stated, and the direct axis always *leads* the quadrature axis by 90 electrical degrees in the *counterclockwise* direction of rotation regardless of the number of poles or the type of rotor construction.

3.2 THREE-PHASE GENERATION

The field and armature windings of the synchronous machine described in Sec. 3.1 are distributed in slots around the periphery of the air gap. Section A.1 of the Appendix shows that these distributed windings can be replaced along their axes by *concentrated* coils with appropriate self- and mutual inductances. Figure 3.4 shows three such coils—a, b, and c—which represent the three armature windings on the stator of the round-rotor machine, and a concentrated coil f, which represents the distributed field winding on the rotor. The three stationary armature coils are identical in every respect and each has one of its two terminals connected to a common point o. The other three terminals are

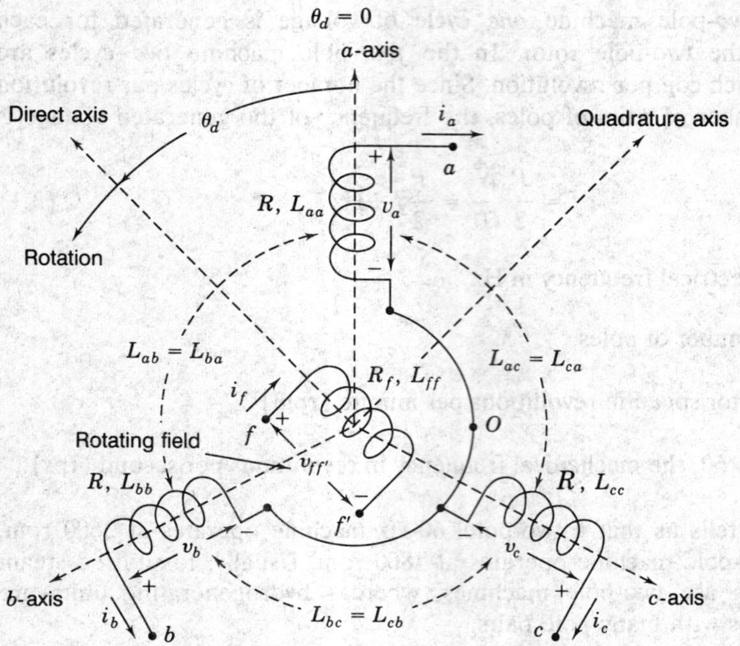

FIGURE 3.4
Idealized three-phase generator showing identical armature coils a, b, and c, and field coil f. Direct axis leads quadrature axis by 90° in the anticlockwise direction of rotation.

marked a, b, and c. The axis of coil a is chosen at $\theta_d = 0°$, and counterclockwise around the air gap are the axes of the b-coil at $\theta_d = 120°$ and of the c-coil at $\theta_d = 240°$. For the round-rotor machine it is shown in Sec. A.1 of the Appendix that:

- Each of the concentrated coils a, b, and c has self-inductance L_s, which is equal to the self-inductances L_{aa}, L_{bb}, and L_{cc} of the distributed armature windings which the coils represent so that

$$L_s = L_{aa} = L_{bb} = L_{cc} \tag{3.2}$$

- The mutual inductances L_{ab}, L_{bc}, and L_{ca} between each adjacent pair of concentrated coils are negative constants denoted by $-M_s$ so that

$$-M_s = L_{ab} = L_{bc} = L_{ca} \tag{3.3}$$

- The mutual inductance between the field coil f and each of the stator coils varies with the rotor position θ_d as a cosinusoidal function with maximum value M_f so that

$$L_{af} = M_f \cos \theta_d$$

$$L_{bf} = M_f \cos(\theta_d - 120°) \tag{3.4}$$

$$L_{cf} = M_f \cos(\theta_d - 240°)$$

The field coil has a constant self-inductance L_{ff}. This is because in the round-rotor machine (and, indeed, in the salient-pole machine also), the field winding on the d-axis produces flux through a similar magnetic path in the stator for all positions of the rotor (neglecting the small effect of armature slots).

Flux linkages with each of the coils a, b, c, and f are due to its own current and the currents in the three other coils. Flux-linkage equations are therefore written for all four coils as follows:

Armature:

$$\lambda_a = L_{aa}i_a + L_{ab}i_b + L_{ac}i_c + L_{af}i_f = L_s i_a - M_s(i_b + i_c) + L_{af}i_f$$

$$\lambda_b = L_{ba}i_a + L_{bb}i_b + L_{bc}i_c + L_{bf}i_f = L_s i_b - M_s(i_a + i_c) + L_{bf}i_f \tag{3.5}$$

$$\lambda_c = L_{ca}i_a + L_{cb}i_b + L_{cc}i_c + L_{cf}i_f = L_s i_c - M_s(i_a + i_b) + L_{cf}i_f$$

Field:

$$\lambda_f = L_{af}i_a + L_{bf}i_b + L_{cf}i_c + L_{ff}i_f \tag{3.6}$$

If i_a, i_b, and i_c are a *balanced* three-phase set of currents, then

$$i_a + i_b + i_c = 0. \tag{3.7}$$

Setting $i_a = -(i_b + i_c)$, $i_b = -(i_a + i_c)$, and $i_c = -(i_a + i_b)$ in Eqs. (3.5) gives

$$\lambda_a = (L_s + M_s)i_a + L_{af}i_f$$

$$\lambda_b = (L_s + M_s)i_b + L_{bf}i_f \tag{3.8}$$

$$\lambda_c = (L_s + M_s)i_c + L_{cf}i_f$$

For now we are interested in steady-state conditions. We assume, therefore, that current i_f is dc with a constant value I_f and that the field rotates at constant angular velocity ω so that for the two-pole machine

$$\frac{d\theta_d}{dt} = \omega \qquad \text{and} \qquad \theta_d = \omega t + \theta_{d0} \tag{3.9}$$

The initial position of the field winding is given by the angle θ_{d0}, which can be arbitrarily chosen at $t = 0$. Equations (3.4) give the expressions for L_{af}, L_{bf}, and L_{cf} in terms of θ_d. Substituting $(\omega t + \theta_{d0})$ for θ_d and using the results along with $i_f = I_f$ in Eqs. (3.8), we obtain

$$\begin{aligned}
\lambda_a &= (L_s + M_s)i_a + M_f I_f \cos(\omega t + \theta_{d0}) \\
\lambda_b &= (L_s + M_s)i_b + M_f I_f \cos(\omega t + \theta_{d0} - 120^\circ) \\
\lambda_c &= (L_s + M_s)i_c + M_f I_f \cos(\omega t + \theta_{d0} - 240^\circ)
\end{aligned} \tag{3.10}$$

The first of these equations shows that λ_a has two flux-linkage components—one due to the field current I_f and the other due to the armature current i_a, which is flowing *out* of the machine for generator action. If coil a has resistance R, then the voltage drop v_a across the coil from terminal a to terminal o in Fig. 3.4 is given by

$$v_a = -Ri_a - \frac{d\lambda_a}{dt} = -Ri_a - (L_s + M_s)\frac{di_a}{dt} + \omega M_f I_f \sin(\omega t + \theta_{d0}) \tag{3.11}$$

The negative signs apply, as discussed in Sec. 2.2, because the machine is being treated as a generator. The last term of Eq. (3.11) represents an internal emf, which we now call $e_{a'}$. This emf can be written

$$e_{a'} = \sqrt{2}\,|E_i| \sin(\omega t + \theta_{d0}) \tag{3.12}$$

where the rms magnitude $|E_i|$, proportional to the field current, is defined by

$$|E_i| = \frac{\omega M_f I_f}{\sqrt{2}} \tag{3.13}$$

The action of the field current causes $e_{a'}$ to appear across the terminals of the a-phase when i_a is zero, and so it is called by various names such as the *no-load voltage*, the *open-circuit voltage*, the *synchronous internal voltage*, or *generated emf* of phase a. The angle θ_{d0} indicates the position of the field winding (and the d-axis) relative to the a-phase at $t = 0$. Hence, $\delta \triangleq \theta_{d0} - 90^\circ$ indicates the position of the q-axis, which is 90° *behind* the d-axis in Fig. 3.4. For later convenience we now set $\theta_{d0} = \delta + 90^\circ$, and then we have

$$\theta_d = (\omega t + \theta_{d0}) = (\omega t + \delta + 90^\circ) \tag{3.14}$$

where θ_d, ω, and δ have consistent units of angular measurement. Substituting from Eq. (3.14) into Eq. (3.12) and noting that $\sin(\alpha + 90°) = \cos\alpha$, we obtain for the open-circuit voltage of phase a

$$e_{a'} = \sqrt{2}\,|E_i|\cos(\omega t + \delta) \tag{3.15}$$

The terminal voltage v_a of Eq. (3.11) is then given by

$$v_a = -Ri_a - (L_s + M_s)\frac{di_a}{dt} + \underbrace{\sqrt{2}\,|E_i|\cos(\omega t + \delta)}_{e_{a'}} \tag{3.16}$$

This equation corresponds to the a-phase circuit of Fig. 3.5 in which the no-load voltage $e_{a'}$ is the source and the external load is balanced across all three phases.

The flux linkages λ_b and λ_c given by Eq. (3.10) can be treated in the same way as λ_a. Since the armature windings are identical, results similar to Eqs. (3.15) and (3.16) can be found for the no-load voltages $e_{b'}$ and $e_{c'}$ which lag $e_{a'}$ by 120° and 240°, respectively, in Fig. 3.5. Hence, $e_{a'}$, $e_{b'}$, and $e_{c'}$ constitute a balanced three-phase set of emfs which give rise to balanced three-phase line

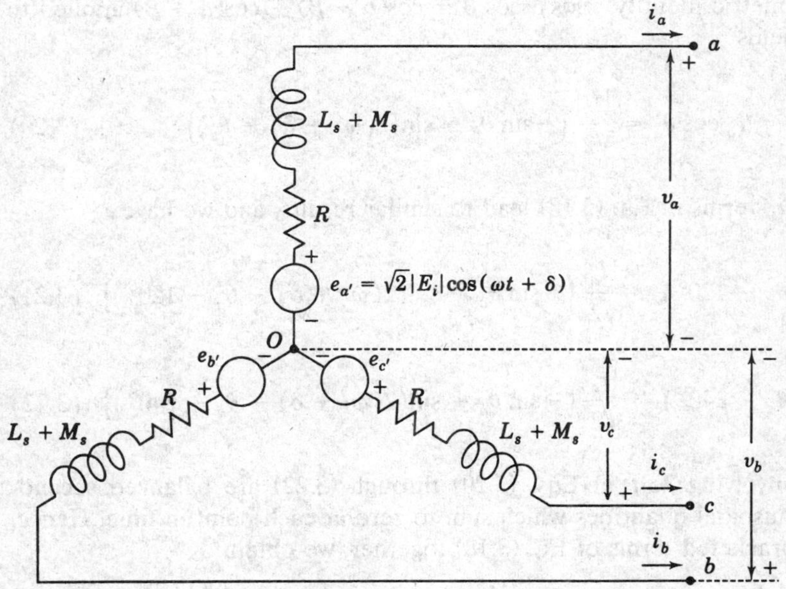

FIGURE 3.5
Armature equivalent circuit of the idealized three-phase generator showing balanced no-load voltages $e_{a'}$, $e_{b'}$, and $e_{c'}$ in the steady state.

currents, say,

$$
\begin{aligned}
i_a &= \sqrt{2}\,|I_a|\cos(\omega t + \delta - \theta_a) \\
i_b &= \sqrt{2}\,|I_a|\cos(\omega t + \delta - \theta_a - 120^\circ) \\
i_c &= \sqrt{2}\,|I_a|\cos(\omega t + \delta - \theta_a - 240^\circ)
\end{aligned}
\tag{3.17}
$$

where $|I_a|$ is the rms value and θ_a is the phase angle of *lag* of the current i_a *with respect to* $e_{a'}$. When the emfs and the currents are expressed as phasors, Fig. 3.5 becomes very much like the equivalent circuit introduced in Fig. 1.11. Before employing the equivalent circuit, let us consider the flux linkages λ_f of the field winding.

The expressions for L_{af}, L_{bf}, and L_{cf} in Eqs. (3.4) can be substituted into Eq. (3.6) to yield

$$\lambda_f = L_{ff}I_f + M_f[i_a \cos\theta_d + i_b\cos(\theta_d - 120^\circ) + i_c\cos(\theta_d - 240^\circ)] \tag{3.18}$$

The first term within the brackets can be expanded according to Eqs. (3.14) and (3.17) as follows:

$$i_a\cos\theta_d = \sqrt{2}\,|I_a|\cos(\omega t + \delta - \theta_a)\cos(\omega t + \delta + 90^\circ) \tag{3.19}$$

The trigonometric identity $2\cos\alpha\cos\beta = \cos(\alpha - \beta) + \cos(\alpha + \beta)$ applied to Eq. (3.19) yields

$$i_a\cos\theta_d = \frac{|I_a|}{\sqrt{2}}\{-\sin\theta_a - \sin(2(\omega t + \delta) - \theta_a)\} \tag{3.20}$$

The i_b and i_c terms in Eq. (3.18) lead to similar results, and we have

$$i_b\cos(\theta_d - 120^\circ) = \frac{|I_a|}{\sqrt{2}}\{-\sin\theta_a - \sin(2(\omega t + \delta) - \theta_a - 120^\circ)\} \tag{3.21}$$

$$i_c\cos(\theta_d - 240^\circ) = \frac{|I_a|}{\sqrt{2}}\{-\sin\theta_a - \sin(2(\omega t + \delta) - \theta_a - 240^\circ)\} \tag{3.22}$$

The terms involving $2\omega t$ in Eqs. (3.20) through (3.22) are balanced second-harmonic sinusoidal quantities which sum to zero at each point in time. Hence, adding the bracketed terms of Eq. (3.18) together, we obtain

$$[i_a\cos\theta_d + i_b\cos(\theta_d - 120^\circ) + i_c\cos(\theta_d - 240^\circ)] = -\frac{3|I_a|}{\sqrt{2}}\sin\theta_a \tag{3.23}$$

and the expression for λ_f takes on the simpler form

$$\lambda_f = L_{ff}I_f - \frac{3M_f|I_a|}{\sqrt{2}}\sin\theta_a = L_{ff}I_f + \sqrt{\frac{3}{2}}M_f i_d \tag{3.24}$$

where *dc current* $i_d = \sqrt{\frac{2}{3}}[i_a\cos\theta_d + i_b\cos(\theta_d - 120°) + i_c(\cos\theta_d - 240°)]$ or by Eq. (3.23)

$$i_d = -\sqrt{3}\,|I_a|\sin\theta_a \tag{3.25}$$

which is useful later in this chapter. For now let us observe from Eq. (3.24) that the flux linkages with the field winding due to the combination of i_a, i_b, and i_c do not vary with time. We can, therefore, regard those flux linkages as coming from the steady dc current i_d in a fictitious dc circuit coincident with the d-axis and thus stationary with respect to the field circuit. The two circuits rotate together in synchronism and have a mutual inductance $(\sqrt{3/2})\,M_f$ between them, as shown in Fig. 3.6. In general, the field winding with resistance R_f and entering current i_f has terminal voltage $v_{ff'}$ given by

$$v_{ff'} = R_f i_f + \frac{d\lambda_f}{dt} \tag{3.26}$$

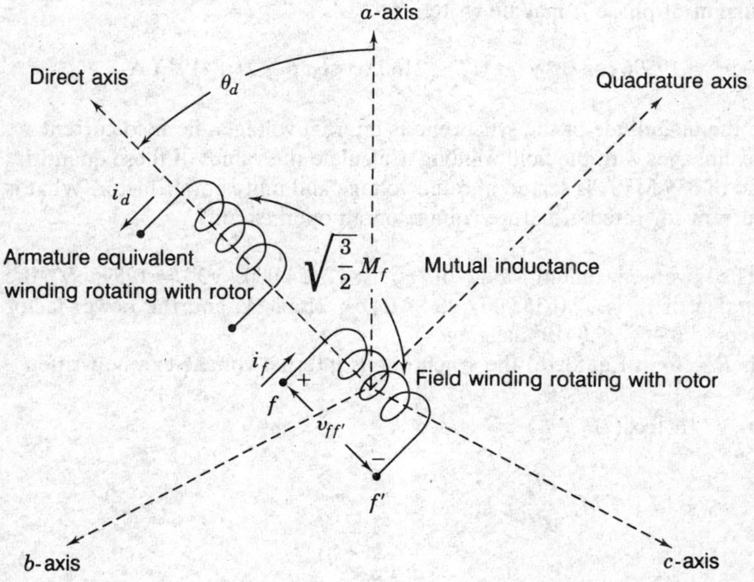

FIGURE 3.6
Representing the armature of the synchronous machine by a direct-axis winding of mutual inductance $\sqrt{3/2}\,M_f$ with the field winding. Both windings rotate together in synchronism.

Because λ_f is not varying with time in the steady state, the field voltage becomes $v_{ff'} = R_f I_f$ and $i_f = I_f$ can be supplied by a dc source.

Equation (3.25) shows that the numerical value of i_d depends on the magnitude of the armature current $|I_a|$ and its phase angle of lag θ_a relative to the internal voltage $e_{a'}$. For lagging power factors θ_a is positive and so i_d is negative, which means that the combined effect of the armature currents i_a, i_b, and i_c is demagnetizing; that is, i_d opposes the magnetizing influence of the field current I_f. To overcome this demagnetizing influence, I_f has to be increased by the excitation system of the generator. At leading power factors θ_a takes on smaller values, which means that the demagnetizing influence of the armature currents (represented by $i_d = -\sqrt{3}\,|I_a|\sin\theta_a$) is reduced and I_f can then be lowered by the excitation system. In an actual machine the effect of the currents i_a, i_b, and i_c is called *armature reaction* and the control of the field current is called *excitation system control*, which is discussed in Sec. 3.4.

Example 3.1. A 60-Hz three-phase synchronous generator with negligible armature resistance has the following inductance parameters:

$$L_{aa} = L_s = 2.7656 \text{ mH} \qquad M_f = 31.6950 \text{ mH}$$

$$L_{ab} = M_s = 1.3828 \text{ mH} \qquad L_{ff} = 433.6569 \text{ mH}$$

The machine is rated at 635 MVA, 0.90 power-factor lagging, 3600 rpm, 24 kV. When operating under rated load conditions, the line-to-neutral terminal voltage and line current of phase a may be written

$$v_a = 19596 \cos \omega t \text{ V} \qquad i_a = 21603 \cos(\omega t - 25.8419^\circ) \text{ A}$$

Determine the magnitude of the synchronous internal voltage, the field current I_f, and the flux linkages with the field winding. Calculate the values of these quantities when a load of 635 MVA is served at rated voltage and unity power factor. What is the field current for rated armature voltage on an open circuit?

Solution. The given maximum value of v_a is $\sqrt{2}(24{,}000/\sqrt{3}) = 19596$ V, the maximum value of i_a is $\sqrt{2}(635{,}000/\sqrt{3} \times 24) = 21603$ A, and the power-factor angle $\theta = \cos^{-1} 0.9 = 25.8419^\circ$ lagging.

With $R = 0$, in Eq. (3.16) the synchronous internal voltage can be written

$$\begin{aligned} e_{a'} &= \sqrt{2}\,|E_i|\cos(\omega t + \delta) \\ &= v_a + (L_s + M_s)\frac{di_a}{dt} \\ &= v_a + (2.7656 + 1.3828)10^{-3}\frac{di_a}{dt} \\ &= 19596 \cos \omega t - (4.1484)10^{-3} \times \omega \times 21603 \sin(\omega t - 25.8419^\circ) \end{aligned}$$

Setting $\omega = 120\pi$, we obtain

$$e_{a'} = \sqrt{2}\,|E_i|\cos(\omega t + \delta) = 19596\cos\omega t - 33785\sin(\omega t - 25.8419^\circ)\text{ V}$$

and expanding the second term according to $\sin(\alpha - \beta) = \sin\alpha\cos\beta - \cos\alpha\sin\beta$ gives

$$e_{a'} = \sqrt{2}\,|E_i|\cos(\omega t + \delta) = 34323\cos\omega t - 30407\sin\omega t$$

$$= 45855\cos(\omega t + 41.5384^\circ)\text{ V}$$

Hence, the synchronous internal voltage has magnitude $\sqrt{2}\,|E_i| = 45855$ V and angle $\delta = 41.5384^\circ$. From Eq. (3.13) we find

$$I_f = \frac{\sqrt{2}|E_i|}{\omega M_f} = \frac{45855}{120\pi \times 31.695 \times 10^{-3}} = 3838\text{ A}$$

The flux linkages with the field windings are given by Eq. (3.24),

$$\lambda_f = L_{ff}I_f - \frac{3M_f}{\sqrt{2}}|I_a|\sin\theta_a$$

where θ_a is the angle of lag of i_a measured with respect to $e_{a'}$. Since i_a lags 25.8419° behind v_a, which lags 41.5384° behind $e_{a'}$, it follows that

$$\theta_a = 25.8419^\circ + 41.5384^\circ = 67.3803^\circ$$

$$|I_a|\sin\theta_a = \frac{21603}{\sqrt{2}}\sin 67.3803^\circ = 14100.6\text{ A}$$

and substituting in the above expression for λ_f yields

$$\lambda_f = (433.6569 \times 10^{-3})3838 - \frac{3 \times 31.695 \times 10^{-3}}{\sqrt{2}} \times 14100.6$$

$$= 1664.38 - 948.06 = 716.32\text{ Wb-turns}$$

Repeating the above sequence of calculations at unity power factor, we obtain

$$e_{a'} = \sqrt{2}\,|E_i|\cos(\omega t + \delta) = 19596\cos\omega t - 33785\sin\omega t$$

$$= 39057\cos(\omega t + 59.8854^\circ)$$

Because $|E_i|$ is directly proportional to I_f, we have from previous calculations

$$I_f = \frac{39057}{45855} \times 3838 = 3269 \text{ A}$$

Current i_a is in phase with v_a and lags $e_{a'}$ by 59.8854°. Therefore,

$$|I_a| \sin \theta_a = 15276 \sin 59.8854° = 13214 \text{ A}$$

$$\lambda_f = (433.6569 \times 10^{-3})3269 - \frac{3 \times 31.695 \times 10^{-3}}{\sqrt{2}} \times 13214$$

$$= 1417.62 - 888.43 = 529.19 \text{ Wb-turns}$$

Thus, when the power factor of the load goes from 0.9 lagging to 1.0 under rated megavoltamperes loading and voltage conditions, the field current is reduced from 3838 to 3269 A. Also, the net air-gap flux linking the field winding of the generator is reduced along with the demagnetizing influence of armature reaction.

The field current required to maintain rated terminal voltage in the machine under open-circuit conditions is found from Eq. (3.13), and Eq. (3.16) with $i_a = 0$,

$$I_f = \frac{\sqrt{2}|E_i|}{\omega M_f} = \frac{19596 \times 10^3}{120\pi \times 31.695} = 1640 \text{ A}$$

3.3 SYNCHRONOUS REACTANCE AND EQUIVALENT CIRCUITS

The coupled-circuit model in Fig. 3.4 represents the idealized Y-connected round-rotor synchronous machine. Let us assume that the machine is rotating at synchronous speed ω and that the field current I_f is steady dc. Under these conditions the balanced three-phase circuit of Fig. 3.5 gives the steady-state operation of the machine. The no-load voltages are the emfs $e_{a'}$, $e_{b'}$, and $e_{c'}$. Choosing a-phase as the reference phase for the machine, we obtain the per-phase equivalent circuit of Fig. 3.7(a) with steady-state sinusoidal currents and voltages which lead the corresponding currents and voltages of phases b and c by 120° and 240°, respectively.

We recall that the phase angle of the current i_a in Eq. (3.17) is chosen with respect to the no-load voltage $e_{a'}$ of the a-phase. In practice, $e_{a'}$ cannot be measured under load, and so it is preferable to choose the terminal voltage v_a as reference and to measure the phase angle of the current i_a with respect to

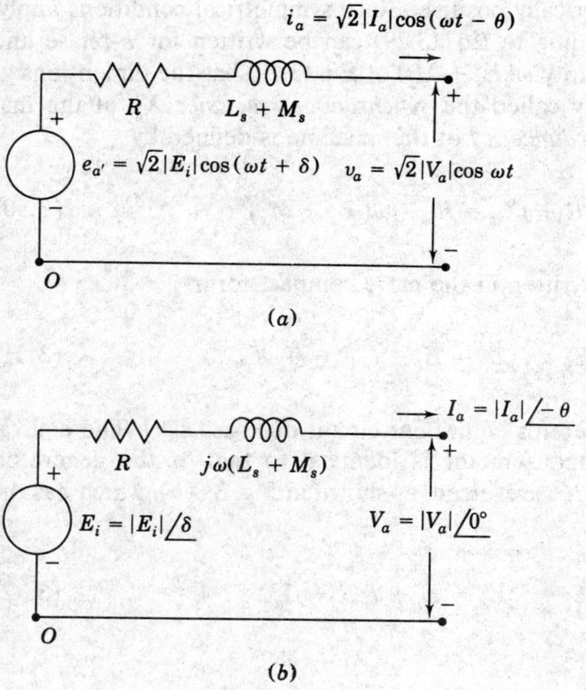

FIGURE 3.7
Equivalent circuit for reference phase a of the synchronous machine showing voltages and currents as (a) cosinusoidal and (b) phasor quantities.

v_a. Therefore, we define

$$v_a = \sqrt{2}\,|V_a|\cos \omega t; \qquad e_{a'} = \sqrt{2}\,|E_i|\cos(\omega t + \delta); \qquad i_a = \sqrt{2}\,|I_a|\cos(\omega t - \theta) \tag{3.27}$$

Note that $e_{a'}$ corresponds to Eq. (3.15) and that i_a differs from Eq. (3.17) only in the respect that the phase angle $\theta = \theta_a - \delta$ is now the angle of lag of i_a measured with respect to terminal voltage v_a. The phasor equivalents of Eqs. (3.27) are

$$V_a = |V_a|\angle 0° \;; \qquad E_{a'} = |E_i|\angle \delta \;; \qquad I_a = |I_a|\angle -\theta \tag{3.28}$$

and these are marked on the equivalent circuit of Fig. 3.7(b) for which the phasor-voltage equation is

$$V_a = \underbrace{E_i}_{\substack{\text{Generated}\\ \text{at no load}}} - \underbrace{RI_a}_{\substack{\text{Due to armature}\\ \text{resistance}}} - \underbrace{j\omega L_s I_a}_{\substack{\text{Due to armature}\\ \text{self-reactance}}} - \underbrace{j\omega M_s I_a}_{\substack{\text{Due to armature}\\ \text{mutual reactance}}} \tag{3.29}$$

When the current I_a leads V_a, the angle θ is numerically negative; and when I_a

lags V_a, the angle θ is numerically positive. Since symmetrical conditions apply, phasor equations corresponding to Eq. (3.29) can be written for *b*-phase and *c*-phase. The combined quantity $\omega(L_s + M_s)$ of Eq. (3.29) has the dimensions of reactance and is customarily called the *synchronous reactance* X_d of the machine. The *synchronous impedance* Z_d of the machine is defined by

$$Z_d = R + jX_d = R + j\omega(L_s + M_s) \tag{3.30}$$

and Eq. (3.29) then can be written in the more compact form

$$V_a = E_i - I_a Z_d = E_i - I_a R - jI_a X_d \tag{3.31}$$

from which follows the generator equivalent circuit of Fig. 3.8(*a*). The equivalent circuit for the synchronous motor is identical to that of the generator, except that the direction of I_a is reversed, as shown in Fig. 3.8(*b*), which has the equation

$$V_a = E_i + I_a Z_d = E_i + I_a R + jI_a X_d \tag{3.32}$$

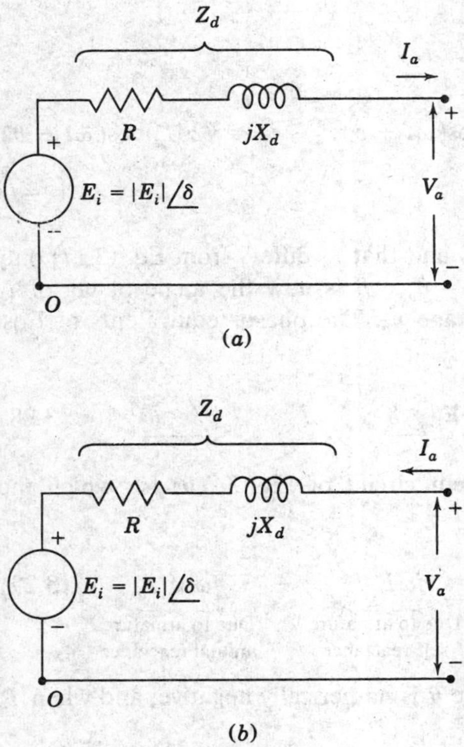

FIGURE 3.8
Equivalent circuits for (*a*) the synchronous generator and (*b*) the synchronous motor with constant synchronous impedance $Z_d = R + jX_d$.

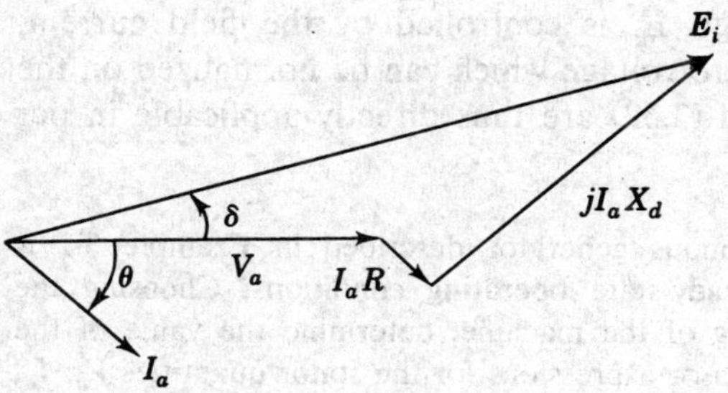

$$E_i = V_a + I_aR + jI_aX_d$$

(*a*)

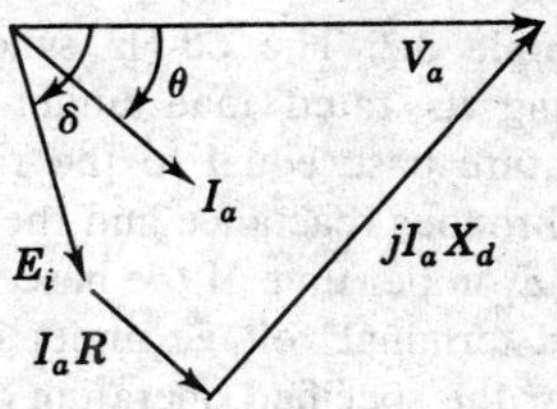

$$V_a = E_i + I_aR + jI_aX_d$$

(*b*)

FIGURE 3.9
Phasor diagrams of: (*a*) overexcited generator delivering lagging current I_a; (*b*) underexcited motor drawing lagging current I_a.

Phasor diagrams for Eqs. (3.31) and (3.32) are shown in Fig. 3.9 for the case of lagging power-factor angle θ measured with respect to the terminal voltage. In Fig. 3.9(*a*) for the generator note that E_i always leads V_a, and in Fig. 3.9(*b*) for the motor E_i always lags V_a.

Except for the case of an isolated generator supplying its own load, most synchronous machines are connected to large interconnected power systems such that the terminal voltage V_a (soon to be called V_t for emphasis) is not altered by machine loading. In that case the point of connection is called an *infinite bus*, which means that its voltage remains constant and no frequency change occurs regardless of changes made in operating the synchronous machine.

Synchronous machine parameters and operating quantities such as voltage and current are normally represented in per unit or normalized values using bases corresponding to the nameplate data of the machine. Such parameters are provided by the manufacturer. Machines of similar design have normalized parameters which fall in a very narrow range regardless of size, and this is very useful when data for a particular machine are not available (see Table A.2 in the Appendix). In the armature of the three-phase machine usually the kilovoltampere base corresponds to the three-phase rating of the machine and base voltage in kilovolts corresponds to the rated line-to-line voltage in kilovolts. Accordingly, the per-phase equivalent circuit of Fig. 3.8 has a kVA base equal to the kilovoltampere rating of one phase and a voltage base equal to the rated line-to-neutral voltage of the machine. Base armature impedance is therefore calculated from Eq. (1.54) in the usual way.

Although the generated voltage E_i is controlled by the field current, nonetheless, it is a per-phase armature voltage which can be normalized on the armature base. Equations (3.31) and (3.32) are thus directly applicable in per unit on the armature base.

Example 3.2. The 60-Hz synchronous generator described in Example 3.1 is serving its rated load under steady-state operating conditions. Choosing the armature base equal to the rating of the machine, determine the value of the synchronous reactance and the phasor expressions for the stator quantities V_a, I_a, and E_i in per unit. If the base field current equals that value of I_f which produces rated terminal voltage under open-circuit conditions, determine the value of I_f under the specified operating conditions.

Solution. From Example 3.1 we find for the armature that

$$\text{Base kVA} = 635{,}000 \text{ kVA}$$

$$\text{Base k}V_{LL} = 24 \text{ kV}$$

$$\text{Base current} = \frac{635{,}000}{\sqrt{3} \times 24} = 15275.726 \text{ A}$$

$$\text{Base impedance} = \frac{24^2}{635} = 0.9071\ \Omega$$

Using the values given for the inductance parameters L_s and M_s of the armature, we compute

$$X_d = \omega(L_s + M_s) = 120\pi(2.7656 + 1.3828)10^{-3} = 1.5639\ \Omega$$

which in per unit is

$$X_d = \frac{1.5639}{0.9071} = 1.7241$$

The load is to be served at rated voltage equal to the specified base, and so if we use the terminal voltage V_a as the reference phasor, we obtain

$$V_a = 1.0\angle 0^\circ \text{ per unit}$$

The load current has the rms magnitude $|I_a| = 635{,}000/(\sqrt{3} \times 24)$ A, which is also the base armature current. Hence, $|I_a| = 1.0$ per unit, and since the power-factor angle of the load is $\theta = \cos^{-1} 0.9 = 25.8419^\circ$ lagging, the phasor form of the lagging current I_a is

$$I_a = |I_a|\angle -\theta = 1.0\angle -25.8419^\circ \text{ per unit}$$

Synchronous internal voltage E_i can be calculated from Eq. (3.31) with $R = 0$,

$$E_i = V_a + jX_d I_a$$

$$= 1.0\angle 0^\circ + j1.7241 \times 1.0\angle -25.8419^\circ$$

$$= 1.7515 + j1.5517 = 2.340\angle 41.5384^\circ \text{ per unit}$$

In Example 3.1 the base field current (which is required to produce 1.0 per-unit open-circuit armature voltage) is 1640 A. Therefore, since $|E_i|$ is directly proportional to I_f, we have an excitation current of $2.34 \times 1640 = 3838$ A under the specified operating conditions.

The interested reader may wish to draw a phasor diagram for the results of this example and compare the phasor method of solution with the time-domain approach of Example 3.1.

3.4 REAL AND REACTIVE POWER CONTROL

When the synchronous machine is connected to an infinite bus, its speed and terminal voltage are fixed and unalterable. Two controllable variables, however, are the field current and the mechanical torque on the shaft. The variation of the field current I_f, referred to as *excitation system control*, is applied to either a generator or a motor to supply or absorb a variable amount of reactive power. Because the synchronous machine runs at constant speed, the only means of varying the real power is through control of the torque imposed on the shaft by either the prime mover in the case of a generator or the mechanical load in the case of a motor.

It is convenient to neglect resistance as we consider reactive power control of the round-rotor generator. Assume that the generator is delivering power so that a certain angle δ exists between the terminal voltage V_t and the generated voltage E_i of the machine [see Fig. 3.10(a)]. The complex power delivered to the system by the generator is given in per unit by

$$S = P + jQ = V_t I_a^* = |V_t|\,|I_a|(\cos\theta + j\sin\theta) \tag{3.33}$$

Equating real and imaginary quantities in this equation, we obtain

$$P = |V_t|\,|I_a|\cos\theta \qquad Q = |V_t|\,|I_a|\sin\theta \tag{3.34}$$

We note that Q is positive for lagging power factors since the angle θ is numerically positive. If we decide to maintain a certain power delivery P from the generator to the constant voltage system, it is clear from Eq. (3.34) that $|I_a|\cos\theta$ must remain constant. As we vary the dc field current I_f under these

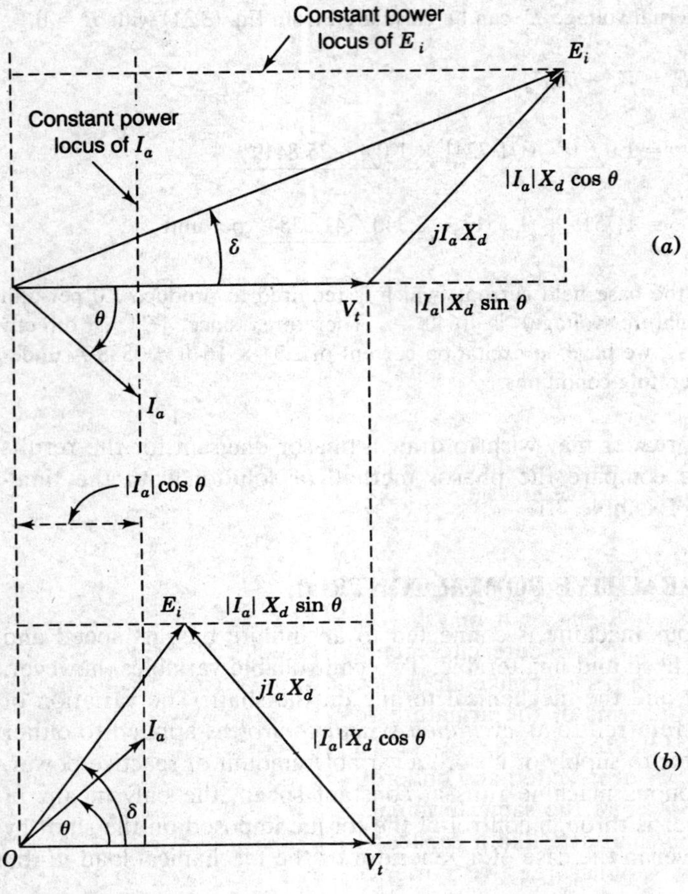

FIGURE 3.10
Phasor diagrams showing constant-power loci of an (*a*) overexcited generator delivering reactive power to the system; (*b*) underexcited generator receiving reactive power from the system. The power delivered by the generator is the same in both cases.

conditions, the generated voltage E_i varies proportionally but always so as to keep $|I_a|\cos\theta$ constant, as shown by the loci of Fig. 3.10(*a*). *Normal excitation* is defined as the condition when

$$|E_i|\cos\delta = |V_t| \tag{3.35}$$

and the machine is said to be either *overexcited* or *underexcited* according to whether $|E_i|\cos\delta > |V_t|$ or $|E_i|\cos\delta < |V_t|$. For the condition of Fig. 3.10(*a*) the generator is overexcited and supplies reactive power Q to the system. Thus, from the system viewpoint the machine is acting like a capacitor. Figure 3.10(*b*)

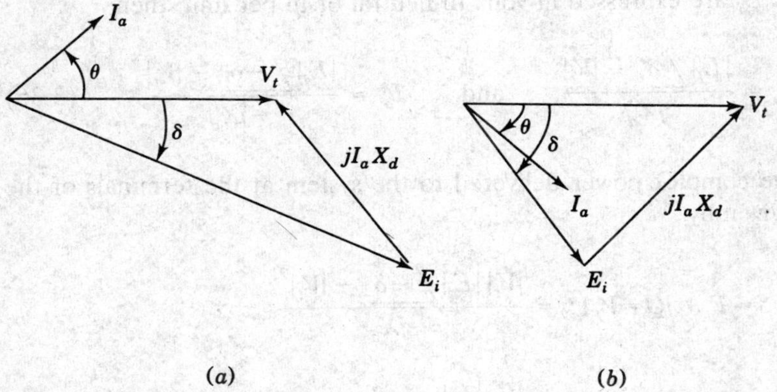

FIGURE 3.11
Phasor diagrams of an (*a*) overexcited and (*b*) underexcited synchronous motor drawing current I_a and constant power at constant terminal voltage.

is for an underexcited generator supplying the same amount of real power and a leading current to the system, or it may be considered to be drawing lagging current from the system. The underexcited generator draws reactive power from the system and in this respect acts like an inductor. The reader is encouraged to explain this action in terms of the armature reaction discussed in connection with Eqs. (3.24) and (3.25).

Figure 3.11 shows overexcited and underexcited synchronous motors drawing the same real power at the same terminal voltage. The overexcited motor draws leading current and acts like a capacitive circuit when viewed from the network to which it supplies reactive power. The underexcited motor draws lagging current, absorbs reactive power, and is acting like an inductive circuit when viewed from the network. Briefly then, Figs. 3.10 and 3.11 show that *overexcited generators and motors supply reactive power to the system* and *underexcited generators and motors absorb reactive power from the system*.

Now we turn our attention to real power P, which is controlled by opening or closing the valves through which steam (or water) enters a turbine. If the power input to the generator is increased, the rotor speed will start to increase, and if the field current I_f and hence $|E_i|$ are held constant, the angle δ between E_i and V_t will increase. Increasing δ results in a larger $|I_a|\cos\theta$, as may be seen by rotating the phasor E_i counterclockwise in Figs. 3.10(*a*) and 3.10(*b*). The generator with a larger δ, therefore, delivers more power to the network; exerts a higher countertorque on the prime mover; and hence, the input from the prime mover is reestablished at the speed corresponding to the frequency of the infinite bus. Similar reasoning applies also to a motor.

The dependence of P on the power angle δ is also shown as follows. If

$$V_t = |V_t|\underline{/0^\circ} \qquad \text{and} \qquad E_i = |E_i|\underline{/\delta}$$

where V_t and E_i are expressed in volts to neutral or in per unit, then

$$I_a = \frac{|E_i|\angle\delta - |V_t|}{jX_d} \quad \text{and} \quad I_a^* = \frac{|E_i|\angle-\delta - |V_t|}{-jX_d} \tag{3.36}$$

Therefore, the complex power delivered to the system at the terminals of the generator is given by

$$S = P + jQ = V_t I_a^* = \frac{|V_t||E_i|\angle-\delta - |V_t|^2}{-jX_d}$$

$$= \frac{|V_t||E_i|(\cos\delta - j\sin\delta) - |V_t|^2}{-jX_d} \tag{3.37}$$

The real and imaginary parts of Eq. (3.37) are

$$P = \frac{|V_t||E_i|}{X_d}\sin\delta \qquad Q = \frac{|V_t|}{X_d}(|E_i|\cos\delta - |V_t|) \tag{3.38}$$

When volts rather than per-unit values are substituted for V_t and E_i in Eqs. (3.38), we must be careful to note that V_t and E_i are line-to-neutral voltages and P and Q will be per-phase quantities. However, line-to-line voltage values substituted for V_t and E_i will yield total three-phase values for P and Q. The per-unit P and Q of Eqs. (3.38) are multiplied by base three-phase megavoltamperes or base megavoltamperes per phase depending on whether total three-phase power or power per phase is wanted.

Equation (3.38) shows very clearly the dependence of P on the power angle δ if $|E_i|$ and $|V_t|$ are constant. However, if P and V_t are constant, Eq. (3.38) shows that δ must decrease if $|E_i|$ is increased by boosting the dc field excitation. With P constant in Eq. (3.38), both an increase in $|E_i|$ and a decrease in δ mean that Q will increase if it is already positive, or it will decrease in magnitude and perhaps become positive if Q is already negative before the field excitation is boosted. These operating characteristics of the generator are made graphically evident in Sec. 3.5.

Example 3.3. The generator of Example 3.1 has synchronous reactance $X_d = 1.7241$ per unit and is connected to a very large system. The terminal voltage is $1.0\angle 0°$ per unit and the generator is supplying to the system a current of 0.8 per unit at 0.9 power-factor lagging. All per-unit values are on the machine base. Neglecting resistance, find the magnitude and angle of the synchronous internal voltage E_i, and P and Q delivered to the infinite bus. If the real power output of the generator remains constant but the excitation of the generator is (*a*) increased by 20% and (*b*) decreased by 20%, find the angle δ between E_i and the terminal bus voltage, and Q delivered to the bus by the generator.

Solution. The power-factor angle is $\theta = \cos^{-1} 0.9 = 25.8419°$ lagging, and so the synchronous internal voltage given by Eq. (3.31) is

$$E_i = |E_i|\underline{/\delta°} = V_t + jX_d I_a$$

$$= 1.0\underline{/0°} + j1.7241 \times 0.8\underline{/-25.8419°}$$

$$= 1.6012 + j1.2414 = 2.0261\underline{/37.7862°} \text{ per unit}$$

Equations (3.38) give the P and Q output of the generator,

$$P = \frac{|V_t||E_i|}{X_d}\sin\delta = \frac{1.0 \times 2.0261}{1.7241}\sin 37.7862° = 0.7200 \text{ per unit}$$

$$Q = \frac{|V_t|}{X_d}(|E_i|\cos\delta - |V_t|) = \frac{1.0}{1.7241}(1.6012 - 1.0) = 0.3487 \text{ per unit}$$

(*a*) Increasing excitation by 20% with P and V_t constant gives

$$\frac{|V_t||E_i|}{X_d}\sin\delta = \frac{1.0 \times 1.2 \times 2.0261}{1.7241}\sin\delta = 0.72$$

$$\delta = \sin^{-1}\left(\frac{0.72 \times 1.7241}{1.20 \times 2.0261}\right) = 30.7016°$$

and the new value of Q supplied by the generator is

$$Q = \frac{1.0}{1.7241}[1.20 \times 2.0261\cos(30.7016°) - 1.0] = 0.6325 \text{ per unit}$$

(*b*) With excitation decreased 20%, we obtain

$$\frac{|V_t||E_i|}{X_d}\sin\delta = \frac{1.0 \times 0.80 \times 2.0261}{1.7241}\sin\delta = 0.72$$

$$\delta = \sin^{-1}\left(\frac{0.72 \times 1.7241}{0.80 \times 2.0261}\right) = 49.9827°$$

and the value of Q now supplied by the generator is

$$Q = \frac{1.0}{1.7241}[0.80 \times 2.0261 \cos(49.9827^\circ) - 1.0] = 0.0245 \text{ per unit}$$

Thus, we see how excitation controls the reactive power output of the generator.

3.5 LOADING CAPABILITY DIAGRAM

All the normal operating conditions of the round-rotor generator connected to an infinite bus can be shown on a single diagram, usually called the *loading capability diagram* or *operation chart* of the machine. The chart is important to the power-plant operators who are responsible for proper loading and operation of the generator.

The chart is constructed on the assumption that the generator has fixed terminal voltage V_t and negligible armature resistance. Construction begins with the phasor diagram of the machine having V_t as the reference phasor, as shown in Fig. 3.10(a). The mirror image of Fig. 3.10(a) can be rotated to give the phasor diagram of Fig. 3.12, which shows five loci passing through the operating point m. These loci correspond to five possible operating modes, in each of which one parameter of the generating unit is kept constant.

CONSTANT EXCITATION. The constant excitation circle has point n as center and a radius of length n-m equal to the internal voltage magnitude $|E_i|$, which can be maintained constant by holding the dc current I_f in the field winding constant according to Eq. (3.13).

CONSTANT $|I_a|$. The circle for constant armature current has point o as center and a radius of length o-m proportional to a fixed value of $|I_a|$. Because $|V_t|$ is fixed, the operating points on this locus correspond to constant megavoltampere output ($|V_t|\,|I_a|$) from the generator.

CONSTANT POWER. Active power output of the machine is given by $P = |V_t|\,|I_a|\cos\theta$ in per unit. Since $|V_t|$ is constant, vertical line m-p at the fixed distance $X_d|I_a|\cos\theta$ from the vertical axis n-o represents a locus of operating points for constant P. The megawatt output of the generator is always positive regardless of the power factor of the output.

CONSTANT REACTIVE POWER. The reactive power output of the machine is given by $Q = |V_t|\,|I_a|\sin\theta$ in per unit when the angle θ is defined positive for lagging power factors. When $|V_t|$ is constant, horizontal line q-m at the fixed distance $X_d|I_a|\,|\sin\theta|$ from the horizontal axis represents a locus of operating points for constant Q. For unity power-factor operation the Q output of the

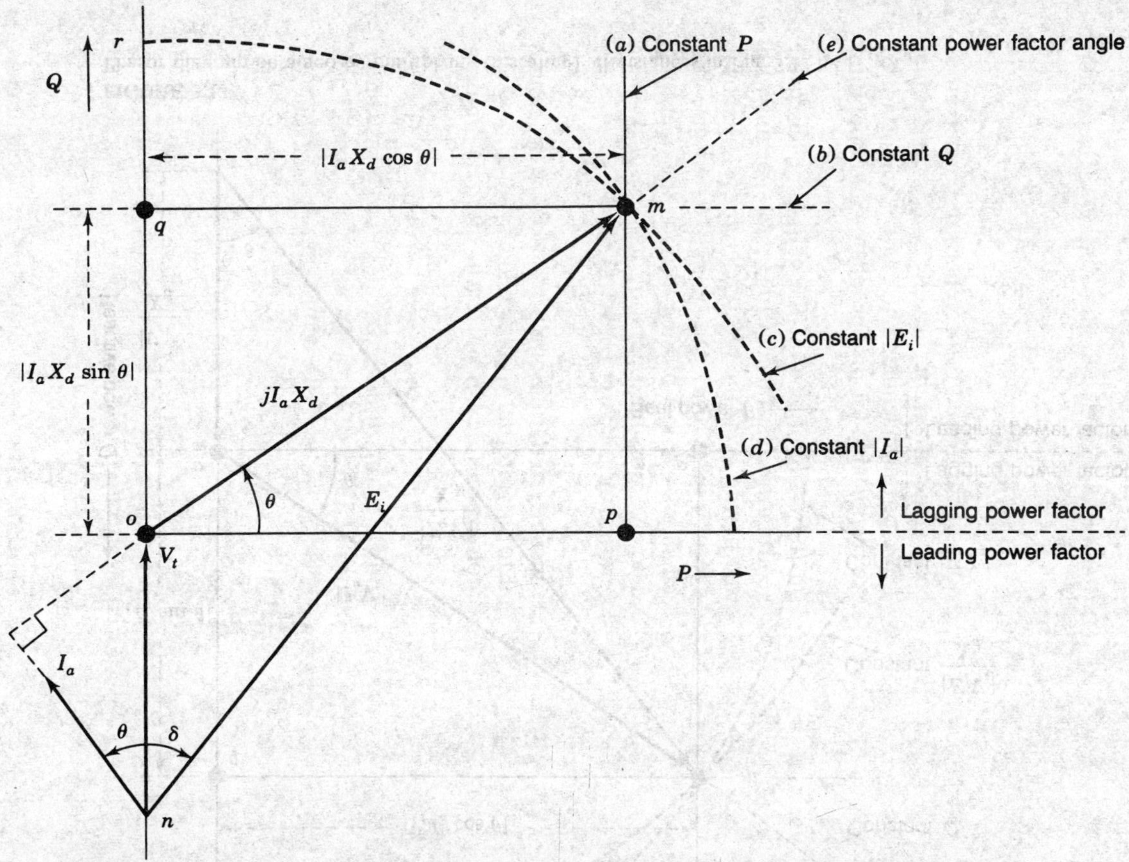

FIGURE 3.12
Phasor diagram obtained from mirror image of Fig. 3.10(*a*) showing five loci through point m corresponding to: (*a*) constant power P; (*b*) constant reactive power Q; (*c*) constant internal voltage $|E_i|$; (*d*) constant armature current $|I_a|$; (*e*) constant power-factor angle θ.

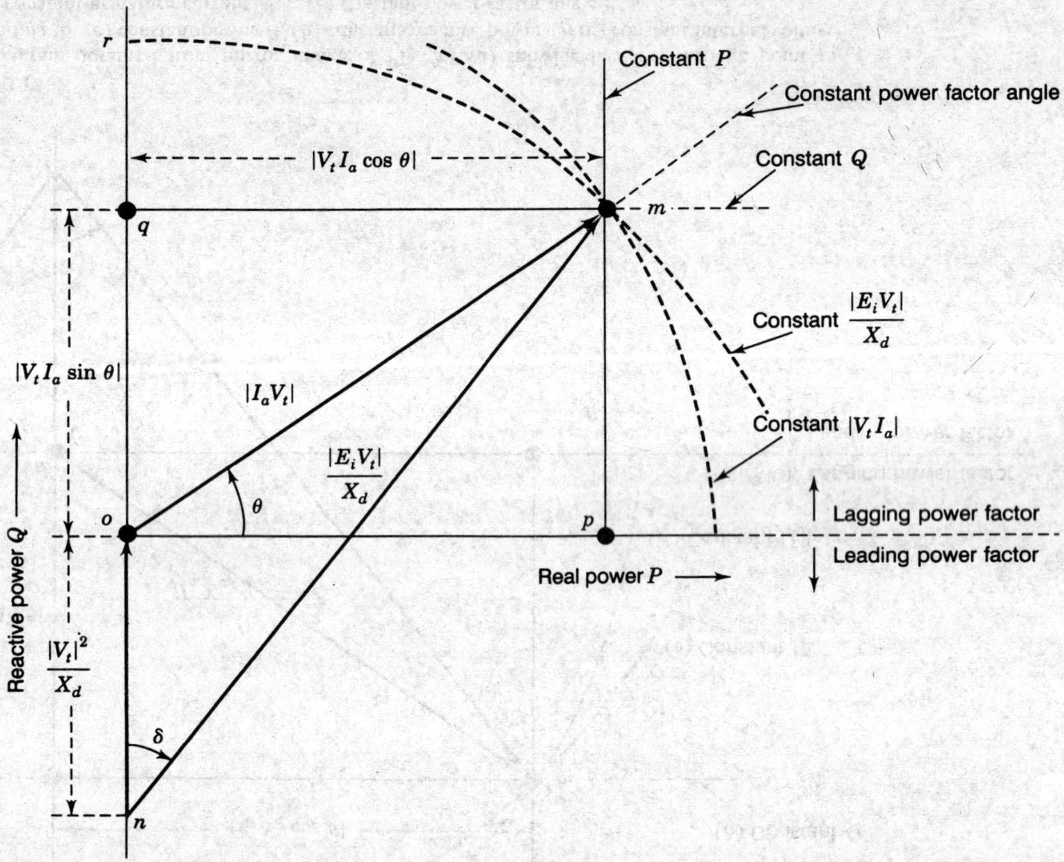

FIGURE 3.13
Phasor diagram obtained by multiplying (rescaling) all distances in Fig. 3.12 by $|V_t/X_d|$.

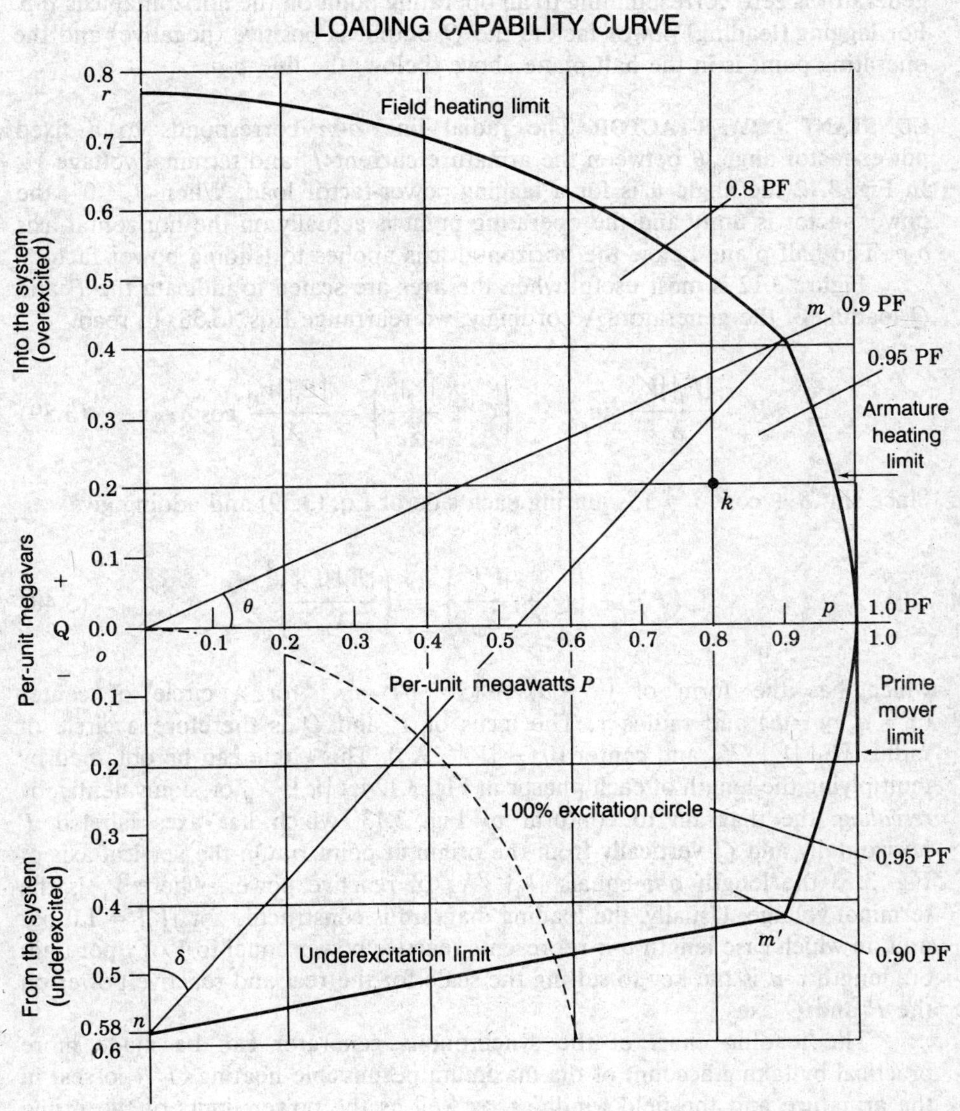

FIGURE 3.14
Loading capability curve for a cylindrical-rotor turbogenerator, 635 MVA, 24 kV, 0.9 power factor, X_d = 172.4% with maximum turbine output = 635 MW. Point k relates to Example 3.4.

generator is zero corresponding to an operating point on the horizontal axis *o-p*. For lagging (leading) power factors the Q output is positive (negative) and the operating point is in the half-plane above (below) the line *o-p*.

CONSTANT POWER-FACTOR. The radial line *o-m* corresponds to a fixed power-factor angle θ between the armature current I_a and terminal voltage V_t. In Fig. 3.12 the angle θ is for a lagging power-factor load. When $\theta = 0°$, the power factor is unity and the operating point is actually on the horizontal axis *o-p*. The half plane below the horizontal axis applies to leading power factors.

Figure 3.12 is most useful when the axes are scaled to indicate the P and Q loading of the generator. Accordingly, we rearrange Eqs. (3.38) to read

$$P = \frac{|E_i||V_t|}{X_d} \sin\delta \qquad \left(Q + \frac{|V_t|^2}{X_d}\right) = \frac{|E_i||V_t|}{X_d} \cos\delta \tag{3.39}$$

Since $\sin^2\delta + \cos^2\delta = 1$, squaring each side of Eq. (3.39) and adding give

$$(P)^2 + \left(Q + \frac{|V_t|^2}{X_d}\right)^2 = \left(\frac{|E_i||V_t|}{X_d}\right)^2 \tag{3.40}$$

which has the form of $(x - a)^2 + (y - b)^2 = r^2$ for a circle of center $(x = a,\ y = b)$ and radius r. The locus of P and Q is therefore a circle of radius $|E_i|\ |V_t|/X_d$ and center $(0, -\ |V_t|^2/X_d)$. This circle can be obtained by multiplying the length of each phasor in Fig. 3.12 by $|V_t|/X_d$ or, equivalently, by *rescaling* the diagram to conform to Fig. 3.13, which has axes labeled P horizontally and Q vertically from the origin at point *o*. On the vertical axis of Fig. 3.13 the length *o-n* equals $|V_t|^2/X_d$ of reactive power, where V_t is the terminal voltage. Usually, the loading diagram is constructed for $|V_t| = 1.0$ per unit, in which case length *o-n* represents reactive power equal to $1/X_d$ per unit. So, length *o-n* is the key to setting the scale for the real and reactive power on the P and Q axes.

The loading chart of the synchronous generator can be made more practical by taking account of the maximum permissible heating (I^2R losses) in the armature and the field windings, as well as the power limits of the prime mover and heating in the armature core. Using the example of a cylindrical-rotor turbogenerating unit rated 635 MVA, 24 kV, 0.9 power factor, $X_d =$ 172.41%, let us demonstrate the procedure for constructing the loading capability diagram of Fig. 3.14 as follows:

- Take $|V_t| = 1.0$ per unit on the rated-voltage base of the machine.
- Using a convenient voltampere scale, mark the point n on the vertical axis so that length *o-n* equals $1/X_d$ in per unit on the rated base of the machine. In

our example $X_d = 1.7241$ per unit, and so the length *o-n* in Fig. 3.14 corresponds to $1/X_d = 0.58$ per unit on the vertical *Q*-axis. The same scale obviously applies to active power *P* in per unit on the horizontal axis.

- Along the *P*-axis, mark the distance corresponding to the maximum power output of the prime mover. For present purposes the megawatt limit of the turbine is assumed in Fig. 3.14 to be 1.00 per unit on the rated megavoltampere base of the machine. Draw the vertical line for $P = 1.00$ per unit.
- Mark the length $o\text{-}m = 1.0$ per unit on the radial line from the origin at the rated power-factor angle θ, which in this case equals $\cos^{-1} 0.90$. With *o* as center and length *o-m* as radius, draw the per-unit megavoltampere circular arc corresponding to the armature-current limit.
- Construct the arc *m-r* of maximum permissible excitation using *n* as center and distance *n-m* as radius. This circular arc corresponds to the maximum field-current limit. The constant excitation circle with radius of length *o-n* usually defines 100% or 1.0 per-unit excitation, and so Fig. 3.14 shows the field-current limit occurring at 2.340 per-unit excitation, that is, (length *r-n*)/(length *o-n*) on the *Q*-axis.
- An underexcitation limit also applies at low levels of excitation when vars are being imported from the system to the machine. It is determined by the manufacturer's design as discussed below.

In Fig. 3.14 point *m* corresponds to the megavoltampere rating of the generator at rated power-factor lagging. The machine designer has to arrange sufficient field current to support *overexcited* operation of the generator at rated point *m*. The level of the field current is limited to this maximum value along the circular arc *m-r*, and the capability of the generator to deliver *Q* to the system is thereby reduced. In actuality, machine saturation decreases the value of the synchronous reactance X_d, and for this reason most manufacturers' curves depart from the theoretical field-heating limits described here.

The mirror image of *m* is the operating point *m'* in the *underexcited* region. Power-plant operators try to avoid operating conditions in the underexcited region of the capability curve for two different reasons. The first relates to steady-state stability of the *system* and the second relates to overheating of the *machine* itself.

Theoretically, the so-called *steady-state stability limit* occurs when the angle δ between E_i and V_t in Figs. 3.12 and 3.13 reaches 90°. In practice, however, system dynamics enter into the picture to complicate the determination of the actual stability limit. For this reason power-plant operators prefer to avoid underexcited machine operation whenever possible.

As the machine enters into the underexcited region of operation eddy currents induced by the system in iron parts of the armature begin to increase. The accompanying I^2R heating also increases in the end region of the armature. To limit such heating, the machine manufacturers prepare capability curves specific to their own designs and recommend limits within which to operate. In Fig. 3.14 the line *m′-n* is therefore drawn for illustrative purposes only.

To obtain megawatt and megavar values for any operating point in Fig. 3.14, the per unit values of P and Q as read from the chart are multiplied by the megavoltampere rating of the machine, which in this case is 635 MVA. Also, distance *n-m* of Fig. 3.14 is the per-unit mega voltampere value of the quantity $|E_i V_t|/X_d$ at operating point m, as shown in Fig. 3.13. Therefore, we can calculate the value of $|E_i|$ in per unit on the rated voltage base (24 kV in this case) by multiplying length *n-m* (expressed in per-unit voltamperes) by the per-unit ratio $X_d/\,|V_t|$, or simply by X_d since $|V_t| = 1.0$ per unit in Fig. 3.14. Conversion to kilovolts then requires multiplication by the voltage rating of the machine in kilovolts.

If the actual terminal voltage $|V_t|$ is not 1.0 per unit, then the per-unit value $1/X_d$ assigned to distance *o-n* of Fig. 3.14 has to be changed to $|V_t|^2/X_d$ in per unit, as shown in Fig. 3.13. This change alters the scale of Fig. 3.14 by $|V_t|^2$, and so the per-unit P and Q readings from the chart must be first multiplied by $|V_t|^2$ in per unit and then by the megavoltampere base (635 MVA in this case) in order to give correct megawatt and megavar values for the actual operating conditions. For instance, if the actual terminal voltage is 1.05 per unit, then the point n on the Q-axis of Fig. 3.14 corresponds to the actual value $0.58 \times (1.05)^2 = 0.63945$ per unit or 406 Mvar, and the point shown as 0.9 per unit on the P-axis has an actual value of $0.9 \times (1.05)^2 = 0.99225$ per unit or 630 MW.

To calculate the correct excitation voltage E_i corresponding to an operating point m when the terminal voltage is not exactly equal to its rated voltage, we could first multiply length *n-m* obtained directly from Fig. 3.14 by $|V_t|^2$ in per unit to correct the scale and then by the ratio $X_d/\,|V_t|$ in per unit to convert to $|E_i|$, as already discussed. The net result is that length *n-m* obtained directly from Fig. 3.14 when multiplied by the actual per-unit value of the *product* $X_d \times |V_t|$ yields the correct per-unit value of $|E_i|$. Then, if physical units of kilovolts are desired, multiplication by the rated kilovolt base of the machine follows. It is important to note that the power-factor angle θ and internal angle δ are the same before and after the rescaling since the geometry of Figs. 3.12 and 3.13 is preserved. The reader should note, however, that the operating constraints forming the boundary of the operating region of the chart are physical limits. So, the boundary of the operating region may be affected once the scale is altered.

The following example illustrates the procedures.

Example 3.4. A 60-Hz three-phase generator rated at 635 MVA, 0.90 power factor, 24 kV, 3600 rpm has the operating chart shown in Fig. 3.14. The generator is delivering 458.47 MW and 114.62 Mvar at 22.8 kV to an infinite bus. Calculate the excitation voltage E_i using (a) the equivalent circuit of Fig. 3.8(a) and (b) the loading diagram of Fig. 3.14. The synchronous reactance is $X_d = 1.7241$ per unit on the machine base and resistance is negligible.

Solution. In the calculations which follow all per-unit values are based on the megavoltampere and kilovolt ratings of the machine.

(*a*) Choosing the terminal voltage as the reference phasor, we have

$$V_t = \frac{22.8}{24.0}\angle 0^\circ = 0.95\angle 0^\circ \text{ per unit}$$

$$P + jQ = \frac{458.47 + j114.62}{635} = 0.722 + j0.1805 \text{ per unit}$$

$$I_a = \frac{0.722 - j0.1805}{0.95\angle 0^\circ} = 0.76 - j0.19 \text{ per unit}$$

$$E_i = V_t + jX_d I_a = 0.95\angle 0^\circ + j1.7241(0.76 - j0.19)$$

$$= 1.2776 + j1.3103 = 1.830\angle 45.7239^\circ \text{ per unit}$$

$$= 43.920\angle 45.7239^\circ \text{ kV}$$

(*b*) The point *k* corresponding to the actual operating conditions can be located on the chart of Fig. 3.14 as follows:

$$P_k + jQ_k = \frac{P + jQ}{0.95^2} = \frac{0.722 + j0.1805}{0.95^2} = 0.8 + j0.2 \text{ per unit}$$

The distance *n-k* equals $\sqrt{0.8^2 + 0.78^2} = 1.1173$ per unit when calculated or measured on the scale of the chart of Fig. 3.14. The actual value of $|E_i|$ is then computed as

$$|E_i| = (1.1173 \times 0.95^2)\frac{1.7241}{0.95} = 1.830 \text{ per unit}$$

which is the same as obtained above. The angle $\delta = 45^\circ$ can be easily measured.

3.6 THE TWO-AXIS MACHINE MODEL

The round-rotor theory already developed in this chapter gives good results for the steady-state performance of the synchronous machine. However, for *transient* analysis we need to consider a *two-axis* model. In this section we introduce the two-axis model by means of the equations of the salient-pole machine in which the air gap is much narrower along the direct axis than along the quadrature axis between poles. The largest generating units are steam-turbine-driven alternators of round-rotor construction; fossil-fired units have two poles and nuclear units have four poles for reasons of economical design and operational efficiency. Hydroelectric generators usually have more pole-pairs and are of salient-pole construction. These units run at lower speeds so as to avoid mechanical damage due to centrifugal forces.

The three-phase salient-pole machine, like its round-rotor counterpart, has three symmetrically distributed armature windings a, b, and c, and a field winding f on the rotor which produces a sinusoidal flux distribution around the air gap.[2] In both types of machines the field sees, so to speak, the same air gap and magnetizing paths in the stator regardless of the rotor position. Consequently, the field winding has constant self-inductance L_{ff}. Moreover, both machine types have the same cosinusoidal mutual inductances L_{af}, L_{bf}, and L_{cf} with the armature phases as given by Eqs. (3.4). Additionally, throughout each revolution of the rotor the self-inductances L_{aa}, L_{bb}, and L_{cc} of the stator windings, and the mutual inductances L_{ab}, L_{bc}, and L_{ca} between them, are not constant in the salient-pole machine but also vary as a function of the rotor angular displacement θ_d. The flux linkages of phases a, b, and c are related to the currents by the inductances so that

$$\begin{aligned} \lambda_a &= L_{aa}i_a + L_{ab}i_b + L_{ac}i_c + L_{af}i_f \\ \lambda_b &= L_{ba}i_a + L_{bb}i_b + L_{bc}i_c + L_{bf}i_f \\ \lambda_c &= L_{ca}i_a + L_{cb}i_b + L_{cc}i_c + L_{cf}i_f \end{aligned} \tag{3.41}$$

These equations look similar to Eqs. (3.5) for the round-rotor machine but all the coefficients are variable, as summarized in Table 3.1.[3] As a result, the equations for the flux linkages λ_a, λ_b, and λ_c of the salient-pole machine are more difficult to use than their round-rotor counterparts. Fortunately, the equations of the salient-pole machine can be expressed in a simple form by transforming the a, b, and c variables of the stator into corresponding sets of new variables, called the *direct-axis*, *quadrature-axis*, and *zero-sequence* quantities which are distinguished by the subscripts d, q, and 0, respectively. For example, the three stator currents i_a, i_b, and i_c can be transformed into three equivalent currents, called the *direct-axis current* i_d, the *quadrature-axis current* i_q and the *zero-sequence current* i_0. The transformation is made by the matrix **P**,

[2] For further discussion on the salient-pole machine, see P. M. Anderson and A. A. Fouad, *Power System Control and Stability*, Chap. 4, The Iowa State University Press, Ames, Iowa, 1977.

[3] The D- and Q-damper windings referred to in Table 3.1 are discussed in Sec. 3.8.

TABLE 3.1
Expressions for the inductances of three-phase salient-pole synchronous generator with field. *D*-damper and *Q*-damper windings on the rotor.

Stator	Self-inductances ($L_s > L_m > 0$)	$L_{aa} = L_s + L_m \cos 2\theta_d$ $L_{bb} = L_s + L_m \cos 2(\theta_d - 2\pi/3)$ $L_{cc} = L_s + L_m \cos 2(\theta_d + 2\pi/3)$
	Mutual-inductances ($M_s > L_m > 0$)	$L_{ab} = L_{ba} = -M_s - L_m \cos 2(\theta_d + \pi/6)$ $L_{bc} = L_{cb} = -M_s - L_m \cos 2(\theta_d - \pi/2)$ $L_{ca} = L_{ac} = -M_s - L_m \cos 2(\theta_d + 5\pi/6)$
Rotor	Self-inductances	Field winding: L_{ff} *D*-damper winding: L_D *Q*-damper winding: L_Q
	Mutual-inductances	Field/*D*-winding: M_r Field/*Q*-winding: 0 *D*-winding/*Q*-winding: 0
Stator-rotor mutual inductances	Armature/field	$L_{af} = L_{fa} = M_f \cos \theta_d$ $L_{bf} = L_{fb} = M_f \cos(\theta_d - 2\pi/3)$ $L_{cf} = L_{fc} = M_f \cos(\theta_d - 4\pi/3)$
	Armature/*D*-winding	$L_{aD} = L_{Da} = M_D \cos \theta_d$ $L_{bD} = L_{Db} = M_D \cos(\theta_d - 2\pi/3)$ $L_{cD} = L_{Dc} = M_D \cos(\theta_d - 4\pi/3)$
	Armature/*Q*-winding	$L_{aQ} = L_{Qa} = M_Q \cos \theta_d$ $L_{bQ} = L_{Qb} = M_Q \cos(\theta_d - 2\pi/3)$ $L_{cQ} = L_{Qc} = M_Q \cos(\theta_d - 4\pi/3)$

called *Park's transformation*, where

$$\mathbf{P} = \sqrt{\frac{2}{3}} \begin{array}{c} \\ ⓐ \\ ⓑ \\ ⓒ \end{array} \begin{array}{c} \begin{array}{ccc} \quad ⓐ \quad & \quad ⓑ \quad & \quad ⓒ \quad \end{array} \\ \begin{bmatrix} \cos\theta_d & \cos(\theta_d - 120°) & \cos(\theta_d - 240°) \\ \sin\theta_d & \sin(\theta_d - 120°) & \sin(\theta_d - 240°) \\ \frac{1}{\sqrt{2}} & \frac{1}{\sqrt{2}} & \frac{1}{\sqrt{2}} \end{bmatrix} \end{array} \tag{3.42}$$

which was introduced by R. H. Park in slightly different form from that shown here. The matrix **P** has the convenient property (called *orthogonality*) that its inverse $\mathbf{P}^{-1}$ equals its transpose $\mathbf{P}^T$, which is found simply by interchanging rows and columns in Eq. (3.42). This property is most important as it ensures that power in the a, b, and c variables is not altered by **P**, as discussed in Sec. 8.9. The currents, voltages, and flux linkages of phases a, b, and c are transformed by **P** to d, q, and 0 variables as follows:

$$\begin{bmatrix} i_d \\ i_q \\ i_0 \end{bmatrix} = \mathbf{P} \begin{bmatrix} i_a \\ i_b \\ i_c \end{bmatrix} \qquad \begin{bmatrix} v_d \\ v_q \\ v_0 \end{bmatrix} = \mathbf{P} \begin{bmatrix} v_a \\ v_b \\ v_c \end{bmatrix} \qquad \begin{bmatrix} \lambda_d \\ \lambda_q \\ \lambda_0 \end{bmatrix} = \mathbf{P} \begin{bmatrix} \lambda_a \\ \lambda_b \\ \lambda_c \end{bmatrix} \tag{3.43}$$

The **P**-transformation defines a set of currents, voltages, and flux linkages for three fictitious coils, one of which is the stationary 0-coil. The other two coils are the d-coil and the q-coil, which rotate in synchronism with the rotor. The d- and q-coils have constant flux linkages with the field and any other windings which may exist on the rotor. Section A.2 of the Appendix illustrates the detailed manipulations which transform the currents, voltages, and flux linkages of phases a, b, and c into d-q-0 quantities according to Eqs. (3.43). The resulting d, q, and 0 flux-linkage equations are

$$\begin{aligned} \lambda_d &= L_d i_d + \sqrt{\frac{3}{2}}\, M_f i_f \\ \lambda_q &= L_q i_q \\ \lambda_0 &= L_0 i_0 \end{aligned} \tag{3.44}$$

in which i_f is the actual field current, and the inductances are defined by

$$L_d = L_s + M_s + \frac{3}{2} L_m; \qquad L_q = L_s + M_s - \frac{3}{2} L_m; \qquad L_0 = L_s - 2M_s \tag{3.45}$$

Parameters L_s and M_s have the same meanings as before and L_m is a positive number. The inductance L_d is called the *direct-axis inductance*, L_q is called the *quadrature-axis inductance*, and L_0 is known as the *zero-sequence inductance*. The flux linkages of the field are still given by Eq. (3.24), which is repeated here in the form

$$\lambda_f = \sqrt{\frac{3}{2}}\, M_f i_d + L_{ff} I_f \tag{3.46}$$

Equations (3.44) and (3.46) have constant inductance coefficients, and thus are quite simple to use. Physically interpreted, these simpler flux-linkage equations show that L_d is the self-inductance of an equivalent d-axis armature winding which rotates at the same speed as the field and which carries current i_d to

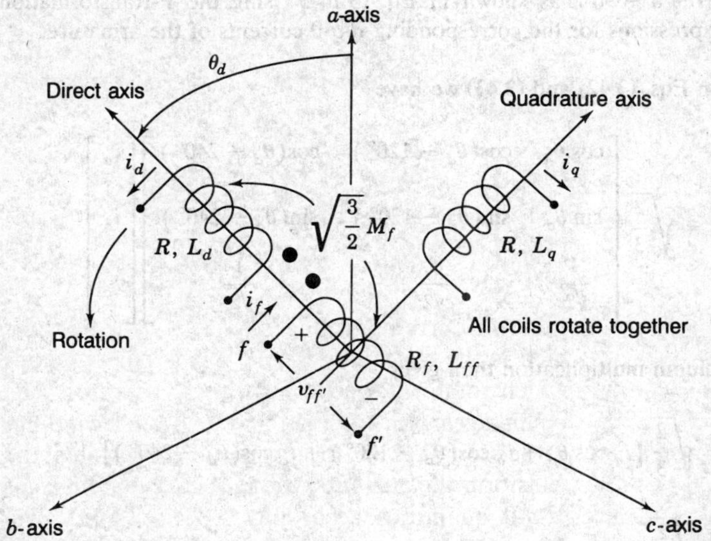

FIGURE 3.15
Representation of the salient-pole synchronous generator by armature-equivalent direct-axis and quadrature-axis coils rotating in synchronism with the field winding on the rotor.

produce the same mmf on the d-axis as do the actual stator currents i_a, i_b, and i_c. Similarly, L_q and i_q apply to the q-axis. Accordingly, i_d and i_q give rise to mmfs which are stationary with respect to the rotor. The fictitious d-axis winding and the f winding representing the physical field can be considered to act like two coupled coils which are stationary with respect to each other as they rotate together sharing the mutual inductance kM_f ($k = \sqrt{3/2}$) between them, as shown by Eqs. (3.44) and (3.46). Furthermore, the field and the d-axis coil do not couple magnetically with the fictitious q winding on the q-axis, which lags the d-axis in space by 90°. The zero-sequence inductance L_0 is associated with a stationary fictitious armature coil with no coupling to any other coils. Under balanced conditions this coil carries no current, and therefore we omit it from further discussion.

The d-axis and q-axis coils representing the stator windings are shown in Fig. 3.15, which should be compared with the single-axis diagram for the round-rotor machine in Fig. 3.6.

Example 3.5. Under steady-state operating conditions the armature of the salient-pole synchronous generator carries symmetrical sinusoidal three-phase currents

$$i_a = \sqrt{2}\,|I_a|\sin(\theta_d - \theta_a)$$

$$i_b = \sqrt{2}\,|I_a|\sin(\theta_d - 120^\circ - \theta_a)$$

$$i_c = \sqrt{2}\,|I_a|\sin(\theta_d - 240^\circ - \theta_a)$$

where $\theta_d = \omega t + \delta + 90°$, as shown in Eq. (3.14). Using the **P**-transformation matrix, find expressions for the corresponding *d-q-0* currents of the armature.

Solution. From Eqs. (3.42) and (3.43) we have

$$\begin{bmatrix} i_d \\ i_q \\ i_0 \end{bmatrix} = \sqrt{\frac{2}{3}} \begin{bmatrix} \cos\theta_d & \cos(\theta_d - 120°) & \cos(\theta_d - 240°) \\ \sin\theta_d & \sin(\theta_d - 120°) & \sin(\theta_d - 240°) \\ \frac{1}{\sqrt{2}} & \frac{1}{\sqrt{2}} & \frac{1}{\sqrt{2}} \end{bmatrix} \begin{bmatrix} i_a \\ i_b \\ i_c \end{bmatrix}$$

and row-by-column multiplication then gives

$$i_d = \sqrt{\frac{2}{3}}\,[i_a \cos\theta_d + i_b \cos(\theta_d - 120°) + i_c \cos(\theta_d - 240°)]$$

$$i_q = \sqrt{\frac{2}{3}}\,[i_a \sin\theta_d + i_b \sin(\theta_d - 120°) + i_c \sin(\theta_d - 240°)]$$

$$i_0 = \sqrt{\frac{2}{3}}\left[\frac{1}{\sqrt{2}}(i_a + i_b + i_c)\right]$$

Under balanced conditions $i_a + i_b + i_c = 0$, and so $i_0 = 0$. By means of the trigonometric identity $2\sin\alpha\cos\beta = \sin(\alpha + \beta) + \sin(\alpha - \beta)$, we obtain

$$\begin{aligned} i_a \cos\theta_d &= \sqrt{2}\,|I_a|\sin(\theta_d - \theta_a)\cos\theta_d \\ &= \frac{|I_a|}{\sqrt{2}}[\sin(2\theta_d - \theta_a) - \sin\theta_a] \end{aligned}$$

Likewise, we have

$$\begin{aligned} i_b \cos(\theta_d - 120°) &= \sqrt{2}\,|I_a|\sin(\theta_d - 120° - \theta_a)\cos(\theta_d - 120°) \\ &= \frac{|I_a|}{\sqrt{2}}[\sin(2\theta_d - 240° - \theta_a) - \sin\theta_a] \end{aligned}$$

$$\begin{aligned} i_c \cos(\theta_d - 240°) &= \sqrt{2}\,|I_a|\sin(\theta_d - 240° - \theta_a)\cos(\theta_d - 240°) \\ &= \frac{|I_a|}{\sqrt{2}}[\sin(2\theta_d - 480° - \theta_a) - \sin\theta_a] \end{aligned}$$

In the three preceding trigonometric expansions the first terms inside the brackets are second harmonic sinusoidal quantities which sum to zero at every instant of time as in Sec. 3.2, and so we obtain

$$i_d = \sqrt{\frac{2}{3}}\,\frac{|I_a|}{\sqrt{2}}[-3\sin\theta_a] = -\sqrt{3}\,|I_a|\sin\theta_a$$

We recall from Sec. 3.3 that $\theta_a = \theta + \delta$, where θ is the phase angle of lag of i_a measured with respect to the terminal voltage and θ_a is the phase angle of lag of i_a with respect to the internal voltage of the machine.

Accordingly,

$$i_d = -\sqrt{3}\,|I_a|\sin\theta_a = -\sqrt{3}\,|I_a|\sin(\theta + \delta)$$

We can show in a similar manner that the quadrature-axis current

$$i_q = \sqrt{3}\,|I_a|\cos\theta_a = \sqrt{3}\,|I_a|\cos(\theta + \delta)$$

Thus, the expression for i_d is exactly the same for the salient-pole and the round-rotor machines. The flux linkages in the field winding are given by Eq. (3.46), which shows that the direct-axis current i_d is directly opposing the magnetizing influence of the field when $\theta_a = \pi/2$, and the quadrature-axis current i_q is then zero.

3.7 VOLTAGE EQUATIONS: SALIENT-POLE MACHINE

In Sec. 3.6 the flux-linkage equations for the salient-pole machine are remarkably simple when expressed in terms of the d, q, and 0 variables. We now consider other important simplifications which occur when the **P**-transformation is also applied to the voltage equations of the armature.

Using the voltage polarities and current directions of Fig. 3.4, let us write the terminal-voltage equations for the armature windings of the salient-pole machine in the form

$$v_a = -Ri_a - \frac{d\lambda_a}{dt}; \qquad v_b = -Ri_b - \frac{d\lambda_b}{dt}; \qquad v_c = -Ri_c - \frac{d\lambda_c}{dt} \qquad (3.47)$$

In these equations the voltages v_a, v_b, and v_c are the line-to-neutral terminal voltages for the armature phases; the negative signs of the coefficients arise because currents i_a, i_b, and i_c are directed out of the generator. While simple in format, Eqs. (3.47) are in fact very difficult to handle if left in terms of λ_a, λ_b, and λ_c. Again, a much simpler set of equations for the voltages v_d, v_q, and v_0 is found by employing the **P**-transformation. The calculations leading to the new voltage equations are straightforward but tedious, as shown in Sec. A.2 of the Appendix, which yields

$$
\begin{aligned}
v_d &= -Ri_d - \frac{d\lambda_d}{dt} - \omega\lambda_q \\
v_q &= -Ri_q - \frac{d\lambda_q}{dt} + \omega\lambda_d \\
v_0 &= -Ri_0 - \frac{d\lambda_0}{dt}
\end{aligned}
\tag{3.48}
$$

where ω is the rotational speed $d\theta_d/dt$. Equation (3.26) for the field winding is not subject to **P**-transformation, and so arranging the d-q-0 flux-linkage and voltage equations according to their axes gives

d-axis:

$$
\begin{aligned}
\lambda_d &= L_d i_d + \mathrm{k}M_f i_f \\
\lambda_f &= \mathrm{k}M_f i_d + L_{ff}\, i_f
\end{aligned}
\tag{3.49}
$$

$$
\begin{aligned}
v_d &= -Ri_d - \frac{d\lambda_d}{dt} - \omega\lambda_q \\
v_{ff'} &= R_f i_f + \frac{d\lambda_f}{dt}
\end{aligned}
\tag{3.50}
$$

q-axis:

$$
\begin{aligned}
\lambda_q &= L_q i_q \\
v_q &= -Ri_q - \frac{d\lambda_q}{dt} + \omega\lambda_d
\end{aligned}
\tag{3.51}
$$

where $\mathrm{k} = \sqrt{3/2}$. Equations involving i_0 and λ_0 stand alone and are not of interest under balanced conditions. Equations (3.49) through (3.51) are much simpler to solve than their corresponding voltage and flux-linkage equations in a-b-c variables. Furthermore, a set of equivalent circuits may be drawn to satisfy these simpler equations, as shown in Fig. 3.16. The f-circuit represents the actual field since the **P**-transformation affects only the armature phases, which are replaced by the d- and q-coils. We see that the f-coil is mutually coupled to the d-coil on the d-axis, and so flux-linkage and voltage equations can be written to agree with Eqs. (3.49) and (3.50). The fictitious q-coil is shown magnetically uncoupled from the other two windings since the d-axis and the q-axis are spatially in quadrature with one another. However, there is interaction between the two axes by means of the voltage sources $-\omega\lambda_q$ and $\omega\lambda_d$, which are *rotational emfs* or *speed voltages* internal to the machine due to the rotation of the rotor. We note that the speed voltage in the d-axis depends on λ_q, and similarly, the speed voltage in the q-axis depends on λ_d. These sources represent ongoing electromechanical energy conversion. No such energy conversion could occur at standstill ($\omega = 0$) since the field and the other d-axis circuit would then act like a stationary transformer and the q-axis circuit like an ordinary inductance coil.

To summarize, Park's transformation replaces the physical stationary windings of the armature by:

1. A direct-axis circuit which rotates with the field circuit and is mutually coupled to it,
2. A quadrature-axis circuit which is displaced 90° from the d-axis, and thus has no mutual inductance with the field or other d-axis circuits although it rotates in synchronism with them, and
3. A stationary stand-alone 0-coil with no coupling to any other circuit, and thus is not shown in Fig. 3.16.

Figure 3.16 is most useful in analyzing the performance of the synchronous machine under short-circuit conditions, which we consider in the next section.

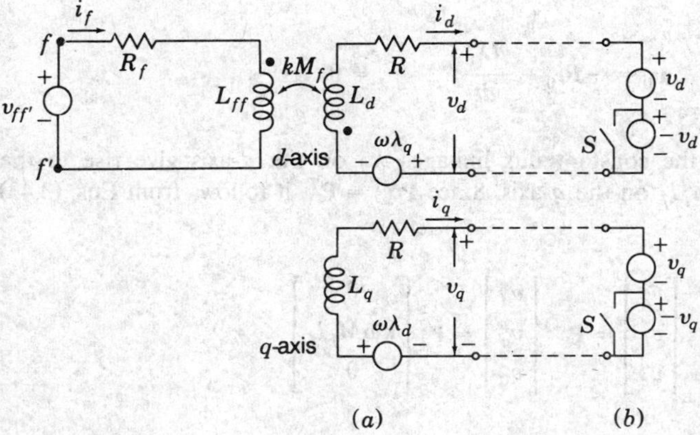

FIGURE 3.16
Equivalent circuit for the salient-pole synchronous generator: (*a*) with terminal voltages v_d and v_q; (*b*) with armature short-circuited.

Example 3.6. Direct current I_f is supplied to the field winding of an unloaded salient-pole synchronous generator rotating with constant angular velocity ω. Determine the form of the open-circuit armature voltages and their *d-q*-0 components.

Solution. The armature currents i_a, i_b, and i_c are zero because of the armature open circuit, and so

$$\begin{bmatrix} i_d \\ i_q \\ i_0 \end{bmatrix} = \mathbf{P} \begin{bmatrix} i_a \\ i_b \\ i_c \end{bmatrix} = \begin{bmatrix} 0 \\ 0 \\ 0 \end{bmatrix}$$

Substituting these values in Eqs. (3.49) through (3.51), we obtain for $k = \sqrt{3/2}$

$$\lambda_d = L_d i_d + \mathrm{k} M_f I_f = \mathrm{k} M_f I_f$$

$$\lambda_q = L_q i_q \qquad = 0$$

$$\lambda_0 = L_0 i_0 \qquad = 0$$

and from Eqs. (3.48) we then find

$$v_d = -Ri_d - \frac{d\lambda_d}{dt} - \omega\lambda_q = 0$$

$$v_q = -Ri_q - \frac{d\lambda_q}{dt} + \omega\lambda_d = k\omega M_f I_f$$

$$v_0 = -Ri_0 - \frac{d\lambda_0}{dt} \qquad = 0$$

Thus, we see that the constant flux linkages λ_d on the *d*-axis give rise to the rotational emf $\mathrm{k}\omega M_f I_f$ on the *q*-axis. Since $\mathbf{P}^{-1} = \mathbf{P}^T$, it follows from Eqs. (3.43) that

$$\begin{bmatrix} v_a \\ v_b \\ v_c \end{bmatrix} = \mathbf{P}^{-1} \begin{bmatrix} v_d \\ v_q \\ v_0 \end{bmatrix} = \mathbf{P}^T \begin{bmatrix} 0 \\ \mathrm{k}\omega M_f I_f \\ 0 \end{bmatrix}$$

$$= \sqrt{\frac{2}{3}} \begin{bmatrix} \sin\theta_d \\ \sin(\theta_d - 120°) \\ \sin(\theta_d - 240°) \end{bmatrix} \mathrm{k}\omega M_f I_f$$

Therefore, the steady-state open-circuit armature voltages in the idealized salient-pole machine are balanced sinusoidal quantities of amplitude $\sqrt{2}\,|E_i| = \omega M_f I_f$, as obtained previously for the round-rotor machine.

3.8 TRANSIENT AND SUBTRANSIENT EFFECTS

When a fault occurs in a power network, the current flowing is determined by the internal emfs of the machines in the network, by their impedances, and by the impedances in the network between the machines and the fault. The current flowing in a synchronous machine immediately after the occurrence of a fault differs from that flowing a few cycles later and from the sustained, or steady-state, value of the fault current. This is because of the effect of the fault current in the armature on the flux generating the voltage in the machine. The current changes relatively slowly from its initial value to its steady-state value owing to the changes in reactance of the synchronous machine.

Our immediate interest is in the inductance effective in the armature of the synchronous machine when a three-phase short circuit suddenly occurs at its terminals. Before the fault occurs, suppose that the armature voltages are v_a, v_b, and v_c and that these give rise to the voltages v_d, v_q, and v_0 according to Eq. (3.43). Figure 3.16(a) shows the voltages v_d and v_q at the terminals of the d-axis and q-axis equivalent circuits. The short circuit of phases a, b, and c imposes the conditions $v_a = v_b = v_c = 0$, which lead to the conditions $v_d = v_q = 0$. Thus, to simulate short-circuit conditions, the terminals of the d-axis and q-axis circuits in Fig. 3.16(a) must also be shorted. Each of these circuits has a net terminal voltage of zero when equal but opposite voltage sources are connected in series, as shown in Fig. 3.16(b). In that figure the switches S should be interpreted in a *symbolic* sense; namely, when the switches are both open, the sources $-v_d$ and $-v_q$ are in the circuit, and when the switches are closed, those two sources are removed from the circuit.

The principle of superposition can be applied to the series-connected voltage sources, provided we assume that the rotor speed ω remains at its prefault steady-state value—for Eqs. (3.49) through (3.51) are then linear. With both switches closed in Fig. 3.16(b), we have the steady-state operation of the machine since the sources v_d and v_q then match perfectly the d-axis and q-axis voltages at the terminals just before the fault occurs. Suddenly opening the switches S adds the voltage source $-v_d$ in series with the source v_d and $-v_q$ in series with the source v_q to produce the required short circuits. Thus, the sources $-v_d$ and $-v_q$ are those determining the instantaneous *changes* from the steady state due to the sudden short-circuit fault. By superposition, we can calculate the fault-induced changes of all variables by setting the *external* sources $v_{ff'}$, v_d, and v_q of Fig. 3.16(b) equal to zero and suddenly applying the voltages $-v_d$ and $-v_q$ to the unexcited rotating machine, as shown in Fig. 3.17. The internal speed voltages $-\omega\lambda_q$ and $\omega\lambda_d$ are initially zero because flux linkages with all coils are zero in Fig. 3.17 before applying the $-v_d$ and $-v_q$

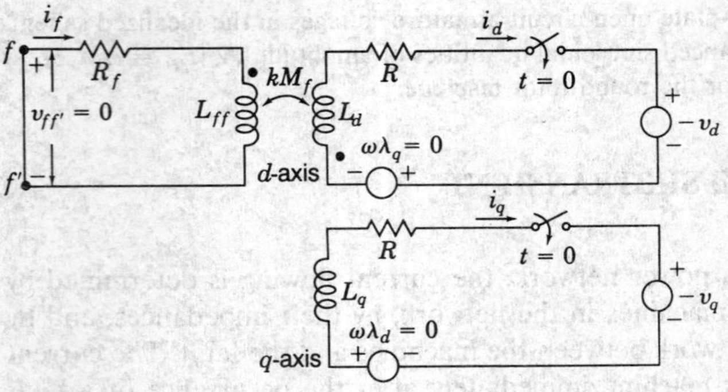

FIGURE 3.17
Equivalent circuit of salient-pole synchronous generator rotating at constant speed with field short-circuited. Closing switches at $t = 0$ corresponds to sudden application of short circuit to machine terminals.

sources. The flux-linkage changes on the d-axis of the machine are governed by Eq. (3.49), which gives

$$\begin{aligned} \Delta\lambda_d &= L_d\,\Delta i_d + \mathrm{k}M_f\,\Delta i_f \\ \Delta\lambda_f &= \mathrm{k}M_f\,\Delta i_d + L_{ff}\,\Delta i_f \end{aligned} \tag{3.52}$$

where Δ denotes incremental changes. Since the field winding is a closed physical winding, its flux linkages cannot change instantaneously according to the principle of constant flux linkages. Therefore, setting $\Delta\lambda_f$ equal to zero in Eq. (3.52) gives

$$\Delta i_f = -(\mathrm{k}M_f/L_{ff})\,\Delta i_d$$

and substituting for Δi_f in the equation for $\Delta\lambda_d$ yields

$$\Delta\lambda_d = \left[L_d - \frac{(\mathrm{k}M_f)^2}{L_{ff}}\right]\Delta i_d \tag{3.53}$$

The flux linkage per unit current in Eq. (3.53) defines the *d-axis transient inductance* L'_d, where

$$L'_d = \frac{\Delta\lambda_d}{\Delta i_d} = L_d - \frac{(\mathrm{k}M_f)^2}{L_{ff}} \tag{3.54}$$

Since $(\mathrm{k}M_f)^2/L_{ff}$ is positive, Eq. (3.54) shows that the *direct-axis transient reactance* $X'_d = \omega L'_d$ is always less than the direct-axis synchronous reactance

$X_d = \omega L_d$. Thus, following abrupt changes at its terminals, the synchronous machine reflects in its armature the transient reactance X'_d, which is less than its steady-state reactance X_d.

In defining X'_d, we assume that the field is the only physical rotor winding. In fact, most salient-pole machines of practical importance have *damper windings* consisting of shorted copper bars through the pole faces of the rotor; and even in a round-rotor machine, under short-circuit conditions eddy currents are induced in the solid rotor as if in damper windings. The effects of the eddy-current damping circuits are represented by direct-axis and quadrature-axis *closed* coils, which are treated in very much the same way as the field winding except that they have no applied voltage. To account for the addition of damper windings, we need only add to Fig. 3.16 the closed D-circuit and Q-circuit of Fig. 3.18, which have self-inductances L_D and L_Q and mutual inductances with the other windings as shown. In the steady state the flux linkages are constant between all circuits on the same rotor axis. The D- and Q-circuits are then passive (having neither induced nor applied voltages) and do not enter into steady-state analysis. Under short-circuit conditions, however, we can determine from Fig. 3.18 the initial d-axis flux-linkage changes resulting from sudden shorting of the synchronous machine with damper-winding effects. The procedure is the same as already discussed. The field and D-damper circuits representing closed physical windings are mutually coupled to each other and to the d-coil representing the armature along the direct axis. There cannot be sudden change in the flux linkages of the closed windings, and so we can write for the flux-linkage *changes* along the d-axis

$$\begin{aligned}\Delta\lambda_d &= L_d\,\Delta i_d + \mathrm{k}M_f\,\Delta i_f + \mathrm{k}M_D\,\Delta i_D\\ \Delta\lambda_f &= \mathrm{k}M_f\,\Delta i_d + L_{ff}\,\Delta i_f + M_r\,\Delta i_D = 0\\ \Delta\lambda_D &= \mathrm{k}M_D\,\Delta i_d + M_r\,\Delta i_f + L_D\,\Delta i_D = 0\end{aligned} \tag{3.55}$$

These equations are similar to Eqs. (3.52), but they have extra terms because of the additional self- and mutual inductances associated with the D-damper circuit. Coefficients reflecting stator-to-rotor mutual coupling have the multiplier $\mathrm{k} = \sqrt{3/2}$. M_r relates to mutual coupling between rotor-based windings on the d-axis and thus has no k multiplier. Solving Eqs. (3.55) for Δi_f and Δi_D in terms of Δi_d yields

$$\Delta i_f = -\left[\frac{(\mathrm{k}M_f)L_D - (\mathrm{k}M_D)M_r}{L_{ff}L_D - M_r^2}\right]\Delta i_d$$

$$\Delta i_D = -\left[\frac{(\mathrm{k}M_D)L_{ff} - (\mathrm{k}M_f)M_r}{L_{ff}L_D - M_r^2}\right]\Delta i_d$$

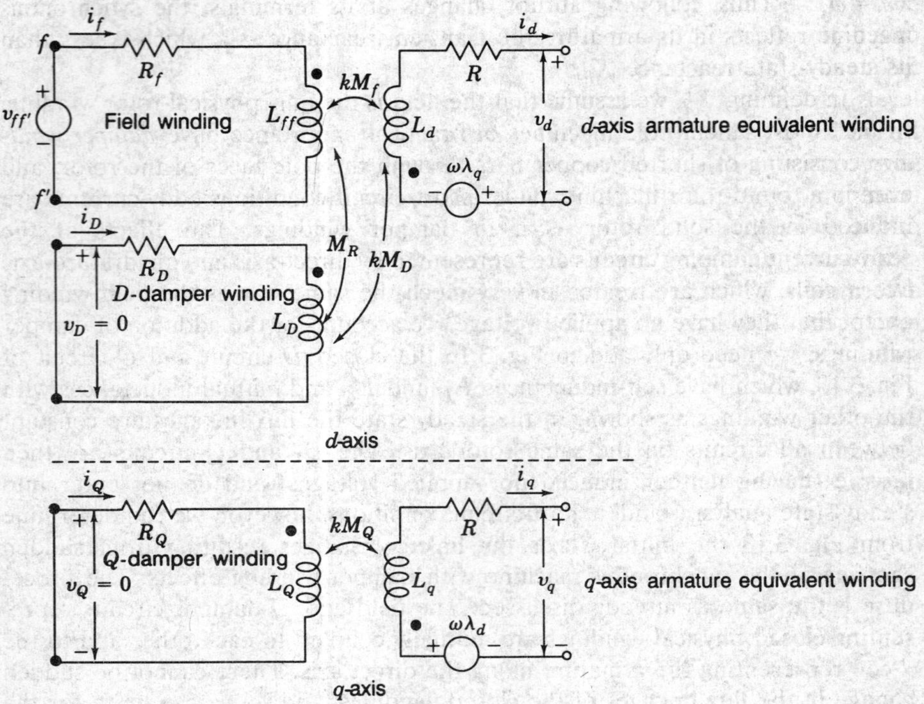

FIGURE 3.18
Equivalent circuit of the salient-pole synchronous generator with one field winding and two damper windings on the rotor.

and substituting these results into the $\Delta\lambda_d$ expression of Eq. (3.55) yields the *direct-axis subtransient inductance* L''_d, defined by

$$\frac{\Delta\lambda_d}{\Delta i_d} = L''_d = L_d - \mathrm{k}^2\left(\frac{M_f^2 L_D + M_D^2 L_{ff} - 2M_f M_D M_r}{L_{ff}L_D - M_r^2}\right) \tag{3.56}$$

The *direct-axis subtransient reactance* X''_d, defined as $X''_d = \omega L''_d$, is considerably smaller than X'_d, which means that $X''_d < X'_d < X_d$. The reader should check the numerical data given by the machine manufacturers in Table A.2 in the Appendix to confirm these inequalities. We should note that similar reactances can be defined for the *q*-axis.

We have shown that the synchronous machine has different reactances when it is subjected to short-circuit faults at its terminals. Immediately upon occurrence of the short circuit, the armature of the machine behaves with an effective reactance X''_d, which combines with an effective resistance determined by the damping circuits to define a *direct-axis, short-circuit subtransient time-*

constant T_d'' in the range of 0.03 s. The period over which X_d'' is effective is called the *subtransient period*, and this is typically 3 to 4 cycles of system frequency in duration. When the damper-winding currents decay to negligible levels, the D- and Q-circuits are no longer needed and Fig. 3.18 reverts to Fig. 3.16. The machine currents then decay more slowly with a *direct-axis, short-circuit transient time-constant* T_d' determined by X_d' and a machine resistance which depends on R_f of the field. The period of effectiveness of X_d' is called the *transient period* and T_d' is of the order of 1 s. Finally, for sustained steady-state conditions the d- and q-axis reactances $X_d = \omega L_d$ and $X_q = \omega L_q$ determine the performance of the salient-pole machine, just as the synchronous reactance X_d applies to the round-rotor synchronous machine in the steady state.

The various reactances supplied by the machine manufacturers are usually expressed in per unit based on the nameplate rating of the machine while time constants are given in seconds. Table A.2 in the Appendix sets forth a summary of typical parameters for the synchronous machines of practical importance.

Example 3.7. Calculate the per-unit value of X_d' for the 60-Hz synchronous generator of Example 3.1. Use the machine rating of 635 MVA, 24 kV as base.

Solution. Values for the inductances of the armature and field windings are given in Example 3.1, and Fig. 3.10 shows $L_d = L_s + M_s$. Therefore,

$$L_d = L_s + M_s = 2.7656 + 1.3828 = 4.1484 \text{ mH}$$

The transient inductance L'_d is now calculated from Eq. (3.54):

$$L'_d = L_d - \frac{(\text{k}M_f)^2}{L_{ff}}$$

$$= 4.1484 - \frac{\left(\sqrt{3/2} \times 31.6950\right)^2}{433.6569} = 0.6736 \text{ mH}$$

and the transient reactance is

$$X_d' = \omega L'_d = 120\pi \times 0.6736 \times 10^{-3} = 0.2540\ \Omega$$

The impedance base on the machine rating equals $(24^2/635)\ \Omega$ so that

$$X_d' = \frac{0.2540 \times 635}{24^2} = 0.28 \text{ per unit}$$

Thus, X_d' is much less than the synchronous reactance $X_d = 1.7341$ per unit.

3.9 SHORT-CIRCUIT CURRENTS

When an ac voltage is applied suddenly across a series $R - L$ circuit, the current which flows generally has two components—a dc component, which decays according to the time constant L/R of the circuit, and a steady-state sinusoidally varying component of constant amplitude. A similar but more complex phenomenon occurs when a short circuit appears suddenly across the terminals of a synchronous machine. The resulting phase currents in the machine will have dc components, which cause them to be offset or *asymmetrical* when plotted as a function of time. In Chap. 10 we shall discuss how the *symmetrical* portion of these short-circuit currents is used in the ratings of circuit breakers. For now let us consider how short circuits affect the reactances of the machine.

A good way to analyze the effect of a three-phase short circuit at the terminals of a previously unloaded generator is to take an oscillogram of the current in one of the phases upon the occurrence of such a fault. Since the voltages generated in the phases of a three-phase machine are displaced 120 electrical degrees from each other, the short circuit occurs at different points on the voltage wave of each phase. For this reason the unidirectional or dc transient component of current is different in each phase.[4] If the dc component of current is eliminated from the current of each phase, the amplitude of the ac component of each phase current plotted versus time, shown in Fig. 3.19, varies approximately according to

$$I(t) = |E_i|\frac{1}{X_d} + |E_i|\left(\frac{1}{X'_d} - \frac{1}{X_d}\right)\varepsilon^{-t/T'_d} + |E_i|\left(\frac{1}{X''_d} - \frac{1}{X'_d}\right)\varepsilon^{-t/T''_d} \quad (3.57)$$

where $e_i = \sqrt{2}\,|E_i|\cos\omega t$ is the synchronous internal or no-load voltage of the machine. Equation (3.57) clearly shows that the armature phase current, with the dc removed, has three components, two of which decay at different rates over the subtransient and transient periods. Neglecting the comparatively small resistance of the armature, the distance *o-a* in Fig. 3.19 is the maximum value of the sustained short-circuit current, with the rms value $|I|$ given by

$$|I| = \frac{o\text{-}a}{\sqrt{2}} = \frac{|E_i|}{X_d} \quad (3.58)$$

If the envelope of the current wave is extended back to zero time and the first few cycles where the decrement appears to be very rapid are neglected, the

[4]For further discussion of the dc components see A. E. Fitzgerald et al., *Electric Machinery*, 4th ed., McGraw-Hill, Inc., New York, 1983 and Chap. 10 of this book.

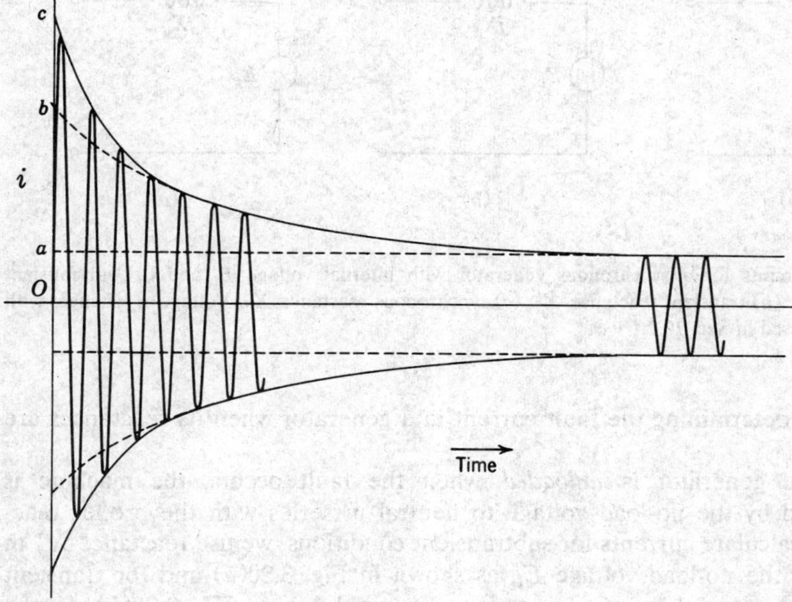

FIGURE 3.19
Current as a function of time for a synchronous generator short-circuited while running at no load. The unidirectional transient component of current has been eliminated in redrawing the oscillogram.

intercept is the distance *o-b*. The rms value of the current represented by this intercept is known as the *transient current* $|I'|$, defined by

$$|I'| = \frac{o\text{-}b}{\sqrt{2}} = \frac{|E_i|}{X'_d} \tag{3.59}$$

The rms value of the current determined by the distance *o-c* in Fig. 3.19 is called the *subtransient current* $|I''|$, given by

$$|I''| = \frac{o\text{-}c}{\sqrt{2}} = \frac{|E_i|}{X''_d} \tag{3.60}$$

Subtransient current is often called the *initial symmetrical rms current*, which is more descriptive because it conveys the idea of neglecting the dc component and taking the rms value of the ac component of current immediately after the occurrence of the fault. Equations (3.59) and (3.60) can be used to calculate the parameters X'_d and X''_d of the machine when an oscillographic record such as Fig. 3.19 is available. On the other hand, Eqs. (3.59) and (3.60) also indicate the

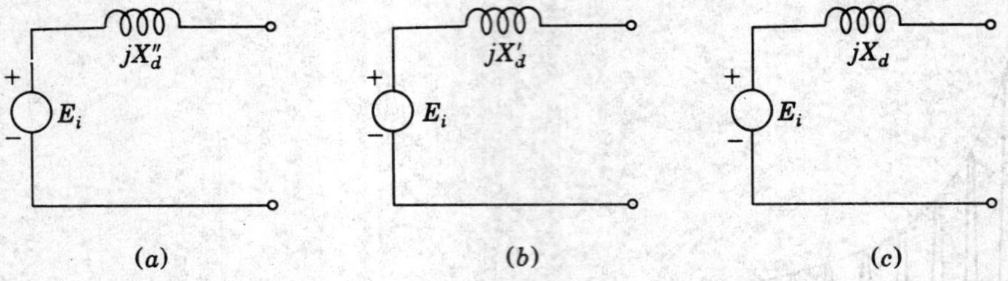

FIGURE 3.20
Equivalent circuits for a synchronous generator with internal voltage E_i and (*a*) subtransient reactance X''_d; (*b*) transient reactance X'_d; (*c*) synchronous reactance X_d. Voltage E_i changes with load as discussed in Sec. 10.2.

method of determining the fault current in a generator when its reactances are known.

If the generator is *unloaded* when the fault occurs, the machine is represented by the no-load voltage to neutral in series with the proper reactance. To calculate currents for subtransient conditions, we use reactance X''_d in series with the no-load voltage E_i, as shown in Fig. 3.20(*a*) and for transient conditions we use the series reactance X'_d, as shown in Fig. 3.20(*b*). In the steady state X_d is used, as shown in Fig. 3.20(*c*). The subtransient current $|I''|$ is much larger than the steady-state current $|I|$ because X''_d is much smaller than X_d. The internal voltage E_i is the same in each circuit of Fig. 3.20 because the generator is assumed to be initially unloaded. In Chap. 10 we shall consider how the equivalent circuits are altered to account for loading on the machine when the short circuit occurs.

Example 3.8. Two generators are connected in parallel to the low-voltage side of a three-phase Δ-*Y* transformer, as shown in Fig. 3.21(*a*). Generator 1 is rated 50,000 kVA, 13.8 kV. Generator 2 is rated 25,000 kVA, 13.8 kV. Each generator has a subtransient reactance of 25% on its own base. The transformer is rated 75,000 kVA, 13.8Δ/69Y kV, with a reactance of 10%. Before the fault occurs, the voltage on the high-voltage side of the transformer is 66 kV. The transformer is unloaded and there is no circulating current between the generators. Find the subtransient current in each generator when a three-phase short circuit occurs on the high-voltage side of the transformer.

Solution. Select 69 kV, 75,000 kVA as base in the high-voltage circuit. Then, the base voltage on the low-voltage side is 13.8 kV.

Generator 1

$$X''_{d1} = 0.25\frac{75{,}000}{50{,}000} = 0.375 \text{ per unit}$$

$$E_{i1} = \frac{66}{69} = 0.957 \text{ per unit}$$

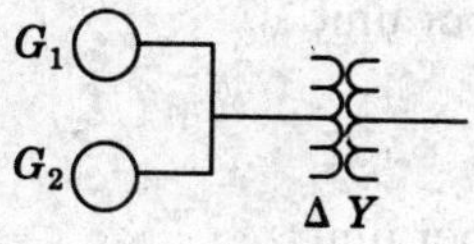

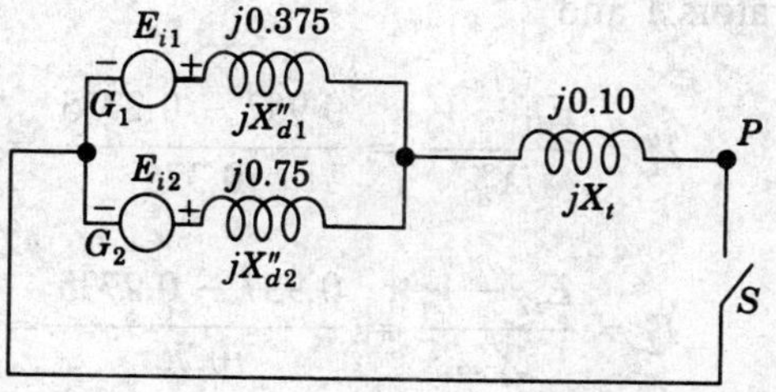

FIGURE 3.21
(*a*) Single-line diagram; (*b*) reactance diagram for Example 3.8.

Generator 2

$$X''_{d2} = 0.25 \frac{75{,}000}{25{,}000} = 0.750 \text{ per unit}$$

$$E_{i2} = \frac{66}{69} = 0.957 \text{ per unit}$$

Transformer

$$X_t = 0.10 \text{ per unit}$$

Figure 3.21(b) shows the reactance diagram before the fault. A three-phase fault at P is simulated by closing switch S. The internal voltages of the two machines may be considered to be in parallel since they are identical in magnitude and phase and no circulating current flows between them. The equivalent parallel subtransient reactance is

$$X''_d = \frac{X''_{d1}X''_{d2}}{X''_{d1} + X''_{d2}} = \frac{0.375 \times 0.75}{0.375 + 0.75} = 0.25 \text{ per unit}$$

Therefore, as a phasor with $E_i \triangleq E_{i1} = E_{i2}$ as reference, the subtransient current in the short circuit is

$$I'' = \frac{E_i}{jX''_d + jX_t} = \frac{0.957}{j0.25 + j0.10} = -j2.735 \text{ per unit}$$

The voltage V_t on the Δ side of the transformer is

$$V_t = I'' \times jX_t = (-j2.735)(j0.10) = 0.2735 \text{ per unit}$$

In generators 1 and 2

$$I_1'' = \frac{E_{i1} - V_t}{jX_{d1}''} = \frac{0.957 - 0.2735}{j0.375} = -j1.823 \text{ per unit}$$

$$I_2'' = \frac{E_{i2} - V_t}{jX_{d2}''} = \frac{0.957 - 0.2735}{j0.75} = -j0.912 \text{ per unit}$$

3.10 SUMMARY

Simplified equivalent circuits for the synchronous generator are developed in this chapter for use throughout the remainder of the text.

We have seen that the steady-state performance of the synchronous machine relies on the concept of synchronous reactance X_d, which is the basis of the steady-state equivalent circuit of the machine. In steady-state operation we have observed that the synchronous generator delivers an increasing amount of reactive power to the system to which it is connected as its excitation is increased. Conversely, as its excitation is reduced, the synchronous generator furnishes less reactive power, and when underexcited, it draws reactive power from the system. All of these normal steady-state operating conditions of the round-rotor generator, connected to a large system as if to an infinite bus, are shown by the loading capability diagram of the machine.

Transient analysis of the synchronous generator requires a two-axis machine model. We have seen that the corresponding equations involving physical *a-b-c* phase variables can be simplified by Park's transformation, which introduces *d-q-*0 currents, voltages, and flux linkages. Simplified equivalent circuits which follow from the *d-q-*0 equations of the machine allow definitions of the subtransient reactance X_d'' and transient reactance X_d'. Subtransient reactance X_d'' is important in calculating currents resulting from short-circuit faults at or near synchronous generators, as discussed in Chap. 10. The transient reactance X_d' is used in stability studies, as demonstrated in Chap. 16.

PROBLEMS

3.1. Determine the highest speed at which two generators mounted on the same shaft can be driven so that the frequency of one generator is 60 Hz and the frequency of the other is 25 Hz. How many poles does each machine have?

3.2. The three-phase synchronous generator described in Example 3.1 is operated at 3600 rpm and supplies a unity power-factor load. If the terminal voltage of the machine is 22 kV and the field current is 2500 A, determine the line current and the total power consumption of the load.

3.3. A three-phase round-rotor synchronous generator has negligible armature resistance and a synchronous reactance X_d of 1.65 per unit. The machine is connected directly to an infinite bus of voltage $1.0\angle 0^\circ$ per unit. Find the internal voltage E_i

of the machine when it delivers a current of (*a*) $1.0\angle 30°$ per unit, (*b*) $1.0\angle 0°$ per unit, and (*c*) $1.0\angle -30°$ per unit to the infinite bus. Draw phasor diagrams depicting the operation of the machine in each case.

3.4. A three-phase round-rotor synchronous generator, rated 10 kV, 50 MVA has armature resistance R of 0.1 per unit and synchronous reactance X_d of 1.65 per unit. The machine operates on a 10-kV infinite bus delivering 2000 A at 0.9 power-factor leading.

(*a*) Determine the internal voltage E_i and the power angle δ of the machine. Draw a phasor diagram depicting its operation.

(*b*) What is the open-circuit voltage of the machine at the same level of excitation?

(*c*) What is the *steady-state* short-circuit current at the same level of excitation? Neglect all saturation effects.

3.5. A three-phase round-rotor synchronous generator, rated 16 kV and 200 MVA, has negligible losses and synchronous reactance of 1.65 per unit. It is operated on an infinite bus having a voltage of 15 kV. The internal emf E_i and the power angle δ of the machine are found to be 24 kV (line to line) and 27.4°, respectively.

(*a*) Determine the line current and the three-phase real and reactive power being delivered to the system.

(*b*) If the mechanical power input and the field current of the generator are now changed so that the line current of the machine is reduced by 25% at the power factor of part (*a*), find the new internal emf E_i and the power angle δ.

(*c*) While delivering the reduced line current of part (*b*), the mechanical power input and the excitation are further adjusted so that the machine operates at unity power factor at its terminals. Calculate the new values of E_i and δ.

3.6. The three-phase synchronous generator of Prob. 3.5 is operated on an infinite bus of voltage 15 kV and delivers 100 MVA at 0.8 power-factor lagging.

(*a*) Determine the internal voltage E_i, the power angle δ, and the line current of the machine.

(*b*) If the field current of the machine is reduced by 10%, while the mechanical power input to the machine is maintained constant, determine the new value of δ and the reactive power delivered to the system.

(*c*) The prime mover power is next adjusted without changing the excitation so that the machine delivers zero reactive power to the system. Determine the new power angle δ and the real power being delivered to the system.

(*d*) What is the maximum reactive power that the machine can deliver if the level of excitation is maintained as in parts (*b*) and (*c*)?

Draw a phasor diagram for the operation of the machine in parts (*a*), (*b*), and (*c*).

3.7. Starting with Eq. (3.31), modify Eq. (3.38) to show that

$$P = \frac{|V_t|}{R^2 + X_d^2}\{|E_i|(R\cos\delta + X_d\sin\delta) - |V_t|R\}$$

$$Q = \frac{|V_t|}{R^2 + X_d^2}\{X_d(|E_i|\cos\delta - |V_t|) - R|E_i|\sin\delta\}$$

when the synchronous generator has nonzero armature resistance R.

3.8. The three-phase synchronous generator described in Example 3.4 is now operated on a 25.2-kV infinite bus. It is found that the internal voltage magnitude $|E_i| =$ 49.5 kV and that the power angle $\delta = 38.5°$. Using the loading capability diagram of Fig. 3.14, determine graphically the real and reactive power delivered to the system by the machine. Verify your answers using Eqs. (3.38).

3.9. A three-phase salient-pole synchronous generator with negligible armature resistance has the following values for the inductance parameters specified in Table 3.1;

$$L_s = 2.7656 \text{ mH} \qquad M_f = 31.6950 \text{ mH} \qquad L_m = 0.3771 \text{ mH}$$

$$M_s = 1.3828 \text{ mH} \qquad L_{ff} = 433.6569 \text{ mH}$$

During balanced steady-state operation the field current and a-phase armature current of the machine have the respective values

$$i_f = 4000 \text{ A} \qquad i_a = 20{,}000 \sin(\theta_d - 30°) \text{ A}$$

(*a*) Using Eq. (3.41), determine the instantaneous values of the flux linkages λ_a, λ_b, λ_c, and λ_f when $\theta_d = 60°$.
(*b*) Using Park's transformation given by Eqs. (3.42) and (3.43), determine the instantaneous values of the flux linkages λ_d, λ_q, and λ_0, and the currents i_d, i_q, and i_0 when $\theta_d = 60°$.
(*c*) Verify results using Eqs. (3.45) and (3.46)

3.10. The armature of a three-phase salient-pole generator carries the currents

$$i_a = \sqrt{2} \times 1000 \sin(\theta_d - \theta_a) \text{ A}$$

$$i_b = \sqrt{2} \times 1000 \sin(\theta_d - 120° - \theta_a) \text{ A}$$

$$i_c = \sqrt{2} \times 1000 \sin(\theta_d - 240° - \theta_a) \text{ A}$$

(*a*) Using the **P**-transformation matrix of Eq. (3.42), find the direct-axis current i_d and the quadrature-axis current i_q. What is the zero-sequence current i_0?
(*b*) Suppose that the armature currents are

$$i_a = \sqrt{2} \times 1000 \sin(\theta_d - \theta_a) \text{ A}$$

$$i_b = i_c = 0$$

Determine i_d, i_q, and i_0.

3.11. Calculate the direct-axis synchronous reactance X_d, the direct-axis transient reactance X_d', and the direct-axis subtransient reactance X_d'' of the 60-Hz salient-pole synchronous machine with the following parameters:

$$L_s = 2.7656 \text{ mH} \qquad L_{ff} = 433.6569 \text{ mH} \qquad L_D = 4.2898 \text{ mH}$$

$$M_s = 1.3828 \text{ mH} \qquad M_f = 31.6950 \text{ mH} \qquad M_D = 3.1523 \text{ mH}$$

$$L_m = 0.3771 \text{ mH} \qquad M_r = 37.0281 \text{ mH}$$

3.12. The single-line diagram of an unloaded power system is shown in Fig. 3.22. Reactances of the two sections of the transmission line are shown on the diagram. The generators and transformers are rated as follows:

Generator 1:	20 MVA, 13.8 kV, $X_d'' = 0.20$ per unit
Generator 2:	30 MVA, 18 kV, $X_d'' = 0.20$ per unit
Generator 3:	30 MVA, 20 kV, $X_d'' = 0.20$ per unit
Transformer T_1:	25 MVA, 220Y/13.8Δ kV, $X = 10\%$
Transformer T_2:	single-phase units, each rated 10 MVA, 127/18 kV, $X = 10\%$
Transformer T_3:	35 MVA, 220Y/22Y kV, $X = 10\%$

(*a*) Draw the impedance diagram with all reactances marked in per unit and with letters to indicate points corresponding to the single-line diagram. Choose a base of 50 MVA, 13.8 kV in the circuit of generator 1.

(*b*) Suppose that the system is unloaded and that the voltage throughout the system is 1.0 per unit on bases chosen in part *a*. If a three-phase short circuit occurs from bus *C* to ground, find the phasor value of the short-circuit current (in amperes) if each generator is represented by its subtransient reactance.

(*c*) Find the megavoltamperes supplied by each generator under the conditions of part (*b*).

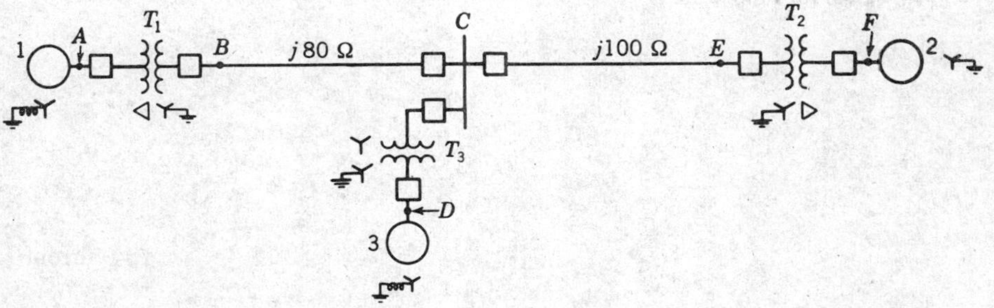

FIGURE 3.22
One-line diagram for Prob. 3.12.

3.13. The ratings of the generators, motors, and transformers of Fig. 3.23 are

Generator 1:	20 MVA, 18 kV, $X_d'' = 20\%$
Generator 2:	20 MVA, 18 kV, $X_d'' = 20\%$
Synchronous motor 3:	30 MVA, 13.8 kV, $X_d'' = 20\%$
Three-phase Y-Y transformers:	20 MVA, 138Y/20Y kV, $X = 10\%$
Three-phase Y-Δ transformers:	15 MVA, 138Y/13.8Δ kV, $X = 10\%$

(*a*) Draw the impedance diagram for the power system. Mark impedances in per unit. Neglect resistance and use a base of 50 MVA, 138 kV in the 40-Ω line.

(*b*) Suppose that the system is unloaded and that the voltage throughout the system is 1.0 per unit on bases chosen in part (*a*). If a three-phase short circuit occurs from bus *C* to ground, find the phasor value of the short-circuit current (in amperes) if each generator is represented by its subtransient reactance.

(*c*) Find the megavoltamperes supplied by each synchronous machine under the conditions of part (*b*).

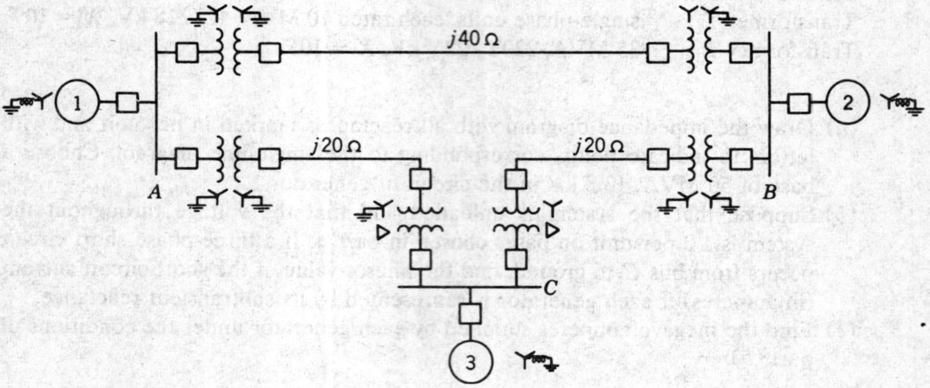

FIGURE 3.23
One-line diagram for Prob. 3.13.

CHAPTER 4

SERIES IMPEDANCE OF TRANSMISSION LINES

An electric transmission line has four parameters which affect its ability to fulfill its function as part of a power system: *resistance*, *inductance*, *capacitance*, *and conductance*. In this chapter we discuss the first two of these parameters, and we shall consider capacitance in the next chapter. The fourth parameter, conductance, exists between conductors or between conductors and the ground. Conductance accounts for the leakage current at the insulators of overhead lines and through the insulation of cables. Since leakage at insulators of overhead lines is negligible, the conductance between conductors of an overhead line is usually neglected.

Another reason for neglecting conductance is that since it is quite variable, there is no good way of taking it into account. Leakage at insulators, the principal source of conductance, changes appreciably with atmospheric conditions and with the conducting properties of dirt that collects on the insulators. Corona, which results in leakage between lines, is also quite variable with atmospheric conditions. It is fortunate that the effect of conductance is such a negligible component of shunt admittance.

Some of the properties of an electric circuit can be explained by the electric and magnetic fields which accompany its current flow. Figure 4.1 shows a single-phase line and its associated magnetic and electric fields. The lines of magnetic flux form closed loops linking the circuit, and the lines of electric flux originate on the positive charges on one conductor and terminate on the

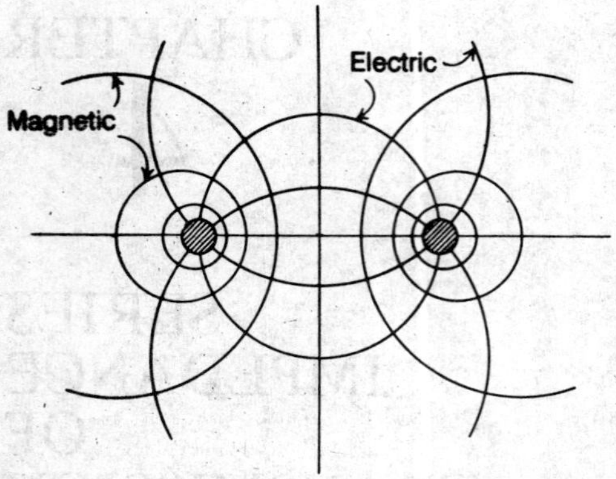

FIGURE 4.1
Magnetic and electric fields associated with a two-wire line.

negative charges on the other conductor. Variation of the current in the conductors causes a change in the number of lines of magnetic flux linking the circuit. Any change in the flux linking a circuit induces a voltage in the circuit which is proportional to the rate of change of flux. The inductance of the circuit relates the voltage induced by changing flux to the rate of change of current.

The capacitance which exists between the conductors is defined as the charge on the conductors per unit of potential difference between them.

The resistance and inductance uniformly distributed along the line form the series impedance. The conductance and capacitance existing between conductors of a single-phase line or from a conductor to neutral of a three-phase line form the shunt admittance. Although the resistance, inductance, and capacitance are distributed, the equivalent circuit of a line is made up of lumped parameters, as we shall see when we discuss them.

4.1 TYPES OF CONDUCTORS

In the early days of the transmission of electric power conductors were usually copper, but aluminum conductors have completely replaced copper for overhead lines because of the much lower cost and lighter weight of an aluminum conductor compared with a copper conductor of the same resistance. The fact that an aluminum conductor has a larger diameter than a copper conductor of the same resistance is also an advantage. With a larger diameter, the lines of electric flux originating on the conductor will be farther apart at the conductor surface for the same voltage. This means there is a lower voltage gradient at the conductor surface and less tendency to ionize the air around the conductor. Ionization produces the undesirable effect called *corona*.

Symbols identifying different types of aluminum conductors are as follows:

AAC	all-aluminum conductors
AAAC	all-aluminum-alloy conductors
ACSR	aluminum conductor, steel-reinforced
ACAR	aluminum conductor, alloy-reinforced

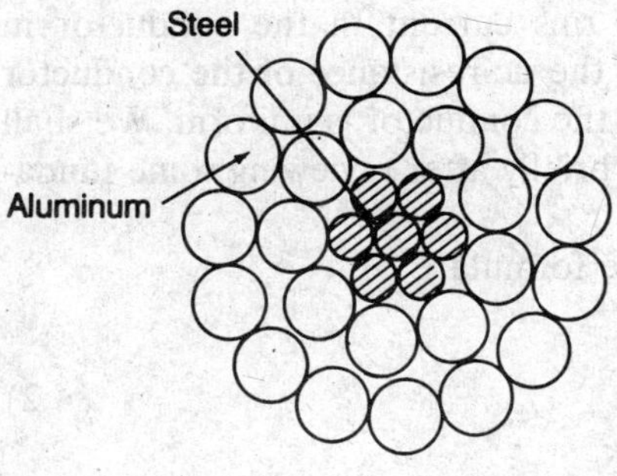

FIGURE 4.2
Cross section of a steel-reinforced conductor, 7 steel strands, and 24 aluminum strands.

Aluminum-alloy conductors have higher tensile strength than the ordinary electrical-conductor grade of aluminum. ACSR consists of a central core of steel strands surrounded by layers of aluminum strands. ACAR has a central core of higher-strength aluminum surrounded by layers of electrical-conductor-grade aluminum.

Alternate layers of wire of a stranded conductor are spiraled in opposite directions to prevent unwinding and to make the outer radius of one layer coincide with the inner radius of the next layer. Stranding provides flexibility for a large cross-sectional area. The number of strands depends on the number of layers and on whether all the strands are of the same diameter. The total number of strands in concentrically stranded cables, where the total annular space is filled with strands of uniform diameter, is 7, 19, 37, 61, 91, or more.

Figure 4.2 shows the cross section of a typical steel-reinforced aluminum cable (ACSR). The conductor shown has 7 steel strands forming a central core, around which there are two layers of aluminum strands. There are 24 aluminum strands in the two outer layers. The conductor stranding is specified as 24 A1/7 St, or simply 24/7. Various tensile strengths, current capacities, and conductor sizes are obtained by using different combinations of steel and aluminum.

Appendix Table A.3 gives some electrical characteristics of ACSR. Code names, uniform throughout the aluminum industry, have been assigned to each conductor for easy reference.

A type of conductor known as *expanded* ACSR has a filler such as paper separating the inner steel strands from the outer aluminum strands. The paper gives a larger diameter (and hence, lower corona) for a given conductivity and tensile strength. Expanded ACSR is used for some extra-high-voltage (EHV) lines.

4.2 RESISTANCE

The resistance of transmission-line conductors is the most important cause of power loss in a transmission line. The term "resistance," unless specifically qualified, means *effective resistance*. The effective resistance of a conductor is

$$R = \frac{\text{power loss in conductor}}{|I|^2}\ \Omega \tag{4.1}$$

where the power is in watts and I is the rms current in the conductor in amperes. The effective resistance is equal to the dc resistance of the conductor only if the distribution of current throughout the conductor is uniform. We shall discuss nonuniformity of current distribution briefly after reviewing some fundamental concepts of dc resistance.

Direct-current resistance is given by the formula

$$R_0 = \frac{\rho l}{A}\,\Omega \tag{4.2}$$

where ρ = resistivity of conductor

l = length

A = cross-sectional area

Any consistent set of units may be used. In power work in the United States l is usually given in feet, A in circular mils (cmil), and ρ in ohm-circular mils per foot, sometimes called ohms per circular mil-foot. In SI units l is in meters, A in square meters and ρ in ohm-meters.[1]

A circular mil is the area of a circle having a diameter of 1 mil. A mil is equal to 10^{-3} in. The cross-sectional area of a solid cylindrical conductor in circular mils is equal to the square of the diameter of the conductor expressed in mils. The number of circular mils multiplied by $\pi/4$ equals the number of square mils. Since manufacturers in the United States identify conductors by their cross-sectional area in circular mils, we must use this unit occasionally. The area in square millimeters equals the area in circular mils multiplied by 5.067×10^{-4}.

The international standard of conductivity is that of annealed copper. Commercial hard-drawn copper wire has 97.3% and aluminum has 61% of the conductivity of standard annealed copper. At 20°C for hard-drawn copper ρ is $1.77 \times 10^{-8}\ \Omega \cdot \text{m}$ ($10.66\ \Omega \cdot \text{cmil/ft}$). For aluminum at 20°C ρ is $2.83 \times 10^{-8}\ \Omega \cdot \text{m}$ ($17.00\ \Omega \cdot \text{cmil/ft}$).

The dc resistance of stranded conductors is greater than the value computed by Eq. (4.2) because spiraling of the strands makes them longer than the conductor itself. For each mile of conductor the current in all strands except the one in the center flows in more than a mile of wire. The increased resistance due to spiraling is estimated as 1% for three-strand conductors and 2% for concentrically stranded conductors.

The variation of resistance of metallic conductors with temperature is practically linear over the normal range of operation. If temperature is plotted on the vertical axis and resistance on the horizontal axis, as in Fig. 4.3, extension

[1]SI is the official designation for the International System of Units.

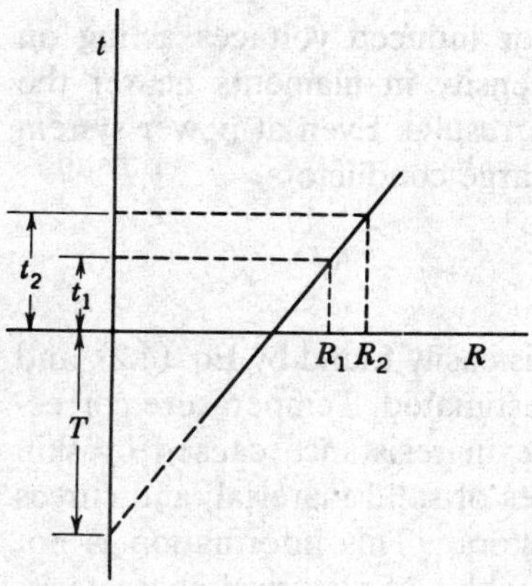

FIGURE 4.3
Resistance of a metallic conductor as a function of temperature.

of the straight-line portion of the graph provides a convenient method of correcting resistance for changes in temperature. The point of intersection of the extended line with the temperature axis at zero resistance is a constant of the material. From the geometry of Fig. 4.3

$$\frac{R_2}{R_1} = \frac{T + t_2}{T + t_1} \tag{4.3}$$

where R_1 and R_2 are the resistances of the conductor at temperatures t_1 and t_2, respectively, in degrees Celsius and T is the constant determined from the graph. Values of the constant T in degrees Celsius are as follows:

$$T = \begin{cases} 234.5 & \text{for annealed copper of 100\% conductivity} \\ 241 & \text{for hard-drawn copper of 97.3\% conductivity} \\ 228 & \text{for hard-drawn aluminum of 61\% conductivity} \end{cases}$$

Uniform distribution of current throughout the cross section of a conductor exists only for direct current. As the frequency of alternating current increases, the nonuniformity of distribution becomes more pronounced. An increase in frequency causes nonuniform current density. This phenomenon is called *skin effect*. In a circular conductor the current density *usually* increases from the interior toward the surface. For conductors of sufficiently large radius, however, a current density oscillatory with respect to radial distance from the center may result.

As we shall see when discussing inductance, some lines of magnetic flux exist inside a conductor. Filaments on the surface of a conductor are not linked by internal flux, and the flux linking a filament near the surface is less than the flux linking a filament in the interior. The alternating flux induces higher voltages acting on the interior filaments than are induced on filaments near the surface of the conductor. By Lenz's law the induced voltage opposes the changes of current producing it, and the higher induced voltages acting on the inner filaments cause the higher current density in filaments nearer the surface, and therefore higher effective resistance results. Even at power system frequencies, skin effect is a significant factor in large conductors.

4.3 TABULATED RESISTANCE VALUES

The dc resistance of various types of conductors is easily found by Eq. (4.2), and the increased resistance due to spiraling can be estimated. Temperature corrections are determined by Eq. (4.3). The increase in resistance caused by skin effect can be calculated for round wires and tubes of solid material, and curves of R/R_0 are available for these simple conductors.[2] This information is not necessary, however, since manufacturers supply tables of electrical characteristics of their conductors. Table A.3 is an example of some of the data available.

Example 4.1. Tables of electrical characteristics of all-aluminum *Marigold* stranded conductor list a dc resistance of 0.01558 Ω per 1000 ft at 20°C and an ac resistance of 0.0956 Ω/mi at 50°C. The conductor has 61 strands and its size is 1,113,000 cmil. Verify the dc resistance and find the ratio of ac to dc resistance.

Solution. At 20°C from Eq. (4.2) with an increase of 2% for spiraling

$$R_0 = \frac{17.0 \times 1000}{1113 \times 10^3} \times 1.02 = 0.01558\ \Omega \text{ per 1000 ft}$$

At a temperature of 50°C from Eq. (4.3)

$$R_0 = 0.01558\frac{228 + 50}{228 + 20} = 0.01746\ \Omega \text{ per 1000 ft}$$

$$\frac{R}{R_0} = \frac{0.0956}{0.01746 \times 5.280} = 1.037$$

Skin effect causes a 3.7% increase in resistance.

4.4 INDUCTANCE OF A CONDUCTOR DUE TO INTERNAL FLUX

The inductance of a transmission line is calculated as flux linkages per ampere. If permeability μ is constant, sinusoidal current produces sinusoidally varying flux in phase with the current. The resulting flux linkages can then be expressed

[2] See The Aluminum Association, *Aluminum Electrical Conductor Handbook*, 2d ed., Washington, DC, 1982.

as a phasor λ, and

$$L = \frac{\lambda}{I} \tag{4.4}$$

If i, the instantaneous value of current, is substituted for the phasor I in Eq. (4.4), then λ should be the value of the instantaneous flux linkages produced by i. Flux linkages are measured in weber-turns, Wbt.

Only flux lines external to the conductors are shown in Fig. 4.1. Some of the magnetic field, however, exists inside the conductors, as we mentioned when considering skin effect. The changing lines of flux inside the conductors also contribute to the induced voltage of the circuit and therefore to the inductance. The correct value of inductance due to internal flux can be computed as the ratio of flux linkages to current by taking into account the fact that each line of internal flux links only a fraction of the total current.

To obtain an accurate value for the inductance of a transmission line, it is necessary to consider the flux inside each conductor as well as the external flux. Let us consider the long cylindrical conductor whose cross section is shown in Fig. 4.4. We assume that the return path for the current in this conductor is so far away that it does not appreciably affect the magnetic field of the conductor shown. Then, the lines of flux are concentric with the conductor.

By Ampere's law the magnetomotive force (mmf) in ampere-turns around any closed path is equal to the net current in amperes enclosed by the path, as discussed in Sec. 2.1. The mmf equals the line integral around the closed path of the component of the magnetic field intensity tangent to the path and is given by Eq. (2.4), now written as Eq. (4.5):

$$\text{mmf} = \oint H \cdot ds = I \text{ At} \tag{4.5}$$

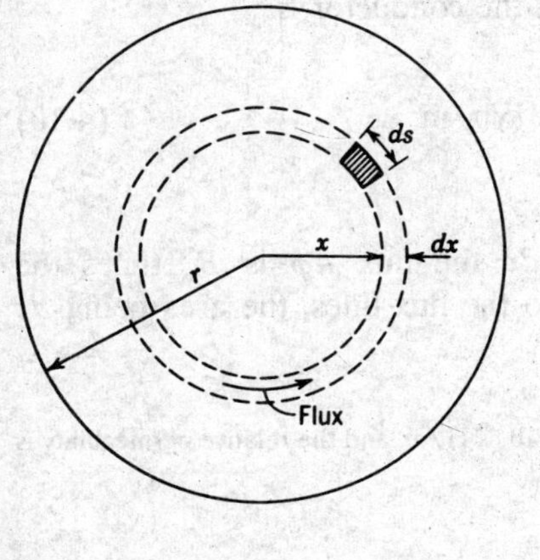

FIGURE 4.4
Cross section of a cylindrical conductor.

where H = magnetic field intensity, At/m

s = distance along path, m

I = current enclosed, A

Note that H and I are shown as phasors to represent sinusoidally alternating quantities since our work here applies equally to alternating and direct current. For simplicity the current I could be interpreted as a direct current and H as a real number. We recall that the dot between H and ds indicates that the value of H is the component of the field intensity tangent to ds.

Let the field intensity at a distance x meters from the center of the conductor be designated H_x. Since the field is symmetrical, H_x is constant at all points equidistant from the center of the conductor. If the integration indicated in Eq. (4.5) is performed around a circular path concentric with the conductor at x meters from the center, H_x is constant over the path and tangent to it. Equation (4.5) becomes

$$\oint H_x \, ds = I_x \tag{4.6}$$

and

$$2\pi x H_x = I_x \tag{4.7}$$

where I_x is the current enclosed. Then, assuming uniform current density,

$$I_x = \frac{\pi x^2}{\pi r^2} I \tag{4.8}$$

where I is the total current in the conductor. Then, substituting Eq. (4.8) in Eq. (4.7) and solving for H_x, we obtain

$$H_x = \frac{x}{2\pi r^2} I \text{ At/m} \tag{4.9}$$

The flux density x meters from the center of the conductor is

$$B_x = \mu H_x = \frac{\mu x I}{2\pi r^2} \text{ Wb/m}^2 \tag{4.10}$$

where μ is the permeability of the conductor.[3]

In the tubular element of thickness dx the flux $d\phi$ is B_x times the cross-sectional area of the element normal to the flux lines, the area being dx

[3] In SI units the permeability of free space is $\mu_0 = 4\pi \times 10^{-7}$ H/m, and the relative permeability is $\mu_r = \mu / \mu_0$.

times the axial length. The flux per meter of length is

$$d\phi = \frac{\mu x I}{2\pi r^2} dx \text{ Wb/m} \tag{4.11}$$

The flux linkages $d\lambda$ per meter of length, which are caused by the flux in the tubular element, are the product of the flux per meter of length and the fraction of the current linked. Thus,

$$d\lambda = \frac{\pi x^2}{\pi r^2} d\phi = \frac{\mu I x^3}{2\pi r^4} dx \text{ Wbt/m} \tag{4.12}$$

Integrating from the center of the conductor to its outside edge to find λ_{int}, the total flux linkages inside the conductor, we obtain

$$\lambda_{\text{int}} = \int_0^r \frac{\mu I x^3}{2\pi r^4} dx = \frac{\mu I}{8\pi} \text{ Wbt/m} \tag{4.13}$$

For a relative permeability of 1, $\mu = 4\pi \times 10^{-7}$ H/m, and

$$\lambda_{\text{int}} = \frac{I}{2} \times 10^{-7} \text{ Wbt/m} \tag{4.14}$$

$$L_{\text{int}} = \frac{1}{2} \times 10^{-7} \text{ H/m} \tag{4.15}$$

We have computed the inductance per unit length (henrys per meter) of a round conductor attributed only to the flux inside the conductor. Hereafter, for convenience, we refer to *inductance per unit length* simply as *inductance*, but we must be careful to use the correct dimensional units.

The validity of computing the internal inductance of a solid round wire by the method of partial flux linkages can be demonstrated by deriving the internal inductance in an entirely different manner. Equating energy stored in the magnetic field within the conductor per unit length at any instant to $L_{\text{int}} i^2/2$ and solving for L_{int} will yield Eq. (4.15).

4.5 FLUX LINKAGES BETWEEN TWO POINTS EXTERNAL TO AN ISOLATED CONDUCTOR

As a step in computing inductance due to flux external to a conductor, let us derive an expression for the flux linkages of an isolated conductor due only to that portion of the external flux which lies between two points at D_1 and D_2 meters from the center of the conductor. In Fig. 4.5 P_1 and P_2 are two such points. The conductor carries a current of I A. Since the flux paths are

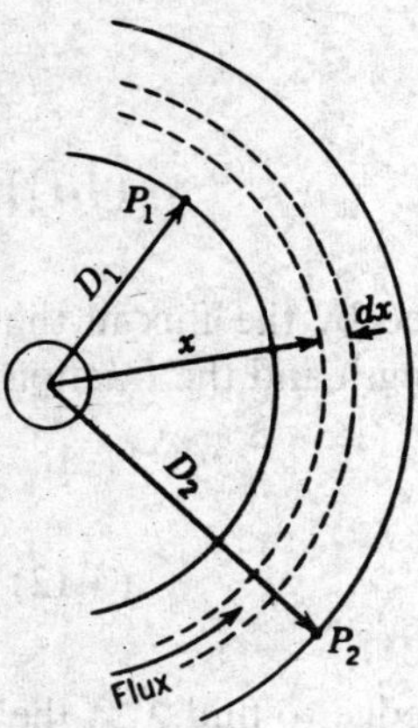

FIGURE 4.5
A conductor and external points P_1 and P_2.

concentric circles around the conductor, all the flux between P_1 and P_2 lies within the concentric cylindrical surfaces (indicated by solid circular lines) which pass through P_1 and P_2. At the tubular element which is x meters from the center of the conductor the field intensity is H_x. The mmf around the element is

$$2\pi x H_x = I \tag{4.16}$$

Solving for H_x and multiplying by μ yield the flux density B_x in the element so that

$$B_x = \frac{\mu I}{2\pi x} \text{ Wb/m}^2 \tag{4.17}$$

The flux $d\phi$ in the tubular element of thickness dx is

$$d\phi = \frac{\mu I}{2\pi x} dx \text{ Wb/m} \tag{4.18}$$

The flux linkages $d\lambda$ per meter are numerically equal to the flux $d\phi$ since flux external to the conductor links all the current in the conductor only once. So, between P_1 and P_2 the flux linkages are

$$\lambda_{12} = \int_{D_1}^{D_2} \frac{\mu I}{2\pi x} dx = \frac{\mu I}{2\pi} \ln \frac{D_2}{D_1} \text{ Wbt/m} \tag{4.19}$$

or for a relative permeability of 1

$$\lambda_{12} = 2 \times 10^{-7} I \ln \frac{D_2}{D_1} \text{ Wbt/m} \tag{4.20}$$

The inductance due only to the flux included between P_1 and P_2 is

$$L_{12} = 2 \times 10^{-7} \ln \frac{D_2}{D_1} \text{ H/m} \tag{4.21}$$

4.6 INDUCTANCE OF A SINGLE-PHASE TWO-WIRE LINE

We can now determine the inductance of a simple two-wire line composed of solid round conductors. Figure 4.6 shows such a line having two conductors of radii r_1 and r_2. One conductor is the return circuit for the other. First, consider only the flux linkages of the circuit caused by the current in conductor 1. A line of flux set up by current in conductor 1 at a distance equal to or greater than $D + r_2$ from the center of conductor 1 does not link the circuit. At a distance less than $D - r_2$ the fraction of the total current linked by a line of flux is 1.0. Therefore, it is logical when D is much greater than r_1 and r_2 to assume that D can be used instead of $D - r_2$ or $D + r_2$. In fact, it can be shown that calculations made with this assumption are correct even when D is small.

We add inductance due to internal flux linkages determined by Eq. (4.15) to inductance due to external flux linkages determined by Eq. (4.21) with r_1 replacing D_1 and D replacing D_2 to obtain

$$L_1 = \left(\frac{1}{2} + 2 \ln \frac{D}{r_1} \right) \times 10^{-7} \text{ H/m} \tag{4.22}$$

which is the inductance of the circuit due to the current in conductor 1 only.

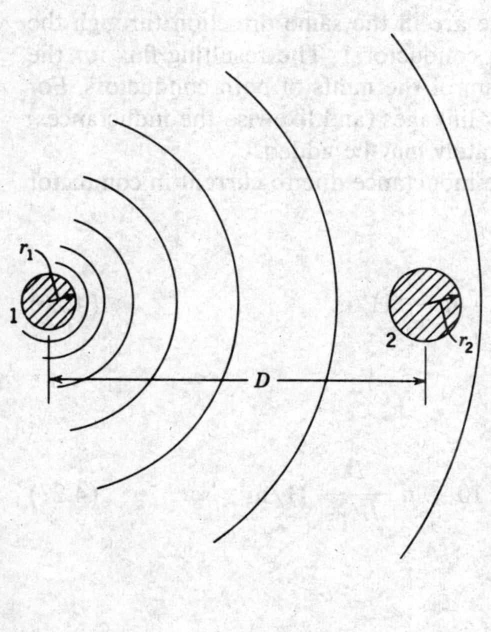

FIGURE 4.6
Conductors of different radii and the magnetic field due to current in conductor 1 only.

The expression for inductance may be put in a more concise form by factoring Eq. (4.22) and by noting that $\ln \varepsilon^{1/4} = 1/4$, whence

$$L_1 = 2 \times 10^{-7}\left(\ln \varepsilon^{1/4} + \ln \frac{D}{r_1}\right) \tag{4.23}$$

Upon combining terms, we obtain

$$L_1 = 2 \times 10^{-7} \ln \frac{D}{r_1 \varepsilon^{-1/4}} \tag{4.24}$$

If we substitute r_1' for $r_1 \varepsilon^{-1/4}$,

$$L_1 = 2 \times 10^{-7} \ln \frac{D}{r_1'} \text{ H/m} \tag{4.25}$$

The radius r_1' is that of a fictitious conductor assumed to have no internal flux but with the same inductance as the actual conductor of radius r_1. The quantity $\varepsilon^{-1/4}$ is equal to 0.7788. Equation (4.25) omits the term accounting for internal flux but compensates for it by using an adjusted value for the radius of the conductor. The multiplying factor of 0.7788, which adjusts the radius in order to account for internal flux, applies only to solid round conductors. We consider other conductors later.

Since the current in conductor 2 flows in the direction opposite to that in conductor 1 (or is 180° out of phase with it), the flux linkages produced by current in conductor 2 considered alone are in the same direction through the circuit as those produced by current in conductor 1. The resulting flux for the two conductors is determined by the sum of the mmfs of both conductors. For constant permeability, however, the flux linkages (and likewise the inductances) of the two conductors considered separately may be added.

By comparison with Eq. (4.25), the inductance due to current in conductor 2 is

$$L_2 = 2 \times 10^{-7} \ln \frac{D}{r_2'} \text{ H/m} \tag{4.26}$$

and for the complete circuit

$$L = L_1 + L_2 = 4 \times 10^{-7} \ln \frac{D}{\sqrt{r_1' r_2'}} \text{ H/m} \tag{4.27}$$

If $r_1' = r_2' = r'$, the total inductance reduces to

$$L = 4 \times 10^{-7} \ln \frac{D}{r'} \text{ H/m} \tag{4.28}$$

This value of inductance is sometimes called the *inductance per loop meter* or *per loop mile* to distinguish it from that component of the inductance of the circuit attributed to the current in one conductor only. The latter, as given by Eq. (4.25), is one-half the total inductance of a single-phase line and is called the *inductance per conductor*.

4.7 FLUX LINKAGES OF ONE CONDUCTOR IN A GROUP

A more general problem than that of the two-wire line is presented by one conductor in a group of conductors where the sum of the currents in all the conductors is zero. Such a group of conductors is shown in Fig. 4.7. Conductors $1, 2, 3, \ldots, n$ carry the phasor currents $I_1, I_2, I_3, \ldots, I_n$. The distances of these conductors from a remote point P are indicated on the figure as $D_{1P}, D_{2P}, D_{3P}, \ldots, D_{nP}$. Let us determine λ_{1P1}, the flux linkages of conductor 1 due to I_1 including internal flux linkages but excluding all the flux beyond the point P. By Eqs. (4.14) and (4.20)

$$\lambda_{1P1} = \left(\frac{I_1}{2} + 2I_1 \ln \frac{D_{1P}}{r_1}\right) 10^{-7} \tag{4.29}$$

$$\lambda_{1P1} = 2 \times 10^{-7} I_1 \ln \frac{D_{1P}}{r_1'} \text{ Wbt/m} \tag{4.30}$$

The flux linkages λ_{1P2} with conductor 1 *due to* I_2 but excluding flux beyond point P is equal to the flux produced by I_2 between the point P and conductor

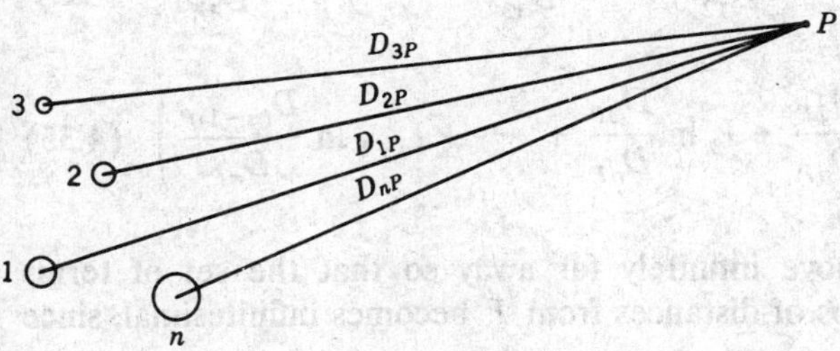

FIGURE 4.7
Cross-sectional view of a group of n conductors carrying currents whose sum is zero. Point P is remote from the conductors.

1 (that is, within the limiting distances D_{2P} and D_{12} from conductor 2), and so

$$\lambda_{1P2} = 2 \times 10^{-7} I_2 \ln \frac{D_{2P}}{D_{12}} \tag{4.31}$$

The flux linkages λ_{1P} with conductor 1 *due to all the conductors* in the group but excluding flux beyond point P is

$$\lambda_{1P} = 2 \times 10^{-7}\left(I_1 \ln \frac{D_{1P}}{r_1'} + I_2 \ln \frac{D_{2P}}{D_{12}} + I_3 \ln \frac{D_{3P}}{D_{13}} + \cdots + I_n \ln \frac{D_{nP}}{D_{1n}}\right) \tag{4.32}$$

which becomes, by expanding the logarithmic terms and regrouping,

$$\lambda_{1P} = 2 \times 10^{-7}\left(I_1 \ln \frac{1}{r_1'} + I_2 \ln \frac{1}{D_{12}} + I_3 \ln \frac{1}{D_{13}} + \cdots + I_n \ln \frac{1}{D_{1n}} \right.$$
$$\left. + I_1 \ln D_{1P} + I_2 \ln D_{2P} + I_3 \ln D_{3P} + \cdots + I_n \ln D_{nP}\right) \tag{4.33}$$

Since the sum of all the currents in the group is zero,

$$I_1 + I_2 + I_3 + \cdots + I_n = 0$$

and solving for I_n, we obtain

$$I_n = -(I_1 + I_2 + I_3 + \cdots + I_{n-1}) \tag{4.34}$$

Substituting Eq. (4.34) in the second term containing I_n in Eq. (4.33) and recombining some logarithmic terms, we have

$$\lambda_{1P} = 2 \times 10^{-7}\left(I_1 \ln \frac{1}{r_1'} + I_2 \ln \frac{1}{D_{12}} + I_3 \ln \frac{1}{D_{13}} + \cdots + I_n \ln \frac{1}{D_{1n}} \right.$$
$$\left. + I_1 \ln \frac{D_{1P}}{D_{np}} + I_2 \ln \frac{D_{2P}}{D_{nP}} + I_3 \ln \frac{D_{3P}}{D_{nP}} + \cdots + I_{n-1} \ln \frac{D_{(n-1)P}}{D_{nP}}\right) \tag{4.35}$$

Now letting the point P move infinitely far away so that the set of terms containing logarithms of ratios of distances from P becomes infinitesimal, since the ratios of the distances approach 1, we obtain

$$\lambda_1 = 2 \times 10^{-7}\left(I_1 \ln \frac{1}{r_1'} + I_2 \ln \frac{1}{D_{12}} + I_3 \ln \frac{1}{D_{13}} + \cdots + I_n \ln \frac{1}{D_{1n}}\right) \text{ Wbt/m} \tag{4.36}$$

By letting point P move infinitely far away, we have included all the flux linkages of conductor 1 in our derivation. Therefore, Eq. (4.36) expresses all the flux linkages of conductor 1 in a group of conductors, provided the sum of all the currents is zero. If the currents are alternating, they must be expressed as instantaneous currents to obtain instantaneous flux linkages or as complex rms values to obtain the rms value of flux linkages as a complex number.

4.8 INDUCTANCE OF COMPOSITE-CONDUCTOR LINES

Stranded conductors come under the general classification of *composite* conductors, which means conductors composed of two or more elements or strands electrically in parallel. We limit ourselves to the case where all the strands are identical and share the current equally. The values of internal inductance of specific conductors are generally available from the various manufacturers and can be found in handbooks. The method to be developed indicates the approach to the more complicated problems of nonhomogeneous conductors and unequal division of current between strands. The method is applicable to the determination of inductance of lines consisting of circuits electrically in parallel since two conductors in parallel can be treated as strands of a single composite conductor.

Figure 4.8 shows a single-phase line composed of two conductors. In order to be more general, each conductor forming one side of the line is shown as an arbitrary arrangement of an indefinite number of conductors. The only restrictions are that the parallel filaments are cylindrical and share the current equally. Conductor X is composed of n identical, parallel filaments, each of which carries the current I/n. Conductor Y, which is the return circuit for the current in conductor X, is composed of m identical, parallel filaments, each of which carries the current $-I/m$. Distances between the elements will be designated by the letter D with appropriate subscripts. Applying Eq. (4.36) to

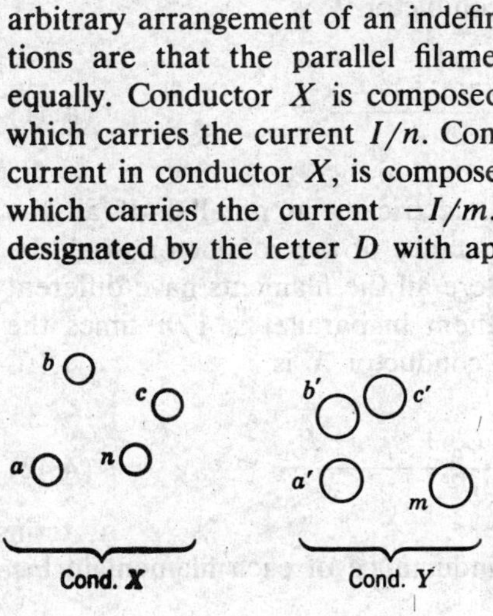

FIGURE 4.8
Single-phase line consisting of two composite conductors.

filament a of conductor X, we obtain for flux linkages of filament a

$$\lambda_a = 2 \times 10^{-7} \frac{I}{n} \left(\ln \frac{1}{r'_a} + \ln \frac{1}{D_{ab}} + \ln \frac{1}{D_{ac}} + \cdots + \ln \frac{1}{D_{an}} \right)$$

$$- 2 \times 10^{-7} \frac{I}{m} \left(\ln \frac{1}{D_{aa'}} + \ln \frac{1}{D_{ab'}} + \ln \frac{1}{D_{ac'}} + \cdots + \ln \frac{1}{D_{am}} \right) \quad (4.37)$$

from which

$$\lambda_a = 2 \times 10^{-7} I \ln \frac{\sqrt[m]{D_{aa'} D_{ab'} D_{ac'} \cdots D_{am}}}{\sqrt[n]{r'_a D_{ab} D_{ac} \cdots D_{an}}} \text{ Wbt/m} \quad (4.38)$$

Dividing Eq. (4.38) by the current I/n, we find that the inductance of filament a is

$$L_a = \frac{\lambda_a}{I/n} = 2n \times 10^{-7} \ln \frac{\sqrt[m]{D_{aa'} D_{ab'} D_{ac'} \cdots D_{am}}}{\sqrt[n]{r'_a D_{ab} D_{ac} \cdots D_{an}}} \text{ H/m} \quad (4.39)$$

Similarly, the inductance of filament b is

$$L_b = \frac{\lambda_b}{I/n} = 2n \times 10^{-7} \ln \frac{\sqrt[m]{D_{ba'} D_{bb'} D_{bc'} \cdots D_{bm}}}{\sqrt[n]{D_{ba} r'_b D_{bc} \cdots D_{bn}}} \text{ H/m} \quad (4.40)$$

The average inductance of the filaments of conductor X is

$$L_{av} = \frac{L_a + L_b + L_c + \cdots + L_n}{n} \quad (4.41)$$

Conductor X is composed of n filaments electrically in parallel. If all the filaments had the same inductance, the inductance of the conductor would be $1/n$ times the inductance of one filament. Here all the filaments have different inductances, but the inductance of all of them in parallel is $1/n$ times the average inductance. Thus, the inductance of conductor X is

$$L_X = \frac{L_{av}}{n} = \frac{L_a + L_b + L_c + \cdots + L_n}{n^2} \quad (4.42)$$

Substituting the logarithmic expression for inductance of each filament in Eq.

(4.42) and combining terms, we obtain

$$L_X = 2 \times 10^{-7}$$

$$\times \ln \frac{\sqrt[mn]{(D_{aa'}D_{ab'}D_{ac'}\cdots D_{am})(D_{ba'}D_{bb'}D_{bc'}\cdots D_{bm})\cdots(D_{na'}D_{nb'}D_{nc'}\cdots D_{nm})}}{\sqrt[n^2]{(D_{aa}D_{ab}D_{ac}\cdots D_{an})(D_{ba}D_{bb}D_{bc}\cdots D_{bn})\cdots(D_{na}D_{nb}D_{nc}\cdots D_{nn})}} \text{ H/m}$$

$$(4.43)$$

where r'_a, r'_b, and r'_n have been replaced by D_{aa}, D_{bb}, and D_{nn}, respectively, to make the expression appear more symmetrical.

Note that the numerator of the argument of the logarithm in Eq. (4.43) is the *mn*th root of *mn* terms, which are the products of the distances from all the *n* filaments of conductor *X* to all the *m* filaments of conductor *Y*. For *each* filament in conductor *X* there are *m* distances to filaments in conductor *Y*, and there are *n* filaments in conductor *X*. The product of *m* distances for each of *n* filaments results in *mn* terms. The *mn*th root of the product of the *mn* distances is called the *geometric mean distance* between conductor *X* and conductor *Y*. It is abbreviated D_m or GMD and is also called the *mutual* GMD between the two conductors.

The denominator of the argument of the logarithm in Eq. (4.43) is the n^2 root of n^2 terms. There are *n* filaments, and for each filament there are *n* terms consisting of r' for that filament times the distances from that filament to every other filament in conductor *X*. Thus, we account for n^2 terms. Sometimes r'_a is called the distance from filament *a* to itself, especially when it is designated as D_{aa}. With this in mind, the terms under the radical in the denominator may be described as the product of the distances from every filament in the conductor to itself and to every other filament. The n^2 root of these terms is called the *self* GMD of conductor *X*, and the r' of a separate filament is called the self GMD of the filament. Self GMD is also called *geometric mean radius*, or GMR. The correct mathematical expression is self GMD, but common practice has made GMR more prevalent. We use GMR in order to conform to this practice and identify it by D_s.

In terms of D_m and D_s Eq. (4.43) becomes

$$L_X = 2 \times 10^{-7} \ln \frac{D_m}{D_s} \text{ H/m} \tag{4.44}$$

The reader should compare Eqs. (4.44) and (4.25).

The inductance of conductor *Y* is determined in a similar manner, and the inductance of the line is

$$L = L_X + L_Y$$

Example 4.2. One circuit of a single-phase transmission line is composed of three solid 0.25-cm-radius wires. The return circuit is composed of two 0.5-cm-radius wires. The arrangement of conductors is shown in Fig. 4.9. Find the inductance due to the current in each side of the line and the inductance of the complete line in henrys per meter (and in millihenrys per mile).

Solution. Find the GMD between sides X and Y:

$$D_m = \sqrt[6]{D_{ad}D_{ae}D_{bd}D_{be}D_{cd}D_{ce}}$$

$$D_{ad} = D_{be} = 9 \text{ m}$$

$$D_{ae} = D_{bd} = D_{ce} = \sqrt{6^2 + 9^2} = \sqrt{117}$$

$$D_{cd} = \sqrt{9^2 + 12^2} = 15 \text{ m}$$

$$D_m = \sqrt[6]{9^2 \times 15 \times 117^{3/2}} = 10.743 \text{ m}$$

Then, find the GMR for side X

$$D_s = \sqrt[9]{D_{aa}D_{ab}D_{ac}D_{ba}D_{bb}D_{bc}D_{ca}D_{cb}D_{cc}}$$

$$= \sqrt[9]{(0.25 \times 0.7788 \times 10^{-2})^3 \times 6^4 \times 12^2} = 0.481 \text{ m}$$

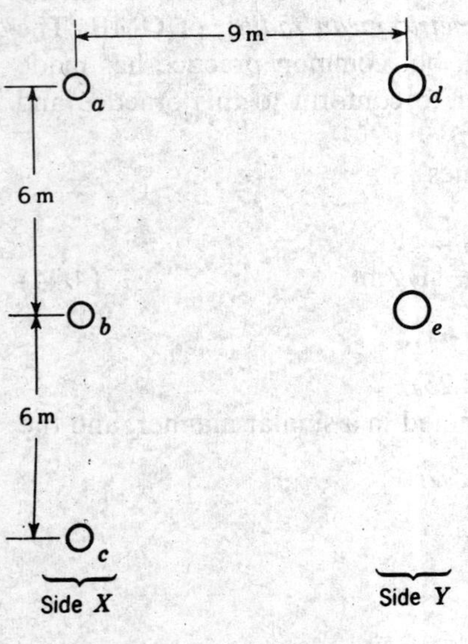

FIGURE 4.9
Arrangement of conductors for Example 4.2.

and for side Y

$$D_s = \sqrt[4]{(0.5 \times 0.7788 \times 10^{-2})^2 \times 6^2} = 0.153 \text{ m}$$

$$L_X = 2 \times 10^{-7} \ln \frac{10.743}{0.481} = 6.212 \times 10^{-7} \text{ H/m}$$

$$L_Y = 2 \times 10^{-7} \ln \frac{10.743}{0.153} = 8.503 \times 10^{-7} \text{ H/m}$$

$$L = L_X + L_Y = 14.715 \times 10^{-7} \text{ H/m}$$

$$(L = 14.715 \times 10^{-7} \times 1609 \times 10^3 = 2.37 \text{ mH/mi})$$

In Example 4.2 the conductors in parallel on one side of the line are separated by 6 m, and the distance between the two sides of the line is 9 m. Here the calculation of mutual GMD is important. For stranded conductors the distance between sides of a line composed of one conductor per side is usually so great that the mutual GMD can be taken as equal to the center-to-center distance with negligible error.

If the effect of the steel core of ACSR is neglected in calculating inductance, a high degree of accuracy results, provided the aluminum strands are in an even number of layers. The effect of the core is more apparent for an odd number of layers of aluminum strands, but the accuracy is good when the calculations are based on the aluminum strands alone.

4.9 THE USE OF TABLES

Tables listing values of GMR are generally available for standard conductors and provide other information for calculating inductive reactance as well as shunt capacitive reactance and resistance. Since industry in the United States continues to use units of inches, feet, and miles, so do these tables. Therefore, some of our examples will use feet and miles, but others will use meters and kilometers.

Inductive reactance rather than inductance is usually desired. The inductive reactance of one conductor of a single-phase two-conductor line is

$$X_L = 2\pi f L = 2\pi f \times 2 \times 10^{-7} \ln \frac{D_m}{D_s}$$

$$= 4\pi f \times 10^{-7} \ln \frac{D_m}{D_s} \ \Omega/\text{m} \tag{4.45}$$

or

$$X_L = 2.022 \times 10^{-3} f \ln \frac{D_m}{D_s} \ \Omega/\text{mi} \tag{4.46}$$

where D_m is the distance between conductors. Both D_m and D_s must be in the same units, usually either meters or feet. The GMR found in tables is an equivalent D_s, which accounts for skin effect where it is appreciable enough to affect inductance. Of course, skin effect is greater at higher frequencies for a conductor of a given diameter. Values of D_s listed in Table A.3 of the Appendix are for a frequency of 60 Hz.

Some tables give values of inductive reactance in addition to GMR. One method is to expand the logarithmic term of Eq. (4.46), as follows:

$$X_L = \underbrace{2.022 \times 10^{-3} f \ln \frac{1}{D_s}}_{X_a} + \underbrace{2.022 \times 10^{-3} f \ln D_m}_{X_d} \ \Omega/\text{mi} \tag{4.47}$$

If both D_s and D_m are in feet, the first term in Eq. (4.47) is the inductive reactance of one conductor of a two-conductor line having a distance of 1 ft between conductors, as may be seen by comparing Eq. (4.47) with Eq. (4.46). Therefore, the first term of Eq. (4.47) is called the *inductive reactance at 1-ft spacing* X_a. It depends on the GMR of the conductor and the frequency. The second term of Eq. (4.47) is called the *inductive reactance spacing factor* X_d. This second term is independent of the type of conductor and depends on frequency and spacing only. Table A.3 includes values of inductive reactance at 1-ft spacing, and Table A.4 lists values of the inductive reactance spacing factor.

Example 4.3. Find the inductive reactance per mile of a single-phase line operating at 60 Hz. The conductor is *Partridge*, and spacing is 20 ft between centers.

Solution. For this conductor Table A.3 lists $D_s = 0.0217$ ft. From Eq. (4.46) for one conductor

$$X_L = 2.022 \times 10^{-3} \times 60 \ln \frac{20}{0.0217}$$

$$= 0.828 \ \Omega/\text{mi}$$

The above calculation is used only if D_s is known. Table A.3, however, lists inductive reactance at 1-ft spacing $X_a = 0.465 \ \Omega/\text{mi}$. From Table A.4 the inductive reactance spacing factor is $X_d = 0.3635 \ \Omega/\text{mi}$, and so the inductive reactance of one conductor is

$$0.465 + 0.3635 = 0.8285 \ \Omega/\text{mi}$$

Since the conductors composing the two sides of the line are identical, the inductive reactance of the line is

$$2X_L = 2 \times 0.8285 = 1.657 \ \Omega/\text{mi}$$

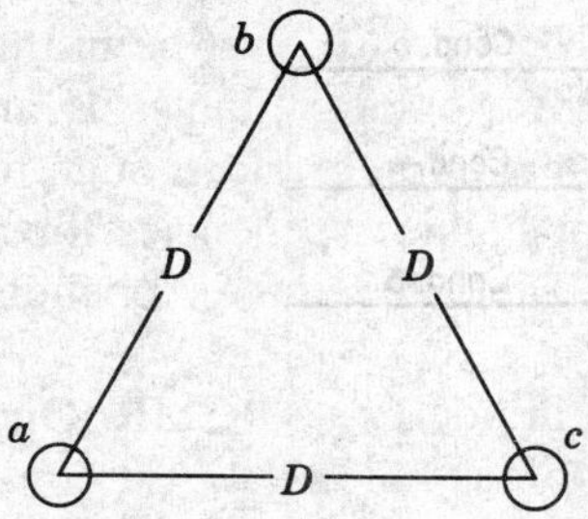

FIGURE 4.10
Cross-sectional view of the equilaterally spaced conductors of a three-phase line.

4.10 INDUCTANCE OF THREE-PHASE LINES WITH EQUILATERAL SPACING

So far in our discussion we have considered only single-phase lines. The equations we have developed are quite easily adapted, however, to the calculation of the inductance of three-phase lines. Figure 4.10 shows the conductors of a three-phase line spaced at the corners of an equilateral triangle. If we assume that there is no neutral wire, or if we assume balanced three-phase phasor currents, $I_a + I_b + I_c = 0$. Equation (4.36) determines the flux linkages of conductor a:

$$\lambda_a = 2 \times 10^{-7}\left(I_a \ln \frac{1}{D_s} + I_b \ln \frac{1}{D} + I_c \ln \frac{1}{D}\right) \text{ Wbt/m} \tag{4.48}$$

Since $I_a = -(I_b + I_c)$, Eq. (4.48) becomes

$$\lambda_a = 2 \times 10^{-7}\left(I_a \ln \frac{1}{D_s} - I_a \ln \frac{1}{D}\right) = 2 \times 10^{-7} I_a \ln \frac{D}{D_s} \text{ Wbt/m} \tag{4.49}$$

$$\text{and } L_a = 2 \times 10^{-7} \ln \frac{D}{D_s} \text{ H/m} \tag{4.50}$$

Equation (4.50) is the same in form as Eq. (4.25) for a single-phase line except that D_s replaces r'. Because of symmetry, the inductances of conductors b and c are the same as the inductance of conductor a. Since each phase consists of only one conductor, Eq. (4.50) gives the inductance per phase of the three-phase line.

4.11 INDUCTANCE OF THREE-PHASE LINES WITH UNSYMMETRICAL SPACING

When the conductors of a three-phase line are not spaced equilaterally, the problem of finding the inductance becomes more difficult. The flux linkages and inductance of each phase are not the same. A different inductance in each

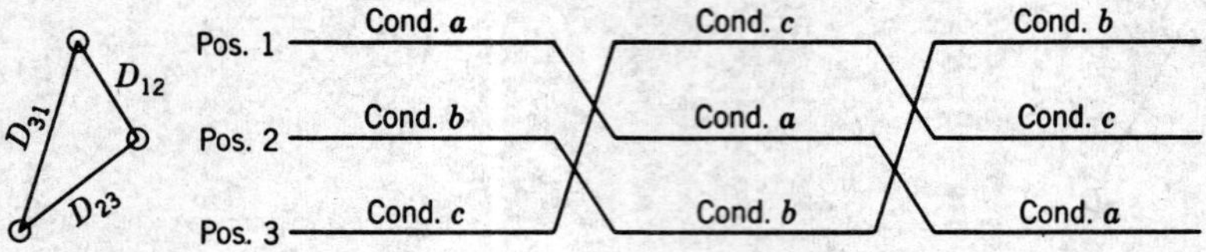

FIGURE 4.11
Transposition cycle.

phase results in an unbalanced circuit. Balance of the three phases can be restored by exchanging the positions of the conductors at regular intervals along the line so that each conductor occupies the original position of every other conductor over an equal distance. Such an exchange of conductor positions is called *transposition*. A complete transposition cycle is shown in Fig. 4.11. The phase conductors are designated a, b, and c, and the positions occupied are numbered 1, 2, and 3, respectively. Transposition results in each conductor having the same average inductance over the whole cycle.

Modern power lines are usually not transposed at regular intervals although an interchange in the positions of the conductors may be made at switching stations in order to balance the inductance of the phases more closely. Fortunately, the dissymmetry between the phases of an untransposed line is small and neglected in most calculations of inductance. If the dissymmetry is neglected, the inductance of the untransposed line is taken as equal to the average value of the inductive reactance of one phase of the same line correctly transposed. The derivations to follow are for transposed lines.

To find the average inductance of one conductor of a transposed line, we first determine the flux linkages of a conductor for each position it occupies in the transposition cycle and then determine the average flux linkages. Applying Eq. (4.36) to conductor a of Fig. 4.11 to find the phasor expression for the flux linkages of a in position 1 when b is in position 2 and c is in position 3, we obtain

$$\lambda_{a1} = 2 \times 10^{-7}\left(I_a \ln \frac{1}{D_s} + I_b \ln \frac{1}{D_{12}} + I_c \ln \frac{1}{D_{31}}\right) \text{ Wbt/m} \tag{4.51}$$

With a in position 2, b in position 3, and c in position 1,

$$\lambda_{a2} = 2 \times 10^{-7}\left(I_a \ln \frac{1}{D_s} + I_b \ln \frac{1}{D_{23}} + I_c \ln \frac{1}{D_{12}}\right) \text{ Wbt/m} \tag{4.52}$$

and, with a in position 3, b in position 1, and c in position 2,

$$\lambda_{a3} = 2 \times 10^{-7}\left(I_a \ln \frac{1}{D_s} + I_b \ln \frac{1}{D_{31}} + I_c \ln \frac{1}{D_{23}}\right) \text{ Wbt/m} \tag{4.53}$$

The average value of the flux linkages of a is

$$\lambda_a = \frac{\lambda_{a1} + \lambda_{a2} + \lambda_{a3}}{3}$$

$$= \frac{2 \times 10^{-7}}{3}\left(3I_a \ln \frac{1}{D_s} + I_b \ln \frac{1}{D_{12}D_{23}D_{31}} + I_c \ln \frac{1}{D_{12}D_{23}D_{31}}\right) \quad (4.54)$$

With the restriction that $I_a = -(I_b + I_c)$,

$$\lambda_a = \frac{2 \times 10^{-7}}{3}\left(3I_a \ln \frac{1}{D_s} - I_a \ln \frac{1}{D_{12}D_{23}D_{31}}\right)$$

$$= 2 \times 10^{-7} I_a \ln \frac{\sqrt[3]{D_{12}D_{23}D_{31}}}{D_s} \text{ Wbt/m} \quad (4.55)$$

and the *average* inductance per phase is

$$L_a = 2 \times 10^{-7} \ln \frac{D_{eq}}{D_s} \text{ H/m} \quad (4.56)$$

where $$D_{eq} = \sqrt[3]{D_{12}D_{23}D_{31}} \quad (4.57)$$

and D_s is the GMR of the conductor. D_{eq}, the geometric mean of the three distances of the unsymmetrical line, is the equivalent equilateral spacing, as may be seen by a comparison of Eq. (4.56) with Eq. (4.50). We should note the similarity between all the equations for the inductance of a conductor. If the inductance is in henrys per meter, the factor 2×10^{-7} appears in all the equations, and the denominator of the logarithmic term is always the GMR of the conductor. The numerator is the distance between wires of a two-wire line, the mutual GMD between sides of a composite-conductor single-phase line, the distance between conductors of an equilaterally spaced line, or the equivalent equilateral spacing of an unsymmetrical line.

Example 4.4. A single-circuit three-phase line operated at 60 Hz is arranged, as shown in Fig. 4.12. The conductors are ACSR *Drake*. Find the inductive reactance per mile per phase.

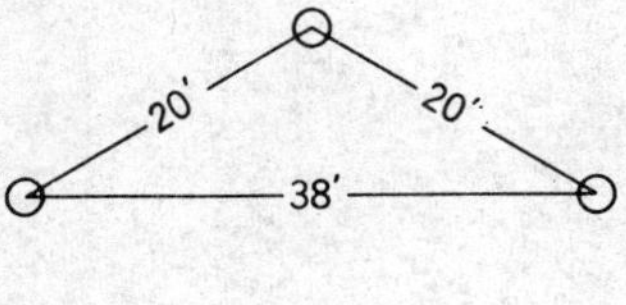

FIGURE 4.12
Arrangement of conductors for Example 4.4.

Solution. From Table A.3

$$D_s = 0.0373 \text{ ft} \qquad D_{eq} = \sqrt[3]{20 \times 20 \times 38} = 24.8 \text{ ft}$$

$$L = 2 \times 10^{-7} \ln \frac{24.8}{0.0373} = 13.00 \times 10^{-7} \text{ H/m}$$

$$X_L = 2\pi 60 \times 1609 \times 13.00 \times 10^{-7} = 0.788 \ \Omega/\text{mi per phase}$$

Equation (4.46) may be used also, or from Tables A.3 and A.4

$$X_a = 0.399$$

and by interpolation for 24.8 ft

$$X_d = 0.3896$$

$$X_L = 0.399 + 0.3896 = 0.7886 \ \Omega/\text{mi per phase}$$

4.12 INDUCTANCE CALCULATIONS FOR BUNDLED CONDUCTORS

At extra-high voltages (EHV), that is, voltages above 230 kV, corona with its resultant power loss and particularly its interference with communications is excessive if the circuit has only one conductor per phase. The high-voltage gradient at the conductor in the EHV range is reduced considerably by having two or more conductors per phase in close proximity compared with the spacing between phases. Such a line is said to be composed of *bundled* conductors. The bundle consists of two, three, or four conductors. Figure 4.13 shows the arrangements. The current will not divide exactly between the conductors of the bundle unless there is a transposition of the conductors within the bundle, but the difference is of no practical importance, and the GMD method is accurate for calculations.

Reduced reactance is the other equally important advantage of bundling. Increasing the number of conductors in a bundle reduces the effects of corona and reduces the reactance. The reduction of reactance results from the increased GMR of the bundle. The calculation of GMR is, of course, exactly the same as that of a stranded conductor. Each conductor of a two-conductor bundle, for instance, is treated as one strand of a two-strand conductor. If we let D_s^b indicate the GMR of a bundled conductor and D_s the GMR of the

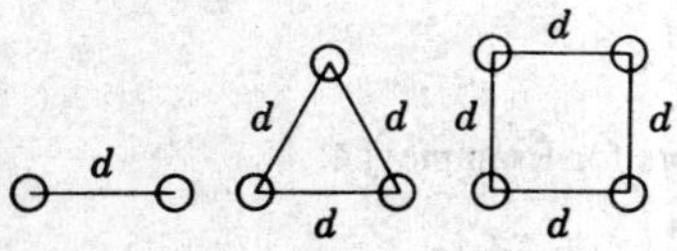

FIGURE 4.13
Bundle arrangements.

individual conductors composing the bundle, we find, referring to Fig. 4.13:

For a two-strand bundle

$$D_s^b = \sqrt[4]{(D_s \times d)^2} = \sqrt{D_s \times d} \tag{4.58}$$

For a three-strand bundle

$$D_s^b = \sqrt[9]{(D_s \times d \times d)^3} = \sqrt[3]{D_s \times d^2} \tag{4.59}$$

For a four-strand bundle

$$D_s^b = \sqrt[16]{\left(D_s \times d \times d \times \sqrt{2}\,d\right)^4} = 1.09\sqrt[4]{D_s \times d^3} \tag{4.60}$$

In computing inductance using Eq. (4.56), D_s^b of the bundle replaces D_s of a single conductor. To compute D_{eq}, the distance from the center of one bundle to the center of another bundle is sufficiently accurate for D_{ab}, D_{bc}, and D_{ca}. Obtaining the actual GMD between conductors of one bundle and those of another would be almost indistinguishable from the center-to-center distances for the usual spacing.

Example 4.5. Each conductor of the bundled-conductor line shown in Fig. 4.14 is ACSR, 1,272,000-cmil *Pheasant*. Find the inductive reactance in ohms per kilometer (and per mile) per phase for $d = 45$ cm. Also, find the per-unit series reactance of the line if its length is 160 km and the base is 100 MVA, 345 kV.

Solution. From Table A.3 $D_s = 0.0466$ ft, and we multiply feet by 0.3048 to convert to meters.

$$D_s^b = \sqrt{0.0466 \times 0.3048 \times 0.45} = 0.080 \text{ m}$$

$$D_{eq} = \sqrt[3]{8 \times 8 \times 16} = 10.08 \text{ m}$$

$$X_L = 2\pi 60 \times 2 \times 10^{-7} \times 10^3 \ln \frac{10.08}{0.08}$$

$$= 0.365\ \Omega/\text{km per phase}$$

$$= 0.365 \times 1.609 = 0.587\ \Omega/\text{mi per phase}$$

$$\text{Base } Z = \frac{(345)^2}{100} = 1190\ \Omega$$

$$X = \frac{0.365 \times 160}{1190} = 0.049 \text{ per unit}$$

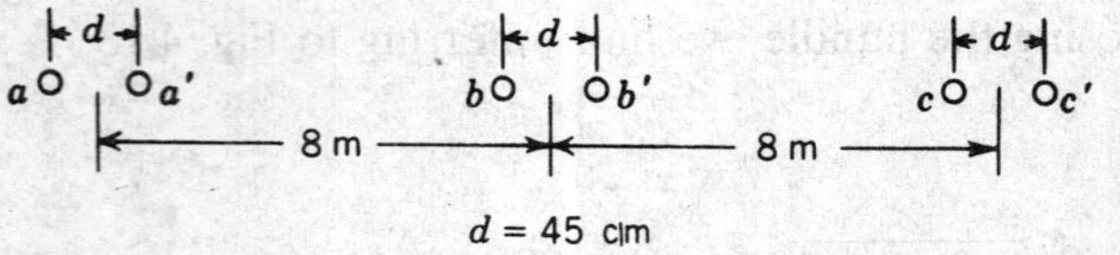

FIGURE 4.14
Spacing of conductors of a bundled-conductor line.

4.13 SUMMARY

Although computer programs are usually available or written rather easily for calculating inductance of all kinds of lines, some understanding of the development of the equations used is rewarding from the standpoint of appreciating the effect of variables in designing a line. However, tabulated values such as those in Tables A.3 and A.4 make the calculations quite simple except for parallel-circuit lines. Table A.3 also lists resistance.

The important equation for inductance per phase of single-circuit three-phase lines is given here for convenience:

$$L = 2 \times 10^{-7} \ln \frac{D_{eq}}{D_s} \text{ H/m per phase} \tag{4.61}$$

Inductive reactance in ohms per kilometer at 60 Hz is found by multiplying inductance in henrys per meter by $2\pi 60 \times 1000$:

$$X_L = 0.0754 \times \ln \frac{D_{eq}}{D_s} \ \Omega\text{/km per phase} \tag{4.62}$$

or

$$X_L = 0.1213 \times \ln \frac{D_{eq}}{D_s} \ \Omega\text{/mi per phase} \tag{4.63}$$

Both D_{eq} and D_s must be in the same units, usually feet. If the line has one conductor per phase, D_s is found directly from tables. For bundled conductors D_s^b, as defined in Sec. 4.12, is substituted for D_s. For both single-conductor and bundled-conductor lines

$$D_{eq} = \sqrt[3]{D_{ab}D_{bc}D_{ca}} \tag{4.64}$$

For bundled-conductor lines D_{ab}, D_{bc}, and D_{ca} are distances between the centers of the bundles of phases a, b, and c.

For lines with one conductor per phase it is convenient to determine X_L from tables by adding X_a for the conductor as found in Table A.3 to X_d as found in Table A.4 corresponding to D_{eq}.

PROBLEMS

4.1. The all-aluminum conductor (AAC) identified by the code word *Bluebell* is composed of 37 strands, each having a diameter of 0.1672 in. Tables of characteristics of AACs list an area of 1,033,500 cmil for this conductor (1 cmil = $(\pi/4) \times 10^{-6}$ in^2). Are these values consistent with each other? Find the overall area of the strands in square millimeters.

4.2. Determine the dc resistance in ohms per km of *Bluebell* at 20°C by Eq. (4.2) and the information in Prob. 4.1, and check the result against the value listed in tables of 0.01678 Ω per 1000 ft. Compute the dc resistance in ohms per kilometer at 50°C and compare the result with the ac 60-Hz resistance of 0.1024 Ω/mi listed in tables for this conductor at 50°C. Explain any difference in values. Assume that the increase in resistance due to spiraling is 2%.

4.3. An AAC is composed of 37 strands, each having a diameter of 0.333 cm. Compute the dc resistance in ohms per kilometer at 75°C. Assume that the increase in resistance due to spiraling is 2%.

4.4. The energy density (that is, the energy per unit volume) at a point in a magnetic field can be shown to be $B^2/2\mu$, where B is the flux density and μ is the permeability. Using this result and Eq. (4.10), show that the total magnetic field energy stored within a unit length of solid circular conductor carrying current I is given by $\mu I^2/16\pi$. Neglect skin effect, and thus verify Eq. (4.15).

4.5. The conductor of a single-phase 60-Hz line is a solid round aluminum wire having a diameter of 0.412 cm. The conductor spacing is 3 m. Determine the inductance of the line in millihenrys per mile. How much of the inductance is due to internal flux linkages? Assume skin effect is negligible.

4.6. A single-phase 60-Hz overhead power line is symmetrically supported on a horizontal crossarm. Spacing between the centers of the conductors (say, a and b) is 2.5 m. A telephone line is also symmetrically supported on a horizontal crossarm 1.8 m directly below the power line. Spacing between the centers of these conductors (say, c and d) is 1.0 m.

(*a*) Using Eq. (4.36), show that the mutual inductance per unit length between circuit a-b and circuit c-d is given by

$$4 \times 10^{-7} \ln \sqrt{\frac{D_{ad}D_{bc}}{D_{ac}D_{bd}}} \quad \text{H/m}$$

where, for example, D_{ad} denotes the distance in meters between conductors a and d.

(*b*) Hence, compute the mutual inductance per kilometer between the power line and the telephone line.

(*c*) Find the 60-Hz voltage per kilometer induced in the telephone line when the power line carries 150 A.

4.7. If the power line and the telephone line described in Prob. 4.6 are in the same horizontal plane and the distance between the nearest conductors of the two lines is 18 m, use the result of Prob. 4.6(*a*) to find the mutual inductance between the power and telephone circuits. Also, find the 60-Hz voltage per kilometer induced in the telephone line when 150 A flows in the power line.

4.8. Find the GMR of a three-strand conductor in terms of r of an individual strand.

4.9. Find the GMR of each of the unconventional conductors shown in Fig. 4.15 in terms of the radius r of an individual strand.

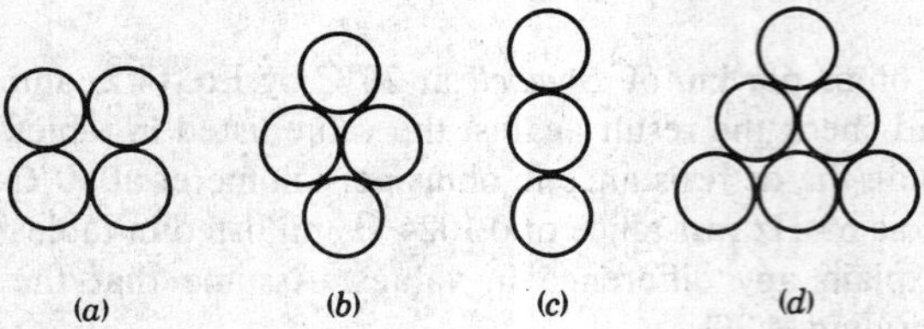

FIGURE 4.15
Cross-sectional view of unconventional conductors for Prob. 4.9.

4.10. The distance between conductors of a single-phase line is 10 ft. Each of its conductors is composed of six strands symmetrically placed around one center strand so that there are seven equal strands. The diameter of each strand is 0.1 in. Show that D_s of each conductor is 2.177 times the radius of each strand. Find the inductance of the line in mH/mi.

4.11. Solve Example 4.2 for the case where side Y of the single-phase line is identical to side X and the two sides are 9 m apart, as shown in Fig. 4.9.

4.12. Find the inductive reactance of ACSR *Rail* in ohms per kilometer at 1-m spacing.

4.13. Which conductor listed in Table A.3 has an inductive reactance at 7-ft spacing of 0.651 Ω/mi?

4.14. A three-phase line has three equilaterally spaced conductors of ACSR *Dove*. If the conductors are 10 ft apart, determine the 60-Hz per-phase reactance of the line in Ω/km.

4.15. A three-phase line is designed with equilateral spacing of 16 ft. It is decided to build the line with horizontal spacing ($D_{13} = 2D_{12} = 2D_{23}$). The conductors are transposed. What should be the spacing between adjacent conductors in order to obtain the same inductance as in the original design?

4.16. A three-phase 60-Hz transmission line has its conductors arranged in a triangular formation so that two of the distances between conductors are 25 ft and the third distance is 42 ft. The conductors are ACSR *Osprey*. Determine the inductance and inductive reactance per phase per mile.

4.17. A three-phase 60-Hz line has flat horizontal spacing. The conductors have a GMR of 0.0133 m with 10 m between adjacent conductors. Determine the inductive reactance per phase in ohms per kilometer. What is the name of this conductor?

4.18. For short transmission lines if resistance is neglected, the maximum power which can be transmitted per phase is equal to

$$\frac{|V_S| \times |V_R|}{|X|}$$

where V_S and V_R are the line-to-neutral voltages at the sending and receiving ends of the line and X is the inductive reactance of the line. This relationship will become apparent in the study of Chap. 6. If the magnitudes of V_S and V_R are held constant, and if the cost of a conductor is proportional to its cross-sectional area, find the conductor in Table A.3 which has the maximum power-handling capacity per cost of conductor at a given geometric mean spacing.

4.19. A three-phase underground distribution line is operated at 23 kV. The three conductors are insulated with 0.5-cm solid black polyethylene insulation and lie flat, side by side, directly next to each other in a dirt trench. The conductor is circular in cross section and has 33 strands of aluminum. The diameter of the conductor is 1.46 cm. The manufacturer gives the GMR as 0.561 cm and the cross section of the conductor as 1.267 cm^2. The thermal rating of the line buried in normal soil whose maximum temperature is 30°C is 350 A. Find the dc and ac resistance at 50°C and the inductive reactance in ohms per kilometer. To decide whether to consider skin effect in calculating resistance, determine the percent skin effect at 50°C in the ACSR conductor of the size nearest that of the underground conductor. Note that the series impedance of the distribution line is dominated by R rather than X_L because of the very low inductance due to the close spacing of the conductors.

4.20. The single-phase power line of Prob. 4.6 is replaced by a three-phase line on a horizontal crossarm in the same position as that of the original single-phase line. Spacing of the conductors of the power line is $D_{13} = 2D_{12} = 2D_{23}$, and equivalent equilateral spacing is 3 m. The telephone line remains in the position described in Prob. 4.6. If the current in the power line is 150 A, find the voltage per kilometer induced in the telephone line. Discuss the phase relation of the induced voltage with respect to the power-line current.

4.21. A 60-Hz three-phase line composed of one ACSR *Bluejay* conductor per phase has flat horizontal spacing of 11 m between adjacent conductors. Compare the inductive reactance in ohms per kilometer per phase of this line with that of a line using a two-conductor bundle of ACSR 26/7 conductors having the same total cross-sectional area of aluminum as the single-conductor line and 11-m spacing measured from the center of the bundles. The spacing between conductors in the bundle is 40 cm.

4.22. Calculate the inductive reactance in ohms per kilometer of a bundled 60-Hz three-phase line having three ACSR *Rail* conductors per bundle with 45 cm between conductors of the bundle. The spacing between bundle centers is 9, 9, and 18 m.

CHAPTER 5

CAPACITANCE OF TRANSMISSION LINES

As we discussed briefly at the beginning of Chap. 4, the shunt admittance of a transmission line consists of conductance and capacitive reactance. We have also mentioned that conductance is usually neglected because its contribution to shunt admittance is very small. For this reason this chapter has been given the title of capacitance rather than shunt admittance.

Capacitance of a transmission line is the result of the potential difference between the conductors; it causes them to be charged in the same manner as the plates of a capacitor when there is a potential difference between them. The capacitance between conductors is the charge per unit of potential difference. Capacitance between parallel conductors is a constant depending on the size and spacing of the conductors. For power lines less than about 80 km (50 mi) long, the effect of capacitance can be slight and is often neglected. For longer lines of higher voltage capacitance becomes increasingly important.

An alternating voltage impressed on a transmission line causes the charge on the conductors at any point to increase and decrease with the increase and decrease of the instantaneous value of the voltage between conductors at the point. The flow of charge is current, and the current caused by the alternate charging and discharging of a line due to an alternating voltage is called the *charging current* of the line. Since capacitance is a shunt between conductors, charging current flows in a transmission line even when it is open-circuited. It affects the voltage drop along the lines as well as efficiency and power factor of the line and the stability of the system of which the line is a part.

The basis of our analysis of capacitance is Gauss's law for electric fields. The law states that the total electric charge within a closed surface equals the total electric flux emerging from the surface. In other words, the total charge within the closed surface equals the integral over the surface of the normal component of the electric flux density.

The lines of electric flux originate on positive charges and terminate on negative charges. Charge density normal to a surface is designated D_f and equals kE, where k is the permittivity of the material surrounding the surface and E is the electric field intensity.[1]

5.1 ELECTRIC FIELD OF A LONG, STRAIGHT CONDUCTOR

If a long, straight cylindrical conductor lies in a uniform medium such as air and is isolated from other charges so that the charge is uniformly distributed around its periphery, the flux is radial. All points equidistant from such a conductor are points of equipotential and have the same electric flux density. Figure 5.1 shows such an isolated conductor. The electric flux density at x meters from the conductor can be computed by imagining a cylindrical surface concentric with the conductor and x meters in radius. Since all parts of the surface are equidistant from the conductor, the cylindrical surface is a surface of equipotential and the electric flux density on the surface is equal to the flux leaving the conductor per meter of length divided by the area of the surface in an axial length of 1 m. The electric flux density is

$$D_f = \frac{q}{2\pi x} \ \mathrm{C/m^2} \tag{5.1}$$

where q is the charge on the conductor in coulombs per meter of length and x is the distance in meters from the conductor to the point where the electric flux density is computed. The electric field intensity, or the negative of the potential gradient, is equal to the electric flux density divided by the permittivity of the medium. Therefore, the electric field intensity is

$$E = \frac{q}{2\pi x k} \ \mathrm{V/m} \tag{5.2}$$

E and q both may be instantaneous, phasor, or dc expressions.

---

[1] In SI units the permittivity of free space k_0 is 8.85×10^{-12} F/m (farads per meter). Relative permittivity k_r is the ratio of the actual permittivity k of a material of the permittivity of free space. Thus, $k_r = k/k_0$. For dry air k_r is 1.00054 and is assumed equal to 1.0 in calculations for overhead lines.

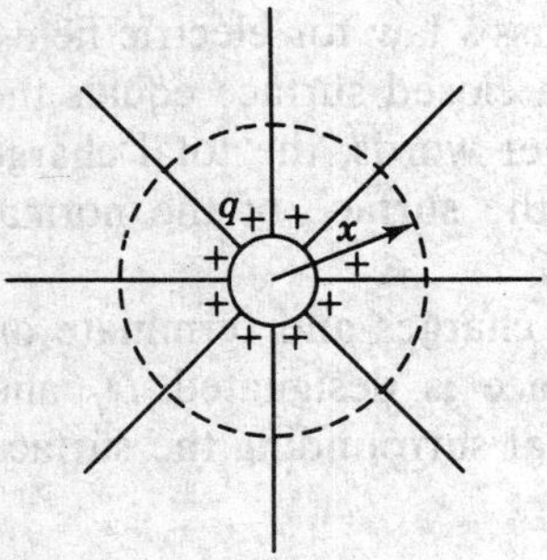

FIGURE 5.1
Lines of electric flux originating on the positive charges uniformly distributed over the surface of an isolated cylindrical conductor.

5.2 THE POTENTIAL DIFFERENCE BETWEEN TWO POINTS DUE TO A CHARGE

The potential difference between two points in volts is numerically equal to the work in joules per coulomb necessary to move a coulomb of charge between the two points. The electric field intensity is a measure of the force on a charge in the field. The electric field intensity in volts per meter is equal to the force in newtons per coulomb on a coulomb of charge at the point considered. Between two points the line integral of the force in newtons acting on a coulomb of positive charge is the work done in moving the charge from the point of lower potential to the point of higher potential and is numerically equal to the potential difference between the two points.

Consider a long, straight wire carrying a positive charge of q C/m, as shown in Fig. 5.2. Points P_1 and P_2 are located at distances D_1 and D_2 meters, respectively, from the center of the wire. The wire is an equipotential surface and the uniformly distributed charge on the wire is equivalent to a charge concentrated at the center of the wire for calculating flux external to the wire. The positive charge on the wire will exert a repelling force on a positive charge placed in the field. For this reason and because D_2 in this case is greater than

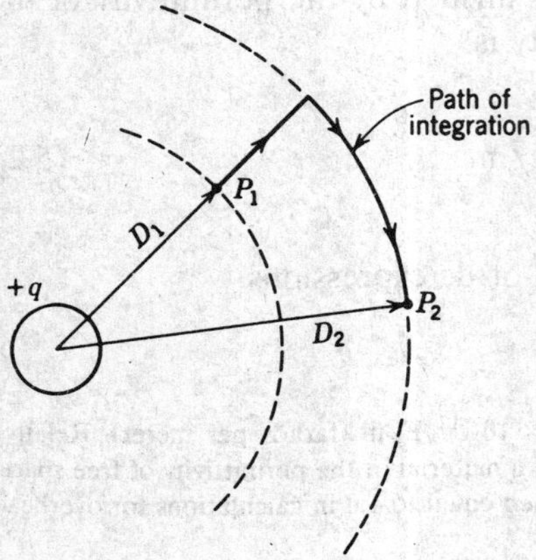

FIGURE 5.2
Path of integration between two points external to a cylindrical conductor having a uniformly distributed positive charge.

D_1, work must be done on a positive charge to move it from P_2 to P_1, and P_1 is at a higher potential than P_2. The difference in potential is the amount of work done per coulomb of charge moved. On the other hand, if the one coulomb of charge moves from P_1 to P_2, it expends energy, and the amount of work, or energy, in newton-meters is the voltage *drop* from P_1 to P_2. The potential difference is independent of the path followed. The simplest way to compute the voltage drop between two points is to compute the voltage between the equipotential surfaces passing through P_1 and P_2 by integrating the field intensity over a *radial* path between the equipotential surfaces. Thus, the instantaneous voltage drop between P_1 and P_2 is

$$v_{12} = \int_{D_1}^{D_2} E\,dx = \int_{D_1}^{D_2} \frac{q}{2\pi kx}\,dx = \frac{q}{2\pi k} \ln \frac{D_2}{D_1} \text{ V} \tag{5.3}$$

where q is the instantaneous charge on the wire in coulombs per meter of length. Note that the voltage drop between two points, as given by Eq. (5.3), may be positive or negative depending on whether the charge causing the potential difference is positive or negative and on whether the voltage drop is computed from a point near the conductor to a point farther away, or vice versa. The sign of q may be either positive or negative, and the logarithmic term is either positive or negative depending on whether D_2 is greater or less than D_1.

5.3 CAPACITANCE OF A TWO-WIRE LINE

Capacitance between the conductors of a two-wire line is defined as the charge on the conductors per unit of potential difference between them. In the form of an equation capacitance per unit length of the line is

$$C = \frac{q}{v} \text{ F/m} \tag{5.4}$$

where q is the charge on the line in coulombs per meter and v is the potential difference between the conductors in volts. Hereafter, for convenience, we refer to *capacitance per unit length* as *capacitance* and indicate the correct dimensions for the equations derived. The capacitance between two conductors can be found by substituting in Eq. (5.4) the expression for v in terms of q from Eq. (5.3). The voltage v_{ab} between the two conductors of the two-wire line shown in Fig. 5.3 can be found by determining the potential difference between the two conductors of the line, first by computing the voltage drop due to the charge q_a on conductor a and then by computing the voltage drop due to the charge q_b on conductor b. By the principle of superposition the voltage drop from conductor a to conductor b due to the charges on both conductors is the sum of the voltage drops caused by each charge alone.

The charge q_a on conductor a of Fig. 5.3 causes surfaces of equipotential in the vicinity of conductor b, which are shown in Fig. 5.4. We avoid the

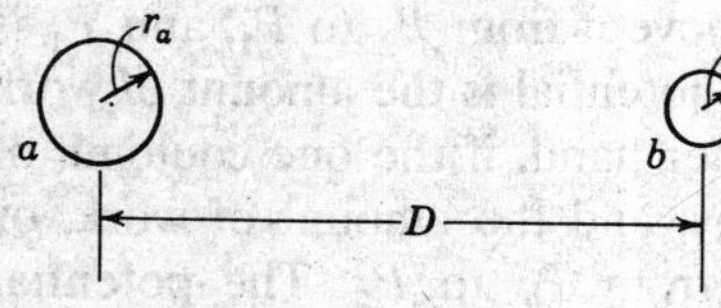

FIGURE 5.3
Cross section of a parallel-wire line.

distorted equipotential surfaces by integrating Eq. (5.3) along the alternate rather than the direct path of Fig. 5.4. In determining v_{ab} due to q_a, we follow the path through the undistorted region and see that distance D_1 of Eq. (5.3) is the radius r_a of conductor a and distance D_2 is the center-to-center distance between conductors a and b. Similarly, in determining v_{ab} due to q_b, we find that the distances D_2 and D_1 are r_b and D, respectively. Converting to phasor notation (q_a and q_b become phasors), we obtain

$$V_{ab} = \underbrace{\frac{q_a}{2\pi k} \ln \frac{D}{r_a}}_{\text{due to } q_a} + \underbrace{\frac{q_b}{2\pi k} \ln \frac{r_b}{D}}_{\text{due to } q_b} \text{ V} \tag{5.5}$$

and since $q_a = -q_b$ for a two-wire line,

$$V_{ab} = \frac{q_a}{2\pi k}\left(\ln \frac{D}{r_a} - \ln \frac{r_b}{D}\right) \text{ V} \tag{5.6}$$

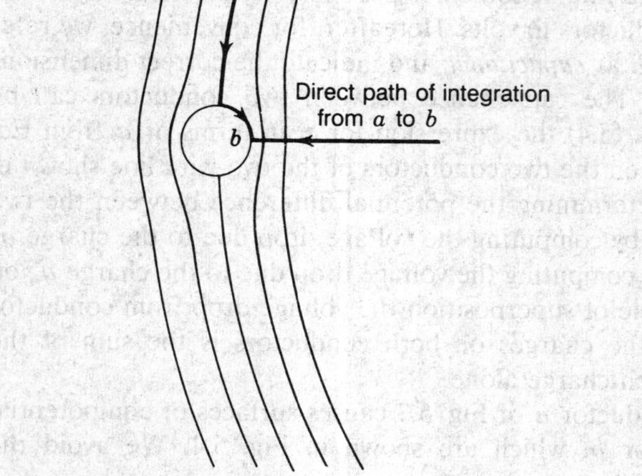

FIGURE 5.4
Equipotential surfaces of a portion of the electric field caused by a charged conductor a (not shown). Conductor b causes the equipotential surfaces to become distorted. Arrows indicate optional paths of integration between a point on the equipotential surface of conductor b and the conductor a, whose charge q_a creates the equipotential surfaces shown.

or by combining the logarithmic terms, we obtain

$$V_{ab} = \frac{q_a}{2\pi k} \ln \frac{D^2}{r_a r_b} \text{ V} \tag{5.7}$$

The capacitance between conductors is

$$C_{ab} = \frac{q_a}{V_{ab}} = \frac{2\pi k}{\ln(D^2/r_a r_b)} \text{ F/m} \tag{5.8}$$

If $r_a = r_b = r$,

$$C_{ab} = \frac{\pi k}{\ln(D/r)} \text{ F/m} \tag{5.9}$$

Equation (5.9) gives the capacitance between the conductors of a two-wire line. If the line is supplied by a transformer having a grounded center tap, the potential difference between each conductor and ground is half the potential difference between the two conductors and the *capacitance to ground*, or *capacitance to neutral*, is

$$C_n = C_{an} = C_{bn} = \frac{q_a}{V_{ab}/2} = \frac{2\pi k}{\ln(D/r)} \text{ F/m to neutral} \tag{5.10}$$

The concept of capacitance to neutral is illustrated in Fig. 5.5.

Equation (5.10) corresponds to Eq. (4.25) for inductance. One difference between the equations for capacitance and inductance should be noted carefully. The radius in the equation for capacitance is the *actual outside radius* of the conductor and not the geometric mean ratio (GMR) of the conductor, as in the inductance formula.

Equation (5.3), from which Eqs. (5.5) through (5.10) were derived, is based on the assumption of uniform charge distribution over the surface of the conductor. When other charges are present, the distribution of charge on the surface of the conductor is not uniform and the equations derived from Eq. (5.3) are not strictly correct. The nonuniformity of charge distribution, however, can

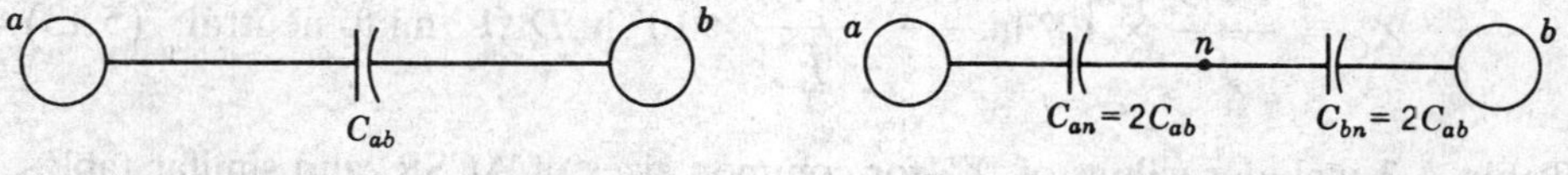

(*a*) Representation of line-to-line capacitance (*b*) Representation of line-to-neutral capacitance

FIGURE 5.5
Relationship between the concepts of line-to-line capacitance and line-to-neutral capacitance.

be neglected entirely in overhead lines since the error in Eq. (5.10) is only 0.01%, even for such a close spacing as that where the ratio $D/r = 50$.

A question arises about the value to be used in the denominator of the argument of the logarithm in Eq. (5.10) when the conductor is a stranded cable because the equation was derived for a solid round conductor. Since electric flux is perpendicular to the surface of a perfect conductor, the electric field at the surface of a stranded conductor is not the same as the field at the surface of a cylindrical conductor. Therefore, the capacitance calculated for a stranded conductor by substituting the outside radius of the conductor for r in Eq. (5.10) will be slightly in error because of the difference between the field in the neighborhood of such a conductor and the field near a solid conductor for which Eq. (5.10) was derived. The error is very small, however, since only the field very close to the surface of the conductor is affected. The outside radius of the stranded conductor is used in calculating the capacitance.

After the capacitance to neutral has been determined, the capacitive reactance existing between one conductor and neutral for relative permittivity $k_r = 1$ is found by using the expression for C given in Eq. (5.10) to yield

$$X_C = \frac{1}{2\pi f C} = \frac{2.862}{f} \times 10^9 \ln \frac{D}{r} \ \Omega \cdot \text{m to neutral} \tag{5.11}$$

Since C in Eq. (5.11) is in farads per meter, the proper units for X_C must be ohm-meters. We should also note that Eq. (5.11) expresses the reactance from line to neutral for 1 m of line. Since capacitance reactance is in parallel along the line, X_C in ohm-meters must be *divided* by the length of the line in meters to obtain the capacitive reactance in ohms to neutral for the entire length of the line.

When Eq. (5.11) is divided by 1609 to convert to ohm-miles, we obtain

$$X_C = \frac{1.779}{f} \times 10^6 \ln \frac{D}{r} \ \Omega \cdot \text{mi to neutral} \tag{5.12}$$

Table A.3 lists the outside diameters of the most widely used sizes of ACSR. If D and r in Eq. (5.12) are in feet, *capacitive reactance at 1-ft spacing* X'_a is the first term and *capacitive reactance spacing factor* X'_d is the second term when the equation is expanded as follows:

$$X_C = \frac{1.779}{f} \times 10^6 \ln \frac{1}{r} + \frac{1.779}{f} \times 10^6 \ln D \ \Omega \cdot \text{mi to neutral} \tag{5.13}$$

Table A.3 includes values of X'_a for common sizes of ACSR, and similar tables are readily available for other types and sizes of conductors. Table A.5 in the Appendix lists values of X'_d which, of course, is different from the synchronous machine transient reactance bearing the same symbol.

Example 5.1. Find the capacitive susceptance per mile of a single-phase line operating at 60 Hz. The conductor is *Partridge*, and spacing is 20 ft between centers.

Solution. For this conductor Table A.3 lists an outside diameter of 0.642 in, and so

$$r = \frac{0.642}{2 \times 12} = 0.0268 \text{ ft}$$

and from Eq. (5.12)

$$X_C = \frac{1.779}{60} \times 10^6 \ln \frac{20}{0.0268} = 0.1961 \times 10^6 \ \Omega \cdot \text{mi to neutral}$$

$$B_C = \frac{1}{X_C} = 5.10 \times 10^{-6} \text{ S/mi to neutral}$$

or in terms of capacitive reactance at 1-ft spacing and capacitive reactance spacing factor from Tables A.3 and A.5

$$X'_a = 0.1074 \text{ M}\Omega \cdot \text{mi}$$

$$X'_d = 0.0889 \text{ M}\Omega \cdot \text{mi}$$

$$X'_C = 0.1074 + 0.0889 = 0.1963 \text{ M}\Omega \cdot \text{mi per conductor}$$

Line-to-line capacitive reactance and susceptance are

$$X_C = 2 \times 0.1963 \times 10^6 = 0.3926 \times 10^6 \ \Omega \cdot \text{mi}$$

$$B_C = \frac{1}{X_C} = 2.55 \times 10^{-6} \text{ S/mi}$$

5.4 CAPACITANCE OF A THREE-PHASE LINE WITH EQUILATERAL SPACING

The three identical conductors of radius r of a three-phase line with equilateral spacing are shown in Fig. 5.6. Equation (5.5) expresses the voltage between two conductors due to the charges on each one if the charge distribution on the conductors can be assumed to be uniform. Thus, the voltage V_{ab} of the three-phase line due only to the charges on conductors a and b is

$$V_{ab} = \underbrace{\frac{1}{2\pi k}\left(q_a \ln \frac{D}{r} + q_b \ln \frac{r}{D}\right)}_{\text{due to } q_a \text{ and } q_b} \text{ V} \tag{5.14}$$

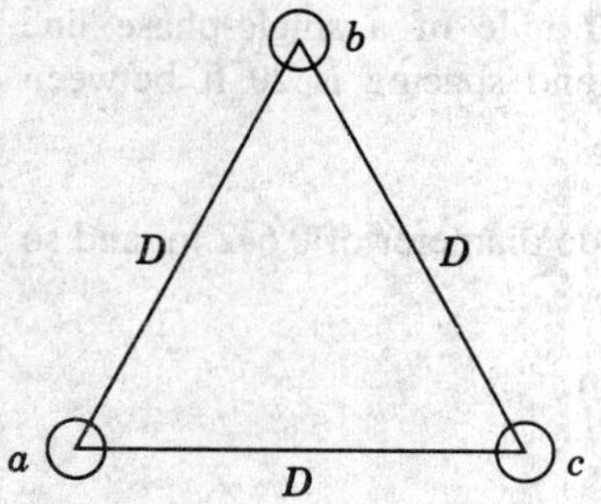

FIGURE 5.6
Cross section of a three-phase line with equilateral spacing.

Equation (5.3) enables us to include the effect of q_c since uniform charge distribution over the surface of a conductor is equivalent to a concentrated charge at the center of the conductor. Therefore, due only to the charge q_c,

$$V_{ab} = \frac{q_c}{2\pi k} \ln \frac{D}{D} \text{ V}$$

which is zero since q_c is equidistant from a and b. However, to show that we are considering all three charges, we can write

$$V_{ab} = \frac{1}{2\pi k}\left(q_a \ln \frac{D}{r} + q_b \ln \frac{r}{D} + q_c \ln \frac{D}{D}\right) \text{ V} \tag{5.15}$$

$$V_{ac} = \frac{1}{2\pi k}\left(q_a \ln \frac{D}{r} + q_b \ln \frac{D}{D} + q_c \ln \frac{r}{D}\right) \text{ V} \tag{5.16}$$

Adding Eqs. (5.15) and (5.16) gives

$$V_{ab} + V_{ac} = \frac{1}{2\pi k}\left[2q_a \ln \frac{D}{r} + (q_b + q_c)\ln \frac{r}{D}\right] \text{ V} \tag{5.17}$$

In deriving these equations, we have assumed that ground is far enough away to have negligible effect. Since the voltages are assumed to be sinusoidal and expressed as phasors, the charges are sinusoidal and expressed as phasors. If there are no other charges in the vicinity, the sum of the charges on the three conductors is zero and we can substitute $-q_a$ in Eq. (5.17) for $q_b + q_c$ and obtain

$$V_{ab} + V_{ac} = \frac{3q_a}{2\pi k} \ln \frac{D}{r} \text{ V} \tag{5.18}$$

Figure 5.7 is the phasor diagram of voltages. From this figure we obtain the following relations between the line voltages V_{ab} and V_{ac} and the voltage V_{an}

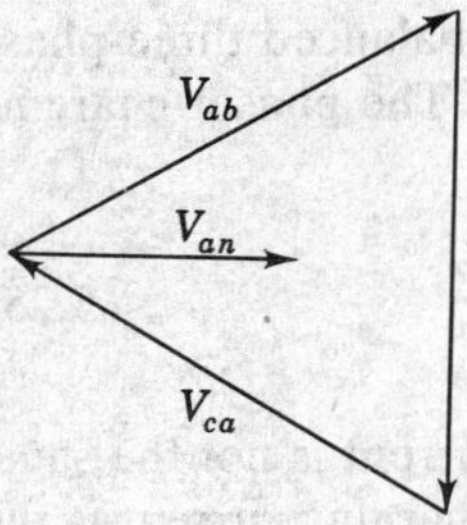

FIGURE 5.7
Phasor diagram of the balanced voltages of a three-phase line.

from line a to the neutral of the three-phase circuit:

$$V_{ab} = \sqrt{3}\,V_{an}\angle 30^\circ = \sqrt{3}\,V_{an}(0.866 + j0.5) \tag{5.19}$$

$$V_{ac} = -V_{ca} = \sqrt{3}\,V_{an}\angle -30^\circ = \sqrt{3}\,V_{an}(0.866 - j0.5) \tag{5.20}$$

Adding Eqs. (5.19) and (5.20) gives

$$V_{ab} + V_{ac} = 3V_{an} \tag{5.21}$$

Substituting $3V_{an}$ for $V_{ab} + V_{ac}$ in Eq. (5.18), we obtain

$$V_{an} = \frac{q_a}{2\pi k}\ln\frac{D}{r}\ \text{V} \tag{5.22}$$

Since capacitance to neutral is the ratio of the charge on a conductor to the voltage between that conductor and neutral,

$$C_n = \frac{q_a}{V_{an}} = \frac{2\pi k}{\ln(D/r)}\ \text{F/m to neutral} \tag{5.23}$$

Comparison of Eqs. (5.23) and (5.10) shows that the two are identical. These equations express the capacitance to neutral for single-phase and equilaterally spaced three-phase lines, respectively. Similarly, we recall that the equations for inductance per conductor are the same for single-phase and equilaterally spaced three-phase lines.

The term *charging current* is applied to the current associated with the capacitance of a line. For a *single-phase* circuit the charging current is the product of the line-to-line voltage and the line-to-line susceptance, or as a phasor,

$$I_{\text{chg}} = j\omega C_{ab} V_{ab} \tag{5.24}$$

For a three-phase line the charging current is found by multiplying the voltage to neutral by the capacitive susceptance to neutral. This gives the charging current per phase and is in accord with the calculation of balanced three-phase circuits on the basis of a single phase with neutral return. The phasor charging current in phase a is

$$I_{\text{chg}} = j\omega C_n V_{an} \text{ A/mi} \tag{5.25}$$

Since the rms voltage varies along the line, the charging current is not the same everywhere. Often the voltage used to obtain a value for charging current is the normal voltage for which the line is designed, such as 220 or 500 kV, which is probably not the actual voltage at either a generating station or a load.

5.5 CAPACITANCE OF A THREE-PHASE LINE WITH UNSYMMETRICAL SPACING

When the conductors of a three-phase line are not equilaterally spaced, the problem of calculating capacitance becomes more difficult. In the usual untransposed line the capacitances of each phase to neutral are unequal. In a transposed line the average capacitance to neutral of any phase for the complete transposition cycle is the same as the average capacitance to neutral of any other phase since each phase conductor occupies the same position as every other phase conductor over an equal distance along the transposition cycle. The dissymmetry of the untransposed line is slight for the usual configuration, and capacitance calculations are carried out as though all lines were transposed.

For the line shown in Fig. 5.8 three equations are found for V_{ab} for the three different parts of the transposition cycle. With phase a in position 1, b in position 2, and c in position 3,

$$V_{ab} = \frac{1}{2\pi k}\left(q_a \ln \frac{D_{12}}{r} + q_b \ln \frac{r}{D_{12}} + q_c \ln \frac{D_{23}}{D_{31}} \right) \text{ V} \tag{5.26}$$

With phase a in position 2, b in position 3, and c in position 1,

$$V_{ab} = \frac{1}{2\pi k}\left(q_a \ln \frac{D_{23}}{r} + q_b \ln \frac{r}{D_{23}} + q_c \ln \frac{D_{31}}{D_{12}} \right) \text{ V} \tag{5.27}$$

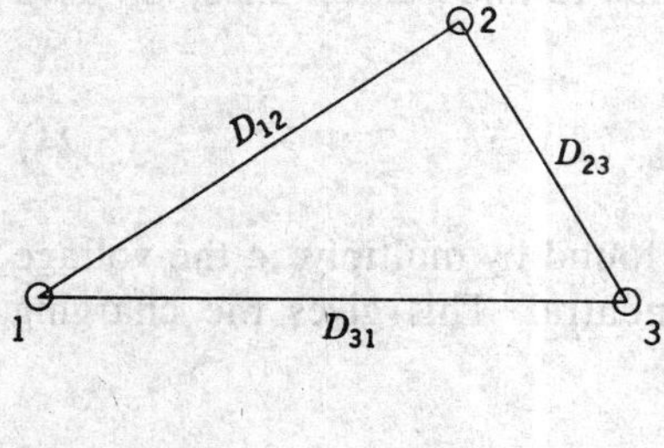

FIGURE 5.8
Cross section of a three-phase line with unsymmetrical spacing.

and with a in position 3, b in position 1, and c in position 2,

$$V_{ab} = \frac{1}{2\pi k}\left(q_a \ln \frac{D_{31}}{r} + q_b \ln \frac{r}{D_{31}} + q_c \ln \frac{D_{12}}{D_{23}}\right) \text{ V} \tag{5.28}$$

Equations (5.26) through (5.28) are similar to Eqs. (4.51) through (4.53) for the magnetic flux linkages of one conductor of a transposed line. However, in the equations for magnetic flux linkages we note that the current in any phase is the same in every part of the transposition cycle. In Eqs. (5.26) through (5.28), if we disregard the voltage drop along the line, the voltage to neutral of a phase in one part of a transposition cycle is equal to the voltage to neutral of that phase in any part of the cycle. Hence, the voltage between any two conductors is the same in all parts of the transposition cycle. It follows that the charge on a conductor must be different when the position of the conductor changes with respect to other conductors. A treatment of Eqs. (5.26) through (5.28) analogous to that of Eqs. (4.51) through (4.53) is not rigorous.

The rigorous solution for capacitances is too involved to be practical except perhaps for flat spacing with equal distances between adjacent conductors. With the usual spacings and conductors, sufficient accuracy is obtained by assuming that the charge per unit length on a conductor is the same in every part of the transposition cycle. When the above assumption is made with regard to charge, the voltage between a pair of conductors is different for each part of the transposition cycle. Then an average value of voltage between the conductors can be found and the capacitance calculated from the average voltage. We obtain the average voltage by adding Eqs. (5.26) through (5.28) and by dividing the result by 3. The average voltage between conductors a and b, assuming the same charge on a conductor regardless of its position in the transposition cycle, is

$$V_{ab} = \frac{1}{6\pi k}\left(q_a \ln \frac{D_{12}D_{23}D_{31}}{r^3} + q_b \ln \frac{r^3}{D_{12}D_{23}D_{31}} + q_c \ln \frac{D_{12}D_{23}D_{31}}{D_{12}D_{23}D_{31}}\right)$$

$$= \frac{1}{2\pi k}\left(q_a \ln \frac{D_{eq}}{r} + q_b \ln \frac{r}{D_{eq}}\right) \tag{5.29}$$

where $$D_{eq} = \sqrt[3]{D_{12}D_{23}D_{31}} \tag{5.30}$$

Similarly, the average voltage drop from conductor a to conductor c is

$$V_{ac} = \frac{1}{2\pi k}\left(q_a \ln \frac{D_{eq}}{r} + q_c \ln \frac{r}{D_{eq}}\right) \text{ V} \tag{5.31}$$

Applying Eq. (5.21) to find the voltage to neutral, we have

$$3V_{an} = V_{ab} + V_{ac} = \frac{1}{2\pi k}\left(2q_a \ln\frac{D_{eq}}{r} + q_b \ln\frac{r}{D_{eq}} + q_c \ln\frac{r}{D_{eq}}\right) \text{ V} \quad (5.32)$$

Since $q_a + q_b + q_c = 0$,

$$3V_{an} = \frac{3}{2\pi k} q_a \ln\frac{D_{eq}}{r} \text{ V} \quad (5.33)$$

and

$$C_n = \frac{q_a}{V_{an}} = \frac{2\pi k}{\ln(D_{eq}/r)} \text{ F/m to neutral} \quad (5.34)$$

Equation (5.34) for capacitance to neutral of a transposed three-phase line corresponds to Eq. (4.56) for the inductance per phase of a similar line. In finding capacitive reactance to neutral corresponding to C_n, we can split the reactance into components of capacitive reactance to neutral at 1-ft spacing X'_a and capacitive reactance spacing factor X'_d, as defined by Eq. (5.13).

Example 5.2. Find the capacitance and the capacitive reactance for 1 mi of the line described in Example 4.4. If the length of the line is 175 mi and the normal operating voltage is 220 kV, find capacitive reactance to neutral for the entire length of the line, the charging current per mile, and the total charging megavoltamperes.

Solution

$$r = \frac{1.108}{2 \times 12} = 0.0462 \text{ ft}$$

$$D_{eq} = 24.8 \text{ ft}$$

$$C_n = \frac{2\pi \times 8.85 \times 10^{-12}}{\ln(24.8/0.0462)} = 8.8466 \times 10^{-12} \text{ F/m}$$

$$X_C = \frac{10^{12}}{2\pi \times 60 \times 8.8466 \times 1609} = 0.1864 \times 10^6 \ \Omega \cdot \text{mi}$$

or from tables

$$X'_a = 0.0912 \times 10^6 \qquad X'_d = 0.0953 \times 10^6$$

$$X_C = (0.0912 + 0.0953) \times 10^6 = 0.1865 \times 10^6 \ \Omega \cdot \text{mi to neutral}$$

For a length of 175 mi

$$\text{Capacitive reactance} = \frac{0.1865 \times 10^6}{175} = 1066\ \Omega \text{ to neutral}$$

$$|I_{chg}| = \frac{220{,}000}{\sqrt{3}} \frac{1}{X_C} = \frac{220{,}000 \times 10^{-6}}{\sqrt{3} \times 0.1865} = 0.681\ \text{A/mi}$$

or $0.681 \times 175 = 119$ A for the line. Reactive power is $Q = \sqrt{3} \times 220 \times 119 \times 10^{-3} = 43.5$ Mvar. This amount of reactive power absorbed by the distributed capacitance is negative in keeping with the convention discussed in Chap. 1. In other words, positive reactive power is being *generated* by the distributed capacitance of the line.

5.6 EFFECT OF EARTH ON THE CAPACITANCE OF THREE-PHASE TRANSMISSION LINES

Earth affects the capacitance of a transmission line because its presence alters the electric field of the line. If we assume that the earth is a perfect conductor in the form of a horizontal plane of infinite extent, we realize that the electric field of charged conductors above the earth is not the same as it would be if the equipotential surface of the earth were not present. The electric field of the charged conductors is forced to conform to the presence of the earth's surface. The assumption of a flat, equipotential surface is, of course, limited by the irregularity of terrain and the type of surface of the earth. The assumption enables us, however, to understand the effect of a conducting earth on capacitance calculations.

Consider a circuit consisting of a single overhead conductor with a return path through the earth. In charging the conductor, charges come from the earth to reside on the conductor, and a potential difference exists between the conductor and the earth. The earth has a charge equal in magnitude to that on the conductor but of opposite sign. The electric flux from the charges on the conductor to the charges on the earth is perpendicular to the earth's equipotential surface since the surface is assumed to be a perfect conductor. Let us imagine a fictitious conductor of the same size and shape as the overhead conductor lying directly below the original conductor at a distance equal to twice the distance of the conductor above the plane of the ground. The fictitious conductor is below the surface of the earth by a distance equal to the distance of the overhead conductor above the earth. If the earth is removed and a charge equal and opposite to that on the overhead conductor is assumed on the fictitious conductor, the plane midway between the original conductor and the fictitious conductor is an equipotential surface and occupies the same position as the equipotential surface of the earth. The electric flux between the overhead conductor and this equipotential surface is the same as that which existed between the conductor and the earth. Thus, for purposes of calculation of capacitance the earth may be replaced by a fictitious charged conductor below the surface of the earth by a distance equal to that of the overhead conductor above the earth. Such a conductor has a charge equal in magnitude and opposite in sign to that of the original conductor and is called the *image conductor*.

The method of calculating capacitance by replacing the earth by the image of an overhead conductor can be extended to more than one conductor. If we locate an image conductor for each overhead conductor, the flux between the original conductors and their images is perpendicular to the plane which replaces the earth, and that plane is an equipotential surface. The flux above the plane is the same as it is when the earth is present instead of the image conductors.

To apply the method of images to the calculation of capacitance for a three-phase line, refer to Fig. 5.9. We assume that the line is transposed and

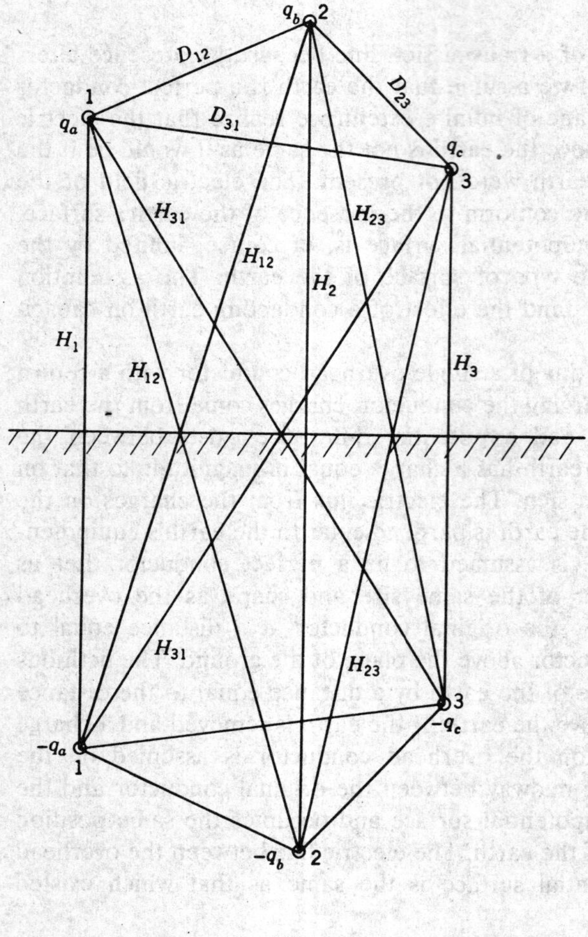

FIGURE 5.9
Three-phase line and its image.

that conductors, a, b, and c carry the charges q_a, q_b, and q_c and occupy positions 1, 2, and 3, respectively, in the first part of the transposition cycle. The plane of the earth is shown, and below it are the conductors with the image charges $-q_a$, $-q_b$, and $-q_c$. Equations for the three parts of the transposition cycle can be written for the voltage drop from conductor a to conductor b as determined by the three charged conductors and their images. With conductor a in position 1, b in position 2, and c in position 3, by Eq. (5.3)

$$V_{ab} = \frac{1}{2\pi k}\left[q_a\left(\ln\frac{D_{12}}{r} - \ln\frac{H_{12}}{H_1}\right) + q_b\left(\ln\frac{r}{D_{12}} - \ln\frac{H_2}{H_{12}}\right)\right.$$
$$\left. + q_c\left(\ln\frac{D_{23}}{D_{31}} - \ln\frac{H_{23}}{H_{31}}\right)\right] \tag{5.35}$$

Similar equations for V_{ab} are written for the other parts of the transposition cycle. Accepting the approximately correct assumption of constant charge per unit length of each conductor throughout the transposition cycle allows us to obtain an average value of the phasor V_{ab}. The equation for the average value of the phasor V_{ac} is found in a similar manner, and $3V_{an}$ is obtained by adding the average values of V_{ab} and V_{ac}. Knowing that the sum of the charges is zero, we then find

$$C_n = \frac{2\pi k}{\ln\left(\dfrac{D_{eq}}{r}\right) - \ln\left(\dfrac{\sqrt[3]{H_{12}H_{23}H_{31}}}{\sqrt[3]{H_1 H_2 H_3}}\right)} \text{ F/m to neutral} \tag{5.36}$$

Comparison of Eqs. (5.34) and (5.36) shows that the effect of the earth is to increase the capacitance of a line. To account for the earth, the denominator of Eq. (5.34) must have subtracted from it the term

$$\ln\left(\frac{\sqrt[3]{H_{12}H_{23}H_{31}}}{\sqrt[3]{H_1 H_2 H_3}}\right)$$

If the conductors are high above ground compared with the distances between them, the diagonal distances in the numerator of the correction term are nearly equal to the vertical distances in the denominator, and the term is very small. This is the usual case, and the effect of ground is generally neglected for three-phase lines except for calculations by symmetrical components when the sum of the three line currents is not zero.

5.7 CAPACITANCE CALCULATIONS FOR BUNDLED CONDUCTORS

Figure 5.10 shows a bundled-conductor line for which we can write an equation for the voltage from conductor a to conductor b as we did in deriving Eq. (5.26), except that now we must consider the charges on all six individual conductors. The conductors of any one bundle are in parallel, and we can assume the charge per bundle divides equally between the conductors of the bundle since the separation between bundles is usually more than 15 times the spacing between the conductors of the bundle. Also, since D_{12} is much greater than d, we can use D_{12} in place of the distances $D_{12} - d$ and $D_{12} + d$ and make other similar substitutions of bundle separation distances instead of using the more exact expressions that occur in finding V_{ab}. The difference due to this approximation cannot be detected in the final result for usual spacings even when the calculation is carried to five or six significant figures.

If charge on phase a is q_a, each of conductors a and a' has the charge $q_a/2$; similar division of charge is assumed for phases b and c. Then,

$$V_{ab} = \frac{1}{2\pi k}\left[\frac{q_a}{2}\left(\underbrace{\ln\frac{D_{12}}{r}}_{a} + \underbrace{\ln\frac{D_{12}}{d}}_{a'}\right) + \frac{q_b}{2}\left(\underbrace{\ln\frac{r}{D_{12}}}_{b} + \underbrace{\ln\frac{d}{D_{12}}}_{b'}\right) + \frac{q_c}{2}\left(\underbrace{\ln\frac{D_{23}}{D_{31}}}_{c} + \underbrace{\ln\frac{D_{23}}{D_{31}}}_{c'}\right)\right] \tag{5.37}$$

The letters under each logarithmic term indicate the conductor whose charge is accounted for by that term. Combining terms gives

$$V_{ab} = \frac{1}{2\pi k}\left(q_a \ln\frac{D_{12}}{\sqrt{rd}} + q_b \ln\frac{\sqrt{rd}}{D_{12}} + q_c \ln\frac{D_{23}}{D_{31}}\right) \tag{5.38}$$

FIGURE 5.10
Cross section of a bundled-conductor three-phase line.

Equation (5.38) is the same as Eq. (5.26), except that $\sqrt{rd}$ has replaced r. It therefore follows that if we consider the line to be transposed, we find

$$C_n = \frac{2\pi k}{\ln\left(\dfrac{D_{eq}}{\sqrt{rd}}\right)} \text{ F/m to neutral} \tag{5.39}$$

The $\sqrt{rd}$ is the same as D_s^b for a two-conductor bundle, except that r has replaced D_s. This leads us to the very important conclusion that a modified geometric mean distance (GMD) method applies to the calculation of capacitance of a bundled-conductor three-phase line having two conductors per bundle. The modification is that we are using outside radius in place of the GMR of a single conductor.

It is logical to conclude that the modified GMD method applies to other bundling configurations. If we let D_{sC}^b stand for the modified GMR to be used in capacitance calculations to distinguish it from D_s^b used in inductance calculations, we have

$$C_n = \frac{2\pi k}{\ln\left(\dfrac{D_{eq}}{D_{sC}^b}\right)} \text{ F/m to neutral} \tag{5.40}$$

Then, for a two-strand bundle

$$D_{sC}^b = \sqrt[4]{(r \times d)^2} = \sqrt{rd} \tag{5.41}$$

for a three-strand bundle

$$D_{sC}^b = \sqrt[9]{(r \times d \times d)^3} = \sqrt[3]{rd^2} \tag{5.42}$$

and for a four-strand bundle

$$D_{sC}^b = \sqrt[16]{(r \times d \times d \times d \times \sqrt{2})^4} = 1.09\sqrt[4]{rd^3} \tag{5.43}$$

Example 5.3. Find the capacitive reactance to neutral of the line described in Example 4.5 in ohm-kilometers (and in ohm-miles) per phase.

Solution. Computed from the diameter given in Table A.3

$$r = \frac{1.382 \times 0.3048}{2 \times 12} = 0.01755 \text{ m}$$

$$D_{sC}^{b} = \sqrt{0.01755 \times 0.45} = 0.0889 \text{ m}$$

$$D_{eq} = \sqrt[3]{8 \times 8 \times 16} = 10.08 \text{ m}$$

$$C_m = \frac{2\pi \times 8.85 \times 10^{-12}}{\ln\left(\dfrac{10.08}{0.0889}\right)} = 11.754 \times 10^{-12} \text{ F/m}$$

$$X_C = \frac{10^{12} \times 10^{-3}}{2\pi 60 \times 11.754} = 0.2257 \times 10^{6}\ \Omega \cdot \text{km per phase to neutral}$$

$$\left(X_C = \frac{0.2257 \times 10^{6}}{1.609} = 0.1403 \times 10^{6}\ \Omega \cdot \text{mi per phase to neutral}\right)$$

5.8 PARALLEL-CIRCUIT THREE-PHASE LINES

If two three-phase circuits that are identical in construction and operating in parallel are so close together that coupling exists between them, the GMD method can be used to calculate the inductive and capacitive reactances of their equivalent circuit.

Figure 5.11 shows a typical arrangement of parallel-circuit three-phase lines on the same tower. Although the line will probably not be transposed, we obtain practical values for inductive and capacitive reactances if transposition is assumed. Conductors a and a' are in parallel to compose phase a. Phases b and

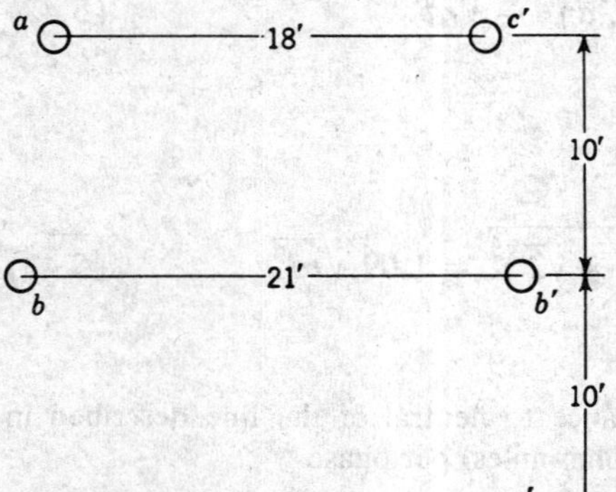

FIGURE 5.11
Typical arrangement of conductors of a parallel-circuit three-phase line.

c are similar. We assume that a and a' take the positions of b and b' and then of c and c' as those conductors are rotated similarly in the transposition cycle.

To calculate D_{eq} the GMD method requires that we use D_{ab}^p, D_{bc}^p, and D_{ca}^p, where the superscript indicates that these quantities are for parallel lines and where D_{ab}^p means the GMD between the conductors of phase a and those of phase b.

For *inductance* calculations D_s of Eq. (4.56) is replaced by D_s^p, which is the geometric mean of the GMR values of the two conductors occupying first the positions of a and a', then the positions of b and b', and finally the positions of c and c'.

Because of the similarity between inductance and capacitance calculations, we can assume that the D_{sC}^p for capacitance is the same as D_s^p for inductance, except that r is used instead of D_s of the individual conductor.

Following each step of Example 5.4 is possibly the best means of understanding the procedure.

Example 5.4. A three-phase double-circuit line is composed of 300,000-cmil 26/7 *Ostrich* conductors arranged as shown in Fig. 5.11. Find the 60-Hz inductive reactance and capacitive susceptance in ohms per mile per phase and siemens per mile per phase, respectively.

Solution. From Table A.3 for *Ostrich*

$$D_s = 0.0229 \text{ ft}$$

$$\text{Distance } a \text{ to } b\text{: original position} = \sqrt{10^2 + 1.5^2} = 10.1 \text{ ft}$$

$$\text{Distance } a \text{ to } b'\text{: original position} = \sqrt{10^2 + 19.5^2} = 21.9 \text{ ft}$$

The GMDs between phases are

$$D_{ab}^p = D_{bc}^p = \sqrt[4]{(10.1 \times 21.9)^2} = 14.88 \text{ ft}$$

$$D_{ca}^p = \sqrt[4]{(20 \times 18)^2} = 18.97 \text{ ft}$$

$$D_{eq} = \sqrt[3]{14.88 \times 14.88 \times 18.97} = 16.1 \text{ ft}$$

For inductance calculations the GMR for the parallel-circuit line is found after first obtaining the GMR values for the three positions. The actual distance from a

to a' is $\sqrt{20^2 + 18^2} = 26.9$ ft. Then, GMR of each phase is

$$\text{In position } a - a'\text{:} \quad \sqrt{26.9 \times 0.0229} = 0.785 \text{ ft}$$

$$\text{In position } b - b'\text{:} \quad \sqrt{21 \times 0.0229} = 0.693 \text{ ft}$$

$$\text{In position } c - c'\text{:} \quad \sqrt{26.9 \times 0.0229} = 0.785 \text{ ft}$$

Therefore,

$$D_s^p = \sqrt[3]{0.785 \times 0.693 \times 0.785} = 0.753 \text{ ft}$$

$$L = 2 \times 10^{-7} \ln \frac{16.1}{0.753} = 6.13 \times 10^{-7} \text{ H/m per phase}$$

$$X_L = 2\pi 60 \times 1609 \times 6.13 \times 10^{-7} = 0.372\ \Omega/\text{mi per phase}$$

For capacitive calculations D_{sC}^p is the same as that of D_s^p, except that the outside radius of the *Ostrich* conductor is used instead of its GMR. The outside diameter of *Ostrich* is 0.680 in:

$$r = \frac{0.680}{2 \times 12} = 0.0283 \text{ ft}$$

$$D_{sC}^p = \left(\sqrt{26.9 \times 0.0283}\,\sqrt{21 \times 0.0283}\,\sqrt{26.9 \times 0.0283}\right)^{1/3}$$

$$= \sqrt{0.0283}\,(26.9 \times 21 \times 26.9)^{1/6} = 0.837 \text{ ft}$$

$$C_n = \frac{2\pi \times 8.85 \times 10^{-12}}{\ln \dfrac{16.1}{0.837}} = 18.807 \times 10^{-12} \text{ F/m}$$

$$B_c = 2\pi \times 60 \times 18.807 \times 1609$$

$$= 11.41 \times 10^{-6} \text{ S/mi per phase to neutral}$$

5.9 SUMMARY

The similarity between inductance and capacitance calculations has been emphasized throughout our discussions. As in inductance calculations, computer programs are recommended if a large number of calculations of capacitance is required. Tables like A.3 and A.5 make the calculations quite simple, however, except for parallel-circuit lines.

The important equation for capacitance to neutral for a single-circuit, three-phase line is

$$C_n = \frac{2\pi k}{\ln \dfrac{D_{eq}}{D_{sC}}} \text{ F/m to neutral} \tag{5.44}$$

D_{sC} is the outside radius r of the conductor for a line consisting of one conductor per phase. For overhead lines k is 8.854×10^{-12} since k_r for air is 1.0. Capacitive reactance is ohm-meters is $1/2\pi f C$, where C is in farads per meter. So, at 60 Hz

$$X_C = 4.77 \times 10^4 \ln \frac{D_{eq}}{D_{sC}} \ \Omega \cdot \text{km to neutral} \tag{5.45}$$

or upon dividing by 1.609 km/mi, we have

$$X_C = 2.965 \times 10^4 \ln \frac{D_{eq}}{D_{sc}} \ \Omega \cdot \text{mi to neutral} \tag{5.46}$$

Values for capacitive susceptance in siemens per kilometer and siemens per mile are the reciprocals of Eqs. (5.45) and (5.46), respectively.

Both D_{eq} and D_{sC} must be in the same units, usually feet. For bundled conductors D_{sC}^b is substituted for D_{sC}. For both single- and bundled-conductor lines

$$D_{eq} = \sqrt[3]{D_{ab} D_{bc} D_{ca}} \tag{5.47}$$

For bundled-conductor lines D_{ab}, D_{bc}, and D_{ca} are distances between the centers of the bundles of phases a, b, and c.

For lines with one conductor per phase it is convenient to determine X_C by adding X_a' for the conductor as found in Table A.3 to X_d' as found in Table A.5 corresponding to D_{eq}.

Inductance, capacitance, and the associated reactances of parallel-circuit lines are found by following the procedure of Example 5.4.

PROBLEMS

5.1. A three-phase transmission line has flat horizontal spacing with 2 m between adjacent conductors. At a certain instant the charge on one of the outside conductors is 60 μC/km, and the charge on the center conductor and on the other outside conductor is -30 μC/km. The radius of each conductor is 0.8 cm. Neglect the effect of the ground and find the voltage drop between the two identically charged conductors at the instant specified.

5.2. The 60-Hz capacitive reactance to neutral of a solid conductor, which is one conductor of a single-phase line with 5-ft spacing, is 196.1 kΩ-mi. What value of reactance would be specified in a table listing the capacitive reactance in ohm-miles to neutral of the conductor at 1-ft spacing for 25 Hz? What is the cross-sectional area of the conductor in circular mils?

5.3. Solve Example 5.1 for 50-Hz operation and 10-ft spacing.

5.4. Using Eq. (5.23), determine the capacitance to neutral (in μF/km) of a three-phase line with three *Cardinal* ACSR conductors equilaterally spaced 20 ft apart. What is the charging current of the line (in A/km) at 60 Hz and 100 kV line to line?

5.5. A three-phase 60-Hz transmission line has its conductors arranged in a triangular formation so that two of the distances between conductors are 25 ft and the third is 42 ft. The conductors are ACSR *Osprey*. Determine the capacitance to neutral in microfarads per mile and the capacitive reactance to neutral in ohm-miles. If the line is 150 mi long, find the capacitance to neutral and capacitive reactance of the line.

5.6. A three-phase 60-Hz line has flat horizontal spacing. The conductors have an outside diameter of 3.28 cm with 12 m between conductors. Determine the capacitive reactance to neutral in ohm-meters and the capacitive reactance of the line in ohms if its length is 125 mi.

5.7. (*a*) Derive an equation for the capacitance to neutral in farads per meter of a single-phase line, taking into account the effect of ground. Use the same nomenclature as in the equation derived for the capacitance of a three-phase line where the effect of ground is represented by image charges.
(*b*) Using the derived equation, calculate the capacitance to neutral in farads per meter of a single-phase line composed of two solid circular conductors, each having a diameter of 0.229 in. The conductors are 10 ft apart and 25 ft above ground. Compare the result with the value obtained by applying Eq. (5.10).

5.8. Solve Prob. 5.6 while taking into account the effect of ground. Assume that the conductors are horizontally placed 20 m above ground.

5.9. A 60-Hz three-phase line composed of one ACSR *Bluejay* conductor per phase has flat horizontal spacing of 11 m between adjacent conductors. Compare the capacitive reactance in ohm-kilometers per phase of this line with that of a line using a two-conductor bundle of ACSR 26/7 conductors having the same total cross-sectional area of aluminum as the single-conductor line and the 11-m spacing measured between bundles. The spacing between conductors in the bundle is 40 cm.

5.10. Calculate the capacitive reactance in ohm-kilometers of a bundled 60-Hz three-phase line having three ACSR *Rail* conductors per bundle with 45 cm between conductors of the bundle. The spacing between bundle centers is 9, 9, and 18 m.

5.11. Six conductors of ACSR *Drake* constitute a 60-Hz double-circuit three-phase line arranged as shown in Fig. 5.11. The vertical spacing, however, is 14 ft; the longer horizontal distance is 32 ft; and the shorter horizontal distances are 25 ft. Find
(*a*) The inductance per phase (in H/mi) and the inductive reactance (in Ω/mi).
(*b*) The capacitive reactance to neutral (in Ω · mi) and the charging current in A/mi per phase and per conductor at 138 kV.

CHAPTER 6

CURRENT AND VOLTAGE RELATIONS ON A TRANSMISSION LINE

We have examined the parameters of a transmission line and are ready to consider the line as an element of a power system. Figure 6.1 shows a 500-kV line having bundled conductors. In overhead lines the conductors are suspended from the tower and insulated from it and from each other by insulators, the number of which is determined by the voltage of the line. Each insulator string in Fig. 6.1 has 22 insulators. The two shorter arms above the phase conductors support wires usually made of steel. These wires, much smaller in diameter than the phase conductors, are not visible in the picture, but they are electrically connected to the tower and are therefore at ground potential. These wires are referred to as *shield* or *ground wires* and shield the phase conductors from lightning strokes.

A very important problem in the design and operation of a power system is the maintenance of the voltage within specified limits at various points in the system. In this chapter we develop formulas by which we can calculate the voltage, current, and power at any point on a transmission line, provided we know these values at one point, usually at one end of the line.

The purpose of this chapter, however, is not merely to develop the pertinent equations, but also to provide an opportunity to understand the effects of the parameters of the line on bus voltages and the flow of power. In

FIGURE 6.1
A 500-kV transmission line. Conductors are 76/19 ACSR with aluminum cross section of 2,515,000 cmil. Spacing between phases is 30 ft 3 in and the two conductors per bundle are 18 in apart. (*Courtesy Carolina Power and Light Company.*)

this way we can see the importance of the design of the line and better understand the developments to come in later chapters. This chapter also provides an introduction to the study of transients on lossless lines in order to indicate how problems arise due to surges caused by lightning and switching.

In the modern power system data from all over the system are being fed continuously into on-line computers for control and information purposes. Power-flow studies performed by a computer readily supply answers to questions concerning the effect of switching lines into and out of the system or of changes in line parameters. Equations derived in this chapter remain important, however, in developing an overall understanding of what is occurring on a system and in calculating efficiency of transmission, losses, and limits of power flow over a line for both steady-state and transient conditions.

6.1 REPRESENTATION OF LINES

The general equations relating voltage and current on a transmission line recognize the fact that all four of the parameters of a transmission line discussed in the two preceding chapters are uniformly distributed along the line. We derive these general equations later, but first we use lumped parameters which give good accuracy for short lines and for lines of medium length. If an overhead line is classified as short, shunt capacitance is so small that it can be omitted entirely with little loss of accuracy, and we need to consider only the series resistance R and the series inductance L for the total length of the line.

A medium-length line can be represented sufficiently well by R and L as lumped parameters, as shown in Fig. 6.2, with half the capacitance to neutral of the line lumped at each end of the equivalent circuit. Shunt conductance G, as mentioned previously, is usually neglected in overhead power transmission lines when calculating voltage and current. The same circuit represents the short line if capacitors are omitted.

Insofar as the handling of capacitance is concerned, open-wire 60-Hz lines less than about 80 km (50 mi) long are short lines. Medium-length lines are roughly between 80 km (50 mi) and 240 km (150 mi) long. Lines longer than 240 km (150 mi) require calculations in terms of distributed constants if a high degree of accuracy is required, although for some purposes a lumped-parameter representation can be used for lines up to 320 km (200 mi) long.

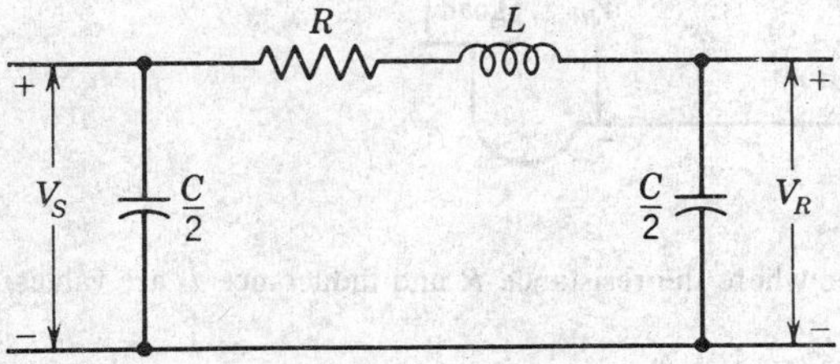

FIGURE 6.2
Single-phase equivalent of a medium-length line. The capacitors are omitted for a short line.

Normally, transmission lines are operated with balanced three-phase loads. Although the lines are not spaced equilaterally and not transposed, the resulting dissymmetry is slight and the phases are considered to be balanced.

In order to distinguish between the total series impedance of a line and the series impedance per unit length, the following nomenclature is adopted:

z = series impedance per unit length per phase

y = shunt admittance per unit length per phase to neutral

l = length of line

$Z = zl$ = total series impedance per phase

$Y = yl$ = total shunt admittance per phase to neutral

6.2 THE SHORT TRANSMISSION LINE

The equivalent circuit of a short transmission line is shown in Fig. 6.3, where I_S and I_R are the sending- and receiving-end currents, respectively, and V_S and V_R are the sending- and receiving-end line-to-neutral voltages.

The circuit is solved as a simple series ac circuit. So,

$$I_S = I_R \tag{6.1}$$

$$V_S = V_R + I_R Z \tag{6.2}$$

where Z is zl, the total series impedance of the line.

The effect of the variation of the power factor of the load on the voltage regulation of a line is most easily understood for the short line and therefore will be considered at this time. Voltage regulation of a transmission line is the rise in voltage at the receiving end, expressed in percent of full-load voltage, when full load at a specified power factor is removed while the sending-end

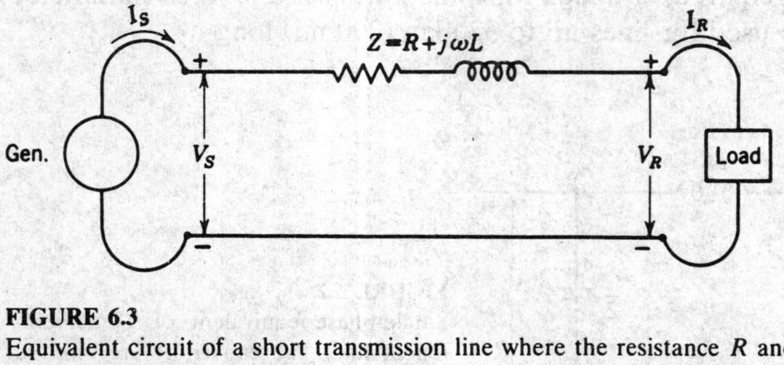

FIGURE 6.3
Equivalent circuit of a short transmission line where the resistance R and inductance L are values for the entire length of the line.

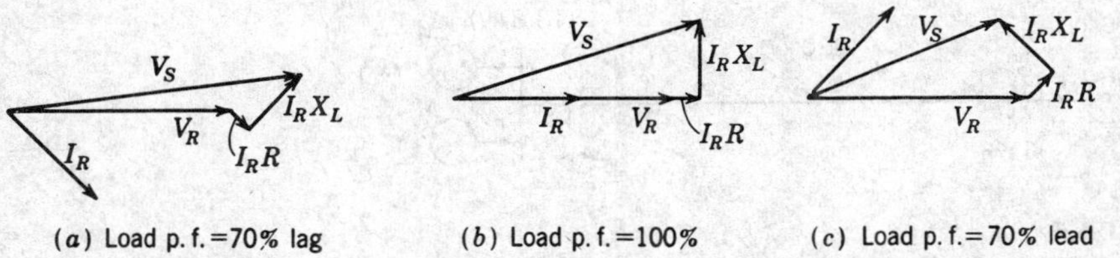

FIGURE 6.4
Phasor diagrams of a short transmission line. All diagrams are drawn for the same magnitudes of V_R and I_R.

voltage is held constant. Corresponding to Eq. (2.33) we can write

$$\text{Percent regulation} = \frac{|V_{R,NL}| - |V_{R,FL}|}{|V_{R,FL}|} \times 100 \tag{6.3}$$

where $|V_{R,NL}|$ is the magnitude of receiving-end voltage at no load and $|V_{R,FL}|$ is the magnitude of receiving-end voltage at full load with $|V_S|$ constant. After the load on a short transmission line, represented by the circuit of Fig. 6.3, is removed, the voltage at the receiving end is equal to the voltage at the sending end. In Fig. 6.3, with the load connected, the receiving-end voltage is designated by V_R, and $|V_R| = |V_{R,FL}|$. The sending-end voltage is V_S, and $|V_S| = |V_{R,NL}|$. The phasor diagrams of Fig. 6.4 are drawn for the same magnitudes of the receiving-end voltage and current and show that a larger value of the sending-end voltage is required to maintain a given receiving-end voltage when the receiving-end current is lagging the voltage than when the same current and voltage are in phase. A still smaller sending-end voltage is required to maintain the given receiving-end voltage when the receiving-end current leads the voltage. The voltage drop is the same in the series impedance of the line in all cases; because of the different power factors, however, the voltage drop is added to the receiving-end voltage at a different angle in each case. The regulation is greatest for lagging power factors and least, or even negative, for leading power factors. The inductive reactance of a transmission line is larger than the resistance, and the principle of regulation illustrated in Fig. 6.4 is true for any load supplied by a predominantly inductive circuit. The magnitudes of the voltage drops $I_R R$ and $I_R X_L$ for a short line have been exaggerated with respect to V_R in drawing the phasor diagrams in order to illustrate the point more clearly. The relation between power factor and regulation for longer lines is similar to that for short lines but is not visualized so easily.

Example 6.1. A 300-MVA 20-kV three-phase generator has a subtransient reactance of 20%. The generator supplies a number of synchronous motors over a 64-km transmission line having transformers at both ends, as shown on the one-line diagram of Fig. 6.5. The motors, all rated 13.2 kV, are represented by just

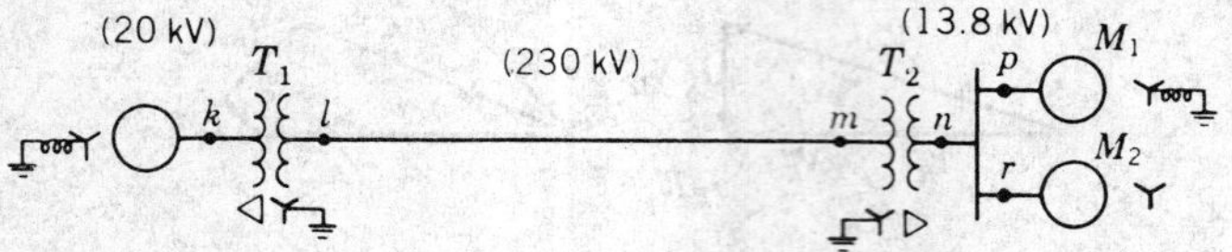

FIGURE 6.5
One-line diagram for Example 6.1.

two equivalent motors. The neutral of one motor M_1 is grounded through reactance. The neutral of the second motor M_2 is not connected to ground (an unusual condition). Rated inputs to the motors are 200 MVA and 100 kVA for M_1 and M_2, respectively. For both motors $X_d'' = 20\%$. The three-phase transformer T_1 is rated 350 MVA, 230/20 kV with leakage reactance of 10%. Transformer T_2 is composed of three single-phase transformers, each rated 127/13.2 kV, 100 MVA with leakage reactance of 10%. Series reactance of the transmission line is 0.5 Ω/km. Draw the reactance diagram with all reactances marked in per unit. Select the generator rating as base in the generator circuit.

Solution. The three-phase rating of transformer T_2 is

$$3 \times 100 = 300 \text{ kVA}$$

and its line-to-line voltage ratio is

$$\sqrt{3} \times \frac{127}{13.2} = \frac{220}{13.2} \text{ kV}$$

A base of 300 MVA, 20 kV in the generator circuit requires a 300-MVA base in all parts of the system and the following voltage bases:

In the transmission line: 230 kV (since T_1 is rated 230/20 kV)

In the motor circuit: $230 \dfrac{13.2}{220} = 13.8$ kV

These bases are shown in parentheses on the one-line diagram of Fig. 6.5. The reactances of the transformers converted to the proper base are

Transformer T_1: $X = 0.1 \times \dfrac{300}{350} = 0.0857$ per unit

Transformer T_2: $X = 0.1\left(\dfrac{13.2}{13.8}\right)^2 = 0.0915$ per unit

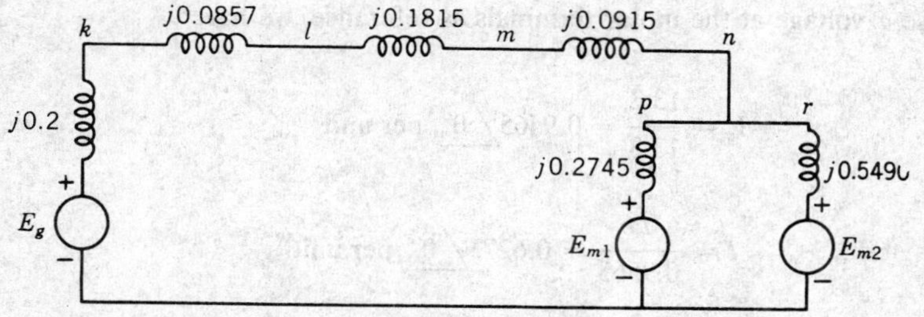

FIGURE 6.6
Reactance diagram for Example 6.1. Reactances are in per unit on the specified base.

The base impedance of the transmission line is

$$\frac{(230)^2}{300} = 176.3\Omega$$

and the reactance of the line is

$$\frac{0.5 \times 64}{176.3} = 0.1815 \text{ per unit}$$

$$\text{Reactance } X_d'' \text{ of motor } M_1 = 0.2\left(\frac{300}{200}\right)\left(\frac{13.2}{13.8}\right)^2 = 0.2745 \text{ per unit}$$

$$\text{Reactance } X_d'' \text{ of motor } M_2 = 0.2\left(\frac{300}{100}\right)\left(\frac{13.2}{13.8}\right)^2 = 0.5490 \text{ per unit}$$

Figure 6.6 is the required reactance diagram when transformer phase shifts are omitted.

Example 6.2. If the motors M_1 and M_2 of Example 6.1 have inputs of 120 and 60 MW, respectively, at 13.2 kV, and both operate at unity power factor, find the voltage at the terminals of the generator and the voltage regulation of the line.

Solution. Together the motors take 180 MW, or

$$\frac{180}{300} = 0.6 \text{ per unit}$$

Therefore, with V and I at the motors in per unit,

$$|V| \times |I| = 0.6 \text{ per unit}$$

With phase-*a* voltage at the motor terminals as reference, we have

$$V = \frac{13.2}{13.8} = 0.9565\angle 0^\circ \text{ per unit}$$

$$I = \frac{0.6}{0.9565} = 0.6273\angle 0^\circ \text{ per unit}$$

Phase-*a* per-unit voltages at other points of Fig. 6.6 are

At m: $V = 0.9565 + 0.6273(j0.0915)$

$$0.9565 + j0.0574 = 0.9582\angle 3.434^\circ \text{ per unit}$$

At l: $V = 0.9565 + 0.6273(j0.0915 + j0.1815)$

$$0.9565 + j0.1713 = 0.9717\angle 10.154^\circ \text{ per unit}$$

At k: $V = 0.9565 + 0.6273(j0.0915 + j0.1815 + j0.0857)$

$$0.9565 + j0.2250 = 0.9826\angle 13.237^\circ \text{ per unit}$$

The voltage regulation of the line is

$$\text{Percent regulation} = \frac{0.9826 - 0.9582}{0.9582} \times 100 = 2.55\%$$

and the magnitude of the voltage at the generator terminals is

$$0.9826 \times 20 = 19.652 \text{ kV}$$

If it is desired to show the phase shifts due to the $Y - \Delta$ transformers, the angles of the phase-*a* voltages at m and l should be increased by 30°. Then the angle of the phase-*a* current in the line should also be increased by 30° from 0°.

6.3 THE MEDIUM-LENGTH LINE

The shunt admittance, usually pure capacitance, is included in the calculations for a line of medium length. If the total shunt admittance of the line is divided into two equal parts placed at the sending and receiving ends of the line, the circuit is called a *nominal* π. We refer to Fig. 6.7 to derive equations. To obtain an expression for V_S, we note that the current in the capacitance at the

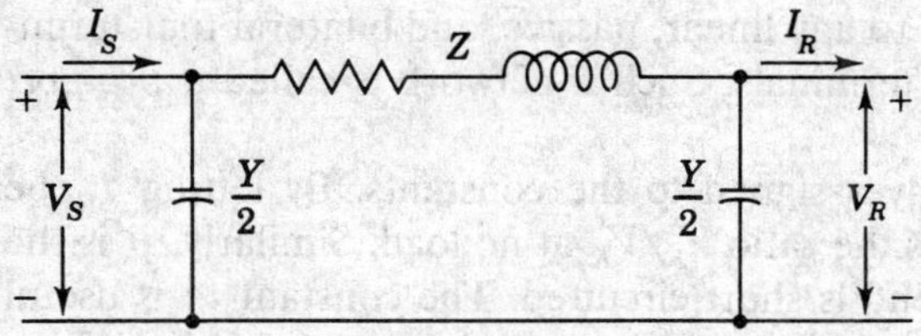

FIGURE 6.7
Nominal-π circuit of a medium-length transmission line.

receiving end is $V_R Y/2$ and the current in the series arm is $I_R + V_R Y/2$. Then,

$$V_S = \left(V_R \frac{Y}{2} + I_R\right)Z + V_R \tag{6.4}$$

$$V_S = \left(\frac{ZY}{2} + 1\right)V_R + ZI_R \tag{6.5}$$

To derive I_S, we note that the current in the shunt capacitance at the sending end is $V_S Y/2$, which added to the current in the series arm gives

$$I_S = V_S \frac{Y}{2} + V_R \frac{Y}{2} + I_R \tag{6.6}$$

Substituting V_S, as given by Eq. (6.5), in Eq. (6.6) yields

$$I_S = V_R Y\left(1 + \frac{ZY}{4}\right) + \left(\frac{ZY}{2} + 1\right)I_R \tag{6.7}$$

Equations (6.5) and (6.7) may be expressed in the general form

$$V_S = AV_R + BI_R \tag{6.8}$$

$$I_S = CV_R + DI_R \tag{6.9}$$

where

$$A = D = \frac{ZY}{2} + 1$$
$$B = Z \qquad C = Y\left(1 + \frac{ZY}{4}\right) \tag{6.10}$$

These *ABCD* constants are sometimes called the *generalized circuit constants* of the transmission line. In general, they are complex numbers. *A* and *D* are dimensionless and equal each other if the line is the same when viewed from either end. The dimensions of *B* and *C* are ohms and mhos or siemens, respectively. The constants apply to any linear, passive, and bilateral four-terminal network having two pairs of terminals. Such a network is called a *two-port network*.

A physical meaning is easily assigned to the constants. By letting I_R be zero in Eq. (6.8), we see that A is the ratio V_S/V_R at no load. Similarly, B is the ratio V_S/I_R when the receiving end is short-circuited. The constant A is useful in computing regulation. If $V_{R,FL}$ is the receiving-end voltage at full load for a sending-end voltage of V_S, Eq. (6.3) becomes

$$\text{Percent regulation} = \frac{|V_S|/|A| - |V_{R,FL}|}{|V_{R,FL}|} \times 100 \tag{6.11}$$

Table A.6 in the Appendix lists $ABCD$ constants for various networks and combinations of networks.

6.4 THE LONG TRANSMISSION LINE: SOLUTION OF THE DIFFERENTIAL EQUATIONS

The exact solution of any transmission line and the one required for a high degree of accuracy in calculating 60-Hz lines more than approximately 150 mi long must consider the fact that the parameters of the lines are not lumped but, rather, are distributed uniformly throughout the length of the line.

Figure 6.8 shows one phase and the neutral connection of a three-phase line. Lumped parameters are not shown because we are ready to consider the solution of the line with the impedance and admittance uniformly distributed. In Fig. 6.8 we consider a differential element of length dx in the line at a distance x from the receiving end of the line. Then $z\,dx$ and $y\,dx$ are, respectively, the series impedance and shunt admittance of the elemental section. V and I are phasors which vary with x.

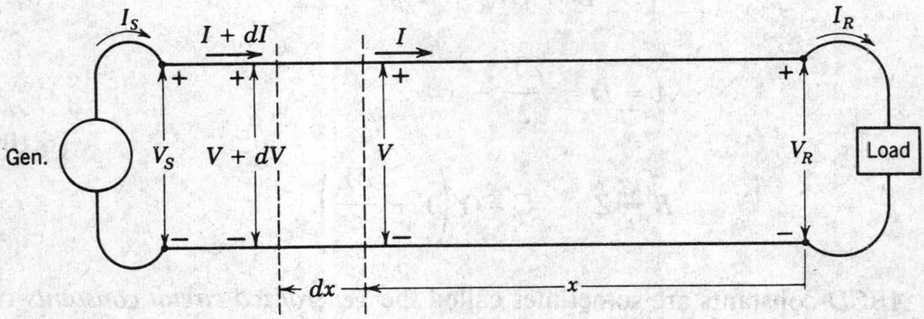

FIGURE 6.8
Schematic diagram of a transmission line showing one phase and the neutral return. Nomenclature for the line and the elemental length are indicated.

Average line current in the element is $(I + I + dI)/2$, and the increase of V in the distance dx is quite accurately expressed as

$$dV = \frac{I + I + dI}{2} z\,dx = Iz\,dx \tag{6.12}$$

when products of the differential quantities are neglected. Similarly,

$$dI = \frac{V + V + dV}{2} y\,dx = Vy\,dx \tag{6.13}$$

Then, from Eqs. (6.12) and (6.13) we have

$$\frac{dV}{dx} = Iz \tag{6.14}$$

and

$$\frac{dI}{dx} = Vy \tag{6.15}$$

Let us differentiate Eqs. (6.14) and (6.15) with respect to x, and we obtain

$$\frac{d^2V}{dx^2} = z\frac{dI}{dx} \tag{6.16}$$

and

$$\frac{d^2I}{dx^2} = y\frac{dV}{dx} \tag{6.17}$$

If we substitute the values of dI/dx and dV/dx from Eqs. (6.15) and (6.14) in Eqs. (6.16) and (6.17), respectively, we obtain

$$\frac{d^2V}{dx^2} = yzV \tag{6.18}$$

and

$$\frac{d^2I}{dx^2} = yzI \tag{6.19}$$

Now we have Eq. (6.18) in which the only variables are V and x and Eq. (6.19) in which the only variables are I and x. The solutions of those equations for V and I, respectively, must be expressions which when differentiated twice with respect to x yield the original expression times the constant yz. For instance, the solution for V when differentiated twice with respect to x must yield yzV.

This suggests an exponential form of solution. Assume that the solution of Eq. (6.18) is

$$V = A_1 \varepsilon^{\sqrt{yz}x} + A_2 \varepsilon^{-\sqrt{yz}x} \tag{6.20}$$

Taking the second derivative of V with respect to x in Eq. (6.20) yields

$$\frac{d^2V}{dx^2} = yz\left[A_1 \varepsilon^{\sqrt{yz}x} + A_2 \varepsilon^{-\sqrt{yz}x}\right] \tag{6.21}$$

which is yz times the assumed solution for V. Therefore, Eq. (6.20) is the solution of Eq. (6.18). When we substitute the value given by Eq. (6.20) for V in Eq. (6.14), we obtain

$$I = \frac{1}{\sqrt{z/y}} A_1 \varepsilon^{\sqrt{yz}x} - \frac{1}{\sqrt{z/y}} A_2 \varepsilon^{-\sqrt{yz}x} \tag{6.22}$$

The constants A_1 and A_2 can be evaluated by using the conditions at the receiving end of the line; namely, when $x = 0$, $V = V_R$ and $I = I_R$. Substitution of these values in Eqs. (6.20) and (6.22) yields

$$V_R = A_1 + A_2 \qquad \text{and} \qquad I_R = \frac{1}{\sqrt{z/y}}(A_1 - A_2)$$

Substituting $Z_c = \sqrt{z/y}$ and solving for A_1 give

$$A_1 = \frac{V_R + I_R Z_c}{2} \qquad \text{and} \qquad A_2 = \frac{V_R - I_R Z_c}{2}$$

Then, substituting the values found for A_1 and A_2 in Eqs. (6.20) and (6.22) and letting $\gamma = \sqrt{yz}$, we obtain

$$V = \frac{V_R + I_R Z_c}{2} \varepsilon^{\gamma x} + \frac{V_R - I_R Z_c}{2} \varepsilon^{-\gamma x} \tag{6.23}$$

$$I = \frac{V_R/Z_c + I_R}{2} \varepsilon^{\gamma x} - \frac{V_R/Z_c - I_R}{2} \varepsilon^{-\gamma x} \tag{6.24}$$

where $Z_c = \sqrt{z/y}$ and is called the *characteristic impedance* of the line, and $\gamma = \sqrt{zy}$ and is called the *propagation constant*.

Equations (6.23) and (6.24) give the rms values of V and I and their phase angles at any specified point along the line in terms of the distance x from the receiving end to the specified point, provided V_R, I_R, and the parameters of the line are known.

6.5 THE LONG TRANSMISSION LINE: INTERPRETATION OF THE EQUATIONS

Both γ and Z_c are complex quantities. The real part of the propagation constant γ is called the *attenuation constant* α and is measured in nepers per unit length. The quadrature part of γ is called the *phase constant* β and is measured in radians per unit length. Thus,

$$\gamma = \alpha + j\beta \tag{6.25}$$

and Eqs. (6.23) and (6.24) become

$$V = \frac{V_R + I_R Z_c}{2}\varepsilon^{\alpha x}\varepsilon^{j\beta x} + \frac{V_R - I_R Z_c}{2}\varepsilon^{-\alpha x}\varepsilon^{-j\beta x} \tag{6.26}$$

and

$$I = \frac{V_R/Z_c + I_R}{2}\varepsilon^{\alpha x}\varepsilon^{j\beta x} - \frac{V_R/Z_c - I_R}{2}\varepsilon^{-\alpha x}\varepsilon^{-j\beta x} \tag{6.27}$$

The properties of $\varepsilon^{\alpha x}$ and $\varepsilon^{j\beta x}$ help to explain the variation of the phasor values of voltage and current as a function of distance along the line. The term $\varepsilon^{\alpha x}$ changes in magnitude as x changes, but $\varepsilon^{j\beta x}$ (identical to $\cos \beta x + j \sin \beta x$) always has a magnitude of 1 and causes a shift in phase of β radians per unit length of line.

The first term in Eq. (6.26), $[(V_R + I_R Z_c)/2]\varepsilon^{\alpha x}\varepsilon^{j\beta x}$, increases in magnitude and advances in phase as distance x from the receiving end increases. Conversely, as progress along the line from the sending end toward the receiving end is considered, the term diminishes in magnitude and is retarded in phase. This is the characteristic of a traveling wave and is similar to the behavior of a wave in water, which varies in magnitude with time at any point, whereas its phase is retarded and its maximum value diminishes with distance from the origin. The variation in instantaneous value is not expressed in the term but is understood since V_R and I_R are phasors. The first term in Eq. (6.26) is called the *incident voltage*.

The second term in Eq. (6.26), $[(V_R - I_R Z_c)/2]\varepsilon^{-\alpha x}\varepsilon^{-j\beta x}$, diminishes in magnitude and is retarded in phase from the receiving end toward the sending end. It is called the *reflected voltage*. At any point along the line the voltage is the sum of the component incident and reflected voltages at that point.

Since the equation for current is similar to the equation for voltage, the current may be considered to be composed of incident and reflected currents.

If a line is terminated in its characteristic impedance Z_c, receiving-end voltage V_R is equal to $I_R Z_c$ and there is no reflected wave of either voltage or current, as may be seen by substituting $I_R Z_c$ for V_R in Eqs. (6.26) and (6.27). A line terminated in its characteristic impedance is called a *flat line* or an *infinite line*. The latter term arises from the fact that a line of infinite length cannot have a reflected wave. Usually, power lines are not terminated in their characteristic impedance, but communication lines are frequently so terminated in order to eliminate the reflected wave. A typical value of Z_c is 400 Ω for a single-circuit overhead line and 200 Ω for two circuits in parallel. The phase angle of Z_c is usually between 0 and $-15°$. Bundled-conductor lines have lower values of Z_c since such lines have lower L and higher C than lines with a single conductor per phase.

In power system work characteristic impedance is sometimes called *surge impedance*. The term "surge impedance," however, is usually reserved for the special case of a lossless line. If a line is lossless, its series resistance and shunt conductance are zero and the characteristic impedance reduces to the real number $\sqrt{L/C}$, which has the dimensions of ohms when L is the series inductance of the line in henrys and C is the shunt capacitance in farads. Also, the propagation constant $\gamma = \sqrt{zy}$ for the line of length l reduces to the imaginary number $j\beta = j\omega\sqrt{LC}/l$ since the attenuation constant α resulting from line losses is zero. When dealing with high frequencies or with surges due to lightning, losses are often neglected and the surge impedance becomes important. Surge-impedance loading (SIL) of a line is the power delivered by a line to a purely resistive load equal to its surge impedance. When so loaded, the line supplies a current of

$$|I_L| = \frac{|V_L|}{\sqrt{3} \times \sqrt{L/C}} \text{ A}$$

where $|V_L|$ is the line-to-line voltage at the load. Since the load is pure resistance,

$$\text{SIL} = \sqrt{3}|V_L|\frac{|V_L|}{\sqrt{3} \times \sqrt{L/C}} \text{ W}$$

or with $|V_L|$ in kilovolts,

$$\text{SIL} = \frac{|V_L|^2}{\sqrt{L/C}} \text{ MW} \tag{6.28}$$

Power system engineers sometimes find it convenient to express the power transmitted by a line in terms of per unit of SIL, that is, as the ratio of the power transmitted to the surge-impedance loading. For instance, the permissible loading of a transmission line may be expressed as a fraction of its SIL, and SIL provides a comparison of load-carrying capabilities of lines.[1]

A *wavelength* λ is the distance along a line between two points of a wave which differ in phase by 360°, or 2π rad. If β is the phase shift in radians per

[1]See R. D. Dunlop, R. Gutman, and P. P. Marchenko, "Analytical Development of Loadability Characteristics for EHV and UHV Transmission Lines," *IEEE Transactions on Power Apparatus and Systems*, vol. PAS-98, no. 2, 1979, pp. 606–617.

mile, the wavelength in miles is

$$\lambda = \frac{2\pi}{\beta} \tag{6.29}$$

The velocity of propagation of a wave in miles per second is the product of the wavelength in miles and the frequency in hertz, or

$$\text{Velocity} = \lambda f = \frac{2\pi f}{\beta} \tag{6.30}$$

For the lossless line of length l meters $\beta = 2\pi f\sqrt{LC}/l$ and Eqs. (6.29) and (6.30) become

$$\lambda = \frac{l}{f\sqrt{LC}}\,\text{m} \qquad \text{velocity} = \frac{l}{\sqrt{LC}}\,\text{m/s}$$

When values of L and C for low-loss overhead lines are substituted in these equations, it is found that the wavelength is approximately 3000 mi at a frequency of 60 Hz and the velocity of propagation is very nearly the speed of light in air (approximately 186,000 mi/s or 3×10^8 m/s).

If there is no load on a line, I_R is equal to zero, and as determined by Eqs. (6.26) and (6.27), the incident and reflected voltages are equal in magnitude and in phase at the receiving end. In this case the incident and reflected currents are equal in magnitude but are 180° out of phase at the receiving end. Thus, the incident and reflected currents cancel each other at the receiving end of an open line but not at any other point unless the line is entirely lossless so that the attenuation α is zero.

6.6 THE LONG TRANSMISSION LINE: HYPERBOLIC FORM OF THE EQUATIONS

The incident and reflected waves of voltage are seldom found when calculating the voltage of a power line. The reason for discussing the voltage and the current of a line in terms of the incident and reflected components is that such an analysis is helpful in obtaining a better understanding of some of the phenomena of transmission lines. A more convenient form of the equations for computing current and voltage of a power line is found by introducing hyperbolic functions. Hyperbolic functions are defined in exponential form

$$\sinh\theta = \frac{\varepsilon^{\theta} - \varepsilon^{-\theta}}{2} \tag{6.31}$$

$$\cosh\theta = \frac{\varepsilon^{\theta} + \varepsilon^{-\theta}}{2} \tag{6.32}$$

By rearranging Eqs. (6.23) and (6.24) and substituting hyperbolic functions for the exponential terms, we find a new set of equations. The new equations, giving voltage and current anywhere along the line, are

$$V = V_R \cosh \gamma x + I_R Z_c \sinh \gamma x \tag{6.33}$$

$$I = I_R \cosh \gamma x + \frac{V_R}{Z_c} \sinh \gamma x \tag{6.34}$$

Letting $x = l$ to obtain the voltage and the current at the sending end, we have

$$V_S = V_R \cosh \gamma l + I_R Z_c \sinh \gamma l \tag{6.35}$$

$$I_S = I_R \cosh \gamma l + \frac{V_R}{Z_c} \sinh \gamma l \tag{6.36}$$

From examination of these equations we see that the generalized circuit constants for a long line are

$$\begin{aligned} A &= \cosh \gamma l \qquad & C &= \frac{\sinh \gamma l}{Z_c} \\ B &= Z_c \sinh \gamma l \qquad & D &= \cosh \gamma l \end{aligned} \tag{6.37}$$

Solving Eqs. (6.35) and (6.36) for V_R and I_R in terms of V_S and I_S, we obtain

$$V_R = V_S \cosh \gamma l - I_S Z_c \sinh \gamma l \tag{6.38}$$

$$I_R = I_S \cosh \gamma l - \frac{V_S}{Z_c} \sinh \gamma l \tag{6.39}$$

For balanced three-phase lines the currents in the above equations are line currents and the voltages are line-to-neutral voltages, that is, line voltages divided by $\sqrt{3}$. In order to solve the equations, the hyperbolic functions must be evaluated. Since γl is usually complex, the hyperbolic functions are also complex and can be evaluated with the assistance of a calculator or computer.

For solving an occasional problem without resorting to a computer there are several choices. The following equations give the expansions of hyperbolic sines and cosines of complex arguments in terms of circular and hyperbolic functions of real arguments:

$$\cosh(\alpha l + j\beta l) = \cosh \alpha l \cos \beta l + j \sinh \alpha l \sin \beta l \tag{6.40}$$

$$\sinh(\alpha l + j\beta l) = \sinh \alpha l \cos \beta l + j \cosh \alpha l \sin \beta l \tag{6.41}$$

Equations (6.40) and (6.41) make possible the computation of hyperbolic functions of complex arguments. The correct mathematical unit for βl is the radian, and the radian is the unit found for βl by computing the quadrature component of γl. Equations (6.40) and (6.41) can be verified by substituting in them the exponential forms of the hyperbolic functions and the similar exponential forms of the circular functions.

Another method of evaluating complex hyperbolic functions is suggested by Eqs. (6.31) and (6.32). Substituting $\alpha + j\beta$ for θ, we obtain

$$\cosh(\alpha + j\beta) = \frac{\varepsilon^{\alpha}\varepsilon^{j\beta} + \varepsilon^{-\alpha}\varepsilon^{-j\beta}}{2} = \frac{1}{2}\left(\varepsilon^{\alpha}\underline{/\beta} + \varepsilon^{-\alpha}\underline{/-\beta}\right) \tag{6.42}$$

$$\sinh(\alpha + j\beta) = \frac{\varepsilon^{\alpha}\varepsilon^{j\beta} - \varepsilon^{-\alpha}\varepsilon^{-j\beta}}{2} = \frac{1}{2}\left(\varepsilon^{\alpha}\underline{/\beta} - \varepsilon^{-\alpha}\underline{/-\beta}\right) \tag{6.43}$$

Example 6.3. A single-circuit 60-Hz transmission line is 370 km (230 mi) long. The conductors are *Rook* with flat horizontal spacing and 7.25 m (23.8 ft) between conductors. The load on the line is 125 MW at 215 kV with 100% power factor. Find the voltage, current, and power at the sending end and the voltage regulation of the line. Also, determine the wavelength and velocity of propagation of the line.

Solution. Feet and miles rather than meters and kilometers are chosen for the calculations in order to use Tables A.3 through A.5 in the Appendix:

$$D_{eq} = \sqrt[3]{23.8 \times 23.8 \times 47.6} \cong 30.0 \text{ ft}$$

and from the tables for *Rook*

$$z = 0.1603 + j(0.415 + 0.4127) = 0.8431\underline{/79.04^\circ}\ \Omega/\text{mi}$$

$$y = j[1/(0.0950 + 0.1009)] \times 10^{-6} = 5.105 \times 10^{-6}\underline{/90^\circ}\ \text{S/mi}$$

$$\gamma l = \sqrt{yz}\,l = 230\sqrt{0.8431 \times 5.105 \times 10^{-6}}\ \underline{\Big/\frac{79.04^\circ + 90^\circ}{2}}$$

$$= 0.4772\underline{/84.52^\circ} = 0.0456 + j0.4750$$

$$Z_c = \sqrt{\frac{z}{y}} = \sqrt{\frac{0.8431}{5.105 \times 10^{-6}}}\ \underline{\Big/\frac{79.04^\circ - 90^\circ}{2}} = 406.4\underline{/-5.48^\circ}\ \Omega$$

$$V_R = \frac{215{,}000}{\sqrt{3}} = 124{,}130\underline{/0^\circ}\ \text{V to neutral}$$

$$I_R = \frac{125{,}000{,}000}{\sqrt{3} \times 215{,}000} = 335.7\underline{/0^\circ}\ \text{A}$$

From Eqs. (6.42) and (6.43) and noting that 0.4750 rad = 27.22°

$$\cosh \gamma l = \frac{1}{2}\varepsilon^{0.0456}\angle 27.22^\circ + \frac{1}{2}\varepsilon^{-0.0456}\angle -27.22^\circ$$

$$= 0.4654 + j0.2394 + 0.4248 - j0.2185$$

$$= 0.8902 + j0.0209 = 0.8904\angle 1.34^\circ$$

$$\sinh \gamma l = 0.4654 + j0.2394 - 0.4248 + j0.2185$$

$$= 0.0406 + j0.4579 = 0.4597\angle 84.93^\circ$$

Then, from Eq. (6.35)

$$V_S = 124{,}130 \times 0.8904\angle 1.34^\circ + 335.7 \times 406.4\angle -5.48^\circ \times 0.4597\angle 84.93^\circ$$

$$= 110{,}495 + j2{,}585 + 11{,}483 + j61{,}656$$

$$= 137{,}860\angle 27.77^\circ \text{ V}$$

and from Eq. (6.36)

$$I_S = 335.7 \times 0.8904\angle 1.34^\circ + \frac{124{,}130}{406.4\angle -5.48^\circ} \times 0.4597\angle 84.93^\circ$$

$$= 298.83 + j6.99 - 1.00 + j140.41$$

$$= 332.31\angle 26.33^\circ \text{ A}$$

At the sending end

$$\text{Line voltage} = \sqrt{3} \times 137.86 = 238.8 \text{ kV}$$

$$\text{Line current} = 332.3 \text{ A}$$

$$\text{Power factor} = \cos(27.77^\circ - 26.33^\circ) = 0.9997 \cong 1.0$$

$$\text{Power} = \sqrt{3} \times 238.8 \times 332.3 \times 1.0 = 137{,}443 \text{ kW}$$

From Eq. (6.35) we see that at no load ($I_R = 0$)

$$V_R = \frac{V_S}{\cosh \gamma l}$$

So, the voltage regulation is

$$\frac{137.86/0.8904 - 124.13}{124.13} \times 100 = 24.7\%$$

The wavelength and velocity of propagation are computed as follows:

$$\beta = \frac{0.4750}{230} = 0.002065 \text{ rad/mi}$$

$$\lambda = \frac{2\pi}{\beta} = \frac{2\pi}{0.002065} = 3043 \text{ mi}$$

$$\text{Velocity} = f\lambda = 60 \times 3043 = 182{,}580 \text{ mi/s}$$

We note particularly in this example that in the equations for V_S and I_S the value of voltage must be expressed in volts and must be the line-to-neutral voltage.

Example 6.4. Solve for the sending-end voltage and the current found in Example 6.3 using per-unit calculations.

Solution. We choose a base of 125 MVA, 215 kV to achieve the simplest per-unit values and to compute base impedance and base current as follows:

$$\text{Base impedance} = \frac{215^2}{125} = 370\ \Omega$$

$$\text{Base current} = \frac{125{,}000}{\sqrt{3} \times 215} = 335.7 \text{ A}$$

So,
$$Z_c = \frac{406.4\underline{/-5.48^\circ}}{370} = 1.098\underline{/-5.48^\circ} \text{ per unit}$$

$$V_R = \frac{215}{215} = \frac{215/\sqrt{3}}{215/\sqrt{3}} = 1.0 \text{ per unit}$$

For use in Eq. (6.35) we chose V_R as the reference voltage. So,

$$V_R = 1.0\underline{/0^\circ} \text{ per unit (as a line-to-neutral voltage)}$$

and since the load is at unity power factor,

$$I_R = \frac{337.5\underline{/0^\circ}}{337.5} = 1.0\underline{/0^\circ}$$

If the power factor had been less than 100%, I_R would have been greater than 1.0 and would have been at an angle determined by the power factor. By Eq. (6.35)

$$V_S = 1.0 \times 0.8904 + 1.0 \times 1.098\underline{/-5.48^\circ} \times 0.4597\underline{/84.93^\circ}$$

$$= 0.8902 + j0.0208 + 0.0923 + j0.4961$$

$$= 1.1102\underline{/27.75^\circ} \text{ per unit}$$

and by Eq. (6.36)

$$I_S = 1.0 \times 0.8904\underline{/1.34^\circ} + \frac{1.0\underline{/0^\circ}}{1.098\underline{/-5.48^\circ}} \times 0.4597\underline{/84.93^\circ}$$

$$= 0.8902 + j0.0208 - 0.0031 + j0.4186$$

$$= 0.990\underline{/26.35^\circ} \text{ per unit}$$

At the sending end

$$\text{Line voltage} = 1.1102 \times 215 = 238.7 \text{ kV}$$

$$\text{Line current} = 0.990 \times 335.7 = 332.3 \text{ A}$$

Note that we multiply line-to-line voltage base by the per-unit magnitude of the voltage to find the line-to-line voltage magnitude. We could have multiplied the line-to-neutral voltage base by the per-unit voltage to find the line-to-neutral voltage magnitude. The factor $\sqrt{3}$ does not enter the calculations after we have expressed all quantities in per unit.

6.7 THE EQUIVALENT CIRCUIT OF A LONG LINE

The nominal-π circuit does not represent a transmission line exactly because it does not account for the parameters of the line being uniformly distributed. The discrepancy between the nominal π and the actual line becomes larger as the length of line increases. It is possible, however, to find the equivalent circuit of a long transmission line and to represent the line accurately, insofar as measurements at the ends of the line are concerned, by a network of lumped parameters. Let us assume that a π circuit similar to that of Fig. 6.7 is the equivalent circuit of a long line, but let us call the series arm of our equivalent-π circuit Z' and the shunt arms $Y'/2$ to distinguish them from the arms of the nominal-π circuit. Equation (6.5) gives the sending-end voltage of a symmetrical-π circuit in terms of its series and shunt arms and the voltage and current at the receiving end. By substituting Z' and $Y'/2$ for Z and $Y/2$ in Eq. (6.5), we obtain the sending-end voltage of our equivalent circuit in terms of its series and shunt arms and the voltage and current at the receiving end:

$$V_S = \left(\frac{Z'Y'}{2} + 1\right)V_R + Z'I_R \tag{6.44}$$

For our circuit to be equivalent to the long transmission line the coefficients of V_R and I_R in Eq. (6.44) must be identical, respectively, to the coefficients of V_R and I_R in Eq. (6.35). Equating the coefficients of I_R in the two equations yields

$$Z' = Z_c \sinh \gamma l \tag{6.45}$$

$$Z' = \sqrt{\frac{z}{y}} \sinh \gamma l = zl\frac{\sinh \gamma l}{\sqrt{zy}\, l}$$

$$Z' = Z\frac{\sinh \gamma l}{\gamma l} \tag{6.46}$$

where Z is equal to zl, the total series impedance of the line. The term $(\sinh \gamma l)/\gamma l$ is the factor by which the series impedance of the nominal π must be multiplied to convert the nominal π to the equivalent π. For small values of γl, *both* $\sinh \gamma l$ and γl are almost identical, and this fact shows that the nominal π represents the medium-length transmission line quite accurately, insofar as the series arm is concerned.

To investigate the shunt arms of the equivalent-π circuit, we equate the coefficients of V_R in Eqs. (6.35) and (6.44) and obtain

$$\frac{Z'Y'}{2} + 1 = \cosh \gamma l \tag{6.47}$$

Substituting $Z_c \sinh \gamma l$ for Z' gives

$$\frac{Y'Z_c \sinh \gamma l}{2} + 1 = \cosh \gamma l \tag{6.48}$$

$$\frac{Y'}{2} = \frac{1}{Z_c}\frac{\cosh \gamma l - 1}{\sinh \gamma l} \tag{6.49}$$

Another form of the expression for the shunt admittance of the equivalent circuit can be found by substituting in Eq. (6.49) the identity

$$\tanh \frac{\gamma l}{2} = \frac{\cosh \gamma l - 1}{\sinh \gamma l} \tag{6.50}$$

The identity can be verified by substituting the exponential forms of Eqs. (6.31)

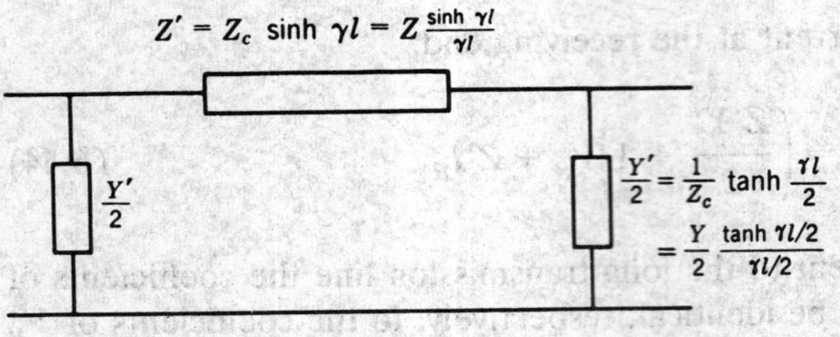

FIGURE 6.9
Equivalent-π circuit of a transmission line.

and (6.32) for the hyperbolic functions and by recalling that $\tanh\theta = \sinh\theta/\cosh\theta$. Now

$$\frac{Y'}{2} = \frac{1}{Z_c} \tanh \frac{\gamma l}{2} \tag{6.51}$$

$$\frac{Y'}{2} = \frac{Y}{2} \frac{\tanh(\gamma l/2)}{\gamma l/2} \tag{6.52}$$

where Y is equal to yl, the total shunt admittance of the line. Equation (6.52) shows the correction factor used to convert the admittance of the shunt arms of the nominal π to that of the equivalent π. Since $\tanh(\gamma l/2)$ and $\gamma l/2$ are very nearly equal for small values of γl, the nominal π represents the medium-length transmission line quite accurately, for we have seen previously that the correction factor for the series arm is negligible for medium-length lines. The equivalent-π circuit is shown in Fig. 6.9. An equivalent-T circuit can also be found for a transmission line.

Example 6.5. Find the equivalent-π circuit for the line described in Example 6.3 and compare it with the nominal-π circuit.

Solution. Since $\sinh \gamma l$ and $\cosh \gamma l$ are already known from Example 6.3, Eqs. (6.45) and (6.49) are now used.

$$Z' = 406.4\underline{/-5.48^\circ} \times 0.4597\underline{/84.93^\circ} = 186.82\underline{/79.45^\circ}\ \Omega \text{ in series arm}$$

$$\frac{Y'}{2} = \frac{0.8902 + j0.0208 - 1}{186.82\underline{/79.45^\circ}} = \frac{0.1118\underline{/169.27^\circ}}{186.82\underline{/79.45^\circ}}$$

$$= 0.000599\underline{/89.82^\circ}\ \text{S} \quad \text{in each shunt arm}$$

Using the values of z and y from Example 6.3, we find for the nominal-π circuit a series impedance of

$$Z = 230 \times 0.8431\underline{/79.04^\circ} = 193.9\underline{/79.04^\circ}$$

and equal shunt arms of

$$\frac{Y}{2} = \frac{5.10510^{-6}\angle 90^\circ}{2} \times 230 = 0.000587\angle 90^\circ \text{ S}$$

For this line the impedance of the series arm of the nominal π exceeds that of the equivalent π by 3.8%. The conductance of the shunt arms of the nominal π is 2.0% less than that of the equivalent π.

We conclude from the preceding example that the nominal π may represent long lines sufficiently well if a high degree of accuracy is not required.

6.8 POWER FLOW THROUGH A TRANSMISSION LINE

Although power flow at any point along a transmission line can always be found if the voltage, current, and power factor are known or can be calculated, very interesting equations for power can be derived in terms of *ABCD* constants. The equations apply to any network of two ports or two terminal pairs. Repeating Eq. (6.8) and solving for the receiving-end current I_R yields

$$V_S = AV_R + BI_R \tag{6.53}$$

$$I_R = \frac{V_S - AV_R}{B} \tag{6.54}$$

Letting

$$A = |A|\angle \alpha \qquad B = |B|\angle \beta$$

$$V_R = |V_R|\angle 0^\circ \qquad V_S = |V_S|\angle \delta$$

we obtain

$$I_R = \frac{|V_S|}{|B|}\angle \delta - \beta - \frac{|A|\,|V_R|}{|B|}\angle \alpha - \beta \tag{6.55}$$

Then, the complex power $V_R I_R^*$ at the receiving end is

$$P_R + jQ_R = \frac{|V_S|\,|V_R|}{|B|}\angle \beta - \delta - \frac{|A|\,|V_R|^2}{|B|}\angle \beta - \alpha \tag{6.56}$$

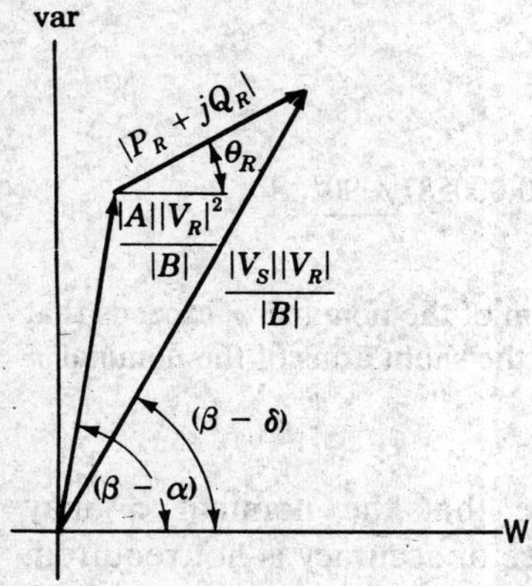

FIGURE 6.10
Phasors of Eq. (6.56) plotted in the complex plane, with magnitudes and angles as indicated.

and real and reactive power at the receiving end are

$$P_R = \frac{|V_S|\,|V_R|}{|B|}\cos(\beta - \delta) - \frac{|A|\,|V_R|^2}{|B|}\cos(\beta - \alpha) \tag{6.57}$$

$$Q_R = \frac{|V_S|\,|V_R|}{|B|}\sin(\beta - \delta) - \frac{|A|\,|V_R|^2}{|B|}\sin(\beta - \alpha) \tag{6.58}$$

Noting that the expression for complex power $P_R + jQ_R$ is shown by Eq. (6.56) to be the resultant of combining two phasors expressed in polar form, we can plot these phasors in the complex plane whose horizontal and vertical coordinates are in power units (watts and vars). Figure 6.10 shows the two complex quantities and their difference as expressed by Eq. (6.56). Figure 6.11 shows the same phasors with the origin of the coordinate axes shifted. This figure is a power diagram with the resultant whose magnitude is $|P_R + jQ_R|$, or $|V_R|\,|I_R|$, at an angle θ_R with the horizontal axis. As expected, the real and imaginary components of $|P_R + jQ_R|$ are

$$P_R = |V_R|\,|I_R|\cos\theta_R \tag{6.59}$$

$$Q_R = |V_R|\,|I_R|\sin\theta_R \tag{6.60}$$

where θ_R is the phase angle by which V_R leads I_R, as discussed in Chap. 1. The sign of Q is consistent with the convention which assigns positive values to Q when current is lagging the voltage.

Now let us determine some points on the power diagram of Fig. 6.11 for various loads with fixed values of $|V_S|$ and $|V_R|$. First, we notice that the position of point n is not dependent on the current I_R and will not change so long as

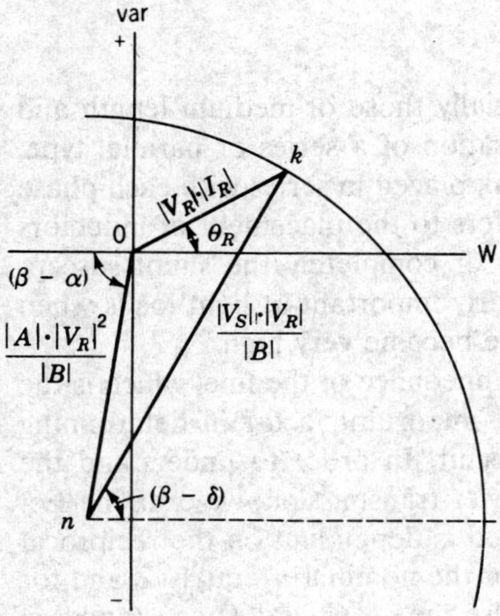

FIGURE 6.11
Power diagram obtained by shifting the origin of the coordinate axes of Fig. 6.10.

$|V_R|$ is constant. We note further that the distance from point n to point k is constant for fixed values of $|V_S|$ and $|V_R|$. Therefore, as the distance 0 to k changes with changing load, the point k, since it must remain at a constant distance from the fixed point n, is constrained to move in a circle whose center is at n. Any change in P_R will require a change in Q_R to keep k on the circle. If a different value of $|V_S|$ is held constant for the same value of $|V_R|$, the location of point n is unchanged but a new circle of radius nk is found.

Examination of Fig. 6.11 shows that there is a limit to the power that can be transmitted to the receiving end of the line for specified magnitudes of sending- and receiving-end voltages. An increase in power delivered means that the point k will move along the circle until the angle $\beta - \delta$ is zero; that is, more power will be delivered until $\delta = \beta$. Further increases in δ result in less power received. The maximum power is

$$P_{R,\max} = \frac{|V_S|\,|V_R|}{|B|} - \frac{|A|\,|V_R|^2}{|B|}\cos(\beta - \alpha) \tag{6.61}$$

The load must draw a large leading current to achieve the condition of maximum power received. Usually, operation is limited by keeping δ less than about 35° and $|V_S|/|V_R|$ equal to or greater than 0.95. For short lines thermal ratings limit the loading.

In Eqs. (6.53) through (6.61) $|V_S|$ and $|V_R|$ are line-to-neutral voltages and coordinates in Fig. 6.11 are watts and vars per phase. However, if $|V_S|$ and $|V_R|$ are line-to-line voltages, each distance in Fig. 6.11 is increased by a factor of 3 and the coordinates on the diagram are total three-phase watts and vars. If the voltages are kilovolts, the coordinates are megawatts and megavars.

6.9 REACTIVE COMPENSATION OF TRANSMISSION LINES

The performance of transmission lines, especially those of medium length and longer, can be improved by reactive compensation of a series or parallel type. *Series compensation* consists of a capacitor bank placed in series with each phase conductor of the line. *Shunt compensation* refers to the placement of inductors from each line to neutral to reduce partially or completely the shunt susceptance of a high-voltage line, which is particularly important at light loads when the voltage at the receiving end may otherwise become very high.

Series compensation reduces the series impedance of the line, which is the principal cause of voltage drop and the most important factor in determining the maximum power which the line can transmit. In order to understand the effect of series impedance Z on maximum power transmission, we examine Eq. (6.61) and see that maximum power transmitted is dependent on the reciprocal of the generalized circuit constant B, which for the nominal-π equals Z and for the equivalent-π equals $Z\,(\sinh \gamma l)/\gamma l$. Because the A, C, and D constants are functions of Z, they will also change in value, but these changes will be small in comparison to the change in B.

The desired reactance of the capacitor bank can be determined by compensating for a specific amount of the total inductive reactance of the line. This leads to the term "compensation factor," which is defined by X_C/X_L, where X_C is the capacitive reactance of the series capacitor bank per phase and X_L is the total inductive reactance of the line per phase.

When the nominal-π circuit is used to represent the line and capacitor bank, the physical location of the capacitor bank along the line is not taken into account. If only the sending- and receiving-end conditions of the line are of interest, this will not create any significant error. However, when the operating conditions along the line are of interest, the physical location of the capacitor bank must be taken into account. This can be accomplished most easily by determining *ABCD* constants of the portions of line on each side of the capacitor bank and by representing the capacitor bank by its *ABCD* constants. The equivalent constants of the combination (actually referred to as a *cascaded* connection) of line-capacitor-line can then be determined by applying the equations found in Table A.6 in the Appendix.

In the southwestern part of the United States series compensation is especially important because large generating plants are located hundreds of miles from load centers and large amounts of power must be transmitted over long distances. The lower voltage drop in the line with series compensation is an additional advantage. Series capacitors are also useful in balancing the voltage drop of two parallel lines.

Example 6.6. In order to show the relative changes in the B constant with respect to the change of the A, C, and D constants of a line as series compensation is applied, find the constants for the line of Example 6.3 when uncompensated and for a series compensation of 70%.

Solution. The equivalent-π circuit and quantities found in Examples 6.3 and 6.5 can be used with Eqs. (6.37) to find, for the uncompensated line

$$A = D = \cosh \gamma l = 0.8904\angle 1.34^\circ$$

$$B = Z' = 186.78\angle 79.46^\circ \ \Omega$$

$$C = \frac{\sinh \gamma l}{Z_c} = \frac{0.4596\angle 84.94^\circ}{406.4\angle -5.48^\circ}$$

$$= 0.001131\angle 90.42^\circ \ \text{S}$$

The series compensation alters only the series arm of the equivalent-π circuit. The new series arm impedance is also the generalized constant B. So,

$$B = 186.78\angle 79.46^\circ - j0.7 \times 230(0.415 + 0.4127)$$

$$= 34.17 + j50.38 = 60.88\angle 55.85^\circ \ \Omega$$

and by Eqs. (6.10)

$$A = 60.88\angle 55.85^\circ \times 0.000599\angle 89.81^\circ + 1 = 0.970\angle 1.24^\circ$$

$$C = 2 \times 0.000599\angle 89.81^\circ + 60.88\angle 55.85^\circ \left(0.000599\angle 89.81^\circ\right)^2$$

$$= 0.001180\angle 90.41^\circ \ \text{S}$$

The example shows that compensation has reduced the constant B to about one-third of its value for the uncompensated line without affecting the A and C constants appreciably. Thus, maximum power which can be transmitted is increased by about 300%.

When a transmission line, with or without series compensation, has the desired load transmission capability, attention is turned to operation under light loads or at no load. Charging current is an important factor to be considered and should not be allowed to exceed the rated full-load current of the line.

Equation (5.25) shows us that the charging current is usually defined as $B_C|V|$ if B_C is the total capacitive susceptance of the line and $|V|$ is the rated voltage to neutral. As noted following Eq. (5.25), this calculation is not an exact determination of charging current because of the variation of $|V|$ along the line. If we connect inductors from line to neutral at various points along the line so that the total inductive susceptance is B_L, the charging current becomes

$$I_{\text{chg}} = (B_C - B_L)|V| = B_C|V|\left(1 - \frac{B_L}{B_C}\right) \tag{6.62}$$

We recognize that the charging current is reduced by the term in parentheses. The shunt compensation factor is B_L/B_C.

The other benefit of shunt compensation is the reduction of the receiving-end voltage of the line which on long high-voltage lines tends to become too high at no load. In the discussion preceding Eq. (6.11) we noted that $|V_S|/|A|$ equals $|V_{R,\,NL}|$. We also have seen that A equals 1.0 when shunt capacitance is neglected. In the medium-length and longer lines, however, the presence of capacitance reduces A. Thus, the reduction of the shunt susceptance to the value of $(B_C - B_L)$ can limit the rise of the no-load voltage at the receiving end of the line if shunt inductors are introduced as load is removed.

By applying both series and shunt compensation to long transmission lines, we can transmit large amounts of power efficiently and within the desired voltage constraints. Ideally, the series and shunt elements should be placed at intervals along the line. Series capacitors can be bypassed and shunt inductors can be switched off when desirable. As with series compensation, *ABCD* constants provide a straightforward method of analysis of shunt compensation.

Example 6.7. Find the voltage regulation of the line of Example 6.3 when a shunt inductor is connected at the receiving end of the line during no-load conditions if the reactor compensates for 70% of the total shunt admittance of the line.

Solution. From Example 6.3 the shunt admittance of the line is

$$y = j5.105 \times 10^{-6} \text{ S/mi}$$

and for the entire line

$$B_C = 5.105 \times 10^{-6} \times 230 = 0.001174 \text{ S}$$

For 70% compensation

$$B_L = 0.7 \times 0.001174 = 0.000822$$

We know the *ABCD* constants of the line from Example 6.6. Table A.6 of the Appendix tells us that the inductor alone is represented by the generalized constants

$$A = D = 1 \qquad B = 0 \qquad C = -jB_L = -j0.000822 \text{ S}$$

The equation in Table A.6 for combining two networks in series tells us that for the line and inductor

$$A_{eq} = 0.8904\underline{/1.34^\circ} + 186.78\underline{/79.46^\circ}\left(0.000822\underline{/-90^\circ}\right)$$

$$= 1.0411\underline{/-0.4}$$

The voltage regulation with the shunt reactor connected at no load becomes

$$\frac{137.86/1.0411 - 124.13}{124.13} = 6.67\%$$

which is a considerable reduction from the value of 24.7% for the regulation of the uncompensated line.

6.10 TRANSMISSION-LINE TRANSIENTS

The transient overvoltages which occur on a power system are either of external origin (for example, a lightning discharge) or generated internally by switching operations. In general, the transients on transmission systems are caused by any sudden change in the operating condition or configuration of the systems. Lightning is always a potential hazard to power system equipment, but switching operations can also cause equipment damage. At voltages up to about 230 kV, the insulation level of the lines and equipment is dictated by the need to protect against lightning. On systems where voltages are above 230 kV but less than 700 kV switching operations as well as lightning are potentially damaging to insulation. At voltages above 700 kV switching surges are the main determinant of the level of insulation.

Of course, underground cables are immune to direct lightning strokes and can be protected against transients originating on overhead lines. However, for economic and technical reasons overhead lines at transmission voltage levels prevail except under unusual circumstances and for short distances such as under a river.

Overhead lines can be protected from direct strokes of lightning in most cases by one or more wires at ground potential strung above the power-line conductors as mentioned in the description of Fig. 6.1. These protecting wires, called *ground wires*, or *shield wires*, are connected to ground through the transmission towers supporting the line. The zone of protection is usually considered to be 3 on each side of vertical beneath a ground wire; that is, the power lines must come within this 6 sector. The ground wires, rather than the power line, receive the lightning strokes in most cases.

Lightning strokes hitting either ground wires or power conductors cause an injection of current, which divides with half the current flowing in one direction and half in the other. The crest value of current along the struck conductor varies widely because of the wide variation in the intensity of the strokes. Values of 10,000 A and upward are typical. In the case where a power line receives a direct stroke the damage to equipment at line terminals is caused by the voltages between the line and the ground resulting from the injected charges which travel along the line as current. These voltages are typically above a million volts. Strokes to the ground wires can also cause high-voltage surges on the power lines by electromagnetic induction.

6.11 TRANSIENT ANALYSIS: TRAVELING WAVES

The study of transmission-line surges, regardless of their origin, is very complex and we can consider here only the case of a lossless line.[2]

A lossless line is a good representation for lines of high frequency where ωL and ωC become very large compared to R and G. For lightning surges on a power transmission line the study of a lossless line is a simplification that enables us to understand some of the phenomena without becoming too involved in complicated theory.

Our approach to the problem is similar to that used earlier for deriving the steady-state voltage and current relations for the long line with distributed constants. We now measure the distance x along the line from the *sending* end (rather than from the receiving end) to the differential element of length Δx shown in Fig. 6.12. The voltage v and the current i are functions of both x and t so that we need to use partial derivatives. The series voltage drop along the elemental length of line is

$$i(R\,\Delta x) + (L\,\Delta x)\frac{\partial i}{\partial t}$$

and we can write

$$\frac{\partial v}{\partial x}\Delta x = -\left(Ri + L\frac{\partial i}{\partial t}\right)\Delta x \tag{6.63}$$

The negative sign is necessary because $v + (\partial v/\partial x)\,\Delta x$ must be less than v for positive values of i and $\partial i/\partial t$. Similarly,

$$\frac{\partial i}{\partial x}\Delta x = -\left(Gv + C\frac{\partial v}{\partial t}\right)\Delta x \tag{6.64}$$

[2] For further study, see A. Greenwood, *Electrical Transients in Power Systems*, 2d ed., Wiley-Interscience, New York, 1991.

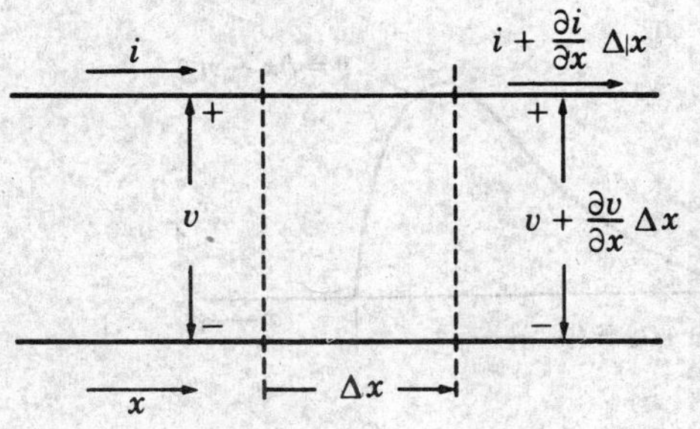

FIGURE 6.12
Schematic diagram of an elemental section of a transmission line showing one phase and neutral return. Voltage v and current i are functions of both x and t. The distance x is measured from the sending end of the line.

We can divide through both Eqs. (6.63) and (6.64) by Δx, and since we are considering only a lossless line, R and G will equal zero to give

$$\frac{\partial v}{\partial x} = -L \frac{\partial i}{\partial t} \tag{6.65}$$

and

$$\frac{\partial i}{\partial x} = -C \frac{\partial v}{\partial t} \tag{6.66}$$

Now we can eliminate i by taking the partial derivative of both terms in Eq. (6.65) with respect to x and the partial derivative of both terms in Eq. (6.66) with respect to t. This procedure yields $\partial^2 i/\partial x\,\partial t$ in both resulting equations, and eliminating this second partial derivative of i between the two equations yields

$$\frac{1}{LC} \frac{\partial^2 v}{\partial x^2} = \frac{\partial^2 v}{\partial t^2} \tag{6.67}$$

Equation (6.67) is the so-called *traveling-wave equation* of a lossless transmission line. A solution of the equation is a function of $(x - \nu t)$, and the voltage is expressed by

$$v = f(x - \nu t) \tag{6.68}$$

The function is undefined but must be single valued. The constant ν must have the dimensions of meters per second if x is in meters and t is in seconds. We can verify this solution by substituting this expression for v into Eq. (6.67) to determine ν. First, we make the change in variable

$$u = x - \nu t \tag{6.69}$$

and write

$$v(x, t) = f(u) \tag{6.70}$$

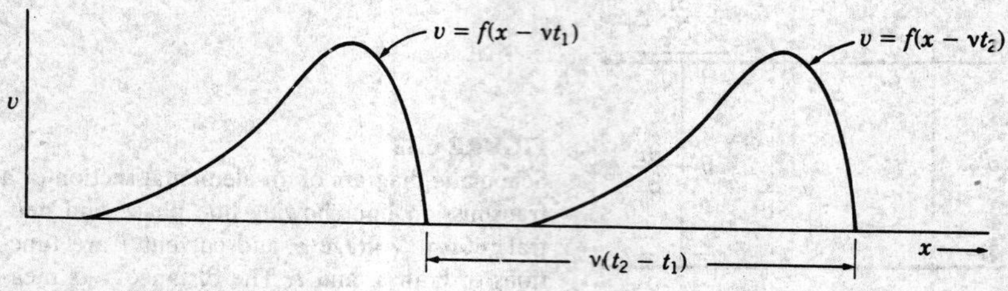

FIGURE 6.13
A voltage wave which is a function of $(x - \nu t)$ is shown for values of t equal to t_1 and t_2.

Then,

$$\frac{\partial v}{\partial t} = \frac{\partial f(u)}{\partial u}\frac{\partial u}{\partial t} = -\nu\frac{\partial f(u)}{\partial u} \tag{6.71}$$

and

$$\frac{\partial^2 v}{\partial t^2} = \nu^2\frac{\partial^2 f(u)}{\partial u^2} \tag{6.72}$$

Similarly, we obtain

$$\frac{\partial^2 v}{\partial x^2} = \frac{\partial^2 f(u)}{\partial u^2} \tag{6.73}$$

Substituting these second partial derivatives of v in Eq. (6.67) yields

$$\frac{1}{LC}\frac{\partial^2 f(u)}{\partial u^2} = \nu^2\frac{\partial^2 f(u)}{\partial u^2} \tag{6.74}$$

and we see that Eq. (6.68) is a solution of Eq. (6.67) if

$$\nu = \frac{1}{\sqrt{LC}} \tag{6.75}$$

The voltage as expressed by Eq. (6.68) is a wave traveling in the positive x direction. Figure 6.13 shows a function of $(x - \nu t)$, which is similar to the shape of a wave of voltage traveling along a line which has been struck by lightning. The function is shown for two values of time t_1 and t_2, where $t_2 > t_1$. An observer traveling with the wave and staying at the same point on the wave sees no change in voltage at that point. To the observer

$$x - \nu t = \text{a constant}$$

from which it follows that

$$\frac{dx}{dt} = \nu = \frac{1}{\sqrt{LC}} \text{ m/s} \tag{6.76}$$

for L and C in henrys per meter and farads per meter, respectively. Thus, the voltage wave travels in the positive x direction with the velocity ν.

A function of $(x + \nu t)$ can also be shown to be a solution of Eq. (6.67) and, by similar reasoning, can be properly interpreted as a wave traveling in the negative x direction. The general solution of Eq. (6.67) is

$$v = f_1(x - \nu t) + f_2(x + \nu t) \tag{6.77}$$

which is a solution for simultaneous occurrence of forward and backward components on the line. Initial conditions and boundary (terminal) conditions determine the particular values for each component.

If we express a forward traveling wave, also called an *incident wave*, as

$$v^+ = f_1(x - \nu t) \tag{6.78}$$

a wave of current will result from the moving charges and will be expressed by

$$i^+ = \frac{1}{\sqrt{L/C}} f_1(x - \nu t) \tag{6.79}$$

which can be verified by substitution of these values of voltage and current in Eq. (6.65) and by the fact that ν is equal to $1/\sqrt{LC}$.

Similarly, for a backward moving wave of voltage where

$$v^- = f_2(x + \nu t) \tag{6.80}$$

the corresponding current is

$$i^- = -\frac{1}{\sqrt{L/C}} f_2(x + \nu t) \tag{6.81}$$

From Eqs. (6.78) and (6.79) we note that

$$\frac{v^+}{i^+} = \sqrt{\frac{L}{C}} \tag{6.82}$$

and from Eqs. (6.80) and (6.81) that

$$\frac{v^-}{i^-} = -\sqrt{\frac{L}{C}} \tag{6.83}$$

If we had decided to assume the positive direction of current for i^- to be in the direction of travel of the backward traveling wave the minus signs would change to plus signs in Eqs. (6.81) and (6.83). We choose, however, to keep the positive x direction as the direction for positive current for both forward and backward traveling waves.

The ratio of v^+ to i^+ is called the characteristic impedance Z_c of the line. We have encountered characteristic impedance previously in the steady-state solution for the long line where Z_c was defined as $\sqrt{z/y}$, which equals $\sqrt{L/C}$ when R and G are zero.

6.12 TRANSIENT ANALYSIS: REFLECTIONS

We now consider what happens when a voltage is first applied to the sending end of a transmission line which is terminated in an impedance Z_R. For our very simple treatment we consider Z_R to be a pure resistance. If the termination is other than a pure resistance, we would resort to Laplace transforms. The transforms of voltage, current, and impedance would be functions of the Laplace transform variable s.

When a switch is closed applying a voltage to a line, a wave of voltage v^+ accompanied by a wave of current i^+ starts to travel along the line. The ratio of the voltage v_R at the end of the line at any time to the current i_R at the end of the line must equal the terminating resistance Z_R. Therefore, the arrival of v^+ and i^+ at the receiving end where their values are v_R^+ and i_R^+ must result in backward traveling or reflected waves v^- and i^- having values v_R^- and i_R^- at the receiving end such that

$$\frac{v_R}{i_R} = \frac{v_R^+ + v_R^-}{i_R^+ + i_R^-} = Z_R \tag{6.84}$$

where v_R^- and i_R^- are the reflected waves v^- and i^- measured at the receiving end.

If we let $Z_c = \sqrt{L/C}$, we find from Eqs. (6.82) and (6.83) that

$$i_R^+ = \frac{v_R^+}{Z_c} \tag{6.85}$$

$$i_R^- = -\frac{v_R^-}{Z_c} \tag{6.86}$$

Then, substituting these values of i_R^+ and i_R^- in Eq. (6.84) yields

$$v_R^- = \frac{Z_R - Z_c}{Z_R + Z_c} v_R^+ \tag{6.87}$$

The voltage v_R^- at the receiving end is evidently the same function of t as v_R^+ (but with diminished magnitude unless Z_R is zero or infinity). The reflection coefficient ρ_R for voltage at the receiving end of the line is defined as v_R^-/v_R^+, so, *for voltage*

$$\rho_R = \frac{Z_R - Z_c}{Z_R + Z_c} \tag{6.88}$$

We note from Eqs. (6.85) and (6.86) that

$$\frac{i_R^+}{i_R^-} = -\frac{v_R^+}{v_R^-} \tag{6.89}$$

and therefore the reflection coefficient for current is always the negative of the reflection coefficient for voltage.

If the line is terminated in its characteristic impedance Z_c, we see that the reflection coefficient for both voltage and current will be zero. There will be no reflected waves, and the line will behave as though it is infinitely long. Only when a reflected wave returns to the sending end does the source sense that the line is neither infinitely long nor terminated in Z_c.

Termination in a short circuit results in a ρ_R for voltage of -1. If the termination is an open circuit, Z_R is infinite and ρ_R is found by dividing the numerator and denominator in Eq. (6.88) by Z_R and by allowing Z_R to approach infinity to yield $\rho_R = 1$ in the limit for voltage.

We should note at this point that waves traveling back toward the sending end will cause new reflections as determined by the reflection coefficient at the sending end ρ_s. For impedance at the sending end equal to Z_s Eq. (6.88) becomes

$$\rho_s = \frac{Z_s - Z_c}{Z_s + Z_c} \tag{6.90}$$

With sending-end impedance of Z_s, the value of the initial voltage impressed across the line will be the source voltage multiplied by $Z_c/(Z_s + Z_c)$. Equation (6.82) shows that the incident wave of voltage experiences a line impedance of Z_c, and at the instant when the source is connected to the line Z_c and Z_s in series act as a voltage divider.

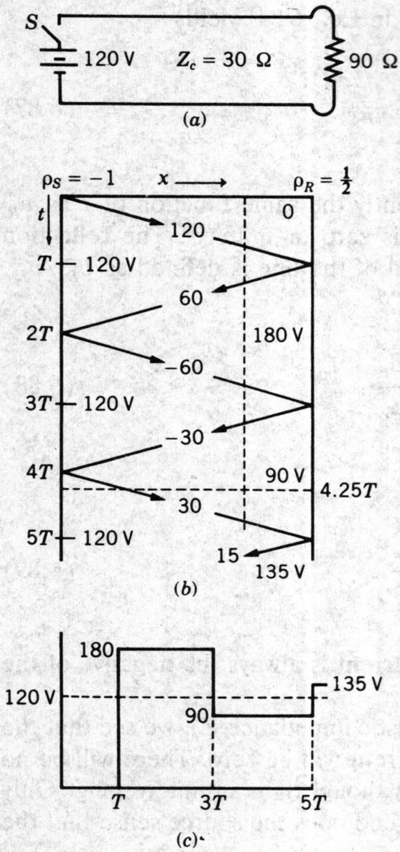

FIGURE 6.14
Circuit diagram, lattice diagram, and plot of voltage versus time for Example 6.8, where the receiving-end resistance is 90 Ω.

Example 6.8. A dc source of 120 V with negligible resistance is connected through a switch S to a lossless transmission line having $Z_c = 30\ \Omega$. The line is terminated in a resistance of 90 Ω. If the switch closes at $t = 0$, plot v_R versus time until $t = 5T$, where T is the time for a voltage wave to travel the length of the line. The circuit is shown in Fig. 6.14(a).

Solution. When switch S is closed, the incident wave of voltage starts to travel along the line and is expressed as

$$v = 120U(\nu t - x)$$

where $U(\nu t - x)$ is the unit step function, which equals zero when $(\nu t - x)$ is negative and equals unity when $(\nu t - x)$ is positive. There can be no reflected wave until the incident wave reaches the end of the line. With impedance to the incident wave of $Z_c = 30\ \Omega$, resistance of the source zero and $v^+ = 120$ V, the reflection coefficient becomes

$$\rho_R = \frac{90 - 30}{90 + 30} = \frac{1}{2}$$

When v^+ reaches the end of the line, a reflected wave originates of value

$$v^- = \left(\frac{1}{2}\right)120 = 60 \text{ V}$$

and so
$$v_R = 120 + 60 = 180 \text{ V}$$

When $t = 2T$, the reflected wave arrives at the sending end where the sending-end reflection coefficient ρ_s is calculated by Eq. (6.90). The line termination for the reflected wave is Z_s, the impedance in series with the source, or zero in this case. So,

$$\rho_R = \frac{0 - 30}{0 + 30} = -1$$

and a reflected wave of -60 V starts toward the receiving end to keep the sending-end voltage equal to 120 V. This new wave reaches the receiving end at $t = 3T$ and reflects toward the sending end a wave of

$$\frac{1}{2}(-60) = -30 \text{ V}$$

and the receiving-end voltage becomes

$$v_R = 180 - 60 - 30 = 90 \text{ V}$$

An excellent method of keeping track of the various reflections as they occur is the *lattice diagram* shown in Fig. 6.14(b). Here time is measured along the vertical axis in intervals of T. On the slant lines there are recorded the values of the incident and reflected waves. In the space between the slant lines there are shown the sum of all the waves above and the current or voltage for a point in that area of the chart. For instance, at x equal to three-fourths of the line length and $t = 4.25T$ the intersection of the dashed lines through these points is within the area which indicates the voltage is 90 V.

Figure 6.14(c) shows the receiving-end voltage plotted against time. The voltage is approaching its steady-state value of 120 V.

Lattice diagrams for current may also be drawn. We must remember, however, that the reflection coefficient for current is always the negative of the reflection coefficient for voltage.

If the resistance at the end of the line of Example 6.8 is reduced to 10 Ω as shown in the circuit of Fig. 6.15(a), the lattice diagram and plot of voltage are as shown in Figs. 6.15(b) and 6.15(c). The resistance of 10 Ω gives a negative value for the reflection coefficient for voltage, which always occurs for resistance Z_R less than Z_c. As we see by comparing Figs. 6.14 and 6.15, the negative ρ_R causes the receiving-end voltage to build up gradually to 120 V, while a positive ρ_R causes an initial jump in voltage to a value greater than that of the voltage originally applied at the sending end.

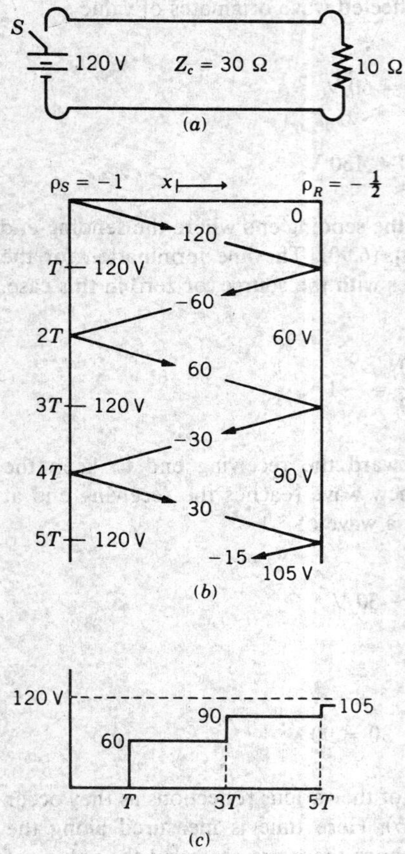

FIGURE 6.15
Circuit diagram, lattice diagram, and plot of voltage versus time when the receiving-end resistance for Example 6.8 is changed to 10 Ω.

Reflections do not necessarily arise only at the end of a line. If one line is joined to a second line of different characteristic impedance, as in the case of an overhead line connected to an underground cable, a wave incident to the junction will behave as though the first line is terminated in the Z_c of the second line. However, the part of the incident wave which is not reflected will travel (as a refracted wave) along the second line at whose termination a reflected wave will occur. Bifurcations of a line will also cause reflected and refracted waves.

It should now be obvious that a thorough study of transmission-line transients in general is a complicated problem. We realize, however, that a voltage surge such as that shown in Fig. 6.13 encountering an impedance at the end of a lossless line (for instance, at a transformer bus) will cause a voltage wave of the same shape to travel back toward the source of the surge. The reflected wave will be reduced in magnitude if the terminal impedance is other than a short or open circuit, but if Z_R is greater than Z_c, our study has shown that the peak terminal voltage will be higher than, often close to double the peak voltage of the surge.

Terminal equipment is protected by *surge arresters*, which are also called *lightning arresters* and *surge diverters*. An ideal arrester connected from the line to a grounded neutral would (1) become conducting at a design voltage above the arrester rating, (2) limit the voltage across its terminals to this design value, and (3) become nonconducting again when the line-to-neutral voltage drops below the design value.

Originally, an arrester was simply an air gap. In this application when the surge voltage reaches a value for which the gap is designed, an arc occurs that causes an ionized path to ground, essentially a short circuit. However, when the surge ends, the 60-Hz current from the generators still flows through the arc to ground. The arc has to be extinguished by the opening of circuit breakers.

Arresters capable of extinguishing a 60-Hz current after conducting surge current to ground were developed later. These arresters are made using nonlinear resistors in series with air gaps to which an arc-quenching capability has been added. The nonlinear resistance decreases rapidly as the voltage across it rises. Typical resistors made of silicon carbide conduct current proportional to approximately the fourth power of the voltage across the resistor. When the gaps arc over as a result of a voltage surge, a low-resistance current path to ground is provided through the nonlinear resistors. After the surge ends and the voltage across the arrester returns to the normal line-to-neutral level the resistance is sufficient to limit the arc current to a value which can be quenched by the series gaps. Quenching is usually accomplished by cooling and deionizing the arc by elongating it magnetically between insulating plates.

The most recent development in surge arresters is the use of zinc oxide in place of silicon carbide. The voltage across the zinc oxide resistor is extremely constant over a very high range of current, which means that its resistance at normal line voltage is so high that a series air gap is not necessary to limit the drain of a 60-Hz current at normal voltage.[3]

6.13 DIRECT-CURRENT TRANSMISSION

The transmission of energy by direct current becomes economical when compared to ac transmission only when the extra cost of the terminal equipment required for dc lines is offset by the lower cost of building the lines. Converters at the two ends of the dc lines operate as both rectifiers to change the generated alternating to direct current and inverters for converting direct to alternating current so that power can flow in either direction.

The year 1954 is generally recognized as the starting date for modern high-voltage dc transmission when a dc line began service at 100 kV from Vastervik on the mainland of Sweden to Visby on the island of Gotland, a

[3]See E. C. Sakshaug, J. S. Kresge, and S. A. Miske, Jr., "A New Concept in Station Arrester Design," *IEEE Transactions on Power Apparatus and Systems*, vol. PAS-96, no. 2, March/April 1977, pp. 647–656.

distance of 100 km (62.5 mi) across the Baltic Sea. Static conversion equipment was in operation much earlier to transfer energy between systems of 25 and 60 Hz, essentially a dc transmission line of zero length. In the United States a dc line operating at 800 kV transfers power generated in the Pacific northwest to the southern part of California. As the cost of conversion equipment decreases with respect to the cost of line construction, the economical minimum length of dc lines also decreases and at this time is about 600 km (375 mi).

Operation of a dc line began in 1977 to transmit power from a mine-mouth generating plant burning lignite at Center, North Dakota, to near Duluth, Minnesota, a distance of 740 km (460 mi). Preliminary studies showed that the dc line including terminal facilities would cost about 30% less than the comparable ac line and auxiliary equipment. This line operates at ± 250 kV (500 kV line to line) and transmits 500 MW.

Direct-current lines usually have one conductor which is at a positive potential with respect to ground and a second conductor operating at an equal negative potential. Such a line is said to be *bipolar*. The line could be operated with one energized conductor with the return path through the earth, which has a much lower resistance to direct current than to alternating current. In this case, or with a grounded return conductor, the line is said to be *monopolar*.

In addition to the lower cost of dc transmission over long distances, there are other advantages. Voltage regulation is less of a problem since at zero frequency the series reactance ωL is no longer a factor, whereas it is the chief contributor to voltage drop in an ac line. Another advantage of direct current is the possibility of monopolar operation in an emergency when one side of a bipolar line becomes grounded.

Due to the fact that underground ac transmission is limited to about 5 km because of excessive charging current at longer distances, direct current was chosen to transfer power under the English Channel between Great Britain and France. The use of direct current for this installation also avoided the difficulty of synchronizing the ac systems of the two countries.

No network of dc lines is possible at this time because no circuit breaker is available for direct current that is comparable to the highly developed ac breakers. The ac breaker can extinguish the arc which is formed when the breaker opens because zero current occurs twice in each cycle. The direction and amount of power in the dc line is controlled by the converters in which grid-controlled mercury-arc devices are being displaced by the semiconductor rectifier (SCR). A rectifier unit will contain perhaps 200 SCRs.

Still another advantage of direct current is the smaller amount of right of way required. The distance between the two conductors of the North Dakota–Duluth 500-kV line is 25 ft. The 500-kV ac line shown in Fig. 6.1 has 60.5 ft between the outside conductors. Another consideration is the peak voltage of the ac line, which is $\sqrt{2} \times 500 = 707$ kV. So, the line requires more insulation between the tower and conductors as well as greater clearance above the earth.

We conclude that dc transmission has many advantages over alternating current, but dc transmission remains very limited in usage except for long lines since there is no dc device which can provide the excellent switching operations and protection of the ac circuit breaker. There is also no simple device to change the voltage level, which the transformer accomplishes for ac systems.

6.14 SUMMARY

The long-line equations given by Eqs. (6.35) and (6.36) are, of course, valid for a line of any length. The approximations for the short- and medium-length lines make analysis easier in the absence of a computer.

Circle diagrams were introduced because of their instructional value in showing the maximum power which can be transmitted by a line and also in showing the effect of the power factor of the load or the addition of capacitors.

ABCD constants provide a straightforward means of writing equations in a more concise form and are very convenient in problems involving network reduction. Their usefulness is apparent in the discussion of series and shunt reactive compensation.

The simple discussion of transients, although confined to lossless lines and dc sources, should give some idea of the complexity of the study of transients which arise from lightning and switching in power systems.

PROBLEMS

6.1. An 18-km, 60-Hz, single-circuit, three-phase line is composed of *Partridge* conductors equilaterally spaced with 1.6 m between centers. The line delivers 2500 kW at 11 kV to a balanced load. Assume a wire temperature of 5C.

(*a*) Determine the per-phase series impedance of the line.

(*b*) What must be the sending-end voltage when the power factor is

(i) 80% lagging,

(ii) unity,

(iii) 90% leading?

(*c*) Determine the percent regulation of the line at the above power factors.

(*d*) Draw phasor diagrams depicting the operation of the line in each case.

6.2. A 100-mi, single-circuit, three-phase transmission line delivers 55 MVA at 0.8 power-factor lagging to the load at 132 kV (line to line). The line is composed of *Drake* conductors with flat horizontal spacing of 11.9 ft between adjacent conductors. Assume a wire temperature of 5C. Determine

(*a*) The series impedance and the shunt admittance of the line.

(*b*) The *ABCD* constants of the line.

(*c*) The sending-end voltage, current, real and reactive powers, and the power factor.

(*d*) The percent regulation of the line.

6.3. Find the *ABCD* constants of a π circuit having a 600-Ω resistor for the shunt branch at the sending end, a 1-kΩ resistor for the shunt branch at the receiving end, and an 80-Ω resistor for the series branch.

6.4. The *ABCD* constants of a three-phase transmission line are

$$A = D = 0.936 + j0.016 = 0.936\angle 0.98^\circ$$

$$B = 33.5 + j138 = 142\angle 76.4^\circ\ \Omega$$

$$C = (-5.18 + j914) \times 10^{-6}\ \text{S}$$

The load at the receiving end is 50 MW at 220 kV with a power factor of 0.9 lagging. Find the magnitude of the sending-end voltage and the voltage regulation. Assume that the magnitude of the sending-end voltage remains constant.

6.5. A 70-mi, single-circuit, three-phase line composed of *Ostrich* conductors is arranged in flat horizontal spacing with 15 ft between adjacent conductors. The line delivers a load of 60 MW at 230 kV with 0.8 power-factor lagging.

(*a*) Using a base of 230 kV, 100 MVA, determine the series impedance and the shunt admittance of the line in per unit. Assume a wire temperature of 50° C. Note that the base admittance must be the reciprocal of base impedance.

(*b*) Find the voltage, current, real and reactive power, and the power factor at the sending end in both per unit and absolute units.

(*c*) What is the percent regulation of the line?

6.6. A single-circuit, three-phase transmission line is composed of *Parakeet* conductors with flat horizontal spacing of 19.85 ft between adjacent conductors. Determine the characteristic impedance and the propagation constant of the line at 60 Hz and 50° C temperature.

6.7. Using Eqs. (6.23) and (6.24), show that if the receiving end of a line is terminated by its characteristic impedance Z_c, then the impedance seen at the sending end of the line is also Z_c regardless of line length.

6.8. A 200-mi transmission line has the following parameters at 60 Hz:

$$\text{Resistance } r = 0.21\ \Omega/\text{mi per phase}$$

$$\text{Series reactance } x = 0.78\ \Omega/\text{mi per phase}$$

$$\text{Shunt susceptance } b = 5.42 \times 10^{-6}\ \text{S/mi per phase}$$

(*a*) Determine the attenuation constant α, wavelength λ, and the velocity of propagation of the line at 60 Hz.

(*b*) If the line is open-circuited at the receiving end and the receiving-end voltage is maintained at 100 kV line to line, use Eqs. (6.26) and (6.27) to determine the incident and reflected components of the sending-end voltage and current.

(*c*) Hence, determine the sending-end voltage and current of the line.

6.9. Evaluate $\cosh\theta$ and $\sinh\theta$ for $\theta = 0.5\angle 82^\circ$.

6.10. Using Eqs. (6.1), (6.2), (6.10), and (6.37), show that the generalized circuit constants of all three transmission-line models satisfy the condition that

$$AD - BC = 1$$

6.11. The sending-end voltage, current, and power factor of the line described in Example 6.3 are found to be 260 kV (line to line), 300 A, and 0.9 lagging, respectively. Find the corresponding receiving-end voltage, current, and power factor.

6.12. A 60-Hz three-phase transmission line is 175 mi long. It has a total series impedance of $35 + j140\ \Omega$ and a shunt admittance of $930 \times 10^{-6} \angle 90^\circ$ S. It delivers 40 MW at 220 kV, with 90% power-factor lagging. Find the voltage at the sending end by (*a*) the short-line approximation, (*b*) the nominal-π approximation, and (*c*) the long-line equation.

6.13. Determine the voltage regulation for the line described in Prob. 6.12. Assume that the sending-end voltage remains constant.

6.14. A three-phase, 60-Hz transmission line is 250 mi long. The voltage at the sending end is 220 kV. The parameters of the line are $R = 0.2\ \Omega/\text{mi}$, $X = 0.8\ \Omega/\text{mi}$, and $Y = 5.3\ \mu\text{S/mi}$. Find the sending-end current when there is no load on the line.

6.15. If the load on the line described in Prob. 6.14 is 80 MW at 220 kV, with unity power factor, calculate the current, voltage, and power at the sending end. Assume that the sending-end voltage is held constant and calculate the voltage regulation of the line for the load specified above.

6.16. A three-phase transmission line is 300 mi long and serves a load of 400 MVA, with 0.8 lagging power factor at 345 kV. The *ABCD* constants of the line are

$$A = D = 0.8180 \angle 1.3^\circ$$

$$B = 172.2 \angle 84.2^\circ\ \Omega$$

$$C = 0.001933 \angle 90.4^\circ\ \text{S}$$

(*a*) Determine the sending-end line-to-neutral voltage, the sending-end current, and the percent voltage drop at full load.
(*b*) Determine the receiving-end line-to-neutral voltage at no load, the sending-end current at no load, and the voltage regulation.

6.17. Justify Eq. (6.50) by substituting for the hyperbolic functions the equivalent exponential expressions.

6.18. Determine the equivalent-π circuit for the line of Prob. 6.12.

6.19. Use Eqs. (6.1) and (6.2) to simplify Eqs. (6.57) and (6.58) for the short transmission line with (*a*) series reactance X and resistance R and (*b*) series reactance X and negligible resistance.

6.20. Rights of way for transmission circuits are difficult to obtain in urban areas, and existing lines are often upgraded by reconductoring the line with larger conductors or by reinsulating the line for operation at higher voltage. Thermal considerations and maximum power which the line can transmit are the important considerations. A 138-kV line is 50 km long and is composed of *Partridge* conductors with flat horizontal spacing of 5 m between adjacent conductors. Neglect resistance and find the percent increase in power which can be transmitted for constant $|V_S|$ and $|V_R|$ while δ is limited to 45°

(a) If the *Partridge* conductor is replaced by *Osprey*, which has more than twice the area of aluminum in square millimeters,

(b) If a second *Partridge* conductor is placed in a two-conductor bundle 40 cm from the original conductor and a center-to-center distance between bundles of 5 m, and

(c) If the voltage of the original line is raised to 230 kV with increased conductor spacing of 8 m.

6.21. Construct a receiving-end power-circle diagram similar to Fig. 6.11 for the line of Prob. 6.12. Locate the point corresponding to the load of Prob. 6.12, and locate the center of circles for various values of $|V_S|$ if $|V_R| = 220$ kV. Draw the circle passing through the load point. From the measured radius of the latter circle determine $|V_S|$ and compare this value with the values calculated for Prob. 6.12.

6.22. A synchronous condenser is connected in parallel with the load described in Prob. 6.12 to improve the overall power factor at the receiving end. The sending-end voltage is always adjusted so as to maintain the receiving-end voltage fixed at 220 kV. Using the power-circle diagram constructed for Prob. 6.21, determine the sending-end voltage and the reactive power supplied by the synchronous condenser when the overall power factor at the receiving end is (a) unity (b) 0.9 leading.

6.23. A series capacitor bank having a reactance of 146.6 Ω is to be installed at the midpoint of the 300-mi line of Prob. 6.16. The *ABCD* constants for each 150-mi portion of line are

$$A = D = 0.9534\angle 0.3^\circ$$

$$B = 90.33\angle 84.1^\circ\ \Omega$$

$$C = 0.001014\angle 90.1^\circ\ \text{S}$$

(a) Determine the equivalent *ABCD* constants for the cascade combination of the line-capacitor-line. (See Table A.6 in the Appendix.)

(b) Solve Prob. 6.16 using these equivalent *ABCD* constants.

6.24. The shunt admittance of a 300-mi transmission line is

$$y_c = 0 + j6.87 \times 10^{-6}\ \text{S/mi}$$

Determine the *ABCD* constants of a shunt reactor that will compensate for 60% of the total shunt admittance.

6.25. A 250-Mvar, 345-kV shunt reactor whose admittance is $0.0021\angle -90^\circ$ S is connected to the receiving end of the 300-mi line of Prob. 6.16 at no load.

(a) Determine the equivalent *ABCD* constants of the line in series with the shunt reactor. (See Table A.6 in the Appendix.)

(b) Rework part (b) of Prob. 6.16 using these equivalent *ABCD* constants and the sending-end voltage found in Prob. 6.16.

6.26. Draw the lattice diagram for current and plot current versus time at the sending end of the line of Example 6.8 for the line terminated in (*a*) an open circuit (*b*) a short circuit.

6.27. Plot voltage versus time for the line of Example 6.8 at a point distant from the sending end equal to one-fourth of the length of the line if the line is terminated in a resistance of 10 Ω.

6.28. Solve Example 6.8 if a resistance of 54 Ω is in series with the source.

6.29. Voltage from a dc source is applied to an overhead transmission line by closing a switch. The end of the overhead line is connected to an underground cable. Assume that both the line and the cable are lossless and that the initial voltage along the line is v^+. If the characteristic impedances of the line and cable are 400 and 50 Ω, respectively, and the end of the cable is open-circuited, find in terms of v^+

(*a*) The voltage at the junction of the line and cable immediately after the arrival of the incident wave and

(*b*) The voltage at the open end of the cable immediately after arrival of the first voltage wave.

6.30. A dc source of voltage V_s and internal resistance R_s is connected through a switch to a lossless line having characteristic impedance R_c. The line is terminated in a resistance R_R. The traveling time of a voltage across the line is T. The switch closes at $t = 0$.

(*a*) Draw a lattice diagram showing the voltage of the line during the period $t = 0$ to $t = 7T$. Indicate the voltage components in terms of V_s and the reflection coefficients ρ_R and ρ_s.

(*b*) Determine the receiving-end voltage at $t = 0$, $2T$, $4T$, and $6T$, and hence at $t = 2nT$ where n is any non-negative integer.

(*c*) Hence, determine the steady-state voltage at the receiving end of the line in terms of V_s, R_s, R_R, and R_c.

(*d*) Verify the result in part (*c*) by analyzing the system as a simple dc circuit in the steady state. (Note that the line is lossless and remember how inductances and capacitances behave as short circuits and open circuits to dc.)

CHAPTER 7

THE ADMITTANCE MODEL AND NETWORK CALCULATIONS

The typical power transmission network spans a large geographic area and involves a large number and variety of network components. The electrical characteristics of the individual components are developed in previous chapters and now we are concerned with the composite representation of those components when they are interconnected to form the network. For large-scale system analysis the network model takes on the form of a *network matrix* with elements determined by the choice of parameter.

There are two choices. The current flow through a network component can be related to the voltage drop across it by either an admittance or an impedance parameter. This chapter treats the admittance representation in the form of a *primitive model* which describes the electrical characteristics of the network components. The primitive model neither requires nor provides any information about how the components are interconnected to form the network. The steady-state behavior of all the components acting together as a system is given by the *nodal admittance matrix* based on nodal analysis of the network equations.

The nodal admittance matrix of the typical power system is large and sparse, and can be constructed in a systematic building-block manner. The building-block approach provides insight for developing algorithms to account for network changes. Because the network matrices are very large, *sparsity techniques* are needed to enhance the computational efficiency of computer programs employed in solving many of the power system problems described in later chapters.

The particular importance of the present chapter, and also Chap. 8, which develops the *nodal impedance matrix*, becomes evident in the course of power-flow and fault analysis of the system.

7.1 BRANCH AND NODE ADMITTANCES

In per-phase analysis the components of the power transmission system are modeled and represented by passive impedances or equivalent admittances accompanied, where appropriate, by active voltage or current sources. In the steady state, for example, a generator can be represented by the circuit of either Fig. 7.1(*a*) or Fig. 7.1(*b*). The circuit having the constant emf E_s, series impedance Z_a, and terminal voltage V has the voltage equation

$$E_s = IZ_a + V \tag{7.1}$$

Dividing across by Z_a gives the current equation for Fig. 7.1(*b*)

$$I_s = \frac{E_s}{Z_a} = I + VY_a \tag{7.2}$$

where $Y_a = 1/Z_a$. Thus, the emf E_s and its series impedance Z_a can be interchanged with the current source I_s and its shunt admittance Y_a, provided

$$I_s = \frac{E_s}{Z_a} \quad \text{and} \quad Y_a = \frac{1}{Z_a} \tag{7.3}$$

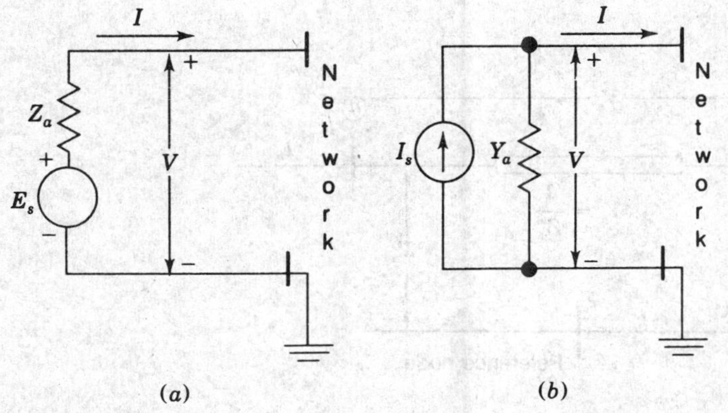

FIGURE 7.1
Circuits illustrating the equivalence of sources when $I_S = E_S/Z_a$ and $Y_a = 1/Z_a$.

Sources such as E_s and I_s may be considered externally applied at the nodes of the transmission network, which then consists of only passive branches. In this chapter subscripts a and b distinguish branch quantities from node quantities which have subscripts m, n, p, and q or else numbers. For network modeling we may then represent the typical branch by either the *branch impedance* Z_a or the *branch admittance* Y_a, whichever is more convenient. The branch impedance Z_a is often called the *primitive impedance*, and likewise, Y_a is called the *primitive admittance*. The equations characterizing the branch are

$$V_a = Z_a I_a \qquad \text{or} \qquad Y_a V_a = I_a \tag{7.4}$$

where Y_a is the reciprocal of Z_a and V_a is the voltage drop across the branch in the direction of the branch current I_a. Regardless of how it is connected into the network, the typical branch has the two associated variables V_a and I_a related by Eqs. (7.4). In this chapter we concentrate on the branch admittance form in order to establish the nodal admittance representation of the power network, and Chap. 8 treats the impedance form.

In Sec. 1.12 rules are given for forming the bus admittance matrix of the network. Review of those rules is recommended since we are about to consider an alternative method for $\mathbf{Y}_{\text{bus}}$ formation. The new method is more general because it is easily extended to networks with mutually coupled elements. Our approach first considers each branch separately and then in combination with other branches of the network.

Suppose that only branch admittance Y_a is connected between nodes ⓜ and ⓝ as part of a larger network of which only the reference node appears in Fig. 7.2. Current injected into the network at any node is considered *positive* and current leaving the network at any node is considered *negative*. In Fig. 7.2 current I_m is that portion of the total current injected into node ⓜ which passes through Y_a. Likewise, I_n is that portion of the current injected into node ⓝ which passes through Y_a. The voltages V_m and V_n are the voltages of nodes

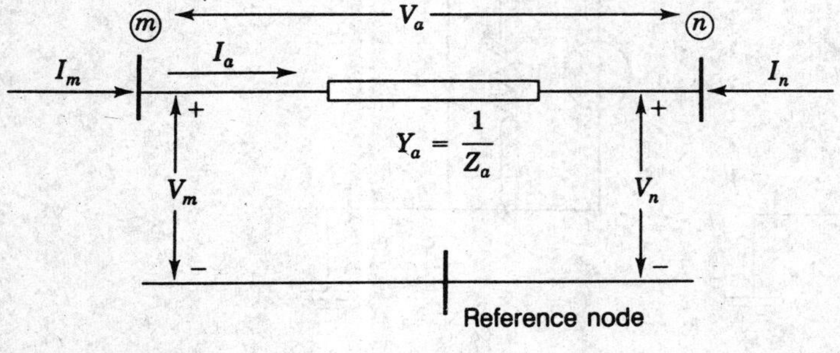

FIGURE 7.2
Primitive branch voltage drop V_a, branch current I_a, injected currents I_m and I_n, node voltages V_m and V_n with respect to network reference.

Ⓜ and Ⓝ, respectively, measured with respect to network reference. By Kirchhoff's law $I_m = I_a$ at node Ⓜ and $I_n = -I_a$ at node Ⓝ. These two current equations arranged in vector form are

$$\begin{bmatrix} I_m \\ I_n \end{bmatrix} = \begin{matrix} ⓜ \\ ⓝ \end{matrix} \begin{bmatrix} 1 \\ -1 \end{bmatrix} I_a \tag{7.5}$$

In Eq. (7.5) the labels or pointers Ⓜ and Ⓝ associate the direction of I_a *from* node Ⓜ *to* node Ⓝ with the entries 1 and -1, which are then said to be in row Ⓜ and row Ⓝ, respectively. In a similar manner, the voltage drop in the direction of I_a has the equation $V_a = V_m - V_n$, or expressed in vector form

$$V_a = \overset{\text{ⓜ}\quad\text{ⓝ}}{[\,1 \quad -1\,]} \begin{bmatrix} V_m \\ V_n \end{bmatrix} \tag{7.6}$$

Substituting this expression for V_a in the admittance equation $Y_a V_a = I_a$ gives

$$Y_a \overset{\text{ⓜ}\quad\text{ⓝ}}{[\,1 \quad -1\,]} \begin{bmatrix} V_m \\ V_n \end{bmatrix} = I_a \tag{7.7}$$

and premultiplying both sides of Eq. (7.7) by the column vector of Eq. (7.5), we obtain

$$\begin{matrix} ⓜ \\ ⓝ \end{matrix} \begin{bmatrix} 1 \\ -1 \end{bmatrix} Y_a \overset{\text{ⓜ}\quad\text{ⓝ}}{[\,1 \quad -1\,]} \begin{bmatrix} V_m \\ V_n \end{bmatrix} = \begin{bmatrix} I_m \\ I_n \end{bmatrix} \tag{7.8}$$

which, simplifies to

$$\begin{matrix} ⓜ \\ ⓝ \end{matrix} \overset{\text{ⓜ}\qquad\text{ⓝ}}{\begin{bmatrix} Y_a & -Y_a \\ -Y_a & Y_a \end{bmatrix}} \begin{bmatrix} V_m \\ V_n \end{bmatrix} = \begin{bmatrix} I_m \\ I_n \end{bmatrix} \tag{7.9}$$

This is the *nodal admittance equation* for branch Y_a and the coefficient matrix is the *nodal admittance matrix*. We note that the off-diagonal elements equal the negative of the branch admittance. The matrix of Eq. (7.9) is singular because neither node Ⓜ nor node Ⓝ connects to the reference. In the particular case where one of the two nodes, say, Ⓝ, is the reference node, then node voltage

V_n is zero and Eq. (7.9) reduces to the 1×1 matrix equation

$$\begin{array}{c} \quad \textcircled{m} \\ \textcircled{m}\ [Y_a]\, V_m = I_m \end{array} \tag{7.10}$$

corresponding to removal of row ⓝ and column ⓝ from the coefficient matrix.

Despite its straightforward derivation, Eq. (7.9) and the procedure leading to it are important in more general situations. We note that the branch voltage V_a is transformed to the node voltages V_m and V_n, and the branch current I_a is likewise represented by the current injections I_m and I_n. The coefficient matrix relating the node voltages and currents of Eq. (7.9) follows from the fact that in Eq. (7.8)

$$\begin{matrix} \textcircled{m} \\ \textcircled{n} \end{matrix}\begin{bmatrix} 1 \\ -1 \end{bmatrix}\overset{\quad\textcircled{m}\quad\ \ \textcircled{n}}{\begin{bmatrix} 1 & -1 \end{bmatrix}} = \begin{matrix} \textcircled{m} \\ \textcircled{n} \end{matrix}\overset{\quad\textcircled{m}\quad\ \ \textcircled{n}}{\begin{bmatrix} 1 & -1 \\ -1 & 1 \end{bmatrix}} \tag{7.11}$$

This 2×2 matrix, also seen in Eq. (1.64), is an important *building block* for representing more general networks, as we shall soon see. The row and column pointers identify each entry in the coefficient matrix by node numbers. For instance, in the first row and second column of Eq. (7.11) the entry -1 is identified with nodes ⓜ and ⓝ of Fig. 7.2 and the other entries are similarly identified.

Thus, the coefficient matrices of Eqs. (7.9) and (7.10) are simply *storage* matrices with row and column labels determined by the end nodes of the branch. Each branch of the network has a similar matrix labeled according to the nodes of the network to which that branch is connected. To obtain the overall nodal admittance matrix of the entire network, we simply combine the individual branch matrices by adding together elements with identical row and column labels. Such addition causes the sum of the branch currents flowing from each node of the network to equal the total current injected into that node, as required by Kirchhoff's current law. In the overall matrix, the off-diagonal element Y_{ij} is the negative sum of the admittances connected between nodes ⓘ and ⓙ and the diagonal entry Y_{ii} is the algebraic sum of the admittances connected to node ⓘ. Provided at least one of the network branches is connected to the reference node, the net result is $\mathbf{Y}_{\text{bus}}$ of the system, as shown in the following example.

Example 7.1. The single-line diagram of a small power system is shown in Fig. 7.3. The corresponding reactance diagram, with reactances specified in per unit, is shown in Fig. 7.4. A generator with emf equal to $1.25\angle 0^\circ$ per unit is connected through a transformer to high-voltage node ③, while a motor with internal voltage equal to $0.85\angle -45^\circ$ is similarly connected to node ④. Develop the nodal

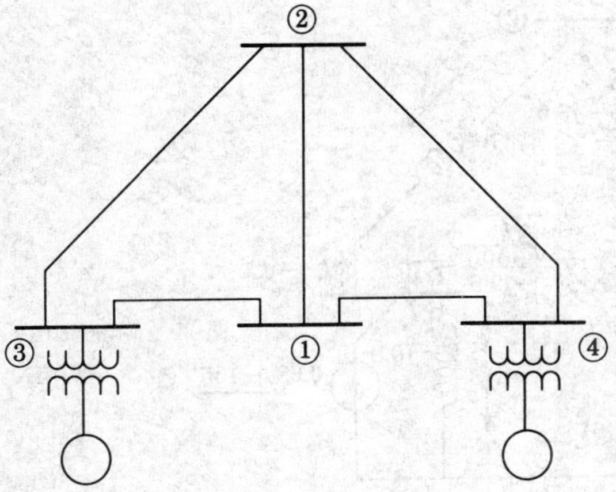

FIGURE 7.3
Single-line diagram of the four-bus system of Example 7.1. Reference node is not shown.

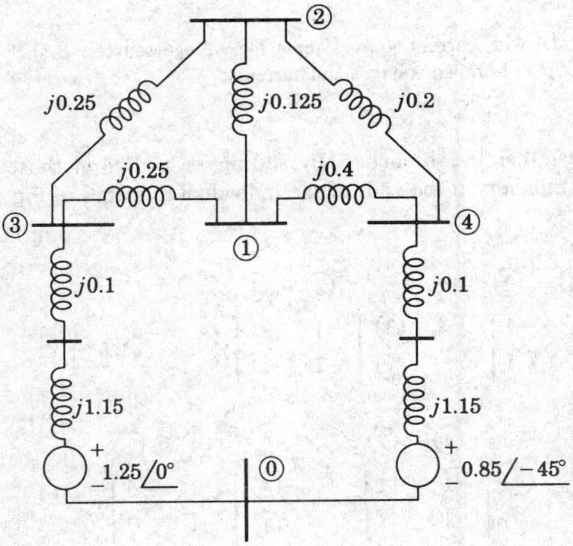

FIGURE 7.4
Reactance diagram for Fig. 7.3. Node ⓪ is reference, reactances and voltages are in per unit.

admittance matrix for each of the network branches and then write the nodal admittance equations of the system.

Solution. The reactances of the generator and the motor may be combined with their respective step-up transformer reactances. Then, by transformation of sources the combined reactances and the generated emfs are replaced by the equivalent current sources and shunt admittances shown in Fig. 7.5. We will treat the current sources as external injections at nodes ③ and ④ and name the seven passive branches according to the subscripts of their currents and voltages. For example, the branch between nodes ① and ③ will be called branch *c*. The admittance of each branch is simply the reciprocal of the branch impedance and Fig. 7.5 shows the resultant admittance diagram with all values in per unit. The two branches *a* and *g* connected to the reference node are characterized by Eq. (7.10), while Eq.

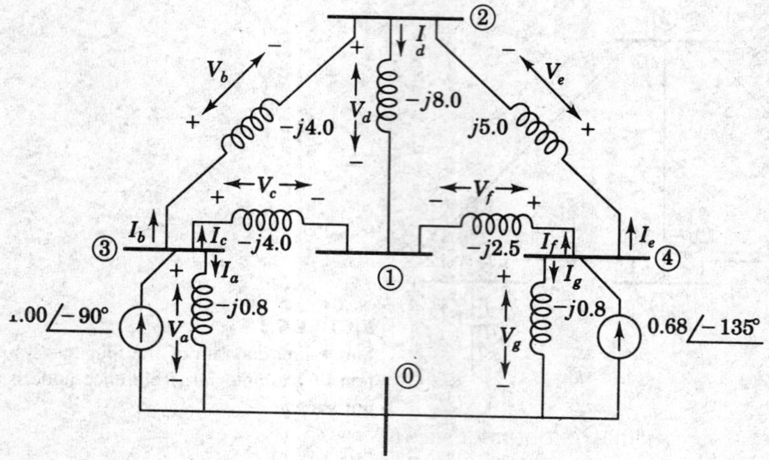

FIGURE 7.5
Per-unit admittance diagram for Fig. 7.4 with current sources replacing voltage sources. Branch names a to g correspond to the subscripts of branch voltages and currents.

(7.9) applies to each of the other five branches. By setting m and n in those equations equal to the node numbers at the ends of the individual branches of Fig. 7.5, we obtain

$$\overset{③}{③\,[1]\,Y_a} \qquad \begin{matrix} & ③ & ② \\ ③ & 1 & -1 \\ ② & -1 & 1 \end{matrix}\,Y_b \qquad \begin{matrix} & ③ & ① \\ ③ & 1 & -1 \\ ① & -1 & 1 \end{matrix}\,Y_c \qquad \overset{④}{④\,[1]\,Y_g}$$

$$\begin{matrix} & ② & ① \\ ② & 1 & -1 \\ ① & -1 & 1 \end{matrix}\,Y_d \qquad \begin{matrix} & ④ & ② \\ ④ & 1 & -1 \\ ② & -1 & 1 \end{matrix}\,Y_e \qquad \begin{matrix} & ④ & ① \\ ④ & 1 & -1 \\ ① & -1 & 1 \end{matrix}\,Y_f$$

The order in which the labels are assigned is not important here, provided the columns and rows follow the *same* order. However, for consistency with later sections let us assign the node numbers in the directions of the branch currents of Fig. 7.5, which also shows the numerical values of the admittances. Combining together those elements of the above matrices having identical row and column labels gives

$$\begin{matrix} & ① & ② & ③ & ④ \\ ① & (Y_c + Y_d + Y_f) & -Y_d & -Y_c & -Y_f \\ ② & -Y_d & (Y_b + Y_d + Y_e) & -Y_b & -Y_e \\ ③ & -Y_c & -Y_b & (Y_a + Y_b + Y_c) & 0 \\ ④ & -Y_f & -Y_e & 0 & (Y_e + Y_f + Y_g) \end{matrix}$$

which is the same as $\mathbf{Y}_{\text{bus}}$ of Eq. (1.62) since Figs. 1.23 and 7.5 are for the same network. Substituting the numerical values of the branch admittances into this matrix, we obtain for the overall network the nodal admittance equations

$$\begin{bmatrix} -j14.5 & j8.0 & j4.0 & j2.5 \\ j8.0 & -j17.0 & j4.0 & j5.0 \\ j4.0 & j4.0 & -j8.8 & 0.0 \\ j2.5 & j5.0 & 0.0 & -j8.3 \end{bmatrix} \begin{bmatrix} V_1 \\ V_2 \\ V_3 \\ V_4 \end{bmatrix} = \begin{bmatrix} 0 \\ 0 \\ 1.00\angle -90^\circ \\ 0.68\angle -135^\circ \end{bmatrix}$$

where V_1, V_2, V_3, and V_4 are the node voltages measured with respect to the reference node and $I_1 = 0$, $I_2 = 0$, $I_3 = 1.00\angle -90^\circ$ and $I_4 = 0.68\angle -135^\circ$ are the external currents injected at the system nodes.

The coefficient matrix obtained in the above example is exactly the same as the bus admittance matrix found in Sec. 1.12 using the usual rules for $\mathbf{Y}_{\text{bus}}$ formation. However, the approach based on the building-block matrix has advantages when extended to networks with mutually coupled branches, as we now demonstrate.

7.2 MUTUALLY COUPLED BRANCHES IN $\mathbf{Y}_{\text{bus}}$

The procedure based on the building-block matrix is now extended to two mutually coupled branches which are part of a larger network but which are not inductively coupled to any other branches. In Sec. 2.2 the primitive equations of such mutually coupled branches are developed in the form of Eq. (2.24) for impedances and Eq. (2.26) for admittances. The notation is different here because we are now using numbers to identify nodes rather than branches.

Assume that branch impedance Z_a connected between nodes ⓜ and ⓝ is coupled through mutual impedance Z_M to branch impedance Z_b connected between nodes ⓟ and ⓠ of Fig. 7.6. The voltage drops V_a and V_b due to the branch currents I_a and I_b are then given by the primitive impedance equation corresponding to Eq. (2.24) in the form

$$\begin{bmatrix} V_a \\ V_b \end{bmatrix} = \begin{bmatrix} Z_a & Z_M \\ Z_M & Z_b \end{bmatrix} \begin{bmatrix} I_a \\ I_b \end{bmatrix} \tag{7.12}$$

in which the coefficient matrix is symmetrical. As noted in Sec. 2.2, the mutual impedance Z_M is considered positive when currents I_a and I_b enter the terminals marked with dots in Fig. 7.6(*a*); the voltage drops V_a and V_b then have the polarities shown. Multiplying Eq. (7.12) by the inverse of the primitive impedance matrix

$$\begin{bmatrix} Z_a & Z_M \\ Z_M & Z_b \end{bmatrix}^{-1} = \frac{1}{Z_a Z_b - Z_M^2} \begin{bmatrix} Z_b & -Z_M \\ -Z_M & Z_a \end{bmatrix} = \begin{bmatrix} Y_a & Y_M \\ Y_M & Y_b \end{bmatrix} \tag{7.13}$$

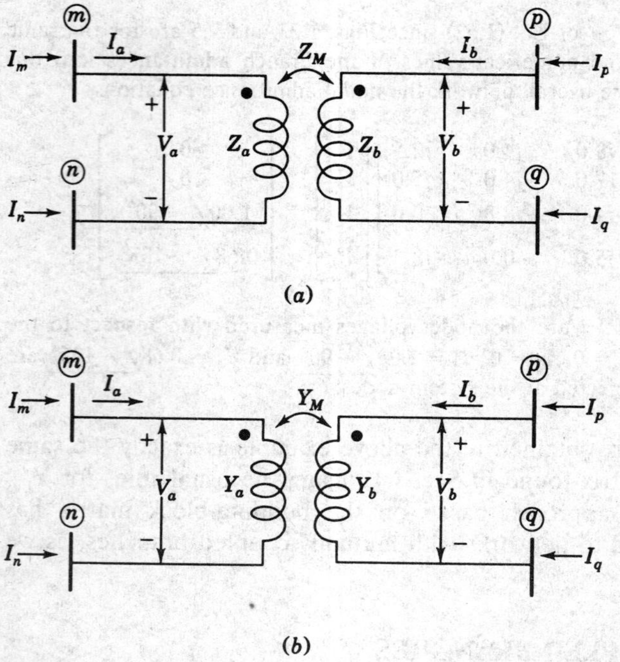

FIGURE 7.6
Two mutually coupled branches with (*a*) impedance parameters and (*b*) corresponding admittances.

we obtain the admittance form of Eq. (2.26) for the two branches

$$\begin{bmatrix} Y_a & Y_M \\ Y_M & Y_b \end{bmatrix} \begin{bmatrix} V_a \\ V_b \end{bmatrix} = \begin{bmatrix} I_a \\ I_b \end{bmatrix} \tag{7.14}$$

which is also symmetrical. The admittance matrix of Eq. (7.14), called the *primitive admittance matrix* of the two coupled branches, corresponds to Fig. 7.6(*b*). The primitive self-admittance Y_a equals $Z_b/(Z_a Z_b - Z_M^2)$ and similar expressions from Eq. (7.13) apply to Y_b and the primitive mutual admittance Y_M. We may write the voltage-drop equations $V_a = V_m - V_n$ and $V_b = V_p - V_q$ of Fig. 7.6 in the matrix form

$$\begin{bmatrix} V_a \\ V_b \end{bmatrix} = \begin{bmatrix} V_m - V_n \\ V_p - V_q \end{bmatrix} = \begin{array}{c} \begin{matrix} \textcircled{m} & \textcircled{n} & \textcircled{p} & \textcircled{q} \end{matrix} \\ \begin{bmatrix} 1 & -1 & 0 & 0 \\ 0 & 0 & 1 & -1 \end{bmatrix} \end{array} \begin{bmatrix} V_m \\ V_n \\ V_p \\ V_q \end{bmatrix} = \mathbf{A} \begin{bmatrix} V_m \\ V_n \\ V_p \\ V_q \end{bmatrix} \tag{7.15}$$

in which the first row of the coefficient matrix **A** is associated with branch admittance Y_a and the second row relates to branch admittance Y_b. The node voltages V_m, V_n, V_p, and V_q are measured with respect to the network reference. In Fig. 7.6 the branch current I_a is related to the injected currents by the two node equations $I_m = I_a$ and $I_n = -I_a$; similarly, branch current I_b is related to the currents I_p and I_q by the two node equations $I_p = I_b$ and $I_q = -I_b$. These four current equations arranged in matrix form become

$$\begin{bmatrix} I_m \\ I_n \\ I_p \\ I_q \end{bmatrix} = \begin{matrix} \textcircled{m} \\ \textcircled{n} \\ \textcircled{p} \\ \textcircled{q} \end{matrix} \begin{bmatrix} 1 & 0 \\ -1 & 0 \\ 0 & 1 \\ 0 & -1 \end{bmatrix} \begin{bmatrix} I_a \\ I_b \end{bmatrix} = \mathbf{A}^T \begin{bmatrix} I_a \\ I_b \end{bmatrix} \tag{7.16}$$

with the coefficient matrix equal to the transpose of that in Eq. (7.15). From Eq. (7.15) we substitute for the voltage drops of Eq. (7.14) to find

$$\begin{bmatrix} Y_a & Y_M \\ Y_M & Y_b \end{bmatrix} \mathbf{A} \begin{bmatrix} V_m \\ V_n \\ V_p \\ V_q \end{bmatrix} = \begin{bmatrix} I_a \\ I_b \end{bmatrix} \tag{7.17}$$

and premultiplying both sides of this equation by the matrix $\mathbf{A}^T$ of Eq. (7.16), we obtain

$$\underbrace{\mathbf{A}^T}_{4\times 2} \underbrace{\begin{bmatrix} Y_a & Y_M \\ Y_M & Y_b \end{bmatrix}}_{2\times 2} \underbrace{\mathbf{A}}_{2\times 4} \begin{bmatrix} V_m \\ V_n \\ V_p \\ V_q \end{bmatrix} = \begin{bmatrix} I_m \\ I_n \\ I_p \\ I_q \end{bmatrix} \tag{7.18}$$

When the multiplications indicated in Eq. (7.18) are performed, the result gives the nodal admittance equations of the two mutually coupled branches in the matrix form

$$\begin{matrix} & \textcircled{m} & \textcircled{n} & \textcircled{p} & \textcircled{q} \\ \textcircled{m} & Y_a & -Y_a & Y_M & -Y_M \\ \textcircled{n} & -Y_a & Y_a & -Y_M & Y_M \\ \textcircled{p} & Y_M & -Y_M & Y_b & -Y_b \\ \textcircled{q} & -Y_M & Y_M & -Y_b & Y_b \end{matrix} \begin{bmatrix} V_m \\ V_n \\ V_p \\ V_q \end{bmatrix} = \begin{bmatrix} I_m \\ I_n \\ I_p \\ I_q \end{bmatrix} \tag{7.19}$$

The two mutually coupled branches are actually part of a larger network, and so the 4 × 4 matrix of Eq. (7.19) forms part of the larger nodal admittance matrix of the overall system. The pointers ⓜ, ⓝ, ⓟ, and ⓠ indicate the rows and columns of the *system* matrix to which the elements of Eq. (7.19) belong. Thus, for example, the quantity entered in row ⓝ and column ⓟ of the system nodal admittance matrix is $-Y_M$ and similar entries are made from the other elements of Eq. (7.19).

The nodal admittance matrix of the two coupled branches may be formed directly by inspection. This becomes clear when we write the coefficient matrix of Eq. (7.19) in the alternative form

$$\left[\begin{array}{c|c} \begin{array}{c} \quad ⓜ \quad ⓝ \\ \begin{matrix} ⓜ \\ ⓝ \end{matrix}\begin{bmatrix} 1 & -1 \\ -1 & 1 \end{bmatrix} Y_a \end{array} & \begin{array}{c} \quad ⓟ \quad ⓠ \\ \begin{matrix} ⓜ \\ ⓝ \end{matrix}\begin{bmatrix} 1 & -1 \\ -1 & 1 \end{bmatrix} Y_M \end{array} \\ \hline \begin{array}{c} \quad ⓜ \quad ⓝ \\ \begin{matrix} ⓟ \\ ⓠ \end{matrix}\begin{bmatrix} 1 & -1 \\ -1 & 1 \end{bmatrix} Y_M \end{array} & \begin{array}{c} \quad ⓟ \quad ⓠ \\ \begin{matrix} ⓟ \\ ⓠ \end{matrix}\begin{bmatrix} 1 & -1 \\ -1 & 1 \end{bmatrix} Y_b \end{array} \end{array}\right] \tag{7.20}$$

To obtain Eq. (7.20), we multiply each element of the primitive admittance matrix by the 2×2 building-block matrix. The labels assigned to the rows and columns of the multipliers in Eq. (7.20) are easily determined. First, we note that the self-admittance Y_a is measured between nodes ⓜ and ⓝ with the dot at node ⓜ. Hence, the 2×2 matrix multiplying Y_a in Eq. (7.20) has rows and columns labeled ⓜ and ⓝ *in that order*. Then, the self-admittance Y_b between nodes ⓟ and ⓠ is multiplied by the 2×2 matrix with labels ⓟ and ⓠ in the order shown since node ⓟ is marked with a dot. Finally, the labels of the matrices multiplying the mutual admittance Y_M are assigned row by row and then column by column so as to align and agree with those already given to the self-inductances. In the nodal admittance matrix of Eqs. (7.19) and (7.20) the sum of the columns (and of the rows) adds up to zero. This is because none of the nodes ⓜ, ⓝ, ⓟ, and ⓠ has been considered as the reference node of the network. In the special case where one of the nodes, say, node ⓝ, is in fact the reference, V_n is zero and column n in Eq. (7.19) does not need to appear; furthermore, I_n does not have to be explicitly represented since the reference node current is not an independent quantity. Consequently, when node ⓝ is the reference, we may eliminate the row and column of that node from Eqs. (7.19) and (7.20).

It is important to note that nodes ⓜ, ⓝ, ⓟ, and ⓠ are often not distinct. For instance, suppose that nodes ⓝ and ⓠ are one and the same node. In that case columns ⓝ and ⓠ of Eq. (7.19) can be combined together since $V_n = V_q$, and the corresponding rows can be added because I_n and I_p are parts of the common injected current. The following example illustrates this situation.

Example 7.2. Two branches having impedances equal to $j0.25$ per unit are coupled through mutual impedance $Z_M = j0.15$ per unit, as shown in Fig. 7.7. Find

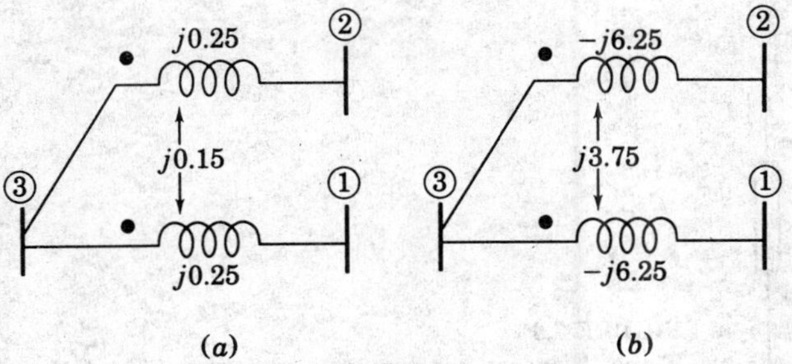

FIGURE 7.7
The two mutually coupled branches of Example 7.2, their (*a*) primitive impedances and (*b*) primitive admittances in per unit.

the nodal admittance matrix for the mutually coupled branches and write the corresponding nodal admittance equations.

Solution. The primitive impedance matrix for the mutually coupled branches of Fig. 7.7(*a*) is inverted as a single entity to yield the primitive admittances of Fig. 7.7(*b*), that is,

$$\begin{bmatrix} j0.25 & j0.15 \\ j0.15 & j0.25 \end{bmatrix}^{-1} = \begin{bmatrix} -j6.25 & j3.75 \\ j3.75 & -j6.25 \end{bmatrix}$$

First, the rows and columns of the building-block matrix which multiplies the primitive self-admittance between nodes ① and ③ are labeled ③ and ① in that order to correspond to the dot marking node ③. Next, the rows and columns of the 2 × 2 matrix multiplying the self-admittance between nodes ② and ③ are labeled ③ and ② in the order shown because node ③ is marked. Finally, the pointers of the matrices multiplying the mutual admittance are aligned with those of the self-admittances to form the 4 × 4 array similar to Eq. (7.20) as follows:

$$\left[\begin{array}{c|c}
\begin{matrix} & \textcircled{3} & \textcircled{1} \\ \textcircled{3} & 1 & -1 \\ \textcircled{1} & -1 & 1 \end{matrix}\ (-j6.25) & \begin{matrix} & \textcircled{3} & \textcircled{2} \\ \textcircled{3} & 1 & -1 \\ \textcircled{1} & -1 & 1 \end{matrix}\ (j3.75) \\
\hline
\begin{matrix} & \textcircled{3} & \textcircled{1} \\ \textcircled{3} & 1 & -1 \\ \textcircled{2} & -1 & 1 \end{matrix}\ (j3.75) & \begin{matrix} & \textcircled{3} & \textcircled{2} \\ \textcircled{3} & 1 & -1 \\ \textcircled{2} & -1 & 1 \end{matrix}\ (-j6.25)
\end{array}\right]$$

Since there are only three nodes in Fig. 7.7, the required 3 × 3 matrix is found by adding the columns and rows of common node ③ to obtain

$$\begin{matrix} & \textcircled{1} & \textcircled{2} & \textcircled{3} \\ \textcircled{1} & -j6.25 & j3.75 & j6.25 - j3.75 \\ \textcircled{2} & j3.75 & -j6.25 & -j3.75 + j6.25 \\ \textcircled{3} & j6.25 - j3.75 & -j3.75 + j6.25 & -j6.25 - j6.25 + 2(j3.75) \end{matrix}$$

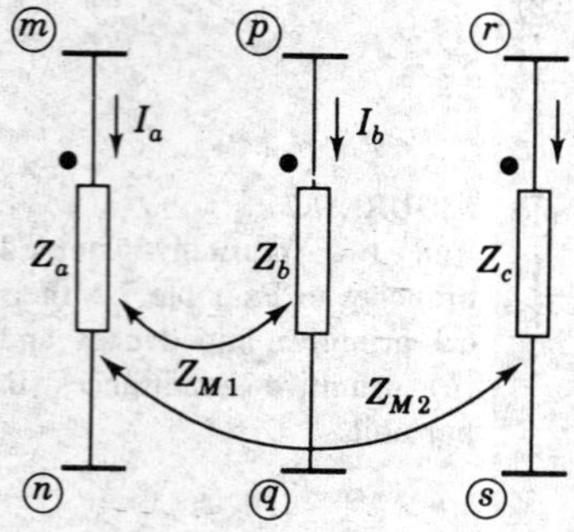

FIGURE 7.8
Three branches with mutual coupling Z_{M1} between branches a and b and Z_{M2} between branches a and c.

The new diagonal element representing node ③, for instance, is the sum of the four elements $(-j6.25 - j6.25 + j3.75 + j3.75)$ in rows ③ and columns ③ of the previous matrix. The three nodal admittance equations in vector-matrix form are then written

$$\begin{bmatrix} -j6.25 & j3.75 & j2.50 \\ j3.75 & -j6.25 & j2.50 \\ j2.50 & j2.50 & -j5.00 \end{bmatrix} \begin{bmatrix} V_1 \\ V_2 \\ V_3 \end{bmatrix} = \begin{bmatrix} I_1 \\ I_2 \\ I_3 \end{bmatrix}$$

where V_1, V_2, and V_3 are the voltages at nodes ①, ②, and ③ measured with respect to reference, while I_1, I_2, and I_3 are the external currents injected at the respective nodes.

As in Sec. 7.1, the coefficient matrix of the last equation can be combined with the nodal admittance matrices of the other branches of the network in order to obtain the nodal admittance matrix of the entire system.

For three or more coupled branches we follow the same procedure as above. For example, the three coupled branches of Fig. 7.8 have primitive impedance and admittance matrices given by

$$\begin{bmatrix} Z_a & Z_{M1} & Z_{M2} \\ Z_{M1} & Z_b & 0 \\ Z_{M2} & 0 & Z_c \end{bmatrix}^{-1} = \begin{bmatrix} Y_a & Y_{M1} & Y_{M2} \\ Y_{M1} & Y_b & Y_{M3} \\ Y_{M2} & Y_{M3} & Y_c \end{bmatrix} \tag{7.21}$$

The zeros in the **Z**-matrix arise because branches b and c are not directly coupled. For all nonzero values of the current I_a of Fig. 7.8, branches b and c are *indirectly* coupled through branch a, as shown by nonzero Y_{M3} of the primitive admittance matrix.

Therefore, to form $\mathbf{Y}_{\text{bus}}$ for a network which has mutually coupled branches, we do the following in sequence:

1. Invert the primitive impedance matrices of the network branches to obtain the corresponding primitive admittance matrices. A single branch has a 1×1 matrix, two mutually coupled branches have a 2×2 matrix, three mutually coupled branches have a 3×3 matrix, and so on.
2. Multiply the elements of each primitive admittance matrix by the 2×2 building-block matrix.
3. Label the two rows and the two columns of each *diagonal* building-block matrix with the end-node numbers of the corresponding self-admittance. For mutually coupled branches it is important to label in the order of the *marked* (*dotted*)—*then*—*unmarked* (*undotted*) node numbers.
4. Label the two rows of each *off-diagonal* building-block matrix with node numbers aligned and consistent with the row labels assigned in (3); then label the columns consistent with the column labels of (3).
5. Combine, by adding together, those elements with identical row and column labels to obtain the nodal admittance matrix of the overall network. If one of the nodes encountered is the reference node, omit its row and column to obtain the system $\mathbf{Y}_{\text{bus}}$.

7.3 AN EQUIVALENT ADMITTANCE NETWORK

We have demonstrated how to write the nodal admittance equations for one branch or a number of mutually coupled branches which are part of a larger network. We now show that such equations can be interpreted as representing an equivalent admittance network with no mutually coupled elements. This may be useful when forming $\mathbf{Y}_{\text{bus}}$ for an original network having mutually coupled elements.

The currents injected into the nodes of Fig. 7.6 are described in terms of node voltages and admittances by Eq. (7.19). For example, the equation for the current I_m at node ③ is given by the first row of Eq. (7.19) as follows:

$$I_m = Y_a V_m - Y_a V_n + Y_M V_p - Y_M V_q \tag{7.22}$$

Adding and subtracting the term $Y_M V_m$ on the right-hand side of Eq. (7.22) and combining terms with common coefficients, we obtain the Kirchhoff's current equation at node ③

$$I_m = \underbrace{Y_a(V_m - V_n)}_{I_{mn}} + \underbrace{(-Y_M)(V_m - V_p)}_{I_{mp}} + \underbrace{Y_M(V_m - V_q)}_{I_{mq}} \tag{7.23}$$

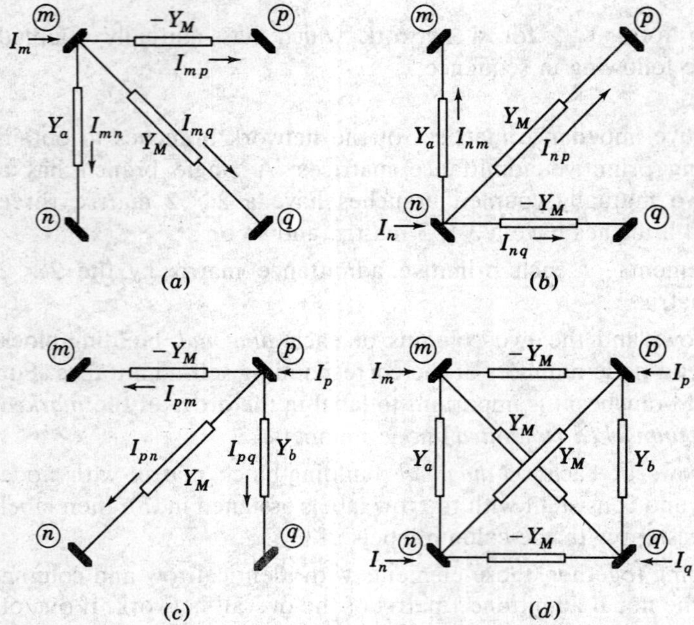

FIGURE 7.9
Developing the nodal admittance network of two mutually coupled branches.

The double subscripts indicate the directions of the currents I_{mn}, I_{mp}, and I_{mq} from node ③ to each of the other nodes ⓝ, ⓟ, and ⓠ, respectively, of Fig. 7.9(*a*). A similar analysis of the second and third rows of Eq. (7.19) leads to equations for the currents I_n and I_p in the form

$$I_n = \underbrace{Y_a(V_n - V_m)}_{I_{nm}} + \underbrace{(-Y_M)(V_n - V_q)}_{I_{nq}} + \underbrace{Y_M(V_n - V_p)}_{I_{np}} \tag{7.24}$$

$$I_p = \underbrace{Y_a(V_p - V_q)}_{I_{pq}} + \underbrace{(-Y_M)(V_p - V_m)}_{I_{pm}} + \underbrace{Y_M(V_p - V_n)}_{I_{pn}} \tag{7.25}$$

and these two equations represent the partial networks of Figs. 7.9(*b*) and (*c*). The fourth row of Eq. (7.19) does not yield a separate partial network because it is not independent of the other rows. Combining the three partial networks without duplicating branches, we obtain an equivalent circuit in the form of the lattice network connected among nodes ⓜ, ⓝ, ⓟ, and ⓠ in Fig. 7.9(*d*). This lattice network has no mutually coupled branches, but it is equivalent in every respect to the two original coupled branches of Fig. 7.6 since it satisfies

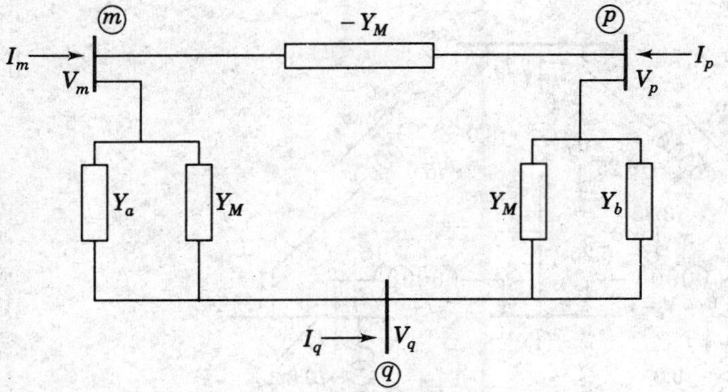

FIGURE 7.10
The nodal admittance network of two mutually coupled branches connected to nodes ⓜ, ⓟ, and ⓠ.

Eq. (7.19). Accordingly, the standard rules of circuit analysis may be applied to the equivalent. For example, if the two coupled branches are physically connected among three independent nodes as in Example 7.2, we may regard nodes ⓝ and ⓠ of Fig. 7.9(*d*) as one and the same node, which we simply join together, as shown in Fig. 7.10. The three-bus equivalent circuit of Fig. 7.10 then yields the nodal equations for the original branches.

Thus, each physical branch or mutually coupled pair gives rise to an equivalent admittance network to which the usual rules of circuit analysis apply. The following example illustrates the role of the equivalent circuit in forming $\mathbf{Y}_{\text{bus}}$.

Example 7.3. Replace branches *b* and *c* between node-pairs ①- ③ and ②- ③ of Fig. 7.5 by the mutually coupled branches of Fig. 7.7. Then, find $\mathbf{Y}_{\text{bus}}$ and the nodal equations of the new network.

Solution. The admittance diagram of the new network including the mutual coupling is shown in Fig. 7.11. From Example 7.2 we know that the mutually coupled branches have the nodal admittance matrix

$$\begin{array}{c} \\ ① \\ ② \\ ③ \end{array} \begin{array}{c} \begin{array}{ccc} ① & ② & ③ \end{array} \\ \begin{bmatrix} -j6.25 & j3.75 & j2.50 \\ j3.75 & -j6.25 & j2.50 \\ j2.50 & j2.50 & -j5.00 \end{bmatrix} \end{array}$$

which corresponds to the equivalent circuit shown encircled in Fig. 7.12. The remaining portion of Fig. 7.12 is drawn from Fig. 7.5. Since mutual coupling is not evident in Fig. 7.12, we may apply the standard rules of $\mathbf{Y}_{\text{bus}}$ formation to the

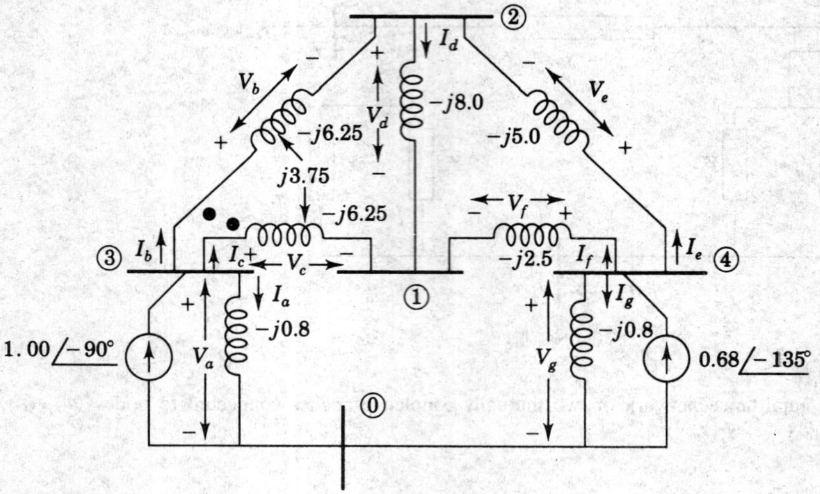

FIGURE 7.11
Per-unit admittance diagram for Example 7.3.

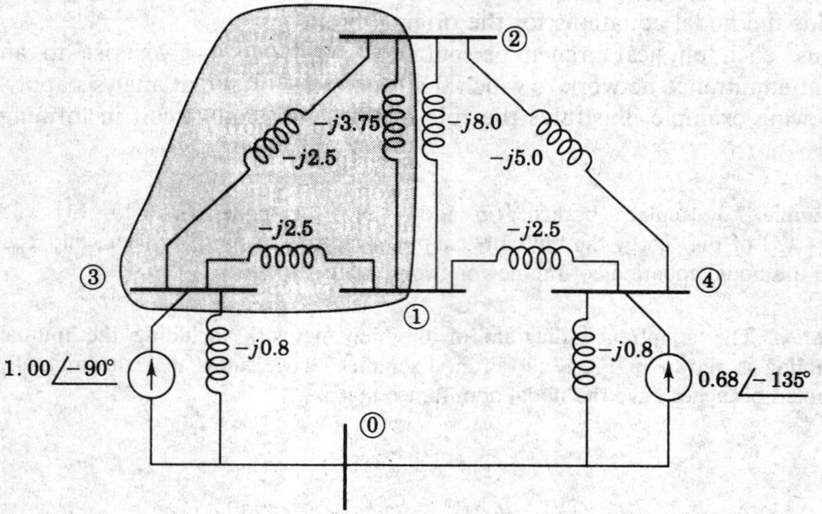

FIGURE 7.12
Nodal admittance network for Example 7.3. The shaded portion represents two mutually coupled branches connected between buses ①, ②, and ③.

overall network, which leads to the nodal admittance equations

$$\begin{array}{c} \textcircled{1} \\ \textcircled{2} \\ \textcircled{3} \\ \textcircled{4} \end{array}\begin{bmatrix} -j16.75 & j11.75 & j2.50 & j2.50 \\ j11.75 & -j19.25 & j2.50 & j5.00 \\ j2.50 & j2.50 & -j5.80 & 0 \\ j2.50 & j5.00 & 0 & -j8.30 \end{bmatrix}\begin{bmatrix} V_1 \\ V_2 \\ V_3 \\ V_4 \end{bmatrix} = \begin{bmatrix} 0 \\ 0 \\ 1.00\angle -90^\circ \\ 0.68\angle -135^\circ \end{bmatrix}$$

(columns labelled ①, ②, ③, ④)

Note that the two admittances between nodes ① and ② combine in parallel to yield

$$Y_{12} = -(-j3.75 - j8.00) = j11.75$$

7.4 MODIFICATION OF $\mathbf{Y}_{\text{bus}}$

The building-block approach and the equivalent circuits of Sec. 7.3 provide important insights into the manner in which each branch self- and mutual admittance contributes to the entries of $\mathbf{Y}_{\text{bus}}$ and the corresponding equivalent network of the overall system. As a result, it is clear that $\mathbf{Y}_{\text{bus}}$ is merely a systematic means of combining the nodal admittance matrices of the various network branches. We simply form a large array with rows and columns ordered according to the sequence in which the nonreference nodes of the network are numbered, and within it we combine entries with matching labels drawn from the nodal admittance matrices of the individual branches. Consequently, we can easily see how to *modify* the system $\mathbf{Y}_{\text{bus}}$ to account for branch additions or other changes to the system network. For instance, to modify $\mathbf{Y}_{\text{bus}}$ of the existing network to reflect the addition of the branch admittance between the nodes Ⓜ and Ⓝ, we simply add Y_a to the elements $\mathbf{Y}_{mm}$ and $\mathbf{Y}_{nn}$ of $\mathbf{Y}_{\text{bus}}$ and subtract Y_a from the symmetrical elements Y_{mn} and Y_{nm}. In other words, to incorporate the new branch admittance Y_a into the network, we add to the existing $\mathbf{Y}_{\text{bus}}$ the change matrix $\Delta\mathbf{Y}_{\text{bus}}$ given by

$$\Delta\mathbf{Y}_{\text{bus}} = \begin{array}{c} \textcircled{m} \\ \textcircled{n} \end{array}\begin{bmatrix} Y_a & -Y_a \\ -Y_a & Y_a \end{bmatrix} \tag{7.26}$$

(columns labelled Ⓜ, Ⓝ)

Again, we recognize $\Delta\mathbf{Y}_{\text{bus}}$ as a storage matrix with rows and columns marked Ⓜ and Ⓝ. Using Eq. (7.26), we may *change* the admittance value of a single branch of the network by adding a new branch between the same end nodes Ⓜ and Ⓝ such that the parallel combination of the old and new branches yields the desired value. Moreover, to *remove* a branch admittance Y_a already connected between nodes Ⓜ and Ⓝ of the network, we simply add the branch admittance $-Y_a$ between the same nodes, which amounts to subtracting the elements of $\Delta\mathbf{Y}_{\text{bus}}$ from the existing $\mathbf{Y}_{\text{bus}}$. Equation (7.20) shows that a *pair* of mutually coupled branches can be removed from the network by subtracting the entries in the change matrix

$$\Delta\mathbf{Y}_{\text{bus}} = \begin{array}{c} \textcircled{m} \\ \textcircled{n} \\ \textcircled{p} \\ \textcircled{q} \end{array} \overset{\begin{array}{cccc} \textcircled{m} & \textcircled{n} & \textcircled{p} & \textcircled{q} \end{array}}{\begin{bmatrix} Y_a & -Y_a & Y_M & -Y_M \\ -Y_a & Y_a & -Y_M & Y_M \\ Y_M & -Y_M & Y_b & -Y_b \\ -Y_M & Y_M & -Y_b & Y_b \end{bmatrix}} \tag{7.27}$$

from the rows and columns of $\mathbf{Y}_{\text{bus}}$ corresponding to the end nodes ⓜ, ⓝ, ⓟ, and ⓠ. Of course, if only one of the two mutually coupled branches is to be removed from the network, we could first remove all the entries for the mutually coupled pair from $\mathbf{Y}_{\text{bus}}$ using Eq. (7.27) and then add the entries for the branch to be retained using Eq. (7.26). Other strategies for modifying $\mathbf{Y}_{\text{bus}}$ to reflect network changes become clear from the insights developed in Secs. 7.1 through 7.3.

Example 7.4. Determine the bus admittance matrix of the network of Fig. 7.5 by removing the effects of mutual coupling from $\mathbf{Y}_{\text{bus}}$ of Fig. 7.11.

Solution. The $\mathbf{Y}_{\text{bus}}$ for the entire system of Fig. 7.11 including mutual coupling is found in Example 7.3 to be

$$\mathbf{Y}_{\text{bus}} = \begin{array}{c} ① \\ ② \\ ③ \\ ④ \end{array} \overset{\begin{array}{cccc} ① & ② & ③ & ④ \end{array}}{\begin{bmatrix} -j16.75 & j11.75 & j2.50 & j2.50 \\ j11.75 & -j19.25 & j2.50 & j5.00 \\ j2.50 & j2.50 & -j5.80 & 0 \\ j2.50 & j5.00 & 0 & -j8.30 \end{bmatrix}}$$

To remove completely the effect of mutual coupling from the network $\mathbf{Y}_{\text{bus}}$, we proceed in two steps by (*a*) first removing the two mutually coupled branches altogether and (*b*) then restoring each of the two branches without mutual coupling between them. (*a*) To remove the two mutually coupled branches from the network, we subtract from the system $\mathbf{Y}_{\text{bus}}$ the entries in

$$\Delta\mathbf{Y}_{\text{bus},1} = \begin{array}{c} ① \\ ② \\ ③ \\ ④ \end{array} \overset{\begin{array}{cccc} ① & ② & ③ & ④ \end{array}}{\begin{bmatrix} -j6.25 & j3.75 & j2.50 & \cdot \\ j3.75 & -j6.25 & j2.50 & \cdot \\ j2.50 & j2.50 & -j5.00 & \cdot \\ \cdot & \cdot & \cdot & \cdot \end{bmatrix}}$$

corresponding to the encircled portion of Fig. 7.12.

(*b*) Now we must reconnect to the network the uncoupled branches, each of which has an admittance $(j0.25)^{-1} = -j4.0$ per unit. Accordingly, to reconnect the branch between nodes ① and ③, we add to $\mathbf{Y}_{bus}$ the change matrix

$$\Delta\mathbf{Y}_{bus,2} = \begin{array}{c} \\ ① \\ ② \\ ③ \\ ④ \end{array}\begin{array}{c} \begin{array}{cccc} ① & ② & ③ & ④ \end{array} \\ \begin{bmatrix} 1 & \cdot & -1 & \cdot \\ \cdot & \cdot & \cdot & \cdot \\ -1 & \cdot & 1 & \cdot \\ \cdot & \cdot & \cdot & \cdot \end{bmatrix} \end{array}(-j4.0)$$

and similarly for the branch between nodes ② and ③ we add

$$\Delta\mathbf{Y}_{bus,3} = \begin{array}{c} \\ ① \\ ② \\ ③ \\ ④ \end{array}\begin{array}{c} \begin{array}{cccc} ① & ② & ③ & ④ \end{array} \\ \begin{bmatrix} \cdot & \cdot & \cdot & \cdot \\ \cdot & 1 & -1 & \cdot \\ \cdot & -1 & 1 & \cdot \\ \cdot & \cdot & \cdot & \cdot \end{bmatrix} \end{array}(-j4.0)$$

Appropriately subtracting and adding the three change matrices and the original $\mathbf{Y}_{bus}$ give the new bus admittance matrix for the uncoupled branches

$$\mathbf{Y}_{bus\,(new)} = \begin{array}{c} \\ ① \\ ② \\ ③ \\ ④ \end{array}\begin{array}{c} \begin{array}{cccc} ① & ② & ③ & ④ \end{array} \\ \begin{bmatrix} -j14.5 & j8.0 & j4.0 & j2.5 \\ j8.0 & -j17.0 & j4.0 & j5.0 \\ j4.0 & j4.0 & -j8.8 & 0 \\ j2.5 & j5.0 & 0 & -j8.3 \end{bmatrix} \end{array}$$

which agrees with Example 7.1.

7.5 THE NETWORK INCIDENCE MATRIX AND $\mathbf{Y}_{bus}$

In Secs. 7.1 and 7.2 nodal admittance equations for each branch and mutually coupled pair of branches are derived independently from those of other branches in the network. The nodal admittance matrices of the individual branches are then combined together in order to build $\mathbf{Y}_{bus}$ of the overall system. Since we now understand the process, we may proceed to the more formal approach which treats all the equations of the system simultaneously rather than separately. We will use the example system of Fig. 7.11 to establish the general procedure.

Two of the seven branches in Fig. 7.11 are mutually coupled as shown. The mutually coupled pair is characterized by Eq. (7.14) and the other five branches by Eq. (7.4). Arranging the seven branch equations into an array format, we obtain

$$\begin{bmatrix} -j0.80 & \cdot & \cdot & \cdot & \cdot & \cdot & \cdot \\ \cdot & -j6.25 & j3.75 & \cdot & \cdot & \cdot & \cdot \\ \cdot & j3.75 & -j6.25 & \cdot & \cdot & \cdot & \cdot \\ \cdot & \cdot & \cdot & -j8.00 & \cdot & \cdot & \cdot \\ \cdot & \cdot & \cdot & \cdot & -j5.00 & \cdot & \cdot \\ \cdot & \cdot & \cdot & \cdot & \cdot & -j2.50 & \cdot \\ \cdot & \cdot & \cdot & \cdot & \cdot & \cdot & -j0.80 \end{bmatrix} \begin{bmatrix} V_a \\ V_b \\ V_c \\ V_d \\ V_e \\ V_f \\ V_g \end{bmatrix} = \begin{bmatrix} I_a \\ I_b \\ I_c \\ I_d \\ I_e \\ I_f \\ I_g \end{bmatrix} \tag{7.28}$$

The coefficient matrix is the primitive admittance matrix formed by inspection of Fig. 7.11. Each branch of the network contributes a diagonal entry equal to the simple reciprocal of its branch impedance except for branches b and c, which are mutually coupled and have entries determined by Eq. (7.13). For the general case Eq. (7.28) may be more compactly written in the form

$$\mathbf{Y}_{\text{pr}}\mathbf{V}_{\text{pr}} = \mathbf{I}_{\text{pr}} \tag{7.29}$$

where $\mathbf{V}_{\text{pr}}$ and $\mathbf{I}_{\text{pr}}$ are the respective column vectors of branch voltages and currents, while $\mathbf{Y}_{\text{pr}}$ represents the primitive admittance matrix of the network. The primitive equations do not tell how the branches are configured within the network. The geometrical configuration of the branches, called the *topology*, is provided by a *directed graph*, as shown in Fig. 7.13(a) in which each branch of the network of Fig. 7.11 is represented between its end nodes by a directed line

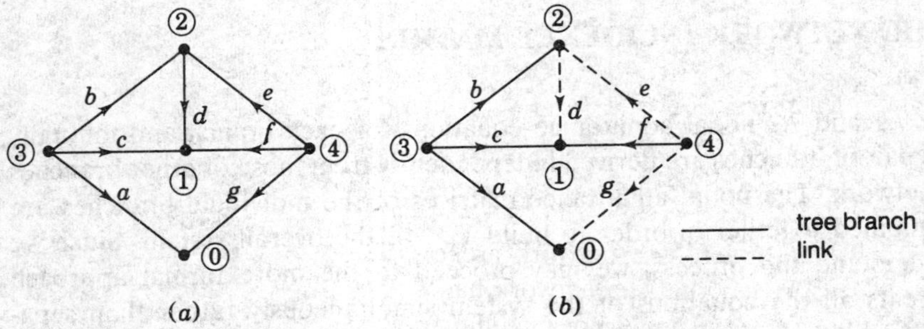

FIGURE 7.13
The linear graph for Fig. 7.11 showing: (a) directed-line segments for branches; (b) branches a, b, c, and f define a tree while branches d, e, and g are links.

segment with an arrow in the direction of the branch current. When a branch connects to a node, the branch and node are said to be *incident*. A *tree* of a graph is formed by those branches of the graph which interconnect or *span* all the nodes of the graph without forming any closed path. In general, there are many possible trees of a network since different combinations of branches can be chosen to span the nodes. Thus, for example, branches a, b, c, and f in Fig. 7.13(b) define a tree. The remaining branches d, e, and g are called *links*, and when a link is added to a tree, a closed path or *loop* is formed.

A graph may be described in terms of a *connection* or *incidence matrix*. Of particular interest is the *branch-to-node incidence matrix* $\hat{\mathbf{A}}$, which has one row for each branch and one column for each node with an entry a_{ij} in row i and column j according to the following rule:

$$a_{ij} = \begin{cases} 0 & \text{if branch } i \text{ is not connected to node } \textcircled{j} \\ 1 & \text{if current in branch } i \text{ is directed away from node } \textcircled{j} \\ -1 & \text{if current in branch } i \text{ is directed toward node } \textcircled{j} \end{cases} \tag{7.30}$$

This rule formalizes for the network as a whole the procedure used to set up the coefficient matrices of Eqs. (7.6) and (7.15) for the individual branches. In network calculations we usually choose a reference node. The column corresponding to the reference node is then omitted from $\hat{\mathbf{A}}$ and the resultant matrix is denoted by $\mathbf{A}$. For example, choosing node ⓪ as the reference in Fig. 7.13 and invoking the rule of Eq. (7.30), we obtain the rectangular branch-to-node matrix

$$\mathbf{A} = \begin{array}{c} \\ a \\ b \\ c \\ d \\ e \\ f \\ g \end{array} \begin{array}{c} \begin{array}{cccc} \textcircled{1} & \textcircled{2} & \textcircled{3} & \textcircled{4} \end{array} \\ \begin{bmatrix} 0 & 0 & 1 & 0 \\ 0 & -1 & 1 & 0 \\ -1 & 0 & 1 & 0 \\ -1 & 1 & 0 & 0 \\ 0 & -1 & 0 & 1 \\ -1 & 0 & 0 & 1 \\ 0 & 0 & 0 & 1 \end{bmatrix} \end{array} \tag{7.31}$$

The nonreference nodes of a network are often called *independent nodes* or *buses*, and when we say that the network has N buses, we generally mean that there are N independent nodes not including the reference node. The $\mathbf{A}$ matrix has the row-column dimension $B \times N$ for any network with B branches and N nodes excluding the reference. We note that each row of Eq. (7.31) has two nonzero entries which add to zero except for rows a and g, each of which has only one nonzero entry. This is because branches a and g of Fig. 7.11 have one end connected to the reference node for which no column is shown.

The voltage across each branch may be expressed as the difference in its end-bus voltages measured with respect to the reference node. For example, in Fig. 7.11 the voltages at buses ①, ②, ③, and ④ with respect to reference node ⓪ are denoted by V_1, V_2, V_3, and V_4, respectively, and so the voltage drops across the branches are given by

$$\begin{aligned} V_a &= V_3 \\ V_b &= V_3 - V_2 \\ V_c &= V_3 - V_1 \\ V_d &= V_2 - V_1 \\ V_e &= V_4 - V_2 \\ V_f &= V_4 - V_1 \\ V_g &= V_4 \end{aligned} \qquad \text{or} \qquad \begin{bmatrix} V_a \\ V_b \\ V_c \\ V_d \\ V_e \\ V_f \\ V_g \end{bmatrix} = \begin{bmatrix} 0 & 0 & 1 & 0 \\ 0 & -1 & 1 & 0 \\ -1 & 0 & 1 & 0 \\ -1 & 1 & 0 & 0 \\ 0 & -1 & 0 & 1 \\ -1 & 0 & 0 & 1 \\ 0 & 0 & 0 & 1 \end{bmatrix} \begin{bmatrix} V_1 \\ V_2 \\ V_3 \\ V_4 \end{bmatrix}$$

in which the coefficient matrix is the $\mathbf{A}$ matrix of Eq. (7.31). This is one illustration of the general result for any N-bus network given by

$$\mathbf{V}_{pr} = \mathbf{A}\mathbf{V} \tag{7.32}$$

where $\mathbf{V}_{pr}$ is the $B \times 1$ column vector of branch voltage drops and $\mathbf{V}$ is the $N \times 1$ column vector of bus voltages measured with respect to the chosen reference node. Equations (7.6) and (7.15) are particular applications of Eq. (7.32) to individual branches. We further note that Kirchhoff's current law at nodes ① to ④ of Fig. 7.11 yields

$$\begin{bmatrix} 0 & 0 & -1 & -1 & 0 & -1 & 0 \\ 0 & -1 & 0 & 1 & -1 & 0 & 0 \\ 1 & 1 & 1 & 0 & 0 & 0 & 0 \\ 0 & 0 & 0 & 0 & 1 & 1 & 1 \end{bmatrix} \begin{bmatrix} I_a \\ I_b \\ I_c \\ I_d \\ I_e \\ I_f \\ I_g \end{bmatrix} = \begin{bmatrix} 0 \\ 0 \\ I_3 \\ I_4 \end{bmatrix}$$

where $I_3 = 1.00\angle{-90^\circ}$ and $I_4 = 0.68\angle{-135^\circ}$ are the external currents injected at nodes ③ and ④, respectively. The coefficient matrix in this equation is $\mathbf{A}^T$. Again, this is illustrative of a general result applicable to every electrical network since it simply states, in accordance with Kirchhoff's current law, that the sum of all the branch currents incident to a node of the network equals the injected current at the node. Accordingly, we may write

$$\mathbf{A}^T \mathbf{I}_{pr} = \mathbf{I} \tag{7.33}$$

where $\mathbf{I}_{pr}$ is the $B \times 1$ column vector of branch currents and $\mathbf{I}$ is the $N \times 1$ column vector with a nonzero entry for each bus with an external current source. Equations (7.5) and (7.16) are particular examples of Eq. (7.33).

The $\mathbf{A}$ matrix fully describes the topology of the network and is independent of the particular values of the branch parameters. The latter are supplied by the primitive admittance matrix. Therefore, two different network configurations employing the same branches will have different $\mathbf{A}$ matrices but the same $\mathbf{Y}_{pr}$. On the other hand, if changes occur in branch parameters while maintaining the same network configuration, only $\mathbf{Y}_{pr}$ is altered but not $\mathbf{A}$.

Multiplying Eq. (7.29) by $\mathbf{A}$, we obtain

$$\mathbf{A}^T \mathbf{Y}_{pr} \mathbf{V}_{pr} = \mathbf{A}^T \mathbf{I}_{pr} \tag{7.34}$$

The right-hand side of Eq. (7.34) equals $\mathbf{I}$ and substituting for $\mathbf{V}_{pr}$ from Eq. (7.32), we find

$$\{\mathbf{A}^T \mathbf{Y}_{pr} \mathbf{A}\} \mathbf{V} = \mathbf{I} \tag{7.35}$$

We may write Eq. (7.35) in the more concise form

$$\mathbf{Y}_{bus} \mathbf{V} = \mathbf{I} \tag{7.36}$$

where the $N \times N$ bus admittance matrix $\mathbf{Y}_{bus}$ for the system is given by

$$\underbrace{\mathbf{Y}_{bus}}_{N \times N} = \underbrace{\mathbf{A}^T}_{N \times B} \; \underbrace{\mathbf{Y}_{pr}}_{B \times B} \; \underbrace{\mathbf{A}}_{B \times N} \tag{7.37}$$

$\mathbf{Y}_{bus}$ has one row and one column for each of the N buses in the network, and so the standard form of the four independent equations of the example system of Fig. 7.11 is

$$\begin{bmatrix} Y_{11} & Y_{12} & Y_{13} & Y_{14} \\ Y_{21} & Y_{22} & Y_{23} & Y_{24} \\ Y_{31} & Y_{32} & Y_{33} & Y_{34} \\ Y_{41} & Y_{42} & Y_{43} & Y_{44} \end{bmatrix} \begin{bmatrix} V_1 \\ V_2 \\ V_3 \\ V_4 \end{bmatrix} = \begin{bmatrix} I_1 \\ I_2 \\ I_3 \\ I_4 \end{bmatrix} \tag{7.38}$$

The four unknowns are the bus voltages V_1, V_2, V_3, and V_4 when the bus-injected currents I_1, I_2, I_3, and I_4 are specified. Generally, $\mathbf{Y}_{pr}$ is symmetrical, in which case taking the transpose of each side of Eq. (7.37) shows that $\mathbf{Y}_{bus}$ is also symmetrical.

Example 7.5. Determine the per-unit bus admittance matrix of the example system of Fig. 7.11 using the tree shown in Fig. 7.13 with reference node ⓪.

Solution. The primitive admittance matrix $\mathbf{Y}_{pr}$ describing the branch admittances is given by Eq. (7.28) and the branch-to-node incidence matrix $\mathbf{A}$ for the specified tree is given by Eq. (7.31). Therefore, performing the row-by-column multiplications for $\mathbf{A}^T\mathbf{Y}_{pr}$ indicated by

$$\begin{bmatrix} 0 & 0 & 1 & 0 \\ 0 & -1 & 1 & 0 \\ -1 & 0 & 1 & 0 \\ -1 & 1 & 0 & 0 \\ 0 & -1 & 0 & 1 \\ -1 & 0 & 0 & 1 \\ 0 & 0 & 0 & 1 \end{bmatrix}^T \begin{bmatrix} -j0.8 & \cdot & \cdot & \cdot & \cdot & \cdot & \cdot \\ \cdot & -j6.25 & j3.75 & \cdot & \cdot & \cdot & \cdot \\ \cdot & j3.75 & -j6.25 & \cdot & \cdot & \cdot & \cdot \\ \cdot & \cdot & \cdot & -j8.0 & \cdot & \cdot & \cdot \\ \cdot & \cdot & \cdot & \cdot & -j5.0 & \cdot & \cdot \\ \cdot & \cdot & \cdot & \cdot & \cdot & -j2.5 & \cdot \\ \cdot & \cdot & \cdot & \cdot & \cdot & \cdot & -j0.8 \end{bmatrix}$$

we obtain the intermediate result

$$\mathbf{A}^T\mathbf{Y}_{pr} = \begin{bmatrix} 0 & -j3.75 & j6.25 & j8.0 & 0 & j2.5 & 0 \\ 0 & j6.25 & -j3.75 & -j8.0 & j5.0 & 0 & 0 \\ -j0.8 & -j2.5 & -j2.5 & 0 & 0 & 0 & 0 \\ 0 & 0 & 0 & 0 & -j5.0 & -j2.5 & -j0.8 \end{bmatrix}$$

which we may now postmultiply by $\mathbf{A}$ to calculate

$$\mathbf{Y}_{bus} = \mathbf{A}^T\mathbf{Y}_{pr}\mathbf{A} = \begin{array}{c} \textcircled{1} \\ \textcircled{2} \\ \textcircled{3} \\ \textcircled{4} \end{array} \overset{\begin{array}{cccc} \textcircled{1} & \textcircled{2} & \textcircled{3} & \textcircled{4} \end{array}}{\begin{bmatrix} -j16.75 & j11.75 & j2.50 & j2.50 \\ j11.75 & -j19.25 & j2.50 & j5.00 \\ j2.50 & j2.50 & -j5.80 & 0 \\ j2.50 & j5.00 & 0 & -j8.30 \end{bmatrix}}$$

Since currents are injected at only buses ③ and ④, the nodal equations in matrix form are written

$$\begin{bmatrix} -j16.75 & j11.75 & j2.50 & j2.50 \\ j11.75 & -j19.25 & j2.50 & j5.00 \\ j2.50 & j2.50 & -j5.80 & 0 \\ j2.50 & j5.00 & 0 & -j8.30 \end{bmatrix} \begin{bmatrix} V_1 \\ V_2 \\ V_3 \\ V_4 \end{bmatrix} = \begin{bmatrix} 0 \\ 0 \\ 1.00\angle -90^\circ \\ 0.68\angle -135^\circ \end{bmatrix}$$

Example 7.6. Solve the node equations of Example 7.5 to find the bus voltages by inverting the bus admittance matrix.

Solution. Premultiplying both sides of the matrix nodal equation by the inverse of the bus admittance matrix (determined by using a standard program on a computer or calculator) yields

$$\begin{bmatrix} V_1 \\ V_2 \\ V_3 \\ V_4 \end{bmatrix} = \begin{bmatrix} j0.73128 & j0.69140 & j0.61323 & j0.63677 \\ j0.69140 & j0.71966 & j0.60822 & j0.64178 \\ j0.61323 & j0.60822 & j0.69890 & j0.55110 \\ j0.63677 & j0.64178 & j0.55110 & j0.69890 \end{bmatrix} \begin{bmatrix} 0 \\ 0 \\ 1.00\angle -90^\circ \\ 0.68\angle -135^\circ \end{bmatrix}$$

Performing the indicated multiplications, we obtain the per-unit results

$$\begin{bmatrix} V_1 \\ V_2 \\ V_3 \\ V_4 \end{bmatrix} = \begin{bmatrix} 0.96903\angle -18.4189^\circ \\ 0.96734\angle -18.6028^\circ \\ 0.99964\angle -15.3718^\circ \\ 0.94866\angle -20.7466^\circ \end{bmatrix} = \begin{bmatrix} 0.91939 - j0.30618 \\ 0.91680 - j0.30859 \\ 0.96388 - j0.26499 \\ 0.88715 - j0.33605 \end{bmatrix}$$

7.6 THE METHOD OF SUCCESSIVE ELIMINATION

In industry-based studies of power systems the networks being solved are geographically extensive and often encompass many hundreds of substations, generating plants and load centers. The $\mathbf{Y}_{\text{bus}}$ matrices for these large networks of thousands of nodes have associated systems of nodal equations to be solved for a correspondingly large number of unknown bus voltages. For such solutions computer-based numerical techniques are required to avoid direct matrix inversion, thereby minimizing computational effort and computer storage. The method of successive elimination, called *gaussian elimination*, underlies many of the numerical methods solving the equations of such large-scale power systems. We now describe this method using the nodal equations of the four-bus system

$$Y_{11}V_1 + Y_{12}V_2 + Y_{13}V_3 + Y_{14}V_4 = I_1 \tag{7.39}$$

$$Y_{21}V_1 + Y_{22}V_2 + Y_{23}V_3 + Y_{24}V_4 = I_2 \tag{7.40}$$

$$Y_{31}V_1 + Y_{32}V_2 + Y_{33}V_3 + Y_{34}V_4 = I_3 \tag{7.41}$$

$$Y_{41}V_1 + Y_{42}V_2 + Y_{43}V_3 + Y_{44}V_4 = I_4 \tag{7.42}$$

Gaussian elimination consists of reducing this system of four equations in the four unknowns V_1, V_2, V_3, and V_4 to a system of three equations in three unknowns and then to a reduced system of two equations in two unknowns, and so on until there remains only one equation in one unknown. The final equation yields a value for the corresponding unknown, which is then substituted back in the reduced sets of equations to calculate the remaining unknowns. The successive elimination of unknowns in the forward direction is called *forward elimination* and the substitution process using the latest calculated values is called *back substitution*. Forward elimination begins by selecting one equation and eliminating from this equation one variable whose coefficient is called the *pivot*. We exemplify the overall procedure by first eliminating V_1 from Eqs. (7.39) through (7.42) as follows:

Step 1

1. Divide Eq. (7.39) by the pivot Y_{11} to obtain

$$V_1 + \frac{Y_{12}}{Y_{11}}V_2 + \frac{Y_{13}}{Y_{11}}V_3 + \frac{Y_{14}}{Y_{11}}V_4 = \frac{1}{Y_{11}}I_1 \tag{7.43}$$

2. Multiply Eq. (7.43) by Y_{21}, Y_{31}, and Y_{41} and subtract the results from Eqs. (7.40) through (7.42), respectively, to get

$$\left(Y_{22} - \frac{Y_{21}Y_{12}}{Y_{11}}\right)V_2 + \left(Y_{23} - \frac{Y_{21}Y_{13}}{Y_{11}}\right)V_3 + \left(Y_{24} - \frac{Y_{21}Y_{14}}{Y_{11}}\right)V_4 = I_2 - \frac{Y_{21}}{Y_{11}}I_1 \tag{7.44}$$

$$\left(Y_{32} - \frac{Y_{31}Y_{12}}{Y_{11}}\right)V_2 + \left(Y_{33} - \frac{Y_{31}Y_{13}}{Y_{11}}\right)V_3 + \left(Y_{34} - \frac{Y_{31}Y_{14}}{Y_{11}}\right)V_4 = I_3 - \frac{Y_{31}}{Y_{11}}I_1 \tag{7.45}$$

$$\left(Y_{42} - \frac{Y_{41}Y_{12}}{Y_{11}}\right)V_2 + \left(Y_{43} - \frac{Y_{41}Y_{13}}{Y_{11}}\right)V_3 + \left(Y_{44} - \frac{Y_{41}Y_{14}}{Y_{11}}\right)V_4 = I_4 - \frac{Y_{41}}{Y_{11}}I_1 \tag{7.46}$$

Equations (7.43) through (7.46) may be written more compactly in the form

$$V_1 + \frac{Y_{12}}{Y_{11}}V_2 + \frac{Y_{13}}{Y_{11}}V_3 + \frac{Y_{14}}{Y_{11}}V_4 = \frac{1}{Y_{11}}I_1 \tag{7.47}$$

$$Y_{22}^{(1)}V_2 + Y_{23}^{(1)}V_3 + Y_{24}^{(1)}V_4 = I_2^{(1)} \tag{7.48}$$

$$Y_{32}^{(1)}V_2 + Y_{33}^{(1)}V_3 + Y_{34}^{(1)}V_4 = I_3^{(1)} \tag{7.49}$$

$$Y_{42}^{(1)}V_2 + Y_{43}^{(1)}V_3 + Y_{44}^{(1)}V_4 = I_4^{(1)} \tag{7.50}$$

where the superscript denotes the Step 1 set of derived coefficients

$$Y_{jk}^{(1)} = Y_{jk} - \frac{Y_{j1}Y_{1k}}{Y_{11}} \qquad \text{for } j \text{ and } k = 2,3,4 \tag{7.51}$$

and the modified right-hand side expressions

$$I_j^{(1)} = I_j - \frac{Y_{j1}}{Y_{11}} I_1 \qquad \text{for } j = 2, 3, 4 \tag{7.52}$$

Note that Eqs. (7.48) through (7.50) may now be solved for V_2, V_3, and V_4 since V_1 has been eliminated; and the coefficients constitute a reduced 3×3 matrix, which can be interpreted as representing a reduced equivalent network with bus ① absent. In this three-bus equivalent the voltages V_2, V_3, and V_4 have exactly the same values as in the original four-bus system. Moreover, the effect of the current injection I_1 on the network is taken into account at buses ②, ③, and ④, as shown by Eq. (7.52). The current I_1 at bus ① is multiplied by the factor $-Y_{j1}/Y_{11}$ before being *distributed*, so to speak, to each bus ⓙ still in the network.

We next consider the elimination of the variable V_2.

Step 2

1. Divide Eq. (7.48) by the new pivot $Y_{22}^{(1)}$ to obtain

$$V_2 + \frac{Y_{23}^{(1)}}{Y_{22}^{(1)}} V_3 + \frac{Y_{24}^{(1)}}{Y_{22}^{(1)}} V_4 = \frac{1}{Y_{22}^{(1)}} I_2^{(1)} \tag{7.53}$$

2. Multiply Eq. (7.53) by $Y_{32}^{(1)}$ and $Y_{42}^{(1)}$ and subtract the results from Eqs. (7.49) and (7.50) to get

$$\left(Y_{33}^{(1)} - \frac{Y_{32}^{(1)} Y_{23}^{(1)}}{Y_{22}^{(1)}}\right) V_3 + \left(Y_{34}^{(1)} - \frac{Y_{32}^{(1)} Y_{24}^{(1)}}{Y_{22}^{(1)}}\right) V_4 = I_3^{(1)} - \frac{Y_{32}^{(1)}}{Y_{22}^{(1)}} I_2^{(1)} \tag{7.54}$$

$$\left(Y_{43}^{(1)} - \frac{Y_{42}^{(1)} Y_{23}^{(1)}}{Y_{22}^{(1)}}\right) V_3 + \left(Y_{44}^{(1)} - \frac{Y_{42}^{(1)} Y_{24}^{(1)}}{Y_{22}^{(1)}}\right) V_4 = I_4^{(1)} - \frac{Y_{42}^{(1)}}{Y_{22}^{(1)}} I_2^{(1)} \tag{7.55}$$

In a manner similar to that of Step 1, we rewrite Eqs. (7.53) through (7.55) in the form

$$V_2 + \frac{Y_{23}^{(1)}}{Y_{22}^{(1)}} V_3 + \frac{Y_{24}^{(1)}}{Y_{22}^{(1)}} V_4 = \frac{1}{Y_{22}^{(1)}} I_2^{(1)} \tag{7.56}$$

$$Y_{33}^{(2)} V_3 + Y_{34}^{(2)} V_4 = I_3^{(2)} \tag{7.57}$$

$$Y_{43}^{(2)} V_3 + Y_{44}^{(2)} V_4 = I_4^{(2)} \tag{7.58}$$

where the second set of calculated coefficients is given by

$$Y_{jk}^{(2)} = Y_{jk}^{(1)} - \frac{Y_{j2}^{(1)} Y_{2k}^{(1)}}{Y_{22}^{(1)}} \quad \text{for } j \text{ and } k = 3, 4 \tag{7.59}$$

and the net currents injected at buses ③ and ④ are

$$I_j^{(2)} = I_j^{(1)} - \frac{Y_{j2}^{(1)}}{Y_{22}^{(1)}} I_2^{(1)} \quad \text{for } j = 3, 4 \tag{7.60}$$

Equations (7.57) and (7.58) describe a further reduced equivalent network having only buses ③ and ④. Voltages V_3 and V_4 are exactly the same as those of the original four-bus network because the current injections $I_3^{(2)}$ and $I_4^{(2)}$ represent the effects of all the original current sources.

We now consider elimination of the variable V_3.

Step 3

1. Divide Eq. (7.57) by the pivot $Y_{33}^{(2)}$ to obtain

$$V_3 + \frac{Y_{34}^{(2)}}{Y_{33}^{(2)}} V_4 = \frac{1}{Y_{33}^{(2)}} I_3^{(2)} \tag{7.61}$$

2. Multiply Eq. (7.61) by $Y_{43}^{(2)}$ and subtract the result from Eq. (7.58) to obtain

$$Y_{44}^{(3)} V_4 = I_4^{(3)} \tag{7.62}$$

in which we have defined

$$Y_{44}^{(3)} = Y_{44}^{(2)} - \frac{Y_{43}^{(2)} Y_{34}^{(2)}}{Y_{33}^{(2)}} \quad \text{and} \quad I_4^{(3)} = I_4^{(2)} - \frac{Y_{43}^{(2)}}{Y_{33}^{(2)}} I_3^{(2)} \tag{7.63}$$

Equation (7.62) describes the single equivalent branch admittance $Y_{44}^{(3)}$ with voltage V_4 from bus ④ to reference caused by the equivalent injected current $I_4^{(3)}$.

The final step in the forward elimination process yields V_4.

Step 4

1. Divide Eq. (7.62) by $Y_{44}^{(3)}$ to obtain

$$V_4 = \frac{1}{Y_{44}^{(3)}} I_4^{(3)} \tag{7.64}$$

At this point we have found a value for bus voltage V_4 which can be substituted back in Eq. (7.61) to obtain a value for V_3. Continuing this process of back substitution using the values of V_3 and V_4 in Eq. (7.56), we obtain V_2 and then solve for V_1 from Eq. (7.47).

Thus, the gaussian-elimination procedure demonstrated here for a four-bus system provides a systematic means of solving large systems of equations without having to invert the coefficient matrix. This is most desirable when a large-scale power system is being analyzed. The following example numerically illustrates the procedure.

Example 7.7. Using gaussian elimination, solve the nodal equations of Example 7.5 to find the bus voltages. At each step of the solution find the equivalent circuit of the reduced coefficient matrix.

Solution. In Example 7.5 the nodal admittance equations in matrix form are found to be

$$\begin{array}{c} \\ ① \\ ② \\ ③ \\ ④ \end{array} \begin{array}{c} \begin{array}{cccc} ① & ② & ③ & ④ \end{array} \\ \begin{bmatrix} -j16.75 & j11.75 & \boxed{j2.50} & j2.50 \\ \boxed{j11.75} & -j19.25 & \underline{j2.50} & j5.00 \\ j2.50 & j2.50 & -j5.80 & 0.00 \\ j2.50 & j5.00 & 0.00 & -j8.30 \end{bmatrix} \end{array} \begin{bmatrix} V_1 \\ V_2 \\ V_3 \\ V_4 \end{bmatrix} = \begin{bmatrix} 0 \\ 0 \\ 1.00\angle -90^\circ \\ 0.68\angle -135^\circ \end{bmatrix}$$

Step 1

To eliminate the variable V_1 from rows 2, 3, and 4, we first divide the first row by the pivot $-j16.75$ to obtain

$$V_1 - 0.70149V_2 - 0.14925V_3 - 0.14925V_4 = 0$$

We now use this equation to eliminate the $j11.75$ entry in (row 2, column 1) of $\mathbf{Y}_{\text{bus}}$, and in the process all the other elements of row 2 are modified. Equation (7.51) shows the procedure. For example, to modify the element $j2.50$ underlined in (row 2, column 3), subtract from it the product of the elements enclosed by rectangles divided by the pivot $-j16.75$; that is,

$$Y_{23}^{(1)} = Y_{23} - \frac{Y_{21}Y_{13}}{Y_{11}} = j2.50 - \frac{j11.75 \times j2.50}{-j16.75} = j4.25373 \text{ per unit}$$

Similarly, the other elements of new row 2 are

$$Y_{22}^{(1)} = -j19.25 - \frac{j11.75 \times j11.75}{-j16.75} = -j11.00746 \text{ per unit}$$

$$Y_{24}^{(1)} = j5.00 - \frac{j11.75 \times j2.50}{-j16.75} = j6.75373 \text{ per unit}$$

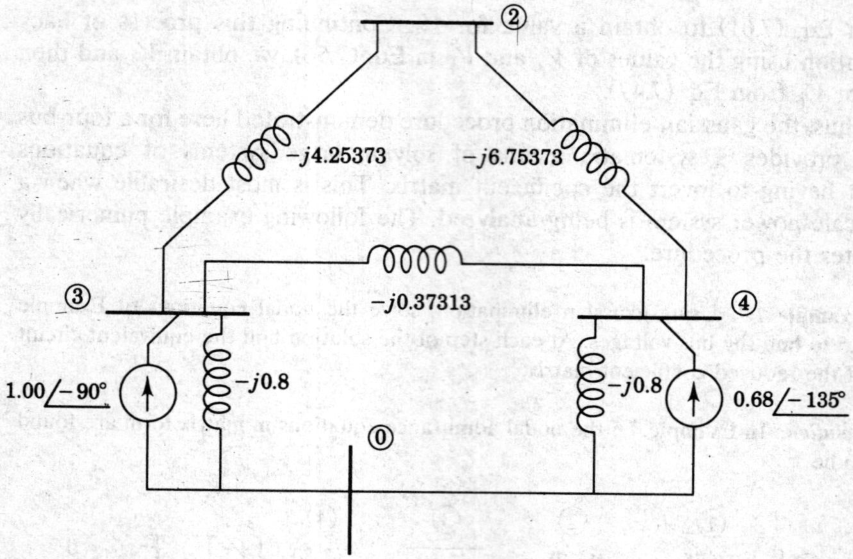

FIGURE 7.14
The equivalent three-bus network following Step 1 of Example 7.7

Modified elements of rows 3 and 4 are likewise found to yield

$$\left[\begin{array}{c|ccc} 1 & -0.70149 & -0.14925 & -0.14925 \\ \hline 0 & -j11.00746 & j4.25373 & j6.75373 \\ 0 & j4.25373 & -j5.42686 & j0.37313 \\ 0 & j6.75373 & j0.37313 & -j7.92686 \end{array}\right]\left[\begin{array}{c} V_1 \\ \hline V_2 \\ V_3 \\ V_4 \end{array}\right] = \left[\begin{array}{c} 0 \\ \hline 0 \\ 1.00\angle -90^\circ \\ 0.68\angle -135^\circ \end{array}\right]$$

Because $I_1 = 0$, no current is distributed from bus ① to the remaining buses ②, ③, and ④, and so the currents I_2^1, I_3^1, and I_4^1 in the right-hand-side vector have the same values before and after Step 1. The partitioned system of equations involving the unknown voltages V_2, V_3, and V_4 corresponds to the three-bus equivalent network constructed in Fig. 7.14 from the reduced coefficient matrix.

Step 2

Forward elimination applied to the partitioned 3×3 system of the last equation proceeds as in Step 1 to yield

$$\left[\begin{array}{cc|cc} 1 & -0.70149 & -0.14925 & -0.14925 \\ 0 & 1 & -0.38644 & -0.61356 \\ \hline 0 & 0 & -j3.78305 & j2.98305 \\ 0 & 0 & j2.98305 & -j3.78305 \end{array}\right]\left[\begin{array}{c} V_1 \\ V_2 \\ \hline V_3 \\ V_4 \end{array}\right] = \left[\begin{array}{c} 0 \\ 0 \\ \hline 1.00\angle -90^\circ \\ 0.68\angle -135^\circ \end{array}\right]$$

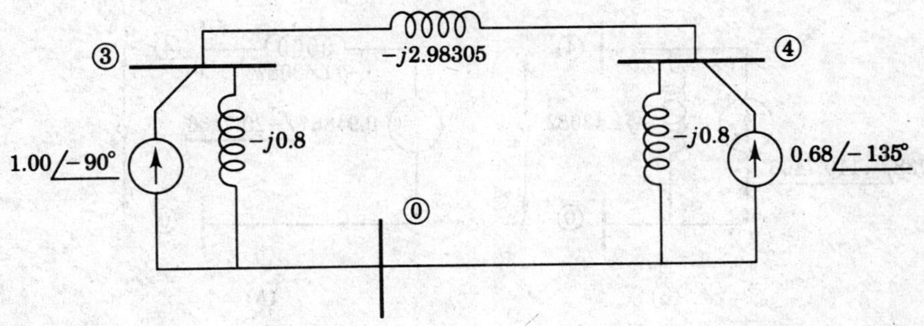

FIGURE 7.15
The equivalent two-bus network following Step 2 of Example 7.7.

in which the injected currents $I_3^{(2)}$ and $I_4^{(2)}$ of Step 2 are also unchanged because $I_2^{(1)} = I_2 = 0$. At this stage we have eliminated V_1 and V_2 from the original 4 × 4 system of equations and there remains the 2 × 2 system in the variables V_3 and V_4 corresponding to $\mathbf{Y}_{\text{bus}}$ of Fig. 7.15. Note that nodes ① and ② are eliminated.

Step 3

Continuing the forward elimination, we find

$$\left[\begin{array}{ccc|c} 1 & -0.70149 & -0.14925 & -0.14925 \\ 0 & 1 & -0.38644 & -0.61356 \\ 0 & 0 & 1 & -0.78853 \\ \hline 0 & 0 & 0 & -j1.43082 \end{array}\right]\left[\begin{array}{c} V_1 \\ V_2 \\ V_3 \\ \hline V_4 \end{array}\right] = \left[\begin{array}{c} 0 \\ 0 \\ 0.26434\angle 0^\circ \\ \hline 1.35738\angle -110.7466^\circ \end{array}\right]$$

in which the entry for bus ③ in the right-hand-side vector is calculated to be

$$\frac{I_3^{(2)}}{Y_{33}^{(2)}} = \frac{1.00\angle -90^\circ}{3.78305\angle -90^\circ} = 0.26434\angle 0^\circ \text{ per unit}$$

and the modified current at bus ④ is given by

$$I_4^{(3)} = I_4^{(2)} - \frac{Y_{43}^{(2)}}{Y_{33}^{(2)}} I_3^{(2)}$$

$$= 0.68\angle -135^\circ - \frac{j2.98305}{-j3.78305} 1.00\angle -90^\circ$$

$$= 1.35738\angle -110.7466^\circ \text{ per unit}$$

Figure 7.16(*a*) shows the single admittance resulting from Step 3.

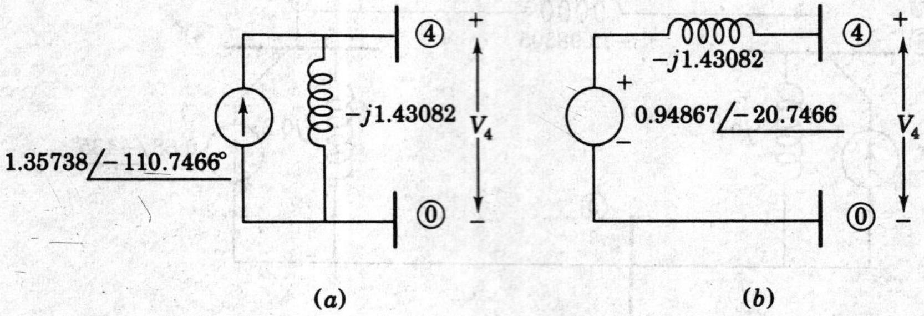

FIGURE 7.16
The equivalent circuits following (*a*) Step 3 and (*b*) Step 4 of Example 7.7.

Step 4

The forward-elimination process terminates with the calculation

$$V_4 = \frac{I_4^{(3)}}{Y_{44}^{(3)}} = \frac{1.35738\angle -110.7466°}{1.43082\angle -90°} = 0.94867\angle -20.7466° \text{ per unit}$$

corresponding to the source transformation of Fig. 7.16(*b*). Therefore, forward elimination leads to the triangular coefficient matrix given by

$$\begin{bmatrix} 1 & -0.70149 & -0.14925 & -0.14925 \\ 0 & 1 & -0.38644 & -0.61356 \\ 0 & 0 & 1 & -0.78853 \\ 0 & 0 & 0 & 1 \end{bmatrix} \begin{bmatrix} V_1 \\ V_2 \\ V_3 \\ V_4 \end{bmatrix} = \begin{bmatrix} 0 \\ 0 \\ 0.26434\angle 0° \\ 0.94867\angle -20.7466° \end{bmatrix}$$

Since $V_4 = 0.94867\angle -20.7466°$, we now begin the back-substitution process to determine V_3 using the third row entries as follows:

$$V_3 - 0.78853V_4 = V_3 - 0.74805\angle -20.7466° = 0.26434\angle 0°$$

which yields

$$V_3 = 0.99965\angle -15.3716° \text{ per unit}$$

Back substitution for V_3 and V_4 in the second-row equation

$$V_2 - 0.38644V_3 - 0.61356V_4 = 0$$

yields

$$V_2 = 0.96734\angle -18.6030° \text{ per unit}$$

When the calculated values of V_2, V_3, and V_4 are substituted in the first-row equation

$$V_1 - 0.70149V_2 - 0.14925V_3 - 0.14925V_4 = 0$$

we obtain

$$V_1 = 0.96903\underline{/-18.4189^\circ} \text{ per unit}$$

and so the per-unit bus voltages are

$$V_1 = 0.96903\underline{/-18.4189^\circ} = 0.91939 - j0.30618$$

$$V_2 = 0.96734\underline{/-18.6030^\circ} = 0.91680 - j0.30859$$

$$V_3 = 0.99964\underline{/-15.3716^\circ} = 0.96388 - j0.26499$$

$$V_4 = 0.94867\underline{/-20.7466^\circ} = 0.88715 - j0.33605$$

which agree almost exactly with the results found in Example 7.6.

7.7 NODE ELIMINATION (KRON REDUCTION)

Section 7.6 shows that gaussian elimination removes the need for matrix inversion when solving the nodal equations of a large-scale power system. At the same time it is also shown that elimination of variables is identical to network reduction since it leads to a sequence of reduced-order network equivalents by node elimination at each step. This may be important in analyzing a large interconnected power system if there is special interest in the voltages at only some of the buses of the overall system. For instance, one electric utility company with interconnections to other companies may wish to confine its study of voltage levels to those substations within its own service territory. By selective numbering of the system buses, we may apply gaussian elimination so as to reduce the $\mathbf{Y}_{bus}$ equations of the overall system to a set which contains only those bus voltages of special interest. The coefficient matrix in the reduced-order set of equations then represents the $\mathbf{Y}_{bus}$ for an equivalent network containing only those buses which are to be retained. All other buses are eliminated in the mathematical sense that their bus voltages and current injections do not appear explicitly. Such reduction in size of the equation set leads to efficiency of computation and helps to focus more directly on that portion of the overall network which is of primary interest.

In gaussian elimination one bus-voltage variable at a time is sequentially removed from the original system of N equations in N unknowns. Following Step 1 of the procedure, the variable V_1 does not explicitly appear in the resultant $(N-1) \times (N-1)$ system, which fully represents the original network if the actual value of the voltage V_1 at bus ① is not of direct interest. If knowledge of V_2 is also not of prime interest, we can interpret the $(N-2) \times (N-2)$ system of equations resulting from Step 2 of the procedure as replacing the actual network by an $(N-2)$ bus equivalent with buses ① and ② removed, and so on. Consequently, in our network calculations if it is advantageous to do so, we may eliminate k nodes from the network representation by employing the first k steps of the gaussian-elimination procedure. Of course, the current injections (if any) at the eliminated nodes are taken into account at the remaining $(N-k)$ nodes by successive application of expressions such as that in Eq. (7.54).

Current injection is always zero at those buses of the network to which there is no external load or generating source connected. At such buses it is usually not necessary to calculate the voltages explicitly, and so we may eliminate them from our representation. For example, when $I_1 = 0$ in the four-bus system, we may write nodal admittance equations in the form

$$\begin{array}{c} \\ ① \\ ② \\ ③ \\ ④ \end{array} \begin{array}{c} \begin{array}{cccc} ① & ② & ③ & ④ \end{array} \\ \begin{bmatrix} Y_{11} & Y_{12} & Y_{13} & Y_{14} \\ Y_{21} & Y_{22} & Y_{23} & Y_{24} \\ Y_{31} & Y_{32} & Y_{33} & Y_{34} \\ Y_{41} & Y_{42} & Y_{43} & Y_{44} \end{bmatrix} \end{array} \begin{bmatrix} V_1 \\ V_2 \\ V_3 \\ V_4 \end{bmatrix} = \begin{bmatrix} 0 \\ I_2 \\ I_3 \\ I_4 \end{bmatrix} \tag{7.65}$$

and following elimination of node ①, we obtain the 3×3 system

$$\begin{array}{c} \\ ② \\ ③ \\ ④ \end{array} \begin{array}{c} \begin{array}{ccc} ② & ③ & ④ \end{array} \\ \begin{bmatrix} Y_{22}^{(1)} & Y_{23}^{(1)} & Y_{24}^{(1)} \\ Y_{32}^{(1)} & Y_{33}^{(1)} & Y_{34}^{(1)} \\ Y_{42}^{(1)} & Y_{43}^{(1)} & Y_{44}^{(1)} \end{bmatrix} \end{array} \begin{bmatrix} V_2 \\ V_3 \\ V_4 \end{bmatrix} = \begin{bmatrix} I_2 \\ I_3 \\ I_4 \end{bmatrix} \tag{7.66}$$

in which the superscripted elements of the reduced coefficient matrix are calculated as before. Systems in which those nodes with zero current injections are eliminated are said to be *Kron*[1] *reduced*. Hence, the system having the particular form of Eq. (7.65) is Kron reduced to Eq. (7.66), and for such systems node elimination and Kron reduction are synonymous terms.

Of course, regardless of which node has the zero current injection, a system can be Kron reduced without having to rearrange the equations as in Eq. (7.65). For example, if $I_p = 0$ in the nodal equations of the N-bus system, we

---

[1]After Dr. Gabriel Kron (1901–1968) of General Electric Company, Schenectady, NY, who contributed greatly to power system analysis.

may directly calculate the elements of the new, reduced bus admittance matrix by choosing Y_{pp} as the pivot and by eliminating bus p using the formula

$$Y_{jk(\text{new})} = Y_{jk} - \frac{Y_{jp}Y_{pk}}{Y_{pp}} \tag{7.67}$$

where j and k take on all the integer values from 1 to N except p since row p and column p are to be eliminated. The subscript (new) distinguishes the elements in the new $\mathbf{Y}_{\text{bus}}$ of dimension $(N-1)\times(N-1)$ from the elements in the original $\mathbf{Y}_{\text{bus}}$.

Example 7.8. Using Y_{22} as the initial pivot, eliminate node ② and the corresponding voltage V_2 from the 4 × 4 system of Example 7.7

Solution. The pivot Y_{22} equals $-j19.25$. With p set equal to 2 in Eq. (7.67), we may eliminate row 2 and column 2 from $\mathbf{Y}_{\text{bus}}$ of Example 7.7 to obtain the new row 1 elements

$$Y_{11(\text{new})} = Y_{11} - \frac{Y_{12}Y_{21}}{Y_{22}} = -j16.75 - \frac{(j11.75)(j11.75)}{-j19.25} = -j9.57792$$

$$Y_{13(\text{new})} = Y_{13} - \frac{Y_{12}Y_{23}}{Y_{22}} = j2.50 - \frac{(j11.75)(j2.50)}{-j19.25} = j4.02597$$

$$Y_{14(\text{new})} = Y_{14} - \frac{Y_{12}Y_{24}}{Y_{22}} = j2.50 - \frac{(j11.75)(j5.00)}{-j19.25} = j5.55195$$

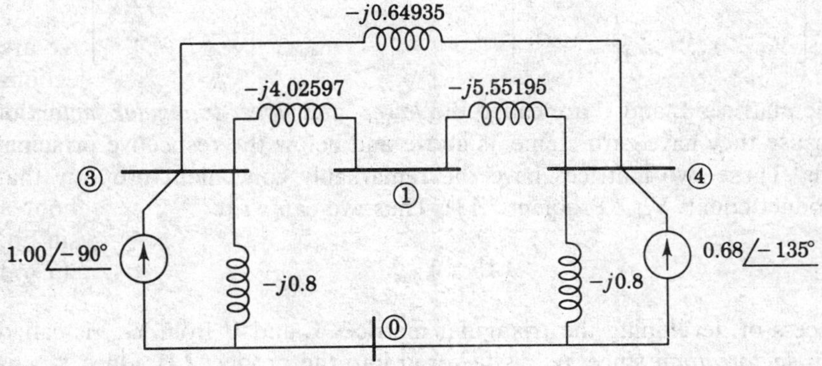

FIGURE 7.17
The Kron-reduced network of Example 7.8.

Similar calculations yield the other elements of the Kron-reduced matrix

$$\begin{matrix} & \textcircled{1} & \textcircled{3} & \textcircled{4} \\ \begin{matrix}\textcircled{1}\\ \textcircled{3}\\ \textcircled{4}\end{matrix} & \multicolumn{3}{c}{\begin{bmatrix} -j9.57791 & j4.02597 & j5.55195 \\ j4.02597 & -j5.47532 & j0.64935 \\ j5.55195 & j0.64935 & -j7.00130 \end{bmatrix}} \end{matrix} \begin{bmatrix} V_1 \\ V_3 \\ V_4 \end{bmatrix} = \begin{bmatrix} 0 \\ 1.00\angle -90^\circ \\ 0.68\angle -135^\circ \end{bmatrix}$$

Because the coefficient matrix is symmetrical, the equivalent circuit of Fig. 7.17 applies. Further use of Eq. (7.67) to eliminate node ① from Fig. 7.17 leads to the Kron-reduced equivalent circuit shown in Fig. 7.15.

7.8 TRIANGULAR FACTORIZATION

In practical studies the nodal admittance equations of a given large-scale power system are solved under different operating conditions. Often in such studies the network configuration and parameters are fixed and the operating conditions differ only because of changes made to the external sources connected at the system buses. In all such cases the same $\mathbf{Y}_{\text{bus}}$ applies and the problem then is to solve repeatedly for the voltages corresponding to different sets of current injections. In finding repeat solutions considerable computational effort is avoided if all the calculations in the forward phase of the gaussian-elimination procedure do not have to be repeated. This may be accomplished by expressing $\mathbf{Y}_{\text{bus}}$ as the product of two matrices $\mathbf{L}$ and $\mathbf{U}$ defined for the four-bus system by

$$\mathbf{L} = \begin{bmatrix} Y_{11} & \cdot & \cdot & \cdot \\ Y_{21} & Y_{22}^{(1)} & \cdot & \cdot \\ Y_{31} & Y_{32}^{(1)} & Y_{33}^{(2)} & \cdot \\ Y_{41} & Y_{42}^{(1)} & Y_{43}^{(2)} & Y_{44}^{(3)} \end{bmatrix} \qquad \mathbf{U} = \begin{bmatrix} 1 & \dfrac{Y_{12}}{Y_{11}} & \dfrac{Y_{13}}{Y_{11}} & \dfrac{Y_{14}}{Y_{11}} \\ \cdot & 1 & \dfrac{Y_{23}^{(1)}}{Y_{22}^{(1)}} & \dfrac{Y_{24}^{(1)}}{Y_{22}^{(1)}} \\ \cdot & \cdot & 1 & \dfrac{Y_{34}^{(2)}}{Y_{33}^{(2)}} \\ \cdot & \cdot & \cdot & 1 \end{bmatrix} \tag{7.68}$$

The matrices $\mathbf{L}$ and $\mathbf{U}$ are called the *lower-* and *upper-triangular factors* of $\mathbf{Y}_{\text{bus}}$ because they have zero elements above and below the respective principal diagonals. These two matrices have the remarkably convenient property that their product equals $\mathbf{Y}_{\text{bus}}$ (Problem 7.13). Thus, we can write

$$\mathbf{LU} = \mathbf{Y}_{\text{bus}} \tag{7.69}$$

The process of developing the triangular matrices $\mathbf{L}$ and $\mathbf{U}$ from $\mathbf{Y}_{\text{bus}}$ is called *triangular factorization* since $\mathbf{Y}_{\text{bus}}$ is factored into the product $\mathbf{LU}$. Once $\mathbf{Y}_{\text{bus}}$ is so factored, the calculations in the forward-elimination phase of gaussian elimination do not have to be repeated since both **L** and **U** are unique and do not change for a given $\mathbf{Y}_{\text{bus}}$. The entries in **L** and **U** are formed by systematically recording the outcome of the calculations at each step of a single pass through the forward-elimination process. Thus, in forming **L** and **U**, no new calculations are involved.

This is now demonstrated for the four-bus system with coefficient matrix

$$\mathbf{Y}_{\text{bus}} = \begin{array}{c} \\ ① \\ ② \\ ③ \\ ④ \end{array} \begin{array}{c} \begin{array}{cccc} ① & ② & ③ & ④ \end{array} \\ \begin{bmatrix} Y_{11} & Y_{12} & Y_{13} & Y_{14} \\ Y_{21} & Y_{22} & Y_{23} & Y_{24} \\ Y_{31} & Y_{32} & Y_{33} & Y_{34} \\ Y_{41} & Y_{42} & Y_{43} & Y_{44} \end{bmatrix} \end{array} \tag{7.70}$$

When gaussian elimination is applied to the four nodal equations corresponding to this $\mathbf{Y}_{\text{bus}}$, we note the following.

Step 1 yields results given by Eqs. (7.47) through (7.50) in which:

1. The coefficients Y_{11}, Y_{21}, Y_{31}, and Y_{41} are eliminated from the *first column* of the original coefficient matrix of Eq. (7.70).
2. New coefficients 1, Y_{12}/Y_{11}, Y_{13}/Y_{11}, and Y_{14}/Y_{11} are generated to replace those in the *first row* of Eq. (7.70).

The coefficients in the other rows and columns are also altered, but we keep a separate record of only those specified in (1) and (2) since these are the only results from Step 1 which are neither used nor altered in Step 2 or subsequent steps of the forward-elimination process. Column 1 of **L** and row 1 of **U** in Eqs. (7.68) show the recorded coefficients.

Step 2 yields results given by Eqs. (7.56) through (7.58) in which:

1. Coefficients $Y_{22}^{(1)}$, $Y_{32}^{(1)}$, and $Y_{42}^{(1)}$ are eliminated from the *second column* of the reduced coefficient matrix corresponding to Eqs. (7.47) through (7.50).
2. New coefficients 1, $Y_{23}^{(1)}/Y_{22}^{(1)}$, and $Y_{24}^{(1)}/Y_{22}^{(1)}$ are generated in Eq. (7.56) for the *second row*.

These coefficients are not needed in the remaining steps of forward elimination, and so we record them as column 2 of **L** and row 2 of **U** to show the Step 2 record. Continuing this record-keeping procedure, we form columns 3 and 4 of **L** and rows 3 and 4 of **U** using the results from Steps 3 and 4 of Sec. 7.6.

Therefore, matrix **L** is simply a record of those columns which are successively eliminated and matrix **U** records those row entries which are successively generated at each step of the forward stage of gaussian elimination.

We may use the triangular factors to solve the original system of equations by substituting the product **LU** for $\mathbf{Y}_{bus}$ in Eq. (7.38) to obtain

$$\underbrace{\mathbf{L}}_{N\times N}\;\underbrace{\mathbf{U}}_{N\times N}\;\underbrace{\mathbf{V}}_{N\times 1} = \underbrace{\mathbf{I}}_{N\times 1} \tag{7.71}$$

As an intermediate step in the solution of Eq. (7.71), we may replace the product **UV** by a new voltage vector **V**′ such that

$$\underbrace{\mathbf{L}}_{N\times N}\;\underbrace{\mathbf{V}'}_{N\times 1} = \underbrace{\mathbf{I}}_{N\times 1} \quad \text{and} \quad \underbrace{\mathbf{U}}_{N\times N}\;\underbrace{\mathbf{V}}_{N\times 1} = \underbrace{\mathbf{V}'}_{N\times 1} \tag{7.72}$$

Expressing Eq. (7.72) in full format shows that the original system of Eq. (7.38) is now replaced by two triangular systems given by

$$\begin{bmatrix} Y_{11} & \cdot & \cdot & \cdot \\ Y_{21} & Y_{22}^{(1)} & \cdot & \cdot \\ Y_{31} & Y_{32}^{(1)} & Y_{33}^{(2)} & \cdot \\ Y_{41} & Y_{42}^{(1)} & Y_{43}^{(2)} & Y_{44}^{(3)} \end{bmatrix} \begin{bmatrix} V_1' \\ V_2' \\ V_3' \\ V_4' \end{bmatrix} = \begin{bmatrix} I_1 \\ I_2 \\ I_3 \\ I_4 \end{bmatrix} \tag{7.73}$$

and

$$\begin{bmatrix} 1 & \dfrac{Y_{12}}{Y_{11}} & \dfrac{Y_{13}}{Y_{11}} & \dfrac{Y_{14}}{Y_{11}} \\ \cdot & 1 & \dfrac{Y_{23}^{(1)}}{Y_{22}^{(1)}} & \dfrac{Y_{24}^{(1)}}{Y_{22}^{(1)}} \\ \cdot & \cdot & 1 & \dfrac{Y_{34}^{(2)}}{Y_{33}^{(2)}} \\ \cdot & \cdot & \cdot & 1 \end{bmatrix} \begin{bmatrix} V_1 \\ V_2 \\ V_3 \\ V_4 \end{bmatrix} = \begin{bmatrix} V_1' \\ V_2' \\ V_3' \\ V_4' \end{bmatrix} \tag{7.74}$$

The lower triangular system of Eq. (7.73) is readily solved by forward substitution beginning with V_1'. We then use the calculated values of V_1', V_2', V_3', and V_4' to solve Eq. (7.74) by back substitution for the actual unknowns V_1, V_2, V_3, and V_4.

Therefore, when changes are made in the current vector **I**, the solution vector **V** is found in two sequential steps; the first involves forward substitution using **L** and the second employs back substitution using **U**.

Example 7.9. Using the triangular factors of $\mathbf{Y}_{bus}$, determine the voltage at bus ③ of Fig. 7.11 when the current source at bus ④ is changed to $I_4 = 0.60\angle -120^\circ$ per unit. All other conditions of Fig. 7.11 are unchanged.

Solution. The $\mathbf{Y}_{bus}$ for the network of Fig. 7.11 is given in Example 7.3. The corresponding matrix **L** may be assembled column by column from Example 7.7 simply by recording the column which is eliminated from the coefficient matrix at each step of the forward-elimination procedure. Then, substituting for **L** and the new current vector **I** in the equation $\mathbf{LV}' = \mathbf{I}$, we obtain

$$\begin{bmatrix} -j16.75 & \cdot & \cdot & \cdot \\ j11.75 & -j11.00746 & \cdot & \cdot \\ j2.50 & j4.25373 & -j3.78305 & \cdot \\ j2.50 & j6.75373 & j2.98305 & -j1.43082 \end{bmatrix} \begin{bmatrix} V_1' \\ V_2' \\ V_3' \\ V_4' \end{bmatrix} = \begin{bmatrix} 0 \\ 0 \\ 1.00\angle -90^\circ \\ 0.60\angle -120^\circ \end{bmatrix}$$

Solution by forward substitution beginning with V_1' yields

$$V_1' = V_2' = 0; \qquad V_3' = \frac{1.00\angle -90^\circ}{3.78305\angle -90^\circ} = 0.26434$$

$$V_4' = \frac{0.60\angle -120^\circ - (j2.98305)V_3'}{-j1.43082} = 0.93800\angle -12.9163^\circ$$

The matrix **U** is shown directly following Step 4 of the forward elimination in Example 7.7. Substituting **U** and the calculated entries of $\mathbf{V}'$ in the equation $\mathbf{UV} = \mathbf{V}'$ gives

$$\begin{bmatrix} 1 & -0.70149 & -0.14925 & -0.14925 \\ \cdot & 1 & -0.38644 & -0.61356 \\ \cdot & \cdot & 1 & -0.78853 \\ \cdot & \cdot & \cdot & 1 \end{bmatrix} \begin{bmatrix} V_1 \\ V_2 \\ V_3 \\ V_4 \end{bmatrix} = \begin{bmatrix} 0 \\ 0 \\ 0.26434 \\ 0.93800\angle -12.9163^\circ \end{bmatrix}$$

which we may solve by back substitution to obtain

$$V_4 = V_4' = 0.93800\angle -12.9163^\circ \text{ per unit}$$

$$V_3 = 0.26434 - (-0.78853)V_4 = 0.99904\angle -9.5257^\circ \text{ per unit}$$

If desired, we may continue the back substitution using the values for V_3 and V_4 to evaluate $V_2 = 0.96118\angle -11.5551^\circ$ per unit and $V_1 = 0.96324\angle -11.4388^\circ$ per unit.

When the coefficient matrix $\mathbf{Y}_{bus}$ is symmetrical, which is almost always the case, an important simplification results. As can be seen by inspection of Eq. (7.68), when the first column of $\mathbf{L}$ is divided by Y_{11}, we obtain the first row of $\mathbf{U}$; when the second column of $\mathbf{L}$ is divided by $Y_{22}^{(1)}$, we obtain the second row of $\mathbf{U}$; and so on for the other columns and rows of Eq. (7.68), provided $Y_{ij} = Y_{ji}$. Therefore, dividing the entries in each column of $\mathbf{L}$ by the principal diagonal element in that column yields $\mathbf{U}^T$ whenever $\mathbf{Y}_{bus}$ is symmetrical. We can then write

$$\mathbf{L} = \mathbf{U}^T\mathbf{D} = \begin{bmatrix} 1 & \cdot & \cdot & \cdot \\ \dfrac{Y_{21}}{Y_{11}} & 1 & \cdot & \cdot \\ \dfrac{Y_{31}}{Y_{11}} & \dfrac{Y_{32}^{(1)}}{Y_{22}^{(1)}} & 1 & \cdot \\ \dfrac{Y_{41}}{Y_{11}} & \dfrac{Y_{42}^{(1)}}{Y_{22}^{(1)}} & \dfrac{Y_{43}^{(2)}}{Y_{33}^{(2)}} & 1 \end{bmatrix} \begin{bmatrix} Y_{11} & \cdot & \cdot & \cdot \\ \cdot & Y_{22}^{(1)} & \cdot & \cdot \\ \cdot & \cdot & Y_{33}^{(2)} & \cdot \\ \cdot & \cdot & \cdot & Y_{44}^{(3)} \end{bmatrix} \tag{7.75}$$

where diagonal matrix $\mathbf{D}$ contains the diagonal elements of $\mathbf{L}$. Substituting in Eq. (7.71) for $\mathbf{L}$ from Eq. (7.75), we obtain the nodal admittance equations in the form

$$\mathbf{Y}_{bus}\mathbf{V} = \mathbf{U}^T\mathbf{D}\mathbf{U}\mathbf{V} = \mathbf{I} \tag{7.76}$$

Equation (7.76) may be solved for the unknown voltages $\mathbf{V}$ in three consecutive steps as follows:

$$\mathbf{U}^T\mathbf{V}'' = \mathbf{I} \tag{7.77}$$

$$\mathbf{D}\mathbf{V}' = \mathbf{V}'' \tag{7.78}$$

$$\mathbf{U}\mathbf{V} = \mathbf{V}' \tag{7.79}$$

These equations will be recognized as an extension of Eqs. (7.72). The intermediate result $\mathbf{V}''$ is first found from Eq. (7.77) by forward substitution. Next, each entry in $\mathbf{V}'$ is calculated from Eq. (7.78) by dividing the corresponding element of $\mathbf{V}''$ by the appropriate diagonal element of $\mathbf{D}$. Finally, the solution $\mathbf{V}$ is obtained from Eq. (7.79) by back substitution as demonstrated in Example 7.9.

Example 7.10. Using Eqs. (7.77) through (7.79), determine the solution vector $\mathbf{V}$ of unknown voltages for the system and operating conditions of Example 7.9.

Solution. Substituting in Eq. (7.77) from the current vector **I** and matrix **U** of Example 7.9, we obtain

$$\begin{bmatrix} 1 & \cdot & \cdot & \cdot \\ -0.70149 & 1 & \cdot & \cdot \\ -0.14925 & -0.38644 & 1 & \cdot \\ -0.14925 & -0.61356 & -0.78853 & 1 \end{bmatrix} \begin{bmatrix} V_1'' \\ V_2'' \\ V_3'' \\ V_4'' \end{bmatrix} = \begin{bmatrix} 0 \\ 0 \\ 1.00\angle -90^\circ \\ 0.60\angle -120^\circ \end{bmatrix}$$

Straightforward solution of this system of equations yields

$$V_1'' = V_2'' = 0 \qquad V_3'' = 1.00\angle -90^\circ \text{ per unit}$$

$$V_4'' = 0.60\angle -120^\circ + 0.78853 V_3'' = 1.34210\angle -102.9164^\circ \text{ per unit}$$

Substituting for $\mathbf{V}''$ in Eq. (7.78) leads to the diagonal system

$$\begin{bmatrix} -j16.75 & \cdot & \cdot & \cdot \\ \cdot & -j11.00746 & \cdot & \cdot \\ \cdot & \cdot & -j3.78305 & \cdot \\ \cdot & \cdot & \cdot & -j1.43082 \end{bmatrix} \begin{bmatrix} V_1' \\ V_2' \\ V_3' \\ V_4' \end{bmatrix} = \begin{bmatrix} 0 \\ 0 \\ 1.00\angle -90^\circ \\ 1.34210\angle -102.9164^\circ \end{bmatrix}$$

the solution of which is exactly equal to $\mathbf{V}'$ of Example 7.9, and so the remaining steps of this example match those of Example 7.9.

7.9 SPARSITY AND NEAR-OPTIMAL ORDERING

Large-scale power systems have only a small number of transmission lines connected to each bulk-power substation. In the network graph for such systems the ratio of the number of branches to the number of nodes is about 1.5 and the corresponding $\mathbf{Y}_{\text{bus}}$ has mainly zero elements. In fact, if there are 750 branches in a 500-node network (excluding the reference node), then since each node has an associated diagonal element and each branch gives rise to two symmetrically placed off-diagonal elements, the total number of nonzero elements is $(500 + 2 \times 750) = 2000$. This compares to a total of 250,000 elements in $\mathbf{Y}_{\text{bus}}$; that is, only 0.8% of the elements in $\mathbf{Y}_{\text{bus}}$ are nonzero. Because of the small number of nonzero elements, such matrices are said to be *sparse*. From the viewpoint of computational speed, accuracy, and storage it is desirable to process only the nonzero entries in $\mathbf{Y}_{\text{bus}}$ and to avoid *filling-in* new nonzero elements in the course of gaussian elimination and triangular factorization. Ordering refers to the sequence in which the equations of a system are processed. When a sparse matrix is triangularized, the order in which the unknown variables are eliminated affects the accumulation of new nonzero entries, called *fill-ins*, in the triangular matrices **L** and **U**. To help minimize such accumulations, we may use ordering schemes as described in Sec. B.1 of the Appendix.

7.10 SUMMARY

Nodal representation of the power transmission network is developed in this chapter. The essential background for understanding the bus admittance matrix and its formation is provided. Incorporation of mutually coupled branches into $\mathbf{Y}_{bus}$ can be handled by the building-block approach described here. Modifications to $\mathbf{Y}_{bus}$ to reflect network changes are thereby facilitated.

Gaussian elimination offers an alternative to matrix inversion for solving large-scale power systems, and triangular factorization of $\mathbf{Y}_{bus}$ enhances computational efficiency and reduces computer memory requirements, especially when the network matrices are symmetric.

These modeling and numerical procedures underlie the solution approaches now being used in daily practice by the electric power industry for power-flow and system analysis.

PROBLEMS

7.1. Using the building-block procedure described in Sec. 7.1, determine $\mathbf{Y}_{bus}$ for the circuit of Fig. 7.18. Assume there is *no* mutual coupling between any of the branches.

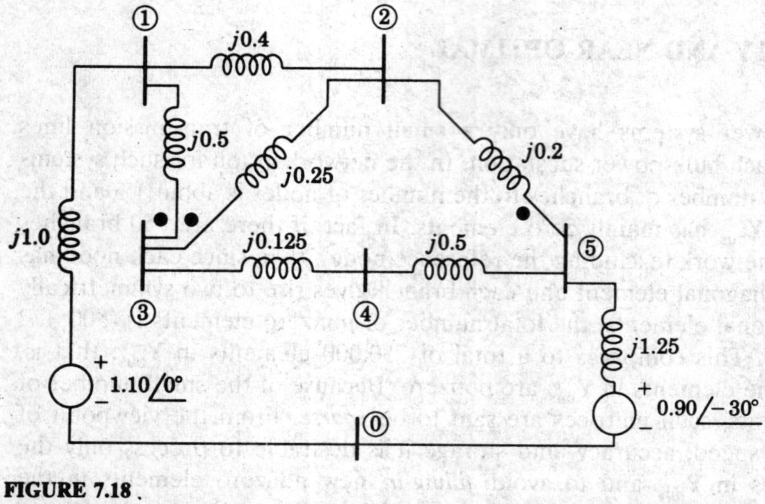

FIGURE 7.18.
Values shown are voltages and impedances in per unit. Dots represent mutual coupling between branches unless otherwise stated in problems.

7.2. Using the $\mathbf{Y}_{bus}$ modification procedure described in Sec. 7.4 and assuming no mutual coupling between branches, modify the $\mathbf{Y}_{bus}$ obtained in Prob. 7.1 to reflect removal of the two branches ①-③ and ②-⑤ from the circuit of Fig. 7.18.

7.3. The circuit of Fig. 7.18 has the linear graph shown in Fig. 7.19, with arrows indicating directions assumed for the branches *a* to *h*. Disregarding all mutual coupling between branches:

(*a*) Determine the branch-to-node incidence matrix **A** for the circuit with node ⓪ as reference.

(*b*) Find the circuit $\mathbf{Y}_{bus}$ using Eq. (7.37).

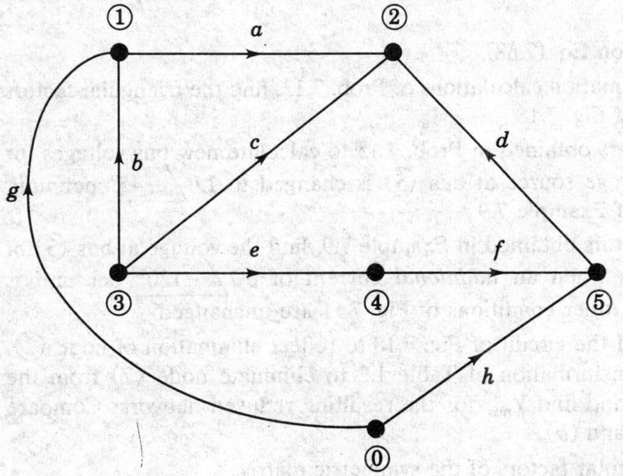

FIGURE 7.19
Linear graph for Prob. 7.3.

7.4. Consider that *only* the two branches ①-③ and ②-③ in the circuit of Fig. 7.18 are mutually coupled as indicated by the dots beside them and that their mutual impedance is $j0.15$ per unit (that is, ignore the dot on branch ②-⑤). Determine the circuit $\mathbf{Y}_{bus}$ by the procedure described in Sec. 7.2.

7.5. Solve Prob. 7.4 using Eq. (7.37). Determine the branch-to-node incidence matrix **A** from the linear graph of Fig. 7.19 with node ⓪ as reference.

7.6. Using the modification procedure of Sec. 7.4, modify the $\mathbf{Y}_{bus}$ solution of Prob. 7.4 (or Prob. 7.5) to reflect removal of the branch ②-③ from the circuit.

7.7. Modify the $\mathbf{Y}_{bus}$ determined in Example 7.3 to reflect removal of the mutually coupled branch ①-③ from the circuit of Fig. 7.11. Use the modification procedure of Sec. 7.4.

7.8. A new branch having a self-impedance of $j0.2$ per unit is added between nodes ② and ③ in the circuit of Fig. 7.11. Mutual impedance of $j0.1$ per unit couples this new branch to the branch already existing between nodes ② and ③. Modify the $\mathbf{Y}_{bus}$ obtained in Example 7.3 to account for the addition of the new branch.

7.9. Suppose that mutual coupling exists pairwise between branches ①-③ and ②-③, and *also* between branches ②-③ and ②-⑤ of Fig. 7.18, as shown by the dots in that figure. The mutual impedance between the former pair of branches is $j0.15$ per unit (the same as in Prob. 7.4) and between the latter pair is $j0.1$ per unit. Use the procedure of Sec. 7.2 to find $\mathbf{Y}_{\text{bus}}$ for the overall circuit including the *three* mutually coupled branches.

7.10. Solve for the $\mathbf{Y}_{\text{bus}}$ of Prob. 7.9 using Eq. (7.37). Use the linear graph of Fig. 7.19 with reference node ⓪ to determine the branch-to-node incidence matrix **A**.

7.11. Suppose that the direction of branch *d* in Fig. 7.19 is reversed so that it is now directed from node ② to node ⑤. Find the branch-to-node incidence matrix **A** of this modified graph and then solve for the $\mathbf{Y}_{\text{bus}}$ of Prob. 7.9 using Eq. (7.37).

7.12. Using the $\mathbf{Y}_{\text{bus}}$ modification procedure described in Sec. 7.4, remove branch ②-③ from the $\mathbf{Y}_{\text{bus}}$ solution obtained in Prob. 7.9 (or Prob. 7.10 or Prob. 7.11).

7.13. Write nodal admittance equations for the circuit of Fig. 7.18 disregarding all mutual coupling. Solve the resultant equations for the bus voltages by the method of gaussian elimination.

7.14. Prove Eq. (7.69) based on Eq. (7.68).

7.15. Using the gaussian-elimination calculations of Prob. 7.13, find the triangular factors of $\mathbf{Y}_{\text{bus}}$ for the circuit of Fig. 7.18.

7.16. Use the triangular factors obtained in Prob. 7.15 to calculate new bus voltages for Fig. 7.18 when the voltage source at bus ⑤ is changed to $1.0\angle -45^\circ$ per unit. Follow the procedure of Example 7.9.

7.17. Using the triangular factors obtained in Example 7.9, find the voltage at bus ③ of the circuit of Fig. 7.11 when an *additional* current of $0.2\angle -120^\circ$ per unit is injected at bus ②. All other conditions of Fig. 7.11 are unchanged.

7.18. (*a*) Kron reduce $\mathbf{Y}_{\text{bus}}$ of the circuit of Fig. 7.18 to reflect elimination of node ②.
(*b*) Use the $Y-\Delta$ transformation of Table 1.2 to eliminate node ② from the circuit of Fig. 7.18 and find $\mathbf{Y}_{\text{bus}}$ for the resulting reduced network. Compare results of parts (*a*) and (*b*).

7.19. Find the **L** and **U** triangular factors of the symmetric matrix

$$\mathbf{M} = \begin{bmatrix} 2 & 1 & 3 \\ 1 & 5 & 4 \\ 3 & 4 & 7 \end{bmatrix}$$

Verify the result using Eq. (7.75).

CHAPTER 8

THE IMPEDANCE MODEL AND NETWORK CALCULATIONS

The bus admittance matrix of a large-scale interconnected power system is typically very sparse with mainly zero elements. In Chap. 7 we saw how $\mathbf{Y}_{\text{bus}}$ is constructed branch by branch from primitive admittances. It is conceptually simple to invert $\mathbf{Y}_{\text{bus}}$ to find the bus impedance matrix $\mathbf{Z}_{\text{bus}}$, but such direct inversion is rarely employed when the systems to be analyzed are large scale. In practice, $\mathbf{Z}_{\text{bus}}$ is rarely explicitly required, and so the triangular factors of $\mathbf{Y}_{\text{bus}}$ are used to generate elements of $\mathbf{Z}_{\text{bus}}$ only as they are needed since this is often the most computationally efficient method. By setting computational considerations aside, however, and regarding $\mathbf{Z}_{\text{bus}}$ as being already constructed and explicitly available, the power system analyst can derive a great deal of insight. This is the approach taken in this chapter.

The bus impedance matrix can be directly constructed element by element using simple algorithms to incorporate one element at a time into the system representation. The work entailed in constructing $\mathbf{Z}_{\text{bus}}$ is much greater than that required to construct $\mathbf{Y}_{\text{bus}}$, but the information content of the bus impedance matrix is far greater than that of $\mathbf{Y}_{\text{bus}}$. We shall see, for example, that each diagonal element of $\mathbf{Z}_{\text{bus}}$ reflects important characteristics of the entire system in the form of the Thévenin impedance at the corresponding bus. Unlike $\mathbf{Y}_{\text{bus}}$, the bus impedance matrix of an interconnected system is never sparse and contains zeros only when the system is regarded as being subdivided into independent parts by open circuits. In Chap. 12, for instance, such open circuits arise in the zero-sequence network of the system.

The bus admittance matrix is widely used for power-flow analysis, as we shall see in Chap. 9. On the other hand, the bus impedance matrix is equally well favored for power system fault analysis. Accordingly, both $\mathbf{Y}_{\text{bus}}$ and $\mathbf{Z}_{\text{bus}}$ have important roles in the analysis of the power system network. In this chapter we study how to construct $\mathbf{Z}_{\text{bus}}$ directly and how to explore some of the conceptual insights which it offers into the characteristics of the power transmission network.

8.1 THE BUS ADMITTANCE AND IMPEDANCE MATRICES

In Example 7.6 we inverted the bus admittance matrix $\mathbf{Y}_{\text{bus}}$ and called the resultant the bus impedance matrix $\mathbf{Z}_{\text{bus}}$. By definition

$$\mathbf{Z}_{\text{bus}} = \mathbf{Y}_{\text{bus}}^{-1} \tag{8.1}$$

and for a network of three independent nodes the standard form is

$$\mathbf{Z}_{\text{bus}} = \begin{matrix} & \begin{matrix} ① & ② & ③ \end{matrix} \\ \begin{matrix} ① \\ ② \\ ③ \end{matrix} & \begin{bmatrix} Z_{11} & Z_{12} & Z_{13} \\ Z_{21} & Z_{22} & Z_{23} \\ Z_{31} & Z_{32} & Z_{33} \end{bmatrix} \end{matrix} \tag{8.2}$$

Since $\mathbf{Y}_{\text{bus}}$ is symmetrical around the principal diagonal, $\mathbf{Z}_{\text{bus}}$ must also be symmetrical. The bus admittance matrix need not be determined in order to obtain $\mathbf{Z}_{\text{bus}}$, and in another section of this chapter we see how $\mathbf{Z}_{\text{bus}}$ may be formulated directly.

The impedance elements of $\mathbf{Z}_{\text{bus}}$ on the principal diagonal are called *driving-point impedances* of the buses, and the off-diagonal elements are called the *transfer impedances* of the buses.

The bus impedance matrix is important and very useful in making fault calculations, as we shall see later. In order to understand the physical significance of the various impedances in the matrix, we compare them with the bus admittances. We can easily do so by looking at the equations at a particular bus. For instance, starting with the node equations expressed as

$$\mathbf{I} = \mathbf{Y}_{\text{bus}}\mathbf{V} \tag{8.3}$$

we have at bus ② of the three independent nodes

$$I_2 = Y_{21}V_1 + Y_{22}V_2 + Y_{23}V_3 \tag{8.4}$$

If V_1 and V_3 are reduced to zero by shorting buses ① and ③ to the reference node, and voltage V_2 is applied at bus ② so that current I_2 enters at bus ②, the self-admittance at bus ② is

$$Y_{22} = \left.\frac{I_2}{V_2}\right|_{V_1=V_3=0} \tag{8.5}$$

Thus, the self-admittance of a particular bus could be measured by shorting all other buses to the reference node and then finding the ratio of the current injected at the bus to the voltage applied at that bus. Figure 8.1 illustrates the method for a three-bus reactive network. The result is obviously equivalent to adding all the admittances directly connected to the bus, which is the procedure up to now when mutually coupled branches are absent. Figure 8.1 also serves to illustrate the off-diagonal admittance terms of $\mathbf{Y}_{\text{bus}}$. At bus ① the equation obtained by expanding Eq. (8.3) is

$$I_1 = Y_{11}V_1 + Y_{12}V_2 + Y_{13}V_3 \tag{8.6}$$

from which we see that

$$Y_{12} = \left.\frac{I_1}{V_2}\right|_{V_1=V_3=0} \tag{8.7}$$

Thus, the mutual admittance term Y_{12} is measured by shorting all buses except bus ② to the reference node and by applying a voltage V_2 at bus ②, as shown in Fig. 8.1. Then, Y_{12} is the ratio of the negative of the current leaving the network in the short circuit at node ① to the voltage V_2. The negative of the current leaving the network at node ① is used since I_1 is defined as the current entering the network. The resultant admittance is the negative

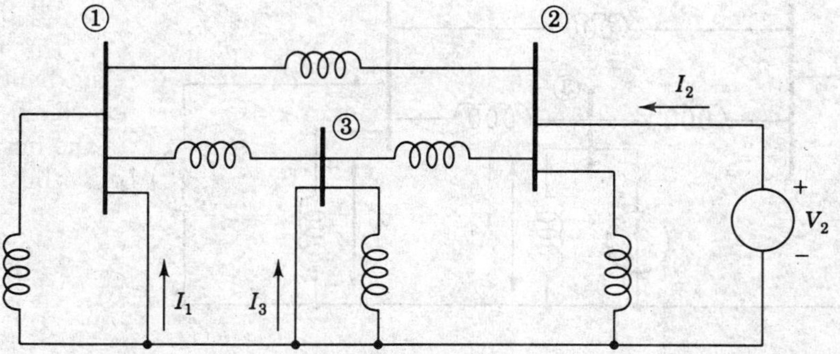

FIGURE 8.1
Circuit for measuring Y_{22}, Y_{12}, and Y_{32}.

of the admittance directly connected between buses ① and ②, as we would expect since mutually coupled branches are absent.

We have made this detailed examination of the bus admittances in order to differentiate them clearly from the impedances of the bus impedance matrix.

Conceptually, we solve Eq. (8.3) by premultiplying both sides of the equation by $\mathbf{Y}_{\text{bus}}^{-1} = \mathbf{Z}_{\text{bus}}$ to yield

$$\mathbf{V} = \mathbf{Z}_{\text{bus}}\mathbf{I} \tag{8.8}$$

and we must remember when dealing with $\mathbf{Z}_{\text{bus}}$ that $\mathbf{V}$ and $\mathbf{I}$ are column vectors of the bus voltages and the currents entering the buses from current sources, respectively. Expanding Eq. (8.8) for a network of three independent nodes yields

$$V_1 = Z_{11}I_1 + Z_{12}I_2 + Z_{13}I_3 \tag{8.9}$$

$$V_2 = Z_{21}I_1 + Z_{22}I_2 + Z_{23}I_3 \tag{8.10}$$

$$V_3 = Z_{31}I_1 + Z_{32}I_2 + Z_{33}I_3 \tag{8.11}$$

From Eq. (8.10) we see that the driving-point impedance Z_{22} is determined by open-circuiting the current sources at buses ① and ③ and by injecting the source current I_2 at bus ②. Then,

$$Z_{22} = \left.\frac{V_2}{I_2}\right|_{I_1 = I_3 = 0} \tag{8.12}$$

Figure 8.2 shows the circuit described. Since Z_{22} is defined by opening the

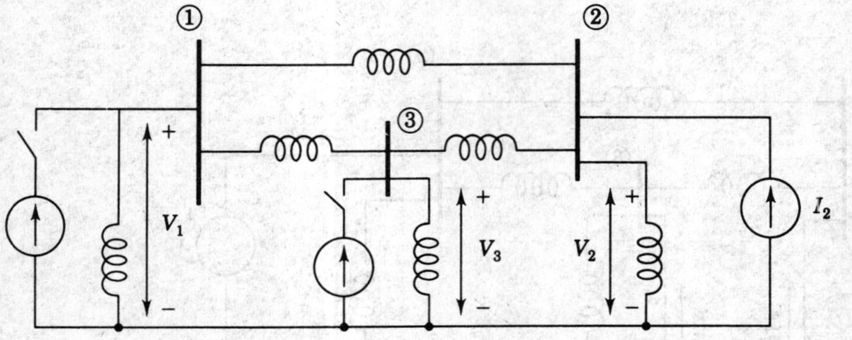

FIGURE 8.2
Circuit for measuring Z_{22}, Z_{12}, and Z_{32}.

current sources connected to the other buses whereas Y_{22} is found with the other buses shorted, we must not expect any reciprocal relation between these two quantities.

The circuit of Fig. 8.2 also enables us to measure some transfer impedances, for we see from Eq. (8.9) that with current sources I_1 and I_3 open-circuited

$$Z_{12} = \left.\frac{V_1}{I_2}\right|_{I_1=I_3=0} \tag{8.13}$$

and from Eq. (8.11)

$$Z_{32} = \left.\frac{V_3}{I_2}\right|_{I_1=I_3=0} \tag{8.14}$$

Thus, we can measure the transfer impedances Z_{12} and Z_{32} by injecting current at bus ② and by finding the ratios of V_1 and V_3 to I_2 with the sources open at all buses except bus ②. We note that a mutual admittance is measured with all but one bus short-circuited and that a transfer impedance is measured with all sources open-circuited except one.

Equation (8.9) tells us that if we inject current into bus ① with current sources at buses ② and ③ open, the only impedance through which I_1 flows is Z_{11}. Under the same conditions, Eqs. (8.10) and (8.11) show that I_1 is causing voltages at buses ② and ③ expressed by

$$V_2 = I_1 Z_{21} \qquad \text{and} \qquad V_3 = I_1 Z_{31} \tag{8.15}$$

It is important to realize the implications of the preceding discussion, for $\mathbf{Z}_{\text{bus}}$ is sometimes used in power-flow studies and is extremely valuable in fault calculations.

8.2 THÉVENIN'S THEOREM AND $\mathbf{Z}_{\text{bus}}$

The bus impedance matrix provides important information regarding the power system network, which we can use to advantage in network calculations. In this section we examine the relationship between the elements of $\mathbf{Z}_{\text{bus}}$ and the Thévenin impedance presented by the network at each of its buses. To establish notation, let us denote the bus voltages corresponding to the initial values $\mathbf{I}^0$ of the bus currents $\mathbf{I}$ by $\mathbf{V}^0 = \mathbf{Z}_{\text{bus}}\mathbf{I}^0$. The voltages V_1^0 to V_N^0 are the effective *open-circuit* voltages, which can be measured by voltmeter between the buses of the network and the reference node. When the bus currents are changed from their initial values to new values $\mathbf{I}^0 = \Delta\mathbf{I}$, the new bus voltages are given by the superposition equation

$$\mathbf{V} = \mathbf{Z}_{\text{bus}}(\mathbf{I}^0 + \Delta\mathbf{I}) = \underbrace{\mathbf{Z}_{\text{bus}}\mathbf{I}^0}_{\mathbf{V}^0} + \underbrace{\mathbf{Z}_{\text{bus}}\,\Delta\mathbf{I}}_{\Delta\mathbf{V}} \tag{8.16}$$

where $\Delta\mathbf{V}$ represents the changes in the bus voltages from their original values.

Figure 8.3(a) shows a large-scale system in schematic form with a representative bus ⓚ extracted along with the reference node of the system. Initially, we consider the circuit not to be energized so that the bus currents $\mathbf{I}^0$ and voltages $\mathbf{V}^0$ are zero. Then, into bus ⓚ a current of ΔI_k amp (or ΔI_k per unit for $\mathbf{Z}_{\text{bus}}$ in per unit) is injected into the system from a current source connected to the reference node. The resulting voltage changes at the buses of

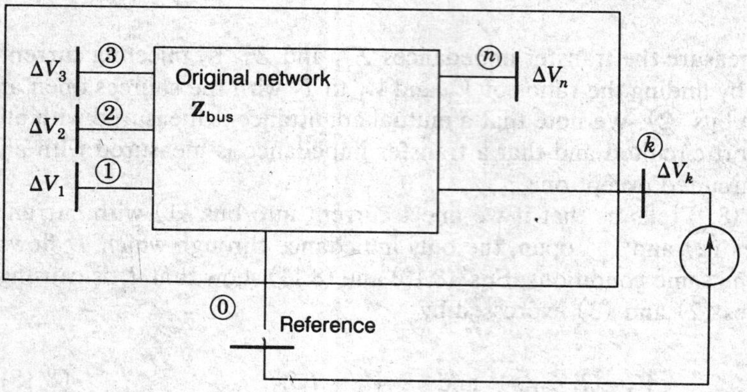

(a)

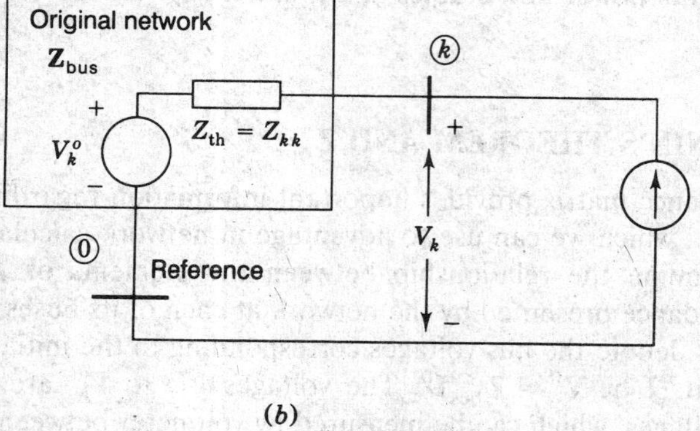

(b)

FIGURE 8.3
(a) Original network with bus ⓚ and reference node extracted. Voltage ΔV_n at bus ⓝ is caused by current ΔI_k entering the network. (b) Thévenin equivalent circuit at node ⓚ.

the network, indicated by the incremental quantities ΔV_1 to ΔV_N, are given by

$$\begin{array}{c} \\ \\ \\ \\ \\ \\ \\ \end{array}\begin{bmatrix} \Delta V_1 \\ \Delta V_2 \\ \vdots \\ \Delta V_k \\ \vdots \\ \Delta V_N \end{bmatrix} = \begin{array}{c} \textcircled{1} \\ \textcircled{2} \\ \\ \textcircled{k} \\ \\ \textcircled{N} \end{array} \overset{\begin{array}{cccccc} \textcircled{1} & \textcircled{2} & & \textcircled{k} & & \textcircled{N} \end{array}}{\begin{bmatrix} Z_{11} & Z_{12} & \cdots & Z_{1k} & \cdots & Z_{1N} \\ Z_{21} & Z_{22} & \cdots & Z_{2k} & \cdots & Z_{2N} \\ \vdots & \vdots & \ddots & \vdots & \ddots & \vdots \\ Z_{k1} & Z_{k2} & \cdots & Z_{kk} & \cdots & Z_{kN} \\ \vdots & \vdots & \ddots & \vdots & \ddots & \vdots \\ Z_{N1} & Z_{N2} & \cdots & Z_{Nk} & \cdots & Z_{NN} \end{bmatrix}} \begin{bmatrix} 0 \\ 0 \\ \vdots \\ \Delta I_k \\ \vdots \\ 0 \end{bmatrix} \tag{8.17}$$

with the only nonzero entry in the current vector equal to ΔI_k in row k. Row-by-column multiplication in Eq. (8.17) yields the incremental bus voltages

$$\begin{bmatrix} \Delta V_1 \\ \Delta V_2 \\ \vdots \\ \Delta V_k \\ \vdots \\ \Delta V_N \end{bmatrix} = \begin{array}{c} \textcircled{1} \\ \textcircled{2} \\ \\ \textcircled{k} \\ \\ \textcircled{N} \end{array} \overset{\textcircled{k}}{\begin{bmatrix} Z_{1k} \\ Z_{2k} \\ \vdots \\ Z_{kk} \\ \vdots \\ Z_{Nk} \end{bmatrix}} \Delta I_k \tag{8.18}$$

which are numerically equal to the entries in column k of $\mathbf{Z}_{\text{bus}}$ multiplied by the current ΔI_k. Adding these voltage changes to the original voltages at the buses according to Eq. (8.16) yields at bus ⓚ

$$V_k = V_k^0 + Z_{kk}\,\Delta I_k \tag{8.19}$$

The circuit corresponding to this equation is shown in Fig. 8.3(b) from which it is evident that the Thévenin impedance Z_{th} at a representative bus ⓚ of the system is given by

$$Z_{\text{th}} = Z_{kk} \tag{8.20}$$

where Z_{kk} is the diagonal entry in row k and column k of $\mathbf{Z}_{\text{bus}}$. With k set equal to 2, this is essentially the same result obtained in Eq. (8.12) for the driving-point impedance at bus ② of Fig. 8.2.

In a similar manner, we can determine the Thévenin impedance between any two buses ⓙ and ⓚ of the network. As shown in Fig. 8.4(a), the otherwise dead network is energized by the current injections ΔI_j at bus ⓙ

and ΔI_k at bus ⓚ. Denoting the changes in the bus voltages resulting from the combination of these two current injections by ΔV_1 to ΔV_N, we obtain

$$\begin{bmatrix} \Delta V_1 \\ \vdots \\ \Delta V_j \\ \Delta V_k \\ \vdots \\ \Delta V_N \end{bmatrix} = \begin{matrix} \textcircled{1} \\ \\ \textcircled{j} \\ \textcircled{k} \\ \\ \textcircled{N} \end{matrix} \overset{\begin{matrix} \textcircled{1} & & \textcircled{j} & \textcircled{k} & & \textcircled{N} \end{matrix}}{\begin{bmatrix} Z_{11} & \cdots & Z_{1j} & Z_{1k} & \cdots & Z_{1N} \\ \vdots & \ddots & \vdots & \vdots & \ddots & \vdots \\ Z_{j1} & \cdots & Z_{jj} & Z_{jk} & \cdots & Z_{jN} \\ Z_{k1} & \cdots & Z_{kj} & Z_{kk} & \cdots & Z_{kN} \\ \vdots & \ddots & \vdots & \vdots & \ddots & \vdots \\ Z_{N1} & \cdots & Z_{Nj} & Z_{Nk} & \cdots & Z_{NN} \end{bmatrix}} \begin{bmatrix} 0 \\ \vdots \\ \Delta I_j \\ \Delta I_k \\ \vdots \\ 0 \end{bmatrix}$$

$$= \begin{bmatrix} Z_{1j}\,\Delta I_j + Z_{1k}\,\Delta I_k \\ \vdots \\ Z_{jj}\,\Delta I_j + Z_{jk}\,\Delta I_k \\ Z_{kj}\,\Delta I_j + Z_{kk}\,\Delta I_k \\ \vdots \\ Z_{Nj}\,\Delta I_j + Z_{Nk}\,\Delta I_k \end{bmatrix} \tag{8.21}$$

in which the right-hand vector is numerically equal to the product of ΔI_j and column j added to the product of ΔI_k and column k of the system $\mathbf{Z}_{\text{bus}}$. Adding these voltage changes to the original bus voltages according to Eq. (8.16), we obtain at buses ⓙ and ⓚ

$$V_j = V_j^0 + Z_{jj}\,\Delta I_j + Z_{jk}\,\Delta I_k \tag{8.22}$$

$$V_k = V_k^0 + Z_{kj}\,\Delta I_j + Z_{kk}\,\Delta I_k \tag{8.23}$$

Adding and subtracting $Z_{jk}\,\Delta I_j$ in Eq. (8.22), and likewise, $Z_{kj}\,\Delta I_k$ in Eq. (8.23), give

$$V_j = V_j^0 + (Z_{jj} - Z_{jk})\,\Delta I_j + Z_{jk}(\Delta I_j + \Delta I_k) \tag{8.24}$$

$$V_k = V_k^0 + Z_{kj}(\Delta I_j + \Delta I_k) + (Z_{kk} - Z_{kj})\,\Delta I_k \tag{8.25}$$

Since $\mathbf{Z}_{\text{bus}}$ is symmetrical, Z_{jk} equals Z_{kj} and the circuit corresponding to these two equations is shown in Fig. 8.4(*b*), which represents the Thévenin equivalent circuit of the system between buses ⓙ and ⓚ. Inspection of Fig. 8.4(*b*) shows that the *open-circuit* voltage from bus ⓚ to bus ⓙ is $V_k^0 - V_j^0$, and the

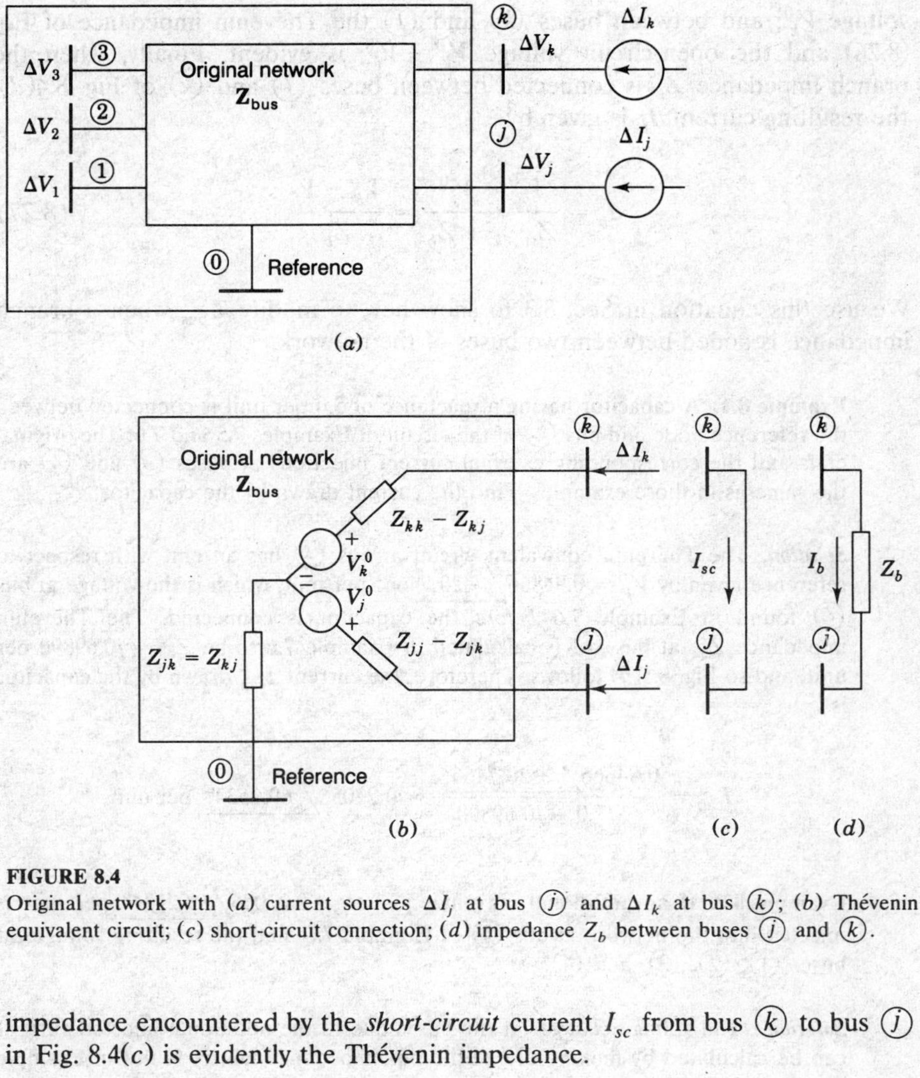

FIGURE 8.4
Original network with (a) current sources ΔI_j at bus ⓙ and ΔI_k at bus ⓚ; (b) Thévenin equivalent circuit; (c) short-circuit connection; (d) impedance Z_b between buses ⓙ and ⓚ.

impedance encountered by the *short-circuit* current I_{sc} from bus ⓚ to bus ⓙ in Fig. 8.4(c) is evidently the Thévenin impedance.

$$Z_{\text{th},\,jk} = Z_{jj} + Z_{kk} - 2Z_{jk} \tag{8.26}$$

This result is readily confirmed by substituting $I_{sc} = \Delta I_j = -\Delta I_k$ in Eqs. (8.24) and (8.25) and by setting the difference $V_j - V_k$ between the resultant equations equal to zero. As far as external connections to buses ⓙ and ⓚ are concerned, Fig. 8.4(b) represents the effect of the original system. From bus ⓙ to the reference node we can trace the Thévenin impedance $Z_{jj} = (Z_{jj} - Z_{jk}) + Z_{jk}$ and the open-circuit voltage V_j^0; from bus ⓚ to the reference node we have the Thévenin impedance $Z_{kk} = (Z_{kk} - Z_{kj}) + Z_{kj}$ and the open-circuit voltage V_k^0; and between buses ⓚ and ⓙ the Thévenin impedance of Eq. (8.26) and the open-circuit voltage $V_k^0 - V_j^0$ is evident. Finally, when the branch impedance Z_b is connected between buses ⓙ and ⓚ of Fig. 8.4(*d*), the resulting current I_b is given by

$$I_b = \frac{V_k^0 - V_j^0}{Z_{\text{th},\,jk} + Z_b} = \frac{V_k - V_j}{Z_b} \tag{8.27}$$

We use this equation in Sec. 8.3 to show how to modify $\mathbf{Z}_{\text{bus}}$ when a branch impedance is added between two buses of the network.

Example 8.1. A capacitor having a reactance of 5.0 per unit is connected between the reference node and bus ④ of the circuit of Examples 7.5 and 7.6. The original emfs and the corresponding external current injections at buses ③ and ④ are the same as in those examples. Find the current drawn by the capacitor.

Solution. The Thévenin equivalent circuit at bus ④ has an emf with respect to reference given by $V_4^0 = 0.94866\underline{/-20.7466^\circ}$ per unit, which is the voltage at bus ④ found in Example 7.6 *before* the capacitor is connected. The Thévenin impedance Z_{44} at bus ④ is calculated in Example 7.6 to be $Z_{44} = j0.69890$ per unit, and so Fig. 8.5(*a*) follows. Therefore, the current I_{cap} drawn by the capacitor is

$$I_{\text{cap}} = \frac{0.94866\underline{/-20.7466^\circ}}{-j5.0 + j0.69890} = 0.22056\underline{/69.2534^\circ} \text{ per unit}$$

Example 8.2. If an additional current equal to $-0.22056\underline{/69.2534^\circ}$ per unit is injected into the network at bus ④ of Example 7.6, find the resulting voltages at buses ①, ②, ③, and ④.

Solution.. The voltage *changes* at the buses due to the additional injected current can be calculated by making use of the bus impedance matrix found in Example 7.6. The required impedances are in column 4 of $\mathbf{Z}_{\text{bus}}$. The voltage changes due to the added current injection at bus ④ in per unit are

$$\Delta V_1 = -I_{\text{cap}} Z_{14} = -0.22056\underline{/69.2534^\circ} \times j0.63677 = 0.14045\underline{/-20.7466^\circ}$$

$$\Delta V_2 = -I_{\text{cap}} Z_{24} = -0.22056\underline{/69.2534^\circ} \times j0.64178 = 0.14155\underline{/-20.7466^\circ}$$

$$\Delta V_3 = -I_{\text{cap}} Z_{34} = -0.22056\underline{/69.2534^\circ} \times j0.55110 = 0.12155\underline{/-20.7466^\circ}$$

$$\Delta V_4 = -I_{\text{cap}} Z_{44} = -0.22056\underline{/69.2534^\circ} \times j0.69890 = 0.15415\underline{/-20.7466^\circ}$$

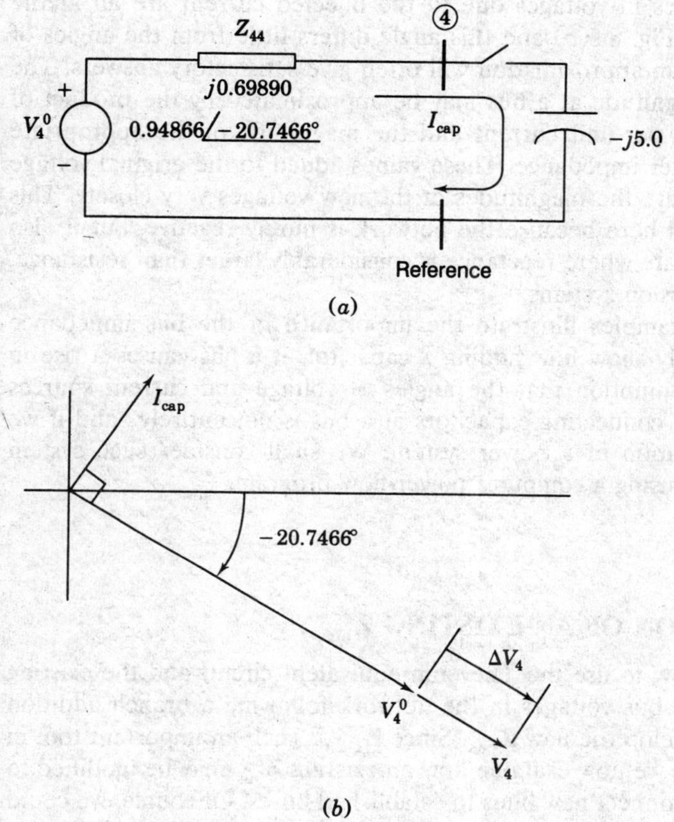

FIGURE 8.5
Circuit for Examples 8.1 and 8.2 showing: (*a*) Thévenin equivalent circuit; (*b*) phasor diagram at bus ④.

By superposition the resulting voltages are determined from Eq. (8.16) by adding these changes to the original bus voltages found in Example 7.6. The new bus voltages in per unit are

$$V_1 = 0.96903\angle -18.4189° + 0.14045\angle -20.7466° = 1.10938\angle -18.7135°$$

$$V_2 = 0.96734\angle -18.6028° + 0.14155\angle -20.7466° = 1.10880\angle -18.8764°$$

$$V_3 = 0.99964\angle -15.3718° + 0.12155\angle -20.7466° = 1.12071\angle -15.9539°$$

$$V_4 = 0.94866\angle -20.7466° + 0.15415\angle -20.7466° = 1.10281\angle -20.7466°$$

Since the changes in voltages due to the injected current are all at the same angle shown in Fig. 8.5(*b*) and this angle differs little from the angles of the original voltages, an approximation will often give satisfactory answers. The change in voltage magnitude at a bus may be approximated by the product of the magnitude of the per-unit current and the magnitude of the appropriate driving-point or transfer impedance. These values added to the original voltage magnitudes approximate the magnitudes of the new voltages very closely. This approximation is valid here because the network is purely reactive, but it also provides a good estimate where reactance is considerably larger than resistance, as is usual in transmission systems.

The last two examples illustrate the importance of the bus impedance matrix and incidentally show how adding a capacitor at a bus causes a rise in bus voltages. The assumption that the angles of voltage and current sources remain constant after connecting capacitors at a bus is not entirely valid if we are considering operation of a power system. We shall consider such system operation in Chap. 9 using a computer power-flow program.

8.3 MODIFICATION OF AN EXISTING $\mathbf{Z}_{\text{bus}}$

In Sec. 8.2 we see how to use the Thévenin equivalent circuit and the existing $\mathbf{Z}_{\text{bus}}$ to solve for new bus voltages in the network following a branch addition without having to develop the new $\mathbf{Z}_{\text{bus}}$. Since $\mathbf{Z}_{\text{bus}}$ is such an important tool in power system analysis we now examine how an existing $\mathbf{Z}_{\text{bus}}$ may be modified to add new buses or to connect new lines to established buses. Of course, we could create a new $\mathbf{Y}_{\text{bus}}$ and invert it, but direct methods of modifying $\mathbf{Z}_{\text{bus}}$ are available and very much simpler than a matrix inversion even for a small number of buses. Also, when we know how to modify $\mathbf{Z}_{\text{bus}}$, we can see how to build it directly.

We recognize several types of modifications in which a branch having impedance Z_b is added to a network with known $\mathbf{Z}_{\text{bus}}$. The original bus impedance matrix is identified as $\mathbf{Z}_{\text{orig}}$, an $N \times N$ matrix.

In the notation to be used in our analysis existing buses will be identified by numbers or the letters h, i, j, and k. The letter p or q will designate a new bus to be added to the network to convert $\mathbf{Z}_{\text{orig}}$ to an $(N + 1) \times (N + 1)$ matrix. At bus ⓚ the original voltage will be denoted by V_k^0, the new voltage after modifying $\mathbf{Z}_{\text{bus}}$ will be V_k, and $\Delta V_k = V_k - V_k^0$ will denote the voltage change at that bus. Four cases are considered in this section.

CASE 1. *Adding Z_b from a new bus* ⓟ *to the reference node.*

The addition of the new bus ⓟ connected to the reference node through Z_b without a connection to any of the buses of the original network cannot alter the original bus voltages when a current I_p is injected at the new bus. The

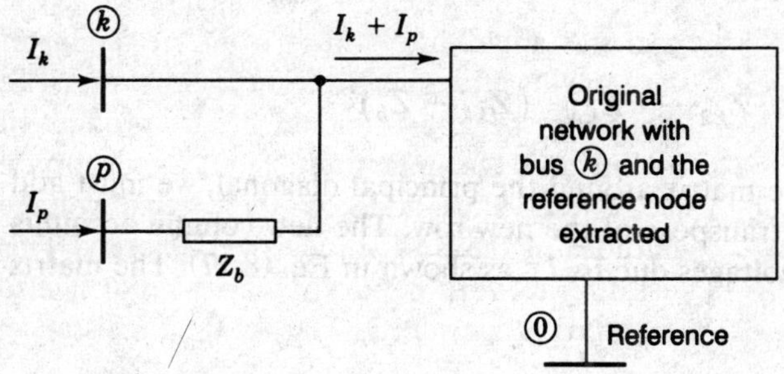

FIGURE 8.6
Addition of new bus ⓟ connected through impedance Z_b to existing bus ⓚ.

voltage V_p at the new bus is equal to $I_p Z_b$. Then,

$$\begin{bmatrix} V_1^0 \\ V_2^0 \\ \vdots \\ V_N^0 \\ \hline V_p \end{bmatrix} = \underbrace{\begin{array}{c} \\ \\ \\ \\ \text{ⓟ} \end{array}\left[\begin{array}{cccc|c} & & & & 0 \\ & & & & 0 \\ & & \mathbf{Z}_{\text{orig}} & & \vdots \\ & & & & 0 \\ \hline 0 & 0 & \cdots & 0 & Z_b \end{array}\right]}_{\mathbf{Z}_{\text{bus(new)}}} \begin{bmatrix} I_1 \\ I_2 \\ \vdots \\ I_N \\ \hline I_p \end{bmatrix} \tag{8.28}$$

We note that the column vector of currents multiplied by the new $\mathbf{Z}_{\text{bus}}$ will not alter the voltages of the original network and will result in the correct voltage at the new bus ⓟ.

CASE 2. *Adding Z_b from a new bus ⓟ to an existing bus ⓚ*

The addition of a new bus ⓟ connected through Z_b to an existing bus ⓚ with I_p injected at bus ⓟ will cause the current entering the original network at bus ⓚ to become the sum of I_k injected at bus ⓚ plus the current I_p coming through Z_b, as shown in Fig. 8.6.

The current I_p flowing into the network at bus ⓚ will increase the original voltage V_k^0 by the voltage $I_p Z_{kk}$ just like in Eq. (8.19); that is,

$$V_k = V_k^0 + I_p Z_{kk} \tag{8.29}$$

and V_p will be larger than the new V_k by the voltage $I_p Z_b$. So,

$$V_p = V_k^0 + I_p Z_{kk} + I_p Z_b \tag{8.30}$$

and substituting for V_k^0, we obtain

$$V_p = \underbrace{I_1 Z_{k1} + I_2 Z_{k2} + \cdots + I_N Z_{kN}}_{V_k^0} + I_p(Z_{kk} + Z_b) \tag{8.31}$$

We now see that the new row which must be added to $\mathbf{Z}_{\text{orig}}$ in order to find V_p is

$$Z_{k1} \quad Z_{k2} \cdots Z_{kN} \quad (Z_{kk} + Z_b)$$

Since $\mathbf{Z}_{\text{bus}}$ must be a square matrix around the principal diagonal, we must add a new column which is the transpose of the new row. The new column accounts for the increase of all bus voltages due to I_p, as shown in Eq. (8.17). The matrix equation is

$$\begin{bmatrix} V_1 \\ V_2 \\ \vdots \\ V_N \\ \hline V_p \end{bmatrix} = \underbrace{\left[\begin{array}{cccc|c} & & & & Z_{1k} \\ & & & & Z_{2k} \\ & \mathbf{Z}_{\text{orig}} & & & \vdots \\ & & & & Z_{Nk} \\ \hline Z_{k1} & Z_{k2} & \cdots & Z_{kN} & Z_{kk} + Z_b \end{array}\right]}_{\mathbf{Z}_{\text{bus(new)}}} \begin{bmatrix} I_1 \\ I_2 \\ \vdots \\ I_N \\ \hline I_p \end{bmatrix} \tag{8.32}$$

(the last row and last column are labelled ⓟ)

Note that the first N elements of the new row are the elements of row k of $\mathbf{Z}_{\text{orig}}$ and the first N elements of the new column are the elements of column k of $\mathbf{Z}_{\text{orig}}$.

CASE 3. *Adding Z_b from existing bus ⓚ to the reference node*

To see how to alter $\mathbf{Z}_{\text{orig}}$ by connecting an impedance Z_b from an existing bus ⓚ to the reference node, we add a new bus ⓟ connected through Z_b to bus ⓚ. Then, we short-circuit bus ⓟ to the reference node by letting V_p equal zero to yield the same matrix equation as Eq. (8.32) except that V_p is zero. So, for the modification we proceed to create a new row and new column exactly the same as in Case 2, but we then eliminate the $(N + 1)$ row and $(N + 1)$ column by Kron reduction, which is possible because of the zero in the column matrix of voltages. We use the method developed in Eq. (7.50) to find each element $Z_{hi(\text{new})}$ in the new matrix, where

$$Z_{hi(\text{new})} = Z_{hi} - \frac{Z_{h(N+1)}Z_{(N+1)i}}{Z_{kk} + Z_b} \tag{8.33}$$

CASE 4. *Adding Z_b between two existing buses ⓙ and ⓚ*

To add a branch impedance Z_b between buses ⓙ and ⓚ already established in $\mathbf{Z}_{\text{orig}}$, we examine Fig. 8.7, which shows these buses extracted from the original network. The current I_b flowing from bus ⓚ to bus ⓙ is similar to that of Fig. 8.4. Hence, from Eq. (8.21) the change in voltage at each

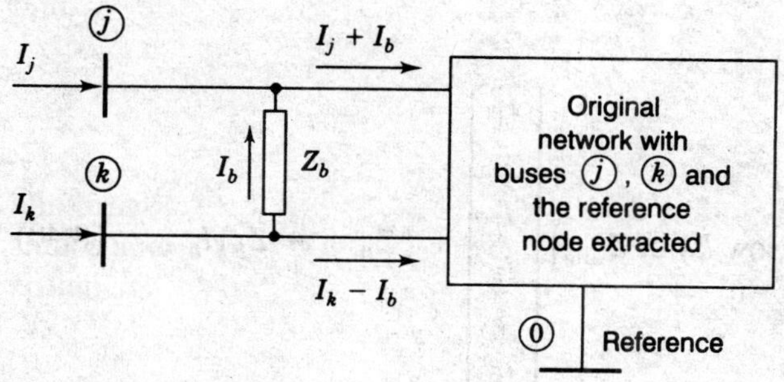

FIGURE 8.7
Addition of impedance Z_b between existing buses ⓙ and ⓚ.

bus ⓗ caused by the injection I_b at bus ⓙ and $-I_b$ at bus ⓚ is given by

$$\Delta V_h = (Z_{hj} - Z_{hk})I_b \tag{8.34}$$

which means that the vector $\mathbf{\Delta V}$ of bus voltage changes is found by subtracting column k from column j of $\mathbf{Z}_{\text{orig}}$ and by multiplying the result by I_b. Based on the definition of voltage change, we now write some equations for the bus voltages as follows:

$$V_1 = V_1^0 + \Delta V_1 \tag{8.35}$$

and using Eq. (8.34) gives

$$V_1 = \underbrace{Z_{11}I_1 + \cdots + Z_{1j}I_j + Z_{1k}I_k + \cdots + Z_{1N}I_N}_{V_1^0} + \underbrace{(Z_{1j} - Z_{1k})I_b}_{\Delta V_1} \tag{8.36}$$

Similarly, at buses ⓙ and ⓚ

$$V_j = \underbrace{Z_{j1}I_1 + \cdots + Z_{jj}I_j + Z_{jk}I_k + \cdots + Z_{jN}I_N}_{V_j^0} + \underbrace{(Z_{jj} - Z_{jk})I_b}_{\Delta V_j} \tag{8.37}$$

$$V_k = \underbrace{Z_{k1}I_1 + \cdots + Z_{kj}I_j + Z_{kk}I_k + \cdots + Z_{kN}I_N}_{V_k^0} + \underbrace{(Z_{kj} - Z_{kk})I_b}_{\Delta V_k} \tag{8.38}$$

We need one more equation since I_b is unknown. This is supplied by Eq. (8.27), which can be rearranged into the form

$$0 = V_j^0 - V_k^0 + (Z_{\text{th},jk} + Z_b)I_b \tag{8.39}$$

From Eq. (8.37) we note that V_j^0 equals the product of row j of $\mathbf{Z}_{\text{orig}}$ and the column of bus currents $\mathbf{I}$; likewise, V_k^0 of Eq. (8.38) equals row k of $\mathbf{Z}_{\text{orig}}$ multiplied by $\mathbf{I}$. Upon substituting the expressions for V_j^0 and V_k^0 in Eq. (8.39), we obtain

$$0 = \left[(\text{row } j - \text{row } k) \text{ of } \mathbf{Z}_{\text{orig}}\right] \begin{bmatrix} I_1 \\ \vdots \\ I_j \\ I_k \\ \vdots \\ I_N \end{bmatrix} + (Z_{\text{th},\,jk} + Z_b) I_b \tag{8.40}$$

By examining the coefficients of Eqs. (8.36) through (8.38) and Eq. (8.40), we can write the matrix equation

$$\begin{bmatrix} V_1 \\ \vdots \\ V_j \\ V_k \\ \vdots \\ V_N \\ \hline 0 \end{bmatrix} = \left[\begin{array}{c|c} \mathbf{Z}_{\text{orig}} & \begin{array}{c}(\text{col. } j - \text{col. } k) \\ \text{of } \mathbf{Z}_{\text{orig}}\end{array} \\ \hline (\text{row } j - \text{row } k) \text{ of } \mathbf{Z}_{\text{orig}} & Z_{bb} \end{array}\right] \begin{bmatrix} I_1 \\ \vdots \\ I_j \\ I_k \\ \vdots \\ I_N \\ \hline I_b \end{bmatrix} \tag{8.41}$$

in which the coefficient of I_b in the last row is denoted by

$$Z_{bb} = Z_{\text{th},\,jk} + Z_b = Z_{jj} + Z_{kk} - 2Z_{jk} + Z_b \tag{8.42}$$

The new column is column j minus column k of $\mathbf{Z}_{\text{orig}}$ with Z_{bb} in the $(N + 1)$ row. The new row is the transpose of the new column. Eliminating the $(N + 1)$ row and $(N + 1)$ column of the square matrix of Eq. (8.41) in the same manner as previously, we see that each element $Z_{hi(\text{new})}$ in the new matrix is

$$Z_{hi(\text{new})} = Z_{hi} - \frac{Z_{h(N+1)} Z_{(N+1)i}}{Z_{jj} + Z_{kk} - 2Z_{jk} + Z_b} \tag{8.43}$$

We need not consider the case of introducing two new buses connected by Z_b because we could always connect one of these new buses through an impedance to an existing bus or to the reference bus before adding the second new bus.

Removing a branch. A single branch of impedance Z_b between two nodes can be removed from the network by adding the *negative* of Z_b between the same terminating nodes. The reason is, of course, that the parallel combination of the existing branch (Z_b) and the added branch $(-Z_b)$ amounts to an effective open circuit.

Table 8.1 summarizes the procedures of Cases 1 to 4.

TABLE 8.1

Modification of existing Z_{bus}

<table>
<tr><th>Case</th><th>Add branch Z_b from</th><th>$Z_{bus\,(new)}$</th></tr>
<tr><td>1</td><td>Reference node to new bus Ⓟ
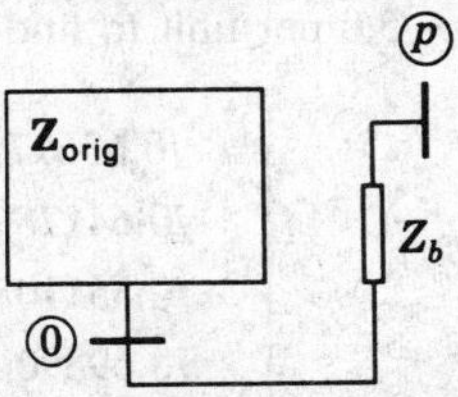
</td><td>$$\begin{array}{c} \\ \\ \\ (p) \end{array}\begin{bmatrix} & & & 0 \\ & Z_{orig} & & \vdots \\ & & & 0 \\ \hline 0 & \cdots & 0 & Z_b \end{bmatrix}$$ (column p at right)</td></tr>
<tr><td>2</td><td>Existing bus Ⓚ to new bus Ⓟ
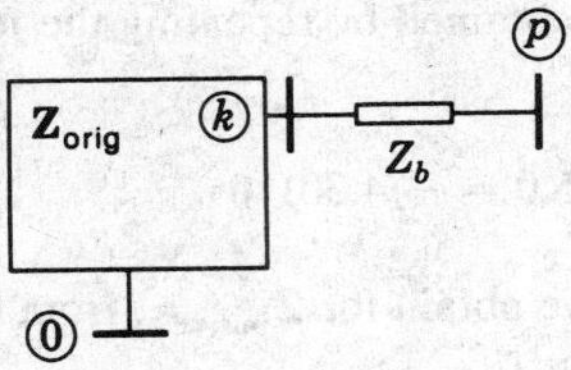
</td><td>$$\begin{array}{c} \\ (k) \\ (p) \end{array}\begin{bmatrix} Z_{orig} & \text{col. } k \\ \hline \text{row } k & Z_{kk} + Z_b \end{bmatrix}$$ (column k copied to column p; row k copied to row p)</td></tr>
<tr><td>3</td><td>Existing bus Ⓚ to reference node
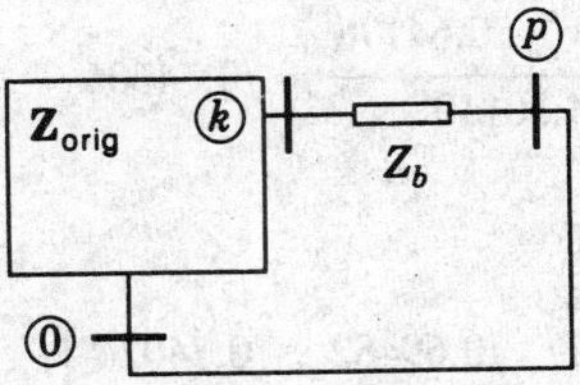

(Node Ⓟ is temporary.)</td><td>• Repeat Case 2 and

• Remove row p and column p by Kron reduction</td></tr>
<tr><td>4</td><td>Existing bus Ⓙ to existing bus Ⓚ
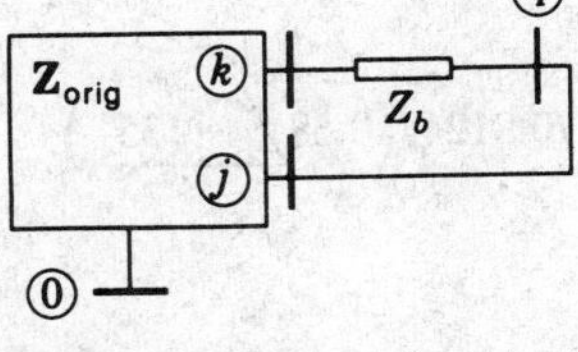

(Node Ⓠ is temporary.)</td><td>• Form the matrix
$$\begin{array}{c} \\ (q) \end{array}\begin{bmatrix} Z_{orig} & \text{col. } j - \text{col. } k \\ \hline \text{row } j - \text{row } k & Z_{th,\,jk} + Z_b \end{bmatrix}$$ (column q at right)

where $Z_{th,\,jk} = Z_{jj} + Z_{kk} - 2Z_{jk}$
and
• Remove row q and column q by Kron reduction</td></tr>
</table>

Example 8.3. Modify the bus impedance matrix of Example 7.6 to account for the connection of a capacitor having a reactance of 5.0 per unit between bus ④ and the reference node of the circuit of Fig. 7.9. Then, find V_4 using the impedances of the new matrix and the current sources of Example 7.6. Compare this value of V_4 with that found in Example 8.2.

Solution. We use Eq. (8.32) and recognize that $\mathbf{Z}_{\text{orig}}$ is the 4×4 matrix of Example 7.6, that subscript $k = 4$, and that $Z_b = -j5.0$ per unit to find

$$\begin{bmatrix} V_1 \\ V_2 \\ V_3 \\ V_4 \\ \hline 0 \end{bmatrix} = \left[\begin{array}{cccc|c} & & & & j0.63677 \\ & \mathbf{Z}_{\text{orig}} & & & j0.64178 \\ & & & & j0.55110 \\ & & & & j0.69890 \\ \hline j0.63677 & j0.64178 & j0.55110 & j0.69890 & -j4.30110 \end{array}\right] \begin{bmatrix} I_1 \\ I_2 \\ I_3 \\ I_4 \\ \hline I_b \end{bmatrix}$$

The terms in the fifth row and column were obtained by repeating the fourth row and column of $\mathbf{Z}_{\text{orig}}$ and noting that

$$Z_{44} + Z_b = j0.69890 - j5.0 = -j4.30110$$

Then, eliminating the fifth row and column, we obtain for $\mathbf{Z}_{\text{bus(new)}}$ from Eq. (8.33)

$$Z_{11(\text{new})} = j0.73128 - \frac{j0.63677 \times j0.63677}{-j4.30110} = j0.82555$$

$$Z_{24(\text{new})} = j0.64178 - \frac{j0.69890 \times j0.64178}{-j4.30110} = j0.74606$$

and other elements in a similar manner to give

$$\mathbf{Z}_{\text{bus(new)}} = \begin{bmatrix} j0.82555 & j0.78641 & j0.69482 & j0.74024 \\ j0.78641 & j0.81542 & j0.69045 & j0.74606 \\ j0.69482 & j0.69045 & j0.76951 & j0.64065 \\ j0.74024 & j0.74606 & j0.64065 & j0.81247 \end{bmatrix}$$

The column matrix of currents by which the new $\mathbf{Z}_{\text{bus}}$ is multiplied to obtain the new bus voltages is the same as in Example 7.6. Since both I_1 and I_2 are zero while I_3 and I_4 are nonzero, we obtain

$$\begin{aligned} V_4 &= j0.64065(1.00\angle -90^\circ) + j0.81247(0.68\angle -135^\circ) \\ &= 1.03131 - j0.39066 \\ &= 1.10281\angle -20.7466^\circ \text{ per unit} \end{aligned}$$

as found in Example 8.2.

It is of interest to note that V_4 may be calculated directly from Eq. (8.27) by setting node ⓙ equal to the reference node. We then obtain for $k = 4$ and $Z_{th} = Z_{44}$

$$V_4 = Z_b \frac{V_4^0}{Z_{th} + Z_b} = -j5.0 I_{cap}$$

$$= 1.10281 \angle -20.7466^\circ \text{ per unit}$$

since I_{cap} is already calculated in Example 8.1.

8.4 DIRECT DETERMINATION OF $\mathbf{Z}_{bus}$

We could determine $\mathbf{Z}_{bus}$ by first finding $\mathbf{Y}_{bus}$ and then inverting it, but this is not convenient for large-scale systems as we have seen. Fortunately, formulation of $\mathbf{Z}_{bus}$ using a direct building algorithm is a straightforward process on the computer.

At the outset we have a list of the branch impedances showing the buses to which they are connected. We start by writing the equation for one bus connected through a branch impedance Z_a to the reference as

$$\begin{array}{cc} & (1) \\ [V_1] = (1) & [Z_a][I_1] \end{array} \tag{8.44}$$

and this can be considered as an equation involving three matrices, each of which has one row and one column. Now we might add a new bus connected to the first bus or to the reference node. For instance, if the second bus is connected to the reference node through Z_b, we have the matrix equation

$$\begin{bmatrix} V_1 \\ V_2 \end{bmatrix} = \begin{array}{c} \\ (1) \\ (2) \end{array} \overset{\begin{array}{cc} (1) & (2) \end{array}}{\begin{bmatrix} Z_a & 0 \\ 0 & Z_b \end{bmatrix}} \begin{bmatrix} I_1 \\ I_2 \end{bmatrix} \tag{8.45}$$

and we proceed to modify the evolving $\mathbf{Z}_{bus}$ matrix by adding other buses and branches following the procedures described in Sec. 8.3. The combination of these procedures constitutes the $\mathbf{Z}_{bus}$ *building algorithm*. Usually, the buses of a network must be renumbered internally by the computer algorithm to agree with the order in which they are to be added to $\mathbf{Z}_{bus}$ as it is built up.

Example 8.4. Determine $\mathbf{Z}_{bus}$ for the network shown in Fig. 8.8, where the impedances labeled 1 through 6 are shown in per unit. Preserve all buses.

Solution. The branches are added in the order of their labels and numbered subscripts on $\mathbf{Z}_{bus}$ will indicate intermediate steps of the solution. We start by

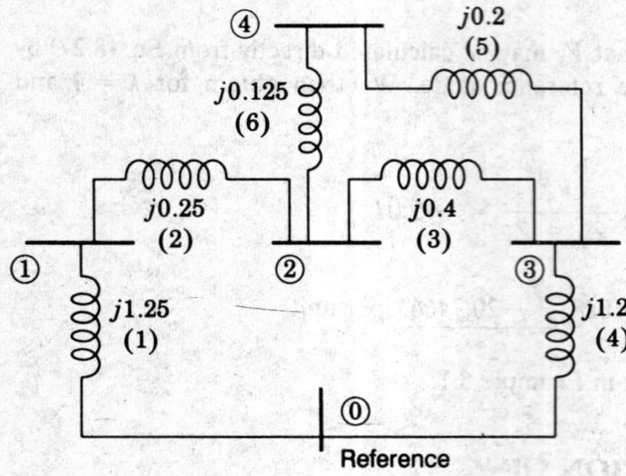

FIGURE 8.8
Network for Examples 8.4 and 8.5. Branch impedances are in per unit and branch numbers are in parentheses.

establishing bus ① with its impedance to the reference node and write

$$[V_1] = \underset{\textcircled{1}}{} \overset{\textcircled{1}}{[\,j1.25\,]}\ [I_1]$$

We then have the 1 × 1 bus impedance matrix

$$\mathbf{Z}_{\text{bus},1} = \textcircled{1}\overset{\textcircled{1}}{[\,j1.25\,]}$$

To establish bus ② with its impedance to bus ①, we follow Eq. (8.32) to write

$$\mathbf{Z}_{\text{bus},2} = \begin{matrix} \\ \textcircled{1} \\ \textcircled{2} \end{matrix} \begin{matrix} \quad\textcircled{1} \qquad \textcircled{2} \\ \begin{bmatrix} j1.25 & j1.25 \\ j1.25 & j1.50 \end{bmatrix} \end{matrix}$$

The term $j1.50$ above is the sum of $j1.25$ and $j0.25$. The elements $j1.25$ in the new row and column are the repetition of the elements of row 1 and column 1 of the matrix being modified.

Bus ③ with the impedance connecting it to bus ② is established by writing

$$\mathbf{Z}_{\text{bus},3} = \begin{matrix} \\ \textcircled{1} \\ \textcircled{2} \\ \textcircled{3} \end{matrix} \begin{matrix} \quad\textcircled{1} \qquad \textcircled{2} \qquad \textcircled{3} \\ \begin{bmatrix} j1.25 & j1.25 & j.125 \\ j1.25 & j1.50 & j1.50 \\ j1.25 & j1.50 & j1.90 \end{bmatrix} \end{matrix}$$

Since the new bus ③ is being connected to bus ②, the term $j1.90$ above is the sum of Z_{22} of the matrix being modified and the impedance Z_b of the branch being connected to bus ② from bus ③. The other elements of the new row and column are the repetition of row 2 and column 2 of the matrix being modified since the new bus is being connected to bus ②.

If we now decide to add the impedance $Z_b = j1.25$ from bus ③ to the reference node, we follow Eq. (8.32) to connect a new bus ⓟ through Z_b and obtain the impedance matrix

$$\mathbf{Z}_{bus,4} = \begin{array}{c} \\ ① \\ ② \\ ③ \\ ⓟ \end{array} \begin{array}{c} \begin{array}{cccc} ① & ② & ③ & ⓟ \end{array} \\ \left[\begin{array}{ccc|c} j1.25 & j1.25 & j.125 & j1.25 \\ j1.25 & j1.50 & j1.50 & j.150 \\ j1.25 & j1.50 & j1.90 & j1.90 \\ \hline j1.25 & j1.50 & j1.90 & j3.15 \end{array}\right] \end{array}$$

where $j3.15$ above is the sum of $Z_{33} + Z_b$. The other elements in the new row and column are the repetition of row 3 and column 3 of the matrix being modified since bus ③ is being connected to the reference node through Z_b.

We now eliminate row p and column p by Kron reduction. Some of the elements of the new matrix from Eq. (8.33) are

$$Z_{11(\text{new})} = j1.25 - \frac{(j1.25)(j1.25)}{j3.15} = j0.75397$$

$$Z_{22(\text{new})} = j1.50 - \frac{(j1.50)(j1.50)}{j3.15} = j0.78571$$

$$Z_{23(\text{new})} = Z_{32(\text{new})} = j1.50 - \frac{(j1.50)(j1.90)}{j3.15} = j0.59524$$

When all the elements are determined, we have

$$\mathbf{Z}_{bus,5} = \begin{array}{c} \\ ① \\ ② \\ ③ \end{array} \begin{array}{c} \begin{array}{ccc} ① & ② & ③ \end{array} \\ \begin{bmatrix} j0.75397 & j0.65476 & j0.49603 \\ j0.65476 & j0.78571 & j0.59524 \\ j0.49603 & j0.59524 & j0.75397 \end{bmatrix} \end{array}$$

We now decide to add the impedance $Z_b = j0.20$ from bus ③ to establish bus ④ using Eq. (8.32), and we obtain

$$\mathbf{Z}_{bus,6} = \begin{array}{c} \\ ① \\ ② \\ ③ \\ ④ \end{array} \begin{array}{c} \begin{array}{cccc} ① & ② & ③ & ④ \end{array} \\ \left[\begin{array}{ccc|c} j0.75397 & j0.65476 & j0.49603 & j0.49603 \\ j0.65476 & j0.78571 & j0.59524 & j0.59524 \\ j0.49603 & j0.59524 & j0.75397 & j0.75397 \\ \hline j0.49603 & j0.59524 & j0.75397 & j0.95397 \end{array}\right] \end{array}$$

The off-diagonal elements of the new row and column are the repetition of row 3 and column 3 of the matrix being modified because the new bus ④ is being connected to bus ③. The new diagonal element is the sum of Z_{33} of the previous matrix and $Z_b = j0.20$.

Finally, we add the impedance $Z_b = j0.125$ between buses ② and ④. If we let j and k in Eq. (8.41) equal 2 and 4, respectively, we obtain the elements for row 5 and column 5

$$Z_{15} = Z_{12} - Z_{14} = j0.65476 - j0.49603 = j0.15873$$

$$Z_{25} = Z_{22} - Z_{24} = j0.78571 - j0.59524 = j0.19047$$

$$Z_{35} = Z_{32} - Z_{34} = j0.59524 - j0.75397 = -j0.15873$$

$$Z_{45} = Z_{42} - Z_{44} = j0.59524 - j0.95397 = -j0.35873$$

and from Eq. (8.42)

$$Z_{55} = Z_{22} + Z_{44} - 2Z_{24} + Z_b$$

$$= j\{0.78571 + 0.95397 - 2(0.59524)\} + j0.125 = j0.67421$$

So, employing $\mathbf{Z}_{\text{bus},6}$ previously found, we write the 5 × 5 matrix

$$\begin{array}{c} \\ \\ \\ \\ \\ (q) \end{array}\left[\begin{array}{cccc|c} & & & & (q) \\ & & & & j0.15873 \\ & & \mathbf{Z}_{\text{bus},6} & & j0.19047 \\ & & & & -j0.15873 \\ & & & & -j0.35873 \\ \hline j0.15873 & j0.19047 & -j0.15873 & -j0.35873 & j0.67421 \end{array}\right]$$

and from Eq. (8.43) we find by Kron reduction

$$\mathbf{Z}_{\text{bus}} = \begin{array}{c} \\ (1) \\ (2) \\ (3) \\ (4) \end{array}\begin{array}{c} \begin{array}{cccc} \quad(1)\quad & \quad(2)\quad & \quad(3)\quad & \quad(4)\quad \end{array} \\ \begin{bmatrix} j0.71660 & j0.60992 & j0.53340 & j0.58049 \\ j0.60992 & j0.73190 & j0.64008 & j0.69659 \\ j0.53340 & j0.64008 & j0.71660 & j0.66951 \\ j0.58049 & j0.69659 & j0.66951 & j0.76310 \end{bmatrix} \end{array}$$

which is the bus impedance matrix to be determined. All calculations have been rounded off to five decimal places.

Since we shall again refer to these results, we note here that the reactance diagram of Fig. 8.8 is derived from Fig. 7.10 by omitting the sources and one of the mutually coupled branches. Also, the buses of Fig. 7.10 have been renumbered in Fig. 8.8 because the $\mathbf{Z}_{bus}$ building algorithm must begin with a bus connected to the reference node, as previously remarked.

The $\mathbf{Z}_{bus}$ building procedures are simple for a computer which first must determine the types of modification involved as each branch impedance is added. However, the operations must follow a sequence such that we avoid connecting an impedance between two new buses.

As a matter of interest, we can check the impedance values of $\mathbf{Z}_{bus}$ by the network calculations of Sec. 8.1.

Example 8.5. Find Z_{11} of the circuit of Example 8.4 by determining the impedance measured between bus ① and the reference node when currents injected at buses ②, ③, and ④ are zero.

Solution. The equation corresponding to Eq. (8.12) is

$$Z_{11} = \left.\frac{V_1}{I_1}\right|_{I_2=I_3=I_4=0}$$

We recognize two parallel paths between buses ② and ③ of the circuit of Fig. 8.8 with the resulting impedance of

$$\frac{(j0.125 + j0.20)(j0.40)}{j(0.125 + 0.20 + 0.40)} = j0.17931$$

This impedance in series with $(j0.25 + j1.25)$ combines in parallel with $j1.25$ to yield

$$Z_{11} = \frac{j1.25(j0.25 + j1.25 + j0.17931)}{j(1.25 + 0.25 + 1.25 + 0.17931)} = j0.71660$$

which is identical with the value found in Example 8.4.

Although the network reduction method of Example 8.5 may appear to be simpler by comparison with other methods of forming $\mathbf{Z}_{bus}$, such is not the case because a different network reduction is required to evaluate each element of the matrix. In Example 8.5 the network reduction to find Z_{44}, for instance, is more difficult than that for finding Z_{11}. The computer could make a network reduction by node elimination but would have to repeat the process for each node.

8.5 CALCULATION OF $\mathbf{Z}_{\text{bus}}$ ELEMENTS FROM $\mathbf{Y}_{\text{bus}}$

When the full numerical form of $\mathbf{Z}_{\text{bus}}$ is not explicitly required for an application, we can readily calculate elements of $\mathbf{Z}_{\text{bus}}$ as needed if the upper- and lower-triangular factors of $\mathbf{Y}_{\text{bus}}$ are available. To see how this can be done, consider postmultiplying $\mathbf{Z}_{\text{bus}}$ by a vector with only one nonzero element $1_m = 1$ in row m and all other elements equal to zero. When $\mathbf{Z}_{\text{bus}}$ is an $N \times N$ matrix, we have

$$\underbrace{\begin{array}{c} \\ ① \\ ② \\ \vdots \\ ⓜ \\ \vdots \\ Ⓝ \end{array}\!\!\begin{array}{c} \begin{array}{ccccccc} ① & ② & & ⓜ & & Ⓝ \end{array} \\ \begin{bmatrix} Z_{11} & Z_{12} & \cdots & Z_{1m} & \cdots & Z_{1N} \\ Z_{21} & Z_{22} & \cdots & Z_{2m} & \cdots & Z_{2N} \\ \vdots & \vdots & \ddots & \vdots & \ddots & \vdots \\ Z_{m1} & Z_{m2} & \cdots & Z_{mm} & \cdots & Z_{mN} \\ \vdots & \vdots & \ddots & \vdots & \ddots & \vdots \\ Z_{N1} & Z_{N2} & \cdots & Z_{Nm} & \cdots & Z_{NN} \end{bmatrix} \end{array}}_{\mathbf{Z}_{\text{bus}}} \begin{bmatrix} 0 \\ 0 \\ \vdots \\ 1_m \\ \vdots \\ 0 \end{bmatrix} = \begin{array}{c} ① \\ ② \\ \vdots \\ ⓜ \\ \vdots \\ Ⓝ \end{array} \underbrace{\overset{ⓜ}{\begin{bmatrix} Z_{1m} \\ Z_{2m} \\ \vdots \\ Z_{mm} \\ \vdots \\ Z_{Nm} \end{bmatrix}}}_{\mathbf{Z}_{\text{bus}}^{(m)}} \tag{8.46}$$

Thus, postmultiplying $\mathbf{Z}_{\text{bus}}$ by the vector shown extracts the mth column, which we have called the vector $\mathbf{Z}_{\text{bus}}^{(m)}$; that is

$$\mathbf{Z}_{\text{bus}}^{(m)} \triangleq \begin{bmatrix} \text{column } m \\ \text{of} \\ \mathbf{Z}_{\text{bus}} \end{bmatrix} = \begin{array}{c} ① \\ ② \\ \vdots \\ ⓜ \\ \vdots \\ Ⓝ \end{array} \overset{ⓜ}{\begin{bmatrix} Z_{1m} \\ Z_{2m} \\ \vdots \\ Z_{mm} \\ \vdots \\ Z_{Nm} \end{bmatrix}}$$

Since the product of $\mathbf{Y}_{\text{bus}}$ and $\mathbf{Z}_{\text{bus}}$ equals the unit matrix, we have

$$\mathbf{Y}_{\text{bus}}\mathbf{Z}_{\text{bus}} \begin{bmatrix} 0 \\ 0 \\ \vdots \\ 1_m \\ \vdots \\ 0 \end{bmatrix} = \mathbf{Y}_{\text{bus}}\mathbf{Z}_{\text{bus}}^{(m)} = \begin{bmatrix} 0 \\ 0 \\ \vdots \\ 1_m \\ \vdots \\ 0 \end{bmatrix} \tag{8.47}$$

If the lower-triangular matrix $\mathbf{L}$ and the upper-triangular matrix $\mathbf{U}$ of $\mathbf{Y}_{\text{bus}}$ are available, we can write Eq. (8.47) in the form

$$\mathbf{LUZ}_{bus}^{(m)} = \begin{bmatrix} 0 \\ 0 \\ \vdots \\ 1_m \\ \vdots \\ 0 \end{bmatrix} \tag{8.48}$$

It is now apparent that the elements in the column vector $\mathbf{Z}_{bus}^{(m)}$ can be found from Eq. (8.48) by forward elimination and back substitution, as explained in Sec. 7.8. If only some of the elements of $\mathbf{Z}_{bus}^{(m)}$ are required, the calculations can be reduced accordingly. For example, suppose that we wish to generate Z_{33} and Z_{43} of $\mathbf{Z}_{bus}$ for a four-bus system. Using convenient notation for the elements of $\mathbf{L}$ and $\mathbf{U}$, we have

$$\begin{bmatrix} l_{11} & \cdot & \cdot & \cdot \\ l_{21} & l_{22} & \cdot & \cdot \\ l_{31} & l_{32} & l_{33} & \cdot \\ l_{41} & l_{42} & l_{43} & l_{44} \end{bmatrix} \begin{bmatrix} 1 & u_{12} & u_{13} & u_{14} \\ \cdot & 1 & u_{23} & u_{24} \\ \cdot & \cdot & 1 & u_{34} \\ \cdot & \cdot & \cdot & 1 \end{bmatrix} \underbrace{\begin{bmatrix} Z_{13} \\ Z_{23} \\ Z_{33} \\ Z_{43} \end{bmatrix}}_{\mathbf{Z}_{bus}^{(3)}} = \begin{bmatrix} 0 \\ 0 \\ 1 \\ 0 \end{bmatrix} \tag{8.49}$$

We can solve this equation for $\mathbf{Z}_{bus}^{(3)}$ in two steps as follows:

$$\begin{bmatrix} l_{11} & \cdot & \cdot & \cdot \\ l_{21} & l_{22} & \cdot & \cdot \\ l_{31} & l_{32} & l_{33} & \cdot \\ l_{41} & l_{42} & l_{43} & l_{44} \end{bmatrix} \begin{bmatrix} x_1 \\ x_2 \\ x_3 \\ x_4 \end{bmatrix} = \begin{bmatrix} 0 \\ 0 \\ 1 \\ 0 \end{bmatrix} \tag{8.50}$$

where

$$\begin{bmatrix} 1 & u_{12} & u_{13} & u_{14} \\ \cdot & 1 & u_{23} & u_{24} \\ \cdot & \cdot & 1 & u_{34} \\ \cdot & \cdot & \cdot & 1 \end{bmatrix} \underbrace{\begin{bmatrix} Z_{13} \\ Z_{23} \\ Z_{33} \\ Z_{43} \end{bmatrix}}_{\mathbf{Z}_{bus}^{(3)}} = \begin{bmatrix} x_1 \\ x_2 \\ x_3 \\ x_4 \end{bmatrix} \tag{8.51}$$

By forward substitution Eq. (8.50) immediately yields

$$x_1 = 0 \qquad x_2 = 0 \qquad x_3 = \frac{1}{l_{33}} \qquad x_4 = -\frac{l_{43}}{l_{44}l_{33}}$$

and by back substitution of these intermediate results in Eq. (8.51) we find the required elements of column 3 of $\mathbf{Z}_{\text{bus}}$,

$$Z_{43} = x_4$$

$$Z_{33} = x_3 - u_{34}Z_{43}$$

If all elements of $\mathbf{Z}_{\text{bus}}^{(3)}$ are required, we can continue the calculations,

$$Z_{23} = x_2 - u_{23}Z_{33} - u_{24}Z_{43}$$

$$Z_{13} = x_1 - u_{12}Z_{23} - u_{13}Z_{33} - u_{14}Z_{43}$$

The computational effort in generating the required elements can be reduced by judiciously choosing the bus numbers.

In later chapters we shall find it necessary to evaluate terms like $(Z_{im} - Z_{in})$ involving *differences* between columns ⓜ and ⓝ of $\mathbf{Z}_{\text{bus}}$. If the elements of $\mathbf{Z}_{\text{bus}}$ are not available explicitly, we can calculate the required differences by solving a system of equations such as

$$\mathbf{LUZ}_{\text{bus}}^{(m-n)} = \begin{bmatrix} 0 \\ \vdots \\ 1_m \\ \vdots \\ -1_n \\ \vdots \\ 0 \end{bmatrix} \tag{8.52}$$

where $\mathbf{Z}_{\text{bus}}^{(m-n)} = \mathbf{Z}_{\text{bus}}^{(m)} - \mathbf{Z}_{\text{bus}}^{(n)}$ is the vector formed by subtracting column n from column m of $\mathbf{Z}_{\text{bus}}$, and $1_m = 1$ in row m and $-1_n = -1$ in row n of the vector shown.

In large-scale system calculations considerable computational efficiency can be realized by solving equations in the triangularized form of Eq. (8.52) while the full $\mathbf{Z}_{\text{bus}}$ need not be developed. Such computational considerations underlie many of the formal developments based on $\mathbf{Z}_{\text{bus}}$ in this text.

Example 8.6. The five-bus system shown in Fig. 8.9 has per-unit *impedances* as marked. The symmetrical bus admittance matrix for the system is given by

$$\mathbf{Y}_{\text{bus}} = \begin{array}{c} \text{①} \\ \text{②} \\ \text{③} \\ \text{④} \\ \text{⑤} \end{array} \overset{\begin{array}{ccccc} \text{①} & \text{②} & \text{③} & \text{④} & \text{⑤} \end{array}}{\begin{bmatrix} -j30.0 & j10.0 & 0 & j20.0 & 0 \\ j10.0 & -j26.2 & j16.0 & 0 & 0 \\ 0 & j16.0 & -j36.0 & 0 & j20.0 \\ j20.0 & 0 & 0 & -j20.0 & 0 \\ 0 & 0 & j20.0 & 0 & -j20.0 \end{bmatrix}}$$

and it is found that the triangular factors of $\mathbf{Y}_{bus}$ are

$$\mathbf{L} = \begin{bmatrix} -j30.0 & \cdot & \cdot & \cdot & \cdot \\ j10.0 & -j22.866667 & \cdot & \cdot & \cdot \\ 0 & j16.000000 & -j24.804666 & \cdot & \cdot \\ j20.0 & j6.666667 & j4.664723 & -j3.845793 & \cdot \\ 0 & 0 & j20.000000 & j3.761164 & -j0.195604 \end{bmatrix}$$

$$\mathbf{U} = \begin{bmatrix} 1 & -0.333333 & 0 & -0.666667 & 0 \\ \cdot & 1 & -0.699708 & -0.291545 & 0 \\ \cdot & \cdot & 1 & -0.188058 & -0.806300 \\ \cdot & \cdot & \cdot & 1 & -0.977995 \\ \cdot & \cdot & \cdot & \cdot & 1 \end{bmatrix}$$

Use the triangular factors to calculate $Z_{th,45} = (Z_{44} - Z_{45}) - (Z_{54} - Z_{55})$, the Thévenin impedance looking into the system between buses ④ and ⑤ of Fig. 8.9.

Solution. Since $\mathbf{Y}_{bus}$ is symmetrical, the reader should check that the row elements of **U** equal the column elements of **L** divided by their corresponding diagonal elements. With l's representing the numerical values of **L**, forward solution of the system of equations

$$\begin{bmatrix} l_{11} & \cdot & \cdot & \cdot & \cdot \\ l_{21} & l_{22} & \cdot & \cdot & \cdot \\ l_{31} & l_{32} & l_{33} & \cdot & \cdot \\ l_{41} & l_{42} & l_{43} & l_{44} & \cdot \\ l_{51} & l_{52} & l_{53} & l_{54} & l_{55} \end{bmatrix} \begin{bmatrix} x_1 \\ x_2 \\ x_3 \\ x_4 \\ x_5 \end{bmatrix} = \begin{bmatrix} 0 \\ 0 \\ 0 \\ 1 \\ -1 \end{bmatrix}$$

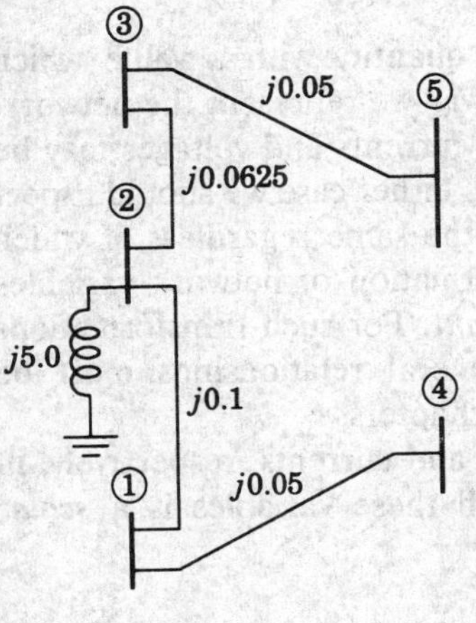

FIGURE 8.9
Reactance diagram for Example 8.6, all values are per-unit impedances.

yields the intermediate values

$$x_1 = x_2 = x_3 = 0$$

$$x_4 = l_{44}^{-1} = (-j3.845793)^{-1} = j0.260024$$

$$x_5 = \frac{-1 - l_{54}x_4}{l_{55}} = \frac{-1 - j3.761164 \times j0.260024}{-j0.195604} = -j0.112500$$

Backsubstituting in the system of equations

$$\begin{bmatrix} 1 & u_{12} & u_{13} & u_{14} & u_{15} \\ \cdot & 1 & u_{23} & u_{24} & u_{25} \\ \cdot & \cdot & 1 & u_{34} & u_{35} \\ \cdot & \cdot & \cdot & 1 & u_{45} \\ \cdot & \cdot & \cdot & \cdot & 1 \end{bmatrix} \left[\mathbf{Z}_{\text{bus}}^{(4-5)}\right] = \begin{bmatrix} 0 \\ 0 \\ 0 \\ j0.260024 \\ -j0.112500 \end{bmatrix}$$

where u's represent the numerical values of $\mathbf{U}$, we find from the last two rows that

$$Z_{54} - Z_{55} = -j0.1125 \text{ per unit}$$

$$Z_{44} - Z_{45} = j0.260024 - u_{45}(Z_{54} - Z_{55}) = j0.260024 - (-0.977995)(-j0.1125)$$

$$= j0.1500 \text{ per unit}$$

The desired Thévenin impedance is therefore calculated as follows:

$$Z_{\text{th},45} = (Z_{44} - Z_{45}) - (Z_{54} - Z_{55}) = j0.1500 - (-j0.1125) = j0.2625 \text{ per unit}$$

Inspection of Fig. 8.9 verifies this result.

8.6 POWER INVARIANT TRANSFORMATIONS

The complex power in a network is a physical quantity with a value which should not change simply because we change the way we represent the network. For example, in Chap. 7 we see that the network currents and voltages may be chosen as branch quantities or as bus quantities. In either case we should expect the power in the branches of the network to be the same regardless of which quantities are used in the calculation. A transformation of network variables which preserves power is said to be *power invariant*. For such transformations involving the bus impedance matrix certain general relationships must be satisfied, which we now establish for use in later chapters.

Let $\mathbf{V}$ and $\mathbf{I}$ describe the set of bus voltages and currents, respectively, in the network. The complex power associated with these variables is a *scalar quantity*, which we may represent by

$$S_L = V_1 I_1^* + V_2 I_2^* + \cdots + V_N I_N^* \tag{8.53}$$

or in matrix form by

$$S_L = [V_1 \;\; V_2 \; \cdots \; V_N] \begin{bmatrix} I_1^* \\ I_2^* \\ \vdots \\ I_N^* \end{bmatrix} = \mathbf{V}^T \mathbf{I}^* \tag{8.54}$$

Suppose that we transform the bus currents $\mathbf{I}$ to a new set of bus currents $\mathbf{I}_{\text{new}}$ using the transformation matrix $\mathbf{C}$ such that

$$\mathbf{I} = \mathbf{C}\mathbf{I}_{\text{new}} \tag{8.55}$$

As we shall see below, such a transformation occurs, for instance, when the reference node of the network is changed and it is required to compute the new bus impedance matrix, which we call $\mathbf{Z}_{\text{bus(new)}}$. The bus voltages in terms of the existing and the new variables are represented by

$$\mathbf{V} = \mathbf{Z}_{\text{bus}}\mathbf{I} \qquad \text{and} \qquad \mathbf{V}_{\text{new}} = \mathbf{Z}_{\text{bus(new)}}\mathbf{I}_{\text{new}} \tag{8.56}$$

and we now seek to establish the conditions to be satisfied by $\mathbf{V}_{\text{new}}$ and $\mathbf{Z}_{\text{bus(new)}}$ so that the power remains invariant when the currents aıe changed according to Eq. (8.55).

Substituting $\mathbf{V}$ from Eq. (8.56) in Eq. (8.54) gives

$$S_L = (\mathbf{Z}_{\text{bus}}\mathbf{I})^T \mathbf{I}^* = \mathbf{I}^T \mathbf{Z}_{\text{bus}} \mathbf{I}^* \tag{8.57}$$

where $\mathbf{Z}_{\text{bus}}$ is symmetrical. From Eq. (8.55) we substitute for $\mathbf{I}$ in Eq. (8.57) to obtain

$$S_L = (\mathbf{C}\mathbf{I}_{\text{new}})^T \mathbf{Z}_{\text{bus}} (\mathbf{C}\mathbf{I}_{\text{new}})^* \tag{8.58}$$

from which it follows that

$$S_L = \mathbf{I}_{\text{new}}^T \underbrace{\mathbf{C}^T \mathbf{Z}_{\text{bus}} \mathbf{C}^*}_{\mathbf{Z}_{\text{bus(new)}}} \mathbf{I}_{\text{new}}^* = \mathbf{I}_{\text{new}}^T \mathbf{Z}_{\text{bus(new)}} \mathbf{I}_{\text{new}}^* \tag{8.59}$$

Comparing Eqs. (8.57) and (8.59), we see that the complex power will be invariant in terms of the new variables, provided the new bus impedance matrix is calculated from the relationship

$$\mathbf{Z}_{\text{bus(new)}} = \mathbf{C}^T \mathbf{Z}_{\text{bus}} \mathbf{C}^* \tag{8.60}$$

This is a fundamental result for constructing the new bus impedance matrix. From Eqs. (8.56) and (8.59) we find that

$$S_L = \mathbf{I}_{\text{new}}^T \mathbf{Z}_{\text{bus(new)}} \mathbf{I}_{\text{new}}^* = \mathbf{V}_{\text{new}}^T \mathbf{I}_{\text{new}}^* \tag{8.61}$$

It also follows from Eq. (8.54) that

$$S_L = \mathbf{V}^T \mathbf{C}^* \mathbf{I}_{\text{new}}^* = (\mathbf{C}^{*T} \mathbf{V})^T \mathbf{I}_{\text{new}}^* \tag{8.62}$$

and we may then conclude from Eqs. (8.61) and (8.62) that the new voltage variables $\mathbf{V}_{\text{new}}$ must be related to the existing voltage variables $\mathbf{V}$ by the fundamental relationship

$$\mathbf{V}_{\text{new}} = \mathbf{C}^{*T} \mathbf{V} \tag{8.63}$$

In many transformations, especially those involving the connection matrices of the network, all the entries in $\mathbf{C}$ are real, and in such cases we drop the complex conjugate superscript of $\mathbf{C}^*$.

Equation (8.53) is the *net* sum of all the real and reactive power entering and leaving the buses of the network. Hence, S_L represents the complex power loss of the system and is a phasor quantity with real and reactive parts given by Eq. (8.59) as

$$S_L = P_L + jQ_L = \mathbf{I}_{\text{new}}^T \mathbf{C}^T \mathbf{Z}_{\text{bus}} \mathbf{C}^* \mathbf{I}_{\text{new}}^* \tag{8.64}$$

The complex conjugate of the transpose of Eq. (8.64) is

$$S_L^* = P_L - jQ_L = \mathbf{I}_{\text{new}}^T \mathbf{C}^T \mathbf{Z}_{\text{bus}}^{T*} \mathbf{C}^* \mathbf{I}_{\text{new}}^* \tag{8.65}$$

Adding Eqs. (8.64) and (8.65) together and solving for P_L yield

$$P_L = \mathbf{I}_{\text{new}}^T \mathbf{C}^T \left[\frac{\mathbf{Z}_{\text{bus}} + \mathbf{Z}_{\text{bus}}^{T*}}{2} \right] \mathbf{C}^* \mathbf{I}_{\text{new}}^* \tag{8.66}$$

When $\mathbf{Z}_{\text{bus}}$ is symmetrical, which is almost always the case, we may write

$$\mathbf{Z}_{\text{bus}} = \mathbf{R}_{\text{bus}} + j\mathbf{X}_{\text{bus}} \tag{8.67}$$

where both $\mathbf{R}_{\text{bus}}$ and $\mathbf{X}_{\text{bus}}$ are symmetrical. We note that $\mathbf{R}_{\text{bus}}$ and $\mathbf{X}_{\text{bus}}$ are available by inspection *after* $\mathbf{Z}_{\text{bus}}$ has been constructed for the network. Substituting from Eq. (8.67) in Eq. (8.66) cancels out the reactance part of $\mathbf{Z}_{\text{bus}}$,

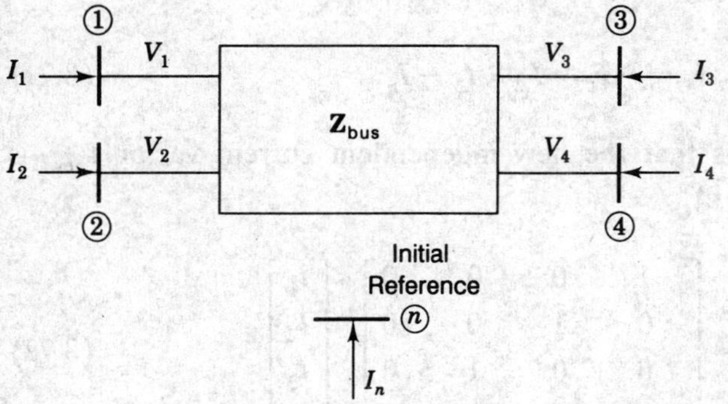

FIGURE 8.10
Changing the reference of $\mathbf{Z}_{bus}$.

and we then find

$$P_L = \mathbf{I}_{new}^T \mathbf{C}^T \mathbf{R}_{bus} \mathbf{C}^* \mathbf{I}_{new}^* \tag{8.68}$$

which simplifies the numerical calculation of P_L since only the resistance portion of $\mathbf{Z}_{bus}$ is involved.

An important application of Eqs. (8.60) and (8.63) arises when the reference node for the $\mathbf{Z}_{bus}$ representation of the system is changed. Of course, we could again use the building algorithm of Sec. 8.4 to rebuild completely the new $\mathbf{Z}_{bus}$ starting with the new reference node. However, this would be computationally inefficient, and we now show how to modify the existing $\mathbf{Z}_{bus}$ to account for the change of reference node. To illustrate, consider that $\mathbf{Z}_{bus}$ has already been constructed for the five-node system of Fig. 8.10 based on node ⓝ as reference. The standard bus equations are then written as

$$\begin{bmatrix} V_1 \\ V_2 \\ V_3 \\ V_4 \end{bmatrix} = \begin{matrix} \textcircled{1} \\ \textcircled{2} \\ \textcircled{3} \\ \textcircled{4} \end{matrix} \overset{\begin{matrix} \textcircled{1} & \textcircled{2} & \textcircled{3} & \textcircled{4} \end{matrix}}{\begin{bmatrix} Z_{11} & Z_{12} & Z_{13} & Z_{14} \\ Z_{21} & Z_{22} & Z_{23} & Z_{24} \\ Z_{31} & Z_{32} & Z_{33} & Z_{34} \\ Z_{41} & Z_{42} & Z_{43} & Z_{44} \end{bmatrix}} \begin{bmatrix} I_1 \\ I_2 \\ I_3 \\ I_4 \end{bmatrix} \tag{8.69}$$

in which the bus voltages V_1, V_2, V_3, and V_4 are measured with respect to node ⓝ as reference, and the current injections I_1, I_2, I_3, and I_4 are independent. Kirchhoff's current law for Fig. 8.10 shows that

$$I_n + I_1 + I_2 + I_3 + I_4 = 0 \tag{8.70}$$

If we now change the reference from node ⓝ to node ④, for instance, then I_4 is no longer independent since it can be expressed in terms of the other four node currents. That is,

$$I_4 = -I_1 - I_2 - I_3 - I_n \tag{8.71}$$

From Eq. (8.71) it follows that the new independent current vector $\mathbf{I}_{\text{new}}$ is related to the old vector $\mathbf{I}$ by

$$\underbrace{\begin{bmatrix} I_1 \\ I_2 \\ I_3 \\ I_4 \end{bmatrix}}_{\mathbf{I}} = \underbrace{\begin{bmatrix} 1 & 0 & 0 & 0 \\ 0 & 1 & 0 & 0 \\ 0 & 0 & 1 & 0 \\ -1 & -1 & -1 & -1 \end{bmatrix}}_{\mathbf{C}} \underbrace{\begin{bmatrix} I_1 \\ I_2 \\ I_3 \\ I_n \end{bmatrix}}_{\mathbf{I}_{\text{new}}} \tag{8.72}$$

Equation (8.72) is merely a statement that I_1, I_2, and I_3 remain as before but that I_4 is replaced by the independent current I_n, which appears in the new current vector $\mathbf{I}_{\text{new}}$ as shown. In the transformation matrix $\mathbf{C}$ of Eq. (8.72) all the entries are real, and so substituting for $\mathbf{C}$ and $\mathbf{Z}_{\text{bus}}$ in Eq. (8.60), we find that

$$\mathbf{Z}_{\text{bus(new)}} = \underbrace{\begin{bmatrix} 1 & 0 & 0 & -1 \\ 0 & 1 & 0 & -1 \\ 0 & 0 & 1 & -1 \\ 0 & 0 & 0 & -1 \end{bmatrix}}_{\mathbf{C}^T} \underbrace{\begin{bmatrix} Z_{11} & Z_{12} & Z_{13} & Z_{14} \\ Z_{21} & Z_{22} & Z_{23} & Z_{24} \\ Z_{31} & Z_{32} & Z_{33} & Z_{34} \\ Z_{41} & Z_{42} & Z_{43} & Z_{44} \end{bmatrix}}_{\mathbf{Z}_{\text{bus}}} \underbrace{\begin{bmatrix} 1 & 0 & 0 & 0 \\ 0 & 1 & 0 & 0 \\ 0 & 0 & 1 & 0 \\ -1 & -1 & -1 & -1 \end{bmatrix}}_{\mathbf{C}} \tag{8.73}$$

The matrix multiplications in Eq. (8.73) are accomplished in two easy stages as follows. First, we compute

$$\mathbf{C}^T\mathbf{Z}_{\text{bus}} = \begin{bmatrix} Z_{11} - Z_{41} & Z_{12} - Z_{42} & Z_{13} - Z_{43} & Z_{14} - Z_{44} \\ Z_{21} - Z_{41} & Z_{22} - Z_{42} & Z_{23} - Z_{43} & Z_{24} - Z_{44} \\ Z_{31} - Z_{41} & Z_{32} - Z_{42} & Z_{33} - Z_{43} & Z_{34} - Z_{44} \\ -Z_{41} & -Z_{42} & -Z_{43} & -Z_{44} \end{bmatrix} \tag{8.74}$$

which for convenience we write in the form

$$\mathbf{C}^T\mathbf{Z}_{\text{bus}} = \begin{bmatrix} Z'_{11} & Z'_{12} & Z'_{13} & Z'_{14} \\ Z'_{21} & Z'_{22} & Z'_{23} & Z'_{24} \\ Z'_{31} & Z'_{32} & Z'_{33} & Z'_{34} \\ Z'_{41} & Z'_{42} & Z'_{43} & Z'_{44} \end{bmatrix} \tag{8.75}$$

It is evident from Eqs. (8.74) and (8.75) that the elements with primed superscripts are found by subtracting the existing row 4 from each of the other rows of $\mathbf{Z}_{\text{bus}}$ and by changing the sign of existing row 4. Second, we postmultiply Eq. (8.75) by $\mathbf{C}$ to obtain

$$\mathbf{Z}_{\text{bus(new)}} = \mathbf{C}^T\mathbf{Z}_{\text{bus}}\mathbf{C}$$

$$= \begin{array}{c} \\ ① \\ ② \\ ③ \\ ⓝ \end{array} \begin{array}{cccc} ① & ② & ③ & ⓝ \\ \left[\begin{array}{c} Z'_{11} - Z'_{14} \\ Z'_{21} - Z'_{24} \\ Z'_{31} - Z'_{34} \\ Z'_{41} - Z'_{44} \end{array}\right. & \begin{array}{c} Z'_{12} - Z'_{14} \\ Z'_{22} - Z_{24} \\ Z'_{32} - Z'_{34} \\ Z'_{42} - Z'_{44} \end{array} & \begin{array}{c} Z'_{13} - Z'_{14} \\ Z'_{23} - Z'_{24} \\ Z'_{33} - Z'_{34} \\ Z'_{43} - Z'_{44} \end{array} & \left.\begin{array}{c} -Z'_{14} \\ -Z'_{24} \\ -Z'_{34} \\ -Z'_{44} \end{array}\right] \end{array} \tag{8.76}$$

which is equally simple to compute from $\mathbf{C}^T\mathbf{Z}_{\text{bus}}$ by subtracting the forth column of Eq. (8.75) from each of its other columns and by changing the sign of the fourth column. It is worthwhile noting that the first diagonal element of $\mathbf{Z}_{\text{bus(new)}}$, expressed in terms of the original $\mathbf{Z}_{\text{bus}}$ entries, has the value $(Z'_{11} - Z'_{14}) = (Z_{11} + Z_{44} - 2Z_{14})$, which is the Thévenin impedance between nodes ① and ④, as we would expect from Eq. (8.26). Similar remarks apply to each of the other diagonal elements of $\mathbf{Z}_{\text{bus(new)}}$.

The bus voltages with respect to the new reference node ④ are given by Eq. (8.63) as follows:

$$\mathbf{V}_{\text{new}} = \begin{bmatrix} V_{1,\text{new}} \\ V_{2,\text{new}} \\ V_{3,\text{new}} \\ V_{n,\text{new}} \end{bmatrix} = \underbrace{\begin{bmatrix} 1 & 0 & 0 & -1 \\ 0 & 1 & 0 & -1 \\ 0 & 0 & 1 & -1 \\ 0 & 0 & 0 & -1 \end{bmatrix}}_{\mathbf{C}^T} \begin{bmatrix} V_1 \\ V_2 \\ V_3 \\ V_4 \end{bmatrix} = \begin{bmatrix} V_1 - V_4 \\ V_2 - V_4 \\ V_3 - V_4 \\ -V_4 \end{bmatrix} \tag{8.77}$$

Therefore, in the general case when bus ⓚ of an existing $\mathbf{Z}_{\text{bus}}$ is chosen as the new reference node, we may determine the new bus impedance matrix $\mathbf{Z}_{\text{bus(new)}}$ in two *consecutive* steps:

1. Subtract the existing row k from each of the other rows in $\mathbf{Z}_{\text{bus}}$ and change the sign of row k. The result is $\mathbf{C}^T\mathbf{Z}_{\text{bus}}$.

2. Subtract column k of the resultant matrix $\mathbf{C}^T\mathbf{Z}_{\text{bus}}$ from each of its other columns and change the sign of column k. The result is $\mathbf{C}^T\mathbf{Z}_{\text{bus}}\mathbf{C} = \mathbf{Z}_{\text{bus(new)}}$ with row k and column k now representing the node which was the previous reference node.

We use these procedures, for example, when studying economic operation in Chap. 13.

8.7 MUTUALLY COUPLED BRANCHES IN $\mathbf{Z}_{\text{bus}}$

So far we have not considered how to incorporate mutually coupled elements of the network into $\mathbf{Z}_{\text{bus}}$. The procedures for doing so are not difficult. But they are somewhat unwieldy as we now demonstrate by extending the $\mathbf{Z}_{\text{bus}}$ building algorithm to provide for addition to the network of one pair of mutually coupled branches.[1] One of the branches may be added to $\mathbf{Z}_{\text{orig}}$ using the appropriate procedure of Sec. 8.3, and the question remaining is how to add the second branch so that it mutually couples with the branch already included in $\mathbf{Z}_{\text{orig}}$. We consider bus ⓟ already established within $\mathbf{Z}_{\text{orig}}$ in the following four cases, which continue the numbering system introduced in Sec. 8.3.

CASE 5. *Adding mutually coupled Z_b from existing bus ⓟ to new bus ⓠ*

Let us assume that the branch impedance Z_a is already added to the energized network between nodes ⓜ and ⓝ of Fig. 8.11. The bus impedance matrix $\mathbf{Z}_{\text{orig}}$ then includes Z_a and the existing buses ⓜ, ⓝ, and ⓟ as shown. Between bus ⓟ and the new bus ⓠ of Fig. 8.11 it is required to add the branch impedance Z_b, which is mutually coupled to Z_a through mutual impedance Z_M. The voltage-drop equations for the two coupled branches are given in Eq. (7.9) and repeated here as

$$V_a = Z_a I_a + Z_M I_b \tag{8.78}$$

$$V_b = Z_M I_a + Z_b I_b \tag{8.79}$$

where I_a is the branch current flow in Z_a from bus ⓜ to bus ⓝ and the current I_b from bus ⓟ to bus ⓠ equals the negative of current injection I_q. Solving Eq. (8.78) for I_a and substituting the result along with $I_b = -I_q$ in Eq. (8.79) give

$$V_b = \frac{Z_M}{Z_a} V_a + \left(\frac{Z_M^2}{Z_a} - Z_b \right) I_q \tag{8.80}$$

In terms of bus voltages the voltage drops across the branches are given by $V_a = (V_m - V_n)$ and $V_b = (V_p - V_q)$, and substituting these relationships in Eq.

[1] For a more general treatment, see G. W. Stagg and A. H. El-Abiad, *Computer Methods in Power System Analysis*, McGraw-Hill, Inc., New York, 1968.

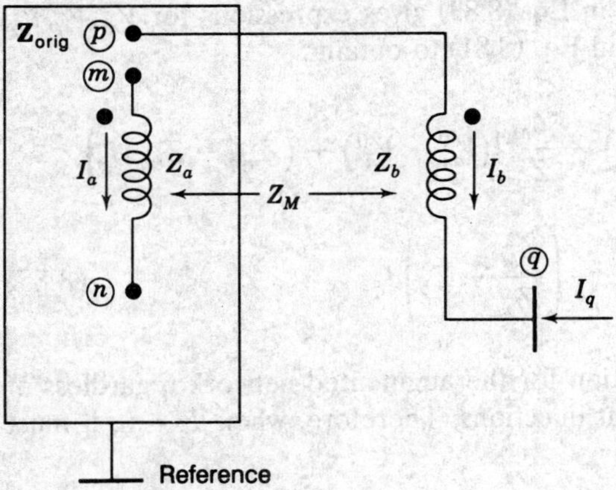

FIGURE 8.11
Adding mutually coupled branch Z_b to $\mathbf{Z}_{orig}$.

(8.80) yields

$$V_q = V_p - \frac{Z_M}{Z_a}(V_m - V_n) - \left(\frac{Z_M^2}{Z_a} - Z_b\right)I_q \tag{8.81}$$

This equation gives the voltage at new bus ⓠ with both mutually coupled branches now included in the network. The bus impedance matrix for the system augmented by new bus ⓠ is given by

$$\begin{bmatrix} V_1 \\ V_2 \\ \vdots \\ V_N \\ \hline V_q \end{bmatrix} = \begin{array}{c} \\ \\ \\ \\ ⓠ \end{array}\left[\begin{array}{cccc|c} & & & & Z_{1q} \\ & & & & Z_{2q} \\ & & \mathbf{Z}_{orig} & & \vdots \\ & & & & Z_{Nq} \\ \hline Z_{q1} & Z_{q2} & \cdots & Z_{qN} & Z_{qq} \end{array}\right] \begin{bmatrix} I_1 \\ I_2 \\ \vdots \\ I_N \\ \hline I_q \end{bmatrix} \tag{8.82}$$

and it is now required to find expressions for the new elements with subscripts q in row q and column q. A typical row i of Eq. (8.82) may be written in the form

$$V_i = \underbrace{Z_{i1}I_1 + Z_{i2}I_2 + \cdots + Z_{iN}I_N}_{V_i^0} + Z_{iq}I_q = V_i^0 + Z_{iq}I_q \tag{8.83}$$

where for convenience we have denoted

$$V_i^0 = \sum_{j=1}^{N} Z_{ij}I_j \tag{8.84}$$

Setting i equal to m, n, p, and q in Eq. (8.83) gives expressions for V_m, V_n, V_p, and V_q, which can be substituted in Eq. (8.81) to obtain

$$V_q^0 + Z_{qq}I_q = \left(V_p^0 + Z_{pq}I_q\right) - \frac{Z_M}{Z_a}\left\{\left(V_m^0 - V_n^0\right) + \left(Z_{mq} - Z_{nq}\right)I_q\right\} - \left(\frac{Z_M^2}{Z_a} - Z_b\right)I_q \tag{8.85}$$

Equation (8.85) is a *general* equation for the augmented network regardless of the particular values of the current injections. Therefore, when $I_q = 0$, it must follow from Eq. (8.85) that

$$V_q^0 = V_p^0 - \frac{Z_M}{Z_a}\left(V_m^0 - V_n^0\right) \tag{8.86}$$

Substituting in Eq. (8.86) for V_m^0, V_n^0, V_p^0, and V_q^0 from Eq. (8.84), collecting terms, and equating coefficients of I_j on both sides of the resultant equation, we find that

$$Z_{qj} = Z_{pj} - \frac{Z_M}{Z_a}\left(Z_{mj} - Z_{nj}\right) \tag{8.87}$$

for all values of j from 1 to N but not including q. Thus, except for Z_{qq}, Eq. (8.87) tells us how to calculate the entries in the new row q of the bus impedance matrix using the known values of Z_M, Z_a, and certain elements of $\mathbf{Z}_{\text{orig}}$. Indeed, to obtain the entry in row q, each element of row n is subtracted from the corresponding element of row m and the difference multiplied by Z_m/Z_a is then subtracted from the corresponding element of row p. Thus, only rows m, n, and p of $\mathbf{Z}_{\text{orig}}$ enter into the calculations of new row q. Because of symmetry, the new column q of Eq. (8.82) is the transpose of the new row q, and so $Z_{qj} = Z_{jq}$. The expression for the diagonal element Z_{qq} is determined by considering all currents except I_q set equal to zero and then equating the coefficients of I_q on both sides of Eq. (8.85), which gives

$$Z_{qq} = Z_{pq} - \frac{Z_M}{Z_a}\left(Z_{mq} - Z_{nq}\right) - \left(\frac{Z_M^2}{Z_a} - Z_b\right) \tag{8.88}$$

This equation shows that there is a sequence to be followed in determining the new elements of the bus impedance matrix. First, we calculate the elements Z_{qj} of the new row q (and thereby Z_{jq} of column q) employing Eq. (8.87), and then we use the newly calculated quantities Z_{mq}, Z_{nq}, and Z_{pq} to find Z_{qq} from Eq. (8.88).

There are three other cases of interest involving mutually coupled branches.

CASE 6. *Adding mutually coupled Z_b from existing bus ⓟ to reference*

For this case the procedure is basically a special application of Case 5. First, between bus ⓟ and new bus ⓠ we add impedance Z_b coupled by mutual impedance Z_M to the impedance Z_a already included in $\mathbf{Z}_{orig}$. Then, we short-circuit bus ⓠ to the reference node by setting V_q equal to zero, which yields the same matrix equation as Eq. (8.82) except that V_q is zero. Thus, for the modified bus impedance matrix we proceed to create a new row q and column q exactly the same as in Case 5. Then, we eliminate the newly formed row and column by the standard technique of Kron reduction since V_q is zero in the column of voltages of Eq. (8.82).

CASE 7. *Adding mutually coupled Z_b between existing buses ⓟ and ⓚ*

The procedure in this case essentially combines Cases 5 and 4. To begin, we follow the procedure of Case 5 to add the mutually coupled branch impedance Z_b from existing bus ⓟ to a new temporary bus ⓠ, recognizing that Z_a is already part of $\mathbf{Z}_{orig}$. The result is the augmented matrix of Eq. (8.82) whose q-elements are given in Eqs. (8.87) and (8.88). Next, we short-circuit bus ⓠ to bus ⓚ by adding a branch of zero impedance between those buses. To do so, we apply Case 4 to Eq. (8.82) as follows. Since $(V_q - V_k)$ is required to be zero, we find an expression for that quantity by subtracting row k from row q in Eq. (8.82) and then use the result to replace the existing row q of Eq. (8.82). By symmetry as in Case 4, a new column follows directly from the transpose of the new row, and we obtain

$$\begin{bmatrix} V_1 \\ V_2 \\ \vdots \\ V_N \\ \hline V_q - V_k \end{bmatrix} = \left[\begin{array}{c|c} \mathbf{Z}_{orig} & \begin{array}{c}(\text{col. } q - \text{col. } k) \\ \text{of} \\ \text{Eq. (8.82)}\end{array} \\ \hline (\text{row } q - \text{row } k) \text{ of Eq. (8.82)} & Z_c \end{array}\right] \begin{bmatrix} I_1 \\ I_2 \\ \vdots \\ I_N \\ \hline I_q \end{bmatrix} \tag{8.89}$$

where Z_c equals the sum $(Z_{qq} + Z_{kk} - 2Z_{qk})$ of elements drawn from Eq. (8.82). Because $(V_q - V_k)$ equals 0, we may eliminate the new row and new column of Eq. (8.89) by Kron reduction to find the final form of the $N \times N$ bus impedance matrix.

To *remove* a mutually coupled branch, we must modify the above procedures as now explained.

CASE 8. *Removing mutually coupled Z_b from existing buses Ⓟ and Ⓚ.*

A single uncoupled branch of impedance Z_b can be removed from the network model by adding the negative of impedance Z_b between the same terminating buses. When the impedance Z_b to be removed is also mutually coupled to a second branch of impedance Z_a, the rule for modifying $\mathbf{Z}_{\text{bus}}$ is to add between the end buses of Z_b a branch, which has negative impedance $-Z_b$ and the *same* mutual coupling to Z_a as the original Z_b. The procedure is illustrated in Fig. 8.12. As shown, the mutually coupled branches Z_a and Z_b with mutual impedance Z_M are already included in $\mathbf{Z}_{\text{orig}}$. In accordance with the above rule, we add the impedance $-Z_b$ having mutual coupling Z_M with the branch Z_a, and the voltage-drop equations for the *three* mutually coupled branches then are

$$V_a = Z_a I_a + Z_M I_b + Z_M I_b' \tag{8.90}$$

$$V_b = Z_M I_a + Z_b I_b + 0 \tag{8.91}$$

$$V_b = Z_M I_a + 0 \quad + (-Z_b) I_b' \tag{8.92}$$

where the branch currents I_a, I_b, and I_b' are as shown in Fig. 8.12. Subtracting Eq. (8.92) from Eq. (8.91) yields

$$0 = Z_b(I_b + I_b') \tag{8.93}$$

which shows that $(I_b + I_b')$ is zero, and substituting this result in Eq. (8.90) gives

$$V_a = Z_a I_a \tag{8.94}$$

Because $(I_b + I_b')$ equals zero, the net coupling effect of the two parallel

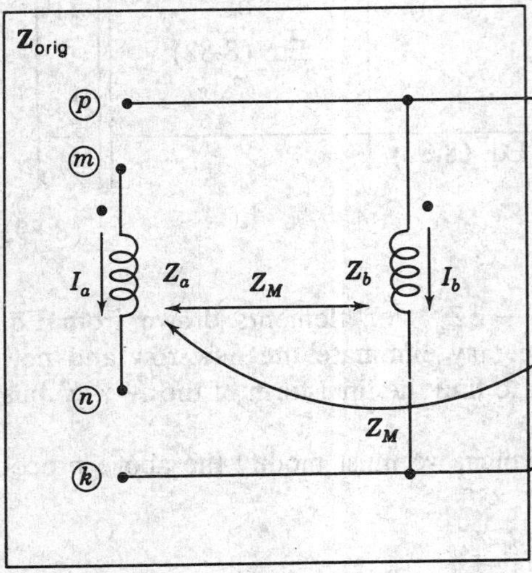

FIGURE 8.12
Removing mutually coupled branch Z_b from $\mathbf{Z}_{\text{orig}}$. Initially, the switch is considered open and the branch $-Z_b$ is added from existing bus Ⓟ to temporary bus Ⓠ. The switch is then closed to connect bus Ⓠ to bus Ⓚ.

branches between buses ⓟ and ⓚ is zero. Consequently, the impedance Z_a between buses ⓜ and ⓝ can be made to stand alone as evidenced by Eq. (8.94). To do so, we follow exactly the same procedure as in Case 7, *except* that the elements of the new row and column of the temporary bus ⓠ are calculated *sequentially* using the modified formulas

$$Z_{jq} = Z_{qj} = Z_{pj} - Z_M[Y_a|Y_M]\left[\frac{Z_{mj} - Z_{nj}}{Z_{pj} - Z_{kj}}\right] \tag{8.95}$$

with index j ranging from 1 to N and

$$Z_{qq} = Z_{pq} - Z_M[Y_a|Y_M]\left[\frac{Z_{mq} - Z_{nq}}{Z_{pq} - Z_{kq}}\right] - (Y_a Z_M^2 + Z_b) \tag{8.96}$$

Using these elements in Eqs. (8.82) and (8.89), and Kron reducing the latter, gives the desired new bus impedance matrix from which the mutually coupled branch between buses ⓟ and ⓚ is omitted. Equations (8.95) and (8.96) (developed in Prob. 8.19) have admittances Y_a and Y_M, which are calculated from the impedance parameters Z_a, Z_b, and Z_M according to Eq. (7.10).

Table 8.2 summarizes the procedures of Cases 5 through 8.

Example 8.7. Between buses ① and ④ of Fig. 8.8 impedance Z_b equal to $j0.25$ per unit is connected so that it couples through mutual impedance $j0.15$ per unit to the branch impedance already connected between buses ① and ②. Modify the bus impedance matrix of Example 8.4 to include the addition of Z_b to Fig. 8.8.

Solution. Connecting Z_b between buses ① and ④ corresponds to the conditions of Case 7 above. Our calculations begin with $\mathbf{Z}_{\text{orig}}$, the 4×4 matrix solution of Example 8.4 which includes the branch between buses ① and ②. To establish row q and column q for temporary bus ⓠ, we follow Eq. (8.87), recognizing that subscript $m = 1$, subscript $n = 2$, subscript $p = 1$, and we find that

$$Z_{qj} = Z_{1j} - \frac{Z_M}{Z_a}(Z_{1j} - Z_{2j}) = \left(1 - \frac{Z_M}{Z_a}\right)Z_{1j} + \frac{Z_M}{Z_a}Z_{2j}$$

The ratio (Z_M/Z_a) is given by $(j0.15/j0.25) = 0.6$, and substituting this value along with the elements of row 1 and row 2 of $\mathbf{Z}_{\text{orig}}$ in the above equation, we obtain

$$Z_{q1} = 0.4Z_{11} + 0.6Z_{21} = 0.4(j0.71660) + 0.6(j0.60992) = j0.65259$$

$$Z_{q2} = 0.4Z_{12} + 0.6Z_{22} = 0.4(j0.60992) + 0.6(j0.73190) = j0.68311$$

$$Z_{q3} = 0.4Z_{13} + 0.6Z_{23} = 0.4(j0.53340) + 0.6(j0.64008) = j0.59741$$

$$Z_{q4} = 0.4Z_{14} + 0.6Z_{24} = 0.4(j0.58049) + 0.6(j0.69659) = j0.65015$$

TABLE 8.2

$\mathbf{Z}_{\text{bus}}$ modifications; mutual coupling

Case	Add mutually coupled $\mathbf{Z}_b$ from	$\mathbf{Z}_{\text{bus(new)}}$
5	Existing bus ⓟ to new bus ⓠ 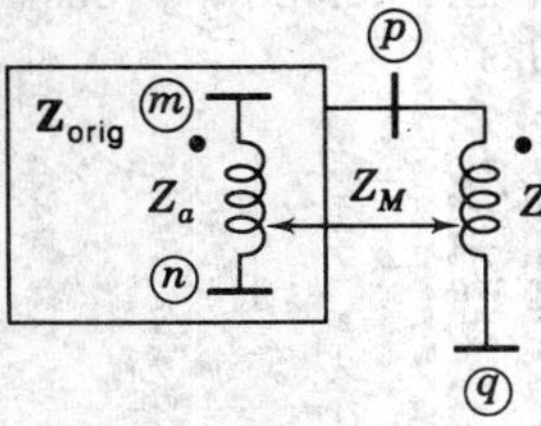	• Form the matrix $\begin{array}{c} \\ ⓠ \end{array}\left[\begin{array}{c\|c} \mathbf{Z}_{\text{orig}} & \text{col. } q \\ \hline \text{row } q & Z_{qq} \end{array}\right]$ (column label ⓠ above col. q) $Z_{jq} = Z_{qj} = Z_{pj} - \dfrac{Z_M}{Z_a}\left(Z_{mj} - Z_{nj}\right)$ $Z_{qq} = Z_{pq} - \dfrac{Z_M}{Z_a}\left(Z_{mq} - Z_{nq}\right) - \left(\dfrac{Z_M^2}{Z_a} - Z_b\right)$
6	Existing bus ⓟ to reference node 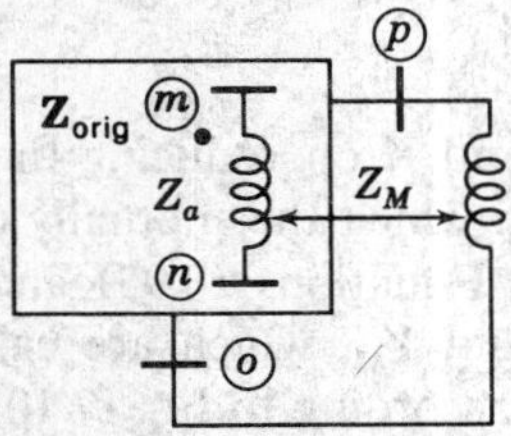	• Repeat Case 5 and • Eliminate row ⓠ and column ⓠ by Kron reduction
7	Existing bus ⓟ to existing bus ⓚ 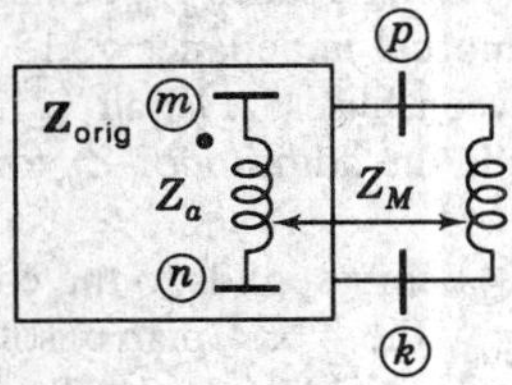	• Repeat Case 5 • Then, form the matrix $\left[\begin{array}{c\|c} \mathbf{Z}_{\text{orig}} & \text{col. } q - \text{col. } k \\ \hline \text{row } q - \text{row } k & Z_{qq} + Z_{kk} - 2Z_{qk} \end{array}\right]$ and • Eliminate last row and column by Kron reduction
8	Removal of mutually coupled line 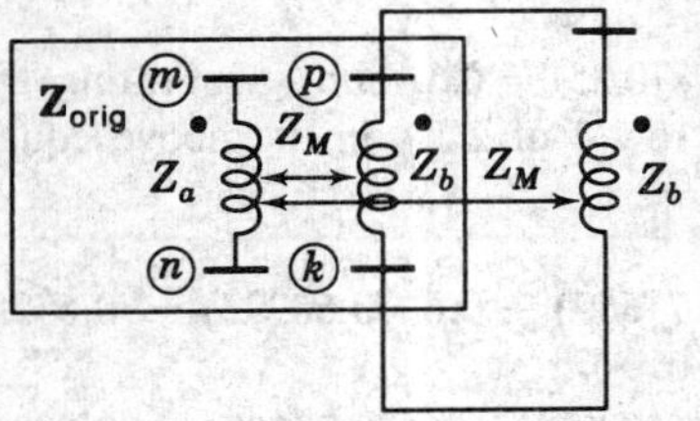	• Form the matrix of Case 5 using Eqs. $Z_{jq} = Z_{qj} = Z_{pj} - Z_M\left[Y_a \quad Y_M\right]\begin{bmatrix} Z_{mj} - Z_{nj} \\ Z_{pj} - Z_{kj} \end{bmatrix}$ $Z_{qq} = Z_{pq} - Z_M\left[Y_a \quad Y_M\right]\begin{bmatrix} Z_{mq} - Z_{nq} \\ Z_{pq} - Z_{kq} \end{bmatrix}$ $\quad - \left(Y_a Z_M^2 + Z_b\right)$ where $\begin{bmatrix} Y_a & Y_M \\ Y_M & Y_b \end{bmatrix} = \begin{bmatrix} Z_a & Z_M \\ Z_M & Z_b \end{bmatrix}^{-1}$ • Continue as in Case 7.

These elements constitute new row q and new column q except for the diagonal element Z_{qq}, which is found from Eq. (8.88) by setting the subscripts m, n, and p as above to yield

$$Z_{qq} = Z_{1q} - \frac{Z_M}{Z_a}(Z_{1q} - Z_{2q}) - \left(\frac{Z_M^2}{Z_a} - Z_b\right)$$

Substituting in this equation for $Z_M = j0.15$, $Z_a = j0.25$ and $Z_b = j0.25$ gives

$$Z_{qq} = 0.4Z_{1q} + 0.6Z_{2q} + j0.16$$

$$= 0.4(j0.65259) + 0.6(j0.68311) + j0.16 = j0.83090$$

Associating the newly calculated elements of row q and column q with $\mathbf{Z}_{\text{orig}}$ from Example 8.4 results in the 5×5 matrix

$$\begin{array}{c} \\ \\ \\ \\ \\ \textcircled{q} \end{array}\overbrace{}^{\mathbf{Z}_{\text{orig}}}\ \textcircled{q}$$

$$\textcircled{q}\quad\left[\begin{array}{cccc|c} j0.71660 & j0.60992 & j0.53340 & j0.58049 & j0.65259 \\ j0.60992 & j0.73190 & j0.64008 & j0.69659 & j0.68311 \\ j0.53340 & j0.64008 & j0.71660 & j0.66951 & j0.59741 \\ j0.58049 & j0.69659 & j0.66951 & j0.76310 & j0.65015 \\ \hline j0.65259 & j0.68311 & j0.59741 & j0.65015 & j0.83090 \end{array}\right]$$

At this point in our solution the mutually coupled branch impedance Z_b has been incorporated into the network between bus ① and bus ⓠ. To complete the connection of Z_b to bus ④, we must find $(V_q - V_4)$ and then set it equal to zero. The first of these steps is accomplished by setting subscript k equal to 4 in Eq. (8.89), subtracting row 4 from row q of the above 5×5 matrix, and using the result to replace the existing row q and column q to obtain

$$\left[\begin{array}{cccc|c} & & & & j0.07210 \\ & & \mathbf{Z}_{\text{orig}} & & -j0.01348 \\ & & & & -j0.07210 \\ & & & & -j0.11295 \\ \hline j0.07210 & -j0.01348 & -j0.07210 & -j0.11295 & j0.29370 \end{array}\right]$$

The new diagonal element is calculated from

$$(Z_{qq} + Z_{44} - 2Z_{q4}) = j0.29370$$

The only remaining calculations involve setting $(V_q - V_4)$ equal to zero in Eq.

(8.89) and eliminating the new row and column by Kron reduction to give

$$\mathbf{Z}_{bus} = \begin{array}{c} \\ ① \\ ② \\ ③ \\ ④ \end{array} \begin{array}{c} \begin{array}{cccc} ① & ② & ③ & ④ \end{array} \\ \begin{bmatrix} j0.69890 & j0.61323 & j0.55110 & j0.60822 \\ j0.61323 & j0.73128 & j0.63677 & j0.69140 \\ j0.55110 & j0.63677 & j0.69890 & j0.64178 \\ j0.60822 & j0.69140 & j0.64178 & j0.71966 \end{bmatrix} \end{array}$$

which is the desired bus impedance matrix. This result is the same as that shown in Example 7.6 except for the change in the bus numbers of Fig. 7.9 to those of Fig. 8.8.

8.8 SUMMARY

This chapter introduces the important $\mathbf{Z}_{bus}$ building algorithm, which starts by choosing a branch tied to the reference from a node and adding to this node a second branch connected from a new node. The starting node and the new node then have one row and one column each in the 2×2 bus impedance matrix which represents them. Next, a third branch connected to one or both of the first two chosen nodes is added to expand the evolving network and its $\mathbf{Z}_{bus}$ representation. In this manner the bus impedance matrix is built up one row and one column at a time until all branches of the physical network have been incorporated into $\mathbf{Z}_{bus}$. Whenever possible at any stage, it is computationally more efficient to select the next branch to be added between two nodes with rows and columns already included in the evolving $\mathbf{Z}_{bus}$. The elements of $\mathbf{Z}_{bus}$ can also be generated as needed using the triangular factors of $\mathbf{Y}_{bus}$, which is often the most attractive method computationally.

The variables used to analyze a power network can take on many different forms. However, regardless of the particular choice of representation, power in the physical network must not be arbitrarily altered when representing currents and voltages in the chosen format. Power invariancy imposes requirements when transforming from one set of currents and voltages to another.

Mutually coupled branches, which do not generally arise except under unbalanced short-circuit (fault) conditions, can be added to and removed from $\mathbf{Z}_{bus}$ by algorithmic methods.

PROBLEMS

8.1. Form $\mathbf{Z}_{bus}$ for the circuit of Fig. 8.13 after removing node ⑤ by converting the voltage source to a current source. Determine the voltages with respect to reference node at each of the four other nodes when $V = 1.2\angle 0°$ and the load currents are $I_{L1} = -j0.1$, $I_{L2} = -j0.1$, $I_{L3} = -j0.2$, and $I_{L4} = -j0.2$, all in per unit.

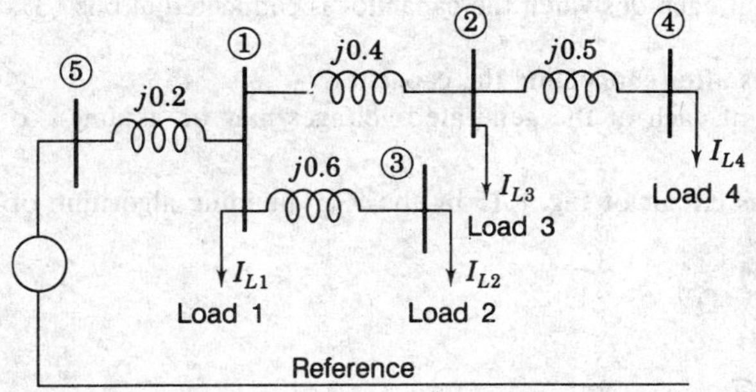

FIGURE 8.13
Circuit diagram showing constant current loads supplied by an ideal voltage source. Circuit parameters are in per unit.

8.2. From the solution of Prob. 8.1, draw the Thévenin equivalent circuit at bus ④ of Fig. 8.13 and use it to determine the current drawn by a capacitor of reactance 5.4 per unit connected between bus ④ and reference. Following the procedure of Example 8.2, calculate the voltage *changes* at each of the buses due to the capacitor.

8.3. Modify $\mathbf{Z}_{\text{bus}}$ of Prob. 8.1 to include a capacitor of reactance 5.4 per unit connected from bus ④ to reference and then calculate the new bus voltages using the modified $\mathbf{Z}_{\text{bus}}$. Check your answers using the results of Probs. 8.1 and 8.2.

8.4. Modify the $\mathbf{Z}_{\text{bus}}$ determined in Example 8.4 for the circuit of Fig. 8.8 by adding a new node connected to bus ③ through an impedance of $j0.5$ per unit.

8.5. Modify the $\mathbf{Z}_{\text{bus}}$ determined in Example 8.4 by adding a branch of impedance $j0.2$ per unit between buses ① and ④ of the circuit of Fig. 8.8.

8.6. Modify the $\mathbf{Z}_{\text{bus}}$ determined in Example 8.4 by removing the impedance connected between buses ② and ③ of the circuit of Fig. 8.8.

8.7. Find $\mathbf{Z}_{\text{bus}}$ for the circuit of Fig. 7.18 by the $\mathbf{Z}_{\text{bus}}$ building algorithm discussed in Sec. 8.4. Assume there is no mutual coupling between branches.

8.8. For the reactance network of Fig. 8.14, find
(*a*) $\mathbf{Z}_{\text{bus}}$ by direct formulation,
(*b*) The voltage at each bus,
(*c*) The current drawn by a capacitor having a reactance of 5.0 per unit connected from bus ③ to neutral,

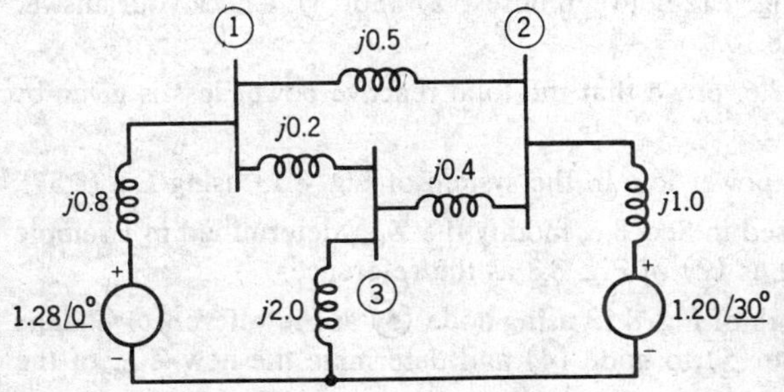

FIGURE 8.14
Circuit for Prob. 8.8. Voltages and impedances are in per unit.

(*d*) The change in voltage at each bus when the capacitor is connected at bus ③, and

(*e*) The voltage at each bus after connecting the capacitor.

The magnitude and angle of each of the generated voltages may be assumed to remain constant.

8.9. Find $\mathbf{Z}_{\text{bus}}$ for the three-bus circuit of Fig. 8.15 by the $\mathbf{Z}_{\text{bus}}$ building algorithm of Sec. 8.4.

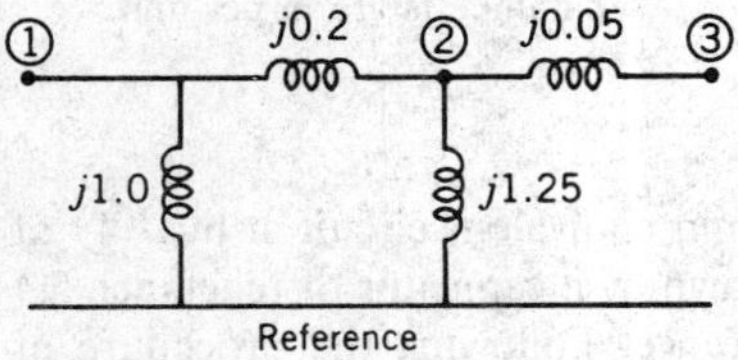

FIGURE 8.15
Circuit for Probs. 8.9, 8.11, and 8.12. Values shown are reactances in per unit.

8.10. Find $\mathbf{Z}_{\text{bus}}$ for the four-bus circuit of Fig. 7.12, which has per-unit *admittances* as marked.

8.11. The three-bus circuit of Fig. 8.15 has per-unit *reactances* as marked. The symmetrical $\mathbf{Y}_{\text{bus}}$ for the circuit has triangular factors

$$\mathbf{L} = \begin{bmatrix} -j6.0 & \cdot & \cdot \\ j5.0 & -j21.633333 & \cdot \\ 0 & j20.0 & -j1.510038 \end{bmatrix} \qquad \mathbf{U} = \begin{bmatrix} 1 & -0.833333 & 0 \\ \cdot & 1 & -0.924499 \\ \cdot & \cdot & 1 \end{bmatrix}$$

Use **L** and **U** to calculate

(*a*) The elements Z_{12}, Z_{23}, and Z_{33} of the system $\mathbf{Z}_{\text{bus}}$ and

(*b*) The Thévenin impedance $Z_{\text{th, 13}}$ looking into the circuit of Fig. 8.15 between buses ① and ③.

8.12. Use the $\mathbf{Y}_{\text{bus}}$ triangular factors of Prob. 8.11 to calculate the Thévenin impedance Z_{22} looking into the circuit of Fig. 8.15 between bus ② and the reference. Check your answer by inspection of Fig. 8.15.

8.13. The $\mathbf{Y}_{\text{bus}}$ for the circuit of Fig. 7.12 has triangular factors **L** and **U** given in Example 7.9. Use the triangular factors to calculate the Thévenin impedance $Z_{\text{th, 24}}$ looking into the circuit of Fig. 7.12 between buses ② and ④. Check your answer using the solution of Prob. 8.10.

8.14. Using the notation of Sec. 8.6, prove that the total reactive power loss is given by the formula $Q_L = \mathbf{I}^T \mathbf{X}_{\text{bus}} \mathbf{I}^*$.

8.15. Calculate the total reactive power loss in the system of Fig. 8.13 using Eq. (8.57).

8.16. Using the procedure discussed in Sec. 8.6, modify the $\mathbf{Z}_{\text{bus}}$ determined in Example 8.4 to reflect the choice of bus ② of Fig. 8.8 as the reference.

8.17. (*a*) Find $\mathbf{Z}_{\text{bus}}$ for the network of Fig. 8.13 using node ⑤ as the reference. Change the reference from node ⑤ to node ④ and determine the new $\mathbf{Z}_{\text{bus}}$ of the network using Eq. (8.60). Use the numerical values of the load currents I_{Li} of Prob. 8.1 to determine $\mathbf{I}_{\text{new}}$ by Eq. (8.55) and $\mathbf{V}_{\text{new}}$ by Eq. (8.56).

(*b*) Change the $\mathbf{Z}_{\text{bus}}$ reference from node ④ back to node ⑤, and using Eq. (8.63), determine the voltages at buses ① and ④ relative to node ⑤. What are the values of these bus voltages with respect to the ground reference of Fig. 8.13?

8.18. A new branch having an impedance of $j0.25$ per unit is connected between nodes ③ and ④ of the circuit of Fig. 8.8 in parallel with the existing impedance of $j0.2$ per unit between the same two nodes. These two branches have mutual impedance of $j0.1$ per unit. Modify the $\mathbf{Z}_{\text{bus}}$ determined in Example 8.4 to account for the addition of the new branch.

8.19. Derive Eqs. (8.95) and (8.96).

8.20. Modify the $\mathbf{Z}_{\text{bus}}$ determined in Example 8.7 to remove the branch between buses ① and ② already coupled by mutual impedance $j0.15$ per unit to the branch between buses ① and ④.

8.21. Assume that the two branches ①–③ and ②–③ in the circuit of Fig. 7.18 are the only mutually coupled branches (as indicated by the dots) with a mutual impedance of $j0.15$ per unit between them. Find $\mathbf{Z}_{\text{bus}}$ for this circuit by the $\mathbf{Z}_{\text{bus}}$ building algorithm.

8.22. Modify the $\mathbf{Z}_{\text{bus}}$ obtained in Prob. 8.21 to remove branch ②–③, which is coupled to branch ①–③ through a mutual impedance of $j0.15$ per unit.

8.23. In Fig. 8.16 a new bus ⓠ is to be connected to an existing bus ⓟ through a new branch *c*. New branch *c* is mutually coupled to branches *a* and *b*, which are already mutually coupled to one another as shown. The primitive impedance matrix defining self- and mutual impedances of these three mutually coupled branches and its reciprocal, the primitive admittance matrix, have the forms

$$\begin{bmatrix} Z_{aa} & Z_{ab} & Z_{ac} \\ Z_{ba} & Z_{bb} & Z_{bc} \\ Z_{ca} & Z_{cb} & Z_{cc} \end{bmatrix}^{-1} = \begin{bmatrix} Y_{aa} & Y_{ab} & Y_{ac} \\ Y_{ba} & Y_{bb} & Y_{bc} \\ Y_{ca} & Y_{cb} & Y_{cc} \end{bmatrix}$$

To account for the addition of new bus ⓠ, prove that the existing bus impedance matrix of the network must be augmented by a new row *q* and column *q* with

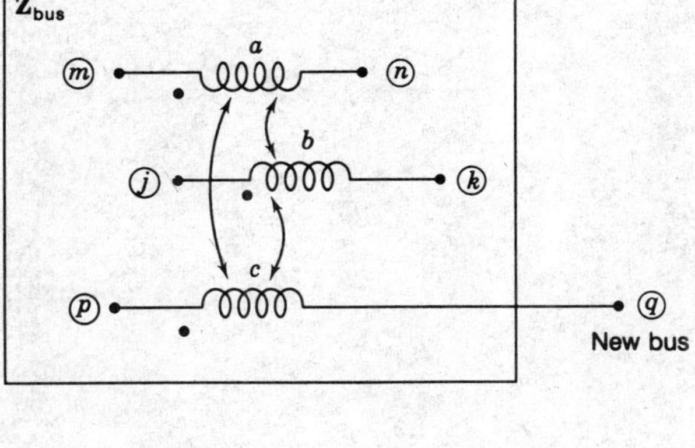

FIGURE 8.16
New branch *c* from new bus ⓠ is mutually coupled to branches *a* and *b*.

elements given by

$$Z_{qi} = Z_{pi} + \frac{1}{Y_{cc}}\begin{bmatrix} Y_{ca} & Y_{cb} \end{bmatrix}\begin{bmatrix} Z_{mi} - Z_{ni} \\ Z_{ji} - Z_{ki} \end{bmatrix} \qquad \text{for } i = 1, \ldots, N$$

$$Z_{qq} = Z_{pq} + \frac{1}{Y_{cc}} + \frac{1}{Y_{cc}}\begin{bmatrix} Y_{ca} & Y_{cb} \end{bmatrix}\begin{bmatrix} Z_{mq} - Z_{nq} \\ Z_{jq} - Z_{kq} \end{bmatrix}$$

Note that these equations are a generalization of Eqs. (8.87) and (8.88).

8.24. Branch ②–③ of the circuit of Fig. 7.18 is mutually coupled to two branches ①–③ and ②–⑤ through mutual impedances of $j0.15$ per unit and $j0.1$ per unit, respectively, as indicated by dots. Using the formulas given in Prob. 8.23, find $\mathbf{Z}_{\text{bus}}$ for the circuit by the $\mathbf{Z}_{\text{bus}}$ building algorithm.

CHAPTER 9

POWER-FLOW SOLUTIONS

Power-flow studies are of great importance in planning and designing the future expansion of power systems as well as in determining the best operation of existing systems. The principal information obtained from a power-flow study is the magnitude and phase angle of the voltage at each bus and the real and reactive power flowing in each line. However, much additional information of value is provided by the printout of the solution from computer programs used by the electric utility companies. Most of these features are made evident in our discussion of power-flow studies in this chapter.

We shall examine some of the methods upon which solutions to the power-flow problem are based. The great value of the power-flow computer program in power system design and operation will become apparent.

9.1 THE POWER-FLOW PROBLEM

Either the bus self- and mutual admittances which compose the bus admittance matrix $\mathbf{Y}_{\text{bus}}$ or the driving-point and transfer impedances which compose $\mathbf{Z}_{\text{bus}}$ may be used in solving the power-flow problem. We confine our study to methods using admittances. The starting point in obtaining the data which must be furnished to the computer is the single-line diagram of the system. Transmission lines are represented by their per-phase nominal-π equivalent circuits like that shown in Fig. 6.7. For each line numerical values for the series impedance Z and the total line-charging admittance Y (usually in terms of line-charging megavars at nominal voltage of the system) are necessary so that the computer can determine all the elements of the $N \times N$ bus admittance matrix of which the typical element Y_{ij} is

$$Y_{ij} = |Y_{ij}|\angle\theta_{ij} = |Y_{ij}|\cos\theta_{ij} + j|Y_{ij}|\sin\theta_{ij} = G_{ij} + jB_{ij} \tag{9.1}$$

Other essential information includes transformer ratings and impedances, shunt capacitor ratings, and transformer tap settings. In advance of each power-flow study certain bus voltages and power injections must be given known values, as discussed below.

The voltage at a typical bus ⓘ of the system is given in polar coordinates by

$$V_i = |V_i|\angle\delta_i = |V_i|(\cos\delta_i + j\sin\delta_i) \tag{9.2}$$

and the voltage at another bus ⓙ is similarly written by changing the subscript from i to j. The net current injected into the network at bus ⓘ in terms of the elements Y_{in} of $\mathbf{Y}_{\text{bus}}$ is given by the summation

$$I_i = Y_{i1}V_1 + Y_{i2}V_2 + \cdots + Y_{iN}V_N = \sum_{n=1}^{N} Y_{in}V_n \tag{9.3}$$

Let P_i and Q_i denote the *net* real and reactive power entering the network at the bus ⓘ. Then, the complex conjugate of the power injected at bus ⓘ is

$$P_i - jQ_i = V_i^* \sum_{n=1}^{N} Y_{in}V_n \tag{9.4}$$

in which we substitute from Eqs. (9.1) and (9.2) to obtain

$$P_i - jQ_i = \sum_{n=1}^{N} |Y_{in}V_iV_n|\angle\theta_{in} + \delta_n - \delta_i \tag{9.5}$$

Expanding this equation and equating real and reactive parts, we obtain

$$P_i = \sum_{n=1}^{N} |Y_{in}V_iV_n|\cos(\theta_{in} + \delta_n - \delta_i) \tag{9.6}$$

$$Q_i = -\sum_{n=1}^{N} |Y_{in}V_iV_n|\sin(\theta_{in} + \delta_n - \delta_i) \tag{9.7}$$

Equations (9.6) and (9.7) constitute the polar form of the *power-flow equations*; they provide *calculated* values for the *net* real power P_i and reactive power Q_i entering the network at typical bus ⓘ. Let P_{gi} denote the scheduled power being generated at bus ⓘ and P_{di} denote the scheduled power demand of the

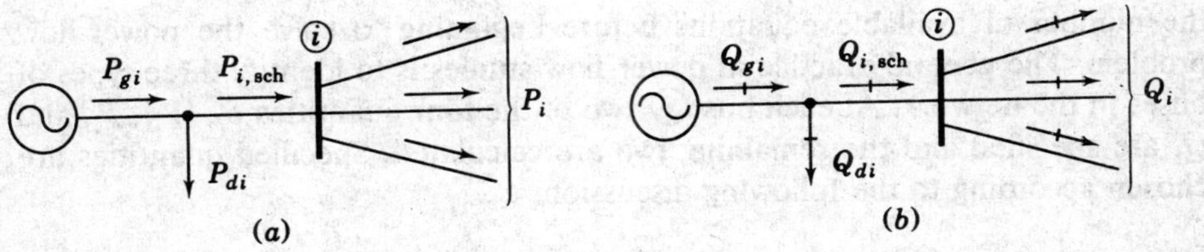

FIGURE 9.1
Notation for (*a*) active and (*b*) reactive power at a typical bus ⓘ in power-flow studies.

load at that bus. Then, $P_{i,\text{sch}} = P_{gi} - P_{di}$ is the net *scheduled* power being injected into the network at bus ⓘ, as illustrated in Fig. 9.1(*a*). Denoting the calculated value of P_i by $P_{i,\text{calc}}$ leads to the definition of *mismatch* ΔP_i as the scheduled value $P_{i,\text{sch}}$ minus the calculated value $P_{i,\text{calc}}$,

$$\Delta P_i = P_{i,\text{sch}} - P_{i,\text{calc}} = (P_{gi} - P_{di}) - P_{i,\text{calc}} \tag{9.8}$$

Likewise, for reactive power at bus ⓘ we have

$$\Delta Q_i = Q_{i,\text{sch}} - Q_{i,\text{calc}} = (Q_{gi} - Q_{di}) - Q_{i,\text{calc}} \tag{9.9}$$

as shown in Fig. 9.1(*b*). Mismatches occur in the course of solving a power-flow problem when calculated values of P_i and Q_i do not coincide with the scheduled values. If the calculated values $P_{i,\text{calc}}$ and $Q_{i,\text{calc}}$ match the scheduled values $P_{i,\text{sch}}$ and $Q_{i,\text{sch}}$ perfectly, then we say that the mismatches ΔP_i and ΔQ_i are zero at bus ⓘ, and we write the *power-balance equations*

$$g_{i'} = P_i - P_{i,\text{sch}} = P_i - (P_{gi} - P_{di}) = 0 \tag{9.10}$$

$$g_{i''} = Q_i - Q_{i,\text{sch}} = Q_i - (Q_{gi} - Q_{di}) = 0 \tag{9.11}$$

As we shall see in Sec. 9.3, the functions $g_{i'}$ and $g_{i''}$ are convenient for writing certain equations involving the mismatches ΔP_i and ΔQ_i. If bus ⓘ has no generation or load, the appropriate terms are set equal to zero in Eqs. (9.10) and (9.11). Each bus of the network has two such equations, and the power-flow problem is to solve Eqs. (9.6) and (9.7) for values of the unknown bus voltages which cause Eqs. (9.10) and (9.11) to be numerically satisfied at each bus. If there is no scheduled value $P_{i,\text{sch}}$ for bus ⓘ, then the mismatch $\Delta P_i = P_{i,\text{sch}} - P_{i,\text{calc}}$ cannot be defined and there is no requirement to satisfy the corresponding Eq. (9.10) in the course of solving the power-flow problem. Similarly, if $Q_{i,\text{sch}}$ is not specified at bus ⓘ, then Eq. (9.11) does not have to be satisfied.

Four potentially unknown quantities associated with each bus ⓘ are P_i, Q_i, voltage angle δ_i, and voltage magnitude $|V_i|$. At most, there are two equations like Eqs. (9.10) and (9.11) available for each node, and so we must consider how the number of unknown quantities can be reduced to agree with the number of available equations before beginning to solve the power-flow problem. The general practice in power-flow studies is to identify three types of buses in the network. At each bus ⓘ two of the four quantities δ_i, $|V_i|$, P_i, and Q_i are specified and the remaining two are calculated. Specified quantities are chosen according to the following discussion:

1. *Load buses*. At each nongenerator bus, called a *load bus*, both P_{gi} and Q_{gi} are zero and the real power P_{di} and reactive power Q_{di} drawn from the system by the load (*negative inputs* into the system) are known from historical record, load forecast, or measurement. Quite often in practice only real power is known and the reactive power is then based on an assumed power factor such as 0.85 or higher. A load bus ⓘ is often called a *P-Q bus* because the scheduled values $P_{i,\text{sch}} = -P_{di}$ and $Q_{i,\text{sch}} = -Q_{di}$ are known and mismatches ΔP_i and ΔQ_i can be defined. The corresponding Eqs. (9.10) and (9.11) are then explicitly included in the statement of the power-flow problem, and the two unknown quantities to be determined for the bus are δ_i and $|V_i|$.
2. *Voltage-controlled buses*. Any bus of the system at which the voltage magnitude is kept constant is said to be *voltage controlled*. At each bus to which there is a generator connected the megawatt generation can be controlled by adjusting the prime mover, and the voltage magnitude can be controlled by adjusting the generator excitation. Therefore, at each generator bus ⓘ we may properly specify P_{gi} and $|V_i|$. With P_{di} also known, we can define mismatch ΔP_i according to Eq. (9.8). Generator reactive power Q_{gi} required to support the scheduled voltage $|V_i|$ cannot be known in advance, and so mismatch ΔQ_i is not defined. Therefore, at a generator bus ⓘ voltage angle δ_i is the unknown quantity to be determined and Eq. (9.10) for P_i is the available equation. *After* the power-flow problem is solved, Q_i can be calculated from Eq. (9.7).

   For obvious reasons a generator bus is usually called a voltage-controlled or *PV* bus. Certain buses without generators may have voltage control capability; such buses are also designated voltage-controlled buses at which the real power generation is simply zero.
3. *Slack bus*. For convenience throughout this chapter bus ① is almost always designated as the *slack bus*. The voltage angle of the slack bus serves as reference for the angles of all other bus voltages. The particular angle assigned to the slack bus voltage is not important because voltage-angle *differences* determine the calculated values of P_i and Q_i in Eqs. (9.6) and (9.7). The usual practice is to set $\delta_1 = 0°$. Mismatches are not defined for the slack bus, as explained below, and so voltage magnitude $|V_1|$ is specified as the other known quantity along with $\delta_1 = 0°$. Then, there is no requirement to include either Eq. (9.10) or Eq. (9.11) for the slack bus in the power-flow problem.

To understand why P_1 and Q_1 are not scheduled at the slack bus, consider that at each of the N buses of the system an equation similar to Eq. (9.10) can be written by letting i range from 1 to N. When the resulting N equations are added together, we obtain

$$\underbrace{P_L}_{\text{Real power loss}} = \sum_{i=1}^{N} P_i = \underbrace{\sum_{i=1}^{N} P_{gi}}_{\text{Total generation}} - \underbrace{\sum_{i=1}^{N} P_{di}}_{\text{Total load}} \tag{9.12}$$

The term P_L in this equation is evidently the total I^2R loss in the transmission lines and transformers of the network. The individual currents in the various transmission lines of the network cannot be calculated until after the voltage magnitude and angle are known at every bus of the system. Therefore, P_L is initially unknown and it is not possible to prespecify *all* the quantities in the summations of Eq. (9.12). In the formulation of the power-flow problem we choose one bus, the slack bus, at which P_g is *not* scheduled or otherwise prespecified. After the power-flow problem has been solved, the difference (slack) between the total specified P going into the system at all the other buses and the total output P plus I^2R losses are assigned to the slack bus. For this reason a generator bus must be selected as the slack bus. The difference between the total megavars supplied by the generators at the buses and the megavars received by the loads is given by

$$\sum_{i=1}^{N} Q_i = \sum_{i=1}^{N} Q_{gi} - \sum_{i=1}^{N} Q_{di} \tag{9.13}$$

This equation is satisfied on an individual bus basis by satisfying Eq. (9.11) at each bus ⓘ in the course of solving the power-flow problem. Individual Q_i can be evaluated from Eq. (9.7) after the power-flow solution becomes available. Thus, the quantity on the left-hand side of Eq. (9.13) accounts for the combined megavars associated with line charging, shunt capacitors and reactors installed at the buses, and the so-called I^2X loss in the series reactances of the transmission lines.

The unscheduled bus-voltage magnitudes and angles in the input data of the power-flow study are called *state variables* or *dependent variables* since their values, which describe the *state* of the system, depend on the quantities specified at all the buses. Hence, the power-flow problem is to determine values for all state variables by solving an equal number of power-flow equations based on the input data specifications. If there are N_g voltage-controlled buses (not counting the slack bus) in the system of N buses, there will be $(2N - N_g - 2)$ equations to be solved for $(2N - N_g - 2)$ state variables, as shown in Table 9.1. Once the state variables have been calculated, the complete *state* of the system is known and all other quantities which depend on the state variables can be

TABLE 9.1
Summary of power-flow problem

Bus type	No. of buses	Quantities specified	No. of available equations	No. of δ_i, $\lvert V_i \rvert$ state variables
Slack: $i = 1$	1	$\delta_1, \lvert V_1 \rvert$	0	0
Voltage controlled ($i = 2, \ldots, N_g + 1$)	N_g	$P_i, \lvert V_i \rvert$	N_g	N_g
Load ($i = N_g + 2, \ldots, N$)	$N - N_g - 1$	P_i, Q_i	$2(N - N_g - 1)$	$2(N - N_g - 1)$
Totals	N	$2N$	$2N - N_g - 2$	$2N - N_g - 2$

determined. Quantities such as P_1 and Q_1 at the slack bus, Q_i at each voltage-controlled bus, and the power loss P_L of the system are examples of dependent functions.

The functions P_i and Q_i of Eqs. (9.6) and (9.7) are nonlinear functions of the state variables δ_i and $|V_i|$. Hence, power-flow calculations usually employ iterative techniques such as the Gauss-Seidel and Newton-Raphson procedures, which are described in this chapter. The Newton-Raphson method solves the polar form of the power-flow equations until the ΔP and ΔQ mismatches at all buses fall within specified tolerances. The Gauss-Seidel method solves the power-flow equations in rectangular (complex variable) coordinates until differences in bus voltages from one iteration to another are sufficiently small. Both methods are based on bus admittance equations.

Example 9.1. Suppose that the *P-Q* load is known at each of the nine buses of a small power system and that synchronous generators are connected to buses ①, ②, ⑤, and ⑦. For a power-flow study, identify the ΔP and ΔQ mismatches and the state variables associated with each bus. Choose bus ① as the slack bus.

Solution. The nine buses of the system are categorized as follows:

P-Q buses: ③, ④, ⑥, ⑧, and ⑨

P-V buses: ②, ⑤, and ⑦

Slack bus: ①

The mismatches corresponding to the specified P and Q are

At *P-Q* buses:

$$\Delta P_3, \Delta Q_3; \quad \Delta P_4, \Delta Q_4; \quad \Delta P_6, \Delta Q_6; \quad \Delta P_8, \Delta Q_8; \quad \Delta P_9, \Delta Q_9$$

At *P-V* buses: $\Delta P_2, \quad \Delta P_5, \quad \Delta P_7$

and the state variables are

$$P\text{-}Q \text{ buses:} \quad \delta_3, |V_3|; \quad \delta_4, |V_4|; \quad \delta_6, |V_6|; \quad \delta_8, |V_8|; \quad \delta_9, |V_9|$$

$$P\text{-}V \text{ buses:} \quad \delta_2, \quad \delta_5, \quad \delta_7$$

Since $N = 9$ and $N_g = 3$, there are $2N - N_g - 2 = 13$ equations to be solved for the 13 state variables shown.

9.2 THE GAUSS-SEIDEL METHOD

The complexity of obtaining a formal solution for power flow in a power system arises because of the differences in the type of data specified for the different kinds of buses. Although the formulation of sufficient equations to match the number of unknown state variables is not difficult, as we have seen, the closed form of solution is not practical. Digital solutions of the power-flow problems follow an iterative process by assigning estimated values to the unknown bus voltages and by calculating a new value for each bus voltage from the estimated values at the other buses and the real and reactive power specified. A new set of values for the voltage at each bus is thus obtained and used to calculate still another set of bus voltages. Each calculation of a new set of voltages if called an *iteration*. The iterative process is repeated until the changes at each bus are less than a specified minimum value.

We derive equations for a four-bus system and write the general equations later. With the slack bus designated as number 1, computations start with bus ②. If $P_{2,\text{sch}}$ and $Q_{2,\text{sch}}$ are the scheduled real and reactive power, respectively, entering the network at bus ②, it follows from Eq. (9.4) with i set equal to 2 and N equal to 4 that

$$\frac{P_{2,\text{sch}} - jQ_{2,\text{sch}}}{V_2^*} = Y_{21}V_1 + Y_{22}V_2 + Y_{23}V_3 + Y_{24}V_4 \tag{9.14}$$

Solving for V_2 gives

$$V_2 = \frac{1}{Y_{22}}\left[\frac{P_{2,\text{sch}} - jQ_{2,\text{sch}}}{V_2^*} - (Y_{21}V_1 + Y_{23}V_3 + Y_{24}V_4)\right] \tag{9.15}$$

For now let us assume that buses ③ and ④ are also load buses with real and reactive power specified. Expressions similar to that in Eq. (9.15) may be written for each bus. At bus ③ we have

$$V_3 = \frac{1}{Y_{33}}\left[\frac{P_{3,\text{sch}} - jQ_{3,\text{sch}}}{V_3^*} - (Y_{31}V_1 + Y_{32}V_2 + Y_{34}V_4)\right] \tag{9.16}$$

If we were to equate real and imaginary parts of Eqs. (9.15), (9.16), and the similar equation of bus ④, we would obtain six equations in the six state variables δ_2 to δ_4 and $|V_2|$ to $|V_4|$. However we solve for the complex voltages directly from the equations as they appear. The solution proceeds by iteration based on the scheduled real and reactive power at buses ②, ③, and ④, the scheduled slack bus voltage $V_1 = |V_1|\angle\delta_1$, and initial voltage estimates $V_2^{(0)}$, $V_3^{(0)}$, and $V_4^{(0)}$ at the other buses.

Solution of Eq. (9.15) gives the corrected voltage $V_2^{(1)}$ calculated from

$$V_2^{(1)} = \frac{1}{Y_{22}}\left[\frac{P_{2,\text{sch}} - jQ_{2,\text{sch}}}{V_2^{(0)*}} - \left(Y_{21}V_1 + Y_{23}V_3^{(0)} + Y_{24}V_4^{(0)}\right)\right] \tag{9.17}$$

in which all quantities in the right-hand-side expression are either fixed specifications or initial estimates. The calculated value $V_2^{(1)}$ and the estimated value $V_2^{(0)}$ will not agree. Agreement would be reached to a good degree of accuracy after several iterations and would be the correct value for V_2 with the estimated voltages but without regard to power at the other buses. This value would *not* be the solution for V_2 for the specific power-flow conditions, however, because the voltages on which this calculation for V_2 depends are the estimated values $V_3^{(0)}$ and $V_4^{(0)}$ at the other buses, and the actual voltages are not yet known.

As the corrected voltage is found at each bus, it is used to calculate the corrected voltage at the next bus. Therefore, substituting $V_2^{(1)}$ in Eq. (9.16), we obtain for the first calculated value at bus ③

$$V_3^{(1)} = \frac{1}{Y_{33}}\left[\frac{P_{3,\text{sch}} - jQ_{3,\text{sch}}}{V_3^{(0)*}} - \left(Y_{31}V_1 + Y_{32}V_2^{(1)} + Y_{34}V_4^{(0)}\right)\right] \tag{9.18}$$

The process is repeated at bus ④ and at each bus consecutively throughout the network (except at the slack bus) to complete the *first iteration* in which calculated values are found for each state variable. Then, the entire process is carried out again and again until the amount of correction in voltage at every bus is less than some predetermined precision index. This process of solving the power-flow equations is known as the *Gauss-Seidel iterative method*.

Convergence upon an erroneous solution is usually avoided if the initial values are of reasonable magnitude and do not differ too widely in phase. It is common practice to set the initial estimates of the unknown voltages at all load buses equal to $1.0\angle 0°$ per unit. Such initialization is called a *flat start* because of the uniform voltage profile assumed.

For a system of N buses the general equation for the calculated voltage at any bus ⓘ where P and Q are scheduled is

$$V_i^{(k)} = \frac{1}{Y_{ii}}\left[\frac{P_{i,\text{sch}} - jQ_{i,\text{sch}}}{V_i^{(k-1)*}} - \sum_{j=1}^{i-1} Y_{ij}V_j^{(k)} - \sum_{j=i+1}^{N} Y_{ij}V_j^{(k-1)}\right] \tag{9.19}$$

The superscript (k) denotes the number of the iteration in which the voltage is currently being calculated and $(k - 1)$ indicates the number of the preceding iteration. Thus, we see that the values for the voltages on the right-hand side of this equation are the most recently calculated values for the corresponding buses (or the estimated voltage if k is 1 and no iteration has yet been made at that particular bus).

Since Eq. (9.19) applies only at load buses where real and reactive power are specified, an additional step is necessary at voltage-controlled buses where voltage magnitude is to remain constant. Before investigating this additional step, let us look at an example of the calculations at a load bus.

Example 9.2. Figure 9.2 shows the one-line diagram of a simple power system. Generators are connected at buses ① and ④ while loads are indicated at all four buses. Base values for the transmission system are 100 MVA, 230 kV. The *line data* of Table 9.2 give per-unit series impedances and line-charging susceptances for the nominal-π equivalents of the four lines identified by the buses at which they terminate. The *bus data* in Table 9.3 list values for P, Q, and V at each bus. The Q values of load are calculated from the corresponding P values assuming a power factor of 0.85. The net scheduled values, $P_{i,\text{sch}}$ and $Q_{i,\text{sch}}$, are negative at the load buses ② and ③. Generated Q_{gi} is not specified where voltage

FIGURE 9.2
One-line diagram for Example 9.2 showing the bus names and numbers.

TABLE 9.2
Line data for Example 9.2†

Line, bus to bus	Series Z: R per unit	Series Z: X per unit	Series $Y = Z^{-1}$: G per unit	Series $Y = Z^{-1}$: B per unit	Shunt Y: Total charging Mvar‡	Shunt Y: Y/2 per unit
1–2	0.01008	0.05040	3.815629	−19.078144	10.25	0.05125
1–3	0.00744	0.03720	5.169561	−25.847809	7.75	0.03875
2–4	0.00744	0.03720	5.169561	−25.847809	7.75	0.03875
3–4	0.01272	0.06360	3.023705	−15.118528	12.75	0.06375

†Base 100MVA, 230 kV.
‡At 230 kV.

TABLE 9.3
Bus data for Example 9.2

	Generation		Load			
Bus	P, MW	Q, Mvar	P, MW	Q, Mvar†	V, per unit	Remarks
1	—	—	50	30.99	$1.00\angle 0°$	Slack bus
2	0	0	170	105.35	$1.00\angle 0°$	Load bus (inductive)
3	0	0	200	123.94	$1.00\angle 0°$	Load bus (inductive)
4	318	—	80	49.58	$1.02\angle 0°$	Voltage controlled

†The Q values of load are calculated from the corresponding P values assuming a power factor of 0.85.

magnitude is constant. In the voltage column the values for the load buses are flat-start estimates. The slack bus voltage magnitude $|V_1|$ and angle δ_1, and also magnitude $|V_4|$ at bus ④, are to be kept constant at the values listed. A power-flow study is to be made by the Gauss-Seidel method. Assuming that the iterative calculations start at bus ②, find the value of V_2 for the first iteration.

Solution. In order to approach the accuracy of a digital computer, we perform the computations indicated below to six decimal places. From the line data given in Table 9.2 we construct the system $\mathbf{Y}_{\text{bus}}$ of Table 9.4. For example, associated with bus ② of Fig. 9.2 are the nonzero off-diagonal elements Y_{21} and Y_{24}, which are equal to the negative of the respective line admittances

$$Y_{21} = -(3.815629 - j19.078144); \qquad Y_{24} = -(5.169561 - j25.847809)$$

Since Y_{22} is the sum of all the admittances connected to bus ②, including the

TABLE 9.4
Bus admittance matrix for Example 9.2†

Bus no.	①	②	③	④
①	8.985190 $-j$44.835953	$-$3.815629 $+j$19.078144	$-$5.169561 $+j$25.847809	0
②	$-$3.815629 $+j$19.078144	8.985190 $-j$44.835953	0	$-$5.169561 $+j$25.847809
③	$-$5.169561 $+j$25.847809	0	8.193267 $-j$40.863838	$-$3.023705 $+j$15.118528
④	0	$-$5.169561 $+j$25.847809	$-$3.023705 $+j$15.118528	8.193267 $-j$40.863838

†Per-unit values rounded to six decimal places

shunt susceptances for line charging of lines ② - ① and ② - ④, we have

$$Y_{22} = (-Y_{21}) + j0.05125 + (-Y_{24}) + j0.03875 = 8.985190 - j44.835953$$

Substitution in Eq. (9.17) yields the per-unit voltage

$$V_2^{(1)} = \frac{1}{Y_{22}}\left[\frac{-1.7 + j1.0535}{1.0 + j0.0} - 1.00(-3.815629 + j19.078144)\right.$$

$$\left. - 1.02(-5.169561 + j25.847809)\right]$$

$$= \frac{1}{Y_{22}}[-1.7 + j1.0535 + 9.088581 - j45.442909]$$

$$= \frac{7.388581 - j44.389409}{8.985190 - j44.835953} = 0.983564 - j0.032316$$

Experience with the Gauss-Seidel method of solution of power-flow problems has shown that the number of iterations required may be reduced considerably if the correction in voltage at each bus is multiplied by some constant that increases the amount of correction to bring the voltage closer to the value it is approaching. The multiplier that accomplishes this improved convergence is called an *acceleration factor*. The difference between the newly calculated voltage and the best previous voltage at the bus is multiplied by the appropriate acceleration factor to obtain a better correction to be added to the previous value. For example, at bus ② in the first iteration we have the accelerated value $V_{2,\text{acc}}^{(1)}$ defined by the straight-line formula

$$V_{2,\text{acc}}^{(1)} = (1 - \alpha)V_2^{(0)} + \alpha V_2^{(1)} = V_2^{(0)} + \alpha\left(V_2^{(1)} - V_2^{(0)}\right) \qquad (9.20)$$

in which α is the acceleration factor. More generally, for bus ⓘ during iteration k the accelerated value is given by

$$V_{i,\text{acc}}^{(k)} = (1 - \alpha)V_{i,\text{acc}}^{(k-1)} + \alpha V_i^{(k)} = V_{i,\text{acc}}^{(k-1)} + \alpha\left(V_i^{(k)} - V_{i,\text{acc}}^{(k-1)}\right) \quad (9.21)$$

If $\alpha = 1$, then the Gauss-Seidel computed value of V_i is stored as the current value. If $0 < \alpha < 1$, then the value to be stored is a weighted average of the Gauss-Seidel value and the value stored from the previous iteration. If $1 < \alpha < 2$, then the value to be stored is essentially an extrapolated value. In power-flow studies α is generally set at about 1.6 and cannot exceed 2 if convergence is to occur.

Substituting the results of Example 9.1 and an acceleration factor of 1.6 in Eq. (9.20), we find that

$$V_{2,\,\mathrm{acc}}^{(1)} = 1 + 1.6[(0.983564 - j0.032316) - 1]$$

$$= 0.973703 - j0.051706 \text{ per unit}$$

Using $V_{2,\,\mathrm{acc}}^{(1)}$ in similar calculations for bus ③ gives first-iteration value

$$V_{3,\,\mathrm{acc}}^{(1)} = 0.953949 - j0.066708 \text{ per unit}$$

Since bus ④ is voltage controlled, we must treat it differently as explained below. The acceleration factor for the real component of the correction may differ from that for the imaginary component. For any system optimum values for acceleration factors exist, and poor choice of factors may result in less rapid convergence or make convergence impossible. An acceleration factor of 1.6 for both the real and imaginary components is usually a good choice but studies may be made to determine the best choice for a particular system.

Voltage-controlled buses. When voltage magnitude rather than reactive power is specified at bus ⓘ, the real and imaginary components of the voltage for each iteration are found by first computing a value for the reactive power. From Eq. (9.4) we have

$$Q_i = -\mathrm{Im}\left\{V_i^* \sum_{j=1}^{N} Y_{ij} V_j\right\} \tag{9.22}$$

which has the equivalent algorithmic expression

$$Q_i^{(k)} = -\mathrm{Im}\left\{V_i^{(k-1)*}\left[\sum_{j=1}^{i-1} Y_{ij} V_j^{(k)} + \sum_{j=i}^{N} Y_{ij} V_j^{(k-1)}\right]\right\} \tag{9.23}$$

where Im means "imaginary part of" and the superscripts indicate the relevant iteration. Reactive power $Q_i^{(k)}$ is evaluated by Eq. (9.23) for the best previous voltage values at the buses, and this value of $Q_i^{(k)}$ is substituted in Eq. (9.19) to find a new value of $V_i^{(k)}$. The components of the new $V_i^{(k)}$ are then multiplied by the ratio of the specified constant magnitude $|V_i|$ to the magnitude of $V_i^{(k)}$ found by Eq. (9.19). The result is the corrected complex voltage of the specified magnitude. In the four-bus example if bus ④ is voltage controlled, Eq. (9.23) yields the calculated value

$$Q_4^{(1)} = -\mathrm{Im}\{V_4^{(0)*}(Y_{41}V_1 + Y_{42}V_{2,\,\mathrm{acc}}^{(1)} + Y_{43}V_{3,\,\mathrm{acc}}^{(1)} + Y_{44}V_4^{(0)})\} \tag{9.24}$$

wherein the calculated voltages of buses ② and ③ are accelerated values of the first iteration. Substituting $Q_4^{(1)}$ for $Q_{4,\,\mathrm{sch}}$ in the equivalent of Eq. (9.19) for bus ④ yields

$$V_4^{(1)} = \frac{1}{Y_{44}}\left[\frac{P_{4,\,\text{sch}} - jQ_4^{(1)}}{V_4^{(0)*}} - \left(Y_{41}V_1 + Y_{42}V_{2,\,\text{acc}}^{(1)} + Y_{43}V_{3,\,\text{acc}}^{(1)}\right)\right] \tag{9.25}$$

and all quantities on the right-hand side are now known. Since $|V_4|$ is specified, we correct the magnitude of $V_4^{(1)}$ as follows:

$$V_{4,\,\text{corr}}^{(1)} = |V_4|\frac{V_4^{(1)}}{|V_4^{(1)}|} \tag{9.26}$$

and proceed to the next step with stored value $V_{4,\,\text{corr}}^{(1)}$ of bus ④ voltage having the specified magnitude in the remaining calculations of the iteration.

As discussed in Sec. 9.1, either voltage magnitude or reactive power must be specified at every bus except the slack bus, where voltage is specified by both voltage magnitude and angle. At buses with generation the voltage magnitude is specified as well as the real power P_g supplied by the generator. The reactive power Q_g entering the network from the generation is then determined by the computer in solving the power-flow problem. From a practical viewpoint the Q_g output of the generator must be within definite limits given by the inequality

$$Q_{\text{min}} \le Q_g \le Q_{\text{max}}$$

where Q_{min} is the minimum and Q_{max} is the maximum limit imposed on the reactive power output of the generator at the bus. In the course of power-flow solution if the calculated value of Q_g is outside either limit, then Q_g is set equal to the limit violated, the originally specified voltage magnitude at the bus is relaxed, and the bus is then treated as a *P-Q* bus for which a new voltage is calculated by the computer program. In subsequent iterations the program endeavors to sustain the originally specified voltage at the bus while ensuring that Q_g is within the permitted range of values. This could well be possible since other changes may occur elsewhere in the system to support the local action of the generator excitation as it adjusts to satisfy the specified terminal voltage.

Example 9.3. To complete the first iteration of the Gauss-Seidel procedure, find the voltage at bus ④ of Example 9.2 computed with the originally estimated voltages at buses ② and ③ replaced by the accelerated values indicated above.

Solution. Table 9.4 shows Y_{41} equal to zero, and so Eq. (9.24) gives

$$Q_4^{(1)} = -\,\text{Im}\left\{V_4^{(0)*}\left[Y_{42}V_{2,\,\text{acc}}^{(1)} + Y_{43}V_{3,\,\text{acc}}^{(1)} + Y_{44}V_4^{(0)}\right]\right\}$$

Substituting values for the indicated quantities in this equation, we obtain

$$Q_4^{(1)} = -\text{Im}\left\{\begin{aligned}1.02[(-5.169561 + j25.847809)(0.973703 - j0.051706)\\ +(-3.023705 + j15.118528)(0.953949 - j0.066708)\\ +(8.193267 - j40.863838)(1.02)]\end{aligned}\right\}$$

$$= -\text{Im}\{1.02[-5.573064 + j40.059396 + (8.193267 - j40.863838)1.02]\}$$

$$= 1.654151 \text{ per unit}$$

This value for $Q_4^{(1)}$ is now substituted into Eq. (9.25) to yield

$$V_4^{(1)} = \frac{1}{Y_{44}}\left[\frac{P_{4,\text{sch}} - jQ_4^{(1)}}{V_4^{(0)*}} - \left(Y_{42}V_{2,\text{acc}}^{(1)} + Y_{43}V_{3,\text{acc}}^{(1)}\right)\right]$$

$$= \frac{1}{Y_{44}}\left[\frac{2.38 - j1.654151}{1.02 - j0.0} - (-5.573066 + j40.059398)\right]$$

$$= \frac{7.906399 - j41.681115}{8.193267 - j40.863838} = 1.017874 - j0.010604 \text{ per unit}$$

Hence, $|V_4^{(1)}|$ equals 1.017929, and so we must correct the magnitude to 1.02,

$$V_{4,\text{corr}}^{(1)} = \frac{1.02}{1.017929}(1.017874 - j0.010604)$$

$$= 1.019945 - j0.010625 \text{ per unit}$$

In this example $Q_4^{(1)}$ is found to be 1.654151 per unit in the first iteration. If the reactive power generation at bus ④ were limited below 1.654151 per unit, then the specified limit value would be used for $Q_4^{(1)}$ and bus ④ in that case would be treated as a load bus within the iteration. The same strategy is used within any other iteration in which the generator Q-limits are violated.

The Gauss-Seidel procedure is one method of solving the power-flow problem. However, today's industry-based studies generally employ the alternative Newton-Raphson iterative method, which is reliable in convergence, computationally faster, and more economical in storage requirements. The converged solution of Examples 9.2 and 9.3 agrees with the results given in Fig. 9.4 found later by the Newton-Raphson method.

9.3 THE NEWTON-RAPHSON METHOD

Taylor's series expansion for a function of two or more variables is the basis for the Newton-Raphson method of solving the power-flow problem. Our study of the method begins by a discussion of the solution of a problem involving only two equations and two variables. Then, we see how to extend the analysis to the solution of power-flow equations.

Let us consider the equation of a function h_1 of two variables x_1 and x_2 equal to a constant b_1 expressed as

$$g_1(x_1, x_2, u) = h_1(x_1, x_2, u) - b_1 = 0 \tag{9.27}$$

and a second equation involving another function h_2 such that

$$g_2(x_1, x_2, u) = h_2(x_1, x_2, u) - b_2 = 0 \tag{9.28}$$

where b_2 is also a constant. The symbol u represents an independent *control*, which is considered constant in this chapter. As in Eqs. (9.9) and (9.10), the functions g_1 and g_2 are introduced for convenience to allow us to discuss the differences between calculated values of h_1 and h_2 and their respective specified values b_1 and b_2.

For a specified value of u let us estimate the solutions of these equations to be $x_1^{(0)}$ and $x_2^{(0)}$. The zero superscripts indicate that these values are initial estimates and not the actual solutions x_1^* and x_2^*. We designate the corrections $\Delta x_1^{(0)}$ and $\Delta x_2^{(0)}$ as the values to be added to $x_1^{(0)}$ and $x_2^{(0)}$ to yield the correct solutions x_1^* and x_2^*. So, we can write

$$g_1(x_1^*, x_2^*, u) = g_1\left(x_1^{(0)} + \Delta x_1^{(0)}, x_2^{(0)} + \Delta x_2^{(0)}, u\right) = 0 \tag{9.29}$$

$$g_2(x_1^*, x_2^*, u) = g_2\left(x_1^{(0)} + \Delta x_1^{(0)}, x_2^{(0)} + \Delta x_2^{(0)}, u\right) = 0 \tag{9.30}$$

Our problem now is to solve for $\Delta x_1^{(0)}$ and $\Delta x_2^{(0)}$, which we do by expanding Eqs. (9.29) and (9.30) in Taylor's series about the assumed solution to give

$$g_1(x_1^*, x_2^*, u) = g_1\left(x_1^{(0)}, x_2^{(0)}, u\right) + \Delta x_1^{(0)} \frac{\partial g_1}{\partial x_1}\bigg|^{(0)} + \Delta x_2^{(0)} \frac{\partial g_1}{\partial x_2}\bigg|^{(0)} + \cdots = 0 \tag{9.31}$$

$$g_2(x_1^*, x_2^*, u) = g_2\left(x_1^{(0)}, x_2^{(0)}, u\right) + \Delta x_1^{(0)} \frac{\partial g_2}{\partial x_1}\bigg|^{(0)} + \Delta x_2^{(0)} \frac{\partial g_2}{\partial x_2}\bigg|^{(0)} + \cdots = 0 \tag{9.32}$$

where the partial derivatives of order greater than 1 in the series of terms of the expansion have not been listed. The term $\partial g_1/\partial x_1|^{(0)}$ indicates that the partial derivative is evaluated for the estimated values of $x_1^{(0)}$ and $x_2^{(0)}$. Other such terms are evaluated similarly.

If we neglect the partial derivatives of order greater than 1, we can rewrite Eqs. (9.31) and (9.32) in matrix form. We then have

$$\underbrace{\begin{bmatrix} \dfrac{\partial g_1}{\partial x_1} & \dfrac{\partial g_1}{\partial x_2} \\ \dfrac{\partial g_2}{\partial x_1} & \dfrac{\partial g_2}{\partial x_2} \end{bmatrix}^{(0)}}_{\mathbf{J}^{(0)}} \begin{bmatrix} \Delta x_1^{(0)} \\ \Delta x_2^{(0)} \end{bmatrix} = \begin{bmatrix} 0 - g_1(x_1^{(0)}, x_2^{(0)}, u) \\ 0 - g_2(x_1^{(0)}, x_2^{(0)}, u) \end{bmatrix} = \begin{bmatrix} b_1 - h_1(x_1^{(0)}, x_2^{(0)}, u) \\ b_2 - h_2(x_1^{(0)}, x_2^{(0)}, u) \end{bmatrix} \tag{9.33}$$

where the square matrix of partial derivatives is called the *jacobian* **J** or in this case $\mathbf{J}^{(0)}$ to indicate that the initial estimates $x_1^{(0)}$ and $x_2^{(0)}$ have been used to compute the numerical values of the partial derivatives. We note that $g_1(x_1^{(0)}, x_2^{(0)}, u)$ is the calculated value of g_1 based on the estimated values of $x_1^{(0)}$ and $x_2^{(0)}$, but this calculated value is not the zero value specified by Eq. (9.27) unless our estimated values $x_1^{(0)}$ and $x_2^{(0)}$ are correct. As before, we designate the specified value of g_1 minus the calculated value of g_1 as the mismatch $\Delta g_1^{(0)}$ and define the mismatch $\Delta g_2^{(0)}$ similarly. We then have the linear system of ***mismatch equations***

$$\mathbf{J}^{(0)} \begin{bmatrix} \Delta x_1^{(0)} \\ \Delta x_2^{(0)} \end{bmatrix} = \begin{bmatrix} \Delta g_1^{(0)} \\ \Delta g_2^{(0)} \end{bmatrix} \tag{9.34}$$

By solving the mismatch equations, either by triangular factorization of the jacobian or (for very small problems) by finding its inverse, we can determine $\Delta x_1^{(0)}$ and $\Delta x_2^{(0)}$. However, since we truncated the series expansion, these values added to our initial guess do not determine for us the correct solution and we must try again by assuming new estimates $x_1^{(1)}$ and $x_2^{(1)}$, where

$$x_1^{(1)} = x_1^{(0)} + \Delta x_1^{(0)}; \qquad x_2^{(1)} = x_2^{(0)} + \Delta x_2^{(0)} \tag{9.35}$$

We repeat the process until the corrections become so small in magnitude that they satisfy a chosen precision index $\varepsilon > 0$; that is, until $|\Delta x_1|$ and $|\Delta x_2|$ are both less than ε. The concepts underlying the Newton-Raphson method are now exemplified numerically.

Example 9.4. Using the Newton-Raphson method, solve for x_1 and x_2 of the nonlinear equations

$$g_1(x_1, x_2, u) = h_1(x_1, x_2, u) - b_1 = 4ux_2 \sin x_1 + 0.6 = 0$$

$$g_2(x_1, x_2, u) = h_2(x_1, x_2, u) - b_2 = 4x_2^2 - 4ux_2 \cos x_1 + 0.3 = 0$$

Treat the parameter u as a fixed number equal to 1.0, and choose the initial conditions $x_1^{(0)} = 0$ rad and $x_2^{(0)} = 1.0$. The precision index ε is 10^{-5}.

Solution. Partial differentiation with respect to the x's yields

$$\mathbf{J} = \begin{bmatrix} \dfrac{\partial g_1}{\partial x_1} & \dfrac{\partial g_1}{\partial x_2} \\ \dfrac{\partial g_2}{\partial x_1} & \dfrac{\partial g_2}{\partial x_2} \end{bmatrix} = \begin{bmatrix} 4ux_2 \cos x_1 & 4u \sin x_1 \\ 4ux_2 \sin x_1 & 8x_2 - 4u \cos x_1 \end{bmatrix}$$

Here the parameter u has a fixed value equal to 1.0, but in some studies it could be treated as a specifiable or control variable.

First iteration. Setting $u = 1.0$ and using the initial estimates of x_1 and x_2, we calculate the mismatches

$$\Delta g_1^{(0)} = 0 - g_{1,\text{calc}} = b_1 - h_1^{(0)} = -0.6 - 4\sin(0) = -0.6$$

$$\Delta g_2^{(0)} = 0 - g_{2,\text{calc}} = b_2 - h_2^{(0)} = -0.3 - 4 \times (1.0)^2 + 4\cos(0) = -0.3$$

which we use in Eq. (9.34) to yield the mismatch equations

$$\begin{bmatrix} 4\cos(0) & 4\sin(0) \\ 4\sin(0) & 8x_2 - 4\cos(0) \end{bmatrix} \begin{bmatrix} \Delta x_1^{(0)} \\ \Delta x_2^{(0)} \end{bmatrix} = \begin{bmatrix} -0.6 \\ -0.3 \end{bmatrix}$$

Inverting this simple 2×2 matrix, we determine the initial corrections

$$\begin{bmatrix} \Delta x_1^{(0)} \\ \Delta x_2^{(0)} \end{bmatrix} = \begin{bmatrix} 4 & 0 \\ 0 & 4 \end{bmatrix}^{-1} \begin{bmatrix} -0.6 \\ -0.3 \end{bmatrix} = \begin{bmatrix} -0.150 \\ -0.075 \end{bmatrix}$$

which provide first iteration values of x_1 and x_2 as follows:

$$x_1^{(1)} = x_1^{(0)} + \Delta x_1^{(0)} = 0.0 + (-0.150) = -0.150 \text{ rad}$$

$$x_2^{(1)} = x_2^{(0)} + \Delta x_2^{(0)} = 1.0 + (-0.075) = 0.925$$

The corrections exceed the specified tolerance, and so we continue.

Second iteration. The new mismatches are

$$\begin{bmatrix} \Delta g_1^{(1)} \\ \Delta g_2^{(1)} \end{bmatrix} = \begin{bmatrix} -0.6 - 4(0.925)\sin(-0.15) \\ -0.3 - 4(0.925)^2 + 4(0.925)\cos(-0.15) \end{bmatrix} = \begin{bmatrix} -0.047079 \\ -0.064047 \end{bmatrix}$$

and updating the jacobian, we compute the new corrections

$$\begin{bmatrix} \Delta x_1^{(1)} \\ \Delta x_2^{(1)} \end{bmatrix} = \begin{bmatrix} 3.658453 & -0.597753 \\ -0.552921 & 3.444916 \end{bmatrix}^{-1} \begin{bmatrix} -0.047079 \\ -0.064047 \end{bmatrix} = \begin{bmatrix} -0.016335 \\ -0.021214 \end{bmatrix}$$

These corrections also exceed the precision index, and so we move on to the next iteration with the new corrected values

$$x_1^{(2)} = -0.150 + (-0.016335) = -0.166335 \text{ rad}$$

$$x_2^{(2)} = 0.925 + (-0.021214) = 0.903786$$

Continuing on to the third iteration, we find that the corrections $\Delta x_1^{(3)}$ and $\Delta x_2^{(3)}$ are each smaller in magnitude than the stipulated tolerance of 10^{-5}. Accordingly, we calculate the solution

$$x_1^{(4)} = -0.166876 \text{ rad}; \qquad x_2^{(4)} = 0.903057$$

The resultant mismatches are insignificant as may be easily checked.

In this example we have actually solved our first power-flow problem by the Newton-Raphson method. This is because the two nonlinear equations of the example are the power-flow equations for the simple system shown in Fig. 9.3,

$$\begin{aligned} g_1(x_1, x_2, u) &= P_2(x_1, x_2, u) - (P_{g2} - P_{d2}) \\ &= 4|V_1|\,|V_2|\sin\delta_2 + 0.6 = 0 \end{aligned} \tag{9.36}$$

$$\begin{aligned} g_2(x_1, x_2, u) &= Q_2(x_1, x_2, u) - (Q_{g2} - Q_{d2}) \\ &= 4|V_2|^2 - 4|V_1|\,|V_2|\cos\delta_2 + 0.3 = 0 \end{aligned} \tag{9.37}$$

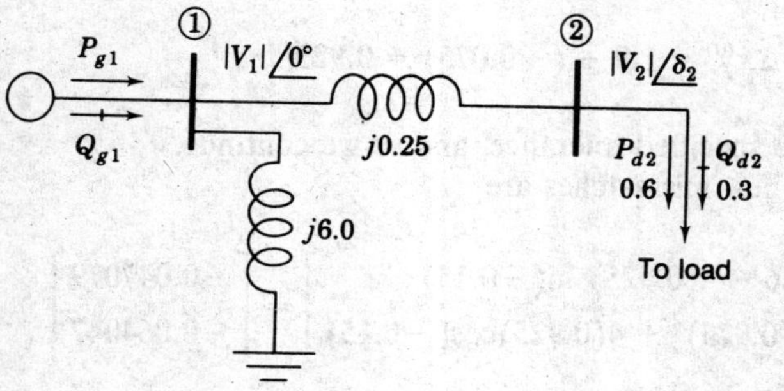

FIGURE 9.3
The system with power-flow equations corresponding to those of Example 9.4.

where x_1 represents the angle δ_2 and x_2 represents the voltage magnitude $|V_2|$ at bus ②. The control u denotes the voltage magnitude $|V_1|$ of the slack bus, and by changing its value from the specified value of 1.0 per unit, we may *control* the solution to the problem. In this textbook we do not investigate this control characteristic but, rather, concentrate on the application of the Newton-Raphson procedure in power-flow studies.

9.4 THE NEWTON-RAPHSON POWER-FLOW SOLUTION

To apply the Newton-Raphson method to the solution of the power-flow equations, we express bus voltages and line admittances in polar form. When n is set equal to i in Eqs. (9.6) and (9.7) and the corresponding terms are separated from the summations, we obtain

$$P_i = |V_i|^2 G_{ii} + \sum_{\substack{n=1 \\ n \neq i}}^{N} |V_i V_n Y_{in}| \cos(\theta_{in} + \delta_n - \delta_i) \tag{9.38}$$

$$Q_i = -|V_i|^2 B_{ii} - \sum_{\substack{n=1 \\ n \neq i}}^{N} |V_i V_n Y_{in}| \sin(\theta_{in} + \delta_n - \delta_i) \tag{9.39}$$

These equations can be readily differentiated with respect to voltage angles and magnitudes. The terms involving G_{ii} and B_{ii} come from the definition of Y_{ij} in Eq. (9.1) and the fact that the angle $(\delta_n - \delta_i)$ is zero when $n = i$.

Let us postpone consideration of voltage-controlled buses for now and regard all buses (except the slack bus) as load buses with known demands P_{di} and Q_{di}. The slack bus has specified values for δ_1 and $|V_1|$, and each of the other buses in the network has the two state variables δ_i and $|V_i|$ to be calculated in the power-flow solution. The known values of P_{di} and Q_{di} correspond to the negative of the b constants shown in Eqs. (9.27) and (9.28), as demonstrated in Example 9.4. At each of the nonslack buses estimated values of δ_i and $|V_i|$ correspond to the estimates $x_1^{(0)}$ and $x_2^{(0)}$ in the preceding section. Correspondence to the Δg mismatches of Eq. (9.34) follows from Eqs. (9.8) and (9.9) by writing the power mismatches for the typical load bus ⓘ,

$$\Delta P_i = P_{i,\text{sch}} - P_{i,\text{calc}} \tag{9.40}$$

$$\Delta Q_i = Q_{i,\text{sch}} - Q_{i,\text{calc}} \tag{9.41}$$

For simplicity sake, we now write mismatch equations for a four-bus system, and it will become obvious how to extend those equations to systems with more than four buses.

For real power P_i we have

$$\Delta P_i = \frac{\partial P_i}{\partial \delta_2}\Delta\delta_2 + \frac{\partial P_i}{\partial \delta_3}\Delta\delta_3 + \frac{\partial P_i}{\partial \delta_4}\Delta\delta_4 + \frac{\partial P_i}{\partial |V_2|}\Delta|V_2| + \frac{\partial P_i}{\partial |V_3|}\Delta|V_3| + \frac{\partial P_i}{\partial |V_4|}\Delta|V_4| \tag{9.42}$$

The last three terms can be multiplied and divided by their respective voltage magnitudes without altering their values, and so we obtain

$$\Delta P_i = \frac{\partial P_i}{\partial \delta_2}\Delta\delta_2 + \frac{\partial P_i}{\partial \delta_3}\Delta\delta_3 + \frac{\partial P_i}{\partial \delta_4}\Delta\delta_4 + |V_2|\frac{\partial P_i}{\partial |V_2|}\frac{\Delta|V_2|}{|V_2|} + |V_3|\frac{\partial P_i}{\partial |V_3|}\frac{\Delta|V_3|}{|V_3|} + |V_4|\frac{\partial P_i}{\partial |V_4|}\frac{\Delta|V_4|}{|V_4|} \tag{9.43}$$

There are advantages to this form of the equation as we shall soon see. A similar mismatch equation can be written for reactive power Q_i,

$$\Delta Q_i = \frac{\partial Q_i}{\partial \delta_2}\Delta\delta_2 + \frac{\partial Q_i}{\partial \delta_3}\Delta\delta_3 + \frac{\partial Q_i}{\partial \delta_4}\Delta\delta_4 + |V_2|\frac{\partial Q_i}{\partial |V_2|}\frac{\Delta|V_2|}{|V_2|} + |V_3|\frac{\partial Q_i}{\partial |V_3|}\frac{\Delta|V_3|}{|V_3|} + |V_4|\frac{\partial Q_i}{\partial |V_4|}\frac{\Delta|V_4|}{|V_4|} \tag{9.44}$$

Each nonslack bus of the system has two equations like those for ΔP_i and ΔQ_i. Collecting all the mismatch equations into vector-matrix form yields

$$\underbrace{\left[\begin{array}{ccc|ccc} \frac{\partial P_2}{\partial \delta_2} & \cdots & \frac{\partial P_2}{\partial \delta_4} & |V_2|\frac{\partial P_2}{\partial |V_2|} & \cdots & |V_4|\frac{\partial P_2}{\partial |V_4|} \\ \vdots & \mathbf{J}_{11} & \vdots & \vdots & \mathbf{J}_{12} & \vdots \\ \frac{\partial P_4}{\partial \delta_2} & \cdots & \frac{\partial P_4}{\partial \delta_4} & |V_2|\frac{\partial P_4}{\partial |V_2|} & \cdots & |V_4|\frac{\partial P_4}{\partial |V_4|} \\ \hline \frac{\partial Q_2}{\partial \delta_2} & \cdots & \frac{\partial Q_2}{\partial \delta_4} & |V_2|\frac{\partial Q_2}{\partial |V_2|} & \cdots & |V_4|\frac{\partial Q_2}{\partial |V_4|} \\ \vdots & \mathbf{J}_{21} & \vdots & \vdots & \mathbf{J}_{22} & \vdots \\ \frac{\partial Q_4}{\partial \delta_2} & \cdots & \frac{\partial Q_4}{\partial \delta_4} & |V_2|\frac{\partial Q_4}{\partial |V_2|} & \cdots & |V_4|\frac{\partial Q_4}{\partial |V_4|} \end{array}\right]}_{\text{Jacobian}} \underbrace{\left[\begin{array}{c} \Delta\delta_2 \\ \vdots \\ \Delta\delta_4 \\ \hline \frac{\Delta|V_2|}{|V_2|} \\ \vdots \\ \frac{\Delta|V_4|}{|V_4|} \end{array}\right]}_{\text{Corrections}} = \underbrace{\left[\begin{array}{c} \Delta P_2 \\ \vdots \\ \Delta P_4 \\ \hline \Delta Q_2 \\ \vdots \\ \Delta Q_4 \end{array}\right]}_{\text{Mismatches}} \tag{9.45}$$

We cannot include mismatches for the slack bus since ΔP_1 and ΔQ_1 are undefined when P_1 and Q_1 are not scheduled. We also omit all terms involving $\Delta\delta_1$ and $\Delta|V_1|$ from the equations because those corrections are both zero at the slack bus.

The partitioned form of Eq. (9.45) emphasizes the four different types of partial derivatives which enter into the jacobian **J**. The elements of $\mathbf{J}_{12}$ and $\mathbf{J}_{22}$ have voltage-magnitude multipliers because a simpler and more symmetrical jacobian results. In choosing this format, we have used the identity

$$\underbrace{|V_j|\frac{\partial P_i}{\partial|V_j|}}_{\text{Element of } \mathbf{J}_{12}} \times \underbrace{\frac{\Delta|V_j|}{|V_j|}}_{\text{Correction}} = \frac{\partial P_i}{\partial|V_j|} \times \Delta|V_j| \tag{9.46}$$

and the corrections become $\Delta|V_j|/|V_j|$ as shown rather than $\Delta|V_j|$.

The solution of Eq. (9.45) is found by iteration as follows:

- Estimate values $\delta_i^{(0)}$ and $|V_i|^{(0)}$ for the state variables.
- Use the estimates to calculate:
  $P_{i,\text{calc}}^{(0)}$ and $Q_{i,\text{calc}}^{(0)}$ from Eqs. (9.38) and (9.39),
  mismatches $\Delta P_i^{(0)}$ and $\Delta Q_i^{(0)}$ from Eqs. (9.40) and (9.41), and
  the partial derivative elements of the jacobian **J**.
- Solve Eq. (9.45) for the initial corrections $\Delta\delta_i^{(0)}$ and $\Delta|V_i|^{(0)}/|V_i|^{(0)}$.
- Add the solved corrections to the initial estimates to obtain

$$\delta_i^{(1)} = \delta_i^{(0)} + \Delta\delta_i^{(0)} \tag{9.47}$$

$$|V_i|^{(1)} = |V_i|^{(0)} + \Delta|V_i|^{(0)} = |V_i|^{(0)}\left(1 + \frac{\Delta|V_i|^{(0)}}{|V_i|^{(0)}}\right) \tag{9.48}$$

- Use the new values $\delta_i^{(1)}$ and $|V_i|^{(1)}$ as starting values for iteration 2 and continue.

In more general terms, the update formulas for the starting values of the state variables are

$$\delta_i^{(k+1)} = \delta_i^{(k)} + \Delta\delta_i^{(k)} \tag{9.49}$$

$$|V_i|^{(k+1)} = |V_i|^{(k)} + \Delta|V_i|^{(k)} = |V_i|^{(k)}\left(1 + \frac{\Delta|V_i|^{(k)}}{|V_i|^{(k)}}\right) \tag{9.50}$$

For the four-bus system submatrix $\mathbf{J}_{11}$ has the form

$$\mathbf{J}_{11} = \begin{bmatrix} \frac{\partial P_2}{\partial \delta_2} & \frac{\partial P_2}{\partial \delta_3} & \frac{\partial P_2}{\partial \delta_4} \\ \frac{\partial P_3}{\partial \delta_2} & \frac{\partial P_3}{\partial \delta_3} & \frac{\partial P_3}{\partial \delta_4} \\ \frac{\partial P_4}{\partial \delta_2} & \frac{\partial P_4}{\partial \delta_3} & \frac{\partial P_4}{\partial \delta_4} \end{bmatrix} \tag{9.51}$$

Expressions for the elements of this equation are easily found by differentiating the appropriate number of terms in Eq. (9.38). When the variable n equals the particular value j, only one of the cosine terms in the summation of Eq. (9.38) contains δ_j, and by partial differentiating that single term with respect to δ_j, we obtain the typical off-diagonal element of $\mathbf{J}_{11}$,

$$\frac{\partial P_i}{\partial \delta_j} = -|V_i V_j Y_{ij}| \sin(\theta_{ij} + \delta_j - \delta_i) \tag{9.52}$$

On the other hand, *every* term in the summation of Eq. (9.38) contains δ_i, and so the typical diagonal element of $\mathbf{J}_{11}$ is

$$\frac{\partial P_i}{\partial \delta_i} = \sum_{\substack{n=1 \\ n \neq i}}^{N} |V_i V_n Y_{in}| \sin(\theta_{in} + \delta_n - \delta_i) = -\sum_{\substack{n=1 \\ n \neq i}}^{N} \frac{\partial P_i}{\partial \delta_n} \tag{9.53}$$

By comparing this expression and that for Q_i in Eq. (9.39), we obtain

$$\frac{\partial P_i}{\partial \delta_i} = -Q_i - |V_i|^2 B_{ii} \tag{9.54}$$

In a quite similar manner, we can derive formulas for the elements of submatrix $\mathbf{J}_{21}$ as follows:

$$\frac{\partial Q_i}{\partial \delta_j} = -|V_i V_j Y_{ij}| \cos(\theta_{ij} + \delta_j - \delta_i) \tag{9.55}$$

$$\frac{\partial Q_i}{\partial \delta_i} = \sum_{\substack{n=1 \\ n \neq i}}^{N} |V_i V_n Y_{in}| \cos(\theta_{in} + \delta_n - \delta_i) = -\sum_{\substack{n=1 \\ n \neq i}}^{N} \frac{\partial Q_i}{\partial \delta_n} \tag{9.56}$$

Comparing this equation for $\partial Q_i/\partial \delta_i$ with Eq. (9.38) for P_i, we can show that

$$\frac{\partial Q_i}{\partial \delta_i} = P_i - |V_i|^2 G_{ii} \tag{9.57}$$

The elements of submatrix $\mathbf{J}_{12}$ are easily found by first finding the expression for the derivative $\partial P_i/\partial |V_j|$ and then multiplying by $|V_j|$ to obtain

$$|V_j| \frac{\partial P_i}{\partial |V_j|} = |V_j|\, |V_i Y_{ij}| \cos(\theta_{ij} + \delta_j - \delta_i) \tag{9.58}$$

Comparison with Eq. (9.55) yields

$$|V_j| \frac{\partial P_i}{\partial |V_j|} = -\frac{\partial Q_i}{\partial \delta_j} \tag{9.59}$$

This is a most useful result, for it reduces the computation involved in forming the jacobian since the off-diagonal elements of $\mathbf{J}_{12}$ are now simply the negatives of the corresponding elements in $\mathbf{J}_{21}$. This fact would not have become apparent if we had not multiplied $\partial P_i/\partial |V_j|$ by the magnitude $|V_j|$ in Eq. (9.43). In an analogous manner, the diagonal elements of $\mathbf{J}_{12}$ are found to be

$$|V_i| \frac{\partial P_i}{\partial |V_i|} = |V_i| \left[2|V_i| G_{ii} + \sum_{\substack{n=1 \\ n \neq i}}^{N} |V_n Y_{in}| \cos(\theta_{in} + \delta_n - \delta_i) \right] \tag{9.60}$$

and comparing this result with Eqs. (9.56) and (9.57), we arrive at the formula

$$|V_i| \frac{\partial P_i}{\partial |V_i|} = \frac{\partial Q_i}{\partial \delta_i} + 2|V_i|^2 G_{ii} = P_i + |V_i|^2 G_{ii} \tag{9.61}$$

Finally, the off-diagonal and diagonal elements of submatrix $\mathbf{J}_{22}$ of the jacobian are determined to be

$$|V_j| \frac{\partial Q_i}{\partial |V_j|} = -|V_j|\, |V_i Y_{ij}| \sin(\theta_{ij} + \delta_j - \delta_i) = \frac{\partial P_i}{\partial \delta_j} \tag{9.62}$$

$$|V_i| \frac{\partial Q_i}{\partial |V_i|} = -\frac{\partial P_i}{\partial \delta_i} - 2|V_i|^2 B_{ii} = Q_i - |V_i|^2 B_{ii} \tag{9.63}$$

Let us now bring together the results developed above in the following definitions:

Off-diagonal elements, $i \neq j$

$$M_{ij} \triangleq \frac{\partial P_i}{\partial \delta_j} = |V_j| \frac{\partial Q_i}{\partial |V_j|} \tag{9.64}$$

$$N_{ij} \triangleq \frac{\partial Q_i}{\partial \delta_j} = -|V_j| \frac{\partial P_i}{\partial |V_j|} \tag{9.65}$$

Diagonal elements, $i = j$

$$M_{ii} \triangleq \frac{\partial P_i}{\partial \delta_i} \qquad |V_i| \frac{\partial Q_i}{\partial |V_i|} = -M_{ii} - 2|V_i|^2 B_{ii} \tag{9.66}$$

$$N_{ii} \triangleq \frac{\partial Q_i}{\partial \delta_i} \qquad |V_i| \frac{\partial P_i}{\partial |V_i|} = N_{ii} + 2|V_i|^2 G_{ii} \tag{9.67}$$

Interrelationships among the elements in the four submatrices of the jacobian are more clearly seen if we use the definitions to rewrite Eq. (9.45) in the following form:

$$\left[\begin{array}{ccc|ccc} M_{22} & M_{23} & M_{24} & N_{22} + 2|V_2|^2 G_{22} & -N_{23} & -N_{24} \\ M_{32} & M_{33} & M_{34} & -N_{32} & N_{33} + 2|V_3|^2 G_{33} & -N_{34} \\ M_{42} & M_{43} & M_{44} & -N_{42} & -N_{43} & N_{44} + 2|V_4|^2 G_{44} \\ \hline N_{22} & N_{23} & N_{24} & -M_{22} - 2|V_2|^2 B_{22} & M_{23} & M_{24} \\ N_{32} & N_{33} & N_{34} & M_{32} & -M_{33} - 2|V_3|^2 B_{33} & M_{34} \\ N_{42} & N_{43} & N_{44} & M_{42} & M_{43} & -M_{44} - 2|V_4|^2 B_{44} \end{array}\right]$$

$$\times \left[\begin{array}{c} \Delta\delta_2 \\ \Delta\delta_3 \\ \Delta\delta_4 \\ \hline \Delta|V_2| / |V_2| \\ \Delta|V_3| / |V_3| \\ \Delta|V_4| / |V_4| \end{array}\right] = \left[\begin{array}{c} \Delta P_2 \\ \Delta P_3 \\ \Delta P_4 \\ \hline \Delta Q_2 \\ \Delta Q_3 \\ \Delta Q_4 \end{array}\right] \tag{9.68}$$

So far we have regarded all nonslack buses as load buses. We now consider voltage-controlled buses also.

Voltage-controlled buses. In the polar form of the power-flow equations voltage-controlled buses are easily taken into account. For example, if bus ④ of the four-bus system is voltage controlled, then $|V_4|$ has a specified constant value and the voltage correction $\Delta|V_4|/|V_4|$ must always be zero. Consequently, the sixth column of the jacobian of Eq. (9.68) is always multiplied by zero, and so it may be removed. Furthermore, since Q_4 is not specified, the mismatch ΔQ_4 cannot be defined, and so we must omit the sixth row of Eq. (9.68) corresponding to Q_4. Of course, Q_4 can be calculated after the power-flow solution becomes available.

In the general case if there are N_g voltage-controlled buses *besides* the slack bus, a row and column for each such bus is omitted from the polar form of the system jacobian, which then has $(2N - N_g - 2)$ rows and $(2N - N_g - 2)$ columns consistent with Table 9.1.

Example 9.5. The small power system of Example 9.2 has the line data and bus data given in Tables 9.2 and 9.3. A power-flow study of the system is to be made by the Newton-Raphson method using the polar form of the equations for P and Q. Determine the number of rows and columns in the jacobian. Calculate the initial mismatch $\Delta P_3^{(0)}$ and the initial values of the jacobian elements of the (second row, third column); of the (second row, second column); and of the (fifth row, fifth column). Use the specified values and initial voltage estimates shown in Table 9.3.

Solution. Since the slack bus has no rows or columns in the jacobian, a 6×6 matrix would be necessary if P and Q were specified for the remaining three buses. In fact, however, the voltage magnitude is specified (held constant) at bus ④, and thus the jacobian will be a 5×5 matrix. In order to calculate $P_{3,\text{calc}}$ based on the estimated and the specified voltages of Table 9.3, we need the polar form of the off-diagonal entries of Table 9.4,

$$Y_{31} = 26.359695\angle 101.30993^\circ; \qquad Y_{34} = 15.417934\angle 101.30993^\circ$$

and the diagonal element $Y_{33} = 8.193267 - j40.863838$. Since Y_{32} and the initial values $\delta_3^{(0)}$ and $\delta_4^{(0)}$ are all zero, from Eq. (9.38) we obtain

$$P_{3,\text{calc}}^{(0)} = |V_3|^2 G_{33} + |V_3 V_1 Y_{31}|\cos\theta_{31} + |V_3 V_4 Y_{34}|\cos\theta_{34}$$

$$= (1.0)^2 8.193267 + (1.0 \times 1.0 \times 26.359695)\cos(101.30993^\circ)$$

$$+ (1.0 \times 1.02 \times 15.417934)\cos(101.30993^\circ)$$

$$= -0.06047 \text{ per unit}$$

Scheduled real power into the network at bus ③ is −2.00 per unit, and so the initial mismatch which we are asked to calculate has the value

$$\Delta P_{3,\text{calc}}^{(0)} = -2.00 - (-0.06047) = -1.93953 \text{ per unit}$$

From Eq. (9.52) the (second row, third column) jacobian element is

$$\frac{\partial P_3}{\partial \delta_4} = -|V_3 V_4 Y_{34}| \sin(\theta_{34} + \delta_4 - \delta_3)$$

$$= -(1.0 \times 1.02 \times 15.417934)\sin(101.30993^\circ)$$

$$= -15.420898 \text{ per unit}$$

and from Eq. (9.53) the element of the (second row, second column) is

$$\frac{\partial P_3}{\partial \delta_3} = -\frac{\partial P_3}{\partial \delta_1} - \frac{\partial P_3}{\partial \delta_2} - \frac{\partial P_3}{\partial \delta_4}$$

$$= |V_3 V_1 Y_{31}| \sin(\theta_{31} + \delta_1 - \delta_3) - 0 - (-15.420898)$$

$$= (1.0 \times 1.0 \times 26.359695)\sin(101.30993^\circ) + 15.420898$$

$$= 41.268707 \text{ per unit}$$

For the element of the (fifth row, fifth column) Eq. (9.63) yields

$$|V_3| \frac{\partial Q_3}{\partial |V_3|} = -\frac{\partial P_3}{\partial \delta_3} - 2|V_3|^2 B_{33}$$

$$= -41.268707 - 2(1.0)^2(-40.863838) = 40.458969 \text{ per unit}$$

Using the initial input data, we can similarly calculate initial values of the other elements of the jacobian and of the power mismatches at all the buses of the system.

For the system of the preceding example the numerical values for initialization of the mismatch equations are now shown, for convenience, to only three decimal places as follows:

$$
\begin{array}{c}
\\
\textcircled{2}\\ \textcircled{3}\\ \textcircled{4}\\ \textcircled{2}\\ \textcircled{3}
\end{array}
\begin{array}{c}
\begin{array}{ccccc} \textcircled{2} & \textcircled{3} & \textcircled{4} & \textcircled{2} & \textcircled{3} \end{array}\\
\left[\begin{array}{ccc|cc}
45.443 & 0 & -26.365 & 8.882 & 0\\
0 & 41.269 & -15.421 & 0 & 8.133\\
-26.365 & -15.421 & 41.786 & -5.273 & -3.084\\
\hline
-9.089 & 0 & 5.273 & 44.229 & 0\\
0 & -8.254 & 3.084 & 0 & 40.459
\end{array}\right]
\end{array}
\left[\begin{array}{c}
\Delta\delta_2\\ \Delta\delta_3\\ \Delta\delta_4\\ \hline \dfrac{\Delta|V_2|}{|V_2|}\\ \dfrac{\Delta|V_3|}{|V_3|}
\end{array}\right]
=
\left[\begin{array}{c}
-1.597\\ -1.940\\ 2.213\\ \hline -0.447\\ -0.835
\end{array}\right]
$$

This system of equations yields values for the voltage corrections of the first iteration which are needed to update the state variables according to Eqs. (9.49) and (9.50). At the end of the first iteration the set of updated voltages at the buses is:

Bus no. i =	①	②	③	④
δ_i (deg.)	0	−0.93094	−1.78790	−1.54383
$\lvert V_i\rvert$ (per unit)	1.00	0.98335	0.97095	1.02

These updated voltages are then used to recalculate the jacobian and mismatches of the second iteration, and so on. The iterative procedure continues until either the mismatches ΔP_i and ΔQ_i become less than their stipulated allowable values or all $\Delta\delta_i$ and $\Delta|V_i|$ are less than the chosen precision index. When the solution is completed, we can use Eqs. (9.38) and (9.39) to calculate real and reactive power, P_1 and Q_1, at the slack bus, and the reactive power Q_4 at voltage-controlled bus ④. Line flows can also be computed from the differences in bus voltages and the known parameters of the lines. Solved values of the bus voltages and line flows for the system of Example 9.5 are shown in Figs. 9.4 and 9.5.

The number of iterations required by the Newton-Raphson method using bus admittances is practically independent of the number of buses. The time for the Gauss-Seidel method (employing bus admittances) increases almost directly with the number of buses. On the other hand, computing the elements of the jacobian is time-consuming, and the time per iteration is considerably longer for the Newton-Raphson method. When sparse matrix techniques are employed, the advantage of shorter computer time for a solution of the same accuracy is in favor of the Newton-Raphson method for all but very small systems. Sparsity features of the jacobian are discussed in Sec. B.2 of the Appendix.

9.5 POWER-FLOW STUDIES IN SYSTEM DESIGN AND OPERATION

Electric utility companies use very elaborate programs for power-flow studies aimed at evaluating the adequacy of a complex, interconnected network. Important information is obtained concerning the design and operation of systems that have not yet been built and the effects of changes on existing systems. A power-flow study for a system operating under actual or projected normal operating conditions is called a *base case*. The results from the base case constitute a benchmark for comparison of changes in network flows and voltages under abnormal or *contingency* conditions. The transmission planning engineer can discover system weaknesses such as low voltages, line overloads, or loading conditions deemed excessive. These weaknesses can be removed by making design studies involving changes and/or additions to the base case system. The system model is then subjected to computer-based contingency testing to discover whether weaknesses arise under contingency conditions involving abnormal generation schedules or load levels. Interaction between the system designer and the computer-based power-flow program continues until system performance satisfies local and regional planning or operating criteria.

A typical power-flow program is capable of handling systems of more than 2000 buses, 3000 lines, and 500 transformers. Of course, programs can be expanded to even greater size provided the available computer facilities are sufficiently large.

Data supplied to the computer must include the numerical values of the line and bus data (such as Tables 9.2 and 9.3) and an indication of whether a bus is the slack bus, a regulated bus where voltage magnitude is held constant by generation of reactive power Q, or a load bus with fixed P and Q. Where values are not to be held constant the quantities given in the tables are interpreted as initial estimates. Limits of P and Q generation usually must be specified as well as the limits of line kilovoltamperes. Unless otherwise specified, programs usually assume a base of 100 MVA.

Total line-charging megavars specified for each line account for shunt capacitance and equal $\sqrt{3}$ times the rated line voltage in kilovolts times I_{chg}, divided by 10^3. That is,

$$(\text{Mvar})_{\text{chg}} = \sqrt{3}\,|V|I_{\text{chg}} \times 10^{-3} = \omega C_n |V|^2 \tag{9.69}$$

where $|V|$ is the rated line-to-line voltage in kilovolts, C_n is line-to-neutral capacitance in farads for the entire length of the line, and I_{chg} is defined by Eqs. (5.24) and (5.25). The program creates a nominal-π representation of the line similar to Fig. 6.7 by dividing the capacitance computed from the given value of charging megavars equally between the two ends of the line. It is evident from Eq. (9.69) that line-charging megavars in per unit equals the per-unit shunt susceptance of the line at 1.0 per-unit voltage. For a long line the computer could be programmed to compute the equivalent π for capacitance distributed evenly along the line.

The printout of results provided by the computer consists of a number of tabulations. Usually, the most important information to be considered first is the table which lists each bus number and name, bus-voltage magnitude in per unit and phase angle, generation and load at each bus in megawatts and megavars, and megavars of static capacitors or reactors on the bus. Accompanying the bus information is the flow of megawatts and megavars from that bus over each transmission line connected to the bus. The totals of system generation and loads are listed in megawatts and megavars. The tabulation described is shown in Fig. 9.4 for the four-bus system of Example 9.5.

A system may be divided into areas, or one study may include the systems of several companies with each designated as a different area. The computer program will examine the flow between areas, and deviations from the prescribed flow will be overcome by causing the appropriate change in generation of a selected generator in each area. In actual system operation interchange of power between areas is monitored to determine whether a given area is producing that amount of power which will result in the desired interchange.

Among other information that may be obtained is a listing of all buses where the per-unit voltage magnitude is above 1.05 or below 0.95, or other limits that may be specified. A list of line loadings in megavoltamperes can be obtained. The printout will also list the total megawatt ($|I|^2R$) losses and megavar ($|I|^2X$) requirements of the system, and both P and Q mismatch at each bus. Mismatch is an indication of the preciseness of the solution and is the difference between P (and also usually Q) entering and leaving each bus.

The numerical results in the printout of Fig. 9.4 are from a Newton-Raphson power-flow study of the system described in Example 9.5. The system line data and bus data are provided in Tables 9.2 and 9.3. Three Newton-Raphson iterations were required. Similar studies employing the Gauss-Seidel procedure required many more iterations, and this is a common observation in comparing the two iterative methods. Inspection of the printout reveals that the $|I|^2R$ losses of the system are $(504.81 - 500.0) = 4.81$ MW.

Figure 9.4 may be examined for more information than just the tabulated results, and the information provided can be displayed on a one-line diagram showing the entire system or a portion of the system such as at load bus ③ of Fig. 9.5. Transmission design engineers and system operators usually can call for video display of such selected power-flow results from computer-interactive terminals or workstations. The megawatt loss in any of the lines can be found by comparing the values of P at the two ends of the line. As an example, from

Bus information										Line flow		
				Generation		Load					Line flow	
Bus no.	Name	Volts (p.u.)	Angle (deg.)	(MW)	(Mvar)	(MW)	(Mvar)	Bus type	To Bus	Name	(MW)	(Mvar)
1	Birch	1.000	0.	186.81	114.50	50.00	30.99	SL	2	Elm	38.69	22.30
									3	Pine	98.12	61.21
2	Elm	0.982	−0.976	0.	0.	170.00	105.35	PQ	1	Birch	−38.46	−31.24
									4	Maple	−131.54	−74.11
3	Pine	0.969	−1.872	0.	0.	200.00	123.94	PQ	1	Birch	−97.09	−63.57
									4	Maple	−102.91	−60.37
4	Maple	1.020	1.523	318.00	181.43	80.00	49.58	PV	2	Elm	133.25	74.92
									3	Pine	104.75	56.93
		Area totals		504.81	295.93	500.00	309.86					

FIGURE 9.4
Newton-Raphson power-flow solution for the system of Example 9.5. Base is 230 kV and 100 MVA. Tables 9.2 and 9.3 show the line data and bus data, respectively.

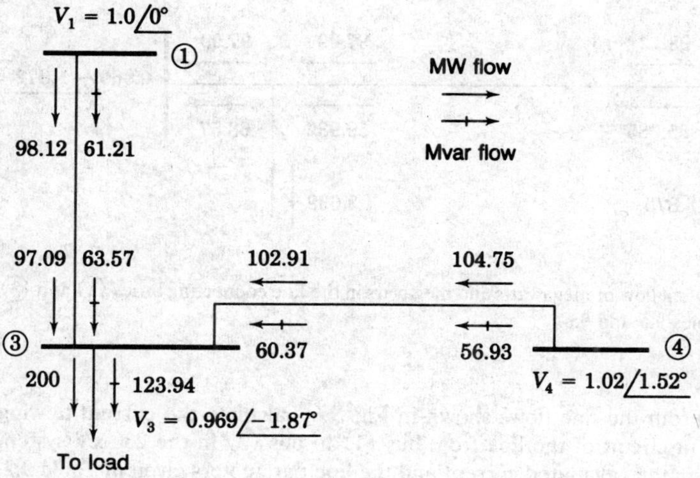

FIGURE 9.5
Flow of P and Q at bus ③ for the system of Example 9.5. Numbers beside the arrows show the flow of P and Q in megawatts and megavars. The bus voltage is shown in per unit.

Fig. 9.5 we see that 98.12 MW flow from bus ① into line ①-③ and 97.09 MW flow into bus ③ from the same line. Evidently, the $|I|^2R$ loss in all three phases of the line is 1.03 MW.

Consideration of the megavar flow on the line between bus ① and bus ③ is slightly more complicated because of the charging megavars. The computer considers the distributed capacitance of the line to be concentrated, half at one end of the line and half at the other end. In the line data given in Table 9.3 the line charging for line ①-③ is 7.75 Mvars, but the computer recognizes this value to be the value when the voltage is 1.0 per unit. Since charging megavars vary as the square of the voltage according to Eq. (9.69), voltages at buses ① and ③ of 1.0 per unit and 0.969 per unit, respectively, make the charging at those buses equal to

$$\frac{7.75}{2} \times (1.0)^2 = 3.875 \text{ Mvars at bus } ①$$

$$\frac{7.75}{2} \times (0.969)^2 = 3.638 \text{ Mvars at bus } ③$$

Figure 9.5 shows 61.21 Mvars going from bus ① into the line to bus ③ and 63.57 Mvars received at bus ③. The increase in megavars is due to the line charging. The three-phase flow of megawatts and megavars in the line is shown in the single-line diagram of Fig. 9.6.

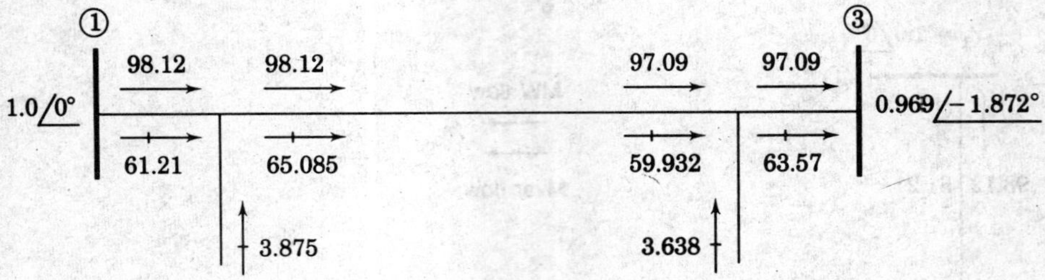

FIGURE 9.6
Single-line diagram showing flow of megawatts and megavars in the line connecting buses ① and ③ of the system of Examples 9.5 and 9.6.

Example 9.6. From the line flows shown in Fig. 9.6, calculate the current flowing in the equivalent circuit of the line from bus ① to bus ③ in the 230-kV system of Fig. 9.2. Using the calculated current and the line parameters given in Table 9.2, calculate the line I^2R loss and compare this value with the difference between the power in the line from bus ① and the power out at bus ③. Similarly, find I^2X in the line and compare with the value which could be found from the data in Fig. 9.6.

Solution. Figure 9.6 shows the megawatt and megavar flow in the per-phase equivalent circuit of line ①–③. The total megavoltampere flow through R and X of all three phases is

$$S = 98.12 + j65.085 = 117.744\angle 33.56^\circ \text{ MVA}$$

or

$$S = 97.09 + j59.932 = 114.098\angle 31.69^\circ \text{ MVA}$$

and

$$|I| = \frac{117{,}744}{\sqrt{3} \times 230 \times 1.0} = 295.56 \text{ A}$$

or

$$|I| = \frac{114{,}098}{\sqrt{3} \times 230 \times 0.969} = 295.57 \text{ A}$$

The magnitude of current I in the series $R + jX$ of line ①–③ can also be calculated using $|I| = |V_1 - V_3| / |R + jX|$. The base impedance is

$$Z_{\text{base}} = \frac{(230)^2}{100} = 529\ \Omega$$

and using the R and X parameters of Table 9.2, we have

$$I^2R \text{ loss} = 3 \times (295.56)^2 \times 0.00744 \times 529 \times 10^{-6} = 1.03 \text{ MW}$$

$$I^2X \text{ of line} = 3 \times (295.56)^2 \times 0.03720 \times 529 \times 10^{-6} = 5.157 \text{ Mvar}$$

These values compare with (98.12 − 97.09) = 1.03 MW and (65.085 − 59.932) = 5.153 Mvar from Fig. 9.6.

9.6 REGULATING TRANSFORMERS

As we have seen in Sec. 2.9, regulating transformers can be used to control the real and reactive power flows in a circuit. We now develop the bus admittance equations to include such transformers in power-flow studies.

Figure 9.7 is a more detailed representation of the regulating transformer of Fig. 2.24(*b*). The admittance Y in per unit is the reciprocal of the per-unit impedance of the transformer which has the transformation ratio 1: t as shown. The admittance Y is shown on the side of the ideal transformer nearest node ⓙ, which is the tap-changing side. This designation is important in using the equations which are to be derived. If we are considering a transformer with off-nominal turns ratio, t may be real or imaginary, such as 1.02 for an approximate 2% boost in voltage magnitude or $\varepsilon^{j\pi/60}$ for an approximate 3° shift per phase.

Figure 9.7 shows currents I_i and I_j entering the two buses, and the voltages are V_i and V_j referred to the reference node. The complex expressions for power into the ideal transformer from bus ⓘ and bus ⓙ are, respectively,

$$S_i = V_i I_i^* \qquad S_j = tV_i I_j^* \tag{9.70}$$

Since we are assuming that we have an ideal transformer with no losses, the power S_i into the ideal transformer from bus ⓘ must equal the power $-S_j$ out of the ideal transformer on the bus ⓙ side, and so from Eqs. (9.70) we obtain

$$I_i = -t^* I_j \tag{9.71}$$

The current I_j can be expressed by

$$I_j = (V_j - tV_i)Y = -tYV_i + YV_j \tag{9.72}$$

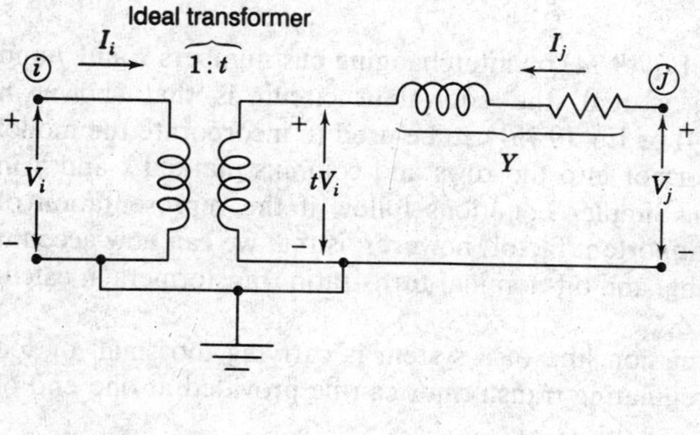

FIGURE 9.7
More detailed per-unit reactance diagram for the transformer of Fig. 2.24(*b*) whose turns ratio is $1/t$.

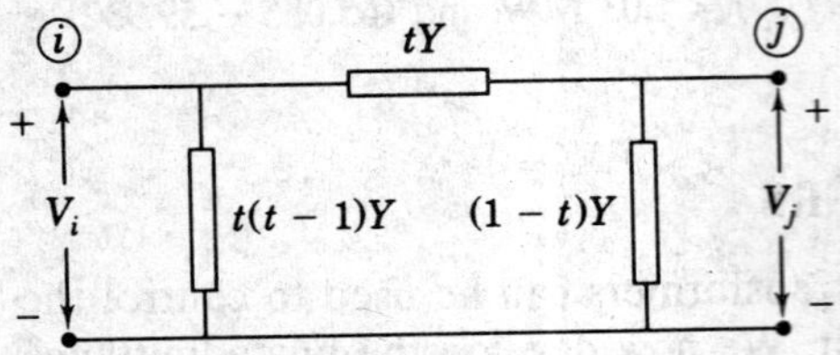

FIGURE 9.8
Circuit having the node admittances of Eq. (9.74) when t is real.

Multiplying by $-t^*$ and substituting I_i for $-t^*I_j$ yield

$$I_i = tt^*YV_i - t^*YV_j \tag{9.73}$$

Setting $tt^* = |t|^2$ and rearranging Eqs. (9.72) and (9.73) into $\mathbf{Y}_{\text{bus}}$ matrix form, we have

$$\begin{matrix} & ⓘ & ⓙ \\ ⓘ & Y_{ii} & Y_{ij} \\ ⓙ & Y_{ji} & Y_{jj} \end{matrix} \begin{bmatrix} V_i \\ V_j \end{bmatrix} = \begin{matrix} & ⓘ & ⓙ \\ ⓘ & |t|^2 Y & -t^*Y \\ ⓙ & -tY & Y \end{matrix} \begin{bmatrix} V_i \\ V_j \end{bmatrix} = \begin{bmatrix} I_i \\ I_j \end{bmatrix} \tag{9.74}$$

The equivalent-π circuit corresponding to these values of node admittances can be found only if t is real because then $Y_{ij} = Y_{ji}$. Otherwise, the coefficient matrix of Eq. (9.74) and the overall system $\mathbf{Y}_{\text{bus}}$ are not symmetrical because of the phase shifter. If the transformer is changing magnitude, not phase shifting, the circuit is that of Fig. 9.8. This circuit cannot be realized if Y has a real component, which would require a negative resistance in the circuit.

Some texts show admittance Y on the side of the transformer opposite to the tap-changing side, and often the transformation ratio is expressed as $1 : a$, as shown in Fig. 9.9(a). Analysis similar to that developed above shows that the bus admittance equations for Fig. 9.9(a) take the form

$$\begin{matrix} & ⓘ & ⓙ \\ ⓘ & Y_{ii} & Y_{ij} \\ ⓙ & Y_{ji} & Y_{jj} \end{matrix} \begin{bmatrix} V_i \\ V_j \end{bmatrix} = \begin{matrix} & ⓘ & ⓙ \\ ⓘ & Y & -Y/a \\ ⓙ & -Y/a^* & Y/|a|^2 \end{matrix} \begin{bmatrix} V_i \\ V_j \end{bmatrix} = \begin{bmatrix} I_i \\ I_j \end{bmatrix} \tag{9.75}$$

which can be verified from Eq. (9.74) by interchanging bus numbers i and j and setting $t = 1/a$. When a is real, the equivalent circuit is that shown in Fig. 9.9(b). Either Eq. (9.74) or Eq. (9.75) can be used to incorporate the model of the tap-changing transformer into the rows and columns marked i and j in $\mathbf{Y}_{\text{bus}}$ of the overall system. Simpler equations follow if the representation of Eq. (9.74) is used, but the important factor, however, is that we can now account for magnitude, phase shifting, and off-nominal-turns-ratio transformers in calculations to obtain $\mathbf{Y}_{\text{bus}}$ and $\mathbf{Z}_{\text{bus}}$.

If a particular transmission line in a system is carrying too small or too large a reactive power, a regulating transformer can be provided at one end of

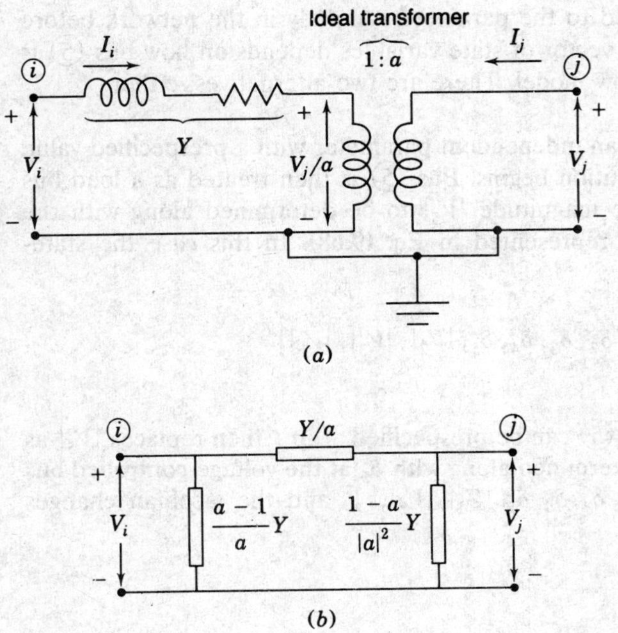

FIGURE 9.9
Per-phase representation of the regulating transformer showing: (*a*) per-unit admittance Y opposite to the tap-changing side; (*b*) per-unit equivalent circuit when a is real.

the line to make the line transmit a larger or smaller reactive power, as demonstrated in Sec. 2.9. And any appreciable drop in voltage on the primary of a transformer due to a change of load may make it desirable to change the tap setting on transformers provided with adjustable taps in order to maintain proper voltage at the load. We can investigate voltage-magnitude adjustment at a bus by means of the automatic tap-changing feature in the power-flow program. For instance, in the four-bus system of Example 9.5 suppose we wish to raise the load voltage at bus ③ by inserting a magnitude-regulating transformer between the load and the bus. With t real, in Eq. (9.74) we set $i = 3$ and assign the number 5 to bus ⓙ from which the load is now to be served. To accommodate the regulator in the power-flow equations, $\mathbf{Y}_{\text{bus}}$ of the network is expanded by one row and one column for bus ⑤, and the elements of buses ③ and ⑤ in the matrix of Eq. (9.74) are then added to the previous bus admittance matrix to give

$$\mathbf{Y}_{\text{bus(new)}} = \begin{array}{c} \text{①} \\ \text{②} \\ \text{③} \\ \text{④} \\ \text{⑤} \end{array} \overset{\begin{array}{ccccc} \text{①} & \text{②} & \text{③} & \text{④} & \text{⑤} \end{array}}{\begin{bmatrix} Y_{11} & Y_{12} & Y_{13} & 0 & 0 \\ Y_{21} & Y_{22} & 0 & Y_{24} & 0 \\ Y_{31} & 0 & Y_{33} + t^2 Y & Y_{34} & -tY \\ 0 & Y_{42} & Y_{43} & Y_{44} & 0 \\ 0 & 0 & -tY & 0 & Y \end{bmatrix}} \tag{9.76}$$

The Y_{ij} elements correspond to the parameters already in the network before the regulator is added. The vector of state variables depends on how bus ⑤ is treated within the power-flow model. There are two alternatives: either

- Tap t can be regarded as an independent parameter with a prespecified value before the power-flow solution begins. Bus ⑤ is then treated as a load bus with angle δ_5 and voltage magnitude $|V_5|$ to be determined along with the other five state variables represented in Eq. (9.68). In this case the state-variable vector is

$$\mathbf{x} = [\delta_2, \delta_3, \delta_4, \delta_5, |V_2|, |V_3|, |V_5|]^T$$

or

- Voltage magnitude at bus ⑤ can be prespecified. Tap t then replaces $|V_5|$ as the state variable to be determined along with δ_5 at the voltage-controlled bus ⑤. In this case $\mathbf{x} = [\delta_2, \delta_3, \delta_4, \delta_5, |V_2|, |V_3|, t]^T$ and the jacobian changes accordingly.

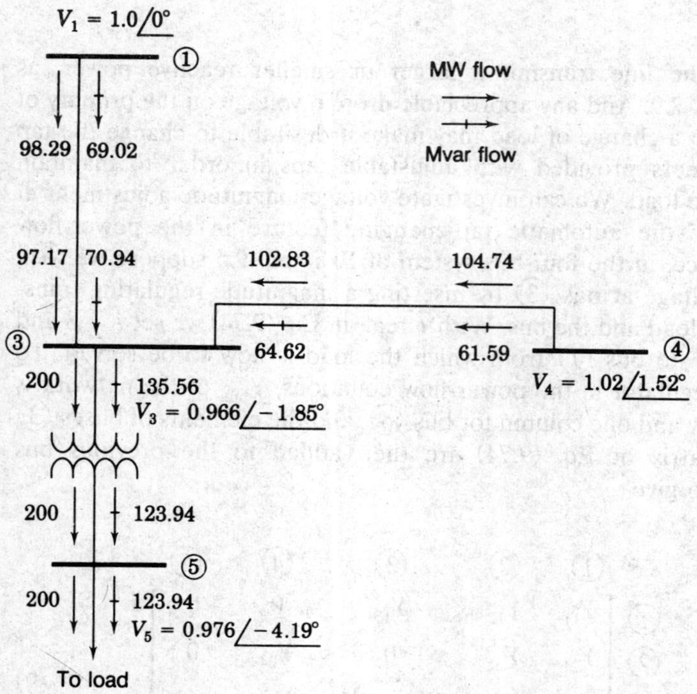

FIGURE 9.10
Flow of P and Q at bus ③ of the system of Fig. 9.5 when a regulating transformer is interposed between the bus and the load.

In some studies the variable tap t is considered as an independent control variable. The reader is encouraged to write the jacobian matrix and the mismatch equations for each of the above alternatives (see Probs. 9.9 and 9.10).

Because there is a definite step between tap settings, discrete control action occurs when regulators are used to boost voltage at a bus. The results of regulating the voltage of the load previously at bus ③ of the system of Fig. 9.5 are shown in the one-line diagram of Fig. 9.10. A per-unit reactance of 0.02 was assumed for the load-tap-changing (LTC) transformer. When the voltage at the load is raised by setting the LTC ratio t equal to 1.0375, the voltage at bus ③ is slightly lowered compared to Fig. 9.5, resulting in slightly higher voltage drops across the lines ①-③ and ④-③. The Q supplied to these lines from buses ① and ④ is increased owing to the reactive power required by the regulator, but the real power flow is relatively unaffected. The increased megavars on the lines cause the losses to increase and the Q contributed by the charging capacitances at bus ③ is decreased.

To determine the effect of phase-shifting transformers, we let t be complex with a magnitude of unity in Eq. (9.74).

Example 9.7. Solve Example 2.13 using the $\mathbf{Y}_{\text{bus}}$ model of Eq. (9.74) for each of the two parallel transformers and compare the solution with the approximate results.

Solution. The admittance Y of each transformer is given by $1/j0.1 = -j10$ per unit. Hence, the currents in transformer T_a of Fig. 9.11 can be determined from

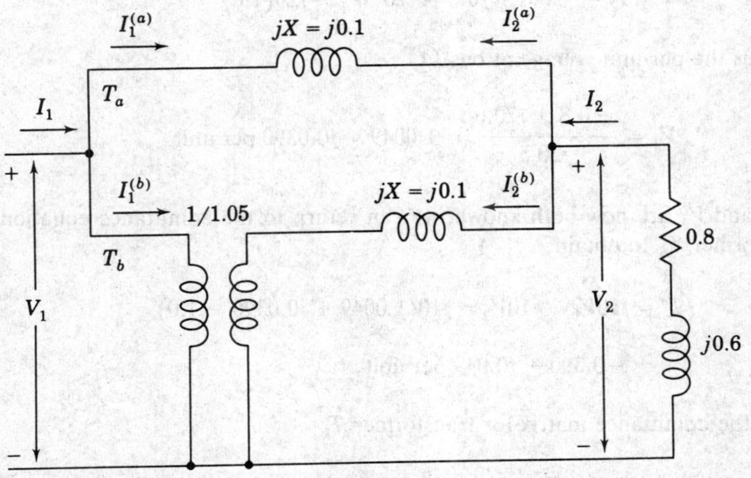

FIGURE 9.11
Circuit for Example 9.7. Values are in per unit.

the bus admittance equation

$$\begin{bmatrix} I_1^{(a)} \\ I_2^{(a)} \end{bmatrix} = \begin{matrix} \\ ① \\ ② \end{matrix} \overset{\quad ① \qquad ②}{\begin{bmatrix} Y & -Y \\ -Y & Y \end{bmatrix}} \begin{bmatrix} V_1 \\ V_2 \end{bmatrix} = \begin{matrix} \\ ① \\ ② \end{matrix} \overset{\quad ① \qquad ②}{\begin{bmatrix} -j10 & j10 \\ j10 & -j10 \end{bmatrix}} \begin{bmatrix} V_1 \\ V_2 \end{bmatrix}$$

and the currents in transformer T_b with $t = 1.05$, as shown in Fig. 9.11, are given by Eq. (9.74), which takes the numerical form

$$\begin{bmatrix} I_1^{(b)} \\ I_2^{(b)} \end{bmatrix} = \begin{matrix} \\ ① \\ ② \end{matrix} \overset{\quad ① \qquad ②}{\begin{bmatrix} t^2 Y & -tY \\ -tY & Y \end{bmatrix}} \begin{bmatrix} V_1 \\ V_2 \end{bmatrix} = \begin{matrix} \\ ① \\ ② \end{matrix} \overset{\quad ① \qquad\qquad ②}{\begin{bmatrix} -j11.025 & j10.500 \\ j10.500 & -j10.000 \end{bmatrix}} \begin{bmatrix} V_1 \\ V_2 \end{bmatrix}$$

In Fig. 9.11 current $I_1 = (I_1^{(a)} + I_1^{(b)})$, and likewise, $I_2 = (I_2^{(a)} + I_2^{(b)})$, which means that the preceding two matrix equations can be directly added (like admittances in parallel) to obtain

$$\begin{bmatrix} I_1 \\ I_2 \end{bmatrix} = \begin{matrix} \\ ① \\ ② \end{matrix} \overset{\quad ① \qquad\qquad ②}{\begin{bmatrix} -j21.025 & j20.500 \\ j20.500 & -j20.000 \end{bmatrix}} \begin{bmatrix} V_1 \\ V_2 \end{bmatrix}$$

In Example 2.13 V_2 is the reference voltage $1.0\angle 0°$ and the current I_2 is calculated to be $-0.8 + j0.6$. Therefore, from the second row of the preceding equation we have

$$I_2 = -0.8 + j0.6 = j20.5V_1 - j20(1.0)$$

which gives the per-unit voltage at bus ①

$$V_1 = \frac{-0.8 + j20.6}{j20.5} = 1.0049 + j0.0390 \text{ per unit}$$

Since V_1 and V_2 are now both known, we can return to the admittance equation for transformer T_a to obtain

$$I_2^{(a)} = j10V_1 - j10V_2 = j10(1.0049 + j0.0390 - 1.0)$$

$$= -0.390 + j0.049 \text{ per unit}$$

and from the admittance matrix for transformer T_b

$$I_2^{(b)} = j10.5V_1 - j10V_2 = j10.5(1.0049 + j0.0390) - j10$$

$$= -0.41 + j0.551 \text{ per unit}$$

Hence, the complex power outputs of the transformers are

$$S_{Ta} = -V_2 I_2^{(a)*} = 0.39 + j0.049 \text{ per unit}$$

$$S_{Tb} = -V_2 I_2^{(b)*} = 0.41 + j0.551 \text{ per unit}$$

The results found by the approximate circulating current method of Example 2.13 compare favorably with the above exact solution.

Example 9.8. Solve the phase-shifter problem of Example 2.14 by the exact $\mathbf{Y}_{\text{bus}}$ model of Eq. (9.74) and compare results.

Solution. The bus admittance equation for the phase-shifting transformer T_b with $t = \varepsilon^{j\pi/60} = 1.0\angle 3°$ is given by Eq. (9.74) as

$$\begin{bmatrix} I_1^{(b)} \\ I_2^{(b)} \end{bmatrix} = \begin{bmatrix} -j10|1.0\angle 3°|^2 & 10\angle 87° \\ 10\angle 93° & -j10 \end{bmatrix} \begin{bmatrix} V_1 \\ V_2 \end{bmatrix}$$

which can be added directly to the admittance equation for transformer T_a given in Example 9.7 to obtain

$$\begin{bmatrix} I_1 \\ I_2 \end{bmatrix} = \begin{bmatrix} -j20.0 & 0.5234 + j19.9863 \\ -0.5234 + j19.9863 & -j20.0 \end{bmatrix} \begin{bmatrix} V_1 \\ V_2 \end{bmatrix}$$

Following the procedure of Example 9.7, we have

$$-0.8 + j0.6 = (-0.5234 + j19.9863)V_1 - j20(1.0)$$

which yields the voltage at bus ①,

$$V_1 = \frac{-0.8 + j20.6}{-0.5234 + j19.9863} = 1.031 + j0.013 \text{ per unit}$$

We then determine the currents

$$I_2^{(a)} = j10(V_1 - V_2) = -0.13 + j0.31 \text{ per unit}$$

$$I_2^{(b)} = I_2 - I_2^{(a)} = -0.8 + j0.6 - (-0.13 + j0.31)$$

$$= -0.67 + j0.29 \text{ per unit}$$

and the complex power outputs are

$$S_{Ta} = -V_2 I_2^{(a)*} = 0.13 + j0.31 \text{ per unit}$$

$$S_{Tb} = -V_2 I_2^{(b)*} = 0.67 + j0.29 \text{ per unit}$$

Again, the approximation values of Example 2.14 are close to the actual values of this example.

9.7 THE DECOUPLED POWER-FLOW METHOD

In the strictest use of the Newton-Raphson procedure, the jacobian is calculated and triangularized in *each iteration* in order to update the **LU** factors. In practice, however, the jacobian is often recalculated only every few iterations and this speeds up the overall solution process. The final solution is determined, of course, by the allowable power mismatches and voltage tolerances at the buses.

When solving large-scale power transmission systems, an alternative strategy for improving computational efficiency and reducing computer storage requirements is the *decoupled power-flow method*, which makes use of an approximate version of the Newton-Raphson procedure. The principle underlying the decoupled approach is based on two observations:

- Change in the voltage angle δ at a bus primarily affects the flow of real power P in the transmission lines and leaves the flow of reactive power Q relatively unchanged.
- Change in the voltage magnitude $|V|$ at a bus primarily affects the flow of reactive power Q in the transmission lines and leaves the flow of real power P relatively unchanged.

We have noted both of these effects in Sec. 9.6 when studying the phase shifter and voltage-magnitude regulator. The first observation states essentially that $\partial P_i/\partial \delta_j$ is much larger than $\partial Q_i/\partial \delta_j$, which we now consider to be approximately zero. The second observation states that $\partial Q_i/\partial |V_j|$ is much larger than $\partial P_i/\partial |V_j|$, which is also considered to be approximately zero.

Incorporation of these approximations into the jacobian of Eq. (9.45) makes the elements of the submatrices $\mathbf{J}_{12}$ and $\mathbf{J}_{21}$ zero. We are then left with two separated systems of equations,

$$\begin{bmatrix} \dfrac{\partial P_2}{\partial \delta_2} & \cdots & \dfrac{\partial P_2}{\partial \delta_4} \\ \vdots & \mathbf{J}_{11} & \vdots \\ \dfrac{\partial P_4}{\partial \delta_2} & \cdots & \dfrac{\partial P_4}{\partial \delta_4} \end{bmatrix} \begin{bmatrix} \Delta \delta_2 \\ \vdots \\ \Delta \delta_4 \end{bmatrix} = \begin{bmatrix} \Delta P_2 \\ \vdots \\ \Delta P_4 \end{bmatrix} \tag{9.77}$$

and

$$\begin{bmatrix} |V_2| \dfrac{\partial Q_2}{\partial |V_2|} & \cdots & |V_4| \dfrac{\partial Q_2}{\partial |V_4|} \\ \vdots & \mathbf{J}_{22} & \vdots \\ |V_2| \dfrac{\partial Q_4}{\partial |V_2|} & \cdots & |V_4| \dfrac{\partial Q_4}{\partial |V_4|} \end{bmatrix} \begin{bmatrix} \dfrac{\Delta V_2}{|V_2|} \\ \vdots \\ \dfrac{\Delta V_4}{|V_4|} \end{bmatrix} = \begin{bmatrix} \Delta Q_2 \\ \vdots \\ \Delta Q_4 \end{bmatrix} \tag{9.78}$$

These equations are *decoupled* in the sense that the voltage-angle corrections $\Delta\delta$ are calculated using only real power mismatches ΔP, while the voltage-magnitude corrections are calculated using only ΔQ mismatches. However, the coefficient matrices $\mathbf{J}_{11}$ and $\mathbf{J}_{22}$ are still interdependent because the elements of $\mathbf{J}_{11}$ depend on the voltage magnitudes being solved in Eq. (9.78), whereas the elements of $\mathbf{J}_{22}$ depend on the angles of Eq. (9.77). Of course, the two sets of equations could be solved alternately, using in one set the most recent solutions from the other set. But this scheme would still require evaluation and factoring of the two coefficient matrices at each iteration. To avoid such computations, we introduce further simplifications, which are justified by the physics of transmission-line power flow, as we now explain.

In a well-designed and properly operated power transmission system:

- The angular differences $(\delta_i - \delta_j)$ between typical buses of the system are usually so small that

$$\cos(\delta_i - \delta_j) = 1; \qquad \sin(\delta_i - \delta_j) \approx (\delta_i - \delta_j) \tag{9.79}$$

- The line susceptances B_{ij} are many times larger than the line conductances G_{ij} so that

$$G_{ij}\sin(\delta_i - \delta_j) \ll B_{ij}\cos(\delta_i - \delta_j) \tag{9.80}$$

- The reactive power Q_i injected into any bus ⓘ of the system during normal operation is much less than the reactive power which would flow if all lines from that bus were short-circuited to reference. That is,

$$Q_i \ll |V_i|^2 B_{ii} \tag{9.81}$$

These approximations can be used to simplify the elements of the jacobian. In Eq. (9.62) the off-diagonal elements of $\mathbf{J}_{11}$ and $\mathbf{J}_{22}$ are given by

$$\frac{\partial P_i}{\partial \delta_j} = |V_j|\frac{\partial Q_i}{\partial |V_j|} = -|V_i V_j Y_{ij}|\sin(\theta_{ij} + \delta_j - \delta_i) \tag{9.82}$$

Using the identity $\sin(\alpha + \beta) = \sin\alpha\cos\beta + \cos\alpha\sin\beta$ in Eq. (9.82) gives us

$$\frac{\partial P_i}{\partial \delta_j} = |V_j|\frac{\partial Q_i}{\partial |V_j|} = -|V_i V_j|\{B_{ij}\cos(\delta_j - \delta_i) + G_{ij}\sin(\delta_j - \delta_i)\} \tag{9.83}$$

where $B_{ij} = |Y_{ij}|\sin\theta_{ij}$ and $G_{ij} = |Y_{ij}|\cos\theta_{ij}$. The approximations listed above then yield the off-diagonal elements

$$\frac{\partial P_i}{\partial \delta_j} = |V_j| \frac{\partial Q_i}{\partial |V_j|} \cong -|V_i V_j| B_{ij} \tag{9.84}$$

The diagonal elements of $\mathbf{J}_{11}$ and $\mathbf{J}_{22}$ have the expressions shown in Eqs. (9.54) and (9.63). Applying the inequality $Q_i \ll |V_i|^2 B_{ii}$ to those expressions yields

$$\frac{\partial P_i}{\partial \delta_i} \cong |V_i| \frac{\partial Q_i}{\partial \delta_i} \cong -|V_i|^2 B_{ii} \tag{9.85}$$

By substituting the approximations of Eqs. (9.84) and (9.85) in the coefficient matrices $\mathbf{J}_{11}$ and $\mathbf{J}_{22}$, we obtain

$$\begin{bmatrix} -|V_2V_2|B_{22} & -|V_2V_3|B_{23} & -|V_2V_4|B_{24} \\ -|V_2V_3|B_{32} & -|V_3V_3|B_{33} & -|V_3V_4|B_{34} \\ -|V_2V_4|B_{42} & -|V_3V_4|B_{43} & -|V_4V_4|B_{44} \end{bmatrix} \begin{bmatrix} \Delta\delta_2 \\ \Delta\delta_3 \\ \Delta\delta_4 \end{bmatrix} = \begin{bmatrix} \Delta P_2 \\ \Delta P_3 \\ \Delta P_4 \end{bmatrix} \tag{9.86}$$

and

$$\begin{bmatrix} -|V_2V_2|B_{22} & -|V_2V_3|B_{23} & -|V_2V_4|B_{24} \\ -|V_2V_3|B_{32} & -|V_3V_3|B_{33} & -|V_3V_4|B_{34} \\ -|V_2V_4|B_{42} & -|V_3V_4|B_{43} & -|V_4V_4|B_{44} \end{bmatrix} \begin{bmatrix} \dfrac{\Delta|V_2|}{|V_2|} \\ \dfrac{\Delta|V_3|}{|V_3|} \\ \dfrac{\Delta|V_4|}{|V_4|} \end{bmatrix} = \begin{bmatrix} \Delta Q_2 \\ \Delta Q_3 \\ \Delta Q_4 \end{bmatrix} \tag{9.87}$$

To show how the voltages are removed from the entries in the coefficient matrix of Eq. (9.87), let us multiply the first row by the correction vector and then divide the resultant equation by $|V_2|$ to obtain

$$-B_{22}\,\Delta|V_2| - B_{23}|V_3| - B_{24}|V_4| = \frac{\Delta Q_2}{|V_2|} \tag{9.88}$$

The coefficients in this equation are constants equal to the negative of the susceptances in the row of $\mathbf{Y}_{\text{bus}}$ corresponding to bus ②. Each row of Eq. (9.87) can be similarly treated by representing the reactive mismatch at bus ⓘ by the quantity $\Delta Q_i / |V_i|$. All the entries in the coefficient matrix of Eq. (9.87) become constants given by the known susceptances of $\mathbf{Y}_{\text{bus}}$. We can also modify Eq. (9.86) by multiplying the first row by the vector of angle corrections and rearranging the result to obtain

$$-|V_2|B_{22}\,\Delta\delta_2 - |V_3|B_{23}\,\Delta\delta_3 - |V_4|B_{24}\,\Delta\delta_4 = \frac{\Delta P_2}{|V_2|} \tag{9.89}$$

The coefficients in this equation can be made the same as those in Eq. (9.88) by setting $|V_2|$, $|V_3|$, and $|V_4|$ equal to 1.0 per unit in the left-hand-side expression. Note that the quantity $\Delta P_2/|V_2|$ represents the real-power mismatch in Eq. (9.89). Treating all the rows of Eq. (9.86) in a similar manner leads to two decoupled systems of equations for the four-bus network

$$\underbrace{\begin{bmatrix} -B_{22} & -B_{23} & -B_{24} \\ -B_{32} & -B_{33} & -B_{34} \\ -B_{42} & -B_{43} & -B_{44} \end{bmatrix}}_{\overline{\mathbf{B}}} \begin{bmatrix} \Delta\delta_2 \\ \Delta\delta_3 \\ \Delta\delta_4 \end{bmatrix} = \begin{bmatrix} \dfrac{\Delta P_2}{|V_2|} \\ \dfrac{\Delta P_3}{|V_3|} \\ \dfrac{\Delta P_4}{|V_4|} \end{bmatrix} \tag{9.90}$$

and

$$\underbrace{\begin{bmatrix} -B_{22} & -B_{23} & -B_{24} \\ -B_{32} & -B_{33} & -B_{34} \\ -B_{42} & -B_{43} & -B_{44} \end{bmatrix}}_{\overline{\mathbf{B}}} \begin{bmatrix} \Delta|V_2| \\ \Delta|V_3| \\ \Delta|V_4| \end{bmatrix} = \begin{bmatrix} \dfrac{\Delta Q_2}{|V_2|} \\ \dfrac{\Delta Q_3}{|V_3|} \\ \dfrac{\Delta Q_4}{|V_4|} \end{bmatrix} \tag{9.91}$$

Matrix $\overline{\mathbf{B}}$ is generally symmetrical and sparse with nonzero elements, which are constant, real numbers exactly equal to the *negative* of the susceptances of $\mathbf{Y}_{\text{bus}}$. Consequently, matrix $\overline{\mathbf{B}}$ is easily formed and its triangular factors, once computed at the beginning of the solution, do not have to be recomputed, which leads to very fast iterations. At voltage-controlled buses Q is not specified and $\Delta|V|$ is zero; the rows and columns corresponding to such buses are omitted from Eq. (9.91).

One typical solution strategy is to:

1. Calculate the initial mismatches $\Delta P/|V|$,
2. Solve Eq. (9.90) for $\Delta\delta$,
3. Update the angles δ and use them to calculate mismatches $\Delta Q/|V|$,
4. Solve Eq. (9.91) for $\Delta|V|$ and update the magnitudes $|V|$, and
5. Return to Eq. (9.90) to repeat the iteration until all mismatches are within specified tolerances.

Using this decoupled version of the Newton-Raphson procedure, faster power-flow solutions may be found within a specified degree of precision.

Example 9.9. Using the decoupled form of the Newton-Raphson method, determine the first-iteration solution to the power-flow problem of Example 9.5.

Solution. The $\overline{\mathbf{B}}$ matrix can be read directly from Table 9.4 and the mismatches corresponding to the initial voltage estimates are already calculated in Example 9.5 so that Eq. (9.90) becomes

$$\begin{bmatrix} 44.835953 & 0 & -25.847809 \\ 0 & 40.863838 & -15.118528 \\ -25.847809 & -15.118528 & 40.863838 \end{bmatrix} \begin{bmatrix} \Delta\delta_2 \\ \Delta\delta_3 \\ \Delta\delta_4 \end{bmatrix} = \begin{bmatrix} -1.59661 \\ -1.93953 \\ 2.21286 \end{bmatrix}$$

Solving this equation gives the angle corrections in radians

$$\Delta\delta_2 = -0.02057; \qquad \Delta\delta_3 = -0.03781; \qquad \Delta\delta_4 = 0.02609$$

Adding these results to the flat-start estimates of Table 9.3 gives the updated values of δ_2, δ_3, and δ_4, which we then use along with the elements of $\mathbf{Y}_{\text{bus}}$ to calculate the reactive mismatches

$$\frac{\Delta Q_2}{|V_2|} = \frac{1}{|V_2|}\{Q_{2,\text{sch}} - Q_{2,\text{calc}}\}$$

$$= \frac{1}{|V_2|}\left\{ Q_{2,\text{sch}} - \left[-|V_2|^2 B_{22} - |Y_{12}V_1V_2|\sin(\theta_{12} + \delta_1 - \delta_2) \right.\right. \\ \left.\left. - |Y_{24}V_2V_4|\sin(\theta_{24} + \delta_4 - \delta_2) \right] \right\}$$

$$= \frac{1}{|1.0|}\left\{ \begin{array}{l} -1.0535 + 1.0^2(-44.835953) + 19.455965 \\ \sin(101.30993 \times \pi/180 + 0 + 0.02057) + 26.359695 \\ \times 1.02\sin(101.30993 \times \pi/180 + 0.02609 + 0.02057) \end{array} \right\}$$

$$= -0.80370 \text{ per unit}$$

$$\frac{\Delta Q_3}{|V_3|} = \frac{1}{|V_3|}\{Q_{3,\text{sch}} - Q_{3,\text{calc}}\}$$

$$= \frac{1}{|V_3|}\left\{ Q_{3,\text{sch}} - \left[-|V_3|^2 B_{33} - |Y_{13}V_1V_3|\sin(\theta_{13} + \delta_1 - \delta_3) \right.\right. \\ \left.\left. - |Y_{34}V_3V_4|\sin(\theta_{34} + \delta_4 - \delta_3) \right] \right\}$$

$$= \frac{1}{|1.0|}\left\{ \begin{array}{l} -1.2394 + 1.0^2(-40.863838) + 26.359695 \\ \sin(101.30993 \times \pi/180 + 0 + 0.03781) + 15.417934 \\ \times 1.02\sin(101.30993 \times \pi/180 + 0.02609 + 0.03781) \end{array} \right\}$$

$$= -1.27684 \text{ per unit}$$

A reactive mismatch calculation is not required for bus ④, which is voltage controlled. Accordingly, in this example Eq. (9.91) becomes

$$\begin{bmatrix} 44.835953 & 0 \\ 0 & 40.863838 \end{bmatrix} \begin{bmatrix} \Delta|V_2| \\ \Delta|V_3| \end{bmatrix} = \begin{bmatrix} -0.80370 \\ -1.27684 \end{bmatrix}$$

which yields the solutions $\Delta|V_2| = -0.01793$ and $\Delta|V_3| = -0.03125$. The new voltage magnitudes at buses ② and ③ are $|V_2| = 0.98207$ and $|V_3| = 0.96875$, which completes the first iteration. Updated mismatches for the second iteration of Eq. (9.90) are calculated using the new voltage values. Repeating the procedure over a number of iterations yields the same solution as tabulated in Fig. 9.4.

Often in industry-based programs certain modifications are made in Eqs. (9.90) and (9.91). The modifications to $\overline{\mathbf{B}}$ in Eq. (9.91) are generally as follows:

- Omit the angle-shifting effects of phase shifters from $\overline{\mathbf{B}}$ by setting $t = 1.0\angle 0°$. When rows and columns for voltage-controlled buses are also omitted as previously indicated, the resulting matrix is called $\mathbf{B}''$.

The coefficient matrix in Eq. (9.90) is generally modified as follows:

- Omit from $\overline{\mathbf{B}}$ those elements that mainly affect megavar flows such as shunt capacitors and reactors, and set taps t of off-nominal transformers equal to 1. Also, ignore series resistances in the equivalent-π circuits of the transmission lines in forming $\mathbf{Y}_{\text{bus}}$ from which $\overline{\mathbf{B}}$ in Eq. (9.90) is obtained. The resulting matrix is called $\mathbf{B}'$.

When $\overline{\mathbf{B}}$ in Eq. (9.90) is replaced by B', the model becomes a lossless network. If, in addition, all bus voltages are assumed to remain constant at nominal values of 1.0 per unit, the so-called *dc power-flow model* results. Under these additional assumptions, Eq. (9.91) is not necessary (since $\Delta|V_i| = 0$ at each bus ⓘ) and Eq. (9.90) for the dc power flow becomes

$$\underbrace{\begin{bmatrix} -B_{22} & -B_{23} & -B_{24} \\ -B_{32} & -B_{33} & -B_{34} \\ -B_{42} & -B_{43} & -B_{44} \end{bmatrix}}_{\mathbf{B}'} \begin{bmatrix} \Delta\delta_2 \\ \Delta\delta_3 \\ \Delta\delta_4 \end{bmatrix} = \begin{bmatrix} \Delta P_2 \\ \Delta P_3 \\ \Delta P_4 \end{bmatrix} \qquad (9.92)$$

where it is understood that the elements of $\mathbf{B}'$ are calculated assuming all lines are lossless. Dc power-flow analysis can be used where approximate solutions are acceptable, as in the contingency studies discussed in Chap. 14.

9.8 SUMMARY

This chapter explains the power-flow problem, which is the determination of the voltage magnitude and angle at each bus of the power network under specified conditions of operation. The Gauss-Seidel and Newton-Raphson iterative procedures for solving the power-flow problem are described and numerically exemplified.

In addition to discussing how power-flow studies are undertaken, some methods of real and reactive power-flow control are presented. The results of paralleling two transformers when the voltage-magnitude ratios are different or when one transformer provides a phase shift are examined. Equations are developed for the nodal admittances of these transformers and equivalent circuits which allow analysis of reactive-power control are introduced.

The computer-based power-flow program can be used to study the application of capacitors at a load bus simply by incorporating the shunt admittance of the capacitor into the system $\mathbf{Y}_{\text{bus}}$. Voltage control at a generator bus can also be investigated by specifying the values of the PV bus voltages.

Fast, approximate methods for solving the power-flow problem are introduced by means of the dc power-flow model, which depends on the linkages between real power P and voltage angle δ, and reactive power Q and voltage magnitude.

Table 9.5 summarizes the equations for each of the methods of power-flow analysis.

TABLE 9.5
Summary of power-flow equations and solution methods

Power-flow equations	$P_i = \sum_{n=1}^{N} \lvert Y_{in}V_iV_n \rvert \cos(\theta_{in} + \delta_n - \delta_i) \qquad Q_i = -\sum_{n=1}^{N} \lvert Y_{in}V_iV_n \rvert \sin(\theta_{in} + \delta_n - \delta_i)$ $\Delta P_i = P_{i,\text{sch}} - P_{i,\text{calc}} \qquad \Delta Q_i = Q_{i,\text{sch}} - Q_{i,\text{calc}}$
Gauss-Seidel method	To obtain V at a bus i with known P and Q: $V_i^{(k)} = \frac{1}{Y_{ii}}\left[\frac{P_{i,\text{sch}} - jQ_{i,\text{sch}}}{V_i^{(k-1)*}} - \sum_{j=1}^{i-1} Y_{ij}V_j^{(k)} - \sum_{j=i+1}^{N} Y_{ij}V_j^{(k-1)}\right]$ To obtain Q at a regulated bus i: $Q_i^{(k)} = -\text{Im}\left\{V_i^{(k-1)*}\left[\sum_{j=1}^{i-1} Y_{ij}V_j^{(k)} + \sum_{j=i}^{N} Y_{ij}V_j^{(k-1)}\right]\right\}$ Use of acceleration factor α at bus i in iteration k. $V_{i,\text{acc}}^{(k)} = (1-\alpha)V_{i,\text{acc}}^{(k-1)} + \alpha V_i^{(k)} = V_{i,\text{acc}}^{(k-1)} + \alpha(V_i^{(k)} - V_{i,\text{acc}}^{(k-1)})$

TABLE 9.5 (*Continued*)

Newton-Raphson method	$\left[\begin{array}{ccc\|ccc} \frac{\partial P_2}{\partial \delta_2} & \cdots & \frac{\partial P_2}{\partial \delta_N} & \|V_2\|\frac{\partial P_2}{\partial \|V_2\|} & \cdots & \|V_N\|\frac{\partial P_2}{\partial \|V_N\|} \\ \vdots & \mathbf{J}_{11} & \vdots & \vdots & \mathbf{J}_{12} & \vdots \\ \frac{\partial P_N}{\partial \delta_2} & \cdots & \frac{\partial P_N}{\partial \delta_N} & \|V_2\|\frac{\partial P_N}{\partial \|V_2\|} & \cdots & \|V_N\|\frac{\partial P_N}{\partial \|V_N\|} \\ \hline \frac{\partial Q_2}{\partial \delta_2} & \cdots & \frac{\partial Q_2}{\partial \delta_N} & \|V_2\|\frac{\partial Q_2}{\partial \|V_2\|} & \cdots & \|V_N\|\frac{\partial Q_2}{\partial \|V_N\|} \\ \vdots & \mathbf{J}_{21} & \vdots & \vdots & \mathbf{J}_{22} & \vdots \\ \frac{\partial Q_N}{\partial \delta_2} & \cdots & \frac{\partial Q_N}{\partial \delta_N} & \|V_2\|\frac{\partial Q_N}{\partial \|V_2\|} & \cdots & \|V_N\|\frac{\partial Q_N}{\partial \|V_N\|} \end{array}\right] \left[\begin{array}{c} \Delta\delta_2 \\ \vdots \\ \Delta\delta_N \\ \hline \frac{\Delta\|V_2\|}{\|V_2\|} \\ \vdots \\ \frac{\Delta\|V_N\|}{\|V_N\|} \end{array}\right] = \left[\begin{array}{c} \Delta P_2 \\ \vdots \\ \Delta P_N \\ \hline \Delta Q_2 \\ \vdots \\ \Delta Q_N \end{array}\right]$ $\frac{\partial P_i}{\partial \delta_j} = -\|V_i V_j Y_{ij}\|\sin(\theta_{ij} + \delta_j - \delta_i)$ $\qquad \frac{\partial P_i}{\partial \delta_i} = -\sum_{\substack{n=1 \\ n \neq i}}^{N} \frac{\partial P_i}{\partial \delta_n} = -Q_i - \|V_i\|^2 B_{ii}$ $\frac{\partial Q_i}{\partial \delta_j} = -\|V_i V_j Y_{ij}\|\cos(\theta_{ij} + \delta_j - \delta_i)$ $\qquad \frac{\partial Q_i}{\partial \delta_i} = -\sum_{\substack{n=1 \\ n \neq i}}^{N} \frac{\partial Q_i}{\partial \delta_n} = P_i - \|V_i\|^2 G_{ii}$ $\|V_j\|\frac{\partial P_i}{\partial \|V_j\|} = -\frac{\partial Q_i}{\partial \delta_j}$ $\qquad \|V_i\|\frac{\partial P_i}{\partial \|V_i\|} = +\frac{\partial Q_i}{\partial \delta_i} + 2\|V_i\|^2 G_{ii}$ $\|V_j\|\frac{\partial Q_i}{\partial \|V_j\|} = +\frac{\partial P_i}{\partial \delta_j}$ $\qquad \|V_i\|\frac{\partial Q_i}{\partial \|V_i\|} = -\frac{\partial P_i}{\partial \delta_j} - 2\|V_i\|^2 B_{ii}$ State variable update formulas: $\delta_i^{(k+1)} = \delta_i^{(k)} + \Delta\delta_i^{(k)}$ $\qquad \|V_i\|^{(k+1)} = \|V_i\|^{(k)} + \Delta\|V_i\|^{(k)} = \|V_i\|^{(k)}\left(1 + \frac{\Delta\|V_i\|^{(k)}}{\|V_i\|^{(k)}}\right)$
Decoupledd power-flow solution technique	$\begin{bmatrix} -B_{22} & -B_{23} & \cdots & -B_{2N} \\ -B_{32} & -B_{33} & \cdots & -B_{3N} \\ \vdots & \vdots & & \vdots \\ -B_{N2} & -B_{N3} & \cdots & -B_{NN} \end{bmatrix} \begin{bmatrix} \Delta\delta_2 \\ \Delta\delta_3 \\ \vdots \\ \Delta\delta_N \end{bmatrix} = \begin{bmatrix} \Delta P_2/\|V_2\| \\ \Delta P_3/\|V_3\| \\ \vdots \\ \Delta P_N/\|V_N\| \end{bmatrix}$ $\begin{bmatrix} -B_{22} & -B_{23} & \cdots & -B_{2N} \\ -B_{32} & -B_{33} & \cdots & -B_{3N} \\ \vdots & \vdots & & \vdots \\ -B_{N2} & -B_{N3} & \cdots & -B_{NN} \end{bmatrix} \begin{bmatrix} \Delta\|V_2\| \\ \Delta\|V_3\| \\ \vdots \\ \Delta\|V_N\| \end{bmatrix} = \begin{bmatrix} \Delta Q_2/\|V_2\| \\ \Delta Q_3/\|V_3\| \\ \vdots \\ \Delta Q_N/\|V_N\| \end{bmatrix}$ B_{ij} are the imaginary parts of the corresponding $\mathbf{Y}_{\text{bus}}$ elements. Voltage-controlled buses are not represented in $\Delta\|V\|$ equations.

PROBLEMS

9.1. In Example 9.3 suppose that the generator's maximum reactive power generation at bus ④ is limited to 125 Mvar. Recompute the first-iteration value of the voltage at bus ④ using the Gauss-Seidel method.

9.2. For the system of Fig. 9.2, complete the second iteration of the Gauss-Seidel procedure using the first-iteration bus voltages obtained in Examples 9.2 and 9.3. Assume an acceleration factor of 1.6.

9.3. A synchronous condenser, whose reactive power capability is assumed to be unlimited, is installed at load bus ② of the system of Example 9.2 to hold the bus-voltage magnitude at 0.99 per unit. Using the Gauss-Seidel method, find the voltages at buses ② and ③ for the first iteration.

9.4. Take Fig. 9.12 as the equivalent-π representation of the transmission line between bus ③ and bus ④ of the system of Fig. 9.2. Using the power-flow solution given in Fig. 9.4, determine and indicate on Fig. 9.12 the values of (*a*) P and Q leaving buses ③ and ④ on line ③- ④, (*b*) charging megavars of the equivalent π of line ③- ④, and (*c*) P and Q at both ends of the series part of the equivalent π of line ③- ④.

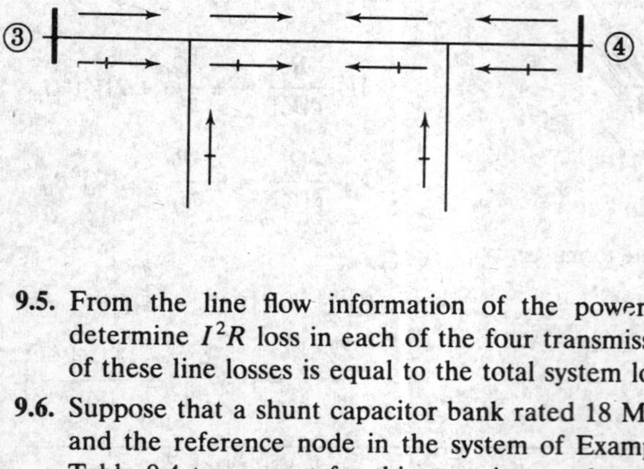

FIGURE 9.12
Diagram for Prob. 9.4.

9.5. From the line flow information of the power-flow solution given in Fig. 9.4, determine I^2R loss in each of the four transmission lines, and verify that the sum of these line losses is equal to the total system loss of 4.81 MW.

9.6. Suppose that a shunt capacitor bank rated 18 Mvar is connected between bus ③ and the reference node in the system of Example 9.5. Modify the $\mathbf{Y}_{\text{bus}}$ given in Table 9.4 to account for this capacitor, and *estimate* the actual megavar reactive power injected into the system from the capacitor.

9.7. For the system of Example 9.5 augmented with a synchronous condenser as described in Prob. 9.3, find the jacobian calculated at the initial estimates. *Hint*: It would be simpler to modify the jacobian matrix shown in Sec. 9.4 following Example 9.5 than to start calculations from the beginning.

9.8. Suppose that in Fig. 9.7 the tap is on the side of node ⓘ so that the transformation ratio is $t:1$. Find elements of $\mathbf{Y}_{\text{bus}}$ similar to those in Eq. (9.74), and draw the equivalent-π representation similar to Fig. 9.8.

9.9. In the four-bus system of Example 9.5 suppose that a magnitude-regulating transformer with 0.2 per-unit reactance is inserted between the load and the bus at bus ③, as shown in Fig. 9.10. The variable tap is on the load side of the transformer. If the voltage magnitude at the new load bus ⑤ is prespecified, and therefore is not a state variable, the variable tap t of the transformer should be regarded as a state variable. The Newton-Raphson method is to be applied to the solution of the power-flow equations.

(*a*) Write mismatch equations for this problem in a symbolic form similar to Eq. (9.45).

(*b*) Write equations of the jacobian elements of the column corresponding to variable t (that is, partial derivatives with respect to t), and evaluate them using the initial voltage estimates shown in Table 9.3 and assuming that the voltage magnitude at bus ⑤ is specified to be 0.97. The initial estimate of δ_5 is 0.

(*c*) Write equations of P and Q mismatches at bus ⑤ and evaluate them for the first iteration. Assume the initial estimate of variable t is 1.0.

9.10. If the tap setting of the transformer of Prob. 9.9 is prespecified instead of the voltage magnitude at bus ⑤, then V_5 should be regarded as a state variable. Suppose that the tap setting t is specified to be 1.05.

(*a*) In this case write mismatch equations in a symbolic form similar to Eq. (9.45).

(*b*) Write equations of the jacobian elements which are partial derivatives with respect to $|V_5|$, and evaluate them using the initial estimates. The initial estimate of V_5 is $1.0\angle 0°$.

(*c*) Write equations for the P and Q mismatches at bus ⑤, and evaluate them for the first iteration.

9.11. Redo Example 9.8 for $t = 1.0\angle -3°$, and compare the two results as to changes in real and reactive power flows.

9.12. The generator at bus ④ of the system of Example 9.5 is to be represented by a generator connected to bus ④ through a generator step-up transformer, as shown in Fig. 9.13. The reactance of this transformer is 0.02 per unit; the tap is on the high-voltage side of the transformer with the off-nominal turns ratio of 1.05. Evaluate the jacobian elements of the rows corresponding to buses ④ and ⑤.

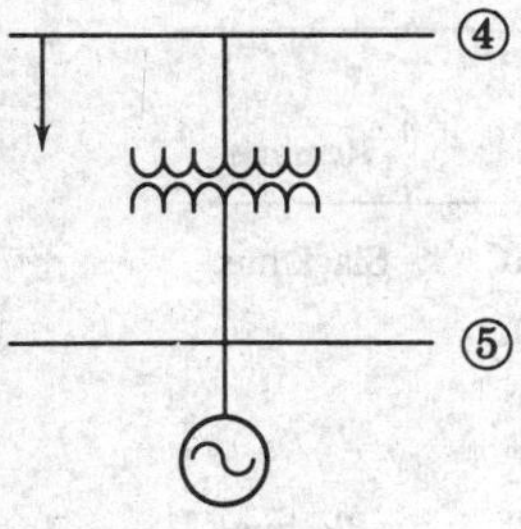

FIGURE 9.13
Generator step-up transformer for Prob. 9.12.

9.13. For the system of Prob. 9.12, find matrices **B**′ and **B**″ for use in the decoupled power-flow method.

9.14. A five-bus power system is shown in Fig. 9.14. The line, bus, transformer, and capacitor data are given in Tables 9.6, 9.7, 9.8, and 9.9, respectively. Use the Gauss-Seidel method to find the bus voltages for the first iteration.

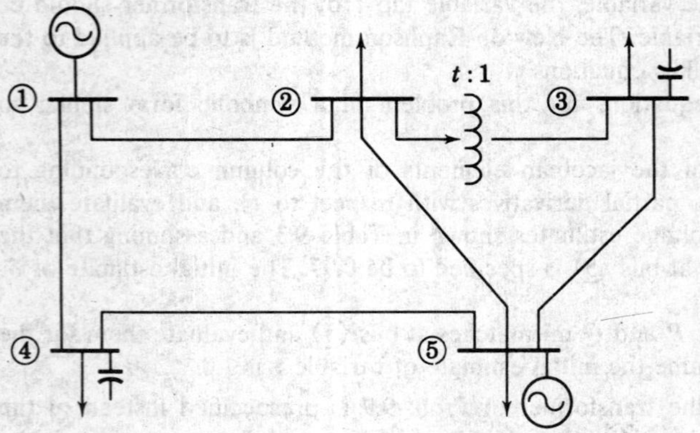

FIGURE 9.14
System for Probs. 9.14 through 9.18. The line and bus data are given in Tables 9.6 through 9.9.

TABLE 9.6
Line data for the system of Figure 9.14

Line bus to bus		Per-unit Series Z R	X	Per-unit Series Y G	B	Charging Mvar
①	②	0.0108	0.0649	2.5	−15	6.6
①	④	0.0235	0.0941	2.5	−10	4.0
②	⑤	0.0118	0.0471	5.0	−20	7.0
③	⑤	0.0147	0.0588	4.0	−16	8.0
④	⑤	0.0118	0.0529	4.0	−18	6.0

TABLE 9.7
Bus data for the system of Figure 9.14

Bus	Generation P (MW)	Generation Q (Mvar)	Load P (MW)	Load Q (Mvar)	V p.u.	Remark
①					$1.01\angle 0°$	Slack bus
②			60	35	$1.0\angle 0°$	
③			70	42	$1.0\angle 0°$	
④			80	50	$1.0\angle 0°$	
⑤	190		65	36	$1.0\angle 0°$	*PV* bus

TABLE 9.8
Transformer data for the system of Figure 9.14

Transformer bus to bus	Per-unit reactance	Tap setting
② - ③	0.04	0.975

TABLE 9.9
Capacitor data for the system of Figure 9.14

Bus	Rating in Mvar
③	18
④	15

9.15. To apply the Newton-Raphson method to the power-flow solution of the system of Fig. 9.14, determine (*a*) $\mathbf{Y}_{bus}$ of the system, (*b*) the mismatch equation at bus ⑤ evaluated at the initial voltage estimates of Table 9.7 for the first iteration, and (*c*) write mismatch equations in a form similar to Eq. (9.45).

9.16. For the system of Fig. 9.14, find matrices $\mathbf{B}'$ and $\mathbf{B}''$ for use in the decoupled power-flow method. Also, determine the first-iteration P and Q mismatch equations at bus ④, and find the voltage magnitude at bus ④ at the end of the first iteration.

9.17. Suppose that in Fig. 9.14 the transformer between buses ② and ③ is a phase shifter where t is now a complex variable and is $1.0\angle -2°$. (*a*) Find $\mathbf{Y}_{bus}$ of this system. (*b*) When compared with the power-flow solution of Prob. 9.15, will the real power flow in the line from bus ⑤ to bus ③ increase or decrease? What about the reactive power flow? Explain why qualitatively.

9.18. To apply the decoupled power-flow method to the system of Prob. 9.17, find matrices $\mathbf{B}'$ and $\mathbf{B}''$.

9.19. Redo Example 9.10 when an 18-Mvar shunt capacitor bank is added to bus ③.

9.20. In applying the Newton-Raphson method, if the amount of reactive power required to maintain the specified voltage at a *PV* bus exceeds the maximum limit of its reactive power generation capability, the reactive power at that bus is set to that limit and the type of bus becomes a load bus. Suppose the maximum reactive power generation at bus ④ is limited to 150 Mvar in the system of Example 9.5. Using the first-iteration result given in Sec. 9.4 following Example 9.5, determine whether or not the type of bus ④ should be converted to a load bus at the start of the second iteration. If so, calculate the reactive power mismatch at bus ④ that should be used in the second-iteration mismatch equation.

CHAPTER 10

SYMMETRICAL FAULTS

A fault in a circuit is any failure which interferes with the normal flow of current. Most faults on transmission lines of 115 kV and higher are caused by lightning, which results in the flashover of insulators. The high voltage between a conductor and the grounded supporting tower causes ionization, which provides a path to ground for the charge induced by the lightning stroke. Once the ionized path to ground is established, the resultant low impedance to ground allows the flow of power current from the conductor to ground and through the ground to the grounded neutral of a transformer or generator, thus completing the circuit. Line-to-line faults not involving ground are less common. Opening circuit breakers to isolate the faulted portion of the line from the rest of the system interrupts the flow of current in the ionized path and allows deionization to take place. After an interval of about 20 cycles to allow deionization, breakers can usually be reclosed without reestablishing the arc. Experience in the operation of transmission lines has shown that ultra-high-speed reclosing breakers successfully reclose after most faults. Of those cases where reclosure is not successful, many are caused by *permanent* faults where reclosure would be impossible regardless of the interval between opening and reclosing. Permanent faults are caused by lines being on the ground, by insulator strings breaking because of ice loads, by permanent damage to towers, and by surge-arrester failures. Experience has shown that between 70 and 80% of transmission-line faults are *single line-to-ground faults*, which arise from the flashover of only one line to the tower and ground. Roughly 5% of all faults involve all three phases. These are the so-called *symmetrical three-phase faults* which are considered in this chapter. Other types of transmission-line faults are *line-to-line faults*, which do not involve ground, and *double line-to-ground faults*. All the above faults except the three-phase type cause an imbalance between the phases, and so they are called *unsymmetrical faults*. These are considered in Chap. 12.

The currents which flow in different parts of a power system immediately after the occurrence of a fault differ from those flowing a few cycles later just before circuit breakers are called upon to open the line on both sides of the fault. And all of these currents differ widely from the currents which would flow under steady-state conditions if the fault were not isolated from the rest of the system by the operation of circuit breakers. Two of the factors on which the proper selection of circuit breakers depends are the current flowing immediately after the fault occurs and the current which the breaker must interrupt. In *fault analysis* values of these currents are calculated for the different types of faults at various locations in the system. The data obtained from fault calculations also serve to determine the settings of relays which control the circuit breakers.

10.1 TRANSIENTS IN *RL* SERIES CIRCUITS

The selection of a circuit breaker for a power system depends not only on the current the breaker is to carry under normal operating conditions, but also on the maximum current it may have to carry *momentarily* and the current it may have to *interrupt* at the voltage of the line in which it is placed.

In order to approach the problem of calculating the initial current when a system is short-circuited, consider what happens when an ac voltage is applied to a circuit containing constant values of resistance and inductance. Let the applied voltage be $V_{\text{max}} \sin(\omega t + \alpha)$, where t is zero at the time of applying the voltage. Then, α determines the magnitude of the voltage when the circuit is closed. If the instantaneous voltage is zero and increasing in a positive direction when it is applied by closing a switch, α is zero. If the voltage is at its positive maximum instantaneous value, α is $\pi/2$. The differential equation is

$$V_{\text{max}} \sin(\omega t + \alpha) = Ri + L\frac{di}{dt} \tag{10.1}$$

The solution of this equation is

$$i = \frac{V_{\text{max}}}{|Z|}\left[\sin(\omega t + \alpha - \theta) - \epsilon^{-Rt/L}\sin(\alpha - \theta)\right] \tag{10.2}$$

where $|Z| = \sqrt{R^2 + (\omega L)^2}$ and $\theta = \tan^{-1}(\omega L/R)$.

The first term of Eq. (10.2) varies sinusoidally with time. The second term is nonperiodic and decays exponentially with a time constant L/R. This nonperiodic term is called the *dc component* of the current. We recognize the sinusoidal term as the steady-state value of the current in an *RL* circuit for the given applied voltage. If the value of the steady-state term is not zero when $t = 0$, the dc component appears in the solution in order to satisfy the physical

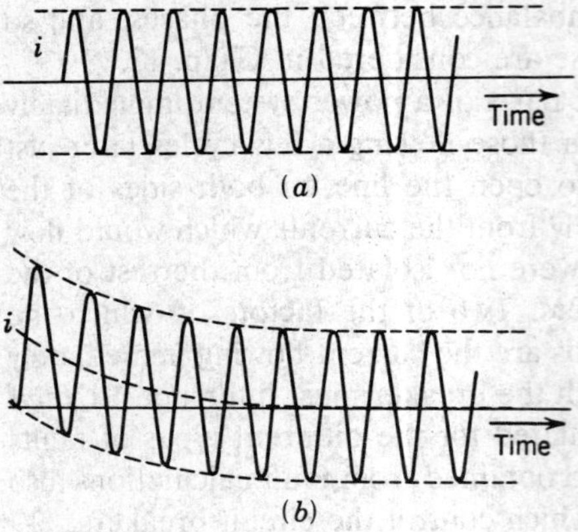

FIGURE 10.1
Current as a function of time in an *RL* circuit for: (*a*) $\alpha - \theta = 0$; (*b*) $\alpha - \theta = -\pi/2$, where $\theta = \tan^{-1}(\omega L/R)$. The voltage is $V_{max}\sin(\omega t + \alpha)$ applied at $t = 0$.

condition of zero current at the instant of closing the switch. Note that the dc term does not exist if the circuit is closed at a point on the voltage wave such that $\alpha - \theta = 0$ or $\alpha - \theta = \pi$. Figure 10.1(*a*) shows the variation of current with time according to Eq. (10.2) when $\alpha - \theta = 0$. If the switch is closed at a point on the voltage wave such that $\alpha - \theta = \pm\pi/2$, the dc component has its maximum initial value, which is equal to the maximum value of the sinusoidal component. Figure 10.1(*b*) shows current versus time when $\alpha - \theta = -\pi/2$. The dc component may have any value from 0 to $V_{max}/|Z|$, depending on the instantaneous value of the voltage when the circuit is closed and on the power factor of the circuit. At the instant of applying the voltage the dc and steady-state components always have the same magnitude but are opposite in sign in order to express the zero value of current then existing.

In Chap. 3 we discussed the principles of operation of a synchronous generator consisting of a rotating magnetic field which generates a voltage in an armature winding having resistance and reactance. The current flowing when a generator is short-circuited is similar to that flowing when an alternating voltage is suddenly applied to a resistance and an inductance in series. There are important differences, however, because the currents in the damper windings and the armature affect the rotating field, as discussed in Secs. 3.8 and 3.9. If the dc component of current is eliminated from the short-circuit current of each armature phase, the resulting plot of each phase current versus time is that shown in Fig. 3.19. Comparison of Figs. 3.19 and 10.1(*a*) shows the difference between applying a voltage to the ordinary *RL* circuit and applying a short circuit to a synchronous machine. There is no dc component in either of these figures, yet the current envelopes are quite different. In a synchronous machine the flux across the air gap is not the same at the instant the short circuit occurs as it is a few cycles later. The change of flux is determined by the combined action of the field, the armature, and the damper windings or iron parts of the round rotor. After a fault occurs, the *subtransient*, *transient*, and *steady-state* periods are characterized by the subtransient reactance X_d'', the transient reactance X_d', and the steady-state reactance X_d, respectively. These reactances have increasing values (that is, $X_d'' < X_d' < X_d$) and the corresponding components of the short-circuit current have decreasing magnitudes ($|I''| > |I'| > |I|$). With the dc component removed, the *initial symmetrical rms current* is the rms value of the ac component of the fault current immediately after the fault occurs.

In analytical work the internal voltage to the machine and the subtransient, transient, and steady-state currents may be expressed as phasors. The voltage induced in the armature windings just after the fault occurs differs from that which exists after steady state is reached. We account for the differences in induced voltage by using the different reactances (X_d'', X_d', and X_d) in series with the internal voltage to calculate currents for subtransient, transient, and steady-state conditions. If a generator is unloaded when the fault occurs, the machine is represented by the no-load voltage to neutral in series with the proper reactance, as shown in Fig. 3.20. The resistance is taken into account if greater accuracy is desired. If there is impedance external to the generator between its terminals and the short circuit, the external impedance must be included in the circuit. We shall examine the transients for machines carrying a load in the next section.

Although machine reactances are not true constants of the machine and depend on the degree of saturation of the magnetic circuit, their values usually lie within certain limits and can be predicted for various types of machines. Table A.2 in the Appendix gives typical values of machine reactances that are needed in making fault calculations and in stability studies. In general, subtransient reactances of generators and motors are used to determine the initial current flowing on the occurrence of a short circuit. For determining the interrupting capacity of circuit breakers, except those which open instantaneously, subtransient reactance is used for generators and transient reactance is used for synchronous motors. In stability studies, where the problem is to determine whether a fault will cause a machine to lose synchronism with the rest of the system if the fault is removed after a certain time interval, transient reactances apply.

10.2 INTERNAL VOLTAGES OF LOADED MACHINES UNDER FAULT CONDITIONS

Let us consider a generator that is loaded when a fault occurs. Figure 10.2(*a*) is the equivalent circuit of a generator that has a balanced three-phase load. Internal voltages and reactances of the generator are now identified by the subscript g since some of the circuits to be considered also have motors. External impedance is shown between the generator terminals and the point P where the fault occurs. The current flowing before the fault occurs at point P is I_L, the voltage at the fault is V_f, and the terminal voltage of the generator is V_t. The steady-state equivalent circuit of the synchronous generator is its no-load voltage E_g in series with its synchronous reactance X_{dg}. If a three-phase fault

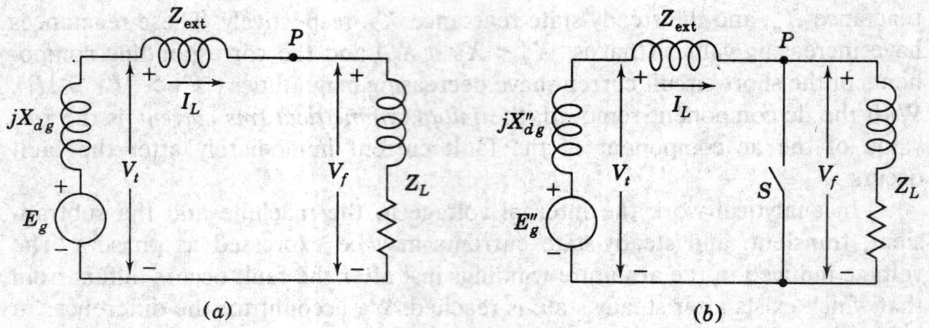

FIGURE 10.2
Equivalent circuit for a generator supplying a balanced three-phase load. Application of a three-phase fault at P is simulated by closing switch S: (a) usual steady-state generator equivalent circuit with load; (b) circuit for calculation of I''.

occurs at point P, we see that a short circuit from P to neutral in the equivalent circuit does not satisfy the conditions for calculating subtransient current, for the reactance of the generator must be X''_{dg} if we are calculating subtransient current I'' or X'_{dg} if we are calculating transient current I'.

The circuit shown in Fig. 10.2(b) gives us the desired result. Here a voltage E''_g in series with X''_{dg} supplies the steady-state current I_L when switch S is open and supplies the current to the short circuit through X''_{dg} and Z_{ext} when switch S is closed. If we can determine E''_g, the current through X''_{dg} will be I''. With switch S open, we see that

$$E''_g = V_t + jX''_{dg}I_L = V_f + (Z_{\text{ext}} + jX''_{dg})I_L \tag{10.3}$$

and this equation *defines* E''_g, which is called the *subtransient internal voltage*. Similarly, when calculating transient current I', which must be supplied through the transient reactance X'_{dg}, the driving voltage is the *transient internal voltage* E'_g, where

$$E'_g = V_t + jX'_{dg}I_L = V_f + (Z_{\text{ext}} + jX'_{dg})I_L \tag{10.4}$$

Thus, the value of the load current I_L determines the values of the voltages E''_g and E'_g, which are both equal to the no-load voltage E_g only when I_L is zero so that E_g is then equal to V_t.

At this point it is important to note that the particular value of E''_g in series with X''_{dg} represents the generator immediately before and immediately after the fault occurs only if the prefault current in the generator has the corresponding value of I_L. On the other hand, E_g in series with the synchronous reactance X_{dg} is the equivalent circuit of the machine under steady-state conditions for any value of the load current. The magnitude of E_g is determined by the field current of the machine, and so for a different value of

I_L in the circuit of Fig. 10.2(a) $|E_g|$ would remain the same but a new value of E_g'' would be required.

Synchronous motors have reactances of the same type as generators. When a motor is short-circuited, it no longer receives electric energy from the power line, but its field remains energized and the inertia of its rotor and connected load keeps it rotating for a short period of time. The internal voltage of a synchronous motor causes it to contribute current to the system, for it is then acting like a generator. By comparison with the corresponding formulas for a generator the subtransient internal voltage E_m'' and transient internal voltage E_m' for a synchronous motor are given by

$$E_m'' = V_t - jX_{dm}''I_L \tag{10.5}$$

$$E_m' = V_t - jX_{dm}'I_L \tag{10.6}$$

where V_t is now the terminal voltage of the motor. Fault currents in systems containing generators and motors under load may be solved in either one of two ways: (1) by calculating the subtransient (or transient) internal voltages of the machines or (2) by using Thévenin's theorem. A simple example will illustrate the two approaches.

Suppose that a synchronous generator is connected to a synchronous motor by a line of external impedance Z_{ext}. The motor is drawing load current I_L from the generator when a symmetrical three-phase fault occurs at the motor terminals. Figure 10.3 shows the equivalent circuits and current flows of the system immediately before and immediately after the fault occurs. By replacing the *synchronous* reactances of the machines by their *subtransient* reactances as shown in Fig. 10.3(a), we can calculate the subtransient internal voltages of the machine immediately before the fault occurs by substituting the values of V_f

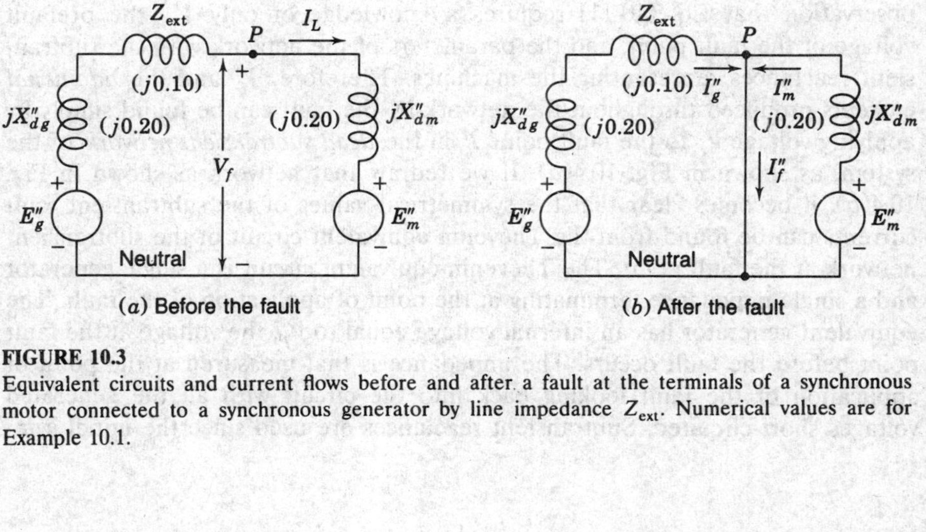

FIGURE 10.3
Equivalent circuits and current flows before and after a fault at the terminals of a synchronous motor connected to a synchronous generator by line impedance Z_{ext}. Numerical values are for Example 10.1.

and I_L in the equations

$$E_g'' = V_f + (Z_{\text{ext}} + jX_{dg}'')I_L \tag{10.7}$$

$$E_m'' = V_f - jX_{dm}''I_L \tag{10.8}$$

When the fault is on the system, as shown in Fig. 10.3(b), the subtransient currents I_g'' out of the generator and I_m'' out of the motor are found from the relations

$$I_g'' = \frac{E_g''}{Z_{\text{ext}} + jX_{dg}''} = \frac{V_f}{Z_{\text{ext}} + jX_{dg}''} + I_L \tag{10.9}$$

$$I_m'' = \frac{E_m''}{jX_{dm}''} = \frac{V_f}{jX_{dm}''} - I_L \tag{10.10}$$

These two currents add together to give the total symmetrical fault current I_f'' shown in Fig. 10.3(b). That is,

$$I_f'' = I_g'' + I_m'' = \underbrace{\frac{V_f}{Z_{\text{ext}} + jX_{dg}''}}_{I_{gf}''} + \underbrace{\frac{V_f}{jX_{dm}''}}_{I_{mf}''} \tag{10.11}$$

where I_{gf}'' and I_{mf}'' are the respective contributions of the generator and motor to the fault current I_f''. Note that the fault current does not include the prefault (load) current.

The alternative approach using Thévenin's theorem is based on the observation that Eq. (10.11) requires a knowledge of only V_f, the prefault voltage of the fault point, and the parameters of the network with the subtransient reactances representing the machines. Therefore, I_f'' and the *additional* currents produced throughout the network by the fault can be found simply by applying voltage V_f to the fault point P in the *dead subtransient network* of the system, as shown in Fig. 10.4(a). If we redraw that network as shown in Fig. 10.4(b), it becomes clear that the symmetrical values of the subtransient fault currents can be found from the Thévenin equivalent circuit of the subtransient network at the fault point. The Thévenin equivalent circuit is a single generator and a single impedance terminating at the point of application of the fault. The equivalent generator has an internal voltage equal to V_f, the voltage at the fault point before the fault occurs. The impedance is that measured at the point of application of the fault looking back into the circuit with all the generated voltages short-circuited. Subtransient reactances are used since the initial sym-

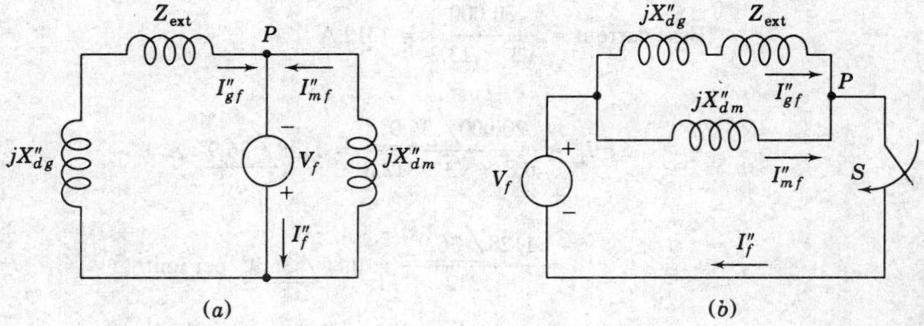

FIGURE 10.4
Circuits illustrating the additional current flows due to the three-phase fault at P: (a) applying V_f to dead network to simulate the fault; (b) Thévenin equivalent looking into the circuit at point P.

metrical fault current is desired. In Fig. 10.4(b) the Thévenin impedance Z_{th} is

$$Z_{th} = \frac{jX''_{dm}(Z_{ext} + jX''_{dg})}{Z_{ext} + j(X''_{dg} + X''_{dm})} \tag{10.12}$$

Upon the occurrence of a three-phase short circuit at P, simulated by closing switch S, the subtransient current in the fault is

$$I''_f = \frac{V_f}{Z_{th}} = \frac{V_f\left[Z_{ext} + j(X''_{dg} + X''_{dm})\right]}{jX''_{dm}(Z_{ext} + jX''_{dg})} \tag{10.13}$$

Thus, three-phase symmetrical faults on systems containing generators and motors under load may be analyzed by either the use of subtransient internal voltages or Thévenin's theorem, as illustrated in the following examples.

Example 10.1. A synchronous generator and motor are rated 30,000 kVA, 13.2 kV, and both have subtransient reactances of 20%. The line connecting them has a reactance of 10% on the base of the machine ratings. The motor is drawing 20,000 kW at 0.8 power-factor leading and a terminal voltage of 12.8 kV when a symmetrical three-phase fault occurs at the motor terminals. Find the subtransient currents in the generator, the motor, and the fault by using the internal voltages of the machines.

Solution. The prefault equivalent circuit of the system corresponds to Fig. 10.3(a). Choosing a base of 30,000 kVA, 13.2 kV and using the voltage V_f at the fault point as the reference phasor, we obtain

$$V_f = \frac{12.8}{13.2} = 0.970\angle 0^\circ \text{ per unit}$$

$$\text{Base current} = \frac{30{,}000}{\sqrt{3} \times 13.2} = 1312 \text{ A}$$

$$I_L = \frac{20{,}000\angle 36.9^\circ}{0.8 \times \sqrt{3} \times 12.8} = 1128\angle 36.9^\circ \text{ A}$$

$$= \frac{1128\angle 36.9^\circ}{1312} = 0.86\angle 36.9^\circ \text{ per unit}$$

$$= 0.86(0.8 + j0.6) = 0.69 + j0.52 \text{ per unit}$$

For the generator

$$V_t = 0.970 + j0.1(0.69 + j0.52) = 0.918 + j0.069 \text{ per unit}$$

$$E''_g = 0.918 + j0.069 + j0.2(0.69 + j0.52) = 0.814 + j0.207 \text{ per unit}$$

$$I''_g = \frac{0.814 + j0.207}{j0.3} = 0.69 - j2.71 \text{ per unit}$$

$$= 1312(0.69 - j2.71) = 905 - j3550 \text{ A}$$

For the motor

$$V_t = V_f = 0.970\angle 0^\circ \text{ per unit}$$

$$E''_m = 0.970 + j0 - j0.2(0.69 + j0.52) = 0.970 - j0.138 + 0.104 \text{ per unit}$$

$$= 1.074 - j0.138 \text{ per unit}$$

$$I''_m = \frac{1.074 - j0.138}{j0.2} = -0.69 - j5.37 \text{ per unit}$$

$$= 1312(-0.69 - j5.37) = -905 - j7050 \text{ A}$$

In the fault

$$I''_f = I''_g + I''_m = 0.69 - j2.71 - 0.69 - j5.37 = -j8.08 \text{ per unit}$$

$$= -j8.08 \times 1312 = -j10{,}600 \text{ A}$$

Figure 10.3(b) shows the paths of I''_g, I''_m, and I''_f

Example 10.2. Solve Example 10.1 by the use of Thévenin's theorem.

Solution. The Thévenin equivalent circuit corresponds to Fig. 10.4.

$$Z_{\text{th}} = \frac{j0.3 \times j0.2}{j0.3 + j0.2} = j0.12 \text{ per unit}$$

$$V_f = 0.970\angle 0^\circ \text{ per unit}$$

In the fault

$$I_f'' = \frac{V_f}{Z_{\text{th}}} = \frac{0.97 + j0}{j0.12} = -j8.08 \text{ per unit}$$

This fault current is divided between the parallel circuits of the machines inversely as their impedances. By simple current division we obtain the fault currents

$$\text{From generator:} \quad I_{gf}'' = -j8.08 \times \frac{j0.2}{j0.5} = -j3.23 \text{ per unit}$$

$$\text{From motor:} \quad I_{mf}'' = -j8.08 \times \frac{j0.3}{j0.5} = -j4.85 \text{ per unit}$$

Neglecting load current gives

$$\text{Fault current from generator} = 3.23 \times 1312 = 4240 \text{ A}$$

$$\text{Fault current from motor} = 4.85 \times 1312 = 6360 \text{ A}$$

$$\text{Current in fault} = 8.08 \times 1312 = 10{,}600 \text{ A}$$

The current in the fault is the same whether or not load current is considered, but the currents in the lines differ. When load current I_L is included, we find from Example 10.1 that

$$I_g'' = I_{gf}'' + I_L = -j3.23 + 0.69 + j0.52 = 0.69 - j2.71 \text{ per unit}$$

$$I_m'' = I_{mf}'' - I_L = -j4.85 - 0.69 - j0.52 = -0.69 - j5.37 \text{ per unit}$$

Note that I_L is in the same direction as I_g'' but opposite to I_m''. The per-unit values found for I_f'', I_g'', and I_m'' are the same as in Example 10.1, and so the ampere values will also be the same.

$$\text{Fault current from generator} = |905 - j3550| = 3600 \text{ A}$$

$$\text{Fault current from motor} = |-905 - j7050| = 7200 \text{ A}$$

The sum of the magnitudes of the generator and motor currents does not equal the fault current because the currents from the generator and motor are not in phase when load current is included.

Usually, load current is omitted in determining the current in each line upon occurrence of a fault. In the Thévenin method neglect of load current means that the prefault current in each line is not added to the component of current flowing toward the fault in the line. The method of Example 10.1 neglects load current if the subtransient internal voltages of all machines are assumed equal to the voltage V_f at the fault before the fault occurs, for such is the case if no current flows anywhere in the network prior to the fault. Resistances, charging capacitances, and off-nominal tap-changing of transformers are also usually omitted in fault studies since they are not likely to influence the *level* of fault current significantly. Calculation of the fault currents is thereby simplified since the network model becomes an interconnection of inductive reactances and all currents throughout the faulted system are then in phase, as demonstrated in Example 10.2.

10.3 FAULT CALCULATIONS USING $\mathbf{Z}_{\text{bus}}$

Our discussion of fault calculations has been confined to simple circuits, but now we extend our study to general networks. We proceed to the general equations by starting with a specific network with which we are already familiar. In the circuit of Fig. 7.4 if the reactances in series with the generated voltages are changed from synchronous to *subtransient* values, and if the generated voltages become *subtransient* internal voltages, we have the network shown in Fig. 10.5. This network can be regarded as the *per-phase* equivalent of a balanced three-phase system. If we choose to study a fault at bus ②, for example, we can follow the notation of Sec. 10.2 and designate V_f as the actual voltage at bus ② before the fault occurs.

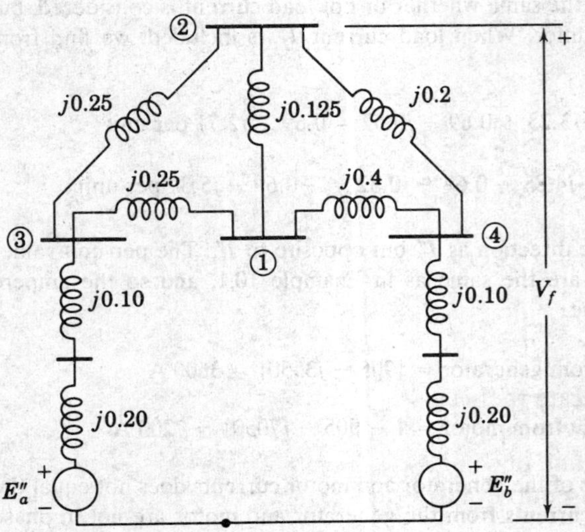

FIGURE 10.5
Reactance diagram obtained from Fig. 7.4 by substituting subtransient values for synchronous reactances and synchronous internal voltages of the machines. Reactance values are marked in per unit.

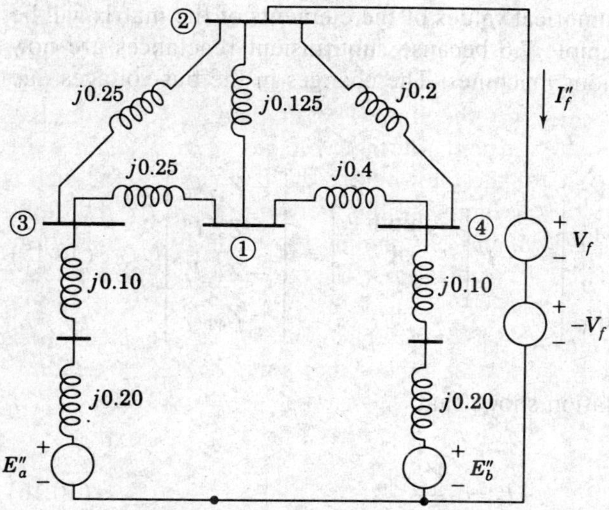

FIGURE 10.6
Circuit of Figure 10.5 with a three-phase fault on bus ② simulated by V_f and $-V_f$ in series.

A three-phase fault at bus ② is simulated by the network of Fig. 10.6, where the source voltages V_f and $-V_f$ in series constitute a short-circuit branch. Source voltage V_f acting alone in this branch would match the prefault voltage already at bus ②, and therefore would not cause current to flow in the branch. With V_f and $-V_f$ in series, the branch becomes a short circuit and the branch current is I_f'' as shown. It is evident, therefore, that I_f'' is caused by the addition of the $-V_f$ source. The current I_f'' distributes itself throughout the system from the reference node before flowing *out* of bus ② through the $-V_f$ source. In doing so, it produces whatever bus voltage *changes* that occur in the system due to the fault. If E_a'', E_b'', and V_f are short-circuited, then $-V_f$ is left to act alone and $-I_f''$ *into* bus ② is then the only current *entering* the network from external sources. With $-V_f$ as the only source, the network has the nodal impedance equations in the $\mathbf{Z}_{bus}$ matrix form

$$\begin{bmatrix} \Delta V_1 \\ \Delta V_2 \\ \Delta V_3 \\ \Delta V_4 \end{bmatrix} = \begin{bmatrix} \Delta V_1 \\ -V_f \\ \Delta V_3 \\ \Delta V_4 \end{bmatrix} = \begin{matrix} \text{①} \\ \text{②} \\ \text{③} \\ \text{④} \end{matrix} \overset{\qquad\text{①}\qquad\text{②}\qquad\text{③}\qquad\text{④}}{\begin{bmatrix} Z_{11} & Z_{12} & Z_{13} & Z_{14} \\ Z_{21} & Z_{22} & Z_{23} & Z_{24} \\ Z_{31} & Z_{32} & Z_{33} & Z_{34} \\ Z_{41} & Z_{42} & Z_{43} & Z_{44} \end{bmatrix}} \begin{bmatrix} 0 \\ -I_f'' \\ 0 \\ 0 \end{bmatrix} \tag{10.14}$$

The prefix Δ is chosen to indicate the *changes* in the voltages at the buses due to the current $-I_f''$ injected into bus ② by the fault.

The $\mathbf{Z}_{bus}$ building algorithm, or some other means such as $\mathbf{Y}_{bus}$ triangularization and inversion, can be used to evaluate the bus impedance matrix for the network of Fig. 10.6. The numerical values of the elements of the matrix will be different from those in Example 7.6 because subtransient reactances are now being used for the synchronous machines. The changes in the bus voltages due to $-I_f''$ are given by

$$\begin{bmatrix} \Delta V_1 \\ \Delta V_2 \\ \Delta V_3 \\ \Delta V_4 \end{bmatrix} = \begin{bmatrix} \Delta V_1 \\ -V_f \\ \Delta V_3 \\ \Delta V_4 \end{bmatrix} = -I_f'' \begin{bmatrix} \text{Column 2} \\ \text{of} \\ \mathbf{Z}_{\text{bus}} \end{bmatrix} = \begin{bmatrix} -Z_{12} I_f'' \\ -Z_{22} I_f'' \\ -Z_{32} I_f'' \\ -Z_{42} I_f'' \end{bmatrix} \tag{10.15}$$

The second row of this equation shows that

$$I_f'' = \frac{V_f}{Z_{22}} \tag{10.16}$$

We recognize Z_{22} as the diagonal element of $\mathbf{Z}_{\text{bus}}$ representing the Thévenin impedance of the network at bus ②. Substituting the expression for I_f'' into Eq. (10.15) gives

$$\begin{bmatrix} \Delta V_1 \\ \Delta V_2 \\ \Delta V_3 \\ \Delta V_4 \end{bmatrix} = \begin{bmatrix} -\dfrac{Z_{12}}{Z_{22}} V_f \\ -V_f \\ -\dfrac{Z_{32}}{Z_{22}} V_f \\ -\dfrac{Z_{42}}{Z_{22}} V_f \end{bmatrix} \tag{10.17}$$

When the generator voltage $-V_f$ is short-circuited in the network of Fig. 10.6 and the sources E_a'', E_b'', and V_f are reinserted into the network, the currents and voltages everywhere in the network will be the same as those existing before the fault. By the principle of superposition these prefault voltages add to the changes given by Eq. (10.17) to yield the total voltages existing after the fault occurs.

The faulted network is usually, but not always, assumed to be without load before the fault occurs. In the absence of loads, as remarked previously, no prefault currents flow and there are no voltage differences across the branch impedances; all bus voltages throughout the network are then the same as V_f, the prefault voltage at the fault point. The assumption of no prefault current simplifies our work considerably, and by applying the principle of superposition, we obtain the bus voltages

$$\begin{bmatrix} V_1 \\ V_2 \\ V_3 \\ V_4 \end{bmatrix} = \begin{bmatrix} V_f \\ V_f \\ V_f \\ V_f \end{bmatrix} + \begin{bmatrix} \Delta V_1 \\ \Delta V_2 \\ \Delta V_3 \\ \Delta V_4 \end{bmatrix} = \begin{bmatrix} V_f - Z_{12} I_f'' \\ V_f - V_f \\ V_f - Z_{32} I_f'' \\ V_f - Z_{42} I_f'' \end{bmatrix} = V_f \begin{bmatrix} 1 - \dfrac{Z_{12}}{Z_{22}} \\ 0 \\ 1 - \dfrac{Z_{32}}{Z_{22}} \\ 1 - \dfrac{Z_{42}}{Z_{22}} \end{bmatrix} \tag{10.18}$$

Thus, the voltages at *all* buses of the network can be calculated using the prefault voltage V_f of the fault bus and the elements in the column of $\mathbf{Z}_{\text{bus}}$ corresponding to the fault bus. The calculated values of the bus voltages will yield the subtransient currents in the branches of the network if the system $\mathbf{Z}_{\text{bus}}$ has been formed with subtransient values for the machine reactances.

In more general terms, when the three-phase fault occurs on bus ⓚ of a large-scale network, we have

$$I_f'' = \frac{V_f}{Z_{kk}} \tag{10.19}$$

and neglecting prefault load currents, we can then write for the voltage at any bus ⓙ during the fault

$$V_j = V_f - Z_{jk} I_f'' = V_f - \frac{Z_{jk}}{Z_{kk}} V_f \tag{10.20}$$

where Z_{jk} and Z_{kk} are elements in column k of the system $\mathbf{Z}_{\text{bus}}$. If the prefault voltage of bus ⓙ is not the same as the prefault voltage of fault bus ⓚ, then we simply replace V_f on the left in Eq. (10.20) by the actual prefault voltage of bus ⓙ. Knowing the bus voltages during the fault, we can calculate the subtransient current I_{ij}'' from bus ⓘ to bus ⓙ in the line of impedance Z_b connecting those two buses,

$$I_{ij}'' = \frac{V_i - V_j}{Z_b} = -I_f''\left(\frac{Z_{ik} - Z_{jk}}{Z_b}\right) = -\frac{V_f}{Z_b}\left(\frac{Z_{ik} - Z_{jk}}{Z_{kk}}\right) \tag{10.21}$$

This equation shows I_{ij}'' as the fraction of the fault current I_f'' appearing as a *line flow* from bus ⓘ to bus ⓙ in the faulted network. If bus ⓙ is directly connected to the faulted bus ⓚ by a line of series impedance Z_b, then the current contributed from bus ⓙ to the current in the fault at bus ⓚ is simply V_j/Z_b, where V_j is given by Eq. (10.20).

The discussion of this section shows that only column k of $\mathbf{Z}_{\text{bus}}$, which we now denote by $\mathbf{Z}_{\text{bus}}^{(k)}$, is required to evaluate the impact on the system of a symmetrical three-phase fault at bus ⓚ. If necessary, the elements of $\mathbf{Z}_{\text{bus}}^{(k)}$ can be generated from the triangular factors of $\mathbf{Y}_{\text{bus}}$, as demonstrated in Sec. 8.5.

Example 10.3. A three-phase fault occurs at bus ② of the network of Fig. 10.5. Determine the initial symmetrical rms current (that is, the subtransient current) in the fault; the voltages at buses ①, ③, and ④ during the fault; the current flow in the line from bus ③ to bus ①; and the current contributions to the fault from lines ③–②, ①–②, and ④–②. Take the prefault voltage V_f at bus ② equal to $1.0\angle 0°$ per unit and neglect all prefault currents.

Solution. Applying the $\mathbf{Z}_{\text{bus}}$ building algorithm to Fig. 10.5, we find that

$$\mathbf{Z}_{\text{bus}} = \begin{array}{c} \\ ① \\ ② \\ ③ \\ ④ \end{array} \begin{array}{c} \begin{array}{cccc} \quad ① \quad & \quad ② \quad & \quad ③ \quad & \quad ④ \quad \end{array} \\ \begin{bmatrix} j0.2436 & j0.1938 & j0.1544 & j0.1456 \\ j0.1938 & j0.2295 & j0.1494 & j0.1506 \\ j0.1544 & j0.1494 & j0.1954 & j0.1046 \\ j0.1456 & j0.1506 & j0.1046 & j0.1954 \end{bmatrix} \end{array}$$

Since load currents are neglected, the prefault voltage at each bus is $1.0\angle 0°$ per unit, the same as V_f at bus ②. When the fault occurs,

$$I_f'' = \frac{1.0}{Z_{22}} = \frac{1.0}{j0.2295} = -j4.3573 \text{ per unit}$$

and from Eq. (10.18) the voltages during the fault are

$$\begin{bmatrix} V_1 \\ V_2 \\ V_3 \\ V_4 \end{bmatrix} = \begin{bmatrix} 1 - \dfrac{j0.1938}{j0.2295} \\ 0 \\ 1 - \dfrac{j0.1494}{j0.2295} \\ 1 - \dfrac{j0.1506}{j0.2295} \end{bmatrix} = \begin{bmatrix} 0.1556 \\ 0 \\ 0.3490 \\ 0.3438 \end{bmatrix} \text{per unit}$$

The current flow in line ③–① is

$$I_{31} = \frac{V_3 - V_1}{Z_b} = \frac{0.3490 - 0.1556}{j0.25} = -j0.7736 \text{ per unit}$$

Fault currents contributed to bus ② by the adjacent unfaulted buses are

$$\text{From bus ①:}\quad \frac{V_1}{Z_{b1}} = \frac{0.1556}{j0.125} = -j1.2448 \text{ per unit}$$

$$\text{From bus ③:}\quad \frac{V_3}{Z_{b3}} = \frac{0.3490}{j0.25} = -j1.3960 \text{ per unit}$$

$$\text{From bus ④:}\quad \frac{V_4}{Z_{b4}} = \frac{0.3438}{j0.20} = -j1.7190 \text{ per unit}$$

Except for round-off errors, the sum of these current contributions equals I_f''.

10.4 FAULT CALCULATIONS USING $\mathbf{Z}_{bus}$ EQUIVALENT CIRCUITS

We cannot devise a physically realizable network which directly incorporates all the individual elements of the bus impedance matrix. However, Fig. 8.4 shows that we can use the matrix elements to construct the Thévenin equivalent circuit between *any pair* of buses in the network that may be of interest. The Thévenin equivalent circuit is very helpful for illustrating the symmetrical fault equations, which are developed in Sec. 10.3.

In the Thévenin equivalent circuit of Fig. 10.7(*a*) bus ⓚ is assumed to be the fault bus and bus ⓙ is unfaulted. The impedances shown correspond directly to the elements of the network $\mathbf{Z}_{bus}$ and all the prefault bus voltages are the same as V_f of the fault bus if load currents are neglected. The two points

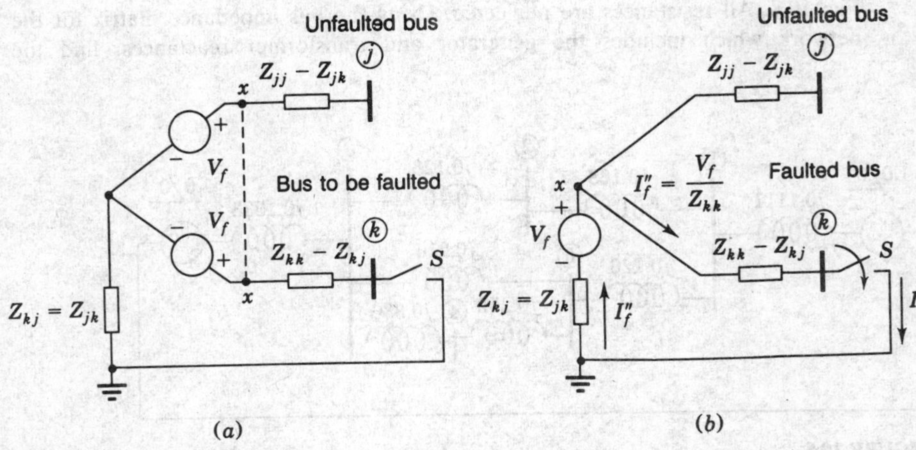

FIGURE 10.7
Thévenin equivalent between buses ⓙ and ⓚ of system with no prefault load currents: (*a*) before the fault (S open); (*b*) during the fault (S closed).

marked x have the same potential, and so they can be joined together to yield the equivalent circuit of Fig. 10.7(b) with a single voltage source V_f as shown. If the switch S is open between bus ⓚ and the reference node, there is no short circuit and no current flows in any branch of the network. When S is closed to represent the fault on bus ⓚ, current flows in the circuit toward bus ⓚ. This current is $I_f'' = V_f/Z_{kk}$, which agrees with Eq. (10.19), and it induces a voltage drop $(Z_{jk}/Z_{kk})V_f$ in the direction *from* the reference node toward bus ⓙ. The voltage from bus ⓙ to the reference *changes* therefore by the amount $-(Z_{jk}/Z_{kk})V_f$ so that the voltage at bus ⓙ during the fault is $V_f - (Z_{jk}/Z_{kk})V_f$, which is consistent with Eq. (10.20).

Thus, by substituting appropriate numerical values for the impedances in the simple equivalent circuit of Fig. 10.7(b), we can calculate the bus voltages of the system before and after the fault occurs. With switch S open in the circuit, the voltages at bus ⓚ and the representative bus ⓙ are equal to V_f. The same uniform voltage profile occurs in Fig. 10.6 if there are no prefault currents so that E_a'' and E_b'' equal V_f. If S is closed in Fig. 10.7(b), the circuit reflects the voltage of representative bus ⓙ with respect to reference while the fault is on bus ⓚ. Therefore, if a three-phase short-circuit fault occurs at bus ⓚ of a large-scale network, we can calculate the current in the fault and the voltage at *any* of the unfaulted buses simply by inserting the proper impedance values into elementary circuits like those in Fig. 10.7. The following example illustrates the procedure.

Example 10.4. A five-bus network has generators at buses ① and ③ rated 270 and 225 MVA, respectively. The generator subtransient reactances plus the reactances of the transformers connecting them to the buses are each 0.30 per unit on the generator rating as base. The turns ratios of the transformers are such that the voltage base in each generator circuit is equal to the voltage rating of the generator. Line impedances in per unit on a 100-MVA system base are shown in Fig. 10.8. All resistances are neglected. Using the bus impedance matrix for the network which includes the generator and transformer reactances, find the

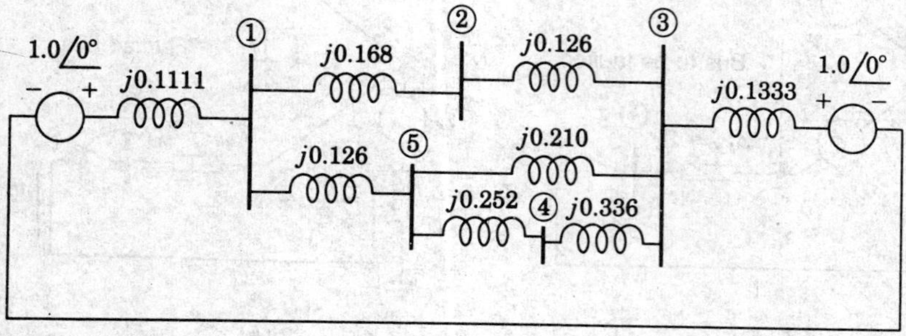

FIGURE 10.8
Impedance diagram for Example 10.4. Generator reactances include subtransient values plus reactances of set-up transformers. All values in per unit on a 100-MVA base.

subtransient current in a three-phase fault at bus ④ and the current coming to the faulted bus over each line. Prefault current is to be neglected and all voltages are assumed to be 1.0 per unit before the fault occurs.

Solution. Converted to the 100-MVA base, the combined generator and transformer reactances are

$$\text{Generator at bus ①:} \quad X = 0.30 \times \frac{100}{270} = 0.1111 \text{ per unit}$$

$$\text{Generator at bus ③:} \quad X = 0.30 \times \frac{100}{225} = 0.1333 \text{ per unit}$$

These values, along with the line impedances, are marked in per unit in Fig. 10.8 from which the bus impedance matrix can be determined by the $\mathbf{Z}_{bus}$ building algorithm to yield

$$\mathbf{Z}_{bus} = \begin{array}{c} \text{①} \\ \text{②} \\ \text{③} \\ \text{④} \\ \text{⑤} \end{array} \overset{\begin{array}{ccccc} \text{①} & \text{②} & \text{③} & \text{④} & \text{⑤} \end{array}}{\begin{bmatrix} j0.0793 & j0.0558 & j0.0382 & j0.0511 & j0.0608 \\ j0.0558 & j0.1338 & j0.0664 & j0.0630 & j0.0605 \\ j0.0382 & j0.0664 & j0.0875 & j0.0720 & j0.0603 \\ j0.0511 & j0.0630 & j0.0720 & j0.2321 & j0.1002 \\ j0.0608 & j0.0605 & j0.0603 & j0.1002 & j0.1301 \end{bmatrix}}$$

Since we are to calculate the currents from buses ③ and ⑤ into the fault at bus

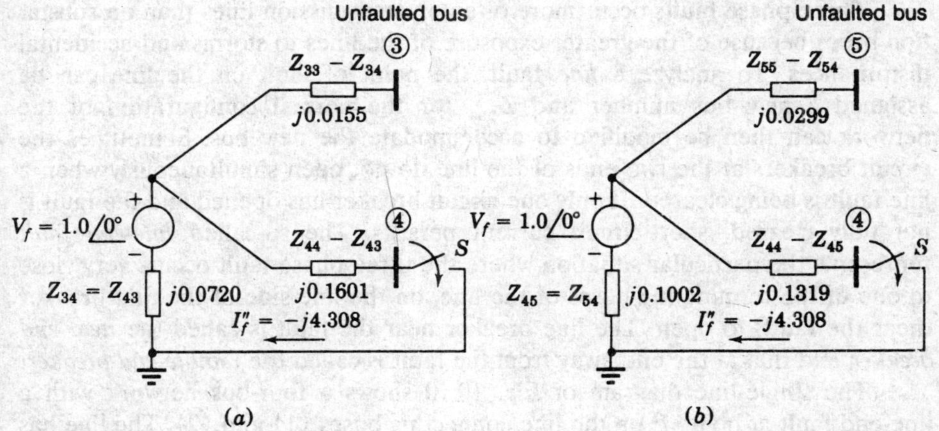

FIGURE 10.9
Use of Thévenin equivalent circuits to calculate voltages at (*a*) bus ③ and (*b*) bus ④ due to fault at bus ⑤.

④, we need to know V_3 and V_5 during the fault. Visualizing equivalent circuits like those of Fig. 10.9 helps in finding the desired currents and voltages.

The subtransient current in the three-phase fault at bus ④ can be calculated from Fig. 10.9(*a*). Simply closing switch S gives

$$I_f'' = \frac{V_f}{Z_{44}} = \frac{1.0}{j0.2321} = -j4.308 \text{ per unit}$$

From Fig. 10.9(*a*) the voltage at bus ③ during the fault is

$$V_3 = V_f - I_f'' Z_{34} = 1.0 - (-j4.308)(j0.0720) = 0.6898 \text{ per unit}$$

From Fig. 10.9(*b*) the voltage at bus ⑤ during the fault is

$$V_5 = V_f - I_f'' Z_{54} = 1.0 - (-j4.308)(j0.1002) = 0.5683 \text{ per unit}$$

Currents into the fault at bus ④ over the line impedances Z_b are

$$\text{From bus ③:} \quad \frac{V_3}{Z_{b3}} = \frac{0.6898}{j0.336} = -j2.053 \text{ per unit}$$

$$\text{From bus ⑤:} \quad \frac{V_5}{Z_{b5}} = \frac{0.5683}{j0.252} = -j2.255 \text{ per unit}$$

Hence, total fault current at bus ④ $= -j4.308$ per unit

Other equivalent circuits based on the given bus impedance matrix can be developed for three-phase faults on any of the other buses or transmission *lines* of the system. A specific application will demonstrate how this is accomplished.

Three-phase faults occur more often on transmission lines than on substation buses because of the greater exposure of the lines to storms and accidental disturbances. To analyze a *line* fault, the point of fault on the line can be assigned a new bus number and $\mathbf{Z}_{\text{bus}}$ for the normal configuration of the network can then be modified to accommodate the new bus. Sometimes the circuit breakers at the two ends of the line do not open simultaneously when a line fault is being cleared. If only one circuit breaker has opened and the fault is not fully cleared, short-circuit current persists. The so-called *line-end fault* represents the particular situation where the three-phase fault occurs very close to one of the terminating buses of the line, on the line side of the first breaker (near the fault) to open. The line breaker near the fault is called the *near-end breaker* and that at the end away from the fault is called the *remote-end breaker*.

The single-line diagram of Fig. 10.10 shows a four-bus network with a line-end fault at point P on the line connecting buses ① and ②. The line has series impedance Z_b. The near-end breaker at bus ② is open and the remote-end breaker is closed, leaving the fault still on at point P, which we now

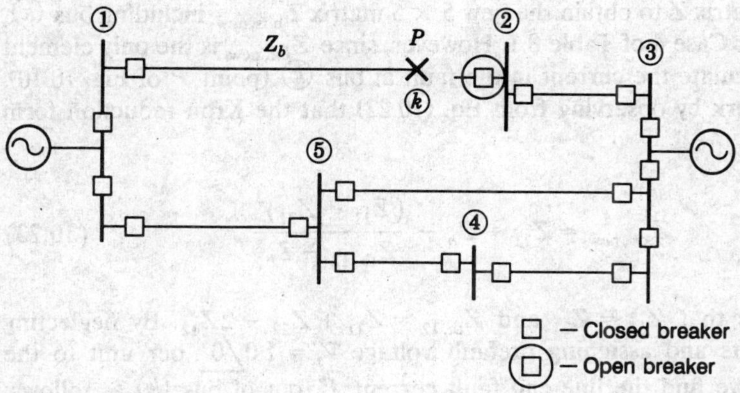

FIGURE 10.10
Line-end fault at point P on line of series impedance Z_b between buses ① and ② of system of Fig. 10.8.

call bus ⓚ. In order to study this fault condition, we need to modify the existing bus impedance matrix $\mathbf{Z}_{orig}$ for the normal configuration of the system to reflect the near-end breaker operation. This is accomplished in two steps:

1. Establish the new bus ⓚ by adding a line of series impedance Z_b between bus ① and bus ⓚ.
2. Remove the line between bus ① and bus ② by adding line impedance $-Z_b$ between those two buses in the manner explained in Sec. 8.4.

The first step follows the procedure given for Case 2 in Table 8.1 and yields, in terms of the elements Z_{ij} of $\mathbf{Z}_{orig}$, the first five rows and columns of the symmetric matrix

$$\mathbf{Z} = \begin{array}{c} \\ (1) \\ (2) \\ (3) \\ (4) \\ (k) \\ (q) \end{array} \left[\begin{array}{cccc|c|c} (1) & (2) & (3) & (4) & (k) & (q) \\ & & & & Z_{11} & Z_{11} - Z_{12} \\ & & & & Z_{21} & Z_{21} - Z_{22} \\ & & \mathbf{Z}_{orig} & & Z_{31} & Z_{31} - Z_{32} \\ & & & & Z_{41} & Z_{41} - Z_{42} \\ \hline Z_{11} & Z_{12} & Z_{13} & Z_{14} & Z_{11} + Z_b & Z_{11} - Z_{12} \\ \hline (Z_{11} - Z_{21}) & (Z_{12} - Z_{22}) & (Z_{13} - Z_{23}) & (Z_{14} - Z_{24}) & (Z_{11} - Z_{21}) & Z_{th,12} - Z_b \end{array} \right] \tag{10.22}$$

where $Z_{th,12} = Z_{11} + Z_{22} - 2Z_{12}$ when $\mathbf{Z}_{orig}$ is symmetric. The second step can be accomplished by forming row ⓠ and column ⓠ as shown and then Kron reducing the matrix $\mathbf{Z}$ to obtain the new 5×5 matrix $\mathbf{Z}_{\text{bus, new}}$ including bus ⓚ, as explained for Case 4 of Table 8.1. However, since $Z_{kk,\text{new}}$ is the only element required to calculate the current in the fault at bus ⓚ (point P of Fig. 10.10), we can save work by observing from Eq. (10.22) that the Kron reduction form gives

$$Z_{kk,\text{new}} = Z_{11} + Z_b - \frac{(Z_{11} - Z_{21})^2}{Z_{\text{th},12} - Z_b} \tag{10.23}$$

Again, we note that $Z_{12} = Z_{21}$ and $Z_{\text{th},12} = Z_{11} + Z_{22} - 2Z_{12}$. By neglecting prefault currents and assigning prefault voltage $V_f = 1.0\angle 0°$ per unit to the fault point P, we find the line-end fault current I_f'' out of bus ⓚ as follows:

$$I_f'' = \frac{1.0}{Z_{kk,\text{new}}} = \frac{1.0}{Z_{11} + Z_b - (Z_{11} - Z_{21})^2/(Z_{\text{th},12} - Z_b)} \tag{10.24}$$

Thus, the only elements of $\mathbf{Z}_{\text{orig}}$ entering into the calculation of I_f'' are Z_{11}, $Z_{12} = Z_{21}$, and Z_{22}.

It is worthwhile observing that the same equation for the line-end fault current can be found directly by inspection of Fig. 10.11(a), which shows the Thévenin equivalent circuit between buses ① and ② of the prefault network. The impedances Z_b and $-Z_b$ are connected as shown in accordance with steps 1 and 2 above. Circuit analysis then shows in a straightforward manner that the impedance looking back into the circuit from the terminals of the open switch S is

$$Z_{kk,\text{new}} = Z_b + \frac{(Z_{11} - Z_{12})(Z_{22} - Z_{21} - Z_b)}{Z_{11} - Z_{12} + Z_{22} - Z_{21} - Z_b} + Z_{12} \tag{10.25}$$

Since $Z_{12} = Z_{21}$ and $Z_{\text{th},12} = Z_{11} + Z_{22} - 2Z_{12}$, Eq. (10.25) can be reduced to give

$$Z_{kk,\text{new}} = Z_b + \frac{(Z_{11} - Z_{12})[(Z_{\text{th},12} - Z_b) - (Z_{11} - Z_{12})]}{Z_{\text{th},12} - Z_b} + Z_{12}$$

$$= Z_{11} + Z_b - \frac{(Z_{11} - Z_{21})^2}{Z_{\text{th},12} - Z_b} \tag{10.26}$$

Thus, simply by closing switch S as shown in Fig. 10.11(b) and using elementary circuit analysis, we can calculate the line-end fault current I_f'' in agreement with Eq. (10.24). Of course, the circuit approach using the Thévenin equivalent must yield the same results as the matrix manipulations of Eq. (10.22)—for the same

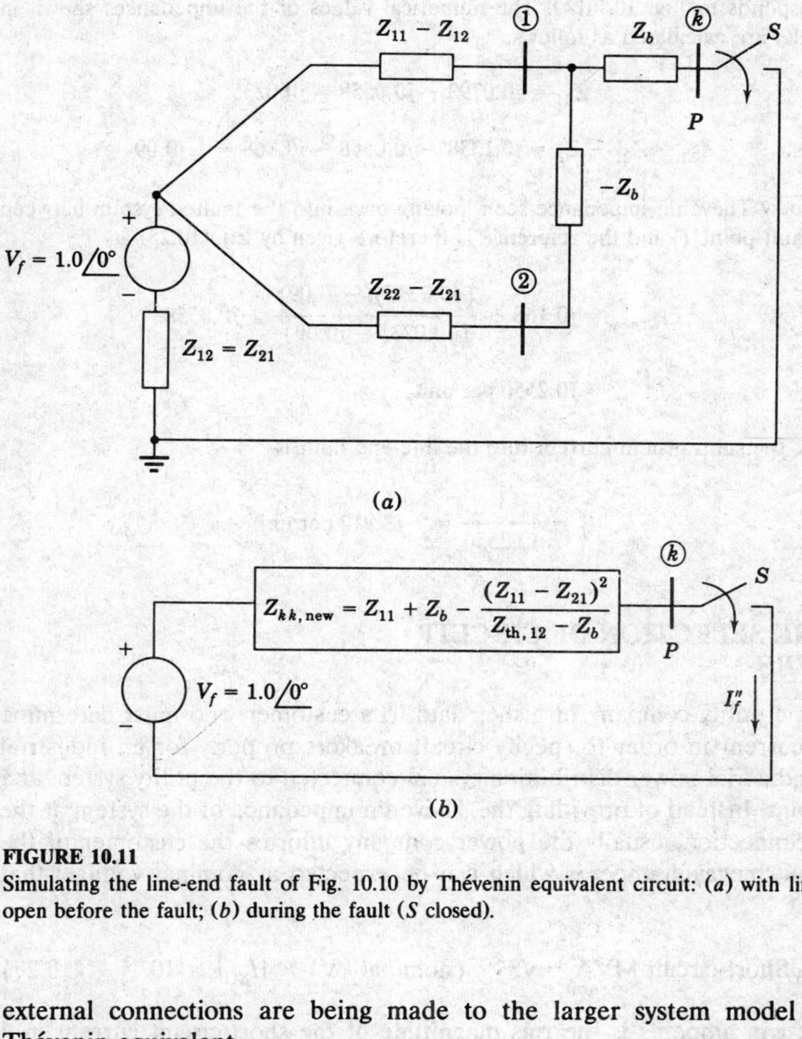

FIGURE 10.11
Simulating the line-end fault of Fig. 10.10 by Thévenin equivalent circuit: (*a*) with line ①- ② open before the fault; (*b*) during the fault (*S* closed).

external connections are being made to the larger system model as to its Thévenin equivalent.

Other uses of the equivalent circuits based on the bus impedance matrix are possible.

Example 10.5. In the five-bus system of Fig. 10.8 a line-end, short-circuit fault occurs on line ①- ②, on the line side of the breaker at bus ②. Neglecting prefault currents and assuming rated system voltage at the fault point, calculate the subtransient current into the fault when only the near-end breaker at bus ② opens.

Solution. Figure 10.8 shows that the impedance of line ①- ② is $Z_b = j0.168$ per unit and the required elements of $\mathbf{Z}_{\text{bus}}$ are given in Example 10.4. The Thévenin equivalent circuit looking into the *intact* system between buses ① and ② corresponds to Fig. 10.11(a). The numerical values of the impedances shown in parallel are calculated as follows:

$$Z_{11} = j0.0793 - j0.0558 = j0.0235$$

$$Z_{22} - Z_{21} - Z_b = j0.1338 - j0.0558 - j0.168 = -j0.09$$

The new Thévenin impedance seen looking back into the faulted system between the fault point P and the reference is therefore given by Eq. (10.25) as

$$Z_{kk,\text{new}} = j0.168 + \frac{(j0.0235)(-j0.09)}{(j0.0235 - j0.09)} + j0.0558$$

$$= j0.2556 \text{ per unit}$$

Thus, the subtransient current into the line-end fault is

$$I_f'' = \frac{1}{j0.2556} = -j3.912 \text{ per unit}$$

10.5 THE SELECTION OF CIRCUIT BREAKERS

The electric utility company furnishes data to a customer who must determine the fault current in order to specify circuit breakers properly for an industrial plant or industrial power distribution system connected to the utility system at a certain point. Instead of providing the Thévenin impedance of the system at the point of connection, usually the power company informs the customer of the short-circuit megavoltamperes which can be expected at nominal voltage; that is,

$$\text{Short-circuit MVA} = \sqrt{3} \times (\text{nominal kV}) \times |I_{SC}| \times 10^{-3} \tag{10.27}$$

where $|I_{SC}|$ in amperes is the rms magnitude of the short-circuit current in a three-phase fault at the connection point. Base megavoltamperes are related to base kilovolts and base amperes $|I_{\text{base}}|$ by

$$\text{Base MVA} = \sqrt{3} \times (\text{base kV}) \times |I_{\text{base}}| \times 10^{-3} \tag{10.28}$$

If base kilovolts equal nominal kilovolts, then dividing Eq. (10.27) by Eq. (10.28) converts the former to per unit, and we obtain

$$\text{Short-circuit MVA in per unit} = |I_{SC}| \text{ in per unit} \tag{10.29}$$

At nominal voltage the Thévenin equivalent circuit looking back into the system from the point of connection is an emf of $1.0\angle 0^\circ$ per unit in series with the per-unit impedance Z_{th}. Therefore, under short-circuit conditions,

$$|Z_{th}| = \frac{1.0}{|I_{SC}|} \text{ per unit} = \frac{1.0}{\text{short-circuit MVA}} \text{ per unit} \tag{10.30}$$

Often resistance and shunt capacitance are neglected, in which case $Z_{th} = X_{th}$. Thus, by specifying short-circuit megavoltamperes at the customer's bus, the electric utility is effectively describing the short-circuit current at nominal voltage and the reciprocal of the Thévenin impedance of the system at the point of connection.

Much study has been given to circuit-breaker ratings and applications, and our discussion here gives some introduction to the subject. The presentation is not intended as a study of breaker applications but, rather, to indicate the importance of understanding fault calculations. For additional guidance in specifying breakers the reader should consult the ANSI publications listed in the footnotes which accompany this section.

From the *current* viewpoint two factors to be considered in selecting circuit breakers are:

- The maximum instantaneous current which the breaker must carry (*withstand*) and
- The total current when the breaker contacts part to *interrupt* the circuit.

Up to this point we have devoted most of our attention to the subtransient current called the *initial symmetrical current*, which does not include the dc component. Inclusion of the dc component results in a rms value of current immediately after the fault, which is higher than the subtransient current. For oil circuit breakers above 5 kV the subtransient current multiplied by 1.6 is considered to be the rms value of the current whose disruptive forces the breaker must withstand during the first half cycle after the fault occurs. This current is called the *momentary current*, and for many years circuit breakers were rated in terms of their momentary current as well as other criteria.[1]

The *interrupting rating* of a circuit breaker was specified in kilovoltamperes or megavoltamperes. The interrupting kilovoltamperes equal $\sqrt{3}$ × (the kilovolts of the bus to which the breaker is connected) × (the current which the breaker must be capable of interrupting when its contacts part). This *interrupting current* is, of course, lower than the momentary current and depends on the speed of the breaker, such as 8, 5, 3, or 2 cycles, which is a measure of the time from the occurrence of the fault to the extinction of the arc. Breakers of different speeds are classified by their *rated interrupting times*. The rated

[1]See G. N. Lester, "High Voltage Circuit Breaker Standards in the USA: Past, Present, and Future," *IEEE Transactions on Power Apparatus and Systems*, vol. 93, 1974, pp. 590–600.

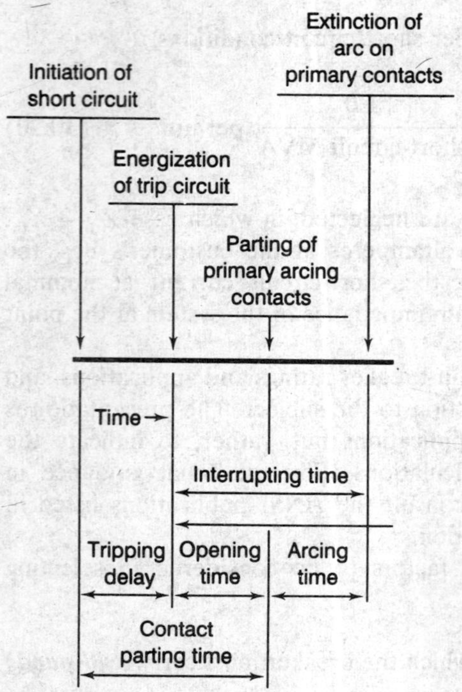

FIGURE 10.12
Definition of interrupting time given in ANSI/IEEE Standard C.37.010-1979, *Application Guide for AC High Voltage Circuit Breakers Rated on a Symmetrical Current Basis.*

interrupting time of a circuit breaker is the period between the instant of energizing the trip circuit and the arc extinction on an opening operation, Fig. 10.12. Preceding this period is the *tripping delay time*, which is usually assumed to be $\frac{1}{2}$ cycle for relays to pick up.

The current which a breaker must interrupt is usually asymmetrical since it still contains some of the decaying dc component. A schedule of preferred ratings for ac high-voltage oil circuit breakers specifies the interrupting current ratings of breakers in terms of the component of the asymmetrical current which is symmetrical about the zero axis. This current is properly called the *required symmetrical interrupting capability* or simply the *rated symmetrical short-circuit current*. Often the adjective *symmetrical* is omitted. Selection of circuit breakers may also be made on the basis of total current (dc component included).[2] We shall limit our discussion to a brief treatment of the symmetrical basis of breaker selection.

---

[2]See *Preferred Ratings and Related Required Capabilities for AC High-Voltage Circuit Breakers Rated on a Symmetrical Current Basis*, ANSI C37.06-1987, and *Guide for Calculation of Fault Currents for Application of AC High-Voltage Circuit Breakers Rated on a Total Current Basis*, ANSI C37.5-1979, American National Standards Institute, New York.

Breakers are identified by nominal-voltage class, such as 69 kV. Among other factors specified are rated continuous current, rated maximum voltage, voltage range factor K, and rated short-circuit current at rated maximum kilovolts. The *rated maximum voltage* of a circuit breaker is the highest rms voltage for which the circuit breaker is designed. The *rated voltage range factor* K is the ratio (rated maximum voltage ÷ the lower limit of the range of operating voltage). K determines the range of voltage over which the product (rated short-circuit current × operating voltage) is constant. In the application of circuit breakers it is important not to exceed the short-circuit capabilities of the breakers. A breaker is required to have a *maximum symmetrical interrupting capability* equal to K × rated short-circuit current. Between the rated maximum voltage and $1/K$ times the rated maximum voltage the *symmetrical interrupting capability* is defined as the product [rated short-circuit current × (rated maximum voltage/operating voltage)].

Example 10.6. A 69-kV circuit breaker having a voltage range factor K of 1.21 and a continuous current rating of 1200 A has a rated short-circuit current of 19,000 A at the maximum rated voltage of 72.5 kV. Determine the maximum symmetrical interrupting capability of the breaker and explain its significance at lower operating voltages.

Solution. The maximum symmetrical interrupting capability is given by

$$K \times \text{rated short-circuit current} = 1.21 \times 19{,}000 = 22{,}990 \text{ A}$$

This value of symmetrical interrupting current must not be exceeded. From the definition of K we have

$$\text{Lower limit of operating voltage} = \frac{\text{rated maximum voltage}}{K} = \frac{72.5}{1.21} \cong 60 \text{ kV}$$

Hence, in the operating voltage range 72.5–60 kV, the symmetrical interrupting current may exceed the rated short-circuit current of 19,000 A, but it is limited to 22,990 A. For example, at 66 kV the interrupting current can be

$$\frac{72.5}{66} \times 19{,}000 = 20{,}871 \text{ A}$$

Breakers of the 115-kV class and higher have a K of 1.0.

A simplified procedure for calculating the symmetrical short-circuit current, called the *E/X method*,[3] disregards all resistance, all static load, and all

[3]See *Application Guide for AC High-Voltage Circuit Breakers Rated on a Symmetrical Current Basis*, ANSI C37.010-1979, American National Standards Institute, New York. This publication is also IEEE Std. 320–1979.

prefault current. Subtransient reactance is used for generators in the E/X method, and for synchronous motors the recommended reactance is the X''_d of the motor times 1.5, which is the approximate value of the transient reactance X'_d of the motor. Induction motors below 50 hp are neglected, and various multiplying factors are applied to the X''_d of larger induction motors depending on their size. If no motors are present, symmetrical short-circuit current equals subtransient current.

The impedance by which the voltage V_f at the fault is divided to find short-circuit current must be examined when the E/X method is used. In specifying a breaker for bus ⓚ, this impedance is Z_{kk} of the bus impedance matrix with the proper machine reactances since the short-circuit current is expressed by Eq. (10.19). If the radio of X/R of this impedance is 15 or less, a breaker of the correct voltage and kilovoltamperes may be used if its interrupting current rating is equal to or exceeds the calculated current. If the X/R ratio is unknown, the calculated current should be no more than 80% of the allowed value for the breaker at the existing bus voltage. The ANSI application guide specifies a corrected method to account for ac and dc time constants for the decay of the current amplitude if the X/R ratio exceeds 15. The corrected method also considers breaker speed.

Example 10.7. A 25,000-kVA 13.8-kV generator with $X''_d = 15\%$ is connected through a transformer to a bus which supplies four identical motors, as shown in Fig. 10.13. The subtransient reactance X''_d of each motor is 20% on a base of 5000 kVA, 6.9 kV. The three-phase rating of the transformer is 25,000 kVA, 13.8/6.9 kV, with a leakage reactance of 10%. The bus voltage at the motors is 6.9 kV when a three-phase fault occurs at point P. For the fault specified, determine (a) the subtransient current in the fault, (b) the subtransient current in breaker A, and (c) the symmetrical short-circuit interrupting current (as defined for circuit-breaker applications) in the fault and in breaker A.

Solution. (a) For a base of 25,000 kVA, 13.8 kV in the generator circuit the base for the motors is 25,000 kVA, 6.9 kV. The subtransient reactance of each motor is

$$X''_d = 0.20\frac{25{,}000}{5000} = 1.0 \text{ per unit}$$

Figure 10.14 is the diagram with subtransient values of reactance marked. For a

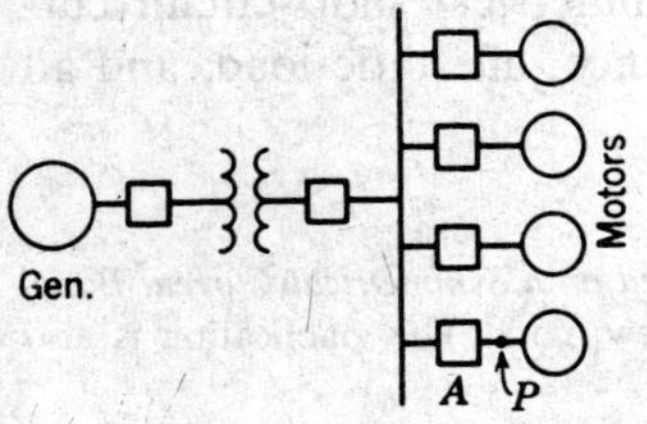

FIGURE 10.13
One-line diagram for Example 10.7.

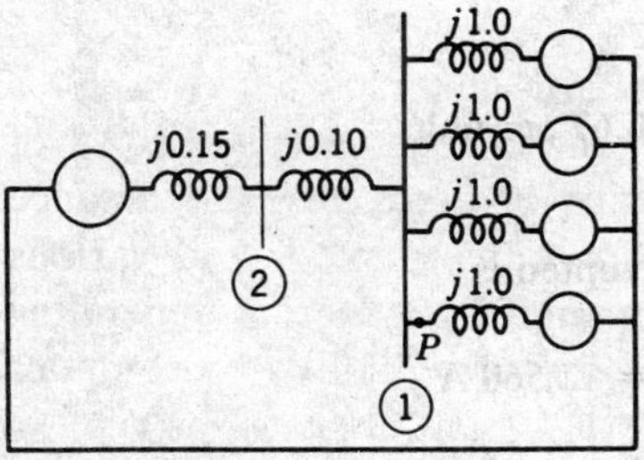

FIGURE 10.14
Reactance diagram for Example 10.7.

fault at P

$$V_f = 1.0\angle 0^\circ \text{ per unit} \qquad Z_{th} = j0.125 \text{ per unit}$$

$$I_f'' = \frac{1.0\angle 0^\circ}{j0.125} = -j8.0 \text{ per unit}$$

The base current in the 6.9 kV circuit is

$$|I_{base}| = \frac{25{,}000}{\sqrt{3} \times 6.9} = 2090 \text{ A}$$

and so $$|I_f''| = 8 \times 2090 = 16{,}720 \text{ A}$$

(b) Through breaker A comes the contribution from the generator and three of the four motors. The generator contributes a current of

$$-j8.0 \times \frac{0.25}{0.50} = -j4.0 \text{ per unit}$$

Each motor contributes 25% of the remaining fault current, or $-j1.0$ per-unit amperes each. breaker A

$$I'' = -j4.0 + 3(-j1.0) = -j7.0 \text{ per unit} \quad \text{or} \quad 7 \times 2090 = 14{,}630 \text{ A}$$

(c) To compute the current to be interrupted by breaker A, replace the subtransient reactance of $j1.0$ by the transient reactance of $j1.5$ in the motor circuits of Fig. 10.14. Then,

$$Z_{th} = j\frac{0.375 \times 0.25}{0.375 + 0.25} = j0.15 \text{ per unit}$$

The generator contributes a current of

$$\frac{1.0}{j0.15} \times \frac{0.375}{0.625} = -j4.0 \text{ per unit}$$

Each motor contributes a current of

$$\frac{1}{4} \times \frac{1.0}{j0.15} \times \frac{0.25}{0.625} = -j0.67 \text{ per unit}$$

The symmetrical short-circuit current to be interrupted is

$$(4.0 + 3 \times 0.67) \times 2090 = 12{,}560 \text{ A}$$

Suppose that all the breakers connected to the bus are rated on the basis of the current into a fault on the bus. In that case the short-circuit current interrupting rating of the breakers connected to the 6.9 kV bus must be at least

$$4 + 4 \times 0.67 = 6.67 \text{ per unit}$$

or

$$6.67 \times 2090 = 13{,}940 \text{ A}$$

A 14.4-kV circuit breaker has a rated maximum voltage of 15.5 kV and a K of 2.67. At 15.5 kV its rated short-circuit interrupting current is 8900 A. This breaker is rated for a symmetrical short-circuit interrupting current of 2.67 × 8900 = 23,760 A, at a voltage of 15.5/2.67 = 5.8 kV. This current is the maximum that can be interrupted even though the breaker may be in a circuit of lower voltage. The short-circuit interrupting current rating at 6.9 kV is

$$\frac{15.5}{6.9} \times 8900 = 20{,}000 \text{ A}$$

The required capability of 13,940 A is well below 80% of 20,000 A, and the breaker is suitable with respect to short-circuit current.

The short-circuit current could have been found by using the bus impedance matrix. For this purpose two buses ① and ② are identified in Fig. 10.14. Bus ① is on the low-voltage side of the transformer and bus ② is on the high-voltage side. For motor reactance of 1.5 per unit

$$Y_{11} = -j10 + \frac{1}{j1.5/4} = -j12.67$$

$$Y_{12} = j10 \qquad Y_{22} = -j10 - j6.67 = -j16.67$$

The node admittance matrix and its inverse are

$$\mathbf{Y}_{\text{bus}} = \begin{matrix} & ① & ② \\ ① & -j12.67 & j10.00 \\ ② & j10.00 & -j16.67 \end{matrix} \qquad \mathbf{Z}_{\text{bus}} = \begin{matrix} & ① & ② \\ ① & j0.150 & j0.090 \\ ② & j0.090 & j0.114 \end{matrix}$$

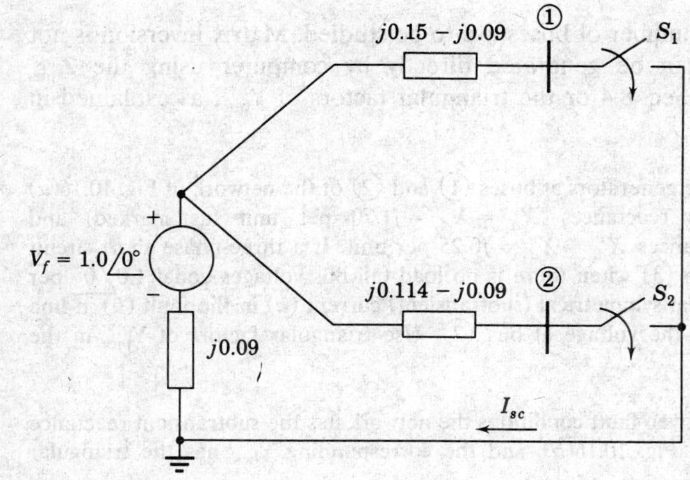

FIGURE 10.15
Bus impedance equivalent circuit for the $\mathbf{Z}_{\text{bus}}$ of Fig. 10.14.

Figure 10.15 is the network corresponding to $\mathbf{Z}_{\text{bus}}$ and $V_f = 1.0$ per unit. Closing S_1 with S_2 open represents a fault on bus ①.

The symmetrical short-circuit interrupting current in a three-phase fault at bus ① is

$$I_{SC} = \frac{1.0}{j0.15} = -j6.67 \text{ per unit}$$

which agrees with our previous calculations. The bus impedance matrix also gives us the voltage at bus ② with the fault on bus ①.

$$V_2 = 1.0 - I_{SC} Z_{21} = 1.0 - (-j6.67)(j0.09) = 0.4$$

and since the admittance between buses ① and ② is $-j10$, the current into the fault from the transformer is

$$(0.4 - 0.0)(-j10) = -j4.0 \text{ per unit}$$

which also agrees with our previous result.

We also know immediately the short-circuit current in a three-phase fault at bus ②, which, by referring to Fig. 10.15 with S_1 open and S_2 closed, is

$$I_{SC} = \frac{1.0}{j0.114} = -j8.77 \text{ per unit}$$

This simple example illustrates the value of the bus impedance matrix where the effects of a fault at a number of buses are to be studied. Matrix inversion is not necessary, for $\mathbf{Z}_{\text{bus}}$ can be generated directly by computer using the $\mathbf{Z}_{\text{bus}}$ building algorithm of Sec. 8.4 or the triangular factors of $\mathbf{Y}_{\text{bus}}$, as explained in Sec. 8.5.

Example 10.8. The generators at buses ① and ② of the network of Fig. 10.16(a) have synchronous reactances $X_{d_1} = X_{d_2} = j1.70$ per unit (as marked) and subtransient reactances $X''_{d_1} = X''_{d_2} = j0.25$ per unit. If a three-phase short-circuit fault occurs at bus ③ when there is no load (all bus voltages equal $1.0\angle 0°$ per unit), find the initial symmetrical (subtransient) current (a) in the fault (b) in line ①–③, and (c) the voltage at bus ②. Use triangular factors of $\mathbf{Y}_{\text{bus}}$ in the calculations.

Solution. For the given fault conditions the network has the subtransient reactance diagram shown in Fig. 10.16(b), and the corresponding $\mathbf{Y}_{\text{bus}}$ has the triangular factors

$$\mathbf{Y}_{\text{bus}} = \underbrace{\begin{bmatrix} -j10 & \cdot & \cdot \\ j1 & -j7.9 & \cdot \\ j5 & j3.5 & -j3.94937 \end{bmatrix}}_{\mathbf{L}} \underbrace{\begin{bmatrix} 1 & -0.1 & -0.5 \\ \cdot & 1 & -0.44304 \\ \cdot & \cdot & 1 \end{bmatrix}}_{\mathbf{U}}$$

Since the fault is at bus ③, Eqs. (10.19) through (10.21) show that the calculations involve column 3, $\mathbf{Z}_{\text{bus}}^{(3)}$, of the subtransient $\mathbf{Z}_{\text{bus}}$, which we now generate as follows:

$$\begin{bmatrix} -j10 & \cdot & \cdot \\ j1 & -j7.9 & \cdot \\ j5 & j3.5 & -j3.94937 \end{bmatrix} \begin{bmatrix} x_1 \\ x_2 \\ x_3 \end{bmatrix} = \begin{bmatrix} 0 \\ 0 \\ 1 \end{bmatrix}$$

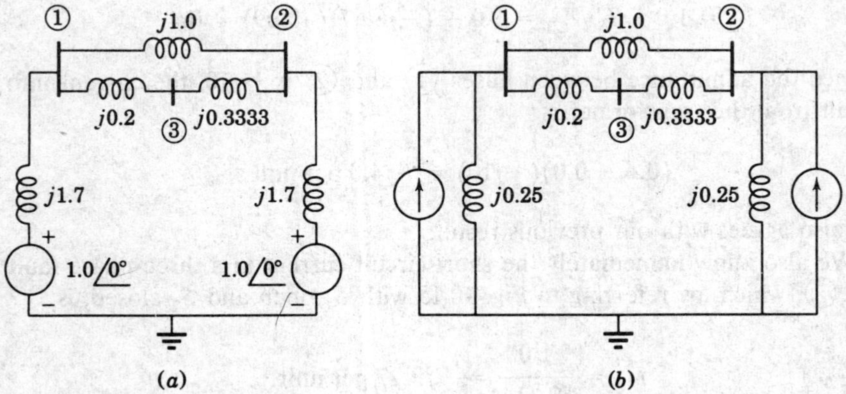

FIGURE 10.16
The reactance diagram for Example 10.8 with generators represented by: (a) series voltage source behind X_d; (b) equivalent current source in parallel with X''_d.

Solving, we obtain

$$x_1 = x_2 = 0; \qquad x_3 = \frac{1}{-j3.94937} = j0.25320 \text{ per unit}$$

and so the elements of $\mathbf{Z}_{\text{bus}}^{(3)}$ are given by

$$\begin{bmatrix} 1 & -0.1 & -0.5 \\ \cdot & 1 & -0.44304 \\ \cdot & \cdot & 1 \end{bmatrix} \begin{bmatrix} Z_{13} \\ Z_{23} \\ Z_{33} \end{bmatrix} = \begin{bmatrix} 0 \\ 0 \\ j0.25320 \end{bmatrix}$$

We find that

$$Z_{33} = j0.25320 \text{ per unit}$$

$$Z_{23} = j0.11218 \text{ per unit}$$

$$Z_{13} = j0.13782 \text{ per unit}$$

(*a*) According to Eq. (10.19), the subtransient current in the fault is

$$I_f'' = \frac{V_f}{Z_{33}} = \frac{1}{j0.25320} = -j3.94937 \text{ per unit}$$

(*b*) From Eq. (10.21) we can write for the current in line ①–③

$$I_{13}'' = -\frac{V_f}{Z_b}\left(\frac{Z_{13} - Z_{33}}{Z_{33}}\right)$$

$$= -\frac{1}{j0.2}\left(\frac{j0.13782 - j0.25320}{j0.25320}\right) = -j2.27844 \text{ per unit}$$

(*c*) During the fault the voltage at bus ② is given by Eq. (10.20) as follows:

$$V_2 = V_f\left(1 - \frac{Z_{23}}{Z_{33}}\right) = 1.0\left(\frac{j0.11218}{j0.25320}\right)$$

$$= 0.55695 \text{ per unit}$$

10.6 SUMMARY

The current flowing immediately after a fault occurs in a power network is determined by the impedances of the network components and the synchronous machines. The initial symmetrical rms fault current can be determined by representing each machine by its subtransient reactance in series with its subtransient internal voltage. Subtransient currents are larger than the transient and steady-state currents. Circuit breakers have ratings determined by the maximum instantaneous current which the breaker must withstand and later interrupt. Interrupting currents depend on the speed of breaker operation. Proper selection and application of circuit breakers should follow the recommendations of ANSI standards, some of which are referenced in this chapter.

Simplifying assumptions usually made in industry-based fault studies are:

- All shunt connections from system buses to the reference node (neutral) can be neglected in the equivalent circuits representing transmission lines and transformers.
- Load impedances are much larger than those of network components, and so they can be neglected in system modeling.
- All buses of the system have rated/nominal voltage of $1.0\angle 0^\circ$ per unit so that no prefault currents flow in the network.
- Synchronous machines can be represented by voltage of $1.0\angle 0^\circ$ per unit behind subtransient or transient reactance, depending on the speed of the circuit breakers and whether the momentary or interrupting fault current is being calculated (ANSI standards should be consulted).
- The voltage-source-plus-series-impedance equivalent circuit of each synchronous machine can be transformed to an equivalent current-source-plus-shunt-impedance model. Then, the shunt impedances of the machine models represent the only shunt connections to the reference node.

The bus impedance matrix is most often used for fault current calculations. The elements of $\mathbf{Z}_{bus}$ can be made available explicitly using the $\mathbf{Z}_{bus}$ building algorithm or they can be generated from the triangular factors of $\mathbf{Y}_{bus}$. Equivalent circuits based on the elements of $\mathbf{Z}_{bus}$ can simplify fault-current calculations as demonstrated in this chapter for the line-end fault.

PROBLEMS

10.1. A 60-Hz alternating voltage having a rms value of 100 V is applied to a series *RL* circuit by closing a switch. The resistance is 15 Ω and the inductance is 0.12 H.

(*a*) Find the value of the dc component of current upon closing the switch if the instantaneous value of the voltage is 50 V at that time.

(*b*) What is the instantaneous value of the voltage which will produce the maximum dc component of current upon closing the switch?

(*c*) What is the instantaneous value of the voltage which will result in the absence of any dc component of current upon closing the switch?

(*d*) If the switch is closed when the instantaneous voltage is zero, find the instantaneous current 0.5, 1.5, and 5.5 cycles later.

10.2. A generator connected through a 5-cycle circuit breaker to a transformer is rated 100 MVA, 18 kV, with reactances of $X_d'' = 19\%$, $X_d' = 26\%$, and $X_d = 130\%$. It is operating at no load and rated voltage when a three-phase short circuit occurs between the breaker and the transformer. Find (*a*) the sustained short-circuit current in the breaker, (*b*) the initial symmetrical rms current in the breaker, and (*c*) the maximum possible dc component of the short-circuit current in the breaker.

10.3. The three-phase transformer connected to the generator described in Prob. 10.2 is rated 100 MVA, 240Y/18Δ kV, X = 10%. If a three-phase short circuit occurs on the high-voltage side of the transformer at rated voltage and no load, find (*a*) the initial symmetrical rms current in the transformer windings on the high-voltage side and (*b*) the initial symmetrical rms current in the line on the low-voltage side.

10.4. A 60-Hz generator is rated 500 MVA, 20 kV, with $X_d'' = 0.20$ per unit. It supplies a purely resistive load of 400 MW at 20 kV. The load is connected directly across the terminals of the generator. If all three phases of the load are short-circuited simultaneously, find the initial symmetrical rms current in the generator in per unit on a base of 500 MVA, 20 kV.

10.5. A generator is connected through a transformer to a synchronous motor. Reduced to the same base, the per-unit subtransient reactances of the generator and motor are 0.15 and 0.35, respectively, and the leakage reactance of the transformer is 0.10 per unit. A three-phase fault occurs at the terminals of the motor when the terminal voltage of the generator is 0.9 per unit and the output current of the generator is 1.0 per unit at 0.8 power factor leading. Find the subtransient current in per unit in the fault, in the generator, and in the motor. Use the terminal voltage of the generator as the reference phasor and obtain the solution (*a*) by computing the voltages behind subtransient reactance in the generator and motor and (*b*) by using Thévenin's theorem.

10.6. Two synchronous motors having subtransient reactances of 0.80 and 0.25 per unit, respectively, on a base of 480 V, 2000 kVA are connected to a bus. This motor is connected by a line having a reactance of 0.023 Ω to a bus of a power system. At the power system bus the short-circuit megavoltamperes of the power system are 9.6 MVA for a nominal voltage of 480 V. When the voltage at the motor bus is 440 V, neglect load current and find the initial symmetrical rms current in a three-phase fault at the motor bus.

10.7. The bus impedance matrix of a four-bus network with values in per unit is

$$\mathbf{Z}_{\text{bus}} = \begin{bmatrix} j0.15 & j0.08 & j0.04 & j0.07 \\ j0.08 & j0.15 & j0.06 & j0.09 \\ j0.04 & j0.06 & j0.13 & j0.05 \\ j0.07 & j0.09 & j0.05 & j0.12 \end{bmatrix}$$

Generators connected to buses ① and ② have their subtransient reactances included in $\mathbf{Z}_{\text{bus}}$. If prefault current is neglected, find the subtransient current in per unit in the fault for a three-phase fault on bus ④. Assume the voltage at the fault is $1.0\angle 0°$ per unit before the fault occurs. Find also the per-unit current from generator 2, whose subtransient reactance is 0.2 per unit.

10.8. For the network shown in Fig. 10.17, find the subtransient current in per unit from generator 1 and in line ①–② and the voltages at buses ① and ③ for a three-phase fault on bus ②. Assume that no current is flowing prior to the fault and that the prefault voltage at bus ② is $1.0\angle 0°$ per unit. Use the bus impedance matrix in the calculations.

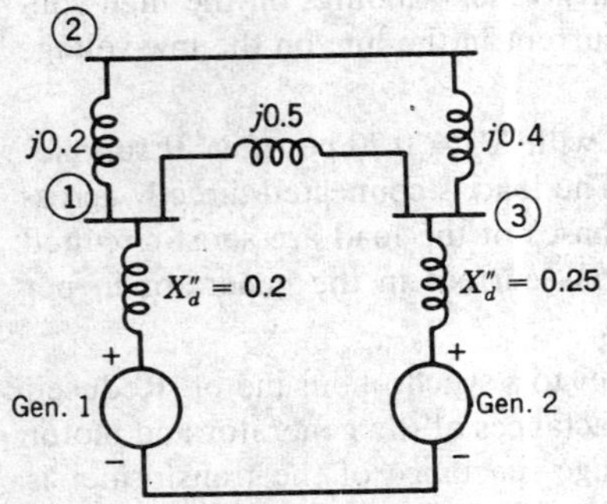

FIGURE 10.17
Network for Probs. 10.8 and 10.9.

10.9. For the network shown in Fig. 10.17, determine $\mathbf{Y}_{bus}$ and its triangular factors. Use the triangular factors to generate the elements of $\mathbf{Z}_{bus}$ needed to solve Prob. 10.8.

10.10. If a three-phase fault occurs at bus ① of the network of Fig. 10.5 when there is no load (all bus voltages equal $1.0\angle 0°$ per unit), find the subtransient current in the fault; the voltages at buses ②, ③, and ④; and the current from the generator connected to bus ④. Use equivalent circuits based on $\mathbf{Z}_{bus}$ of Example 10.3 and similar to those of Fig. 10.7 to illustrate your calculations.

10.11. The network of Fig. 10.8 has the bus impedance matrix given in Example 10.4. If a short-circuit fault occurs at bus ② of the network when there is no load (all bus voltages equal $1.0\angle 0°$ per unit), find the subtransient current in the fault, the voltages at buses ① and ③, and the current from the generator connected to bus ①. Use equivalent circuits based on $\mathbf{Z}_{bus}$ and similar to those of Fig. 10.7 to illustrate your calculations.

10.12. $\mathbf{Z}_{bus}$ for the network of Fig. 10.8 is given in Example 10.4. If a line-end short-circuit fault occurs on line ③–⑤ of the network on the line side of the breaker at bus ③, calculate the subtransient current in the fault when only the near-end breaker at bus ③ has opened. Use the equivalent circuit approach of Fig. 10.11.

10.13. Figure 9.2 shows the one-line diagram of a single power network which has the line data given in Table 9.2. Each generator connected to buses ① and ④ has a subtransient reactance of 0.25 per unit. Making the usual fault study assumptions, summarized in Sec. 10.6, determine for the network (*a*) $\mathbf{Y}_{bus}$, (*b*) $\mathbf{Z}_{bus}$, (*c*) the subtransient current in per unit in a three-phase fault on bus ③ and (*d*) the contributions to the fault current from line ①–③ and from line ④–③.

10.14. A 625-kV generator with $X''_d = 0.20$ per unit is connected to a bus through a circuit breaker, as shown in Fig. 10.18. Connected through circuit breakers to the same bus are three synchronous motors rated 250 hp, 2.4 kV, 1.0 power factor, 90% efficiency, with $X''_d = 0.20$ per unit. The motors are operating at full load, unity power factor, and rated voltage, with the load equally divided among the machines.

(*a*) Draw the impedance diagram with the impedances marked in per unit on a base of 625 kVA, 2.4 kV.

(*b*) Find the symmetrical short-circuit current in amperes, which must be interrupted by breakers *A* and *B* for a three-phase fault at point *P*. Simplify the calculations by neglecting the prefault current.

(*c*) Repeat part (*b*) for a three-phase fault at point *Q*.

(*d*) Repeat part (*b*) for a three-phase fault at point *R*.

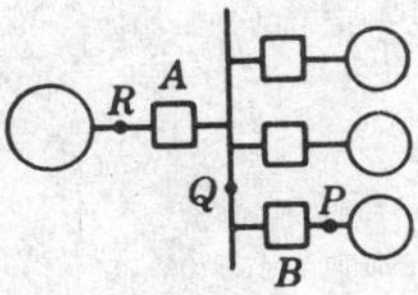

FIGURE 10.18
One-line diagram for Prob. 10.14.

10.15. A circuit breaker having a nominal rating of 34.5 kV and a continuous current rating of 1500 A has a voltage range factor *K* of 1.65. Rated maximum voltage is 38 kV and the rated short-circuit current at that voltage is 22 kA. Find (*a*) the voltage below which rated short-circuit current does not increase as operating voltage decreases and the value of that current and (*b*) rated short-circuit current at 34.5 kV.

CHAPTER 11

SYMMETRICAL COMPONENTS AND SEQUENCE NETWORKS

One of the most powerful tools for dealing with unbalanced polyphase circuits is the *method of symmetrical components* introduced by C. L. Fortescue.[1] Fortescue's work proves that an unbalanced system of *n* related phasors can be resolved into *n* systems of balanced phasors called the *symmetrical components* of the original phasors. The *n* phasors of each set of components are equal in length, and the angles between adjacent phasors of the set are equal. Although the method is applicable to any unbalanced polyphase system, we confine our discussion to three-phase systems.

In a three-phase system which is normally balanced, unbalanced fault conditions generally cause unbalanced currents and voltages to exist in each of the phases. If the currents and voltages are related by constant impedances, the system is said to be *linear* and the principle of superposition applies. The voltage response of the linear system to the unbalanced currents can be determined by considering the separate responses of the individual elements to the symmetrical components of the currents. The system elements of interest are the machines, transformers, transmission lines, and loads connected to Δ or Y configurations.

[1] C. L. Fortescue, "Method of Symmetrical Coordinates Applied to the Solution of Polyphase Networks," *Transactions of AIEE*, vol. 37, 1918, pp. 1027–1140.

In this chapter we study symmetrical components and show that the response of each system element depends, in general, on its connections and the component of the current being considered. Equivalent circuits, called *sequence circuits*, will be developed to reflect the separate responses of the elements to each current component. There are three equivalent circuits for each element of the three-phase system. By organizing the individual equivalent circuits into networks according to the interconnections of the elements, we arrive at the concept of three *sequence networks*. Solving the sequence networks for the fault conditions gives symmetrical current and voltage components which can be combined together to reflect the effects of the original unbalanced fault currents on the overall system.

Analysis by symmetrical components is a powerful tool which makes the calculation of unsymmetrical faults almost as easy as the calculation of three-phase faults. Unsymmetrical faults are studied in Chap. 12.

11.1 SYNTHESIS OF UNSYMMETRICAL PHASORS FROM THEIR SYMMETRICAL COMPONENTS

According to Fortescue's theorem, three unbalanced phasors of a three-phase system can be resolved into *three balanced systems* of phasors. The balanced sets of components are:

1. *Positive-sequence components* consisting of three phasors equal in magnitude, displaced from each other by 120° in phase, and having the same phase sequence as the original phasors,
2. *Negative-sequence components* consisting of three phasors equal in magnitude, displaced from each other by 120° in phase, and having the phase sequence opposite to that of the original phasors, and
3. *Zero-sequence components* consisting of three phasors equal in magnitude and with zero phase displacement from each other.

It is customary when solving a problem by symmetrical components to designate the three phases of the system as a, b, and c in such a manner that the phase sequence of the voltages and currents in the system is *abc*. Thus, the phase sequence of the positive-sequence components of the unbalanced phasors is *abc*, and the phase sequence of the negative-sequence components is *acb*. If the original phasors are voltages, they may be designated V_a, V_b, and V_c. The three sets of symmetrical components are designated by the additional superscript 1 for the positive-sequence components, 2 for the negative-sequence components, and 0 for the zero-sequence components. Superscripts are chosen so as not to confuse bus numbers with sequence indicators later on in this chapter. The positive-sequence components of V_a, V_b, and V_c are $V_a^{(1)}$, $V_b^{(1)}$, and $V_c^{(1)}$, respectively. Similarly, the negative-sequence components are $V_a^{(2)}$, $V_b^{(2)}$,

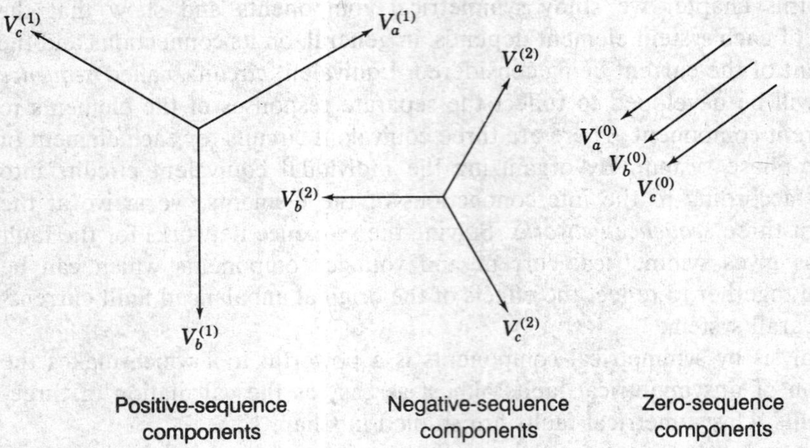

FIGURE 11.1
Three sets of balanced phasors which are the symmetrical components of three unbalanced phasors.

and $V_c^{(2)}$, and the zero-sequence components are $V_a^{(0)}$, $V_b^{(0)}$, and $V_c^{(0)}$, respectively. Figure 11.1 shows three such sets of symmetrical components. Phasors representing currents will be designated by I with superscripts as for voltages.

Since each of the original unbalanced phasors is the sum of its components, the original phasors expressed in terms of their components are

$$V_a = V_a^{(0)} + V_a^{(1)} + V_a^{(2)} \tag{11.1}$$

$$V_b = V_b^{(0)} + V_b^{(1)} + V_b^{(2)} \tag{11.2}$$

$$V_c = V_c^{(0)} + V_c^{(1)} + V_c^{(2)} \tag{11.3}$$

The synthesis of a set of three unbalanced phasors from the three sets of symmetrical components of Fig. 11.1 is shown in Fig. 11.2.

The many advantages of analysis of power systems by the method of symmetrical components will become apparent gradually as we apply the method to the study of unsymmetrical faults on otherwise symmetrical systems. It is sufficient to say here that the method consists in finding the symmetrical components of current at the fault. Then, the values of current and voltage at various points in the system can be found by means of the bus impedance matrix. The method is simple and leads to accurate predictions of system behavior.

11.2 THE SYMMETRICAL COMPONENTS OF UNSYMMETRICAL PHASORS

In Fig. 11.2 we observe the synthesis of three unsymmetrical phasors from three sets of symmetrical phasors. The synthesis is made in accordance with Eqs.

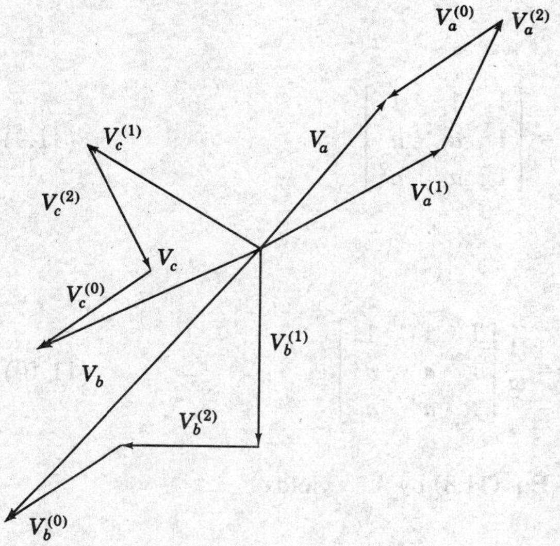

FIGURE 11.2
Graphical addition of the components shown in Fig. 11.1 to obtain three unbalanced phasors.

(11.1) through (11.3). Now let us examine these same equations to determine how to resolve three unsymmetrical phasors into their symmetrical components.

First, we note that the number of unknown quantities can be reduced by expressing each component of V_b and V_c as the product of a component of V_a and some function of the operator $a = 1\angle 120°$, which was introduced in Chap. 1. Reference to Fig. 11.1 verifies the following relations:

$$\begin{aligned} V_b^{(0)} &= V_a^{(0)} & V_c^{(0)} &= V_a^{(0)} \\ V_b^{(1)} &= a^2 V_a^{(1)} & V_c^{(1)} &= a V_a^{(1)} \\ V_b^{(2)} &= a V_a^{(2)} & V_c^{(2)} &= a^2 V_a^{(2)} \end{aligned} \tag{11.4}$$

Repeating Eq. (11.1) and substituting Eqs. (11.4) in Eqs. (11.2) and (11.3) yield

$$V_a = V_a^{(0)} + V_a^{(1)} + V_a^{(2)} \tag{11.5}$$

$$V_b = V_a^{(0)} + a^2 V_a^{(1)} + a V_a^{(2)} \tag{11.6}$$

$$V_c = V_a^{(0)} + a V_a^{(1)} + a^2 V_a^{(2)} \tag{11.7}$$

or in matrix form

$$\begin{bmatrix} V_a \\ V_b \\ V_c \end{bmatrix} = \begin{bmatrix} 1 & 1 & 1 \\ 1 & a^2 & a \\ 1 & a & a^2 \end{bmatrix} \begin{bmatrix} V_a^{(0)} \\ V_a^{(1)} \\ V_a^{(2)} \end{bmatrix} = \mathbf{A} \begin{bmatrix} V_a^{(0)} \\ V_a^{(1)} \\ V_a^{(2)} \end{bmatrix} \tag{11.8}$$

where, for convenience, we let

$$\mathbf{A} = \begin{bmatrix} 1 & 1 & 1 \\ 1 & a^2 & a \\ 1 & a & a^2 \end{bmatrix} \tag{11.9}$$

Then, as may be verified easily,

$$\mathbf{A}^{-1} = \frac{1}{3}\begin{bmatrix} 1 & 1 & 1 \\ 1 & a & a^2 \\ 1 & a^2 & a \end{bmatrix} \tag{11.10}$$

and premultiplying both sides of Eq. (11.8) by $\mathbf{A}^{-1}$ yields

$$\begin{bmatrix} V_a^{(0)} \\ V_a^{(1)} \\ V_a^{(2)} \end{bmatrix} = \frac{1}{3}\begin{bmatrix} 1 & 1 & 1 \\ 1 & a & a^2 \\ 1 & a^2 & a \end{bmatrix}\begin{bmatrix} V_a \\ V_b \\ V_c \end{bmatrix} = \mathbf{A}^{-1}\begin{bmatrix} V_a \\ V_b \\ V_c \end{bmatrix} \tag{11.11}$$

which shows us how to resolve three unsymmetrical phasors into their symmetrical components. These relations are so important that we write the separate equations in the expanded form

$$V_a^{(0)} = \tfrac{1}{3}(V_a + V_b + V_c) \tag{11.12}$$

$$V_a^{(1)} = \tfrac{1}{3}(V_a + aV_b + a^2V_c) \tag{11.13}$$

$$V_a^{(2)} = \tfrac{1}{3}(V_a + a^2V_b + aV_c) \tag{11.14}$$

If required, the components $V_b^{(0)}$, $V_b^{(1)}$, $V_b^{(2)}$, $V_c^{(0)}$, $V_c^{(1)}$, and $V_c^{(2)}$ can be found by Eqs. (11.4). Similar results apply to line-to-line voltages simply by replacing V_a, V_b, and V_c in above equations by V_{ab}, V_{bc}, and V_{ca}, respectively.

Equation (11.12) shows that no zero-sequence components exist if the sum of the unbalanced phasors is zero. Since the sum of the line-to-line voltage phasors in a three-phase system is always zero, zero-sequence components are never present in the line voltages regardless of the degree of unbalance. The sum of the three line-to-line neutral voltage phasors is not necessarily zero, and voltages to neutral may contain zero-sequence components.

The preceding equations could have been written for any set of related phasors, and we might have written them for currents instead of for voltages. They may be solved either analytically or graphically. Because some of the preceding equations are so fundamental, they are summarized for currents:

$$\begin{aligned} I_a &= I_a^{(0)} + I_a^{(1)} + I_a^{(2)} \\ I_b &= I_a^{(0)} + a^2 I_a^{(1)} + aI_a^{(2)} \\ I_c &= I_a^{(0)} + a\, I_a^{(1)} + a^2 I_a^{(2)} \end{aligned} \tag{11.15}$$

$$\begin{aligned} I_a^{(0)} &= \tfrac{1}{3}\left(I_a + I_b + I_c\right) \\ I_a^{(1)} &= \tfrac{1}{3}\left(I_a + aI_b + a^2 I_c\right) \\ I_a^{(2)} &= \tfrac{1}{3}\left(I_a + a^2 I_b + aI_c\right) \end{aligned} \tag{11.16}$$

Finally, these results can be extended to phase currents of a Δ circuit [such as that of Fig. 11.4(*a*)] by replacing I_a, I_b, and I_c by I_{ab}, I_{bc}, and I_{ca}, respectively.

Example 11.1. One conductor of a three-phase line is open. The current flowing to the Δ-connected load through line *a* is 10 A. With the current in line *a* as reference and assuming that line *c* is open, find the symmetrical components of the line currents.

Solution. Figure 11.3 is a diagram of the circuit. The line currents are

$$I_a = 10\angle 0^\circ \text{ A} \qquad I_b = 10\angle 180^\circ \text{ A} \qquad I_c = 0 \text{ A}$$

From Eqs. (11.16)

$$I_a^{(0)} = \tfrac{1}{3}\left(10\angle 0^\circ + 10\angle 180^\circ + 0\right) = 0$$

$$I_a^{(1)} = \tfrac{1}{3}\left(10\angle 0^\circ + 10\angle 180^\circ + 120^\circ + 0\right)$$

$$= 5 - j2.89 = 5.78\angle -30^\circ \text{ A}$$

$$I_a^{(2)} = \tfrac{1}{3}\left(10\angle 0 + 10\angle 180^\circ + 240^\circ + 0\right)$$

$$= 5 + j2.89 = 5.78\angle 30^\circ \text{ A}$$

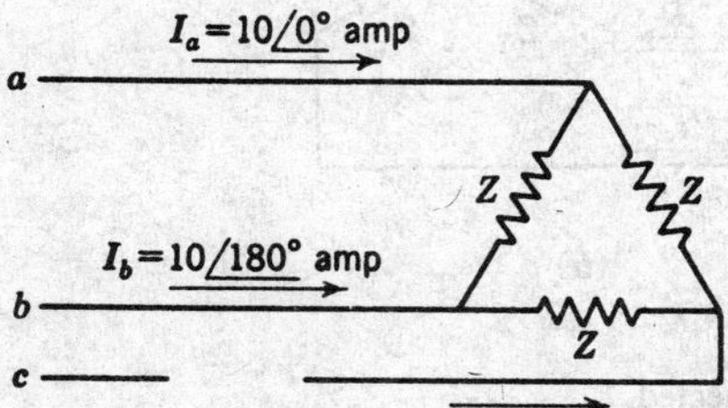

FIGURE 11.3
Circuit for Example 11.1.

From Eqs. (11.4)

$$I_b^{(0)} = 0 \qquad I_c^{(0)} = 0$$

$$I_b^{(1)} = 5.78\angle -150^\circ \text{ A} \qquad I_c^{(1)} = 5.78\angle 90^\circ \text{ A}$$

$$I_b^{(2)} = 5.78\angle 150^\circ \text{ A} \qquad I_c^{(2)} = 5.78\angle -90^\circ \text{ A}$$

The result $I_a^{(0)} = I_b^{(0)} = I_c^{(0)} = 0$ holds for any three-wire system.

In Example 11.1 we note that components $I_c^{(1)}$ and $I_c^{(2)}$ have nonzero values although line c is open and can carry no net current. As is expected, therefore, the sum of the components in line c is zero. Of course, the sum of the components in line a is $10\angle 0^\circ$ A, and the sum of the components in line b is $10\angle 180^\circ$ A.

11.3 SYMMETRICAL Y AND Δ CIRCUITS

In three-phase systems circuit elements are connected between lines a, b, and c in either Y or Δ configuration. Relationships between the symmetrical components of Y and Δ currents and voltages can be established by referring to Fig. 11.4, which shows *symmetrical* impedances connected in Y and Δ. Let us agree that the reference phase for Δ quantities is branch a–b. The particular choice of reference phase is arbitrary and does not affect the results. For currents we have

$$\begin{aligned} I_a &= I_{ab} - I_{ca} \\ I_b &= I_{bc} - I_{ab} \\ I_c &= I_{ca} - I_{bc} \end{aligned} \tag{11.17}$$

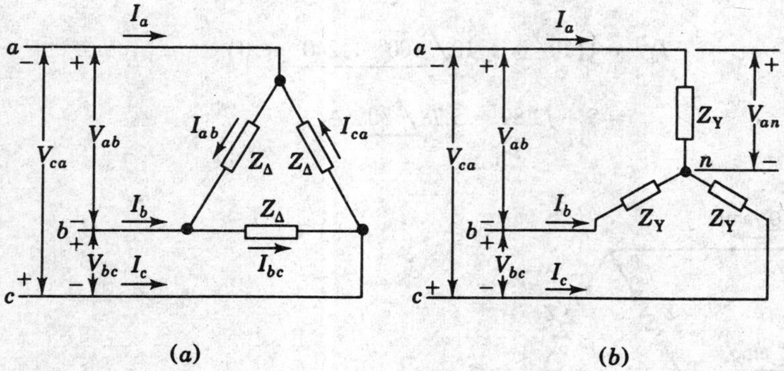

FIGURE 11.4
Symmetrical impedances: (a) Δ-connected; (b) Y-connected.

Adding all three equations together and invoking the definition of zero-sequence current, we obtain $I_a^{(0)} = (I_a + I_b + I_c)/3 = 0$, which means that *line currents into a Δ-connected circuit have no zero-sequence currents*. Substituting components of current in the equation for I_a yields

$$
\begin{aligned}
I_a^{(1)} + I_a^{(2)} &= \left(I_{ab}^{(0)} + I_{ab}^{(1)} + I_{ab}^{(2)}\right) - \left(I_{ca}^{(0)} + I_{ca}^{(1)} + I_{ca}^{(2)}\right) \\
&= \underbrace{\left(I_{ab}^{(0)} - I_{ca}^{(0)}\right)}_{0} + \left(I_{ab}^{(1)} - I_{ca}^{(1)}\right) + \left(I_{ab}^{(2)} - I_{ca}^{(2)}\right) \qquad (11.18)
\end{aligned}
$$

Evidently, if a nonzero value of circulating current $I_{ab}^{(0)}$ exists in the Δ circuit, it cannot be determined from the line currents alone. Noting that $I_{ca}^{(1)} = aI_{ab}^{(1)}$ and $I_{ca}^{(2)} = a^2 I_{ab}^{(2)}$, we now write Eq. (11.18) as follows:

$$
I_a^{(1)} + I_a^{(2)} = (1 - a) I_{ab}^{(1)} + (1 - a^2) I_{ab}^{(2)} \qquad (11.19)
$$

A similar equation for phase b is $I_b^{(1)} + I_b^{(2)} = (1 - a)I_{bc}^{(1)} + (1 - a^2)I_{bc}^{(2)}$, and expressing $I_b^{(1)}$, $I_b^{(2)}$, $I_{bc}^{(1)}$, and $I_{bc}^{(2)}$ in terms of $I_a^{(1)}$, $I_a^{(2)}$, $I_{ab}^{(1)}$, and $I_{ab}^{(2)}$, we obtain a resultant equation which can be solved along with Eq. (11.19) to yield the important results

$$
I_a^{(1)} = \sqrt{3}\,\angle{-30^\circ} \times I_{ab}^{(1)} \qquad I_a^{(2)} = \sqrt{3}\,\angle{30^\circ} \times I_{ab}^{(2)} \qquad (11.20)
$$

These results amount to equating currents of the same sequence in Eq. (11.19). Complete sets of positive- and negative-sequence components of currents are shown in the phasor diagram of Fig. 11.5(a).

In a similar manner, the line-to-line voltages can be written in terms of line-to-neutral voltages of a Y-connected system,

$$
\begin{aligned}
V_{ab} &= V_{an} - V_{bn} \\
V_{bc} &= V_{bn} - V_{cn} \qquad (11.21) \\
V_{ca} &= V_{cn} - V_{an}
\end{aligned}
$$

Adding together all three equations shows that $V_{ab}^{(0)} = (V_{ab} + V_{bc} + V_{ca})/3 = 0$. In words, *line-to-line voltages have no zero-sequence components*. Substituting components of the voltages in the equation for V_{ab} yields

$$
\begin{aligned}
V_{ab}^{(1)} + V_{ab}^{(2)} &= \left(V_{an}^{(0)} + V_{an}^{(1)} + V_{an}^{(2)}\right) - \left(V_{bn}^{(0)} + V_{bn}^{(1)} + V_{bn}^{(2)}\right) \\
&= \underbrace{\left(V_{an}^{(0)} - V_{bn}^{(0)}\right)}_{0} + \left(V_{an}^{(1)} - V_{bn}^{(1)}\right) + \left(V_{an}^{(2)} - V_{bn}^{(2)}\right) \quad (11.22)
\end{aligned}
$$

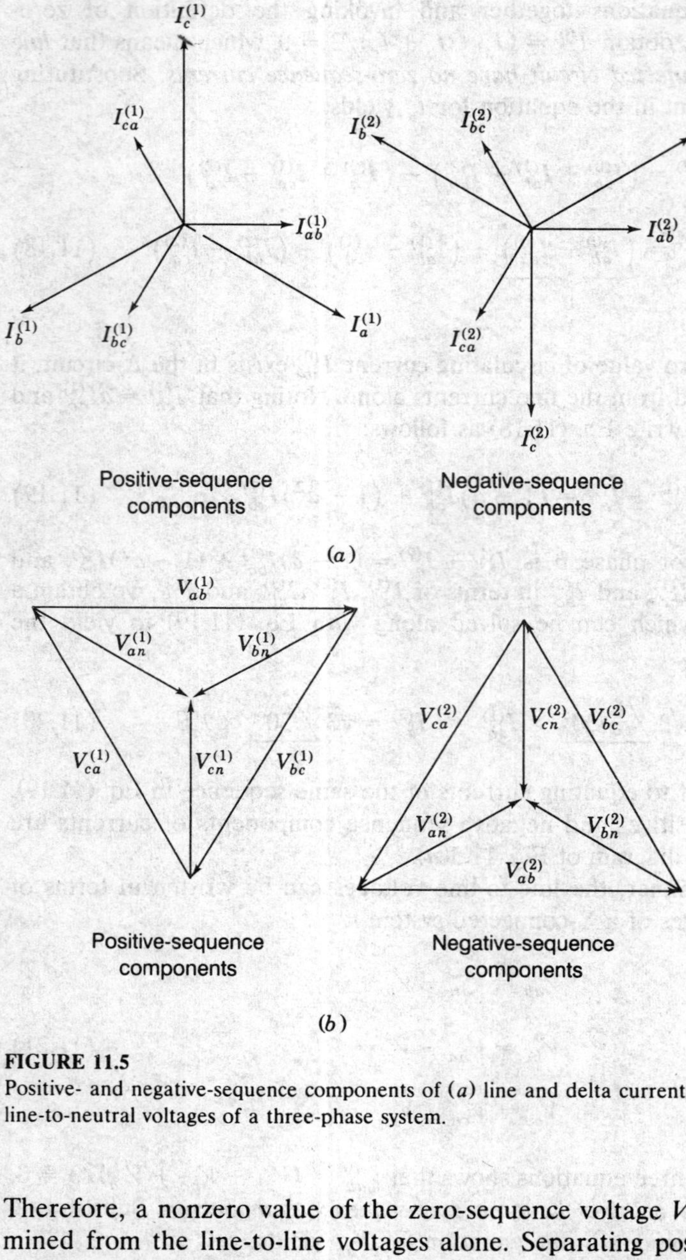

FIGURE 11.5
Positive- and negative-sequence components of (*a*) line and delta currents and (*b*) line-to-line and line-to-neutral voltages of a three-phase system.

Therefore, a nonzero value of the zero-sequence voltage $V_{an}^{(0)}$ cannot be determined from the line-to-line voltages alone. Separating positive- and negative-sequence quantities in the manner explained for Eq. (11.19), we obtain the important voltage relations

$$\begin{aligned} V_{ab}^{(1)} &= (1 - a^2)V_{an}^{(1)} = \sqrt{3}\,\angle 30^\circ \times V_{an}^{(1)} \\ V_{ab}^{(2)} &= (1 - a)V_{an}^{(2)} = \sqrt{3}\,\angle -30^\circ \times V_{an}^{(2)} \end{aligned} \tag{11.23}$$

Complete sets of positive- and negative-sequence components of voltages are shown in the phasor diagrams of Fig. 11.5(*b*). If the voltages to neutral are in per unit referred to the base voltage to neutral and the line voltages are in per unit referred to the base voltage from line to line, the $\sqrt{3}$ multipliers must be omitted from Eqs. (11.23). If both voltages are referred to the *same* base, however, the equations are correct as given. Similarly, when line and Δ currents are expressed in per unit, each on its own base, the $\sqrt{3}$ in Eqs. (11.20) disappears since the two bases are related to one another in the ratio of $\sqrt{3}:1$. When the currents are expressed on the same base, the equation is correct as written.

From Fig. 11.4 we note that $V_{ab}/I_{ab} = Z_\Delta$ when there are no sources or mutual coupling inside the Δ circuit. When positive- and negative-sequence quantities are both present, we have

$$\frac{V_{ab}^{(1)}}{I_{ab}^{(1)}} = Z_\Delta = \frac{V_{ab}^{(2)}}{I_{ab}^{(2)}} \tag{11.24}$$

Substituting from Eqs. (11.20) and (11.23), we obtain

$$\frac{\sqrt{3}\,V_{an}^{(1)}\angle 30^\circ}{\dfrac{I_a^{(1)}}{\sqrt{3}}\angle 30^\circ} = Z_\Delta = \frac{\sqrt{3}\,V_{an}^{(2)}\angle -30^\circ}{\dfrac{I_a^{(2)}}{\sqrt{3}}\angle -30^\circ}$$

so that

$$\frac{V_{an}^{(1)}}{I_a^{(1)}} = \frac{Z_\Delta}{3} = \frac{V_{an}^{(2)}}{I_a^{(2)}} \tag{11.25}$$

which shows that the Δ-connected impedances Z_Δ are equivalent to the *per-phase* or Y-connected impedances $Z_Y = Z_\Delta/3$ of Fig. 11.6(*a*) insofar as posi-

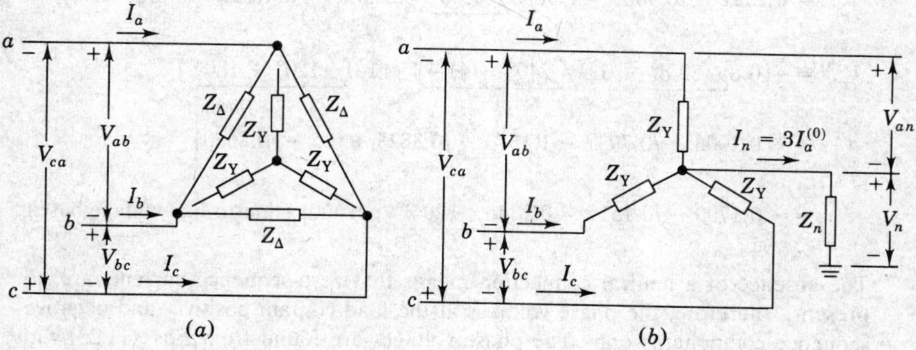

FIGURE 11.6
(*a*) Symmetrical Δ-connected impedances and their Y-connected equivalents related by $Z_Y = Z_\Delta/3$; (*b*) Y-connected impedances with neutral connection to ground.

tive- or negative-sequence currents are concerned. Of course, this result could have been anticipated from the usual Δ-Y transformations of Table 1.2. The relation $Z_Y = Z_\Delta/3$ is correct when the impedances Z_Δ and Z_Y are both expressed in ohms or in per unit on the same kilovoltampere and voltage bases.

Example 11.2. Three identical Y-connected resistors form a load bank with a three-phase rating of 2300 V and 500 kVA. If the load bank has applied voltages

$$|V_{ab}| = 1840 \text{ V} \qquad |V_{bc}| = 2760 \text{ V} \qquad |V_{ca}| = 2300 \text{ V}$$

find the line voltages and currents in per unit into the load. Assume that the neutral of the load is not connected to the neutral of the system and select a base of 2300 V, 500 kVA.

Solution. The rating of the load bank coincides with the specified base, and so the resistance values are 1.0 per unit. On the same base the given line voltages in per unit are

$$|V_{ab}| = 0.8 \qquad |V_{bc}| = 1.2 \qquad |V_{ca}| = 1.0$$

Assuming an angle of 180° for V_{ca} and using the law of cosines to find the angles of the other line voltages, we find the per-unit values

$$V_{ab} = 0.8\angle 82.8^\circ \qquad V_{bc} = 1.2\angle -41.4^\circ \qquad V_{ca} = 1.0\angle 180^\circ$$

The symmetrical components of the line voltages are

$$V_{ab}^{(1)} = \tfrac{1}{3}\left(0.8\angle 82.8^\circ + 1.2\angle 120^\circ - 41.4^\circ + 1.0\angle 240^\circ + 180^\circ\right)$$

$$= \tfrac{1}{3}(0.1003 + j0.7937 + 0.2372 + j1.1763 + 0.5 + j0.8660)$$

$$= 0.2792 + j0.9453 = 0.9857\angle 73.6^\circ \text{ per unit (line-to-line voltage base)}$$

$$V_{ab}^{(2)} = \tfrac{1}{3}\left(0.8\angle 82.8^\circ + 1.2\angle 240^\circ - 41.4^\circ + 1.0\angle 120^\circ + 180^\circ\right)$$

$$= \tfrac{1}{3}(0.1003 + j0.7937 - 1.1373 - j0.3828 + 0.5 - j0.8660)$$

$$= -0.1790 - j0.1517 = 0.2346\angle 220.3^\circ \text{ per unit (line-to-line voltage base)}$$

The absence of a neutral connection means that zero-sequence currents are not present. Therefore, the phase voltages at the load contain positive- and negative-sequence components only. The phase voltages are found from Eqs. (11.23) with the $\sqrt{3}$ factor omitted since the line voltages are expressed in terms of the base voltage from line to line and the phase voltages are desired in per unit of the base voltage to neutral. Thus,

$$V_{an}^{(1)} = 0.9857\angle 73.6^\circ - 30^\circ$$

$$= 0.9857\angle 43.6^\circ \text{ per unit (line-to-neutral voltage base)}$$

$$V_{an}^{(2)} = 0.2346\angle 220.3^\circ + 30^\circ$$

$$= 0.2346\angle 250.3^\circ \text{ per unit (line-to-neutral voltage base)}$$

Since each resistor has an impedance of $1.0\angle 0^\circ$ per unit,

$$I_a^{(1)} = \frac{V_a^{(1)}}{1.0\angle 0^\circ} = 0.9857\angle 43.6^\circ \text{ per unit}$$

$$I_a^{(2)} = \frac{V_a^{(2)}}{1.0\angle 0^\circ} = 0.2346\angle 250.3^\circ \text{ per unit}$$

The positive direction of current is chosen to be from the supply toward the load.

11.4 POWER IN TERMS OF SYMMETRICAL COMPONENTS

If the symmetrical components of current and voltage are known, the power expended in a three-phase circuit can be computed directly from the components. Demonstration of this statement is a good example of the matrix manipulation of symmetrical components.

The total complex power flowing into a three-phase circuit through three lines a, b, and c is

$$S_{3\phi} = P + jQ = V_a I_a^* + V_b I_b^* + V_c I_c^* \qquad (11.26)$$

where V_a, V_b, and V_c are the voltages to reference at the terminals and I_a, I_b, and I_c are the currents flowing into the circuit in the three lines. A neutral connection may or may not be present. If there is impedance in the neutral connection to ground, then the voltages V_a, V_b, and V_c must be interpreted as voltages from the line to ground rather than to neutral. In matrix notation

$$S_{3\phi} = \begin{bmatrix} V_a & V_b & V_c \end{bmatrix} \begin{bmatrix} I_a \\ I_b \\ I_c \end{bmatrix}^* = \begin{bmatrix} V_a \\ V_b \\ V_c \end{bmatrix}^T \begin{bmatrix} I_a \\ I_b \\ I_c \end{bmatrix}^* \qquad (11.27)$$

where the conjugate of a matrix is understood to be composed of elements that are the conjugates of the corresponding elements of the original matrix.

To introduce the symmetrical components of the voltages and currents, we make use of Eq. (11.8) to obtain

$$S_{3\phi} = [\mathbf{A}\mathbf{V}_{012}]^T[\mathbf{A}\mathbf{I}_{012}]^* \tag{11.28}$$

where
$$\mathbf{V}_{012} = \begin{bmatrix} V_a^{(0)} \\ V_a^{(1)} \\ V_a^{(2)} \end{bmatrix} \quad \text{and} \quad \mathbf{I}_{012} = \begin{bmatrix} I_a^{(0)} \\ I_a^{(1)} \\ I_a^{(2)} \end{bmatrix} \tag{11.29}$$

The *reversal rule* of matrix algebra states that the transpose of the product of two matrices is equal to the product of the transposes of the matrices in reverse order. According to this rule,

$$[\mathbf{A}\mathbf{V}_{012}]^T = \mathbf{V}_{012}^T\mathbf{A}^T \tag{11.30}$$

and so
$$S_{3\phi} = \mathbf{V}_{012}^T\mathbf{A}^T[\mathbf{A}\mathbf{I}_{012}]^* = \mathbf{V}_{012}^T\mathbf{A}^T\mathbf{A}^*\mathbf{I}_{012}^* \tag{11.31}$$

Noting that $\mathbf{A}^T = \mathbf{A}$ and that a and a^2 are conjugates, we obtain

$$S_{3\phi} = \begin{bmatrix} V_a^{(0)} & V_a^{(1)} & V_a^{(2)} \end{bmatrix} \begin{bmatrix} 1 & 1 & 1 \\ 1 & a^2 & a \\ 1 & a & a^2 \end{bmatrix} \begin{bmatrix} 1 & 1 & 1 \\ 1 & a & a^2 \\ 1 & a^2 & a \end{bmatrix} \begin{bmatrix} I_a^{(0)} \\ I_a^{(1)} \\ I_a^{(2)} \end{bmatrix}^* \tag{11.32}$$

or since

$$\mathbf{A}^T\mathbf{A}^* = \begin{bmatrix} 1 & 1 & 1 \\ 1 & a^2 & a \\ 1 & a & a^2 \end{bmatrix} \begin{bmatrix} 1 & 1 & 1 \\ 1 & a & a^2 \\ 1 & a^2 & a \end{bmatrix} = 3\begin{bmatrix} 1 & 0 & 0 \\ 0 & 1 & 0 \\ 0 & 0 & 1 \end{bmatrix}$$

$$S_{3\phi} = 3\begin{bmatrix} V_a^{(0)} & V_a^{(1)} & V_a^{(2)} \end{bmatrix} \begin{bmatrix} I_a^{(0)} \\ I_a^{(1)} \\ I_a^{(2)} \end{bmatrix}^* \tag{11.33}$$

So, complex power is

$$S_{3\phi} = V_aI_a^* + V_bI_b^* + V_cI_c^* = 3V_a^{(0)}I_a^{(0)*} + 3V_a^{(1)}I_a^{(1)*} + 3V_a^{(2)}I_a^{(2)*} \tag{11.34}$$

which shows how complex power (in *voltamperes*) can be computed from the symmetrical components of the voltages to reference (in *volts*) and line currents (in *amperes*) of an unbalanced three-phase circuit. It is important to note that the transformation of *a-b-c* voltages and currents to symmetrical components is power-invariant in the sense discussed in Sec. 8.6, only if each product of sequence voltage (in volts) times the complex conjugate of the corresponding sequence current (in amperes) is multiplied by 3, as shown in Eq. (11.34). When the complex power $S_{3\phi}$ is expressed in per unit of a three-phase voltampere base, however, the multiplier 3 disappears.

Example 11.3. Using symmetrical components, calculate the power absorbed in the load of Example 11.2 and check the answer.

Solution. In per unit of the three-phase 500-kVA base, Eq. (11.34) becomes

$$S_{3\phi} = V_a^{(0)} I_a^{(0)*} + V_a^{(1)} I_a^{(1)*} + V_a^{(2)} I_a^{(2)*}$$

Substituting the components of voltages and currents from Example 11.2, we obtain

$$\begin{aligned} S_{3\phi} &= 0 + 0.9857\angle 43.6^\circ \times 0.9857\angle -43.6^\circ + 0.2346\angle 250.3^\circ \times 0.2346\angle -250.3^\circ \\ &= (0.9857)^2 + (0.2346)^2 = 1.02664 \text{ per unit} \\ &= 513.32 \text{ kW} \end{aligned}$$

The per-unit value of the resistors in each phase of the Y-connected load bank is 1.0 per unit. In ohms, therefore,

$$R_Y = \frac{(2300)^2}{500{,}000} = 10.58\ \Omega$$

and the equivalent Δ-connected resistors are

$$R_\Delta = 3R_Y = 31.74\ \Omega$$

From the given line-to-line voltages we calculate directly

$$\begin{aligned} S_{3\phi} &= \frac{|V_{ab}|^2}{R_\Delta} + \frac{|V_{bc}|^2}{R_\Delta} + \frac{|V_{ca}|^2}{R_\Delta} \\ &= \frac{(1840)^2 + (2760)^2 + (2300)^2}{31.74} = 513.33 \text{ kW} \end{aligned}$$

11.5 SEQUENCE CIRCUITS OF Y AND Δ IMPEDANCES

If impedance Z_n is inserted between the neutral and ground of the Y-connected impedances shown in Fig. 11.6(*b*), then the sum of the line currents is equal to the current I_n in the return path through the neutral. That is,

$$I_n = I_a + I_b + I_c \tag{11.35}$$

Expressing the unbalanced line currents in terms of their symmetrical components gives

$$\begin{aligned}
I_n &= \left(I_a^{(0)} + I_a^{(1)} + I_a^{(2)}\right) + \left(I_b^{(0)} + I_b^{(1)} + I_b^{(2)}\right) + \left(I_c^{(0)} + I_c^{(1)} + I_c^{(2)}\right) \\
&= \left(I_a^{(0)} + I_b^{(0)} + I_c^{(0)}\right) + \underbrace{\left(I_a^{(1)} + I_b^{(1)} + I_c^{(1)}\right)}_{0} + \underbrace{\left(I_a^{(2)} + I_b^{(2)} + I_c^{(2)}\right)}_{0} \\
&= 3I_a^{(0)}
\end{aligned} \tag{11.36}$$

Since the positive-sequence and negative-sequence currents add separately to zero at neutral point n, there cannot be any positive-sequence or negative-sequence currents in the connections from neutral to ground regardless of the value of Z_n. Moreover, the zero-sequence currents combining together at n become $3I_a^{(0)}$, which produces the voltage drop $3I_a^{(0)}Z_n$ between neutral and ground. It is important, therefore, to distinguish between voltages to neutral and voltages to ground under unbalanced conditions. Let us designate voltages of phase a with respect to neutral and ground as V_{an} and V_a, respectively. Thus, the voltage of phase a with respect to ground is given by $V_a = V_{an} + V_n$, where $V_n = 3I_a^{(0)}Z_n$. Referring to Fig. 11.6(b), we can write the voltage drops to ground from each of the lines a, b, and c as

$$\begin{bmatrix} V_a \\ V_b \\ V_c \end{bmatrix} = \begin{bmatrix} V_{an} \\ V_{bn} \\ V_{cn} \end{bmatrix} + \begin{bmatrix} V_n \\ V_n \\ V_n \end{bmatrix} = Z_Y \begin{bmatrix} I_a \\ I_b \\ I_c \end{bmatrix} + 3I_a^{(0)}Z_n \begin{bmatrix} 1 \\ 1 \\ 1 \end{bmatrix} \tag{11.37}$$

The a-b-c voltages and currents in this equation can be replaced by their symmetrical components as follows:

$$\mathbf{A} \begin{bmatrix} V_a^{(0)} \\ V_a^{(1)} \\ V_a^{(2)} \end{bmatrix} = Z_Y \mathbf{A} \begin{bmatrix} I_a^{(0)} \\ I_a^{(1)} \\ I_a^{(2)} \end{bmatrix} + 3I_a^{(0)}Z_n \begin{bmatrix} 1 \\ 1 \\ 1 \end{bmatrix} \tag{11.38}$$

Multiplying across by the inverse matrix $\mathbf{A}^{-1}$, we obtain

$$\begin{bmatrix} V_a^{(0)} \\ V_a^{(1)} \\ V_a^{(2)} \end{bmatrix} = Z_Y \begin{bmatrix} I_a^{(0)} \\ I_a^{(1)} \\ I_a^{(2)} \end{bmatrix} + 3I_a^{(0)}Z_n \mathbf{A}^{-1} \begin{bmatrix} 1 \\ 1 \\ 1 \end{bmatrix}$$

Postmultiplying $\mathbf{A}^{-1}$ by $[1 \quad 1 \quad 1]^T$ amounts to adding the elements in each row of $\mathbf{A}^{-1}$,

and so
$$\begin{bmatrix} V_a^{(0)} \\ V_a^{(1)} \\ V_a^{(2)} \end{bmatrix} = Z_Y \begin{bmatrix} I_a^{(0)} \\ I_a^{(1)} \\ I_a^{(2)} \end{bmatrix} + 3I_a^{(0)}Z_n \begin{bmatrix} 1 \\ 0 \\ 0 \end{bmatrix} \tag{11.39}$$

In expanded form, Eq. (11.39) becomes three separate or *decoupled* equations,

$$V_a^{(0)} = (Z_Y + 3Z_n)I_a^{(0)} = Z_0 I_a^{(0)} \tag{11.40}$$

$$V_a^{(1)} = Z_Y I_a^{(1)} = Z_1 I_a^{(1)} \tag{11.41}$$

$$V_a^{(2)} = Z_Y I_a^{(2)} = Z_2 I_a^{(2)} \tag{11.42}$$

It is customary to use the symbols Z_0, Z_1, and Z_2 as shown.

Equations (11.40) through (11.42) could have been easily developed in a less formal manner, but the matrix approach adopted here will be useful in developing other important relations in the sections which follow. Equations (11.24) and (11.25) combine with Eqs. (11.40) through (11.42) to show that currents of one sequence cause voltage drops of only the *same* sequence in Δ- or Y-connected circuits with symmetrical impedances in each phase. This most important result allows us to draw the three single-phase *sequence circuits* shown in Fig. 11.7. These three circuits, considered simultaneously, provide the same information as the actual circuit of Fig. 11.6(*b*), and they are independent of one another because Eqs. (11.40) through (11.42) are decoupled. The circuit of Fig. 11.7(*a*) is called the *zero-sequence circuit* because it relates the zero-sequence voltage $V_a^{(0)}$ to the zero-sequence current $I_a^{(0)}$, and thereby serves to define the *impedance to zero-sequence current* given by

$$\frac{V_a^{(0)}}{I_a^{(0)}} = Z_0 = Z_Y + 3Z_n \tag{11.43}$$

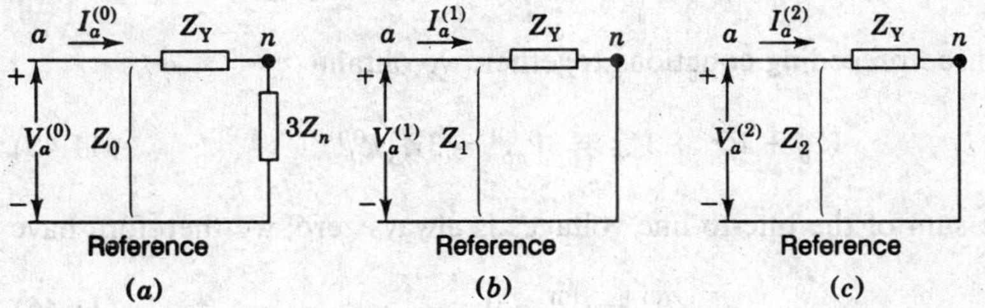

FIGURE 11.7
Zero-, positive-, and negative-sequence circuits for Fig. 11.6(*b*).

Likewise, Fig. 11.7(b) is called the *positive-sequence circuit* and Z_1 is called the *impedance to positive-sequence current*, whereas Fig. 11.7(c) is the *negative-sequence circuit* and Z_2 is the *impedance to negative-sequence current*. The names of the impedances to currents of the different sequences are usually shortened to the less descriptive terms *zero-sequence impedance* Z_0, *positive-sequence impedance* Z_1, and *negative-sequence impedance* Z_2. Here the positive- and negative-sequence impedances Z_1 and Z_2, respectively, are both found to be equal to the usual per-phase impedance Z_Y, which is generally the case for stationary symmetrical circuits. Each of the three sequence circuits represents one phase of the actual three-phase circuit when the latter carries current of only that sequence. When the three sequence currents are simultaneously present, *all three* sequence circuits are needed to fully represent the original circuit.

Voltages in the positive-sequence and negative-sequence circuits can be regarded as voltages measured with respect to either neutral or ground whether or not there is a connection of some finite value of impedance Z_n between neutral and ground. Accordingly, in the positive-sequence circuit there is no difference between $V_a^{(1)}$ and $V_{an}^{(1)}$, and a similar statement applies to $V_a^{(2)}$ and $V_{an}^{(2)}$ in the negative-sequence circuit. However, a voltage difference can exist between the neutral and the reference of the zero-sequence circuit. In the circuit of Fig. 11.7(a) the current $I_a^{(0)}$ flowing through impedance $3Z_n$ produces the same voltage drop from neutral to ground as the current $3I_a^{(0)}$ flowing through impedance Z_n in the actual circuit of Fig. 11.6(b).

If the neutral of the Y-connected circuit is grounded through zero impedance, we set $Z_n = 0$ and a zero-impedance connection then joins the neutral point to the reference node of the zero-sequence circuit. If there is no connection between neutral and ground, there cannot be any zero-sequence current flow, for then $Z_n = \infty$, which is indicated by the open circuit between neutral and the reference node in the zero-sequence circuit of Fig. 11.8(a).

Obviously, a Δ-connected circuit cannot provide a path through neutral, and so line currents flowing into a Δ-connected load or its equivalent Y circuit cannot contain any zero-sequence components. Consider the symmetrical Δ-connected circuit of Fig. 11.4 with

$$V_{ab} = Z_\Delta I_{ab} \qquad V_{bc} = Z_\Delta I_{bc} \qquad V_{ca} = Z_\Delta I_{ca} \tag{11.44}$$

Adding the three preceding equations together, we obtain

$$V_{ab} + V_{bc} + V_{ca} = 3V_{ab}^{(0)} = 3Z_\Delta I_{ab}^{(0)} \tag{11.45}$$

and since the sum of the line-to-line voltages is always zero, we therefore have

$$V_{ab}^{(0)} = I_{ab}^{(0)} = 0 \tag{11.46}$$

Thus, in Δ-connected circuits with impedances only and no sources or mutual coupling there cannot be any circulating currents. Sometimes single-phase

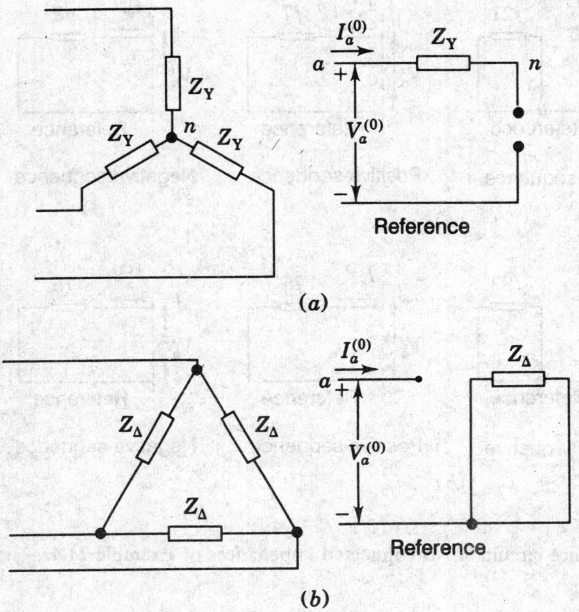

FIGURE 11.8
(*a*) Ungrounded Y-connected and (*b*) Δ-connected circuits and their zero-sequence circuits.

circulating currents can be produced in the Δ circuits of transformers and generators by either induction or zero-sequence generated voltages. A Δ circuit and its zero-sequence circuit are shown in Fig. 11.8(*b*). Note, however, that even if zero-sequence voltages were generated in the phases of the Δ, no zero-sequence voltage could exist between the Δ terminals, for the *rise* in voltage in each phase would then be matched by the voltage *drop* in the zero-sequence impedance of each phase.

Example 11.4. Three equal impedances of $j21\ \Omega$ are connected in Δ. Determine the sequence impedances and circuits of the combination. Repeat the solution for the case where a mutual impedance of $j6\ \Omega$ exists between each pair of adjacent branches in the Δ.

Solution. The line-to-line voltages are related to the Δ currents by

$$\begin{bmatrix} V_{ab} \\ V_{bc} \\ V_{ca} \end{bmatrix} = \begin{bmatrix} j21 & 0 & 0 \\ 0 & j21 & 0 \\ 0 & 0 & j21 \end{bmatrix} \begin{bmatrix} I_{ab} \\ I_{bc} \\ I_{ca} \end{bmatrix}$$

Transforming to symmetrical components of voltages and currents gives

$$\mathbf{A} \begin{bmatrix} V_{ab}^{(0)} \\ V_{ab}^{(1)} \\ V_{ab}^{(2)} \end{bmatrix} = \begin{bmatrix} j21 & 0 & 0 \\ 0 & j21 & 0 \\ 0 & 0 & j21 \end{bmatrix} \mathbf{A} \begin{bmatrix} I_{ab}^{(0)} \\ I_{ab}^{(1)} \\ I_{ab}^{(2)} \end{bmatrix}$$

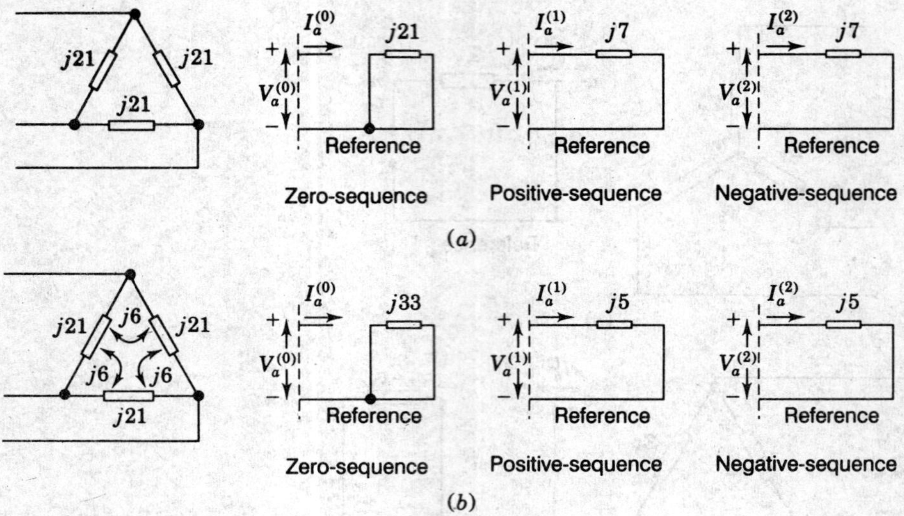

FIGURE 11.9
Zero-, positive-, and negative-sequence circuits for Δ-connected impedances of Example 11.4.

and premultiplying each side by $\mathbf{A}^{-1}$, we obtain

$$\begin{bmatrix} V_{ab}^{(0)} \\ V_{ab}^{(1)} \\ V_{ab}^{(2)} \end{bmatrix} = j21\mathbf{A}^{-1}\mathbf{A}\begin{bmatrix} I_{ab}^{(0)} \\ I_{ab}^{(1)} \\ I_{ab}^{(2)} \end{bmatrix} = \begin{bmatrix} j21 & 0 & 0 \\ 0 & j21 & 0 \\ 0 & 0 & j21 \end{bmatrix}\begin{bmatrix} I_{ab}^{(0)} \\ I_{ab}^{(1)} \\ I_{ab}^{(2)} \end{bmatrix}$$

The positive- and negative-sequence circuits have per-phase impedances $Z_1 = Z_2 = j7\ \Omega$, as shown in Fig. 11.9(*a*), and since $V_{ab}^{(0)} = 0$, the zero-sequence current $I_{ab}^{(0)} = 0$ so that the zero-sequence circuit is an open circuit. The $j21$-Ω resistance in the zero-sequence network has significance only when there is an internal source in the original Δ circuit.

When there is mutual inductance $j6\ \Omega$ between phases,

$$\mathbf{A}\begin{bmatrix} V_{ab}^{(0)} \\ V_{ab}^{(1)} \\ V_{ab}^{(2)} \end{bmatrix} = \begin{bmatrix} j21 & j6 & j6 \\ j6 & j21 & j6 \\ j6 & j6 & j21 \end{bmatrix}\mathbf{A}\begin{bmatrix} I_{ab}^{(0)} \\ I_{ab}^{(1)} \\ I_{ab}^{(2)} \end{bmatrix}$$

The coefficient matrix can be separated into two parts as follows:

$$\begin{bmatrix} j21 & j6 & j6 \\ j6 & j21 & j6 \\ j6 & j6 & j21 \end{bmatrix} = j15\begin{bmatrix} 1 & 0 & 0 \\ 0 & 1 & 0 \\ 0 & 0 & 1 \end{bmatrix} + j6\begin{bmatrix} 1 & 1 & 1 \\ 1 & 1 & 1 \\ 1 & 1 & 1 \end{bmatrix}$$

and substituting into the previous equation, we obtain

$$\begin{bmatrix} V_{ab}^{(0)} \\ V_{ab}^{(1)} \\ V_{ab}^{(2)} \end{bmatrix} = \left\{ j15\mathbf{A}^{-1}\mathbf{A} + j6\mathbf{A}^{-1}\begin{bmatrix} 1 & 1 & 1 \\ 1 & 1 & 1 \\ 1 & 1 & 1 \end{bmatrix}\mathbf{A} \right\} \begin{bmatrix} I_{ab}^{(0)} \\ I_{ab}^{(1)} \\ I_{ab}^{(2)} \end{bmatrix}$$

$$= \left\{ \begin{bmatrix} j15 & 0 & 0 \\ 0 & j15 & 0 \\ 0 & 0 & j15 \end{bmatrix} + j6 \begin{bmatrix} 3 & 0 & 0 \\ 0 & 0 & 0 \\ 0 & 0 & 0 \end{bmatrix} \right\} \begin{bmatrix} I_{ab}^{(0)} \\ I_{ab}^{(1)} \\ I_{ab}^{(2)} \end{bmatrix}$$

$$= \begin{bmatrix} j33 & 0 & 0 \\ 0 & j15 & 0 \\ 0 & 0 & j15 \end{bmatrix} \begin{bmatrix} I_{ab}^{(0)} \\ I_{ab}^{(1)} \\ I_{ab}^{(2)} \end{bmatrix}$$

The positive- and negative-sequence impedances Z_1 and Z_2 now take on the value $j5\ \Omega$, as shown in Fig. 11.9(*b*), and since $V_{ab}^{(0)} = I_{ab}^{(0)} = 0$, the zero-sequence circuit is open. Again, we note that the $j33$-Ω resistance in the zero-sequence network has no significance because there is no internal source in the original Δ circuit.

The matrix manipulations of this example are useful in the sections which follow.

11.6 SEQUENCE CIRCUITS OF A SYMMETRICAL TRANSMISSION LINE

We are concerned primarily with systems that are essentially symmetrically balanced and which become unbalanced only upon the occurrence of an unsymmetrical fault. In practical transmission systems such complete symmetry is more ideal than realized, but since the effect of the departure from symmetry is usually small, perfect balance between phases is often assumed especially if the lines are transposed along their lengths. Let us consider Fig. 11.10, for instance, which shows one section of a three-phase transmission line with a neutral conductor. The self-impedance Z_{aa} is the same for each phase conductor, and the neutral conductor has self-impedance Z_{nn}. When currents I_a, I_b, and I_c in the phase conductors are unbalanced, the neutral conductor serves as a return path. All the currents are assumed positive in the directions shown even though some of their numerical values may be negative under unbalanced conditions caused by faults. Because of mutual coupling, current flow in any one of the phases induces voltages in each of the other adjacent phases and in the neutral conductor. Similarly, I_n in the neutral conductor induces voltages in each of the phases. The coupling between all three phase conductors is regarded as being symmetrical and mutual impedance Z_{ab} is assumed between

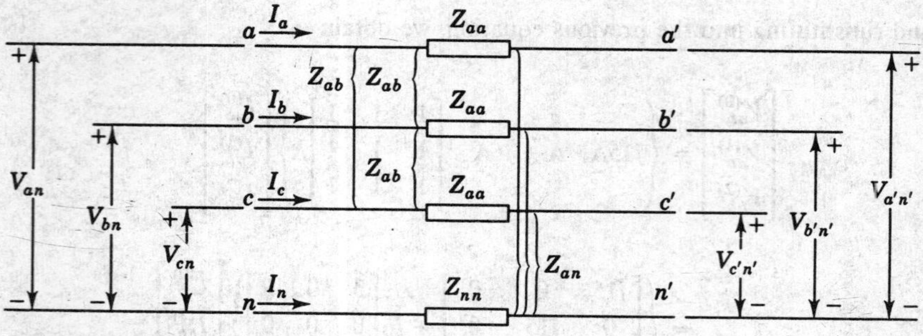

FIGURE 11.10
Flow of unbalanced currents in a symmetrical three-phase-line section with neutral conductor.

each pair. Likewise, the mutual impedance between the neutral conductor and each of the phases is taken to be Z_{an}.

The voltages induced in phase a, for example, by currents in the other two phases and the neutral conductor are shown as sources in the loop circuit of Fig. 11.11, along with the similar voltages induced in the neutral conductor. Applying Kirchhoff's voltage law around the loop circuit gives

$$V_{an} = Z_{aa}I_a + Z_{ab}I_b + Z_{ab}I_c + Z_{an}I_n + V_{a'n'}$$
$$-(Z_{nn}I_n + Z_{an}I_c + Z_{an}I_b + Z_{an}I_a) \tag{11.47}$$

from which voltage drop across the line section is found to be

$$V_{an} - V_{a'n'} = (Z_{aa} - Z_{an})I_a + (Z_{ab} - Z_{an})(I_b + I_c) + (Z_{an} - Z_{nn})I_n \tag{11.48}$$

Similar equations can be written for phases b and c as follows:

$$V_{bn} - V_{b'n'} = (Z_{aa} - Z_{an})I_b + (Z_{ab} - Z_{an})(I_a + I_c) + (Z_{an} - Z_{nn})I_n$$
$$V_{cn} - V_{c'n'} = (Z_{aa} - Z_{an})I_c + (Z_{ab} - Z_{an})(I_a + I_b) + (Z_{an} - Z_{nn})I_n \tag{11.49}$$

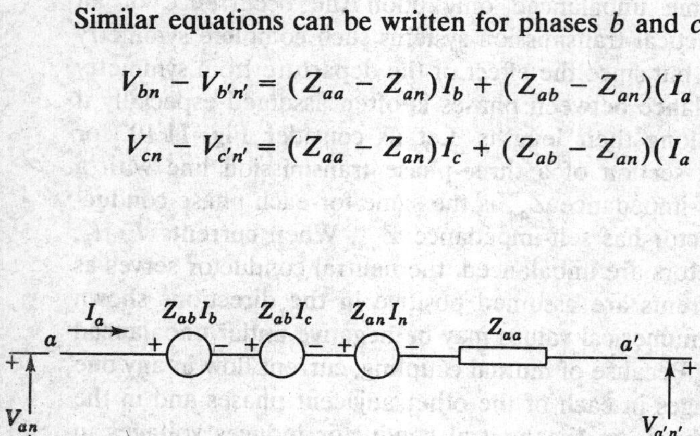

FIGURE 11.11
Writing Kirchhoff's voltage equation around the loop formed by line a and the neutral conductor.

When the line currents I_a, I_b, and I_c return together as I_n in the neutral conductor of Fig. 11.10, we have

$$I_n = -(I_a + I_b + I_c) \tag{11.50}$$

Let us now substitute for I_n in Eqs. (11.48) and (11.49) to obtain

$$\begin{aligned}
V_{an} - V_{a'n'} &= (Z_{aa} + Z_{nn} - 2Z_{an})I_a + (Z_{ab} + Z_{nn} - 2Z_{an})I_b \\
&\quad + (Z_{ab} + Z_{nn} - 2Z_{an})I_c \\
V_{bn} - V_{b'n'} &= (Z_{ab} + Z_{nn} - 2Z_{an})I_a + (Z_{aa} + Z_{nn} - 2Z_{an})I_b \\
&\quad + (Z_{ab} + Z_{nn} - 2Z_{an})I_c \\
V_{cn} - V_{c'n'} &= (Z_{ab} + Z_{nn} - 2Z_{an})I_a + (Z_{ab} + Z_{nn} - 2Z_{an})I_b \\
&\quad + (Z_{aa} + Z_{nn} - 2Z_{an})I_c
\end{aligned} \tag{11.51}$$

The coefficients in these equations show that the presence of the neutral conductor changes the self- and mutual impedances of the phase conductors to the following effective values:

$$\begin{aligned}
Z_s &\triangleq Z_{aa} + Z_{nn} - 2Z_{an} \\
Z_m &\triangleq Z_{ab} + Z_{nn} - 2Z_{an}
\end{aligned} \tag{11.52}$$

Using these definitions, we can rewrite Eqs. (11.51) in the convenient matrix form

$$\begin{bmatrix} V_{aa'} \\ V_{bb'} \\ V_{cc'} \end{bmatrix} = \begin{bmatrix} V_{an} - V_{a'n'} \\ V_{bn} - V_{b'n'} \\ V_{cn} - V_{c'n'} \end{bmatrix} = \begin{bmatrix} Z_s & Z_m & Z_m \\ Z_m & Z_s & Z_m \\ Z_m & Z_m & Z_s \end{bmatrix} \begin{bmatrix} I_a \\ I_b \\ I_c \end{bmatrix} \tag{11.53}$$

where the voltage drops across the phase conductors are now denoted by

$$V_{aa'} \triangleq V_{an} - V_{a'n'} \qquad V_{bb'} \triangleq V_{bn} - V_{b'n'} \qquad V_{cc'} \triangleq V_{cn} - V_{c'n'} \tag{11.54}$$

Since Eq. (11.53) does not *explicitly* include the neutral conductor, Z_s and Z_m can be regarded as parameters of the phase conductors alone, without any self- or mutual inductance being associated with the return path.

The *a-b-c* voltage drops and currents of the line section can be written in terms of their symmetrical components according to Eq. (11.8) so that with phase a as the reference phase, we have

$$\mathbf{A}\begin{bmatrix} V_{aa'}^{(0)} \\ V_{aa'}^{(1)} \\ V_{aa'}^{(2)} \end{bmatrix} = \left\{ \begin{bmatrix} Z_s - Z_m & \cdot & \cdot \\ \cdot & Z_s - Z_m & \cdot \\ \cdot & \cdot & Z_s - Z_m \end{bmatrix} + \begin{bmatrix} Z_m & Z_m & Z_m \\ Z_m & Z_m & Z_m \\ Z_m & Z_m & Z_m \end{bmatrix} \right\} \mathbf{A} \begin{bmatrix} I_a^{(0)} \\ I_a^{(1)} \\ I_a^{(2)} \end{bmatrix} \tag{11.55}$$

This particular form of the equation makes calculations easier, as demonstrated in Example 11.4. Multiplying across by $\mathbf{A}^{-1}$, we obtain

$$\begin{bmatrix} V_{aa'}^{(0)} \\ V_{aa'}^{(1)} \\ V_{aa'}^{(2)} \end{bmatrix} = \mathbf{A}^{-1} \left\{ (Z_s - Z_m) \begin{bmatrix} 1 & \cdot & \cdot \\ \cdot & 1 & \cdot \\ \cdot & \cdot & 1 \end{bmatrix} + Z_m \begin{bmatrix} 1 & 1 & 1 \\ 1 & 1 & 1 \\ 1 & 1 & 1 \end{bmatrix} \right\} \mathbf{A} \begin{bmatrix} I_a^{(0)} \\ I_a^{(1)} \\ I_a^{(2)} \end{bmatrix} \tag{11.56}$$

The matrix multiplication here is the same as in Example 11.4 and yields

$$\begin{bmatrix} V_{aa'}^{(0)} \\ V_{aa'}^{(1)} \\ V_{aa'}^{(2)} \end{bmatrix} = \begin{bmatrix} Z_s + 2Z_m & \cdot & \cdot \\ \cdot & Z_s - Z_m & \cdot \\ \cdot & \cdot & Z_s - Z_m \end{bmatrix} \begin{bmatrix} I_a^{(0)} \\ I_a^{(1)} \\ I_a^{(2)} \end{bmatrix} \tag{11.57}$$

Let us now define zero-, positive-, and negative-sequence impedances in terms of Z_s and Z_m introduced in Eqs. (11.52),

$$\begin{aligned} Z_0 &= Z_s + 2Z_m = Z_{aa} + 2Z_{ab} + 3Z_{nn} - 6Z_{an} \\ Z_1 &= Z_s - \ \ Z_m = Z_{aa} - Z_{ab} \\ Z_2 &= Z_s - \ \ Z_m = Z_{aa} - Z_{ab} \end{aligned} \tag{11.58}$$

From Eqs. (11.57) and (11.58) the sequence components of the voltage drops between the two ends of the line section can be written as three simple equations of the form

$$\begin{aligned} V_{aa'}^{(0)} &= V_{an}^{(0)} - V_{a'n'}^{(0)} = Z_0 I_a^{(0)} \\ V_{aa'}^{(1)} &= V_{an}^{(1)} - V_{a'n'}^{(1)} = Z_1 I_a^{(1)} \\ V_{aa'}^{(2)} &= V_{an}^{(2)} - V_{a'n'}^{(2)} = Z_2 I_a^{(2)} \end{aligned} \tag{11.59}$$

Because of the assumed symmetry of the circuit of Fig. 11.10, once again we see

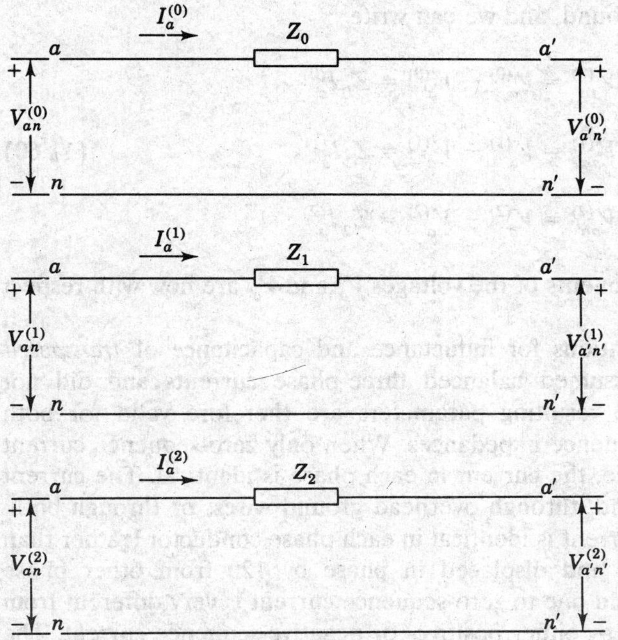

FIGURE 11.12
Sequence circuits for the symmetrical line section of Fig. 11.10.

that the zero-, positive-, and negative-sequence equations decouple from one another, and corresponding zero-, positive-, and negative-sequence circuits can be drawn without any mutual coupling between them, as shown in Fig. 11.12. Despite the simplicity of the line model in Fig. 11.10, the above development has demonstrated important characteristics of the sequence impedances which apply to more elaborate and practical line models. We note, for instance, that the positive- and negative-sequence impedances are equal and that they do not include the neutral-conductor impedances Z_{nn} and Z_{an}, which enter into the calculation of only the zero-sequence impedance Z_0, as shown by Eqs. (11.58). In other words, impedance parameters of the return-path conductors enter into the values of the zero-sequence impedances of transmission lines, but they do not affect either the positive- or negative-sequence impedance.

Most aerial transmission lines have at least two overhead conductors called *ground wires*, which are grounded at uniform intervals along the length of the line. The ground wires combine with the earth return path to constitute an effective neutral conductor with impedance parameters, like Z_{nn} and Z_{an}, which depend on the resistivity of the earth. The more specialized literature shows, as we have demonstrated here, that the parameters of the return path are included in the zero-sequence impedance of the line. By regarding the neutral conductor of Fig. 11.10 as the effective return path for the zero-sequence components of the unbalanced currents and including its parameters in the zero-sequence impedance, we can treat the ground as an ideal conductor. The voltages of Fig. 11.12 are then interpreted as being measured with respect to perfectly conducting ground, and we can write

$$\begin{aligned} V_{aa'}^{(0)} &= V_a^{(0)} - V_{a'}^{(0)} = Z_0 I_a^{(0)} \\ V_{aa'}^{(1)} &= V_a^{(1)} - V_{a'}^{(1)} = Z_1 I_a^{(1)} \\ V_{aa'}^{(2)} &= V_a^{(2)} - V_{a'}^{(2)} = Z_2 I_a^{(2)} \end{aligned} \tag{11.60}$$

where the sequence components of the voltages V_a and $V_{a'}$ are now with respect to *ideal ground*.

In deriving the equations for inductance and capacitance of *transposed* transmission lines, we assumed balanced three-phase currents and did not specify phase order. The resulting parameters are therefore valid for both positive- and negative-sequence impedances. When only zero-sequence current flows in a transmission line, the current in each phase is identical. The current returns through the ground, through overhead ground wires, or through both. Because zero-sequence current is identical in each phase conductor (rather than equal only in magnitude and displaced in phase by 120 from other phase currents), the magnetic field due to zero-sequence current is very different from the magnetic field caused by either positive- or negative-sequence current. The difference in magnetic field results in the zero-sequence inductive reactance of overhead transmission lines being 2 to 3.5 times as large as the positive-sequence reactance. The ratio is toward the higher portion of the specified range for double-circuit lines and lines without ground wires.

Example 11.5. In Fig. 11.10 the terminal voltages at the left-hand and right-hand ends of the line are given by

$$V_{an} = 182.0 + j70.0 \text{ kV} \qquad V_{a'n'} = 154.0 + j28.0 \text{ kV}$$

$$V_{bn} = 72.24 - j32.62 \text{ kV} \qquad V_{b'n'} = 44.24 - j74.62 \text{ kV}$$

$$V_{cn} = -170.24 + j88.62 \text{ kV} \qquad V_{c'n'} = -198.24 + j46.62 \text{ kV}$$

The line impedances in ohms are

$$Z_{aa} = j60 \qquad Z_{ab} = j20 \qquad Z_{nn} = j80 \qquad Z_{an} = 0$$

Determine the line currents I_a, I_b, and I_c using symmetrical components. Repeat the solution without using symmetrical components.

Solution. The sequence impedances have calculated values

$$Z_0 = Z_{aa} + 2Z_{ab} + 3Z_{nn} - 6Z_{an} = j60 + j40 + j240 - j180 = j160\ \Omega$$

$$Z_1 = Z_2 = Z_{aa} - Z_{ab} = j60 - j20 = j40\ \Omega$$

The sequence components of the voltage drops in the line are

$$\begin{bmatrix} V_{aa'}^{(0)} \\ V_{aa'}^{(1)} \\ V_{aa'}^{(2)} \end{bmatrix} = \mathbf{A}^{-1} \begin{bmatrix} V_{an} - V_{a'n'} \\ V_{bn} - V_{b'n'} \\ V_{cn} - V_{c'n'} \end{bmatrix} = \mathbf{A}^{-1} \begin{bmatrix} (182.0 - 154.0) + j(70.0 - 28.0) \\ (72.24 - 44.24) - j(32.62 - 74.62) \\ -(170.24 - 198.24) + j(88.62 - 46.62) \end{bmatrix}$$

$$= \mathbf{A}^{-1} \begin{bmatrix} 28.0 + j42.0 \\ 28.0 + j42.0 \\ 28.0 + j42.0 \end{bmatrix} = \begin{bmatrix} 28.0 + j42.0 \\ 0 \\ 0 \end{bmatrix} \text{kV}$$

Substituting in Eq. (11.59), we obtain

$$V_{aa'}^{(0)} = 28{,}000 + j42{,}000 = j160 I_a^{(0)}$$

$$V_{aa'}^{(1)} = \quad 0 \quad = j40 I_a^{(1)}$$

$$V_{aa'}^{(2)} = \quad 0 \quad = j40 I_a^{(2)}$$

from which we determine the symmetrical components of the currents in phase a,

$$I_a^{(0)} = 262.5 - j175 \text{ A} \qquad I_a^{(1)} = I_a^{(2)} = 0$$

The line currents are therefore

$$I_a = I_b = I_c = 262.5 - j175 \text{ A}$$

The self- and mutual impedances of Eq. (11.52) have values

$$Z_s = Z_{aa} + Z_{nn} - 2Z_{an} = j60 + j80 - j60 = j80\ \Omega$$

$$Z_m = Z_{ab} + Z_{nn} - 2Z_{an} = j20 + j80 - j60 = j40\ \Omega$$

and so line currents can be calculated from Eq. (11.53) without symmetrical components as follows:

$$\begin{bmatrix} V_{aa'} \\ V_{bb'} \\ V_{cc'} \end{bmatrix} = \begin{bmatrix} 28 + j42 \\ 28 + j42 \\ 28 + j42 \end{bmatrix} \times 10^3 = \begin{bmatrix} j80 & j40 & j40 \\ j40 & j80 & j40 \\ j40 & j40 & j80 \end{bmatrix} \begin{bmatrix} I_a \\ I_b \\ I_c \end{bmatrix}$$

$$\begin{bmatrix} I_a \\ I_b \\ I_c \end{bmatrix} = \begin{bmatrix} j80 & j40 & j40 \\ j40 & j80 & j40 \\ j40 & j40 & j80 \end{bmatrix}^{-1} \begin{bmatrix} 28 + j42 \\ 28 + j42 \\ 28 + j42 \end{bmatrix} \times 10^3 = \begin{bmatrix} 262.5 - j175 \\ 262.5 - j175 \\ 262.5 - j175 \end{bmatrix} \text{A}$$

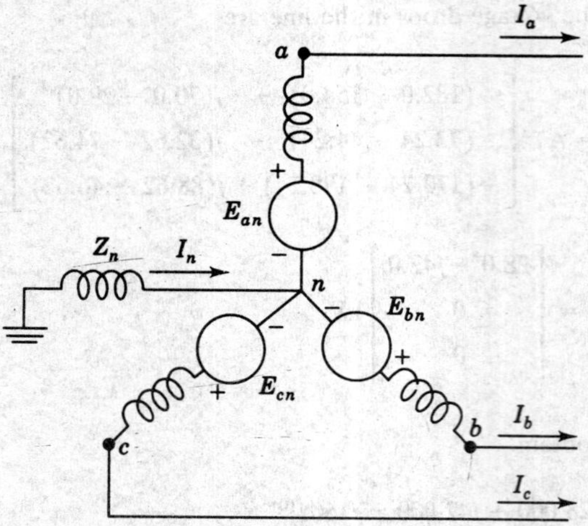

FIGURE 11.13
Circuit diagram of a generator grounded through a reactance. The phase emfs E_{an}, E_{bn}, and E_{cn} are positive sequence.

11.7 SEQUENCE CIRCUITS OF THE SYNCHRONOUS MACHINE

A synchronous generator, grounded through a reactor, is shown in Fig. 11.13. When a fault (not indicated in the figure) occurs at the terminals of the generator, currents I_a, I_b, and I_c flow in the lines. If the fault involves ground, the current flowing into the neutral of the generator is designated I_n and the line currents can be resolved into their symmetrical components regardless of how unbalanced they may be.

The equations developed in Sec. 3.2 for the idealized synchronous machine are all based on the assumption of balanced instantaneous armature currents. In Eq. (3.7) we assumed that $i_a + i_b + i_c = 0$, and then set $i_a = -(i_b + i_c)$ in Eq. (3.5) in order to arrive at Eq. (3.11) for the terminal voltage of phase a in the form

$$v_{an} = -Ri_a - (L_s + M_s)\frac{di_a}{dt} + e_{an} \tag{11.61}$$

The steady-state counterpart of this equation is given in Eq. (3.24) as

$$V_{an} = -RI_a - j\omega(L_s + M_s)I_a + E_{an} \tag{11.62}$$

where E_{an} is the synchronous internal voltage of the machine. The subscripts of the voltages in Eqs. (11.61) and (11.62) differ slightly from those of Chap. 3 so as to emphasize that the voltages are with respect to neutral. If the substitution $i_a = -(i_b + i_c)$ had not been made as indicated, then we would have found

$$v_{an} = -Ri_a - L_s\frac{di_a}{dt} + M_s\frac{d}{dt}(i_b + i_c) + e_{an} \tag{11.63}$$

Assuming for now that steady-state sinusoidal currents and voltages of nominal system frequency ω continue to exist in the armature, we can write Eq. (11.63) in the phasor form

$$V_{an} = -RI_a - j\omega L_s I_a + j\omega M_s(I_b + I_c) + E_{an} \tag{11.64}$$

where E_{an} again designates the phasor equivalent of e_{an}. The armature phases b and c of the idealized machine have similar equations

$$\begin{aligned} V_{bn} &= -RI_b - j\omega L_s I_b + j\omega M_s(I_a + I_c) + E_{bn} \\ V_{cn} &= -RI_c - j\omega L_s I_c + j\omega M_s(I_a + I_b) + E_{cn} \end{aligned} \tag{11.65}$$

We can arrange Eqs. (11.64) and (11.65) in vector-matrix form as follows:

$$\begin{bmatrix} V_{an} \\ V_{bn} \\ V_{cn} \end{bmatrix} = -[R + j\omega(L_s + M_s)] \begin{bmatrix} I_a \\ I_b \\ I_c \end{bmatrix} + j\omega M_s \begin{bmatrix} 1 & 1 & 1 \\ 1 & 1 & 1 \\ 1 & 1 & 1 \end{bmatrix} \begin{bmatrix} I_a \\ I_b \\ I_c \end{bmatrix} + \begin{bmatrix} E_{an} \\ E_{bn} \\ E_{cn} \end{bmatrix} \tag{11.66}$$

Following the procedure demonstrated in the two preceding sections, we now express the *a-b-c* quantities of the machine in terms of symmetrical components of phase a of the armature

$$\begin{bmatrix} V_{an}^{(0)} \\ V_{an}^{(1)} \\ V_{an}^{(2)} \end{bmatrix} = -[R + j\omega(L_s + M_s)] \begin{bmatrix} I_a^{(0)} \\ I_a^{(1)} \\ I_a^{(2)} \end{bmatrix}$$

$$+ j\omega M_s \mathbf{A}^{-1} \begin{bmatrix} 1 & 1 & 1 \\ 1 & 1 & 1 \\ 1 & 1 & 1 \end{bmatrix} \mathbf{A} \begin{bmatrix} I_a^{(0)} \\ I_a^{(1)} \\ I_a^{(2)} \end{bmatrix} + \mathbf{A}^{-1} \begin{bmatrix} E_{an} \\ a^2 E_{an} \\ a E_{an} \end{bmatrix} \tag{11.67}$$

Since the synchronous generator is designed to supply balanced three-phase voltages, we have shown the generated voltages E_{an}, E_{bn}, and E_{cn} as a positive-sequence set of phasors in Eq. (11.67), where the operator $a = 1\angle 120°$ and $a^2 = \angle 240°$. The matrix multiplications of Eq. (11.67) are similar to those of Eqs. (11.56), and so we obtain

$$\begin{bmatrix} V_{an}^{(0)} \\ V_{an}^{(1)} \\ V_{an}^{(2)} \end{bmatrix} = -[R + j\omega(L_s + M_s)] \begin{bmatrix} I_a^{(0)} \\ I_a^{(1)} \\ I_a^{(2)} \end{bmatrix} + j\omega M_s \begin{bmatrix} 3 & \cdot & \cdot \\ \cdot & \cdot & \cdot \\ \cdot & \cdot & \cdot \end{bmatrix} \begin{bmatrix} I_a^{(0)} \\ I_a^{(1)} \\ I_a^{(2)} \end{bmatrix} + \begin{bmatrix} 0 \\ E_{an} \\ 0 \end{bmatrix} \tag{11.68}$$

The zero-, positive-, and negative-sequence equations decouple to give

$$\begin{aligned}
V_{an}^{(0)} &= -RI_a^{(0)} - j\omega(L_s - 2M_s)I_a^{(0)} \\
V_{an}^{(1)} &= -RI_a^{(1)} - j\omega(L_s + M_s)I_a^{(1)} + E_{an} \\
V_{an}^{(2)} &= -RI_a^{(2)} - j\omega(L_s + M_s)I_a^{(2)}
\end{aligned} \tag{11.69}$$

Drawing the corresponding sequence circuits is made simple by writing Eqs. (11.69) in the form

$$\begin{aligned}
V_{an}^{(0)} &= \quad\;\; - I_a^{(0)}[R + j\omega(L_s - 2M_s)] = \quad\;\; - I_a^{(0)}Z_{g0} \\
V_{an}^{(1)} &= E_{an} - I_a^{(1)}[R + j\omega(L_s + M_s)] = E_{an} - I_a^{(1)}Z_1 \\
V_{an}^{(2)} &= \quad\;\; - I_a^{(2)}[R + j\omega(L_s + M_s)] = \quad\;\; - I_a^{(2)}Z_2
\end{aligned} \tag{11.70}$$

where Z_{g0}, Z_1, and Z_2 are the zero-, positive-, and negative-sequence impedances, respectively, of the generator. The sequence circuits shown in Fig. 11.14 are the single-phase equivalent circuits of the balanced three-phase machine through which the symmetrical components of the unbalanced currents are considered to flow. The sequence components of current are flowing through impedances of their own sequence only, as indicated by the appropriate subscripts on the impedances shown in the figure. This is because the machine is symmetrical with respect to phases a, b, and c. The positive-sequence circuit is composed of an emf in series with the positive-sequence impedance of the generator. The negative- and zero-sequence circuits contain no emfs but include the impedances of the generator to negative- and zero-sequence currents, respectively.

The reference node for the positive- and negative-sequence circuits is the neutral of the generator. So far as positive- and negative-sequence components are concerned, the neutral of the generator is at ground potential if there is a connection between neutral and ground having a finite or zero impedance since the connection will carry no positive- or negative-sequence current. Once again, we see that there is no essential difference between $V_a^{(1)}$ and $V_{an}^{(1)}$ in the positive-sequence circuit or between $V_a^{(2)}$ and $V_{an}^{(2)}$ in the negative-sequence circuit. This explains why the positive- and negative-sequence voltages $V_a^{(1)}$ and $V_a^{(2)}$ of Fig. 11.14 are written without subscript n.

The current flowing in the impedance Z_n between neutral and ground is $3I_a^{(0)}$. By referring to Fig. 11.14(e), we see that the voltage *drop* of zero sequence *from point a to ground* is $-3I_a^{(0)}Z_n - I_a^{(0)}Z_{g0}$, where Z_{g0} is the zero-sequence impedance per phase of the generator. The zero-sequence circuit, which is a single-phase circuit assumed to carry only the zero-sequence

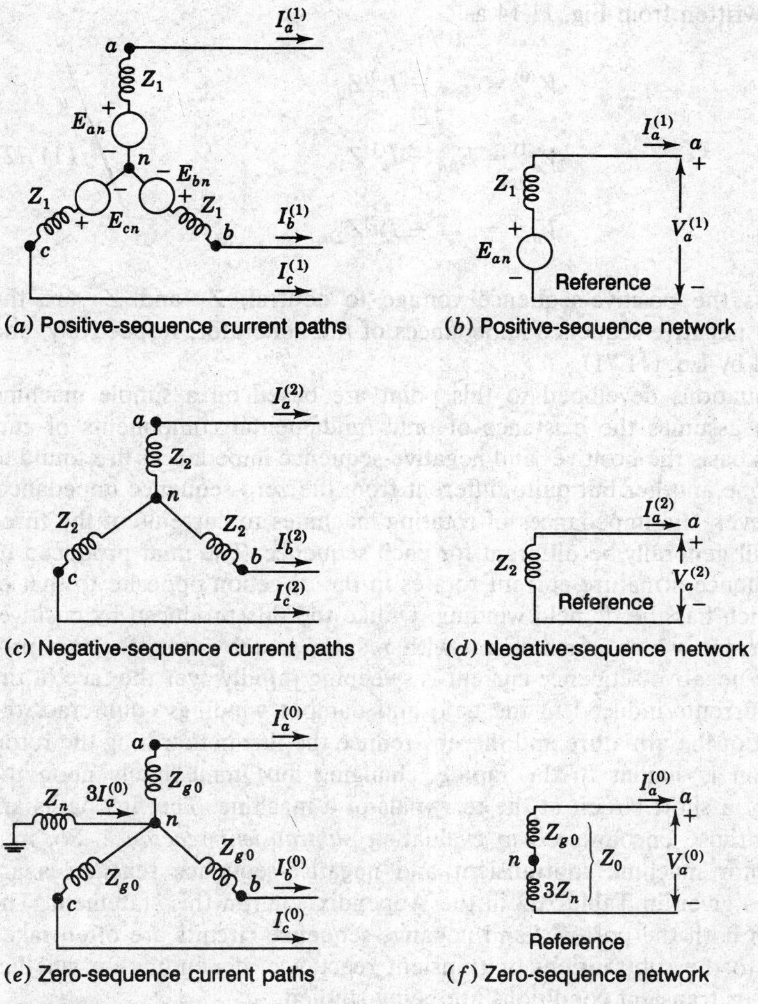

FIGURE 11.14
Paths for current of each sequence in a generator and the corresponding sequence networks.

current of one phase, must therefore have an impedance of $3Z_n + Z_{g0}$, as shown in Fig. 11.14(f). The total zero-sequence impedance through which $I_a^{(0)}$ flows is

$$Z_0 = 3Z_n + Z_{g0} \tag{11.71}$$

Usually, the components of current and voltage for phase a are found from equations determined by the sequence circuits. The equations for the components of voltage drop from point a of phase a to the reference node (or ground) are written from Fig. 11.14 as

$$\begin{aligned} V_a^{(0)} &= \qquad\; - I_a^{(0)} Z_0 \\ V_a^{(1)} &= E_{an} - I_a^{(1)} Z_1 \\ V_a^{(2)} &= \qquad\; - I_a^{(2)} Z_2 \end{aligned} \qquad (11.72)$$

where E_{an} is the positive-sequence voltage to neutral, Z_1 and Z_2 are the positive- and negative-sequence impedances of the generator, respectively, and Z_0 is defined by Eq. (11.71).

The equations developed to this point are based on a simple machine model which assumes the existence of only fundamental components of currents; on this basis the positive- and negative-sequence impedances are found to be equal to one another but quite different from the zero-sequence impedance. In fact, however, the impedances of rotating machines to currents of the three sequences will generally be different for each sequence. The mmf produced by negative-sequence armature current rotates in the direction opposite to that of the rotor which has the dc field winding. Unlike the flux produced by positive-sequence current, which is stationary with respect to the rotor, the flux produced by the negative-sequence current is sweeping rapidly over the face of the rotor. The currents induced in the field and damper windings counteract the rotating mmf of the armature and thereby reduce the flux penetrating the rotor. This condition is similar to the rapidly changing flux immediately upon the occurrence of a short circuit at the terminals of a machine. The flux paths are the same as those encountered in evaluating *subtransient reactance*. So, in a cylindrical-rotor machine subtransient and negative-sequence reactances are equal. Values given in Table A.2 in the Appendix confirm this statement. The reactances in both the positive- and negative-sequence circuits are often taken to be equal to the subtransient or transient reactance, depending on whether subtransient or transient conditions are being studied.

When only zero-sequence current flows in the armature winding of a three-phase machine, the current and mmf of one phase are a maximum at the same time as the current and mmf of each of the other phases. The windings are so distributed around the circumference of the armature that the point of maximum mmf produced by one phase is displaced 120 electrical degrees in space from the point of maximum mmf of each of the other phases. If the mmf produced by the current of each phase had a perfectly sinusoidal distribution in space, a plot of mmf around the armature would result in three sinusoidal curves whose sum would be zero at every point. No flux would be produced across the air gap, and the only reactance of any phase winding would be that due to leakage and end turns. In an actual machine the winding is not distributed to produce perfectly sinusoidal mmf. The flux resulting from the sum of the mmfs is very small, which makes the zero-sequence reactance the smallest of the machine's reactances—just somewhat higher than zero of the ideal case where there is no air-gap flux due to zero-sequence current.

Equations (11.72), which apply to any generator carrying unbalanced currents, are the starting points for the derivation of equations for the components of current for different types of faults. As we shall see, they apply to the Thévenin equivalent circuits at any bus of the system as well as to the case of a loaded generator under steady-state conditions. When computing transient or subtransient conditions, the equations apply to a loaded generator if E' or E'' is substituted for E_{an}.

Example 11.6. A salient-pole generator without dampers is rated 20 MVA, 13.8 kV and has a direct-axis subtransient reactance of 0.25 per unit. The negative- and zero-sequence reactances are, respectively, 0.35 and 0.10 per unit. The neutral of the generator is solidly grounded. With the generator operating unloaded at rated voltage with $E_{an} = 1.0\angle 0°$ per unit, a single line-to-ground fault occurs at the machine terminals, which then have per-unit voltages to ground,

$$V_a = 0 \qquad V_b = 1.013\angle -102.25° \qquad V_c = 1.013\angle 102.25°$$

Determine the subtransient current in the generator and the line-to-line voltages for subtransient conditions due to the fault.

Solution. Figure 11.15 shows the line-to-ground fault on phase a of the machine. In rectangular coordinates V_b and V_c are

$$V_b = -0.215 - j0.990 \text{ per unit}$$

$$V_c = -0.215 + j0.990 \text{ per unit}$$

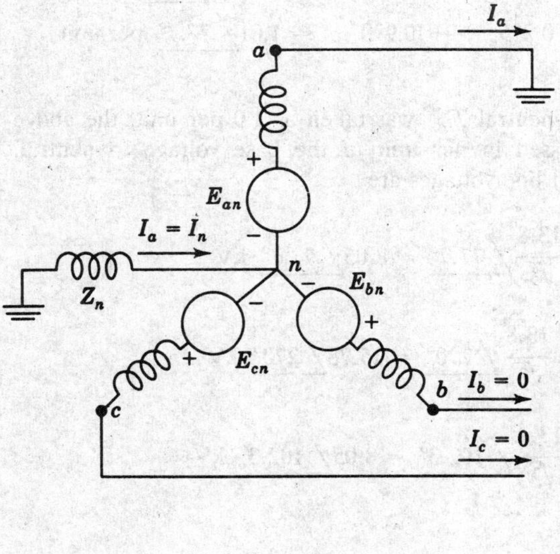

FIGURE 11.15
Circuit diagram for a single line-to-ground fault on phase a at the terminals of an unloaded generator whose neutral is grounded through a reactance.

The symmetrical components of the voltages at the fault point are

$$\begin{bmatrix} V_a^{(0)} \\ V_a^{(1)} \\ V_a^{(2)} \end{bmatrix} = \frac{1}{3}\begin{bmatrix} 1 & 1 & 1 \\ 1 & a & a^2 \\ 1 & a^2 & a \end{bmatrix}\begin{bmatrix} 0 \\ -0.215 - j0.990 \\ -0.215 + j0.990 \end{bmatrix} = \begin{bmatrix} -0.143 + j0 \\ 0.643 + j0 \\ -0.500 + j0 \end{bmatrix} \text{ per unit}$$

From Eqs. (11.72) and Fig. 11.14 with $Z_n = 0$ we calculate

$$I_a^{(0)} = -\frac{V_a^{(0)}}{Z_{g0}} = -\frac{(-0.143 + j0)}{j0.10} = -j1.43 \text{ per unit}$$

$$I_a^{(1)} = \frac{E_{an} - V_a^{(1)}}{Z_1} = \frac{(1.0 + j0) - (0.643 + j0)}{j0.25} = -j1.43 \text{ per unit}$$

$$I_a^{(2)} = -\frac{V_a^{(2)}}{Z_2} = -\frac{(-0.500 + j0)}{j0.35} = -j1.43 \text{ per unit}$$

Therefore, the fault current into the ground is

$$I_a = I_a^{(0)} + I_a^{(1)} + I_a^{(2)} = 3I_a^{(0)} = -j4.29 \text{ per unit}$$

The base current is $20{,}000/(\sqrt{3} \times 13.8) = 837$ A, and so the subtransient current in line a is

$$I_a = -j4.29 \times 837 = -j3{,}590 \text{ A}$$

Line-to-line voltages during the fault are

$$V_{ab} = V_a - V_b = 0.215 + j0.990 = 1.01\angle 77.7^\circ \text{ per unit}$$

$$V_{bc} = V_b - V_c = 0 - j1.980 = 1.980\angle 270^\circ \text{ per unit}$$

$$V_{ca} = V_c - V_a = -0.215 + j0.990 = 1.01\angle 77.7^\circ \text{ per unit}$$

Since the generated voltage-to-neutral E_{an} was taken as 1.0 per unit, the above line-to-line voltages are expressed in per unit of the base voltage to neutral. Expressed in volts, the postfault line voltages are

$$V_{ab} = 1.01 \times \frac{13.8}{\sqrt{3}}\angle 77.7^\circ = 8.05\angle 77.7^\circ \text{ kV}$$

$$V_{bc} = 1.980 \times \frac{13.8}{\sqrt{3}}\angle 270^\circ = 15.78\angle 270^\circ \text{ kV}$$

$$V_{ca} = 1.01 \times \frac{13.8}{\sqrt{3}}\angle 102.3^\circ = 8.05\angle 102.3^\circ \text{ kV}$$

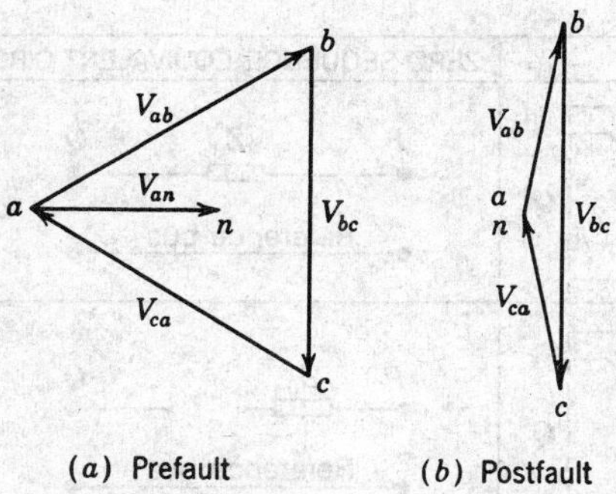

FIGURE 11.16
Phasor diagrams of the line voltages of Example 11.6 before and after the fault.

Before the fault the line voltages were balanced and equal to 13.8 kV. For comparison with the line voltages after the fault occurs, the prefault voltages, with $V_{an} = E_{an}$ as reference, are given as

$$V_{ab} = 13.8\underline{/30°} \text{ kV} \qquad V_{bc} = 13.8\underline{/270°} \text{ kV} \qquad V_{ca} = 13.8\underline{/150°} \text{ kV}$$

Figure 11.16 shows phasor diagrams of prefault and postfault voltages.

The preceding example shows that $I_a^{(0)} = I_a^{(1)} = I_a^{(2)}$ in the case of a single line-to-ground fault. This is a general result, which is established in Sec. 12.2.

11.8 SEQUENCE CIRCUITS OF Y-Δ TRANSFORMERS

The sequence equivalent circuits of three-phase transformers depend on the connections of the primary and secondary windings. The different combinations of Δ and Y windings determine the configurations of the zero-sequence circuits and the phase shift in the positive- and negative-sequence circuits. On this account the reader may wish to review some portions of Chap. 2, notably Secs. 2.5 and 2.6.

We remember that no current flows in the primary of a transformer unless current flows in the secondary, if we neglect the relatively small magnetizing current. We know also that the primary current is determined by the secondary current and the turns ratio of the windings, again with magnetizing current neglected. These principles guide us in the analysis of individual cases. Five possible connections of two-winding transformers will be discussed. These connections are summarized, along with their zero-sequence circuits, in Fig. 11.17. The arrows on the connection diagrams of the figures to follow show the possible paths for the flow of zero-sequence current. Absence of an arrow indicates that the transformer connection is such that zero-sequence current cannot flow. The zero-sequence equivalent circuits are approximate as shown since resistance and the magnetizing-current path are omitted from each circuit. The letters P and Q identify corresponding points on the connection diagram

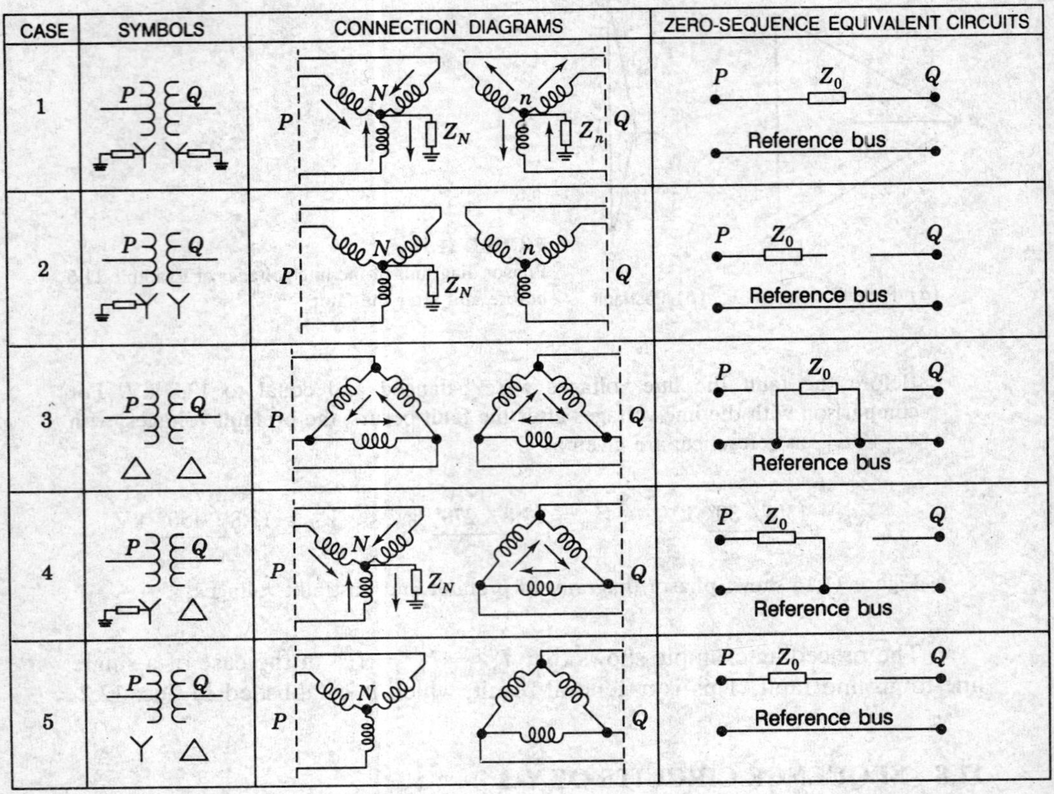

FIGURE 11.17
Zero-sequence equivalent circuits of three-phase transformer banks, together with diagrams of connections and the symbols for one-line diagrams. Impedance Z_0 accounts for the leakage impedance Z and the neutral impedances $3Z_N$ and $3Z_n$ where applicable.

and equivalent circuit. The reasoning to justify the equivalent circuit for each connection follows.

CASE 1. Y-Y *Bank, Both Neutrals Grounded*

Figure 11.18(a) shows the neutrals of a Y-Y bank grounded through impedance Z_N on the high-voltage side and Z_n on the low-voltage side. The arrows on the diagram show the directions chosen for the currents. We first treat the transformers as ideal and add series leakage impedance later when the shunt magnetizing current can also be included if necessary. We continue to designate voltages with respect to ground by a single subscript such as V_A, V_N, and V_a. Voltages with respect to neutral have two subscripts such as V_{AN} and V_{an}. Capital letters are assigned to the high-voltage and lowercase letters are assigned on the other side. As before, windings that are drawn in parallel

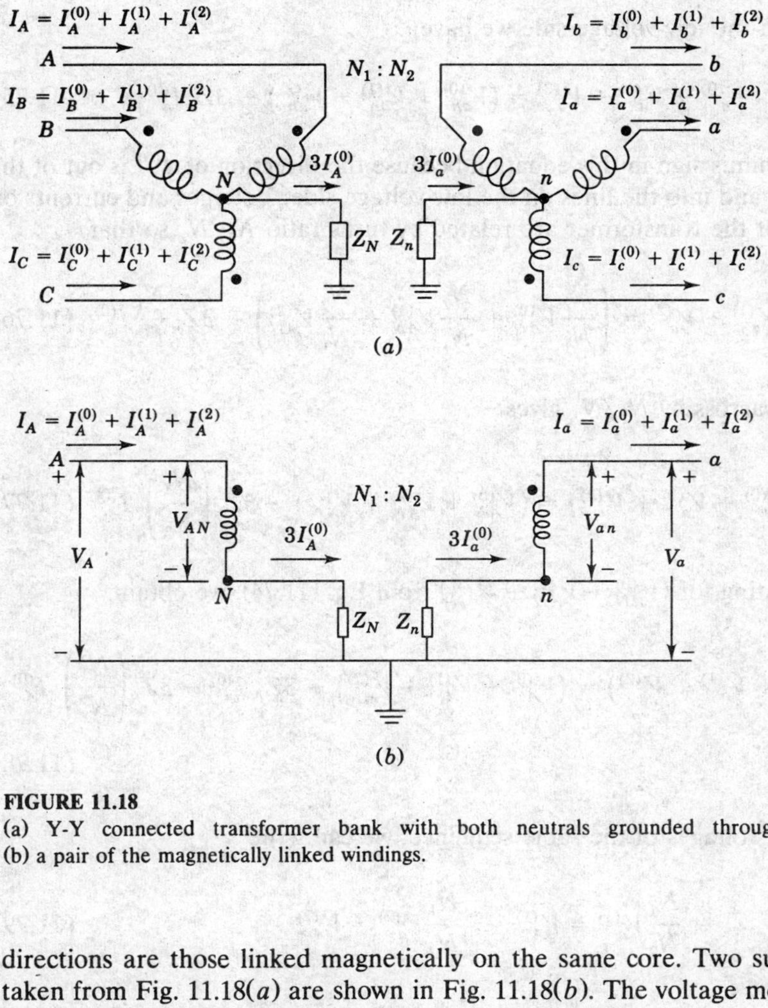

FIGURE 11.18
(a) Y-Y connected transformer bank with both neutrals grounded through impedances; (b) a pair of the magnetically linked windings.

directions are those linked magnetically on the same core. Two such windings taken from Fig. 11.18(a) are shown in Fig. 11.18(b). The voltage measured with respect to ground on the high-voltage side is given by

$$V_A = V_{AN} + V_N \tag{11.73}$$

Substituting the symmetrical components of each voltage gives

$$V_A^{(0)} + V_A^{(1)} + V_A^{(2)} = \left(V_{AN}^{(0)} + V_{AN}^{(1)} + V_{AN}^{(2)}\right) + 3Z_N I_A^{(0)} \tag{11.74}$$

and equating quantities of the same sequence, as explained for Eq. (11.19), again confirms the fact that positive- and negative-sequence voltages to ground are equal to positive- and negative-sequence voltages to neutral. The zero-sequence voltage difference between neutral and ground is equal to $(3Z_N)I_A^{(0)}$.

Similarly, on the low-voltage side we have

$$V_a^{(0)} + V_a^{(1)} + V_a^{(2)} = \left(V_{an}^{(0)} + V_{an}^{(1)} + V_{an}^{(2)}\right) - 3Z_n I_a^{(0)} \qquad (11.75)$$

There is a minus sign in this equation because the direction of $I_a^{(0)}$ is out of the transformer and into the lines on the low-voltage side. Voltages and currents on both sides of the transformer are related by turns ratio N_1/N_2 so that

$$V_a^{(0)} + V_a^{(1)} + V_a^{(2)} = \left(\frac{N_2}{N_1}V_{AN}^{(0)} + \frac{N_2}{N_1}V_{AN}^{(1)} + \frac{N_2}{N_1}V_{AN}^{(2)}\right) - 3Z_n \frac{N_1}{N_2} I_A^{(0)} \qquad (11.76)$$

Multiplying across by N_1/N_2 gives

$$\frac{N_1}{N_2}\left(V_a^{(0)} + V_a^{(1)} + V_a^{(2)}\right) = \left(V_{AN}^{(0)} + V_{AN}^{(1)} + V_{AN}^{(2)}\right) - 3Z_n\left(\frac{N_1}{N_2}\right)^2 I_A^{(0)} \qquad (11.77)$$

and substituting for $(V_{AN}^{(0)} + V_{AN}^{(1)} + V_{AN}^{(2)})$ from Eq. (11.74), we obtain

$$\frac{N_1}{N_2}\left(V_a^{(0)} + V_a^{(1)} + V_a^{(2)}\right) = \left(V_A^{(0)} + V_A^{(1)} + V_A^{(2)}\right) - 3Z_N I_A^{(0)} - 3Z_n\left(\frac{N_1}{N_2}\right)^2 I_A^{(0)} \qquad (11.78)$$

By equating voltages of the same sequence, we can write

$$\frac{N_1}{N_2}V_a^{(1)} = V_A^{(1)} \qquad \frac{N_1}{N_2}V_a^{(2)} = V_A^{(2)} \qquad (11.79)$$

$$\frac{N_1}{N_2}V_a^{(0)} = V_A^{(0)} - \left[3Z_N + 3Z_n\left(\frac{N_1}{N_2}\right)^2\right] I_A^{(0)} \qquad (11.80)$$

The positive- and negative-sequence relations of Eqs. (11.79) are exactly the same as in Chap. 2, and the usual per-phase equivalent circuit of the transformer therefore applies when positive- or negative-sequence voltages and currents are present. The zero-sequence equivalent circuit representing Eq. (11.80) is drawn in Fig. 11.19. We have added the leakage impedance Z of the transformer in series on the high-voltage side as shown so that the total impedance to zero-sequence current is now $Z + 3Z_N + 3(N_1/N_2)^2 Z_n$ referred to the high-voltage side. It is apparent that the shunt magnetizing impedance could also be added to the circuit of Fig. 11.19 if so desired. When voltages on both sides of the transformer are expressed in per unit on kilovolt line-to-line

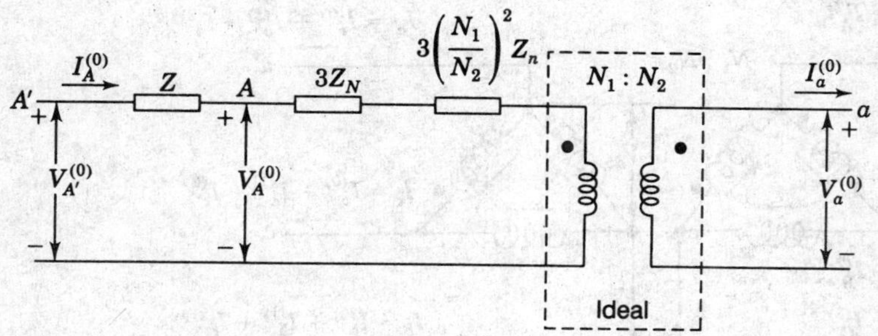

FIGURE 11.19
Zero-sequence circuit of Y-Y connected transformers of Fig. 11.18. Impedance Z is the leakage reactance as measured on the high-voltage side of the transformer.

bases chosen in accordance with the rated voltages, the turns ratio in Fig. 11.19 becomes unity and N_1/N_2 disappears, and we arrive at the zero-sequence circuit shown under Case 1 in Fig. 11.17, where

$$Z_0 = Z + 3Z_N + 3Z_n \text{ per unit} \tag{11.81}$$

Again, we note that impedances connected from neutral to ground in the actual circuit are multiplied by 3 in the zero-sequence circuit. Where both neutrals of a Y-Y bank are grounded directly or through impedance a path through the transformer exists for zero-sequence currents in both windings. Provided the zero-sequence current can follow a complete circuit outside the transformer on both sides, it can flow in both windings of the transformer. In the zero-sequence circuit points on the two sides of the transformer are connected by the zero-sequence impedance of the transformer in the same manner as in the positive- and negative-sequence networks.

CASE 2. Y-Y *Bank, One Neutral Grounded*

If *either one* of the neutrals of a Y-Y bank is ungrounded, zero-sequence current cannot flow in either winding. This can be seen by setting either Z_N or Z_n equal to ∞ in Fig. 11.19. The absence of a path through one winding prevents current in the other and an open circuit exists for zero-sequence current between the two parts of the system connected by the transformer, as shown in Fig. 11.17.

CASE 3. Δ-Δ *Bank*

The phasor sum of the line-to-line voltages equals zero on each side of the Δ-Δ transformer of Fig. 11.20, and so $V_{AB}^{(0)} = V_{ab}^{(0)} = 0$. Applying the rules of the

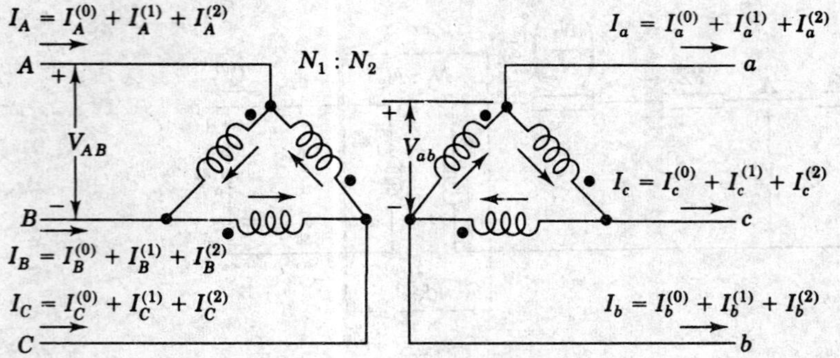

FIGURE 11.20
Wiring diagram of the three-phase Δ-Δ connected transformer.

conventional dot notation for coupled coils to that figure, we have

$$V_{AB} = \frac{N_1}{N_2} V_{ab}$$
$$V_{AB}^{(1)} + V_{AB}^{(2)} = \frac{N_1}{N_2}\left(V_{ab}^{(1)} + V_{ab}^{(2)}\right) \tag{11.82}$$

The line-to-line voltages can be written as line-to-neutral voltages according to Eqs. (11.23) giving

$$\sqrt{3}\, V_{AN}^{(1)} \angle 30^\circ + \sqrt{3}\, V_{AN}^{(2)} \angle -30^\circ = \frac{N_1}{N_2}\left(\sqrt{3}\, V_{an}^{(1)} \angle 30^\circ + \sqrt{3}\, V_{an}^{(2)} \angle -30^\circ\right) \tag{11.83}$$

and so
$$V_{AN}^{(1)} = \frac{N_1}{N_2} V_{an}^{(1)} \qquad V_{AN}^{(2)} = \frac{N_1}{N_2} V_{an}^{(2)} \tag{11.84}$$

Thus, the positive- and negative-sequence equivalent circuits for the Δ-Δ transformer, like those for the Y-Y connection, correspond exactly to the usual per-phase equivalent circuit of Chap. 2. Since a Δ circuit provides no return path for zero-sequence current, no zero-sequence current can flow into either side of a Δ-Δ bank although it can sometimes circulate within the Δ windings. Hence, $I_A^{(0)} = I_a^{(0)} = 0$ in Fig. 11.20, and we obtain the zero-sequence equivalent circuit shown in Fig. 11.17.

CASE 4. Y-Δ *Bank*, *Grounded* Y

If the neutral of a Y-Δ bank is grounded, zero-sequence currents have a path to ground through the Y because corresponding induced currents can

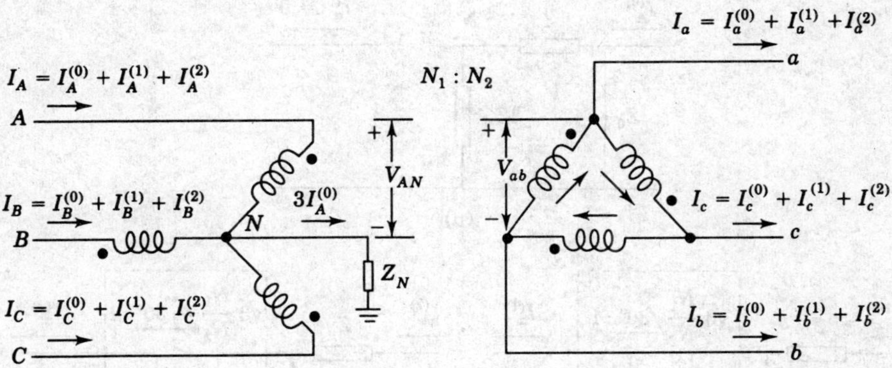

FIGURE 11.21
Wiring diagram for a three-phase Y-Δ transformer bank with neutral grounded through impedance Z_N.

circulate in the Δ. The zero-sequence current circulating in the Δ magnetically balances the zero-sequence current in the Y but cannot flow in the lines connected to the Δ. Hence, $I_a^{(0)} = 0$ in Fig. 11.21. Phase-A voltage on the Y side can be written the same as in Eq. (11.74) from which we obtain

$$V_A^{(0)} + V_A^{(1)} + V_A^{(2)} = \frac{N_1}{N_2}V_{ab}^{(0)} + \frac{N_1}{N_2}V_{ab}^{(1)} + \frac{N_1}{N_2}V_{ab}^{(2)} + 3Z_N I_A^{(0)} \quad (11.85)$$

Equating corresponding sequence components, as explained for Eq. (11.19), gives

$$V_A^{(0)} - 3Z_N I_A^{(0)} = \frac{N_1}{N_2}V_{ab}^{(0)} = 0 \quad (11.86)$$

$$V_A^{(1)} = \frac{N_1}{N_2}V_{ab}^{(1)} = \frac{N_1}{N_2}\sqrt{3}\angle 30^\circ \times V_a^{(1)}$$

$$V_A^{(2)} = \frac{N_1}{N_2}V_{ab}^{(2)} = \frac{N_1}{N_2}\sqrt{3}\angle -30^\circ \times V_a^{(2)} \quad (11.87)$$

Equation (11.86) enables us to draw the zero-sequence circuit shown in Fig. 11.22(a) in which $Z_0 = Z + 3Z_N$ when leakage impedance Z is referred to the high side of the transformer. The equivalent circuit provides for a zero-sequence current path from the line on the Y side through the equivalent resistance and leakage reactance of the transformer to the reference node. An open circuit must exist between the line and the reference node on the Δ side. When the connection from neutral to ground contains impedance Z_N as shown,

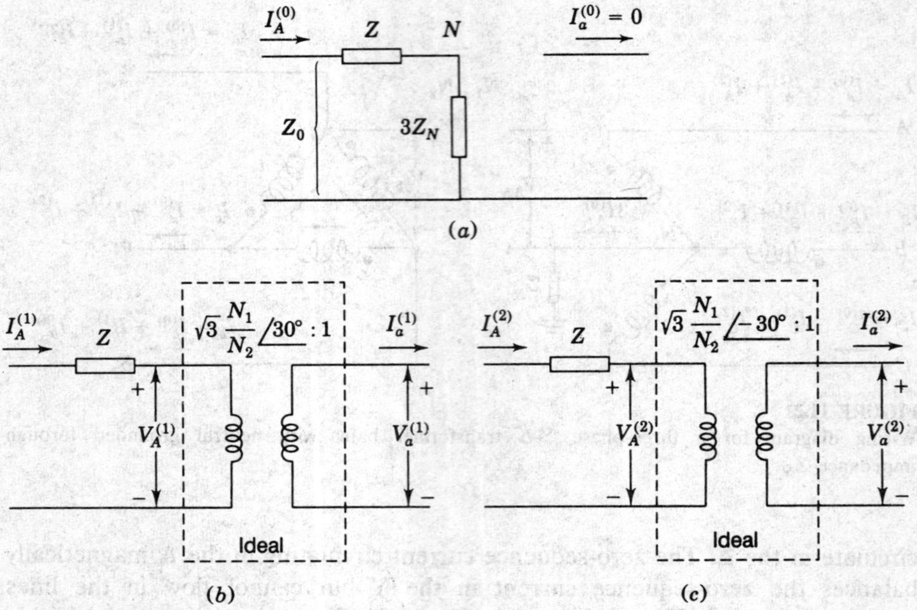

FIGURE 11.22
(*a*) The zero-sequence circuit of Y-Δ transformer bank with grounding impedance Z_N and corresponding (*b*) positive-sequence and (*c*) negative-sequence circuits.

the zero-sequence equivalent circuit must have impedance $3Z_N$ in series with the equivalent resistance and leakage reactance of the transformer to connect the line on the Y side to ground.

CASE 5. Y-Δ *Bank, Ungrounded* Y

An ungrounded Y is a special case where the impedance Z_N between neutral and ground is infinite. The impedance $3Z_n$ in the zero-sequence equivalent circuit of Case 4 becomes infinite and zero-sequence current cannot flow in the transformer windings.

The positive- and negative-sequence equivalent circuits of the Y-Δ transformer shown in Figs. 11.22(*b*) and 11.22(*c*) are based on Eqs. (11.87). We recall from Sec. 2.6 that the multiplier $\sqrt{3}\,N_1/N_2$ in Eqs. (11.87) is the ratio of the two rated line-to-line (and also line-to-neutral) voltages of the Y-Δ transformer. In per-unit calculations, therefore, Eqs. (11.87) become exactly the same as Eqs. (2.35), and again we have the rules

$$V_A^{(1)} = V_a^{(1)} \times 1\angle 30^\circ \qquad I_A^{(1)} = I_a^{(1)} \times 1\angle 30^\circ$$
$$V_A^{(2)} = V_a^{(2)} \times 1\angle -30^\circ \qquad I_A^{(2)} = I_a^{(2)} \times 1\angle -30^\circ \tag{11.88}$$

That is,

> When stepping up from the low-voltage side to the high-voltage side of a Δ-Y or Y-Δ transformer, *advance* the positive-sequence voltages (and currents) by 30° and *retard* the negative-sequence voltages (and currents) by 30°.

The next example shows the numerical application of Eqs. (11.88).

Example 11.7. The resistive Y-connected load bank of Example 11.2 is supplied from the low-voltage Y-side of a Y-Δ transformer. The voltages at the load are the same as in that example. Find the line voltages and currents in per unit on the high-voltage side of the transformer.

Solution. In Example 11.2 we found that the positive- and negative-sequence currents flowing toward the resistive load are

$$I_a^{(1)} = 0.9857\angle 43.6^\circ \text{ per unit}$$

$$I_a^{(2)} = 0.2346\angle 250.3^\circ \text{ per unit}$$

while the corresponding voltages on the low-voltage Y side of the transformer are

$$V_{an}^{(1)} = 0.9857\angle 43.6^\circ \text{ per unit (line-to-neutral voltage base)}$$

$$V_{an}^{(2)} = 0.2346\angle 250.3^\circ \text{ per unit (line-to-neutral voltage base)}$$

Advancing the phase angle of the low-voltage positive-sequence voltage by 30° and retarding the negative-sequence voltage by 30° give on the high-voltage side

$$V_A^{(1)} = 0.9857\angle 43.6 + 30^\circ = 0.9857\angle 73.6^\circ = 0.2783 + j0.9456$$

$$V_A^{(2)} = 0.2346\angle 250.3 - 30^\circ = 0.2346\angle 220.3^\circ = -0.1789 - j0.1517$$

$$V_A = V_A^{(1)} + V_A^{(2)} = 0.0994 + j0.7939 = 0.8\angle 82.8^\circ \text{ per unit}$$

$$V_B^{(1)} = a^2V_A^{(1)} = 0.9857\angle -46.4^\circ = 0.6798 - j0.7138$$

$$V_B^{(2)} = aV_A^{(2)} = 0.2346\angle -19.7^\circ = 0.2209 - j0.0791$$

$$V_B = V_B^{(1)} + V_B^{(2)} = 0.9007 - j0.7929 = 1.20\angle -41.4^\circ \text{ per unit}$$

$$V_C^{(1)} = aV_A^{(1)} = 0.9857\angle 193.6^\circ = -0.9581 - j0.2318$$

$$V_C^{(2)} = a^2V_A^{(2)} = 0.2346\angle 100.3^\circ = -0.0419 + j0.2318$$

$$V_C = V_C^{(1)} + V_C^{(2)} = -1.0 + j0 = 1.0\angle 180^\circ \text{ per unit}$$

Note that the *line-to-neutral* voltages on the high-voltage Δ side of the transformer are equal in per unit to the *line-to-line* voltages found in Example 11.2 for the low-voltage Y side. The line-to-line voltages are

$$V_{AB} = V_A - V_B = 0.0994 + j0.7939 - 0.9007 + j0.7929 = -0.8013 + j1.5868$$

$$= 1.78\underline{/116.8^\circ} \text{ per unit (line-neutral voltage base)}$$

$$= \frac{1.78}{\sqrt{3}}\underline{/116.8^\circ} = 1.028\underline{/116.8^\circ} \text{ per unit (line-line voltage base)}$$

$$V_{BC} = V_B - V_C = 0.9007 - j0.7939 + 1.0 = 1.9007 - j0.7939$$

$$= 2.06\underline{/-22.7^\circ} \text{ per unit (line-neutral voltage base)}$$

$$= \frac{2.06}{\sqrt{3}}\underline{/-22.7^\circ} = 1.19\underline{/-22.7^\circ} \text{ per unit (line-line voltage base)}$$

$$V_{CA} = V_C - V_A = -1.0 - 0.0994 - j0.7939 = 1.0994 - j0.7939$$

$$= 1.356\underline{/215.8^\circ} \text{ per unit (line-neutral voltage base)}$$

$$= \frac{1.356}{\sqrt{3}}\underline{/215.8^\circ} = 0.783\underline{/215.8^\circ} \text{ per unit (line-line voltage base)}$$

Since the load impedance in each phase is a resistance of $1.0\underline{/0^\circ}$ per unit, $I_a^{(1)}$ and $V_a^{(1)}$ are found to have identical per-unit values in this problem. Likewise, $I_a^{(2)}$ and $V_a^{(2)}$ are identical in per unit. Therefore, I_A must be identical to V_A expressed in per unit. Thus,

$$I_A = 0.80\underline{/82.8^\circ} \text{ per unit}$$

$$I_B = 1.20\underline{/-41.4^\circ} \text{ per unit}$$

$$I_C = 1.0\underline{/180^\circ} \text{ per unit}$$

This example emphasizes the fact that in going from one side of the Δ-Y or Y-Δ transformer to the other, the positive-sequence components of currents and voltages on one side must be phase shifted *separately* from the negative-sequence components on the same side before combining them together to form the actual voltage on the other side.

Remarks on Phase Shift. The American National Standards Institute (ANSI) requires connection of Y-Δ and Δ-Y transformers so that the *positive-sequence* voltage to neutral, V_{H_1N}, on the high-voltage side leads the *positive-sequence* voltage to neutral, V_{X_1n}, on the low-voltage side by 30°. The wiring

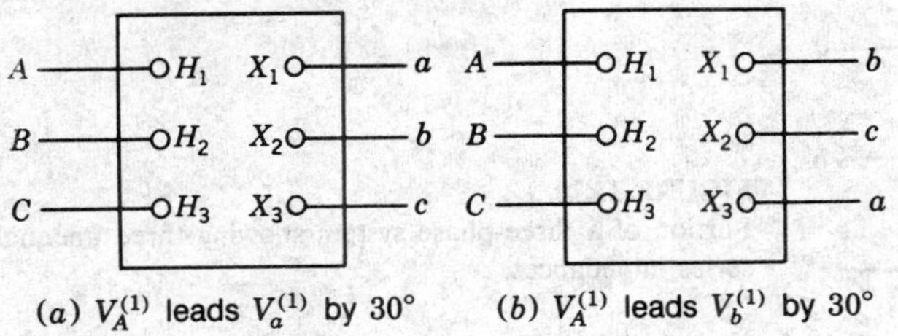

FIGURE 11.23 Labeling of lines connected to a three-phase Y-Δ transformer.

diagram of Fig. 11.21 and the connection diagram of Fig. 11.23(*a*) both satisfy the ANSI requirement; and because the connections of the phases to the transformer terminals H_1, H_2, $H_3 - X_1$, X_2, X_3 are respectively marked $A, B, C - a, b, c$ as shown in those figures, we find that the positive-sequence voltage to neutral $V_{AN}^{(1)}$ leads the positive-sequence voltage to neutral $V_{an}^{(1)}$ by 30°.

It is not absolutely necessary, however, to label lines attached to the transformer terminals X_1, X_2, and X_3 by the letters a, b, and c, respectively, as we have done, since no standards have been adopted for such labeling. In fact, in calculations the designation of lines could be chosen as shown in Fig. 11.23(*b*), which shows the letters b, c, and a associated, respectively, with X_1, X_2, and X_3. If the scheme of Fig. 11.23(*b*) is preferred, it is necessary only to exchange b for a, c for b, and a for c in the wiring and phasor diagrams of Fig. 11.21, which would then show that $V_{an}^{(1)}$ leads $V_{AN}^{(1)}$ by 90° and that $V_{an}^{(2)}$ lags $V_{AN}^{(2)}$ by 90°. It is easy to show that similar statements also apply to the corresponding currents.

We shall continue to follow the labeling scheme of Fig. 11.23(*a*), and Eqs. (11.88) then become identical to the ANSI requirement. When problems involving unsymmetrical faults are solved, positive- and negative-sequence components are found separately and phase shift is taken into account, if necessary, by applying Eqs. (11.88). Computer programs can be written to incorporate the effects of phase shift.

A transformer in a three-phase circuit may consist of three individual single-phase units, or it may be a three-phase transformer. Although the zero-sequence series impedances of three-phase units may differ slightly from the positive- and negative-sequence values, it is customary to assume that series impedances of all sequences are equal regardless of the type of transformer. Table A.1 in the Appendix lists tranformer reactances. Reactance and impedance are almost equal for transformers of 1000 kVA or larger. For simplicity in our calculations we omit the shunt admittance which accounts for exciting current.

11.9 UNSYMMETRICAL SERIES IMPEDANCES

In the previous sections we have been concerned particularly with systems that are normally balanced. Let us look, however, at the equations of a three-phase

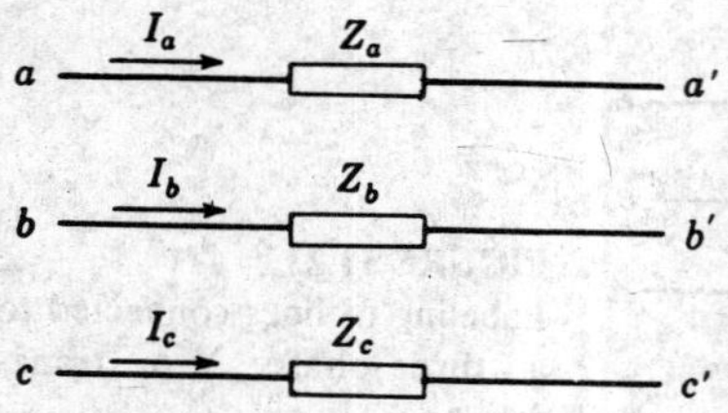

FIGURE 11.24
Portion of a three-phase system showing three unequal series impedances.

circuit when the series impedances are unequal. We shall reach a conclusion that is important in analysis by symmetrical components. Figure 11.24 shows the unsymmetrical part of a system with three unequal series impedances Z_a, Z_b, and Z_c. If we assume no mutual inductance (no coupling) among the three impedances, the voltage drops across the part of the system shown are given by the matrix equation

$$\begin{bmatrix} V_{aa'} \\ V_{bb'} \\ V_{cc'} \end{bmatrix} = \begin{bmatrix} Z_a & 0 & 0 \\ 0 & Z_b & 0 \\ 0 & 0 & Z_c \end{bmatrix} \begin{bmatrix} I_a \\ I_b \\ I_c \end{bmatrix} \tag{11.89}$$

and in terms of the symmetrical components of voltage and current

$$\mathbf{A}\begin{bmatrix} V_{aa'}^{(0)} \\ V_{aa'}^{(1)} \\ V_{aa'}^{(2)} \end{bmatrix} = \begin{bmatrix} Z_a & 0 & 0 \\ 0 & Z_b & 0 \\ 0 & 0 & Z_c \end{bmatrix} \mathbf{A} \begin{bmatrix} I_a^{(0)} \\ I_a^{(1)} \\ I_a^{(2)} \end{bmatrix} \tag{11.90}$$

where $\mathbf{A}$ is the matrix defined by Eq. (11.9). Premultiplying both sides of the equation by $\mathbf{A}^{-1}$ yields the matrix equation from which we obtain

$$\begin{aligned} V_{aa'}^{(0)} &= \tfrac{1}{3}I_a^{(0)}(Z_a + Z_b + Z_c) + \tfrac{1}{3}I_a^{(1)}(Z_a + a^2Z_b + aZ_c) \\ &\quad + \tfrac{1}{3}I_a^{(2)}(Z_a + aZ_b + a^2Z_c) \\ V_{aa'}^{(1)} &= \tfrac{1}{3}I_a^{(0)}(Z_a + aZ_b + a^2Z_c) + \tfrac{1}{3}I_a^{(1)}(Z_a + Z_b + Z_c) \\ &\quad + \tfrac{1}{3}I_a^{(2)}(Z_a + a^2Z_b + aZ_c) \\ V_{aa'}^{(2)} &= \tfrac{1}{3}I_a^{(0)}(Z_a + a^2Z_b + aZ_c) + \tfrac{1}{3}I_a^{(1)}(Z_a + aZ_b + a^2Z_c) \\ &\quad + \tfrac{1}{3}I_a^{(2)}(Z_a + Z_b + Z_c) \end{aligned} \tag{11.91}$$

If the impedances are made equal (that is, if $Z_a = Z_b = Z_c$), Eqs. (11.91) reduce to

$$V_{aa'}^{(0)} = I_a^{(0)} Z_a \qquad V_{aa'}^{(1)} = I_a^{(1)} Z_a \qquad V_{aa'}^{(2)} = I_a^{(2)} Z_a \tag{11.92}$$

If the impedances are unequal, however, Eqs. (11.91) show that the voltage drop of any one sequence is dependent on the currents of all three sequences. Thus, we conclude that the symmetrical components of unbalanced currents flowing in a *balanced* load or in *balanced* series impedances produce voltage drops of like sequence only. If asymmetric coupling (such as unequal mutual inductances) existed among the three impedances of Fig. 11.24, the square matrix of Eqs. (11.89) and (11.90) would contain off-diagonal elements and Eqs. (11.91) would have additional terms.

Although current in any conductor of a three-phase transmission line induces a voltage in the other phases, the way in which reactance is calculated eliminates consideration of coupling. The self-inductance calculated on the basis of complete transposition includes the effect of mutual reactance. The assumption of transposition yields equal series impedances. Thus, the component currents of any one sequence produce only voltage drops of like sequence in a transmission line; that is, positive-sequence currents produce positive-sequence voltage drops only. Likewise, negative-sequence currents produce negative-sequence voltage drops only and zero-sequence currents produce zero-sequence voltage drops only. Equations (11.91) apply to unbalanced Y loads because points a', b', and c' may be connected to form a neutral. We could study variations of these equations for special cases such as single-phase loads where $Z_b = Z_c = 0$, but we continue to confine our discussion to systems that are balanced before a fault occurs.

11.10 SEQUENCE NETWORKS

In the preceding sections of this chapter we have developed single-phase equivalent circuits in the form of zero-, positive-, and negative-sequence circuits for load impedances, transformers, transmission lines, and synchronous machines, all of which constitute the main parts of the three-phase power transmission network. Except for rotating machines, all parts of the network are static and without sources. Each individual part is assumed to be linear and three-phase symmetrical when connected in Y or Δ configuration. On the basis of these assumptions, we have found that:

- In any part of the network voltage drop caused by current of a certain sequence depends on only the impedance of that part of the network to current flow of that sequence.
- The impedances to positive- and negative-sequence currents, Z_1 and Z_2, are equal in any static circuit and may be considered approximately equal in synchronous machines under subtransient conditions.

- In any part of the network impedance to zero-sequence current, Z_0, is generally different from Z_1 and Z_2.
- Only positive-sequence circuits of rotating machines contain sources which are of positive-sequence voltages.
- Neutral is the reference for voltages in positive- and negative-sequence circuits, and such voltages to neutral are the same as voltages to ground if a physical connection of zero or other finite impedance exists between neutral and ground in the actual circuit.
- No positive- or negative-sequence currents flow between neutral points and ground.
- Impedances Z_n in the physical connections between neutral and ground are not included in positive- and negative-sequence circuits but are represented by impedances $3Z_n$ between the points for neutral and ground in the zero-sequence circuits only.

These characteristics of individual sequence circuits guide the construction of corresponding *sequence networks*. The object of obtaining the values of the sequence impedances of the various parts of a power system is to enable us to construct the sequence networks for the complete system. The network of a particular sequence—constructed by joining together all the corresponding sequence circuits of the separate parts—shows all the paths for the flow of current of that sequence in one phase of the actual system.

In a balanced three-phase system the currents flowing in the three phases under normal operating conditions constitute a symmetrical positive-sequence set. These positive-sequence currents cause voltage drops of the same sequence only. Because currents of only *one* sequence occurred in the preceding chapters, we considered them to flow in an independent *per-phase* network which combined the positive-sequence emfs of rotating machines and the impedances of other static circuits to positive-sequence currents only. That same per-phase equivalent network is now called the *positive-sequence network* in order to distinguish it from the networks of the other two sequences.

We have discussed the construction of impedance and admittance representations of some rather complex positive-sequence networks in earlier chapters. Generally, we have not included the phase shift associated with Δ-Y and Y-Δ transformers in positive-sequence networks since practical systems are designed with such phase shifts summing to zero around all loops. In detailed calculations, however, we must remember to *advance* all positive-sequence voltages and currents by 30° when *stepping up* from the low-voltage side to the high-voltage side of a Δ-Y or Y-Δ transformer.

The transition from a positive-sequence network to a *negative-sequence network* is simple. Three-phase synchronous generators and motors have internal voltages of positive sequence only because they are designed to generate balanced voltages. Since the positive- and negative-sequence impedances are the same in a static symmetrical system, conversion of a positive-sequence network to a negative-sequence network is accomplished by changing, if necessary, only the impedances that represent rotating machinery and by omitting the emfs. Electromotive forces are omitted on the assumption of balanced generated voltages and the absence of negative-sequence voltages induced from outside sources. Of course, in using the negative-sequence network for detailed calculations, we must also remember to *retard* the negative-sequence voltages and currents by 30° when *stepping up* from the low-voltage side to the high-voltage side of a Δ-Y or Y-Δ transformer.

Since all the neutral points of a symmetrical three-phase system are at the same potential when balanced three-phase currents are flowing, all the neutral points must be at the same potential for either positive- or negative-sequence currents. Therefore, the neutral of a symmetrical three-phase system is the logical reference potential for specifying positive- and negative-sequence voltage drops and is the reference node of the positive- and negative-sequence networks. Impedance connected between the neutral of a machine and ground is not a part of either the positive- or negative-sequence network because neither positive- nor negative-sequence current can flow in an impedance so connected.

Negative-sequence networks, like the positive-sequence networks of previous chapters, may contain the exact equivalent circuits of parts of the system or be simplified by omitting series resistance and shunt admittance.

Example 11.8. Draw the negative-sequence network for the system described in Example 6.1. Assume that the negative-sequence reactance of each machine is equal to its subtransient reactance. Omit resistance and phase shifts associated with the transformer connections.

Solution. Since all the negative-sequence reactances of the system are equal to the positive-sequence reactances, the negative-sequence network is identical to the positive-sequence network of Fig. 6.6 except for the omission of emfs from the negative-sequence network. The required network is drawn without transformer phase shifts in Fig. 11.25.

Zero-sequence equivalent circuits determined for the various separate parts of the system are readily combined to form the complete *zero-sequence network*. A three-phase system operates single phase insofar as the zero-

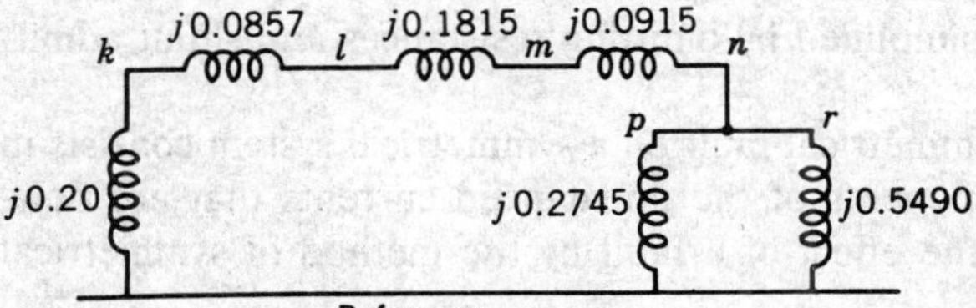

FIGURE 11.25
Negative-sequence network for Example 11.8.

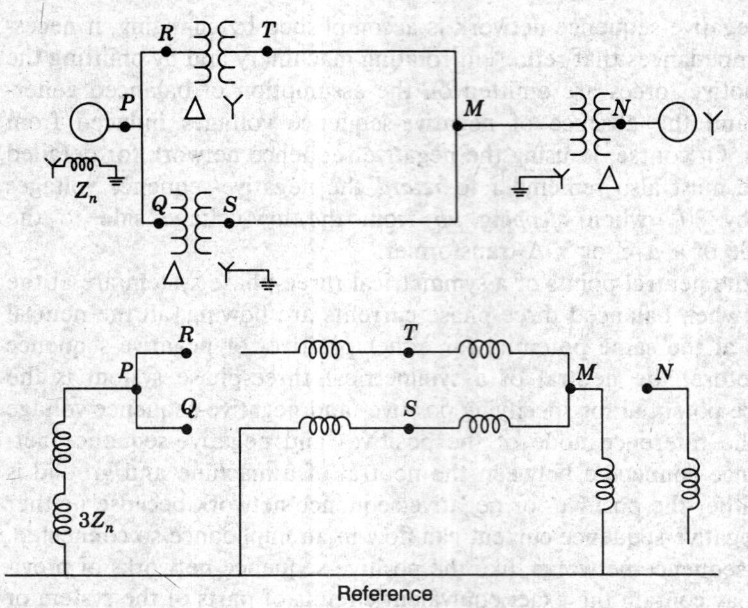

FIGURE 11.26
One-line diagram of a small power system and the corresponding zero-sequence network.

sequence currents are concerned, for the zero-sequence currents are the same in magnitude and phase at any point in all the phases of the system. Therefore, zero-sequence currents will flow only if a return path exists through which a completed circuit is provided. The reference for zero-sequence voltages is the potential of the ground at the point in the system at which any particular voltage is specified. Since zero-sequence currents may be flowing in the ground, the ground is not necessarily at the same potential at all points and the reference node of the zero-sequence network does not represent a ground of uniform potential. We have already discussed the fact that the impedance of the ground and ground wires is included in the zero-sequence impedance of the transmission line, and the return circuit of the zero-sequence network is a conductor of zero impedance, which is the reference node of the system. It is because the impedance of the ground is included in the zero-sequence impedance that voltages measured to the reference node of the zero-sequence network give the correct voltage to equivalent ideal ground. Figures 11.26 and 11.27 show one-line diagrams of two small power systems and their corresponding zero-sequence networks simplified by omitting resistances and shunt admittances.

The analysis of an unsymmetrical fault on a symmetrical system consists in finding the symmetrical components of the unbalanced currents that are flowing. Therefore, to calculate the effect of a fault by the method of symmetrical

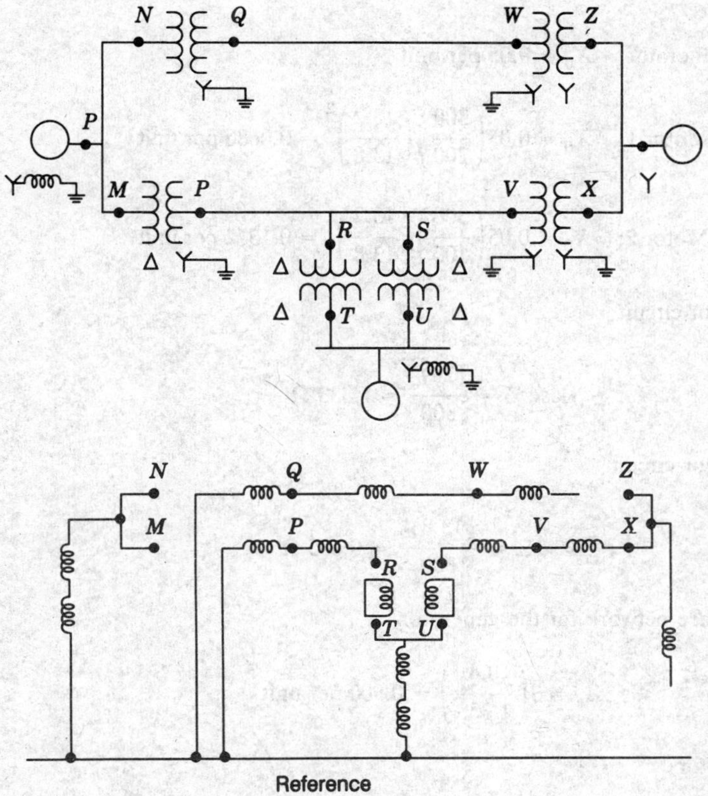

FIGURE 11.27
One-line diagram of a small power system and the corresponding zero-sequence network.

components, it is essential to determine the sequence impedances and to combine them to form the sequence networks. The sequence networks carrying the symmetrical-component currents $I_a^{(0)}$, $I_a^{(1)}$, and $I_a^{(2)}$ are then interconnected to represent various unbalanced fault conditions, as described in Chap. 12.

Example 11.9. Draw the zero-sequence network for the system described in Example 6.1. Assume zero-sequence reactances for the generator and motors of 0.05 per unit. A current-limiting reactor of 0.4 Ω is in each of the neutrals of the generator and the larger motor. The zero-sequence reactance of the transmission line is 1.5 Ω/km.

Solution. The zero-sequence leakage reactance of transformers is equal to the positive-sequence reactance. So, for the transformers $X_0 = 0.0857$ per unit and 0.0915 per unit, as in Example 6.1. Zero-sequence reactances of the generator and motors are

$$\text{Generator:} \quad X_0 = 0.05 \text{ per unit}$$

$$\text{Motor 1:} \quad X_0 = 0.05\left(\frac{300}{200}\right)\left(\frac{13.2}{13.8}\right)^2 = 0.0686 \text{ per unit}$$

$$\text{Motor 2:} \quad X_0 = 0.05\left(\frac{300}{100}\right)\left(\frac{13.2}{13.8}\right)^2 = 0.1372 \text{ per unit}$$

In the generator circuit

$$\text{Base Z} = \frac{(20)^2}{300} = 1.333\ \Omega$$

and in the motor circuit

$$\text{Base Z} = \frac{(13.8)^2}{300} = 0.635\ \Omega$$

In the impedance network for the generator

$$3Z_n = 3\left(\frac{0.4}{1.333}\right) = 0.900 \text{ per unit}$$

and for the motor

$$3Z_n = 3\left(\frac{0.4}{0.635}\right) = 1.890 \text{ per unit}$$

For the transmission line

$$Z_0 = \frac{1.5 \times 64}{176.3} = 0.5445 \text{ per unit}$$

The zero-sequence network is shown in Fig. 11.28.

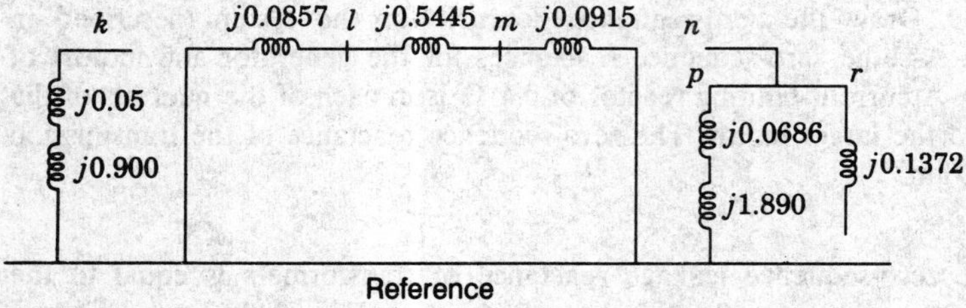

FIGURE 11.28
Zero-sequence network for Example 11.9.

11.11 SUMMARY

Unbalanced voltages and currents can be resolved into their symmetrical components. Problems are solved by treating each set of components separately and superimposing the results.

In balanced networks having strictly symmetrical coupling between phases the currents of one phase sequence induce voltage drops of like sequence only. Impedances of circuit elements to currents of different sequences are not necessarily equal.

A knowledge of the positive-sequence network is necessary for power-flow studies, fault calculations, and stability studies. If the fault calculations or stability studies involve unsymmetrical faults on otherwise symmetrical systems, the negative- and zero-sequence networks are also needed. Synthesis of the zero-sequence network requires particular care because the zero-sequence network may differ from the others considerably.

PROBLEMS

11.1. If $V_{an}^{(1)} = 50\angle 0°$, $V_{an}^{(2)} = 20\angle 90°$, and $V_{an}^{(0)} = 10\angle 180°$ V, determine analytically the voltages to neutral V_{an}, V_{bn}, and V_{cn}, and also show graphically the sum of the given symmetrical components which determine the line-to-neutral voltages.

11.2. When a generator has terminal *a* open and the other two terminals are connected to each other with a short circuit from this connection to ground, typical values for the symmetrical components of current in phase *a* are $I_a^{(1)} = 600\angle -90°$, $I_a^{(2)} = 250\angle 90°$, and $I_a^{(0)} = 350\angle 90°$ A. Find the current into the ground and the current in each phase of the generator.

11.3. Determine the symmetrical components of the three currents $I_a = 10\angle 0°$, $I_b = 10\angle 230°$, and $I_c = 10\angle 130°$ A.

11.4. The currents flowing in the lines toward a balanced load connected in Δ are $I_a = 100\angle 0°$, $I_b = 141.4\angle 225°$, and $I_c = 100\angle 90°$. Find the symmetrical components of the given line currents and draw phasor diagrams of the positive- and negative-sequence line and phase currents. What is I_{ab} in amperes?

11.5. The voltages at the terminals of a balanced load consisting of three 10-Ω resistors connected in Y are $V_{ab} = 100\angle 0°$, $V_{bc} = 80.8\angle -121.44°$, and $V_{ca} = 90\angle 130°$ V. Assuming that there is no connection to the neutral of the load, find the line currents from the symmetrical components of the given line voltages.

11.6. Find the power expended in the three 10-Ω resistors of Prob. 11.5 from the symmetrical components of currents and voltages. Check the answer.

11.7. If there is impedance in the neutral connection to ground of a Y-connected load, then show that the voltages V_a, V_b, and V_c of Eq. (11.26) must be interpreted as voltages with respect to ground.

11.8. A balanced three-phase load consists of Δ-connected impedances Z_Δ in parallel with solidly grounded Y-connected impedances Z_Y.

(*a*) Express the currents I_a, I_b, and I_c flowing in the lines from the supply source toward the load in terms of the source voltages V_a, V_b, and V_c.

(*b*) Transform the expressions of part (*a*) into their symmetrical component equivalents, and thus express $I_a^{(0)}$, $I_a^{(1)}$, and $I_a^{(2)}$ in terms of $V_a^{(0)}$, $V_a^{(1)}$, and $V_a^{(2)}$.

(*c*) Hence, draw the sequence circuit for the combined load.

11.9. The Y-connected impedances in parallel with the Δ-connected impedances Z_Δ of Prob. 11.8 are now grounded through an impedance Z_g.

(*a*) Express the currents I_a, I_b, and I_c flowing in the lines from the supply source toward the load in terms of the source voltages V_a, V_b, and V_c and the voltage V_n of the neutral point.

(*b*) Expressing V_n in terms of $I_a^{(0)}$, $I_a^{(1)}$, $I_a^{(2)}$, and Z_g, find the equations for these currents in terms of $V_a^{(0)}$, $V_a^{(1)}$, and $V_a^{(2)}$.

(*c*) Hence, draw the sequence circuit for the combined load.

11.10. Suppose that the line-to-neutral voltages at the sending end of the line described in Example 11.5 can be maintained constant at 200-kV and that a single-phase inductive load of 420 Ω is connected between phase *a* and neutral at the receiving end.

(*a*) Use Eqs. (11.51) to express numerically the receiving-end sequence voltages $V_{a'n'}^{(0)}$, $V_{a'n'}^{(1)}$, and $V_{a'n'}^{(2)}$ in terms of the load current I_L and the sequence impedances Z_0, Z_1, and Z_2 of the line.

(*b*) Hence, determine the line current I_L in amperes.

(*c*) Determine the open-circuit voltages to neutral of phases *b* and *c* at the receiving end.

(*d*) Verify your answer to part (*c*) without using symmetrical components.

11.11. Solve Prob. 11.10 if the same 420-Ω inductive load is connected between phases *a* and *b* at the receiving end. In part (*c*) find the open-circuit voltage of phase *c* only.

11.12. A Y-connected synchronous generator has sequence reactances $X_0 = 0.09$, $X_1 = 0.22$, and $X_2 = 0.36$, all in per unit. The neutral point of the machine is grounded through a reactance of 0.09 per unit. The machine is running on no load with rated terminal voltage when it suffers an unbalanced fault. The fault currents out of the machine are $I_a = 0$, $I_b = 3.75\angle 150°$, and $I_c = 3.75\angle 30°$, all in per unit with respect to phase *a* line-to-neutral voltage. Determine

(*a*) The terminal voltages in each phase of the machine with respect to ground,

(*b*) The voltage of the neutral point of the machine with respect to ground, and

(*c*) The nature (type) of the fault from the results of part (*a*).

11.13. Solve Prob. 11.12 if the fault currents in per unit are $I_a = 0$, $I_b = -2.986\angle 0°$, and $I_c = 2.986\angle 0°$.

11.14. Assume that the currents specified in Prob. 11.4 are flowing toward a load from lines connected to the Y side of a Δ-Y transformer rated 10 MVA, 13.2Δ/66Y kV. Determine the currents flowing in the lines on the Δ side by converting the symmetrical components of the currents to per unit on the base of the transformer rating and by shifting the components according to Eq. (11.88). Check the results by computing the currents in each phase of the Δ windings in amperes directly from the currents on the Y side by multiplying by the turns ratio of the windings. Complete the check by computing the line currents from the phase currents on the Δ side.

11.15. Three single-phase transformers are connected as shown in Fig. 11.29 to form a Y-Δ transformer. The high-voltage windings are Y-connected with polarity marks as indicated. Magnetically coupled windings are drawn in parallel directions. Determine the correct placement of polarity marks on the low-voltage windings. Identify the numbered terminals on the low-voltage side (*a*) with the letters *a*, *b*, and *c*, where $I_A^{(1)}$ leads $I_a^{(1)}$ by 30, and (*b*) with the letters *a*′, *b*′, and *c*′ so that $I_{a'}^{(1)}$ is 90 out of phase with $I_A^{(1)}$.

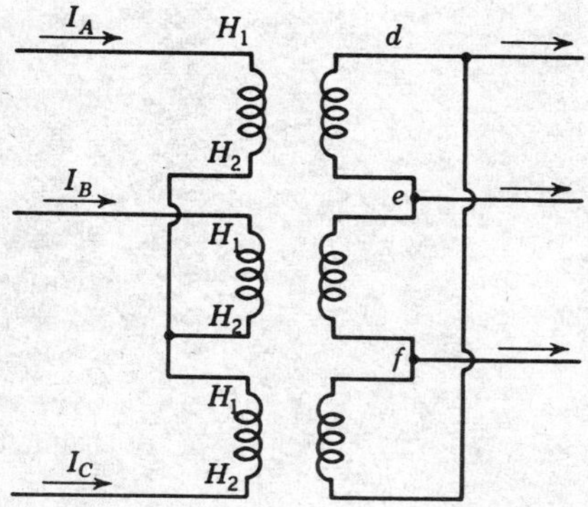

FIGURE 11.29
Circuit for Prob. 11.15.

11.16. Balanced three-phase voltages of 100 V line to line are applied to a Y-connected load consisting of three resistors. The neutral of the load is not grounded. The resistance in phase *a* is 10 Ω, in phase *b* is 20 Ω, and in phase *c* is 30 Ω. Select voltage to neutral of the three-phase line as reference and determine the current in phase *a* and the voltage V_{an}.

11.17. Draw the negative- and zero-sequence impedance networks for the power system of Prob. 3.12. Mark the values of all reactances in per unit on a base of 50 MVA, 13.8 kV in the circuit of generator 1. Letter the networks to correspond to the single-line diagram. The neutrals of generators 1 and 3 are connected to ground through current-limiting reactors having a reactance of 5%, each on the base of the machine to which it is connected. Each generator has negative- and zero-sequence reactances of 20 and 5%, respectively, on its own rating as base. The zero-sequence reactance of the transmission line is 210 Ω from *B* to *C* and 250 Ω from *C* to *E*.

11.18. Draw the negative- and zero-sequence impedance networks for the power system of Prob. 3.13. Choose a base of 50 MVA, 138 kV in the 40-Ω transmission line and mark all reactances in per unit. The negative-sequence reactance of each synchronous machine is equal to its subtransient reactance. The zero-sequence reactance of each machine is 8% based on its own rating. The neutrals of the machines are connected to ground through current-limiting reactors having a reactance of 5%, each on the base of the machine to which it is connected. Assume that the zero-sequence reactances of the transmission lines are 300% of their positive-sequence reactances.

11.19. Determine the zero-sequence Thévenin impedance seen at bus © of the system described in Prob. 11.17 if transformer T_3 has (*a*) one ungrounded and one solidly grounded neutral, as shown in Fig. 3.23, and (*b*) both neutrals are solidly grounded.

CHAPTER 12

UNSYMMETRICAL FAULTS

Most of the faults that occur on power systems are unsymmetrical faults, which may consist of unsymmetrical short circuits, unsymmetrical faults through impedances, or open conductors. Unsymmetrical faults occur as single line-to-ground faults, line-to-line faults, or double line-to-ground faults. The path of the fault current from line to line or line to ground may or may not contain impedance. One or two open conductors result in unsymmetrical faults, through either the breaking of one or two conductors or the action of fuses and other devices that may not open the three phases simultaneously. Since any unsymmetrical fault causes unbalanced currents to flow in the system, the method of symmetrical components is very useful in an analysis to determine the currents and voltages in all parts of the system after the occurrence of the fault. We consider faults on a power system by applying Thévenin's theorem, which allows us to find the current in the fault by replacing the entire system by a single generator and series impedance, and we show how the bus impedance matrix is applied to the analysis of unsymmetrical faults.

12.1 UNSYMMETRICAL FAULTS ON POWER SYSTEMS

In the derivation of equations for the symmetrical components of currents and voltages in a general network the currents flowing *out* of the original balanced

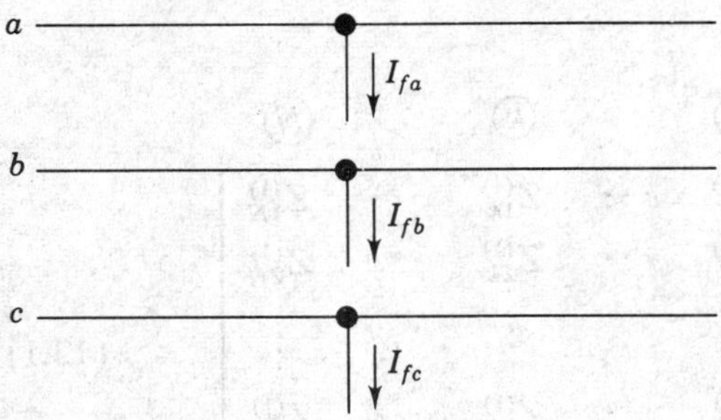

FIGURE 12.1
Three conductors of a three-phase system. The stubs carrying currents I_{fa}, I_{fb}, and I_{fc} may be interconnected to represent different types of faults.

system from phases a, b, and c at the fault point will be designated as I_{fa}, I_{fb}, and I_{fc}, respectively. We can visualize these currents by referring to Fig. 12.1, which shows the three lines a, b, and c of the three-phase system at the part of the network where the fault occurs. The flow of current from each line into the fault is indicated by arrows shown on the diagram beside *hypothetical stubs* connected to each line at the fault location. Appropriate connections of the stubs represent the various types of fault. For instance, direct connection of stubs b and c produces a line-to-line fault through zero impedance. The current in stub a is then zero, and I_{fb} equals $-I_{fc}$.

The line-to-ground voltages at *any* bus Ⓙ of the system *during* the fault will be designated V_{ja}, V_{jb}, and V_{jc}; and we shall continue to use superscripts 1, 2, and 0, respectively, to denote positive-, negative-, and zero-sequence quantities. Thus, for example, $V_{ja}^{(1)}$, $V_{ja}^{(2)}$, and $V_{ja}^{(0)}$ will denote, respectively, the positive-, negative-, and zero-sequence components of the line-to-ground voltage V_{ja} at bus Ⓙ *during* the fault. The line-to-neutral voltage of phase a at the fault point *before* the fault occurs will be designated simply by V_f, which is a *positive-sequence* voltage since the system is balanced. We met the prefault voltage V_f previously in Sec. 10.3 when calculating the currents in a power system with a symmetrical three-phase fault applied.

A single-line diagram of a power system containing two synchronous machines is shown in Fig. 12.2. Such a system is sufficiently general for equations derived therefrom to be applicable to any balanced system regardless of the complexity. Figure 12.2 also shows the sequence networks of the system. The point where a fault is assumed to occur is marked P, and in this particular example it is called bus Ⓚ on the single-line diagram and in the sequence networks. Machines are represented by their subtransient internal voltages in series with their subtransient reactances when subtransient fault conditions are being studied.

In Sec. 10.3 we used the bus impedance matrix composed of positive-sequence impedances to determine currents and voltages upon the occurrence of a symmetrical three-phase fault. The method can be easily extended to apply to *unsymmetrical* faults by realizing that the negative- and zero-sequence networks also can be represented by bus impedance matrices. The bus impedance matrix will now be written symbolically for the positive-sequence network in the following form:

$$
\mathbf{Z}_{\text{bus}}^{(1)} = \begin{array}{c} \\ (1) \\ (2) \\ \vdots \\ (k) \\ \vdots \\ (N) \end{array}
\begin{array}{c}
\begin{array}{cccccc} (1) & (2) & & (k) & & (N) \end{array} \\
\left[\begin{array}{cccccc}
Z_{11}^{(1)} & Z_{12}^{(1)} & \cdots & Z_{1k}^{(1)} & \cdots & Z_{1N}^{(1)} \\
Z_{21}^{(1)} & Z_{22}^{(1)} & \cdots & Z_{2k}^{(1)} & \cdots & Z_{2N}^{(1)} \\
\vdots & \vdots & \ddots & \vdots & \ddots & \vdots \\
Z_{k1}^{(1)} & Z_{k2}^{(1)} & \cdots & Z_{kk}^{(1)} & \cdots & Z_{kN}^{(1)} \\
\vdots & \vdots & \ddots & \vdots & \ddots & \vdots \\
Z_{N1}^{(1)} & Z_{N2}^{(1)} & \cdots & Z_{Nk}^{(1)} & \cdots & Z_{NN}^{(1)}
\end{array} \right]
\end{array} \tag{12.1}
$$

Similarly, the bus impedance matrices for the negative- and zero-sequence networks will be written

$$
\mathbf{Z}_{\text{bus}}^{(2)} = \begin{array}{c} \\ (1) \\ (2) \\ \vdots \\ (k) \\ \vdots \\ (N) \end{array}
\begin{array}{c}
\begin{array}{cccccc} (1) & (2) & & (k) & & (N) \end{array} \\
\left[\begin{array}{cccccc}
Z_{11}^{(2)} & Z_{12}^{(2)} & \cdots & Z_{1k}^{(2)} & \cdots & Z_{1N}^{(2)} \\
Z_{21}^{(2)} & Z_{22}^{(2)} & \cdots & Z_{2k}^{(2)} & \cdots & Z_{2N}^{(2)} \\
\vdots & \vdots & \ddots & \vdots & \ddots & \vdots \\
Z_{k1}^{(2)} & Z_{k2}^{(2)} & \cdots & Z_{kk}^{(2)} & \cdots & Z_{kN}^{(2)} \\
\vdots & \vdots & \ddots & \vdots & \ddots & \vdots \\
Z_{N1}^{(2)} & Z_{N2}^{(2)} & \cdots & Z_{Nk}^{(2)} & \cdots & Z_{NN}^{(2)}
\end{array} \right]
\end{array} \tag{12.2}
$$

$$
\mathbf{Z}_{\text{bus}}^{(0)} = \begin{array}{c} \\ (1) \\ (2) \\ \vdots \\ (k) \\ \vdots \\ (N) \end{array}
\begin{array}{c}
\begin{array}{cccccc} (1) & (2) & & (k) & & (N) \end{array} \\
\left[\begin{array}{cccccc}
Z_{11}^{(0)} & Z_{12}^{(0)} & \cdots & Z_{1k}^{(0)} & \cdots & Z_{1N}^{(0)} \\
Z_{21}^{(0)} & Z_{22}^{(0)} & \cdots & Z_{2k}^{(0)} & \cdots & Z_{2N}^{(0)} \\
\vdots & \vdots & \ddots & \vdots & \ddots & \vdots \\
Z_{k1}^{(0)} & Z_{k2}^{(0)} & \cdots & Z_{kk}^{(0)} & \cdots & Z_{kN}^{(0)} \\
\vdots & \vdots & \ddots & \vdots & \ddots & \vdots \\
Z_{N1}^{(0)} & Z_{N2}^{(0)} & \cdots & Z_{Nk}^{(0)} & \cdots & Z_{NN}^{(0)}
\end{array} \right]
\end{array}
$$

Thus, $Z_{ij}^{(1)}$, $Z_{ij}^{(2)}$, and $Z_{ij}^{(0)}$ denote representative elements of the bus impedance matrices for the positive-, negative-, and zero-sequence networks, respectively. If so desired, each of the networks can be replaced by its Thévenin equivalent between any one of the buses and the reference node.

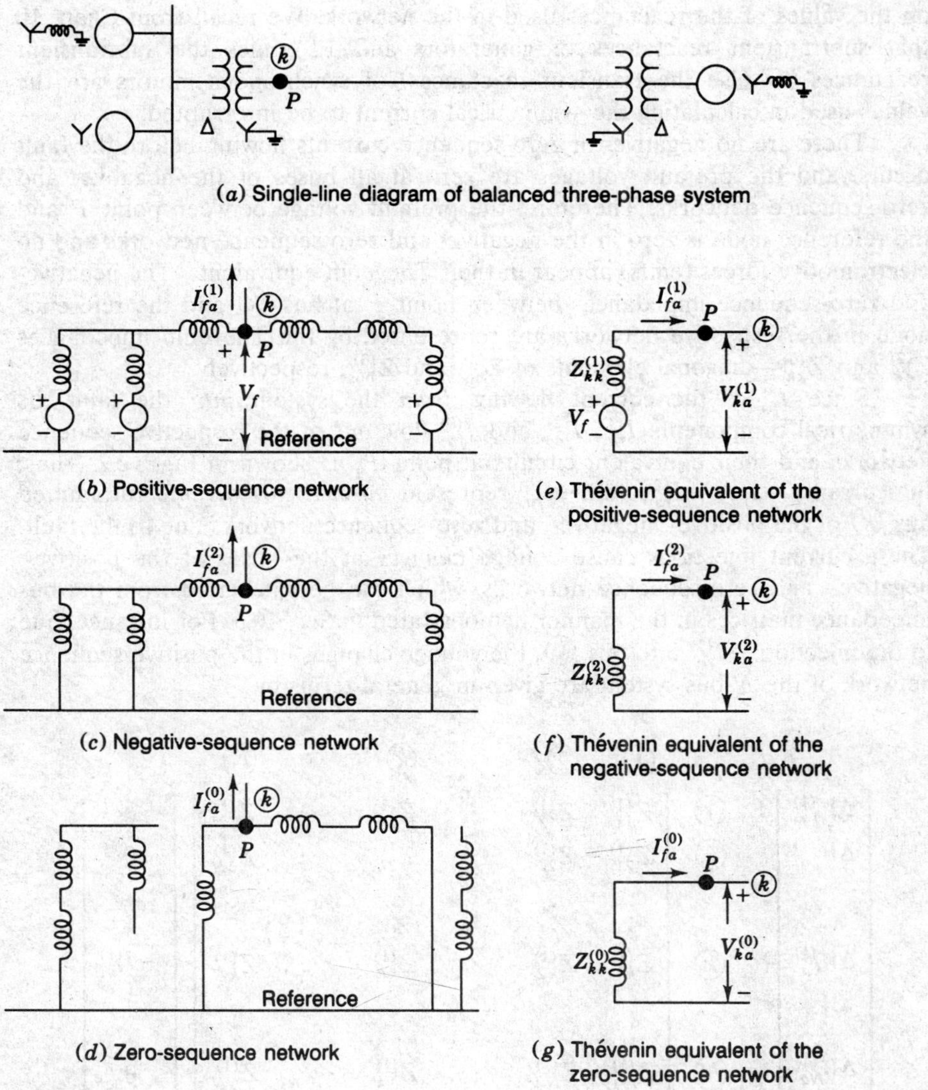

FIGURE 12.2
Single-line diagram of a three-phase system, the three sequence networks of the system, and the Thévenin equivalent of each network for a fault at P, which is called bus Ⓚ.

The Thévenin equivalent circuit between the fault point P and the reference node in each sequence network is shown adjacent to the diagram of the corresponding network in Fig. 12.2. As in Chap. 10, the voltage source in the positive-sequence network and its Thévenin equivalent circuit is V_f, the prefault voltage to neutral at the fault point P, which happens to be bus Ⓚ in this illustration. The Thévenin impedance measured between point P and the reference node of the positive-sequence network is $Z_{kk}^{(1)}$, and its value depends on the values of the reactances used in the network. We recall from Chap. 10 that subtransient reactances of generators and 1.5 times the subtransient reactances (or else the transient reactances) of synchronous motors are the values used in calculating the symmetrical current to be interrupted.

There are no negative- or zero-sequence currents flowing before the fault occurs, and the prefault voltages are zero at all buses of the negative- and zero-sequence networks. Therefore, the prefault voltage between point P and the reference node is zero in the negative- and zero-sequence networks and no electromotive forces (emfs) appear in their Thévenin equivalents. The negative- and zero-sequence impedances between point P at bus ⓚ and the reference node in the respective networks are represented by the Thévenin impedances $Z_{kk}^{(2)}$ and $Z_{kk}^{(0)}$—diagonal elements of $\mathbf{Z}_{\text{bus}}^{(2)}$ and $\mathbf{Z}_{\text{bus}}^{(0)}$, respectively.

Since I_{fa} is the current flowing *from* the system *into* the fault, its symmetrical components $I_{fa}^{(1)}$, $I_{fa}^{(2)}$, and $I_{fa}^{(0)}$ flow *out* of the respective sequence networks and their equivalent circuits at point P, as shown in Fig. 12.2. Thus, the currents $-I_{fa}^{(1)}$, $-I_{fa}^{(2)}$, and $-I_{fa}^{(0)}$ represent *injected currents into* the faulted bus ⓚ of the positive-, negative-, and zero-sequence networks due to the fault. These current injections cause voltage *changes* at the buses of the positive-, negative-, and zero-sequence networks, which can be calculated from the bus impedance matrices in the manner demonstrated in Sec. 10.3. For instance, due to the injection $-I_{fa}^{(1)}$ into bus ⓚ, the voltage changes in the positive-sequence network of the N-bus system are given in general terms by

$$\begin{array}{c} \\ \left[\begin{array}{c} \Delta V_{1a}^{(1)} \\ \Delta V_{2a}^{(1)} \\ \vdots \\ \Delta V_{ka}^{(1)} \\ \vdots \\ \Delta V_{Na}^{(1)} \end{array}\right] \end{array} = \begin{array}{c} \\ \begin{array}{c} \textcircled{1} \\ \textcircled{2} \\ \\ \textcircled{k} \\ \\ \textcircled{N} \end{array} \end{array} \begin{array}{c} \begin{array}{cccccc} \textcircled{1} & \textcircled{2} & & \textcircled{k} & & \textcircled{N} \end{array} \\ \left[\begin{array}{cccccc} Z_{11}^{(1)} & Z_{12}^{(1)} & \cdots & Z_{1k}^{(1)} & \cdots & Z_{1N}^{(1)} \\ Z_{21}^{(1)} & Z_{22}^{(1)} & \cdots & Z_{2k}^{(1)} & \cdots & Z_{2N}^{(1)} \\ \vdots & \vdots & \ddots & \vdots & \ddots & \vdots \\ Z_{k1}^{(1)} & Z_{k2}^{(1)} & \cdots & Z_{kk}^{(1)} & \cdots & Z_{kN}^{(1)} \\ \vdots & \vdots & \ddots & \vdots & \ddots & \vdots \\ Z_{N1}^{(1)} & Z_{N2}^{(1)} & \cdots & Z_{Nk}^{(1)} & \cdots & Z_{NN}^{(1)} \end{array}\right] \end{array} \begin{array}{c} \\ \left[\begin{array}{c} 0 \\ 0 \\ \vdots \\ -I_{fa}^{(1)} \\ \vdots \\ 0 \end{array}\right] \end{array}$$

$$= \left[\begin{array}{c} -Z_{1k}^{(1)} I_{fa}^{(1)} \\ -Z_{2k}^{(1)} I_{fa}^{(1)} \\ \vdots \\ -Z_{kk}^{(1)} I_{fa}^{(1)} \\ \vdots \\ -Z_{Nk}^{(1)} I_{fa}^{(1)} \end{array}\right] \qquad (12.3)$$

This equation is quite similar to Eq. (10.15) for symmetrical faults. Note that only column k of $\mathbf{Z}_{\text{bus}}^{(1)}$ enters into the calculations. In industry practice, it is customary to regard all prefault currents as being zero and to designate the voltage V_f as the positive-sequence voltage at all buses of the system before the fault occurs. Superimposing the changes of Eq. (12.3) on the prefault voltages then yields the total positive-sequence voltage of phase a at each bus during the fault,

$$\begin{bmatrix} V_{1a}^{(1)} \\ V_{2a}^{(1)} \\ \vdots \\ V_{ka}^{(1)} \\ \vdots \\ V_{Na}^{(1)} \end{bmatrix} = \begin{bmatrix} V_f \\ V_f \\ \vdots \\ V_f \\ \vdots \\ V_f \end{bmatrix} + \begin{bmatrix} \Delta V_{1a}^{(1)} \\ \Delta V_{2a}^{(1)} \\ \vdots \\ \Delta V_{ka}^{(1)} \\ \vdots \\ \Delta V_{Na}^{(1)} \end{bmatrix} = \begin{bmatrix} V_f - Z_{1k}^{(1)} I_{fa}^{(1)} \\ V_f - Z_{2k}^{(1)} I_{fa}^{(1)} \\ \vdots \\ V_f - Z_{kk}^{(1)} I_{fa}^{(1)} \\ \vdots \\ V_f - Z_{Nk}^{(1)} I_{fa}^{(1)} \end{bmatrix} \tag{12.4}$$

This equation is similar to Eq. (10.18) for symmetrical faults, the only difference being the added superscripts and subscripts denoting the positive-sequence components of the phase a quantities.

Equations for the negative- and zero-sequence voltage changes due to the fault at bus Ⓚ of the N-bus system are similarly written with the superscripts in Eq. (12.3) changed from 1 to 2 and from 1 to 0, respectively. Because the prefault voltages are zero in the negative- and zero-sequence networks, the voltage changes express the *total* negative- and zero-sequence voltages during the fault, and so we have

$$\begin{bmatrix} V_{1a}^{(2)} \\ V_{2a}^{(2)} \\ \vdots \\ V_{ka}^{(2)} \\ \vdots \\ V_{Na}^{(2)} \end{bmatrix} = \begin{bmatrix} -Z_{1k}^{(2)} I_{fa}^{(2)} \\ -Z_{2k}^{(2)} I_{fa}^{(2)} \\ \vdots \\ -Z_{kk}^{(2)} I_{fa}^{(2)} \\ \vdots \\ -Z_{Nk}^{(2)} I_{fa}^{(2)} \end{bmatrix} \qquad \begin{bmatrix} V_{1a}^{(0)} \\ V_{2a}^{(0)} \\ \vdots \\ V_{ka}^{(0)} \\ \vdots \\ V_{Na}^{(0)} \end{bmatrix} = \begin{bmatrix} -Z_{1k}^{(0)} I_{fa}^{(0)} \\ -Z_{2k}^{(0)} I_{fa}^{(0)} \\ \vdots \\ -Z_{kk}^{(0)} I_{fa}^{(0)} \\ \vdots \\ -Z_{Nk}^{(0)} I_{fa}^{(0)} \end{bmatrix} \tag{12.5}$$

When the fault is at bus Ⓚ, note that only the entries in columns k of $\mathbf{Z}_{\text{bus}}^{(2)}$ and $\mathbf{Z}_{\text{bus}}^{(0)}$ are involved in the calculations of negative- and zero-sequence voltages. Thus, knowing the symmetrical components $I_{fa}^{(0)}$, $I_{fa}^{(1)}$, and $I_{fa}^{(2)}$ of the fault currents at bus Ⓚ, we can determine the sequence voltages at *any* bus Ⓙ of the system from the jth rows of Eqs. (12.4) and (12.5). That is, during the fault at bus Ⓚ the voltages at any bus Ⓙ are

$$\begin{aligned} V_{ja}^{(0)} &= \quad\;\, - Z_{jk}^{(0)} I_{fa}^{(0)} \\ V_{ja}^{(1)} &= V_f - Z_{jk}^{(1)} I_{fa}^{(1)} \\ V_{ja}^{(2)} &= \quad\;\, - Z_{jk}^{(2)} I_{fa}^{(2)} \end{aligned} \tag{12.6}$$

If the prefault voltage at bus Ⓙ is not V_f, then we simply replace V_f in Eq. (12.6) by the actual value of the prefault (positive-sequence) voltage at that bus. Since V_f is *by definition* the actual prefault voltage at the faulted bus Ⓚ, we always have at that bus

$$\begin{aligned} V_{ka}^{(0)} &= \quad\;\, - Z_{kk}^{(0)} I_{fa}^{(0)} \\ V_{ka}^{(1)} &= V_f - Z_{kk}^{(1)} I_{fa}^{(1)} \\ V_{ka}^{(2)} &= \quad\;\, - Z_{kk}^{(2)} I_{fa}^{(2)} \end{aligned} \tag{12.7}$$

and these are the terminal voltage equations for the Thévenin equivalents of the sequence networks shown in Fig. 12.2.

It is important to remember that the currents $I_{fa}^{(0)}$, $I_{fa}^{(1)}$, and $I_{fa}^{(2)}$ are symmetrical-component currents in the stubs hypothetically attached to the system at the fault point. These currents take on values determined by the particular type of fault being studied, and once they have been calculated, they can be regarded as negative injections into the corresponding sequence networks. If the system has Δ-Y transformers, some of the sequence voltages calculated from Eqs. (12.6) may have to be shifted in phase angle before being combined with other components to form the new bus voltages of the faulted system. There are no phase shifts involved in Eq. (12.7) when the voltage V_f at the fault point is chosen as reference, which is customary.

In a system with Δ-Y transformers the open circuits encountered in the zero-sequence network require careful consideration in computer applications of the $\mathbf{Z}_{\text{bus}}$ building algorithm. Consider, for instance, the solidly grounded Y-Δ transformer connected between buses Ⓜ and Ⓝ of Fig. 12.3(*a*). The positive- and zero-sequence circuits are shown in Figs. 12.3(*b*) and 12.3(*c*), respectively. The negative-sequence circuit is the same as the positive-sequence circuit. It is straightforward to include these sequence circuits in the bus impedance matrices $\mathbf{Z}_{\text{bus}}^{(0)}$, $\mathbf{Z}_{\text{bus}}^{(1)}$, and $\mathbf{Z}_{\text{bus}}^{(2)}$ using the *pictorial* representations shown in the figures. This will be done in the sections which follow when Y-Δ transformers are present. Suppose, however, that we wish to represent *removal* of the transformer connections from bus Ⓝ in a computer algorithm which cannot avail of pictorial representations. We can easily undo the connections to bus Ⓝ in the positive- and negative-sequence networks by applying the building algorithm to

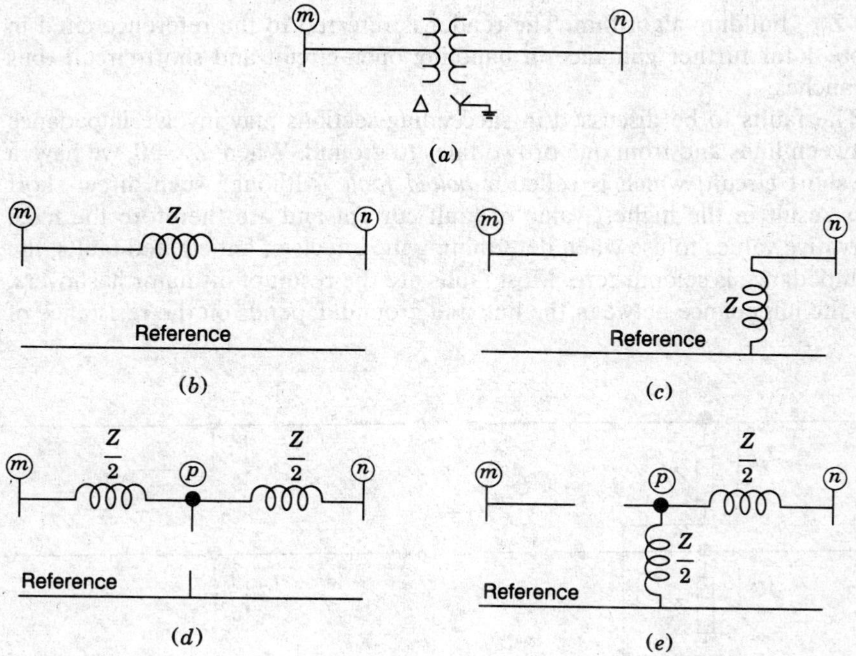

FIGURE 12.3
(*a*) Δ-Y grounded transformer with leakage impedance *Z*; (*b*) positive-sequence circuit; (*c*) zero-sequence circuit; (*d*) positive-sequence circuit with internal node; (*e*) zero-sequence circuit with internal node.

the matrices $\mathbf{Z}_{\text{bus}}^{(1)}$ and $\mathbf{Z}_{\text{bus}}^{(2)}$ in the usual manner—that is, by adding the negative of the leakage impedance Z between buses ⓜ and ⓝ in the positive- and negative-sequence networks. However, a similar strategy does not apply to the zero-sequence matrix $\mathbf{Z}_{\text{bus}}^{(0)}$ if it has been formed directly from the pictorial representation shown in Fig. 12.3(*c*). Adding $-Z$ between buses ⓜ and ⓝ does not remove the zero-sequence connection from bus ⓝ. To permit uniform procedures for all sequence networks, one strategy is to include an internal node ⓟ, as shown in Figs. 12.3(*d*) and 12.3(*e*).[1] Note that the leakage impedance is now subdivided into two parts between node ⓟ and the other nodes as shown. Connecting $-Z/2$ between buses ⓝ and ⓟ in each of the sequence circuits of Figs. 12.3(*d*) and 12.3(*e*) will open the transformer connections to bus ⓝ. Also, the open circuits can be represented within the computer algorithm by branches of arbitrarily large impedances (say, 10^6 per unit). Internal nodes of transformers can be useful in practical computer applications

[1]See H. E. Brown, *Solution of Large Networks by Matrix Methods*, 2d ed., John Wiley & Sons, Inc., New York, 1985.

of the $\mathbf{Z}_{\text{bus}}$ building algorithm. The reader is referred to the reference cited in footnote 1 for further guidance in handling open-circuit and short-circuit (bus tie) branches.

The faults to be discussed in succeeding sections may involve impedance Z_f between lines and from one or two lines to ground. When $Z_f = 0$, we have a direct short circuit, which is called a *bolted fault*. Although such direct short circuits result in the highest value of fault current and are therefore the most conservative values to use when determining the effects of anticipated faults, the fault impedance is seldom zero. Most faults are the result of insulator flashovers, where the impedance between the line and ground depends on the resistance of

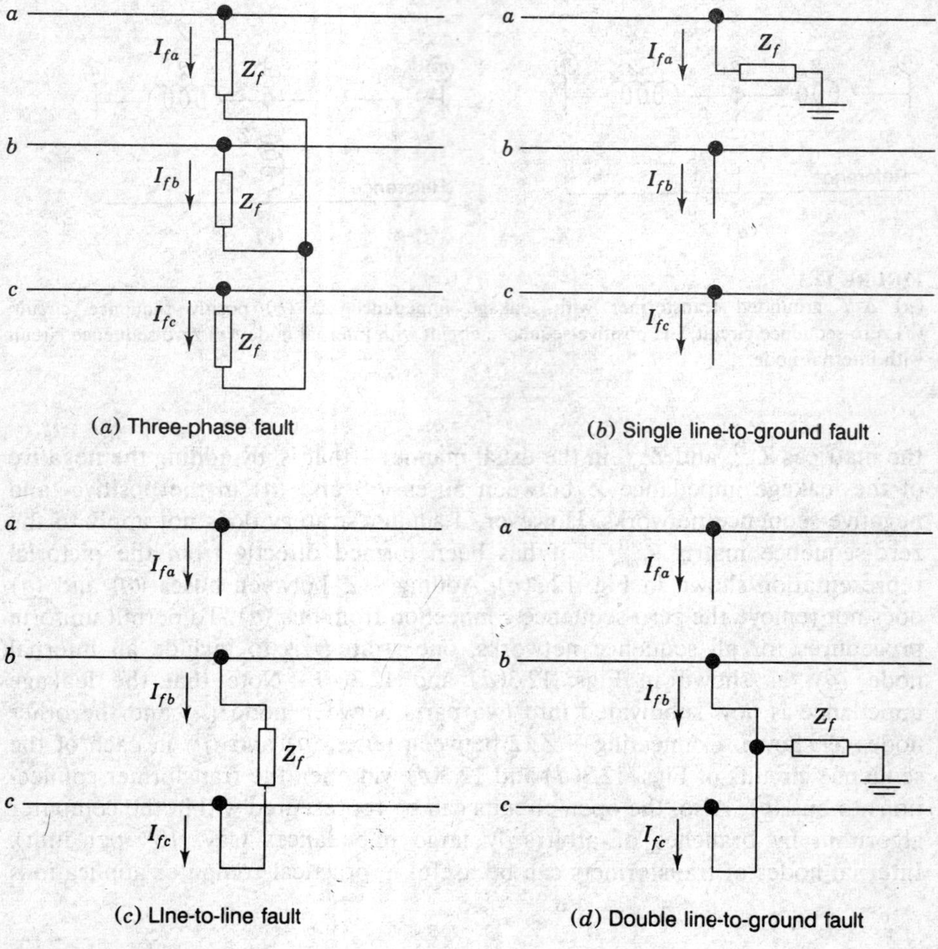

FIGURE 12.4
Connection diagrams of the hypothetical stubs for various faults through impedance.

the arc, of the tower itself, and of the tower footing if ground wires are not used. Tower-footing resistances form the major part of the resistance between line and ground and depend on the soil conditions. The resistance of dry earth is 10 to 100 times the resistance of swampy ground. Connections of the hypothetical stubs for faults through impedance Z_f are shown in Fig. 12.4.

A balanced system remains symmetrical after the occurrence of a *three-phase fault* having the same impedance between each line and a common point. Only positive-sequence currents flow. With the fault impedance Z_f equal in all phases, as shown in Fig. 12.4(*a*), we simply add impedance Z_f to the usual (positive-sequence) Thévenin equivalent circuit of the system at the fault bus ⓚ and calculate the fault current from the equation

$$I_{fa}^{(1)} = \frac{V_f}{Z_{kk}^{(1)} + Z_f} \tag{12.8}$$

For each of the other faults shown in Fig. 12.4, formal derivations of the equations for the symmetrical-component currents $I_{fa}^{(0)}$, $I_{fa}^{(1)}$, and $I_{fa}^{(2)}$ are provided in the sections which follow. In each case the fault point P is designated as bus ⓚ.

Example 12.1. Two synchronous machines are connected through three-phase transformers to the transmission line shown in Fig. 12.5. The ratings and reactances of the machines and transformers are

Machines 1 and 2: 100 MVA, 20 kV; $X_d'' = X_1 = X_2 = 20\%$,

$X_0 = 4\%, \quad X_n = 5\%$

Transformers T_1 and T_2: 100 MVA, 20Δ/345Y kV; $X = 8\%$

On a chosen base of 100 MVA, 345 kV in the transmission-line circuit the line reactances are $X_1 = X_2 = 15\%$ and $X_0 = 50\%$. Draw each of the three sequence networks and find the zero-sequence bus impedance matrix by means of the $\mathbf{Z}_{\text{bus}}$ building algorithm.

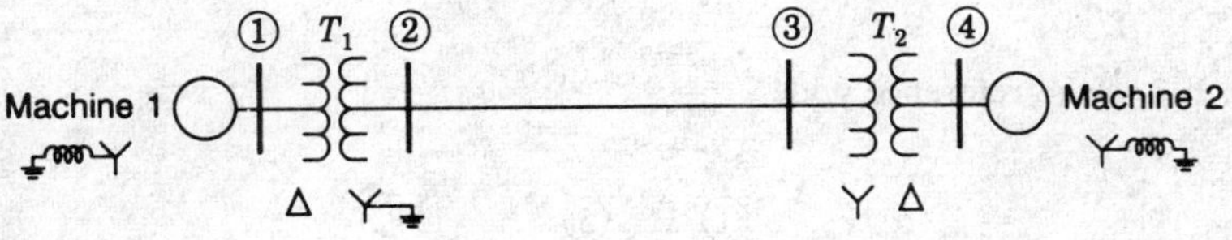

FIGURE 12.5
Single-line diagram of the system of Example 12.1.

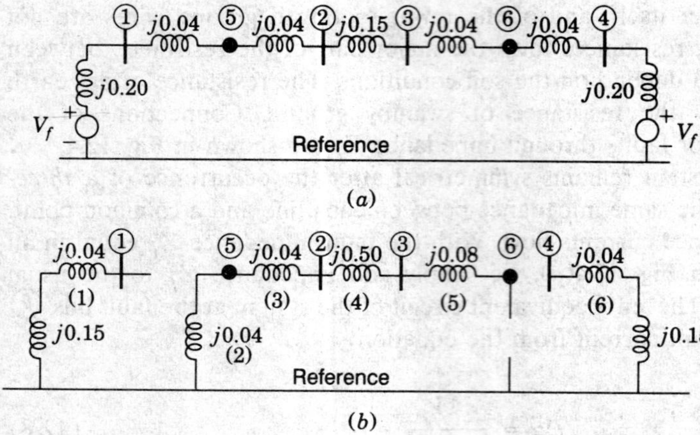

FIGURE 12.6
(a) Positive-sequence and (b) zero-sequence networks of the system of Fig. 12.5. Buses ⑤ and ⑥ are internal nodes of the transformers.

Solution. The given per-unit impedance values correspond to the chosen base, and so they can be used directly to form the sequence networks. Figure 12.6(a) shows the positive-sequence network, which is identical to the negative-sequence network when the emfs are short-circuited; Fig. 12.6(b) shows the zero-sequence network with reactance $3X_n = 0.15$ per unit in the neutral connection of each machine. Note that each transformer is assigned an internal node—bus ⑤ for transformer T_1 and bus ⑥ for transformer T_2. These internal nodes do not have an active role in the analysis of the system. In order to apply the $\mathbf{Z}_{\text{bus}}$ building algorithm, which is particularly simple in this example, let us label the zero-sequence branches from 1 to 7 as shown.

Step 1

Add branch 1 to the reference node

$$\begin{array}{cc} & ① \\ ① & \left[\, j0.19 \,\right] \end{array}$$

Step 2

Add branch 2 to the reference node

$$\begin{array}{cc} & \begin{array}{cc} ① & ⑤ \end{array} \\ \begin{array}{c} ① \\ ⑤ \end{array} & \left[\begin{array}{c|c} j0.19 & 0 \\ \hline 0 & j0.04 \end{array}\right] \end{array}$$

Step 3

Add branch 3 between buses ⑤ and ②

	①	⑤	②
①	$j0.19$	0	0
⑤	0	$j0.04$	$j0.04$
②	0	$j0.04$	$j0.08$

Step 4

Add branch 4 between buses ② and ③

	①	⑤	②	③
①	$j0.19$	0	0	0
⑤	0	$j0.04$	$j0.04$	$j0.04$
②	0	$j0.04$	$j0.08$	$j0.08$
③	0	$j0.04$	$j0.08$	$j0.58$

Step 5

Add branch 5 between buses ③ and ⑥

	①	⑤	②	③	⑥
①	$j0.19$	0	0	0	0
⑤	0	$j0.04$	$j0.04$	$j0.04$	$j0.04$
②	0	$j0.04$	$j0.08$	$j0.08$	$j0.08$
③	0	$j0.04$	$j0.08$	$j0.58$	$j0.58$
⑥	0	$j0.04$	$j0.08$	$j0.58$	$j0.66$

Step 6

Add branch 6 from bus ④ to the reference

	①	⑤	②	③	⑥	④
①	$j0.19$	0	0	0	0	0
⑤	0	$j0.04$	$j0.04$	$j0.04$	$j0.04$	0
②	0	$j0.04$	$j0.08$	$j0.08$	$j0.08$	0
③	0	$j0.04$	$j0.08$	$j0.58$	$j0.58$	0
⑥	0	$j0.04$	$j0.08$	$j0.58$	$j0.66$	0
④	0	0	0	0	0	$j0.19$

Buses ⑤ and ⑥ are the fictitious internal nodes of the transformers which facilitate computer application of the $\mathbf{Z}_{\text{bus}}$ building algorithm. We have not shown calculations for the very high impedance branches representing the open circuits. Let us remove the rows and columns for buses ⑤ and ⑥ from the matrix to obtain the effective working matrix

$$\mathbf{Z}_{\text{bus}}^{(0)} = \begin{array}{c} \text{①} \\ \text{②} \\ \text{③} \\ \text{④} \end{array} \overset{\begin{array}{cccc} \text{①} & \text{②} & \text{③} & \text{④} \end{array}}{\begin{bmatrix} j0.19 & 0 & 0 & 0 \\ 0 & j0.08 & j0.08 & 0 \\ 0 & j0.08 & j0.58 & 0 \\ 0 & 0 & 0 & j0.19 \end{bmatrix}}$$

The zeros in $\mathbf{Z}_{\text{bus}}^{(0)}$ show that zero-sequence current injected into bus ① or bus ④ of Fig. 12.6(*b*) cannot cause voltages at the other buses because of the open circuits introduced by the Δ-Y transformers. Note also that the $j0.08$ per-unit reactance in series with the open circuit between buses ⑥ and ④ does not affect $\mathbf{Z}_{\text{bus}}^{(0)}$ since it cannot carry current.

By applying the $\mathbf{Z}_{\text{bus}}$ building algorithm to the positive- and negative-sequence networks in a similar manner, we obtain

$$\mathbf{Z}_{\text{bus}}^{(1)} = \mathbf{Z}_{\text{bus}}^{(2)} = \begin{array}{c} \text{①} \\ \text{②} \\ \text{③} \\ \text{④} \end{array} \overset{\begin{array}{cccc} \text{①} & \text{②} & \text{③} & \text{④} \end{array}}{\begin{bmatrix} j0.1437 & j0.1211 & j0.0789 & j0.0563 \\ j0.1211 & j0.1696 & j0.1104 & j0.0789 \\ j0.0789 & j0.1104 & j0.1696 & j0.1211 \\ j0.0563 & j0.0789 & j0.1211 & j0.1437 \end{bmatrix}}$$

We use the above matrices in the examples which follow.

12.2 SINGLE LINE-TO-GROUND FAULTS

The single line-to-ground fault, the most common type, is caused by lightning or by conductors making contact with grounded structures. For a single line-to-ground fault through impedance Z_f the hypothetical stubs on the three lines are connected, as shown in Fig. 12.7, where phase a is the one on which the fault occurs. The relations to be developed for this type of fault will apply only when the fault is on phase a, but this should cause no difficulty since the phases are labeled arbitrarily and any phase can be designated as phase a. The conditions at the fault bus ⓚ are expressed by the following equations:

$$I_{fb} = 0 \qquad I_{fc} = 0 \qquad V_{ka} = Z_f I_{fa} \tag{12.9}$$

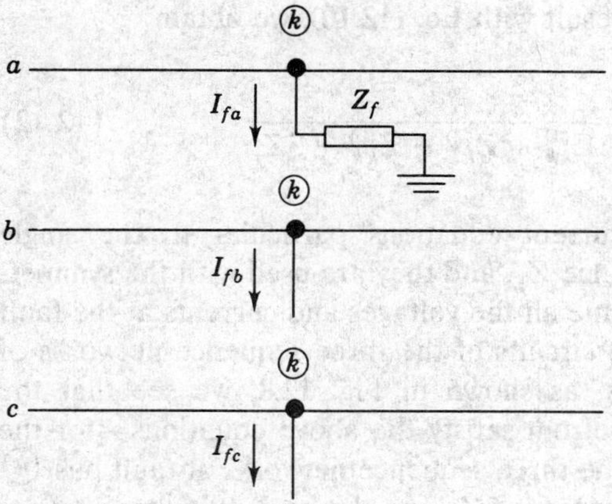

FIGURE 12.7
Connection diagram of the hypothetical stubs for a single line-to-ground fault. The fault point is called bus ⓚ.

With $I_{fb} = I_{fc} = 0$, the symmetrical components of the stub currents are given by

$$\begin{bmatrix} I_{fa}^{(0)} \\ I_{fa}^{(1)} \\ I_{fa}^{(2)} \end{bmatrix} = \frac{1}{3}\begin{bmatrix} 1 & 1 & 1 \\ 1 & a & a^2 \\ 1 & a^2 & a \end{bmatrix}\begin{bmatrix} I_{fa} \\ 0 \\ 0 \end{bmatrix}$$

and performing the multiplication yields

$$I_{fa}^{(0)} = I_{fa}^{(1)} = I_{fa}^{(2)} = \frac{I_{fa}}{3} \tag{12.10}$$

Substituting $I_{fa}^{(0)}$ for $I_{fa}^{(1)}$ and $I_{fa}^{(2)}$ shows that $I_{fa} = 3I_{fa}^{(0)}$, and from Eqs. (12.7) we obtain

$$\begin{aligned} V_{ka}^{(0)} &= \quad\; - Z_{kk}^{(0)} I_{fa}^{(0)} \\ V_{ka}^{(1)} &= V_f - Z_{kk}^{(1)} I_{fa}^{(0)} \\ V_{ka}^{(2)} &= \quad\; - Z_{kk}^{(2)} I_{fa}^{(0)} \end{aligned} \tag{12.11}$$

Summing these equations and noting that $V_{ka} = 3Z_f I_{fa}^{(0)}$ give

$$V_{ka} = V_{ka}^{(0)} + V_{ka}^{(1)} + V_{ka}^{(2)} = V_f - \left(Z_{kk}^{(0)} + Z_{kk}^{(1)} + Z_{kk}^{(2)}\right) I_{fa}^{(0)} = 3Z_f I_{fa}^{(0)}$$

Solving for $I_{fa}^{(0)}$ and combining the result with Eq. (12.10), we obtain

$$I_{fa}^{(0)} = I_{fa}^{(1)} = I_{fa}^{(2)} = \frac{V_f}{Z_{kk}^{(1)} + Z_{kk}^{(2)} + Z_{kk}^{(0)} + 3Z_f} \tag{12.12}$$

Equations (12.12) are the fault current equations particular to the single line-to-ground fault through impedance Z_f, and they are used with the symmetrical-component relations to determine all the voltages and currents at the fault point P. If the Thévenin equivalent circuits of the three sequence networks of the system are connected *in series*, as shown in Fig. 12.8, we see that the currents and voltages resulting therefrom satisfy the above equations—for the Thévenin impedances looking into the three sequence networks at fault bus ⓚ are then in series with the fault impedance $3Z_f$ and the prefault voltage source V_f. With the equivalent circuits so connected, the voltage across each sequence network is the corresponding symmetrical component of the voltage V_{ka} at the fault bus ⓚ, and the current injected into each sequence network at bus ⓚ is the *negative* of the corresponding sequence current in the fault. The series connection of the Thévenin equivalents of the sequence networks, as shown in Fig. 12.8, is a convenient means of remembering the equations for the solution of the single line-to-ground fault, for all the necessary equations for the *fault*

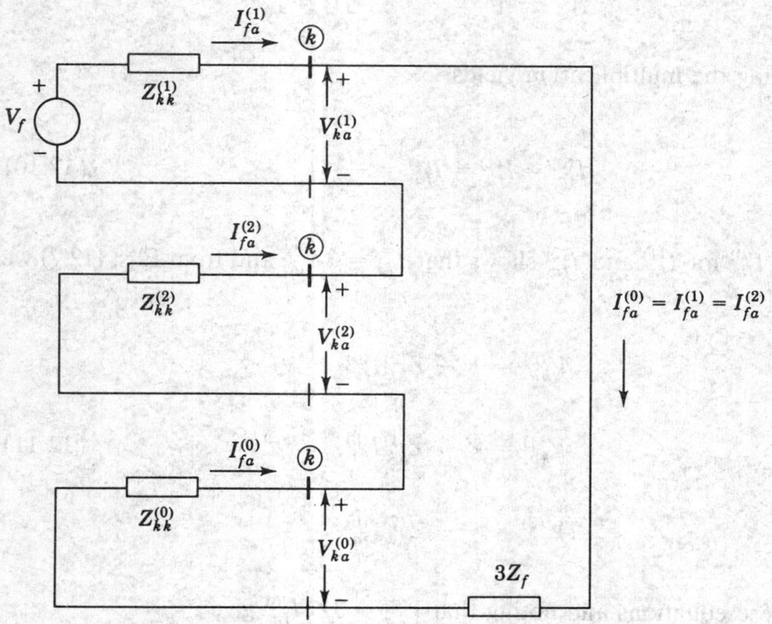

FIGURE 12.8
Connection of the Thévenin equivalents of the sequence networks to simulate a single line-to-ground fault on phase a at bus ⓚ of the system.

point can be determined from the sequence-network connection. Once the currents $I_{fa}^{(0)}$, $I_{fa}^{(1)}$, and $I_{fa}^{(2)}$ are known, the components of voltages at all other buses of the system can be determined from the bus impedance matrices of the sequence networks according to Eqs. (12.6).

Example 12.2. Two synchronous machines are connected through three-phase transformers to the transmission line shown in Fig. 12.9(*a*). The ratings and reactances of the machines and transformers are

Machines 1 and 2: 100 MVA, 20 kV; $X_d'' = X_1 = X_2 = 20\%$,

$X_0 = 4\%$, $X_n = 5\%$

Transformers T_1 and T_2: 100 MVA, 20Y/345Y kV; $X = 8\%$

Both transformers are solidly grounded on two sides. On a chosen base of 100 MVA, 345 kV in the transmission-line circuit the line reactances are $X_1 = X_2 =$ 15% and $X_0 = 50\%$. The system is operating at nominal voltage without prefault currents when a bolted ($Z_f = 0$) single line-to-ground fault occurs on phase A at bus ③. Using the bus impedance matrix for each of the three sequence networks, determine the subtransient current to ground at the fault, the line-to-ground voltages at the terminals of machine 2, and the subtransient current out of phase c of machine 2.

Solution. The system is the same as in Example 12.1, except that the transformers are now Y-Y connected. Therefore, we can continue to use $\mathbf{Z}_{\text{bus}}^{(1)}$ and $\mathbf{Z}_{\text{bus}}^{(2)}$ corresponding to Fig. 12.6(*a*), as given in Example 12.1. However, because the

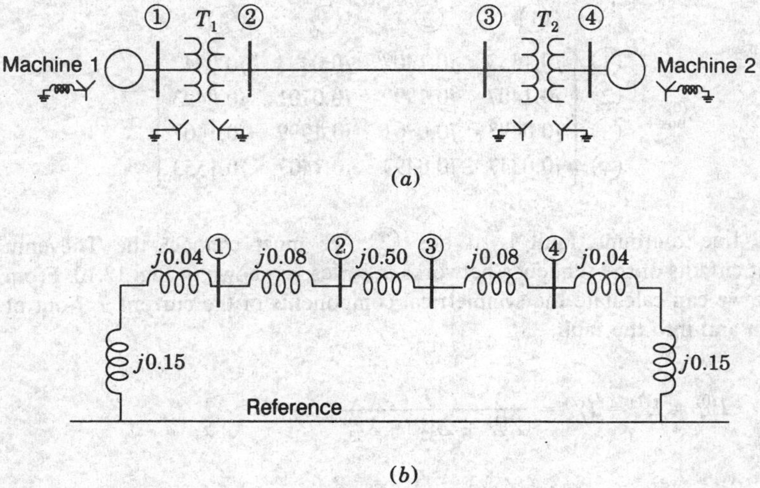

FIGURE 12.9
(*a*) The single-line diagram and (*b*) zero-sequence network of the system of Example 12.2.

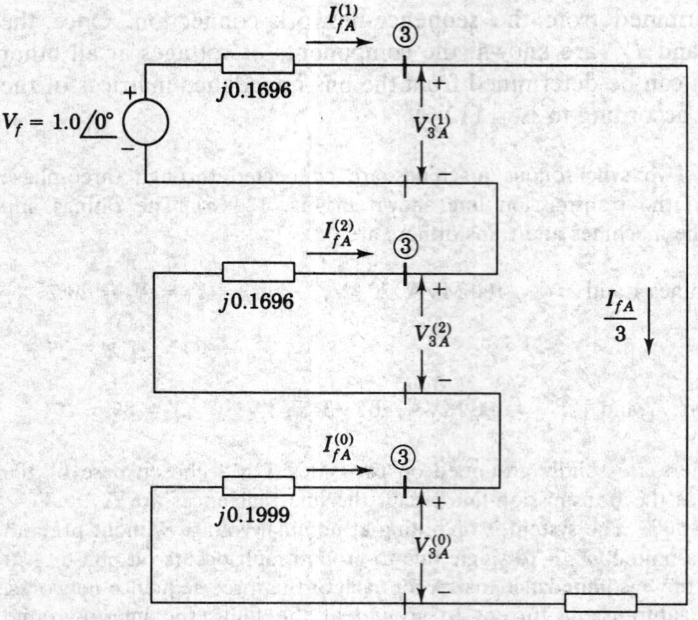

FIGURE 12.10
Series connection of the Thévenin equivalents of the sequence networks for the single line-to-ground fault of Example 12.2.

transformers are solidly grounded on both sides, the zero-sequence network is fully connected, as shown in Fig. 12.9(*b*), and has the bus impedance matrix

$$\mathbf{Z}_{\text{bus}}^{(0)} = \begin{array}{c} \\ ① \\ ② \\ ③ \\ ④ \end{array} \begin{array}{c} \begin{array}{cccc} \quad ① \quad & \quad ② \quad & \quad ③ \quad & \quad ④ \quad \end{array} \\ \begin{bmatrix} j0.1553 & j0.1407 & j0.0493 & j0.0347 \\ j0.1407 & j0.1999 & j0.0701 & j0.0493 \\ j0.0493 & j0.0701 & j0.1999 & j0.1407 \\ j0.0347 & j0.0493 & j0.1407 & j0.1553 \end{bmatrix} \end{array}$$

Since the line-to-ground fault is at bus ③, we must connect the Thévenin equivalent circuits of the sequence networks in series, as shown in Fig. 12.10. From this figure we can calculate the symmetrical components of the current I_{fA} out of the system and into the fault,

$$I_{fA}^{(0)} = I_{fA}^{(1)} = I_{fA}^{(2)} = \frac{V_f}{Z_{33}^{(1)} + Z_{33}^{(2)} + Z_{33}^{(0)}}$$

$$= \frac{1.0\angle 90°}{j(0.1696 + 0.1696 + 0.1999)} = -j1.8549 \text{ per unit}$$

The total current in the fault is

$$I_{fA} = 3I_{fA}^{(0)} = -j5.5648 \text{ per unit}$$

and since the base current in the high-voltage transmission line is $100{,}000/\sqrt{3} \times 345 = 167.35$ A, we have

$$I_{fA} = -j5.5648 \times 167.35 = 931\angle 270^\circ \text{ A}$$

The phase-*a* sequence voltages at bus ④, the terminals of machine 2, are calculated from Eqs. (12.6) with $k = 3$ and $j = 4$,

$$V_{4a}^{(0)} = \quad -Z_{43}^{(0)}I_{fA}^{(0)} = \quad -(j0.1407)(-j1.8549) = -0.2610 \text{ per unit}$$

$$V_{4a}^{(1)} = V_f - Z_{43}^{(1)}I_{fA}^{(1)} = 1 - (j0.1211)(-j1.8549) = \quad 0.7754 \text{ per unit}$$

$$V_{4a}^{(2)} = \quad -Z_{43}^{(2)}I_{fA}^{(2)} = \quad -(j0.1211)(-j1.8549) = -0.2246 \text{ per unit}$$

Note that subscripts *A* and *a* denote voltages and currents in the high-voltage and low-voltage circuits, respectively, of the Y-Y connected transformer. No phase shift is involved. From the above symmetrical components we can calculate *a-b-c* line-to-ground voltages at bus ④ as follows:

$$\begin{bmatrix} V_{4a} \\ V_{4b} \\ V_{4c} \end{bmatrix} = \begin{bmatrix} 1 & 1 & 1 \\ 1 & a^2 & a \\ 1 & a & a^2 \end{bmatrix} \begin{bmatrix} -0.2610 \\ 0.7754 \\ -0.2246 \end{bmatrix} = \begin{bmatrix} 0.2898 + j0.0 \\ -0.5364 - j0.8660 \\ -0.5364 + j0.8660 \end{bmatrix}$$

$$= \begin{bmatrix} 0.2898\angle 0^\circ \\ 1.0187\angle -121.8^\circ \\ 1.0187\angle 121.8^\circ \end{bmatrix}$$

To express the line-to-ground voltages of machine 2 in kilovolts, we multiply by $20/\sqrt{3}$, which gives

$$V_{4a} = 3.346\angle 0^\circ \text{ kV} \qquad V_{4b} = 11.763\angle -121.8^\circ \text{ kV} \qquad V_{4c} = 11.763\angle 121.8^\circ \text{ kV}$$

To determine phase-*c* current out of machine 2, we must first calculate the symmetrical components of the phase-*a* current in the branches representing the machine in the sequence networks. From Fig. 12.9(*b*) the zero-sequence current *out* of the machine is

$$I_a^{(0)} = -\frac{V_{4a}^{(0)}}{jX_0} = \frac{0.2610}{j0.04} = -j6.525 \text{ per unit}$$

and from Fig. 12.6(a) the other sequence currents are calculated as

$$I_a^{(1)} = \frac{V_f - V_{4a}^{(1)}}{jX''} = \frac{1.0 - 0.7754}{j0.20} = -j1.123 \text{ per unit}$$

$$I_a^{(2)} = -\frac{V_{4a}^{(2)}}{jX_2} = \frac{0.2246}{j0.20} = -j1.123 \text{ per unit}$$

Note that the machine currents are shown without subscript f, which is reserved exclusively for the (stub) currents and voltages of the fault point. The phase-c currents in machine 2 are now easily calculated,

$$I_c = I_a^{(0)} + aI_a^{(1)} + a^2I_a^{(2)}$$

$$= -j6.525 + a(-j1.123) + a^2(-j1.123) = -j5.402 \text{ per unit}$$

The base current in the machine circuits is $100{,}000/(\sqrt{3} \times 20) = 2886.751$ A, and so $|I_c| = 15{,}594$ A. Other voltages and currents in the system can be calculated similarly.

12.3 LINE-TO-LINE FAULTS

To represent a line-to-line fault through impedance Z_f, the hypothetical stubs on the three lines at the fault are connected, as shown in Fig. 12.11. Bus ⓚ is again the fault point P, and without any loss of generality, the line-to-line fault is regarded as being on phases b and c. The following relations must be satisfied at the fault point

$$I_{fa} = 0 \qquad I_{fb} = -I_{fc} \qquad V_{kb} - V_{kc} = I_{fb}Z_f \tag{12.13}$$

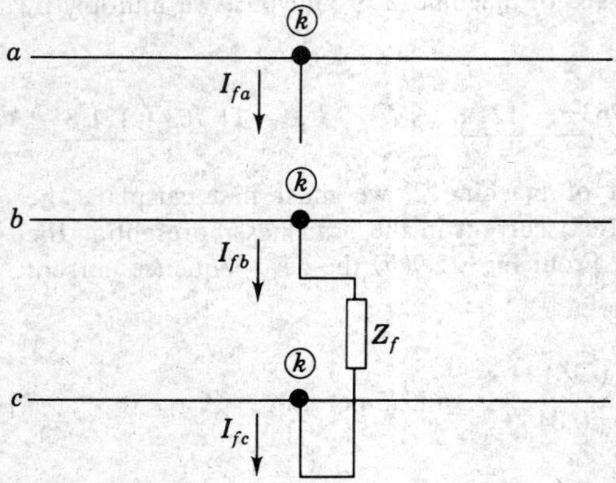

FIGURE 12.11
Connection of the hypothetical stubs for a line-to-line fault. The fault point is called bus ⓚ.

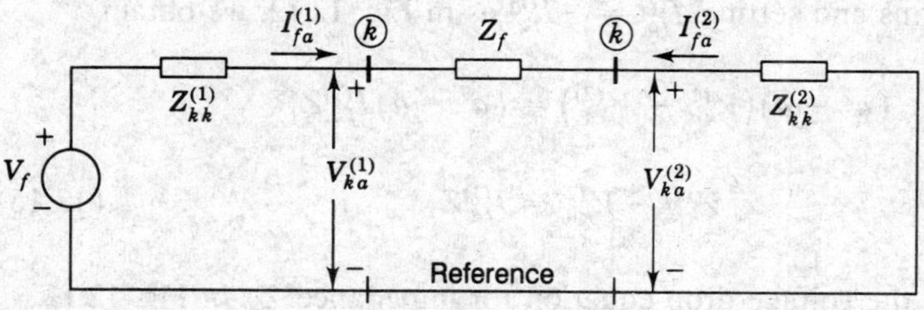

FIGURE 12.12
Connection of the Thévenin equivalents of the positive- and negative-sequence networks for a line-to-line fault between phases b and c at bus ⓚ of the system.

Since $I_{fb} = -I_{fc}$ and $I_{fa} = 0$, the symmetrical components of current are

$$\begin{bmatrix} I_{fa}^{(0)} \\ I_{fa}^{(1)} \\ I_{fa}^{(2)} \end{bmatrix} = \frac{1}{3}\begin{bmatrix} 1 & 1 & 1 \\ 1 & a & a^2 \\ 1 & a^2 & a \end{bmatrix}\begin{bmatrix} 0 \\ I_{fb} \\ -I_{fb} \end{bmatrix}$$

and multiplying through in this equation shows that

$$I_{fa}^{(0)} = 0 \tag{12.14}$$

$$I_{fa}^{(1)} = -I_{fa}^{(2)} \tag{12.15}$$

The voltages throughout the zero-sequence network must be zero since there are no zero-sequence sources, and because $I_{fa}^{(0)} = 0$, current is not being injected into that network due to the fault. Hence, line-to-line fault calculations do not involve the zero-sequence network, which remains the same as before the fault—a dead network.

To satisfy the requirement that $I_{fa}^{(1)} = -I_{fa}^{(2)}$, let us connect the Thévenin equivalents of the positive- and negative-sequence networks *in parallel*, as shown in Fig. 12.12. To show that this connection of the networks also satisfies the voltage equation $V_{kb} - V_{kc} = I_{fb}Z_f$, we now expand each side of that equation separately as follows:

$$V_{kb} - V_{kc} = \left(V_{kb}^{(1)} + V_{kb}^{(2)}\right) - \left(V_{kc}^{(1)} + V_{kc}^{(2)}\right) = \left(V_{kb}^{(1)} - V_{kc}^{(1)}\right) + \left(V_{kb}^{(2)} - V_{kc}^{(2)}\right)$$

$$= (a^2 - a)V_{ka}^{(1)} + (a - a^2)V_{ka}^{(2)} = (a^2 - a)\left(V_{ka}^{(1)} - V_{ka}^{(2)}\right)$$

$$I_{fb}Z_f = \left(I_{fb}^{(1)} + I_{fb}^{(2)}\right)Z_f = \left(a^2 I_{fa}^{(1)} + aI_{fa}^{(2)}\right)Z_f$$

Equating both terms and setting $I_{fa}^{(2)} = -I_{fa}^{(1)}$ as in Fig. 12.12, we obtain

$$(a^2 - a)(V_{ka}^{(1)} - V_{ka}^{(2)}) = (a^2 - a)I_{fa}^{(1)}Z_f$$

or

$$V_{ka}^{(1)} - V_{ka}^{(2)} = I_{fa}^{(1)}Z_f \tag{12.16}$$

which is precisely the voltage-drop equation for impedance Z_f in Fig. 12.12.

Thus, all the fault conditions of Eqs. (12.13) are satisfied by connecting the positive- and negative-sequence networks *in parallel* through impedance Z_f, as shown in Fig. 12.12. The zero-sequence network is inactive and does not enter into the line-to-line fault calculations. The equation for the positive-sequence current in the fault can be determined directly from Fig. 12.12 so that

$$I_{fa}^{(1)} = -I_{fa}^{(2)} = \frac{V_f}{Z_{kk}^{(1)} + Z_{kk}^{(2)} + Z_f} \tag{12.17}$$

For a bolted line-to-line fault we set $Z_f = 0$.

Equations (12.17) are the fault current equations for a line-to-line fault through impedance Z_f. Once $I_{fa}^{(1)}$ and $I_{fa}^{(2)}$ are known, they can be treated as injections $-I_{fa}^{(1)}$ and $-I_{fa}^{(2)}$ into the positive- and negative-sequence networks, respectively, and the changes in the sequence voltages at the buses of the system due to the fault can be obtained from the bus impedance matrices, as previously demonstrated. When Δ-Y transformers are present, phase shift of the positive- and negative-sequence currents and voltages must be taken into account in the calculations. The following example shows how this is accomplished.

Example 12.3. The same system as in Example 12.1 is operating at nominal system voltage without prefault currents when a bolted line-to-line fault occurs at bus ③. Using the bus impedance matrices of the sequence networks for subtransient conditions, determine the currents in the fault, the line-to-line voltages at the fault bus, and the line-to-line voltages at the terminals of machine 2.

Solution. $\mathbf{Z}_{\text{bus}}^{(1)}$ and $\mathbf{Z}_{\text{bus}}^{(2)}$ are already set forth in Example 12.1. Although $\mathbf{Z}_{\text{bus}}^{(0)}$ is also given, we are not concerned with the zero-sequence network in this solution since the fault is line to line.

To simulate the fault, the Thévenin equivalent circuits at bus ③ of the positive- and negative-sequence networks of Example 12.1 are connected in parallel, as shown in Fig. 12.13. From this figure the sequence currents are calculated as follows:

$$I_{fA}^{(1)} = -I_{fA}^{(2)} = \frac{V_f}{Z_{33}^{(1)} + Z_{33}^{(2)}} = \frac{1 + j0}{j0.1696 + j0.1696} = -j2.9481 \text{ per unit}$$

Uppercase A is used here because the fault is in the high-voltage transmission-line

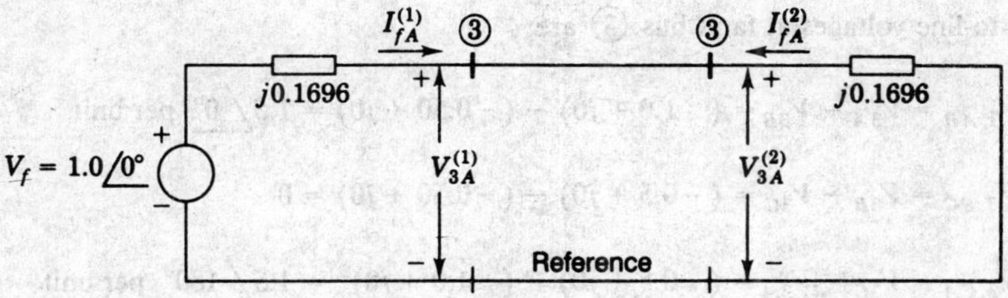

FIGURE 12.13
Connection of the Thévenin equivalent circuits for the line-to-line fault of Example 12.3.

circuit. Since $I_{fA}^{(0)} = 0$, the components of currents in the fault are calculated from

$$I_{fA} = I_{fA}^{(1)} + I_{fA}^{(2)} = -j2.9481 + j2.9481 = 0$$

$$I_{fB} = a^2 I_{fA}^{(1)} + a I_{fA}^{(2)} = -j2.9481(-0.5 - j0.866) + j2.9481(-0.5 + j0.866)$$

$$= -5.1061 + j0 \text{ per unit}$$

$$I_{fC} = -I_{fB} = 5.1061 + j0 \text{ per unit}$$

As in Example 12.2, base current in the transmission line is 167.35 A, and so

$$I_{fA} = 0$$

$$I_{fB} = -5.1061 \times 167.35 = 855\angle 180^\circ \text{ A}$$

$$I_{fC} = -5.1061 \times 167.35 = 855\angle 0^\circ \text{ A}$$

Symmetrical components of phase-A voltage to ground at bus ③ are

$$V_{3A}^{(0)} = 0$$

$$V_{3A}^{(1)} = V_{3A}^{(2)} = 1 - Z_{kk}^{(1)} I_{fA}^{(1)} = 1 - (j0.1696)(-j2.9481) = 0.5 + j0 \text{ per unit}$$

Line-to-ground voltages at fault bus ③ are

$$V_{3A} = V_{3A}^{(0)} + V_{3A}^{(1)} + V_{3A}^{(2)} = 0 + 0.5 + 0.5 = 1.0\angle 0^\circ \text{ per unit}$$

$$V_{3B} = V_{3A}^{(0)} + a^2 V_{3A}^{(1)} + a V_{3A}^{(2)} = 0 + a^2 0.5 + a0.5 = 0.5\angle 180^\circ \text{ per unit}$$

$$V_{3C} = V_{3B} = 0.5\angle 180^\circ \text{ per unit}$$

Line-to-line voltages at fault bus ③ are

$$V_{3,AB} = V_{3A} - V_{3B} = (\ \ 1.0 + j0) - (-0.50 + j0) = 1.5\angle 0^\circ \text{ per unit}$$

$$V_{3,BC} = V_{3B} - V_{3C} = (-0.5 + j0) - (-0.50 + j0) = 0$$

$$V_{3,CA} = V_{3C} - V_{3A} = (-0.5 + j0) - (\ \ 1.0 + j0) = 1.5\angle 180^\circ \text{ per unit}$$

Expressed in volts, these line-to-line voltages are

$$V_{3,AB} = 1.5\angle 0^\circ \times \frac{345}{\sqrt{3}} = 299\angle 0^\circ \text{ kV}$$

$$V_{3,BC} = 0$$

$$V_{3,CA} = 1.5\angle 180^\circ \times \frac{345}{\sqrt{3}} = 299\angle 180^\circ \text{ kV}$$

For the moment, let us avoid phase shifts due to the Δ-Y transformer connected to machine 2 and proceed to calculate the sequence voltages of phase A at bus ④ using the bus impedance matrices of Example 12.1 and Eqs. (12.6) with $k = 3$ and $j = 4$

$$V_{4A}^{(0)} = \quad - Z_{43}^{(0)} I_{fA}^{(0)} = 0$$

$$V_{4A}^{(1)} = V_f - Z_{43}^{(1)} I_{fA}^{(1)} = 1 - (j0.1211)(-j2.9481) = 0.643 \text{ per unit}$$

$$V_{4A}^{(2)} = \quad - Z_{43}^{(2)} I_{fA}^{(2)} = \quad - (j0.1211)(j2.9481) = 0.357 \text{ per unit}$$

To account for phase shifts in stepping *down* from the high-voltage transmission line to the low-voltage terminals of machine 2, we must retard the positive-sequence voltage and advance the negative-sequence voltage by 30°. At machine 2 terminals, indicated by lowercase a, the voltages are

$$V_{4a}^{(0)} = 0$$

$$V_{4a}^{(1)} = V_{4A}^{(1)}\angle -30^\circ = 0.643\angle -30^\circ = 0.5569 - j0.3215 \text{ per unit}$$

$$V_{4a}^{(2)} = V_{4A}^{(2)}\angle 30^\circ = 0.357\angle 30^\circ = 0.3092 + j0.1785 \text{ per unit}$$

$$V_{4a} = V_{4a}^{(0)} + V_{4a}^{(1)} + V_{4a}^{(2)} = 0 + (0.5569 - j0.3215) + (0.3092 + j0.1785)$$

$$= 0.8661 - j0.1430 = 0.8778\angle -9.4^\circ \text{ per unit}$$

Phase-b voltages at terminals of machine 2 are now calculated,

$$V_{4b}^{(0)} = V_{4a}^{(0)} = 0$$

$$V_{4b}^{(1)} = a^2 V_{4a}^{(1)} = \left(1\underline{/240^\circ}\right)\left(0.643\underline{/-30^\circ}\right) = -0.5569 - j0.3215 \text{ per unit}$$

$$V_{4b}^{(2)} = a\, V_{4a}^{(2)} = \left(1\underline{/120^\circ}\right)\left(0.357\underline{/30^\circ}\right) = -0.3092 + j0.1785 \text{ per unit}$$

$$V_{4b} = V_{4b}^{(0)} + V_{4b}^{(1)} + V_{4b}^{(2)} = 0 + (-0.5569 - j0.3215) + (-0.3092 + j0.1785)$$

$$= -0.8661 - j0.143 = 0.8778\underline{/-170.6^\circ} \text{ per unit}$$

and for phase c of machine 2

$$V_{4c}^{(0)} = V_{4a}^{(0)} = 0$$

$$V_{4c}^{(1)} = a\, V_{4a}^{(1)} = \left(1\underline{/120^\circ}\right)\left(0.643\underline{/-30^\circ}\right) = 0.643\underline{/90^\circ} \text{ per unit}$$

$$V_{4c}^{(2)} = a^2 V_{4a}^{(2)} = \left(1\underline{/240^\circ}\right)\left(0.357\underline{/30^\circ}\right) = 0.357\underline{/-90^\circ} \text{ per unit}$$

$$V_{4c} = V_{4c}^{(0)} + V_{4c}^{(1)} + V_{4c}^{(2)} = 0 + (j0.643) + (-j0.357) = 0 + j0.286 \text{ per unit}$$

Line-to-line voltages at the terminals of machine 2 are

$$V_{4,ab} = V_{4a} - V_{4b} = (0.8661 - j0.143) - (0.8661 - j0.143)$$

$$= 1.7322 + j0 \text{ per unit}$$

$$V_{4,bc} = V_{4b} - V_{4c} = (-0.8661 - j0.143) - (0 + j0.286)$$

$$= -0.8661 - j0.429 = 0.9665\underline{/-153.65^\circ} \text{ per unit}$$

$$V_{4,ca} = V_{4c} - V_{4a} = (0 + j0.286) - (0.8661 - j0.143)$$

$$= -0.8661 + j0.429 = 0.9665\underline{/153.65^\circ} \text{ per unit}$$

In volts, line-to-line voltages at machine 2 terminals are

$$V_{4,ab} = 1.7322\underline{/0^\circ} \times \frac{20}{\sqrt{3}} = 20\underline{/0^\circ} \text{ kV}$$

$$V_{4,bc} = 0.9665\underline{/-153.65^\circ} \times \frac{20}{\sqrt{3}} = 11.2\underline{/-153.65^\circ} \text{ kV}$$

$$V_{4,ca} = 0.9665\underline{/153.65^\circ} \times \frac{20}{\sqrt{3}} = 11.2\underline{/153.65^\circ} \text{ kV}$$

Thus, from the currents $I_{fA}^{(0)}$, $I_{fA}^{(1)}$, and $I_{fA}^{(2)}$ of the fault and the bus impedance matrices of the sequence networks we can determine the unbalanced bus voltages and branch currents throughout the system due to the line-to-line fault.

12.4 DOUBLE LINE-TO-GROUND FAULTS

For a double line-to-ground fault the hypothetical stubs are connected, as shown in Fig. 12.14. Again, the fault is taken to be on phases b and c, and the relations now existing at the fault bus ⓚ are

$$I_{fa} = 0 \qquad V_{kb} = V_{kc} = (I_{fb} + I_{fc})Z_f \tag{12.18}$$

Since I_{fa} is zero, the zero-sequence current is given by $I_{fa}^{(0)} = (I_{fb} + I_{fc})/3$, and the voltages of Eq. (12.18) then become

$$V_{kb} = V_{kc} = 3Z_f I_{fa}^{(0)} \tag{12.19}$$

Substituting V_{kb} for V_{kc} in the symmetrical-component transformation, we find that

$$\begin{bmatrix} V_{ka}^{(0)} \\ V_{ka}^{(1)} \\ V_{ka}^{(2)} \end{bmatrix} = \frac{1}{3}\begin{bmatrix} 1 & 1 & 1 \\ 1 & a & a^2 \\ 1 & a^2 & a \end{bmatrix}\begin{bmatrix} V_{ka} \\ V_{kb} \\ V_{kb} \end{bmatrix} \tag{12.20}$$

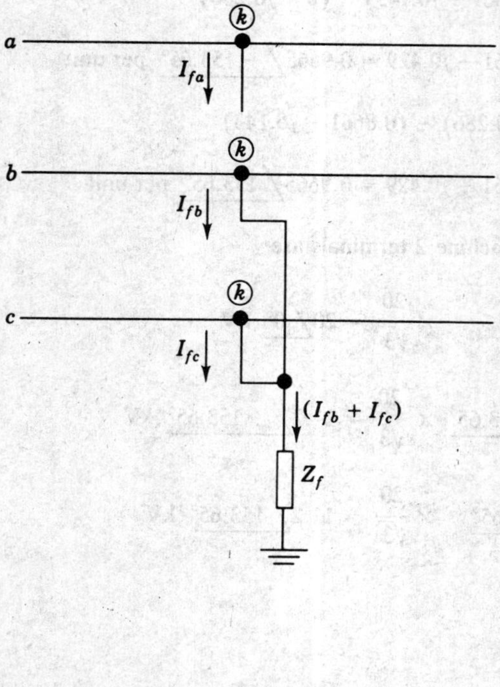

FIGURE 12.14
Connection diagram for the hypothetical stubs for a double line-to-ground fault. The fault point is called bus ⓚ.

The second and third rows of this equation show that

$$V_{ka}^{(1)} = V_{ka}^{(2)} \tag{12.21}$$

while the first row and Eq. (12.19) show that

$$3V_{ka}^{(0)} = V_{ka} + 2V_{kb} = \left(V_{ka}^{(0)} + V_{ka}^{(1)} + V_{ka}^{(2)}\right) + 2\left(3Z_f I_{fa}^{(0)}\right)$$

Collecting zero-sequence terms on one side, setting $V_{ka}^{(2)} = V_{ka}^{(1)}$, and solving for $V_{ka}^{(1)}$, we obtain

$$V_{ka}^{(1)} = V_{ka}^{(0)} - 3Z_f I_{fa}^{(0)} \tag{12.22}$$

Bringing together Eqs. (12.21) and (12.22), and again noting that $I_{fa} = 0$, we arrive at the results

$$V_{ka}^{(1)} = V_{ka}^{(2)} = V_{ka}^{(0)} - 3Z_f I_{fa}^{(0)}$$

$$I_{fa}^{(0)} + I_{fa}^{(1)} + I_{fa}^{(2)} = 0 \tag{12.23}$$

These characterizing equations of the double line-to-ground fault are satisfied when all three of the sequence networks are connected *in parallel*, as shown in Fig. 12.15. The diagram of network connections shows that the positive-sequence current $I_{fa}^{(1)}$ is determined by applying prefault voltage V_f across the total impedance consisting of $Z_{kk}^{(1)}$ in series with the parallel combination of $Z_{kk}^{(2)}$ and $(Z_{kk}^{(0)} + 3Z_f)$. That is,

$$I_{fa}^{(1)} = \frac{V_f}{Z_{kk}^{(1)} + \left[\dfrac{Z_{kk}^{(2)}\left(Z_{kk}^{(0)} + 3Z_f\right)}{Z_{kk}^{(2)} + Z_{kk}^{(0)} + 3Z_f}\right]} \tag{12.24}$$

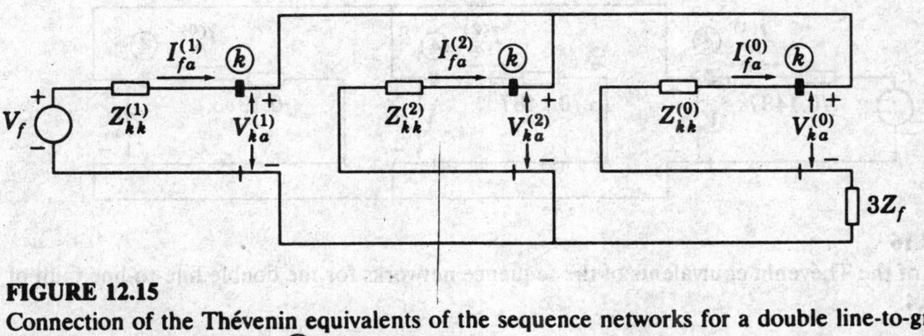

FIGURE 12.15
Connection of the Thévenin equivalents of the sequence networks for a double line-to-ground fault on phases *b* and *c* at bus ⓚ of the system.

The negative- and zero-sequence currents *out* of the system and *into* the fault can be determined from Fig. 12.15 by simple current division so that

$$I_{fa}^{(2)} = -I_{fa}^{(1)}\left[\frac{Z_{kk}^{(0)} + 3Z_f}{Z_{kk}^{(2)} + Z_{kk}^{(0)} + 3Z_f}\right] \tag{12.25}$$

$$I_{fa}^{(0)} = -I_{fa}^{(1)}\left[\frac{Z_{kk}^{(2)}}{Z_{kk}^{(2)} + Z_{kk}^{(0)} + 3Z_f}\right] \tag{12.26}$$

For a bolted fault Z_f is set equal to 0 in the above equations. When $Z_f = \infty$, the zero-sequence circuit becomes an open circuit; no zero-sequence current can then flow and the equations revert back to those for the line-to-line fault discussed in the preceding section.

Again, we observe that the sequence currents $I_{fa}^{(1)}$, $I_{fa}^{(2)}$, and $I_{fa}^{(0)}$, once calculated, can be treated as negative injections into the sequence networks at the fault bus ⓚ and the sequence voltage *changes* at all buses of the system can then be calculated from the bus impedance matrices, as we have done in preceding sections.

Example 12.4. Find the subtransient currents and the line-to-line voltages at the fault under subtransient conditions when a double line-to-ground fault with $Z_f = 0$ occurs at the terminals of machine 2 in the system of Fig. 12.5. Assume that the system is unloaded and operating at rated voltage when the fault occurs. Use the bus impedance matrices and neglect resistance.

Solution. The bus impedance matrices $\mathbf{Z}_{\text{bus}}^{(1)}$, $\mathbf{Z}_{\text{bus}}^{(2)}$, and $\mathbf{Z}_{\text{bus}}^{(0)}$ are the same as in Example 12.1, and so the Thévenin impedances at fault bus ④ are equal in per unit to the diagonal elements $Z_{44}^{(0)} = j0.19$ and $Z_{44}^{(1)} = Z_{44}^{(2)} = j0.1437$. To simulate the double line-to-ground fault at bus ④, we connect the Thévenin equivalents of all three sequence networks in parallel, as shown in Fig. 12.16, from which we

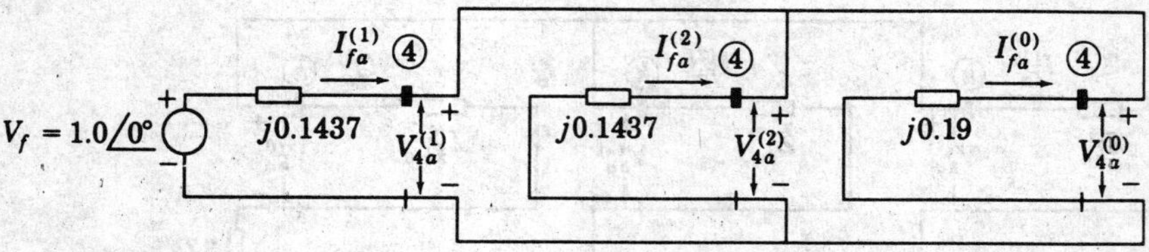

FIGURE 12.16
Connection of the Thévenin equivalents of the sequence networks for the double line-to-line fault of Example 12.4.

obtain

$$I_{fa}^{(1)} = \frac{V_f}{Z_{44}^{(1)} + \left[\dfrac{Z_{44}^{(2)}Z_{44}^{(0)}}{Z_{44}^{(2)} + Z_{44}^{(0)}}\right]} = \frac{1 + j0}{j0.1437 + \left[\dfrac{(j0.1437)(j0.19)}{(j0.1437 + j0.19)}\right]}$$

$$= -j4.4342 \text{ per unit}$$

Therefore, the sequence voltages at the fault are

$$V_{4a}^{(1)} = V_{4a}^{(2)} = V_{4a}^{(0)} = V_f - I_{fa}^{(1)}Z_{44}^{(1)} = 1 - (-j4.4342)(j0.1437) = 0.3628 \text{ per unit}$$

Current injections into the negative- and zero-sequence networks at the fault bus are calculated by current division as follows:

$$I_{fa}^{(2)} = -I_{fa}^{(1)}\left[\frac{Z_{44}^{(0)}}{Z_{44}^{(2)} + Z_{44}^{(0)}}\right] = j4.4342\left[\frac{j0.19}{j(0.1437 + 0.19)}\right] = j2.5247 \text{ per unit}$$

$$I_{fa}^{(0)} = -I_{fa}^{(1)}\left[\frac{Z_{44}^{(2)}}{Z_{44}^{(2)} + Z_{44}^{(0)}}\right] = j4.4342\left[\frac{j0.1437}{j(0.1437 + 0.19)}\right] = j1.9095 \text{ per unit}$$

The currents out of the system at the fault point are

$$I_{fa} = I_{fa}^{(0)} + I_{fa}^{(1)} + I_{fa}^{(2)} = j1.9095 - j4.4342 + j2.5247 = 0$$

$$I_{fb} = I_{fa}^{(0)} + a^2 I_{fa}^{(1)} + a I_{fa}^{(2)}$$

$$= j1.9095 + (1\angle 240^\circ)(4.4342\angle -90^\circ) + (1\angle 120^\circ)(2.5247\angle 90^\circ)$$

$$= -6.0266 + j2.8642 = 6.6726\angle 154.6^\circ \text{ per unit}$$

$$I_{fc} = I_{fa}^{(0)} + a I_{fa}^{(1)} + a^2 I_{fa}^{(2)}$$

$$= j1.9095 + (1\angle 120^\circ)(4.4342\angle -90^\circ) + (1\angle 240^\circ)(2.5247\angle 90^\circ)$$

$$= 6.0266 + j2.8642 = 6.6726\angle 25.4^\circ \text{ per unit}$$

and the current I_f into the ground is

$$I_f = I_{fb} + I_{fc} = 3I_{fa}^{(0)} = j5.7285 \text{ per unit}$$

Calculating *a-b-c* voltages at the fault bus, we find that

$$V_{4a} = V_{4a}^{(0)} + V_{4a}^{(1)} + V_{4a}^{(2)} = 3V_{4a}^{(1)} = 3(0.3628) = 1.0884 \text{ per unit}$$

$$V_{4b} = V_{4c} = 0$$

$$V_{4,ab} = V_{4a} - V_{4b} = 1.0884 \text{ per unit}$$

$$V_{4,bc} = V_{4b} - V_{4c} = 0$$

$$V_{4,ca} = V_{4c} - V_{4a} = -1.0884 \text{ per unit}$$

Base current equals $100 \times 10^3/(\sqrt{3} \times 20) = 2887$ A in the circuit of machine 2, and so we find that

$$I_{fa} = 0$$

$$I_{fb} = 2887 \times 6.6726\underline{/154.6^\circ} = 19{,}262\underline{/154.6^\circ} \text{ A}$$

$$I_{fc} = 2887 \times 6.6726\underline{/25.4^\circ} = 19{,}262\underline{/25.4^\circ} \text{ A}$$

$$I_f = 2887 \times 5.7285\underline{/90^\circ} = 16{,}538\underline{/90^\circ} \text{ A}$$

The base line-to-neutral voltage in machine 2 is $20/\sqrt{3}$ kV, and so

$$V_{4,ab} = 1.0884 \times \frac{20}{\sqrt{3}} = 12.568\underline{/0^\circ} \text{ kV}$$

$$V_{4,bc} = 0$$

$$V_{4,ca} = -1.0884 \times \frac{20}{\sqrt{3}} = 12.568\underline{/180^\circ} \text{ kV}$$

Examples 12.3 and 12.4 show that phase shifts due to Δ-Y transformers do not enter into the calculations of sequence currents and voltages in that part of the system where the fault occurs, provided V_f at the fault point is chosen as the reference voltage for the calculations. However, for those parts of the system which are separated by Δ-Y transformers from the fault point, the sequence currents, and voltages calculated by bus impedance matrix must be shifted in phase before being combined to form the actual voltages. This is because the bus impedance matrices of the sequence networks are formed without consideration of phase shifts, and so they consist of per-unit impedances *referred* to the part of the network which includes the fault point.

Example 12.5. Solve for the subtransient voltages to ground at bus ②, the end of the transmission line remote from the double line-to-ground fault, in the system of Example 12.4.

Solution. Numerical values of the fault-current components are given in the solution of Example 12.4 and the elements of $\mathbf{Z}_{\text{bus}}^{(1)}$, $\mathbf{Z}_{\text{bus}}^{(2)}$, and $\mathbf{Z}_{\text{bus}}^{(0)}$ are provided in the solution of Example 12.1. Neglecting phase shift of the Δ-Y transformers for the moment and substituting the appropriate values in Eq. (12.6), we obtain for the voltages at bus ② due to the fault at bus ④,

$$V_{2a}^{(0)} = -I_{fa}^{(0)}Z_{24}^{(0)} = -(j1.9095)(0) = 0$$

$$V_{2a}^{(1)} = V_f - I_{fa}^{(1)}Z_{24}^{(1)} = 1 - (-j4.4342)(j0.0789) = 0.6501 \text{ per unit}$$

$$V_{2a}^{(2)} = -I_{fa}^{(2)}Z_{24}^{(2)} = -(-j2.5247)(j0.0789) = 0.1992 \text{ per unit}$$

Accounting for phase shift in stepping *up* to the transmission-line circuit from the fault at bus ④, we have

$$V_{2A}^{(0)} = 0$$

$$V_{2A}^{(1)} = V_{2a}^{(1)}\angle 30^\circ = 0.6501\angle 30^\circ = 0.5630 + j0.3251 \text{ per unit}$$

$$V_{2A}^{(2)} = V_{2a}^{(2)}\angle -30^\circ = 0.1992\angle -30^\circ = 0.1725 - j0.0996 \text{ per unit}$$

The required voltages can now be calculated:

$$V_{2A} = V_{2A}^{(0)} + V_{2A}^{(1)} + V_{2A}^{(2)} = (0.5630 + j0.3251) + (0.1725 - j0.0996)$$

$$= 0.7355 + j0.2255 = 0.7693\angle 17.0^\circ \text{ per unit}$$

$$V_{2B} = V_{2A}^{(0)} + a^2V_{2A}^{(1)} + aV_{2A}^{(2)} = (1\angle 240^\circ)(0.6531\angle 30^\circ)$$

$$+(1\angle 120^\circ)(0.1992\angle 30^\circ)$$

$$= -0.1725 - j0.5535 = 0.5798\angle 107.3^\circ \text{ per unit}$$

$$V_{2C} = V_{2A}^{(0)} + aV_{2A}^{(1)} + a^2V_{2A}^{(2)} = (1\angle 120^\circ)(0.6531\angle 30^\circ)$$

$$+(1\angle 240^\circ)(0.1992\angle 30^\circ)$$

$$= -0.5656 + j0.1274 = 0.5798\angle 167.3^\circ \text{ per unit}$$

These per-unit values can be converted to volts by multiplying by the line-to-neutral base voltage $345/\sqrt{3}$ kV of the transmission line.

12.5 DEMONSTRATION PROBLEMS

Large-scale computer programs based on the bus impedance matrices of the sequence networks are generally used to analyze faults on electric utility transmission systems. Three-phase and single line-to-ground faults are usually the only types of fault studied. Since circuit-breaker applications are made according to the symmetrical short-circuit current that must be interrupted, this current is calculated for the two types of fault. The printout includes the total fault current and the contributions from each line. The results also list those quantities when each line connected to the faulted bus is opened in turn while all others are in operation.

The program uses the impedances for the lines as provided in the line data for the power-flow program and includes the appropriate reactance for each machine in forming the positive- and zero-sequence bus impedance matrices. As far as impedances are concerned, the negative-sequence network is taken to be the same as the positive-sequence network. So, for a single line-to-ground fault at bus ⓚ, $I_{fa}^{(1)}$ is calculated in per unit as 1.0 divided by the sum $(2Z_{kk}^{(1)} + Z_{kk}^{(0)} + 3Z_f)$. The bus voltages are included in the computer printout, if called for, as well as the current in lines other than those connected to the faulted bus since this information can easily be found from the bus impedance matrices.

The following numerical examples show the analysis of a single line-to-ground fault on (1) an industrial power system and (2) a small electric utility system. Both of these systems are quite small in extent compared to the large-scale systems normally encountered. The calculations are presented without matrices in order to emphasize the circuit concepts which underlie fault analysis. The presentation should allow the reader to become more familiar with the sequence networks and how they are used to analyze faults. The principles demonstrated here are essentially the same as those employed within the large-scale computer programs used by industry. The same examples are to be solved by the bus impedance matrices in the problems at the end of this chapter.

Example 12.6. A group of identical synchronous motors is connected through a transformer to a 4.16-kV bus at a location remote from the generating plants of a power system. The motors are rated 600 V and operate at 89.5% efficiency when carrying a full load at unity power factor and rated voltage. The sum of their output ratings is 4476 kW (6000 hp). The reactances in per unit of each motor based on its own input kilovoltampere rating are $X_d'' = X_1 = 0.20$, $X_2 = 0.20$, $X_0 = 0.04$, and each is grounded through a reactance of 0.02 per unit. The motors are connected to the 4.16-kV bus through a transformer bank composed of three single-phase units, each of which is rated 2400/600 V, 2500 kVA. The 600-V windings are connected in Δ to the motors and the 2400-V windings are connected in Y. The leakage reactance of each transformer is 10%.

The power system which supplies the 4.16-kV bus is represented by a Thévenin equivalent generator rated 7500 kVA, 4.16 kV with reactances of

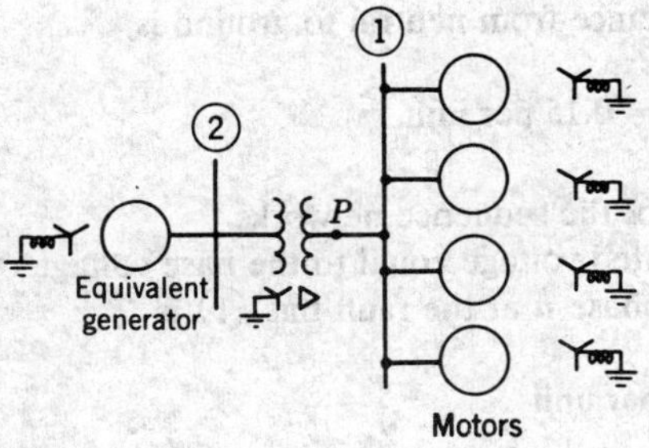

FIGURE 12.17
Single-line diagram of the system of Example 12.6.

$X''_d = X_2 = 0.10$ per unit, $X_0 = 0.05$ per unit, and X_n from neutral to ground equal to 0.05 per unit.

Each of the identical motors is supplying an equal share of a total load of 3730 kW (5000 hp) and is operating at rated voltage, 85% power-factor lag, and 88% efficiency when a single line-to-ground fault occurs on the low-voltage side of the transformer bank. Treat the group of motors as a single equivalent motor. Draw the sequence networks showing values of the impedances. Determine the subtransient line currents in all parts of the system with prefault current neglected.

Solution. The single-line diagram of the system is shown in Fig. 12.17. The 600-V bus and the 4.16-kV bus are numbered ① and ②, respectively. Choose the rating of the equivalent generator as base: 7500 kVA, 4.16 kV at the system bus.

Since

$$\sqrt{3} \times 2400 = 4160 \text{ V} \qquad 3 \times 2500 = 7500 \text{ kVA}$$

the three-phase rating of the transformer is 7500 kVA, 4160Y/600Δ V. So, the base for the motor circuit is 7500 kVA, 600 V.

The input *rating* of the single equivalent motor is

$$\frac{6000 \times 0.746}{0.895} = 5000 \text{ kVA}$$

and the reactances of the equivalent motor in percent are the same on the base of the combined rating as the reactances of the individual motors on the base of the rating of an individual motor. The reactances of the equivalent motor in per unit on the selected base are

$$X''_d = X_1 = X_2 = 0.2\frac{7500}{5000} = 0.3 \qquad X_0 = 0.04\frac{7500}{5000} = 0.06$$

In the zero-sequence network the reactance between neutral and ground of the equivalent motor is

$$3X_n = 3 \times 0.02\frac{7500}{5000} = 0.09 \text{ per unit}$$

and for the equivalent generator the reactance from neutral to ground is

$$3X_n = 3 \times 0.05 = 0.15 \text{ per unit}$$

Figure 12.18 shows the series connection of the sequence networks.

Since the motors are operating at rated voltage equal to the base voltage of the motor circuit, the prefault voltage of phase a at the fault bus ① is

$$V_f = 1.0 \text{ per unit}$$

Base current for the motor circuit is

$$\frac{7{,}500{,}000}{\sqrt{3} \times 600} = 7217 \text{ A}$$

and the actual motor current is

$$\frac{746 \times 5000}{0.88 \times \sqrt{3} \times 600 \times 0.85} = 4798 \text{ A}$$

Current drawn by the motor through line a before the fault occurs is

$$\frac{4798}{7217}\angle -\cos^{-1} 0.85 = 0.665\angle -31.8^\circ = 0.565 - j0.350 \text{ per unit}$$

If prefault current is neglected, E_g'' and E_m'' are made equal to $1.0\angle 0^\circ$ in Fig. 12.18. Thévenin impedances are computed at bus ① in each sequence network as follows:

$$Z_{11}^{(1)} = Z_{11}^{(2)} = \frac{(j0.1 + j0.1)(j0.3)}{j(0.1 + 0.1 + 0.3)} = j0.12 \text{ per unit} \qquad Z_{11}^{(0)} = j0.15 \text{ per unit}$$

Fault current in the series connection of the sequence networks is

$$I_{fa}^{(1)} = \frac{V_f}{Z_{11}^{(1)} + Z_{11}^{(2)} + Z_{11}^{(0)}} = \frac{1.0}{j0.12 + j0.12 + j0.15} = \frac{1.0}{j0.39} = -j2.564$$

$$I_{fa}^{(2)} = I_{fa}^{(0)} = I_{fa}^{(1)} = -j2.564 \text{ per unit}$$

Current in the fault $= 3I_{fa}^{(0)} = 3(-j2.564) = -j7.692$ per unit. In the positive-sequence network the portion of $I_{fa}^{(1)}$ flowing toward P from the transformer is found by current division to be

$$\frac{-j2.564 \times j0.30}{j0.50} = -j1.538 \text{ per unit}$$

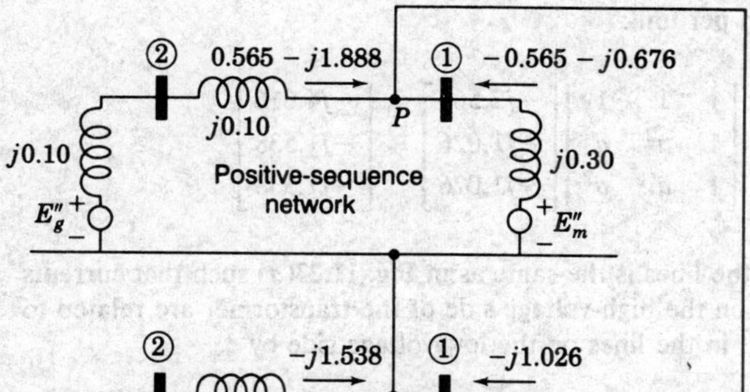

FIGURE 12.18
Connection of the sequence networks of Example 12.6. Subtransient currents are marked in per unit for a single line-to-ground fault at P. Prefault current is included.

and the portion of $I_{fa}^{(1)}$ flowing from the motor toward P is

$$\frac{-j2.564 \times j0.20}{j0.50} = -j1.026 \text{ per unit}$$

Similarly, the portion of $I_{fa}^{(2)}$ from the transformer is $-j1.538$ per unit, and the component of $I_{fa}^{(2)}$ from the motor is $-j1.026$ per unit. All of $I_{fa}^{(0)}$ flows toward P from the motor.

Currents *in the lines* at the fault, shown without subscript f, are:
To P from the transformer in per unit:

$$\begin{bmatrix} I_a \\ I_b \\ I_c \end{bmatrix} = \begin{bmatrix} 1 & 1 & 1 \\ 1 & a^2 & a \\ 1 & a & a^2 \end{bmatrix} \begin{bmatrix} 0 \\ -j1.538 \\ -j1.538 \end{bmatrix} = \begin{bmatrix} -j3.076 \\ j1.538 \\ j1.538 \end{bmatrix}$$

To P from the motors in per unit:

$$\begin{bmatrix} I_a \\ I_b \\ I_c \end{bmatrix} = \begin{bmatrix} 1 & 1 & 1 \\ 1 & a^2 & a \\ 1 & a & a^2 \end{bmatrix} \begin{bmatrix} -j2.564 \\ -j1.026 \\ -j1.026 \end{bmatrix} = \begin{bmatrix} -j4.616 \\ -j1.538 \\ -j1.538 \end{bmatrix}$$

Our method of labeling the lines is the same as in Fig. 11.23(a) such that currents $I_A^{(1)}$ and $I_A^{(2)}$ in the lines on the high-voltage side of the transformer are related to the currents $I_a^{(1)}$ and $I_a^{(2)}$ in the lines on the low-voltage side by

$$I_A^{(1)} = I_a^{(1)}\angle 30^\circ \qquad I_A^{(2)} = I_a^{(2)}\angle -30^\circ$$

Hence,

$$I_A^{(1)} = (-j1.538)\angle 30^\circ = 1.538\angle -60^\circ = 0.769 - j1.332$$

$$I_A^{(2)} = (-j1.538)\angle -30^\circ = 1.538\angle -120^\circ = -0.769 - j1.332$$

and from Fig. 12.18 we note that $I_A^{(0)} = 0$ in the zero-sequence network. Since there are no zero-sequence currents on the high-voltage side of the transformer, we have

$$I_A = I_A^{(1)} + I_A^{(2)} = (0.769 - j1.332) + (-0.769 - j1.332) = -j2.664 \text{ per unit}$$

$$I_B^{(1)} = a^2 I_A^{(1)} = (1\angle 240^\circ)(1.538\angle -60^\circ) = -1.538 + j0$$

$$I_B^{(2)} = a\, I_A^{(2)} = (1\angle 120^\circ)(1.538\angle -120^\circ) = 1.538 + j0$$

$$I_B = I_B^{(1)} + I_B^{(2)} = 0$$

$$I_C^{(1)} = a\, I_A^{(1)} = (1\angle 120^\circ)(1.538\angle -60^\circ) = 0.769 + j1.332$$

$$I_C^{(2)} = a^2 I_A^{(2)} = (1\angle 240^\circ)(1.538\angle -120^\circ) = -0.769 + j1.332$$

$$I_C = I_C^{(1)} + I_C^{(2)} = j2.664 \text{ per unit}$$

If voltages throughout the system are to be found by circuit analysis, their components at any point can be calculated from the currents and reactances of the sequence networks. Components of voltages on the high-voltage side of the transformer are found first without regard for phase shift. Then, the effect of phase shift must be determined.

By evaluating the base currents on the two sides of the transformer, we can convert the above per-unit currents to amperes. Base current for the motor circuit was found previously and equals 7217 A. Base current for the high-voltage circuit is

$$\frac{7{,}500{,}000}{\sqrt{3} \times 4160} = 1041 \text{ A}$$

Current in the fault is

$$7.692 \times 7217 = 55{,}500 \text{ A}$$

Currents in the lines between the transformer and the fault are

In line a: $3.076 \times 7217 = 22{,}200$ A

In line b: $1.538 \times 7217 = 11{,}100$ A

In line c: $1.538 \times 7217 = 11{,}100$ A

Currents in the lines between the motor and the fault are

In line a: $4.616 \times 7271 = 33{,}300$ A

In line b: $1.538 \times 7217 = 11{,}100$ A

In line c: $1.538 \times 7217 = 11{,}100$ A

Currents in the lines between the 4.16 kV bus and the transformer are

In line A: $2.664 \times 1041 = 2773$ A

In line B: 0

In line C: $2.664 \times 1041 = 2773$ A

The currents we have calculated in the above example are those which would flow upon the occurrence of a single line-to-ground fault when there is no load on the motors. These currents are correct only if the motors are drawing no current whatsoever. The statement of the problem specifies the load conditions at the time of the fault, however, and the load can be considered. To account for the load, we add the per-unit current drawn by the motor through line a before the fault occurs to the portion of $I_{fa}^{(1)}$ flowing toward P from the transformer and subtract the same current from the portion of $I_{fa}^{(1)}$ flowing from the motor to P. The new value of positive-sequence current from the transformer to the fault in phase a is

$$0.565 - j0.350 - j1.538 = 0.565 - j1.888$$

and the new value of positive-sequence current from the motor to the fault in

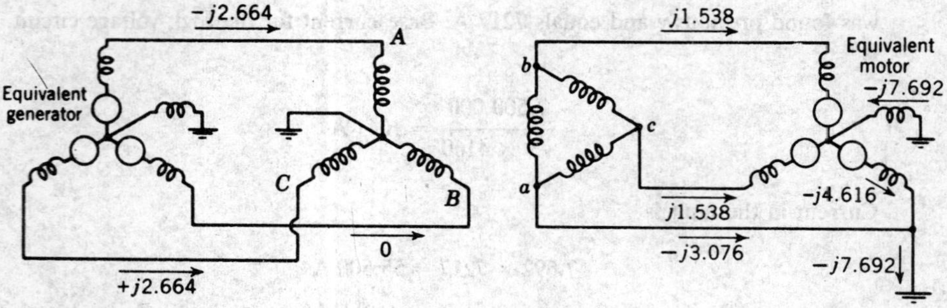

FIGURE 12.19
Per-unit values of subtransient line currents in all parts of the system of Example 12.6, prefault current neglected.

phase a is

$$-0.565 + j0.350 - j1.026 = -0.565 - j0.676$$

These values are shown in Fig. 12.18. The remainder of the calculation, using these new values, proceeds as in the example.

Figure 12.19 gives the per-unit values of subtransient line currents in all parts of the system when the fault occurs at no load. Figure 12.20 shows the values for the fault occurring on the system when the load specified in the example is considered. In a larger system where the fault current is much higher in comparison with the load current the effect of neglecting the load current is less than is indicated by comparing Figs. 12.19 and 12.20. In the large system, however, the prefault currents determined by a power-flow study could simply be added to the fault current found with the load neglected.

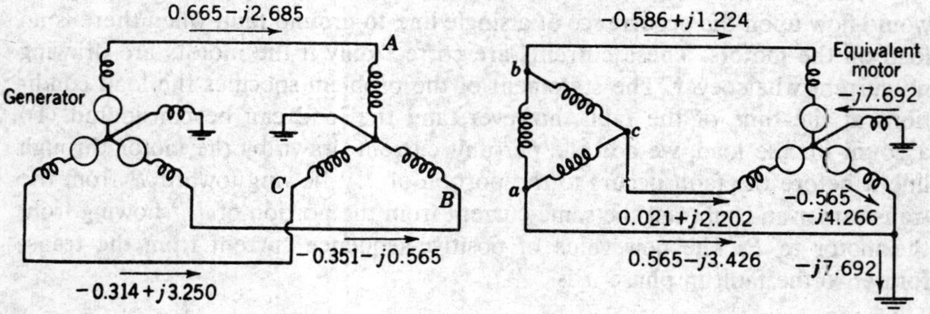

FIGURE 12.20
Per-unit values of subtransient line currents in all parts of the system of Example 12.6, prefault current considered.

Example 12.7. The single-line diagram of a small power system is shown in Fig. 12.21. A bolted single line-to-ground fault at point P is to be analyzed. The ratings and reactances of the generator and the transformers are

Generator: 100 MVA, 20 kV; $X'' = X_2 = 20\%$, $X_0 = 4\%$,

$X_n = 5\%$

Transformers T_1 and T_2: 100 MVA, 20Δ/345Y kV; $X = 10\%$

On a chosen base of 100 MVA, 345 kV in the transmission-line circuit the line reactances are

From T_1 to P: $X_1 = X_2 = 20\%$, $X_0 = 50\%$

From T_2 to P: $X_1 = X_2 = 10\%$, $X_0 = 30\%$

To simulate the fault, the sequence networks of the system with reactances marked in per unit are connected in series, as shown in Fig. 12.22. Verify the values of the currents shown in the figure and draw a complete three-phase circuit diagram with all current flows marked in per unit. Assume that the transformers are lettered so that Eqs. (11.88) apply.

Solution. With switch S open, the prefault currents are zero and the open-circuit voltage of phase A at point P can be taken as the reference voltage $1.0 + j0.0$ per unit. The impedances seen looking into the sequence networks at the fault point are

$$Z_{pp}^{(0)} = \frac{(j0.6)(j0.4)}{j0.6 + j0.4} = j0.24 \text{ per unit}$$

$$Z_{pp}^{(1)} = Z_{pp}^{(2)} = j0.5 \text{ per unit}$$

The sequence currents in the hypothetical stub of phase A at P are

$$I_{fA}^{(0)} = I_{fA}^{(1)} = I_{fA}^{(2)} = \frac{1.0 + j0.0}{j0.5 + j0.5 + j0.24} = -j0.8065 \text{ per unit}$$

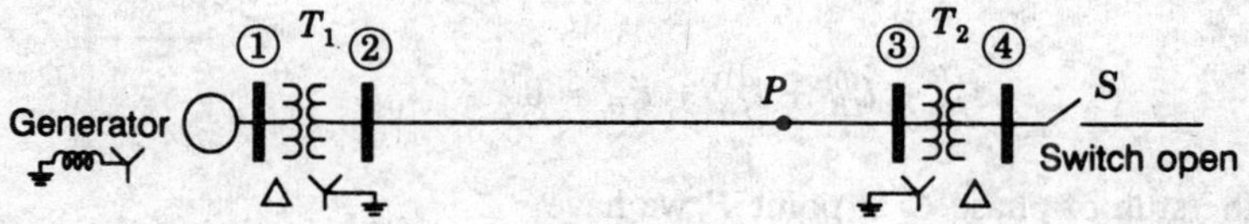

FIGURE 12.21
Single-line diagram of the system of Example 12.7. Single line-to-ground fault is at point P.

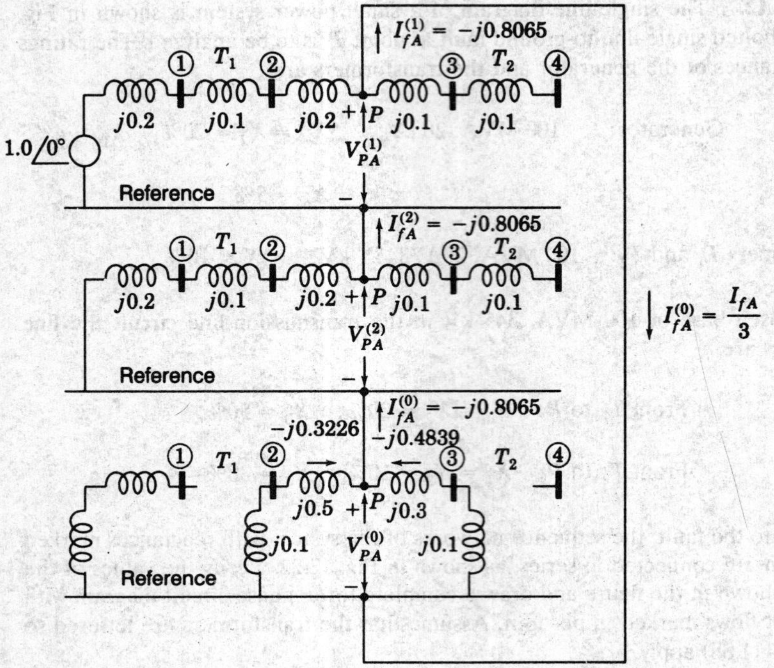

FIGURE 12.22
Connection of the sequence networks for the system of Fig. 12.21 to simulate single line-to-ground fault at point P.

The total current in the fault is

$$I_{fA} = 3I_{fA}^{(0)} = -j2.4195 \text{ per unit}$$

In the stub of phase B at point P we have

$$I_{fB}^{(1)} = a^2 I_{fA}^{(1)} = 0.8065\angle{-90° + 240°} = 0.8065\angle 150°$$

$$I_{fB}^{(2)} = aI_{fA}^{(2)} = 0.8065\angle{-90° + 120°} = 0.8065\angle 30°$$

$$I_{fB}^{(0)} = I_{fA}^{(0)} = 0.8065\angle{-90°}$$

$$I_{fB} = I_{fB}^{(0)} + I_{fB}^{(1)} + I_{fB}^{(2)} = 0$$

Likewise, in the stub of phase C at point P we have

$$I_{fC} = I_{fC}^{(0)} + I_{fC}^{(1)} + I_{fC}^{(2)} = 0$$

In the zero-sequence network the currents are:

Toward P from T_1

$$I_A^{(0)} = \frac{j0.4}{j0.6 + j0.4}\left(0.8065\angle -90^\circ\right)$$

$$= 0.3226\angle -90^\circ \text{ per unit}$$

Toward P from T_2

$$I_A^{(0)} = \frac{j0.6}{j0.6 + j0.4}\left(0.8065\angle -90^\circ\right)$$

$$= 0.4839\angle -90^\circ \text{ per unit}$$

In the transmission line the currents are:

Toward P from T_1

In line A: $\quad 0.3226\angle -90^\circ + 0.8065\angle -90^\circ + 0.8065\angle -90^\circ = -j1.9356$ per unit

In line B: $\quad 0.3226\angle -90^\circ + 0.8065\angle 150^\circ + 0.8065\angle 30^\circ = j0.4839$ per unit

In line C: $\quad 0.3226\angle -90^\circ + 0.8065\angle 30^\circ + 0.8065\angle 150^\circ = j0.4839$ per unit

Toward P from T_2

In line A: $\quad I_A = -j0.4839$ per unit

In line B: $\quad I_B = -j0.4839$ per unit

In line C: $\quad I_C = -j0.4839$ per unit

Note that positive-, negative-, and zero-sequence components of current flow in lines A, B, and C from T_1 but only zero-sequence components flow in these lines from T_2. Kirchhoff's current law is fulfilled, however.

In the generator the currents are

$$I_a = I_a^{(0)} + I_a^{(1)} + I_a^{(2)} = 0 + 0.8065\angle -90^\circ - 30^\circ + 0.8065\angle -90^\circ + 30^\circ$$

$$= -j1.3969$$

$$I_b = I_a^{(0)} + a^2 I_a^{(1)} + a\, I_a^{(2)} = 0 + 0.8065\angle -120^\circ + 240^\circ + 0.8065\angle -60^\circ + 120^\circ$$

$$= j1.3969$$

$$I_c = I_a^{(0)} + a\, I_a^{(1)} + a^2 I_a^{(2)} = 0 + 0.8065\angle -120^\circ + 120^\circ + 0.8065\angle -60^\circ + 240^\circ$$

$$= 0$$

The three-phase circuit diagram of Fig. 12.23 shows all the current flows in per

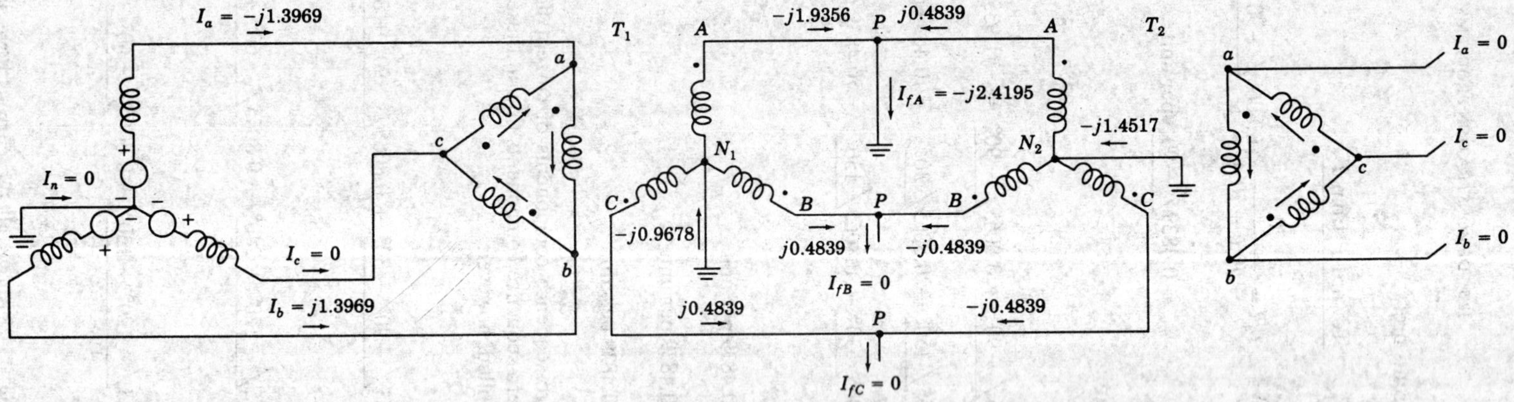

FIGURE 12.23
Current flows in the system of Fig. 12.21 due to single line-to-ground fault at P.

unit. From this diagram note that:

- Lines are lettered and polarity marks are placed so that Eqs. (11.88) are valid.
- Stubs are connected to each line at the fault.
- For a single line-to-ground fault stub currents $I_B = I_C = 0$, but $I_B^{(0)}$, $I_B^{(1)}$, $I_B^{(2)}$, $I_C^{(0)}$, $I_C^{(1)}$, and $I_C^{(2)}$ in the stubs all have nonzero values.
- Fault current flows out of stub A, then partly to T_1 and partly to T_2.
- In the generator only positive- and negative-sequence currents are flowing.
- In the Δ windings of T_2 only zero-sequence currents are flowing.
- In the Δ windings of T_1 each phase winding contains positive-, negative-, and zero-sequence components of current. These components are shown in Fig.

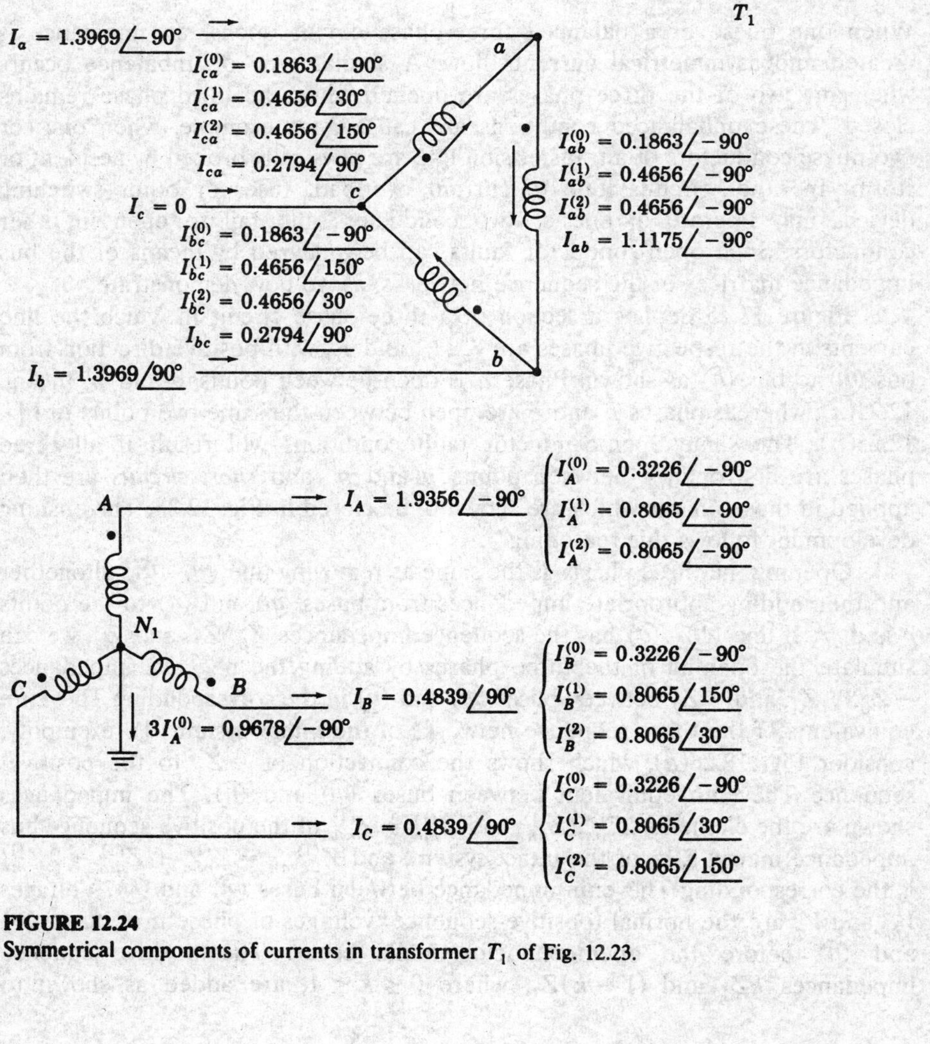

FIGURE 12.24
Symmetrical components of currents in transformer T_1 of Fig. 12.23.

12.24 and yield

$$I_{ab} = \frac{I_A}{\sqrt{3}} = 1.1175\angle{-90^\circ}$$

$$I_{bc} = \frac{I_B}{\sqrt{3}} = 0.2794\angle{90^\circ}$$

$$I_{ca} = \frac{I_C}{\sqrt{3}} = 0.2794\angle{90^\circ}$$

12.6 OPEN-CONDUCTOR FAULTS

When one phase of a balanced three-phase circuit opens, an unbalance is created and asymmetrical currents flow. A similar type of unbalance occurs when any two of the three phases are opened while the third phase remains closed. These unbalanced conditions are caused, for example, when one- or two-phase conductors of a transmission line are physically broken by accident or storm. In other circuits, due to current overload, fuses or other switching devices may operate in one or two conductors and fail to open in other conductors. Such open-conductor faults can be analyzed by means of the bus impedance matrices of the sequence networks, as we now demonstrate.

Figure 12.25 depicts a section of a three-phase circuit in which the line currents in the respective phases are I_a, I_b, and I_c, with positive direction from bus ⓜ to bus ⓝ as shown. Phase a is open between points p and p' in Fig. 12.25(a), whereas phases b and c are open between the same two points in Fig. 12.25(b). The same open-conductor fault conditions will result if all three phases are first opened between points p and p', and *short circuits* are then applied in those phases which are shown to be closed in Fig. 12.25. The ensuing development follows this reasoning.

Opening the three phases is the same as removing line ⓜ–ⓝ altogether and then adding appropriate impedances from buses ⓜ and ⓝ to the points p and p'. If line ⓜ–ⓝ has the sequence impedances Z_0, Z_1, and Z_2, we can simulate the opening of the three phases by adding the negative impedances $-Z_0$, $-Z_1$, and $-Z_2$ between buses ⓜ and ⓝ in the corresponding Thévenin equivalents of the three sequence networks of the *intact* system. To exemplify, consider Fig. 12.26(a), which shows the connection of $-Z_1$ to the positive-sequence Thévenin equivalent between buses ⓜ and ⓝ. The impedances shown are the elements $Z_{mm}^{(1)}$, $Z_{nn}^{(1)}$, and $Z_{mn}^{(1)} = Z_{nm}^{(1)}$ of the positive-sequence bus impedance matrix $\mathbf{Z}_{\text{bus}}^{(1)}$ of the intact system, and $Z_{\text{th},\,mn}^{(1)} = Z_{mm}^{(1)} + Z_{nm}^{(1)} - 2Z_{mn}^{(1)}$ is the corresponding Thévenin impedance between buses ⓜ and ⓝ. Voltages V_m and V_n are the normal (positive-sequence) voltages of phase a at buses ⓜ and ⓝ before the open-conductor faults occur. The positive-sequence impedances kZ_1 and $(1-k)Z_1$, where $0 \le k \le 1$, are added as shown to

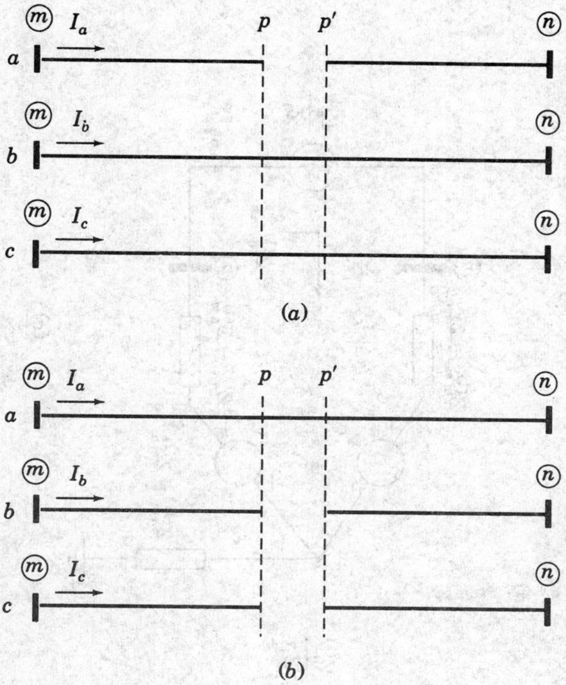

FIGURE 12.25
Open-conductor faults on a section of a three-phase system between buses (m) and (n): (*a*) conductor *a* open; (*b*) conductors *b* and *c* open between points *p* and *p*′.

represent the fractional lengths of the broken line (m)–(n) from bus (m) to point *p* and bus (n) to point *p*′, respectively. To use a convenient notation, let the voltage $V_a^{(1)}$ denote the phase-*a* positive-sequence component of the *voltage drops* $V_{pp',a}$, $V_{pp',b}$, and $V_{pp',c}$ from *p* to *p*′ in the phase conductors. We shall soon see that $V_a^{(1)}$, and the corresponding negative- and zero-sequence components $V_a^{(2)}$ and $V_a^{(0)}$, take on different values depending on which one of the open-conductor faults is being considered.

By source transformation we can replace the voltage drop $V_a^{(1)}$ in series with the impedance $[kZ_1 + (1-k)Z_1]$ in Fig. 12.26(*a*) by the current $V_a^{(1)}/Z_1$ in parallel with the impedance Z_1, as shown in Fig. 12.26(*b*). In this latter figure the parallel combination of $-Z_1$ and Z_1 can be canceled, as shown in Fig. 12.26(*c*).

The above considerations for the positive-sequence network also apply directly to the negative- and zero-sequence networks, but we must remember that the latter networks do not contain any internal sources of their own. In drawing the negative- and zero-sequence equivalent circuits of Fig. 12.27, it is understood that the currents $V_a^{(2)}/Z_2$ and $V_a^{(0)}/Z_0$, like the current $V_a^{(1)}/Z_1$ of Fig. 12.26(*c*), owe their origin to the open-conductor fault between points *p* and *p*′ in the system. If there is no open conductor, the voltages $V_a^{(1)}$, $V_a^{(2)}$, and $V_a^{(0)}$ are all zero and the current sources disappear. It is evident from the figures that each of the sequence currents $V_a^{(0)}/Z_0$, $V_a^{(1)}/Z_1$, and $V_a^{(2)}/Z_2$ can be regarded in turn as a *pair* of injections into buses (m) and (n) of the corresponding

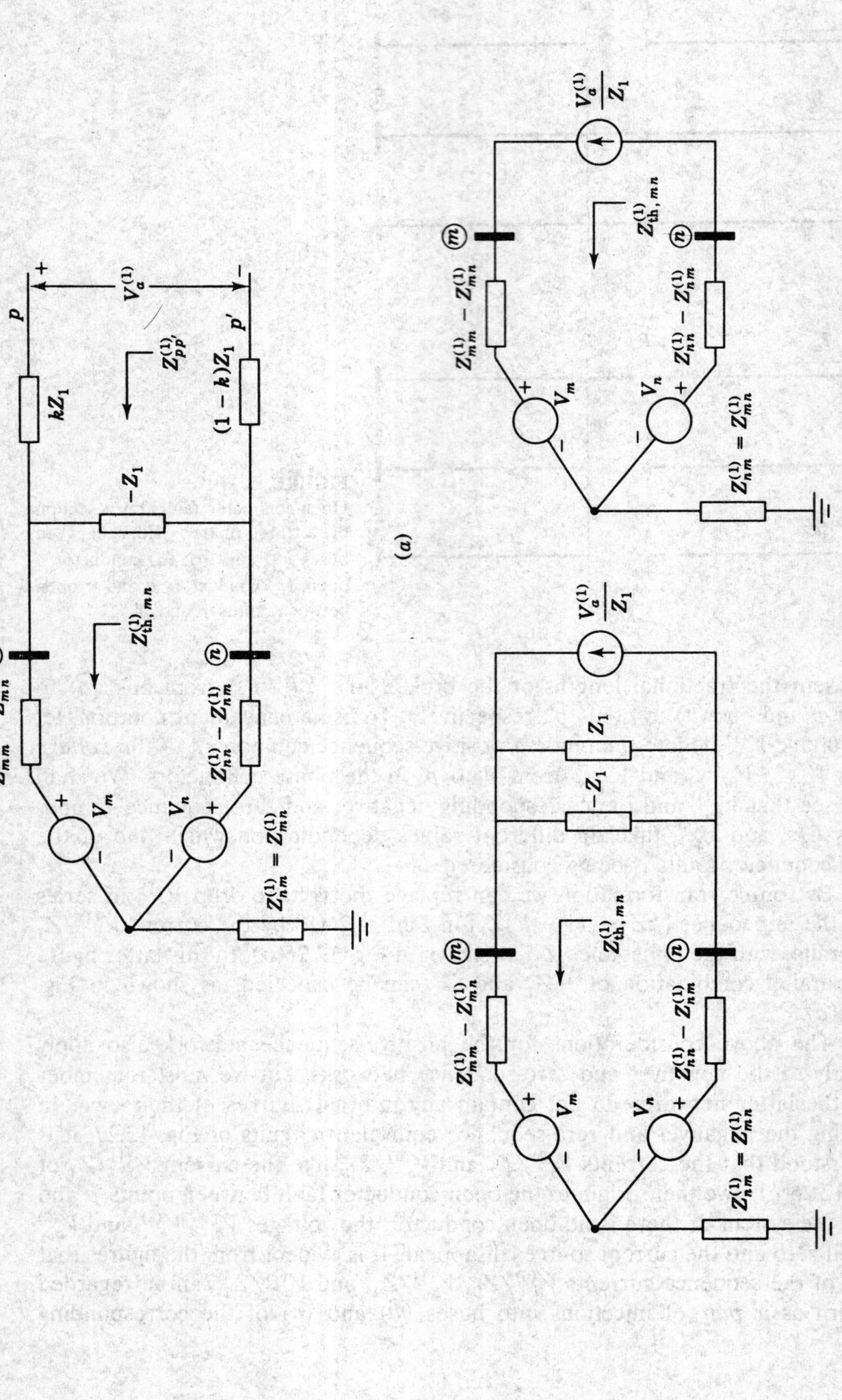

FIGURE 12.26
Simulating the opening of line (m)-(n) between points p and p': (*a*) connections to positive-sequence Thévenin equivalent of the system; (*b*) transformation to current source; (*c*) resultant equivalent circuit.

sequence network of the *intact* system. Hence, we can use the bus impedance matrices $\mathbf{Z}_{\text{bus}}^{(0)}$, $\mathbf{Z}_{\text{bus}}^{(1)}$, and $\mathbf{Z}_{\text{bus}}^{(2)}$ of the *normal* configuration of the system to determine the voltage changes due to the open-conductor faults. But first we must find expressions for the symmetrical components $V_a^{(0)}$, $V_a^{(1)}$, and $V_a^{(2)}$ of the voltage drops across the fault points p and p' for each type of fault shown in Fig. 12.25. These voltage drops can be regarded as giving rise to the following sets of injection currents into the sequence networks of the normal system configuration:

	Positive Sequence	Negative Sequence	Zero Sequence
At bus ⓜ:	$\dfrac{V_a^{(1)}}{Z_1}$	$\dfrac{V_a^{(2)}}{Z_2}$	$\dfrac{V_a^{(0)}}{Z_0}$
At bus ⓝ:	$-\dfrac{V_a^{(1)}}{Z_1}$	$-\dfrac{V_a^{(2)}}{Z_2}$	$-\dfrac{V_a^{(0)}}{Z_0}$

as shown in Figs. 12.26 and 12.27. By multiplying the bus impedance matrices $\mathbf{Z}_{\text{bus}}^{(0)}$, $\mathbf{Z}_{\text{bus}}^{(1)}$, and $\mathbf{Z}_{\text{bus}}^{(2)}$ by current vectors containing only these current injections, we obtain the following *changes* in the symmetrical components of the phase-a voltage of each bus ⓘ:

$$\text{Zero sequence:} \quad \Delta V_i^{(0)} = \frac{Z_{im}^{(0)} - Z_{in}^{(0)}}{Z_0} V_a^{(0)}$$

$$\text{Positive sequence:} \quad \Delta V_i^{(1)} = \frac{Z_{im}^{(1)} - Z_{in}^{(1)}}{Z_1} V_a^{(1)} \tag{12.27}$$

$$\text{Negative sequence:} \quad \Delta V_i^{(2)} = \frac{Z_{im}^{(2)} - Z_{in}^{(2)}}{Z_2} V_a^{(2)}$$

Before developing the equations for $V_a^{(0)}$, $V_a^{(1)}$, and $V_a^{(2)}$ for each type of open-conductor fault, let us derive expressions for the Thévenin equivalent impedances of the sequence networks, as seen from fault points p and p'.

Looking into the positive-sequence network of Fig. 12.26(a) between p and p', we see the impedance $Z_{pp'}^{(1)}$ given by

$$Z_{pp'}^{(1)} = kZ_1 + \frac{Z_{\text{th},mn}^{(1)}(-Z_1)}{Z_{\text{th},mn}^{(1)} - Z_1} + (1-k)Z_1 = \frac{-Z_1^2}{Z_{\text{th},mn}^{(1)} - Z_1} \tag{12.28}$$

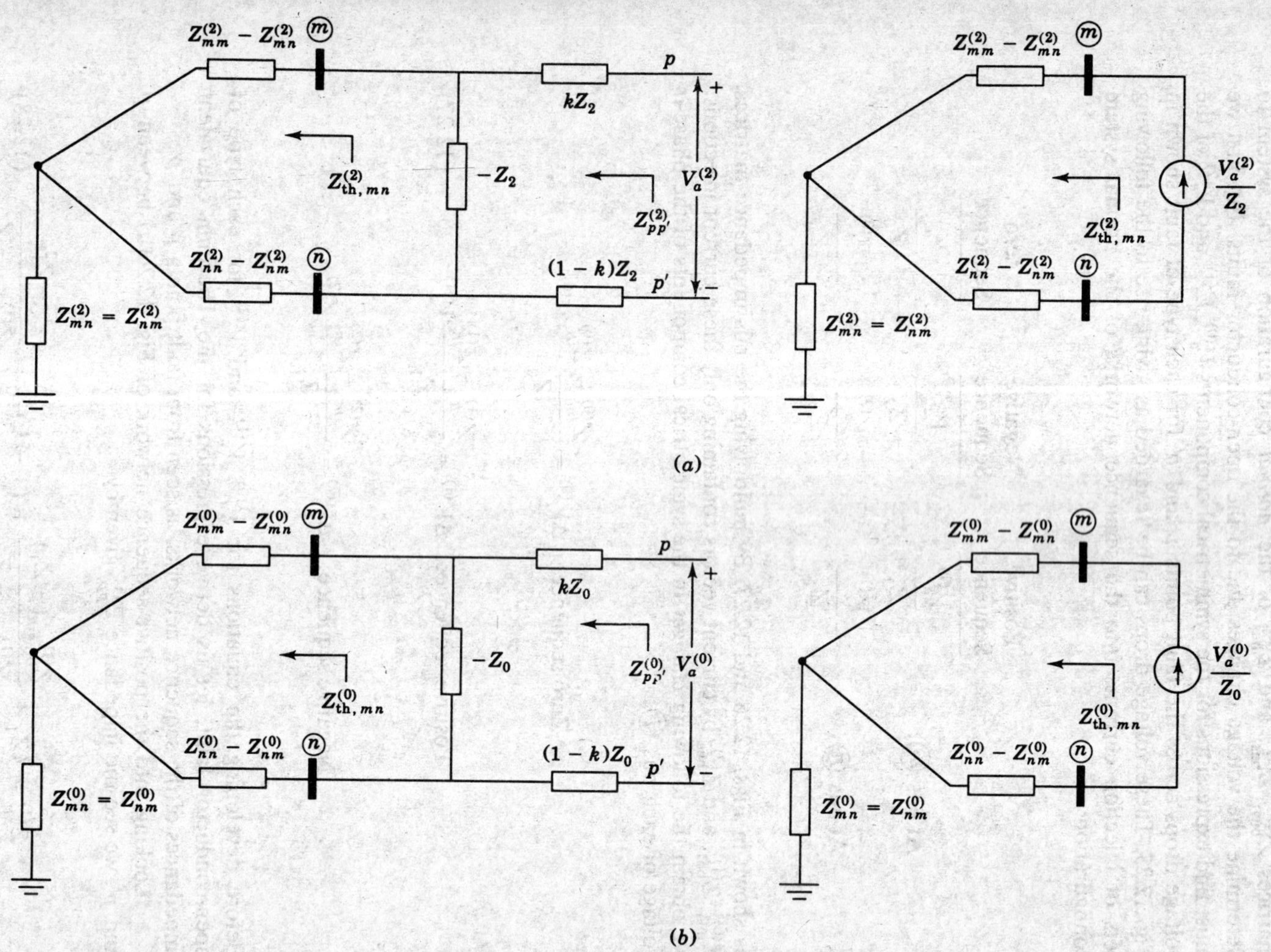

FIGURE 12.27
Simulating the opening of line ⓜ – ⓝ between points p and p': (a) negative-sequence and (b) zero-sequence equivalent circuits.

and from p to p' the open-circuit voltage obtained by voltage division is

$$\text{Open-circuit voltage from } p \text{ to } p' = \frac{-Z_1}{Z_{\text{th},mn}^{(1)} - Z_1}(V_m - V_n) = \frac{Z_{pp'}^{(1)}}{Z_1}(V_m - V_n) \tag{12.29}$$

Before any conductor opens, the current I_{mn} in phase a of the line ⓜ- ⓝ is positive sequence and is given by

$$I_{mn} = \frac{V_m - V_n}{Z_1} \tag{12.30}$$

Substituting this expression for I_{mn} in Eq. (12.29), we obtain

$$\text{Open-circuit voltage from } p \text{ to } p' = I_{mn} Z_{pp'}^{(1)} \tag{12.31}$$

Figure 12.28(a) shows the resulting positive-sequence equivalent circuit between points p and p'. Analogous to Eq. (12.28) we have

$$Z_{pp'}^{(2)} = \frac{-Z_2^2}{Z_{\text{th},mn}^{(2)} - Z_2} \quad \text{and} \quad Z_{pp'}^{(0)} = \frac{-Z_0^2}{Z_{\text{th},mn}^{(0)} - Z_0} \tag{12.32}$$

which are the negative- and zero-sequence impedances, respectively, between p and p' in Figs. 12.28(b) and 12.28(c). We can now proceed to develop expressions for the sequence voltage drops $V_a^{(0)}$, $V_a^{(1)}$, and $V_a^{(2)}$.

One open conductor

Let us consider one open conductor as in Fig. 12.25(a). Owing to the open circuit in phase a, the current $I_a = 0$, and so

$$I_a^{(0)} + I_a^{(1)} + I_a^{(2)} = 0 \tag{12.33}$$

where $I_a^{(0)}$, $I_a^{(1)}$, and $I_a^{(2)}$ are the symmetrical components of the *line* currents I_a, I_b, and I_c from p to p'. Since phases b and c are closed, we also have the voltage drops

$$V_{pp',b} = 0 \qquad V_{pp',c} = 0 \tag{12.34}$$

Resolving the series voltage drops across the fault point into their symmetrical

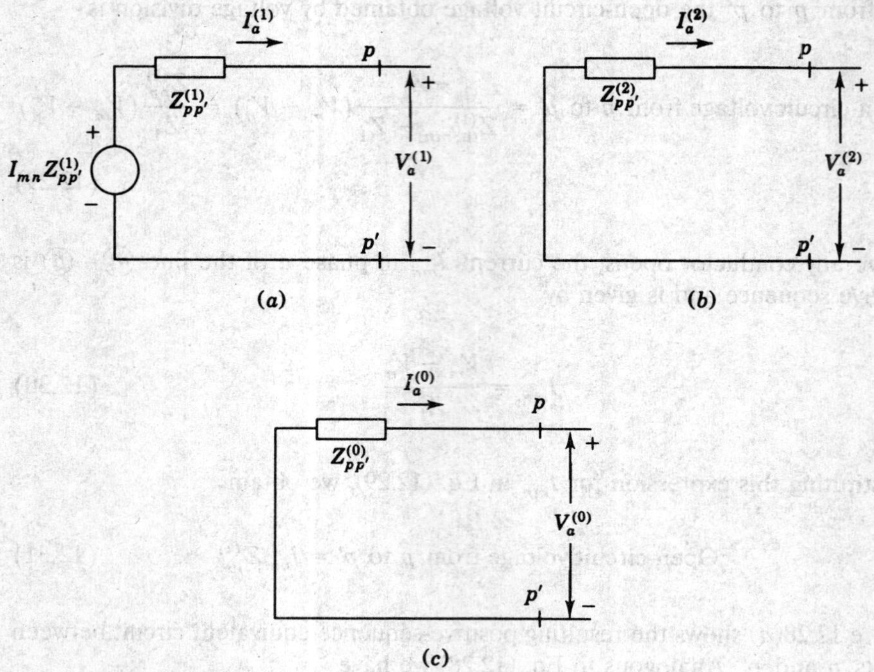

FIGURE 12.28
Looking into the system between points p and p': (a) positive-sequence, (b) negative-sequence, and (c) zero-sequence equivalent circuits.

components, we obtain

$$\begin{bmatrix} V_a^{(0)} \\ V_a^{(1)} \\ V_a^{(2)} \end{bmatrix} = \frac{1}{3}\begin{bmatrix} 1 & 1 & 1 \\ 1 & a & a^2 \\ 1 & a^2 & a \end{bmatrix}\begin{bmatrix} V_{pp',a} \\ 0 \\ 0 \end{bmatrix} = \frac{1}{3}\begin{bmatrix} V_{pp',a} \\ V_{pp',a} \\ V_{pp',a} \end{bmatrix} \tag{12.35}$$

That is,

$$V_a^{(0)} = V_a^{(1)} = V_a^{(2)} = \frac{V_{pp',a}}{3} \tag{12.36}$$

In words, this equation states that the open conductor in phase a causes equal voltage drops to appear from p to p' in each of the sequence networks. We can satisfy this requirement and that of Eq. (12.33) by connecting the Thévenin equivalents of the sequence networks *in parallel* at the points p and p', as shown in Fig. 12.29. From this figure the expression for the positive-sequence

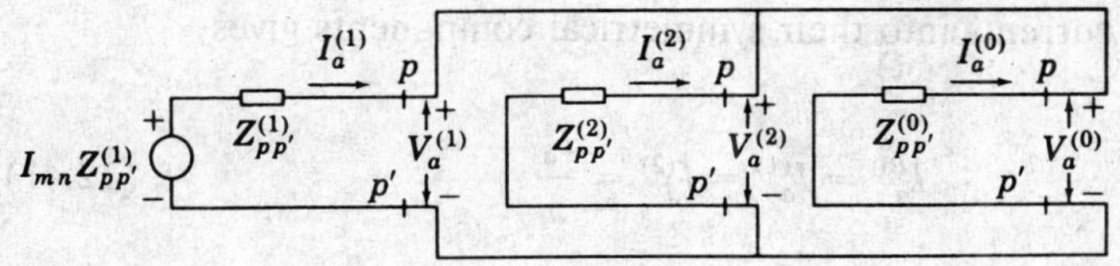

FIGURE 12.29
Connection of the sequence networks of the system to simulate opening phase a between points p and p'.

current $I_a^{(1)}$ is found to be

$$I_a^{(1)} = I_{mn} \frac{Z_{pp'}^{(1)}}{Z_{pp'}^{(1)} + \dfrac{Z_{pp'}^{(2)} Z_{pp'}^{(0)}}{Z_{pp'}^{(2)} + Z_{pp'}^{(0)}}}$$

$$= I_{mn} \frac{Z_{pp'}^{(1)}\left(Z_{pp'}^{(2)} + Z_{pp'}^{(0)}\right)}{Z_{pp'}^{(0)} Z_{pp'}^{(1)} + Z_{pp'}^{(1)} Z_{pp'}^{(2)} + Z_{pp'}^{(2)} Z_{pp'}^{(0)}} \tag{12.37}$$

The sequence voltage drops $V_a^{(0)}$, $V_a^{(1)}$, and $V_a^{(2)}$ are then given by Fig. 12.29 as

$$V_a^{(0)} = V_a^{(2)} = V_a^{(1)} = I_a^{(1)} \frac{Z_{pp'}^{(2)} Z_{pp'}^{(0)}}{Z_{pp'}^{(2)} + Z_{pp'}^{(0)}}$$

$$= I_{mn} \frac{Z_{pp'}^{(0)} Z_{pp'}^{(1)} Z_{pp'}^{(2)}}{Z_{pp'}^{(0)} Z_{pp'}^{(1)} + Z_{pp'}^{(1)} Z_{pp'}^{(2)} + Z_{pp'}^{(2)} Z_{pp'}^{(0)}} \tag{12.38}$$

The quantities on the right-hand side of this equation are known from the impedance parameters of the sequence networks and the prefault current in phase a of the line ⓜ– ⓝ. Thus, the currents $V_a^{(0)}/Z_0$, $V_a^{(1)}/Z_1$, and $V_a^{(2)}/Z_2$ for injection into the corresponding sequence networks can be determined from Eq. (12.38).

Two open conductors

When two conductors are open, as shown in Fig. 12.25(b), we have fault conditions which are the *duals*[2] of those in Eqs. (12.33) and (12.34); namely,

$$V_{pp',a} = V_a^{(0)} + V_a^{(1)} + V_a^{(2)} = 0 \tag{12.39}$$

$$I_b = 0 \qquad I_c = 0 \tag{12.40}$$

[2] *Duality* is treated in many textbooks on electrical circuits.

Resolving the line currents into their symmetrical components gives

$$I_a^{(0)} = I_a^{(1)} = I_a^{(2)} = \frac{I_a}{3} \tag{12.41}$$

Equations (12.39) and (12.41) are both satisfied by connecting the Thévenin equivalent of the negative- and zero-sequence networks *in series* between points p and p', as shown in Fig. 12.30. The sequence currents are now expressed by

$$I_a^{(0)} = I_a^{(2)} = I_a^{(1)} = I_{mn} \frac{Z_{pp'}^{(1)}}{Z_{pp'}^{(0)} + Z_{pp'}^{(1)} + Z_{pp'}^{(2)}} \tag{12.42}$$

where I_{mn} is again the prefault current in phase a of the line Ⓜ– Ⓝ before the open circuits occur in phases b and c. The sequence voltage drops are now given by

$$V_a^{(1)} = I_a^{(1)}\left(Z_{pp'}^{(2)} + Z_{pp'}^{(0)}\right) = I_{mn} \frac{Z_{pp'}^{(1)}\left(Z_{pp'}^{(2)} + Z_{pp'}^{(0)}\right)}{Z_{pp'}^{(1)} + Z_{pp'}^{(2)} + Z_{pp'}^{(0)}}$$

$$V_a^{(2)} = -I_a^{(2)} Z_{pp'}^{(2)} = I_{mn} \frac{-Z_{pp'}^{(1)} Z_{pp'}^{(2)}}{Z_{pp'}^{(1)} + Z_{pp'}^{(2)} + Z_{pp'}^{(0)}} \tag{12.43}$$

$$V_a^{(0)} = -I_a^{(0)} Z_{pp'}^{(0)} = I_{mn} \frac{-Z_{pp'}^{(1)} Z_{pp'}^{(0)}}{Z_{pp'}^{(1)} + Z_{pp'}^{(2)} + Z_{pp'}^{(0)}}$$

In each of these equations the right-hand side quantities are all known before the fault occurs. Therefore, Eq. (12.38) can be used to evaluate the symmetrical components of the voltage drops between the fault points p and p' when an open-conductor fault occurs; and Eq. (12.43) can be similarly used when a fault due to two open conductors occurs.

The net effect of the open conductors on the positive-sequence network is to increase the *transfer impedance* across the line in which the open-conductor fault occurs. For one open conductor this *increase* in impedance equals the *parallel* combination of the negative- and zero-sequence networks between points p and p'; for two open conductors the *increase* in impedance equals the *series* combination of the negative- and zero-sequence networks between points p and p'.

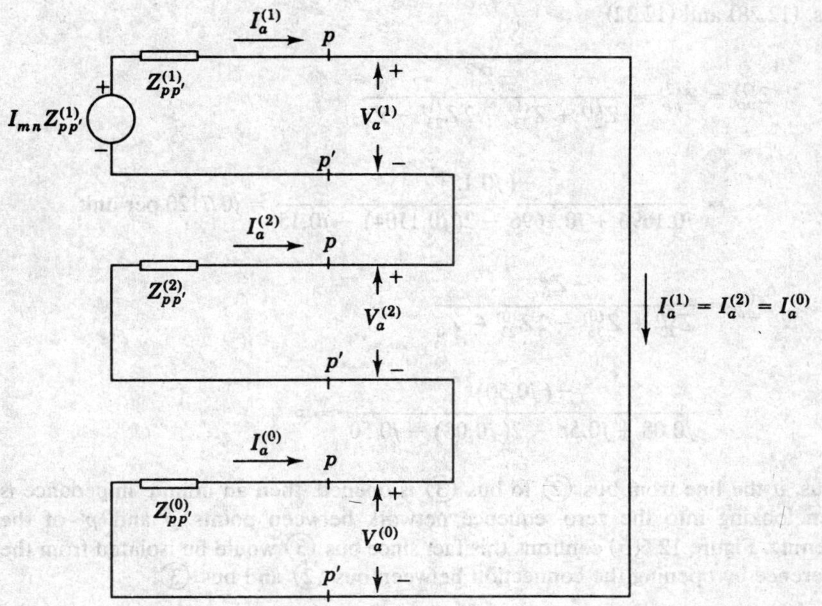

FIGURE 12.30
Connection of the sequence networks of the system to simulate opening phases b and c between points p and p'.

Example 12.8. In the system of Fig. 12.5 consider that machine 2 is a motor drawing a load equivalent to 50 MVA at 0.8 power-factor lagging and nominal system voltage of 345 kV at bus ③. Determine the change in voltage at bus ③ when the transmission line undergoes (a) a one-open-conductor fault and (b) a two-open-conductor fault along its span between buses ② and ③. Choose a base of 100 MVA, 345 kV in the transmission line.

Solution. All the per-unit parameters given in Example 12.1 apply directly here. By choosing the voltage at bus ③ as $1.0 + j0.0$ per unit, we can calculate the prefault current in line ②–③ as follows:

$$I_{23} = \frac{P - jQ}{V_3^*} = \frac{0.5(0.8 - j0.6)}{1.0 + j0.0} = 0.4 - j0.3 \text{ per unit}$$

The sequence networks of Fig. 12.6 show that line ②–③ has parameters

$$Z_1 = Z_2 = j0.15 \text{ per unit} \qquad Z_0 = j0.50 \text{ per unit}$$

The bus impedance matrices $\mathbf{Z}_{\text{bus}}^{(0)}$ and $\mathbf{Z}_{\text{bus}}^{(1)} = \mathbf{Z}_{\text{bus}}^{(2)}$ are also given in Example 12.1. Designating the open-circuit points of the line as p and p', we can calculate from

Eqs. (12.28) and (12.32)

$$Z_{pp'}^{(1)} = Z_{pp'}^{(2)} = \frac{-Z_1^2}{Z_{22}^{(1)} + Z_{33}^{(1)} - 2Z_{23}^{(1)} - Z_1}$$

$$= \frac{-(j0.15)^2}{j0.1696 + j0.1696 - 2(j0.1104) - j0.15} = j0.7120 \text{ per unit}$$

$$Z_{pp'}^{(0)} = \frac{-Z_0^2}{Z_{22}^{(0)} + Z_{33}^{(0)} - 2Z_{23}^{(0)} - Z_0}$$

$$= \frac{-(j0.50)^2}{j0.08 + j0.58 - 2(j0.08) - j0.50} = \infty$$

Thus, if the line from bus ② to bus ③ is opened, then an infinite impedance is seen looking into the zero-sequence network between points p and p' of the opening. Figure 12.6(b) confirms this fact since bus ③ would be isolated from the reference by opening the connection between bus ② and bus ③.

One open conductor

In this example Eq. (12.38) becomes

$$V_a^{(0)} = V_a^{(2)} = V_a^{(1)} = I_{23} \frac{Z_{pp'}^{(1)} Z_{pp'}^{(2)}}{Z_{pp'}^{(1)} + Z_{pp'}^{(2)}}$$

$$= (0.4 - j0.3)\frac{(j0.7120)(j0.7120)}{j0.7120 + j0.7120}$$

$$= 0.1068 + j0.1424 \text{ per unit}$$

and from Eqs. (12.27) we now calculate the symmetrical components of the voltage at bus ③:

$$\Delta V_3^{(1)} = \Delta V_3^{(2)} = \frac{Z_{32}^{(1)} - Z_{33}^{(1)}}{Z_1} V_a^{(1)} = \left(\frac{j0.1104 - j0.1696}{j0.15}\right)(0.1068 + j0.1424)$$

$$= -0.0422 - j0.0562 \text{ per unit}$$

$$\Delta V_3^{(0)} = \frac{Z_{32}^{(0)} - Z_{33}^{(0)}}{Z_0} V_a^{(0)} = \left(\frac{j0.08 - j0.58}{j0.50}\right)(0.1068 + j0.1424)$$

$$= -0.1068 - j0.1424 \text{ per unit}$$

$$\Delta V_3 = \Delta V_3^{(0)} + \Delta V_3^{(1)} + \Delta V_3^{(2)} = -0.1068 - j0.1424 - 2(0.0422 + j0.0562)$$

$$= -0.1912 - j0.2548 \text{ per unit}$$

Since the prefault voltage at bus ③ equals 1.0 + j0.0, the new voltage at bus ③ is

$$V_3' = V_3 + \Delta V_3 = (1.0 + j0.0) + (-0.1912 - j0.2548)$$

$$= 0.8088 - j0.2548 = 0.848\underline{/-17.5^\circ} \text{ per unit}$$

Two open conductors

Inserting the infinite impedance of the zero-sequence network in *series* between points p and p' of the positive-sequence network causes an open circuit in the latter. No power transfer can occur in the system—confirmation of the fact that power cannot be transferred by only one phase conductor of the transmission line in this case since the zero-sequence network offers no return path for current.

12.7 SUMMARY

If the emfs in a positive-sequence network like that shown in Fig. 12.2 are replaced by short circuits, the impedance between the fault bus ⓚ and the reference node is the positive-sequence impedance $Z_{kk}^{(1)}$ in the equation developed for faults on a power system and is the series impedance of the Thévenin equivalent of the circuit between bus ⓚ and the reference node. Thus, we can regard $Z_{kk}^{(1)}$ as a single impedance or the entire positive-sequence network between bus ⓚ and the reference with no emfs present. If the voltage V_f is connected in series with this modified positive-sequence network, the resulting circuit, shown in Fig. 12.2(e), is the Thévenin equivalent of the original positive-sequence network. The circuits shown in Fig. 12.2 are equivalent only in their effect on any external connections made between bus ⓚ and the reference node of the original networks. We can easily see that no current flows in the branches of the equivalent circuit in the absence of an external connection, but current will flow in the branches of the *original* positive-sequence network if any difference exists in the phase or magnitude of the two emfs in the network. In Fig. 12.2(b) the current flowing in the branches in the absence of an external connection is the prefault load current.

When the other sequence networks are interconnected with the positive-sequence network of Fig. 12.2(b) or its equivalent shown in Fig. 12.2(e), the current flowing out of the network or its equivalent is $I_{fa}^{(1)}$ and the voltage between bus ⓚ and the reference is $V_{ka}^{(1)}$. With such an external connection, the current in any branch of the original positive-sequence network of Fig. 12.2(b) is the positive-sequence current in phase a of that branch during the fault. The prefault component of this current is included. The current in any branch of the Thévenin equivalent of Fig. 12.2(e), however, is only that portion of the actual positive-sequence current found by apportioning $I_{fa}^{(1)}$ of the fault among the branches represented by $Z_{kk}^{(1)}$ according to their impedances and does not include the prefault component.

In the preceding sections we have seen that the Thévenin equivalents of the sequence networks of a power system can be interconnected so that solving

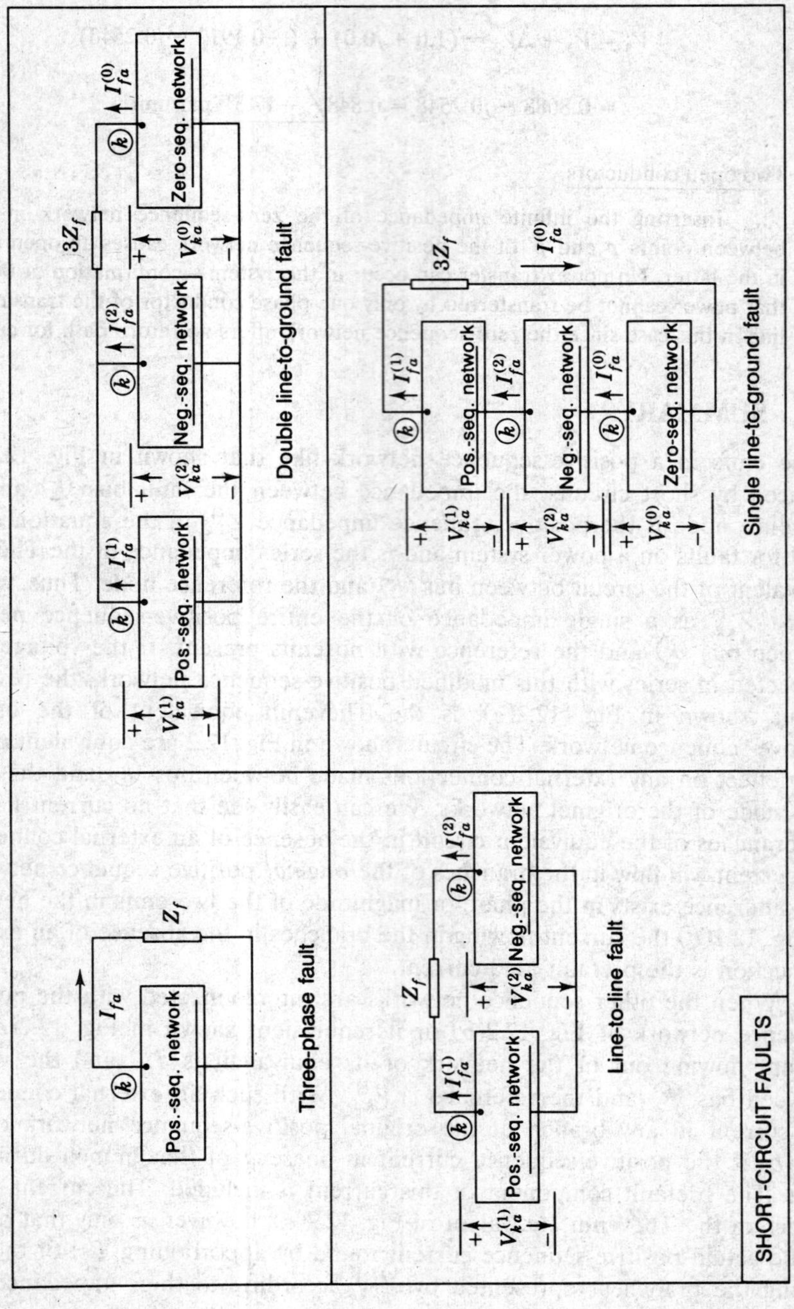
SHORT-CIRCUIT FAULTS
Three-phase fault
Pos.-seq. network
I_{fa}
Z_f
Line-to-line fault
$V_{ka}^{(1)}$
$I_{fa}^{(1)}$
$V_{ka}^{(2)}$
$I_{fa}^{(2)}$
Neg.-seq. network
Double line-to-ground fault
Zero-seq. network
$V_{ka}^{(0)}$
$I_{fa}^{(0)}$
$3Z_f$
Single line-to-ground fault

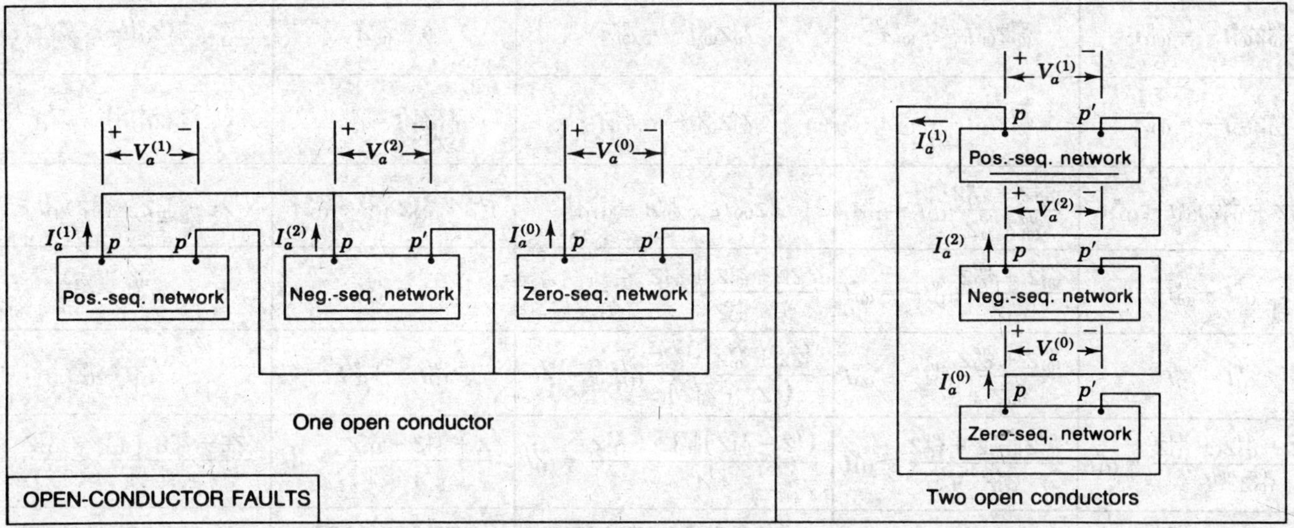

FIGURE 12.31
Summary of the connections of the sequence networks to simulate various types of short-circuit faults through impedance Z_f. $V_{ka}^{(0)}$, $V_{ka}^{(1)}$, and $V_{ka}^{(2)}$ are the symmetrical components of phase a voltage at the fault bus ⓚ with respect to reference. $V_a^{(0)}$, $V_a^{(1)}$, and $V_a^{(2)}$ are the symmetrical components of the phase a voltage drops across the open-circuit points p and p'.

TABLE 12.1
Summary of equations for sequence voltages and currents at the fault point for various types of faults

	Short-circuit faults			Open-circuit faults	
	Line-to-ground fault	Line-to-line fault	Double line-to-ground fault	One open conductor	Two open conductors
Sequence currents	$I_{fa}^{(1)} = \dfrac{V_f}{Z_{kk}^{(1)} + Z_{kk}^{(2)} + Z_{kk}^{(0)} + 3Z_f}$	$I_{fa}^{(1)} = \dfrac{V_f}{Z_{kk}^{(1)} + Z_{kk}^{(2)} + Z_f}$	$I_{fa}^{(1)} = \dfrac{V_f}{Z_{kk}^{(1)} + Z_{kk}^{(2)} \Vert \left(Z_{kk}^{(0)} + Z_f\right)}$	$I_a^{(1)} = \dfrac{I_{mn} Z_{pp'}^{(1)}}{Z_{pp'}^{(1)} + Z_{pp'}^{(2)} \Vert Z_{pp'}^{(0)}}$	$I_a^{(1)} = \dfrac{I_{mn} Z_{pp'}^{(1)}}{Z_{pp'}^{(1)} + Z_{pp'}^{(2)} + Z_{pp'}^{(0)}}$
	$I_{fa}^{(2)} = I_{fa}^{(1)}$	$I_{fa}^{(2)} = -I_{fa}^{(1)}$	$I_{fa}^{(2)} = -I_{fa}^{(1)} \dfrac{\left(Z_{kk}^{(0)} + 3Z_f\right)}{Z_{kk}^{(2)} + Z_{kk}^{(0)} + 3Z_f}$	$I_a^{(2)} = -I_a^{(1)} \dfrac{Z_{pp'}^{(0)}}{Z_{pp'}^{(2)} + Z_{pp'}^{(0)}}$	$I_a^{(2)} = I_a^{(1)}$
	$I_{fa}^{(0)} = I_{fa}^{(1)}$	$I_{fa}^{(0)} = 0$	$I_{fa}^{(0)} = -I_{fa}^{(1)} \dfrac{Z_{kk}^{(2)}}{Z_{kk}^{(2)} + Z_{kk}^{(0)} + 3Z_f}$	$I_a^{(0)} = -I_a^{(1)} \dfrac{Z_{pp'}^{(2)}}{Z_{pp'}^{(2)} + Z_{pp'}^{(0)}}$	$I_a^{(0)} = I_a^{(1)}$
Sequence voltages	$V_{ka}^{(1)} = I_{fa}^{(1)}(Z_{kk}^{(2)} + Z_{kk}^{(0)} + 3Z_f)$	$V_{ka}^{(1)} = I_{fa}^{(1)}(Z_{kk}^{(2)} + Z_f)$	$V_{ka}^{(1)} = V_{ka}^{(0)} - 3I_{fa}^{(0)} Z_f$	$V_a^{(1)} = I_a^{(1)} \dfrac{Z_{pp'}^{(2)} Z_{pp'}^{(0)}}{Z_{pp'}^{(2)} + Z_{pp'}^{(0)}}$	$V_a^{(1)} = I_a^{(1)}(Z_{pp'}^{(2)} + Z_{pp'}^{(0)})$
	$V_{ka}^{(2)} = -I_{fa}^{(1)} Z_{kk}^{(2)}$	$V_{ka}^{(2)} = I_{fa}^{(1)} Z_{kk}^{(2)}$	$V_{ka}^{(2)} = -I_{fa}^{(2)} Z_{kk}^{(2)}$	$V_a^{(2)} = -I_a^{(2)} Z_{pp'}^{(2)}$	$V_a^{(2)} = -I_a^{(2)} Z_{pp'}^{(2)}$
	$V_{ka}^{(0)} = -I_{fa}^{(1)} Z_{kk}^{(0)}$	$V_{ka}^{(0)} = 0$	$V_{ka}^{(0)} = -I_{fa}^{(0)} Z_{kk}^{(0)}$	$V_a^{(0)} = -I_a^{(0)} Z_{pp'}^{(0)}$	$V_a^{(0)} = -I_a^{(0)} Z_{pp'}^{(0)}$

Note: "‖" implies parallel combination of impedances.
$V_{ka}^{(0)}$, $V_{ka}^{(1)}$, and $V_{ka}^{(2)}$ are the symmetrical components of phase a voltage at the fault bus ⓚ with respect to the reference.
$V_a^{(0)}$, $V_a^{(1)}$, and $V_a^{(2)}$ are the symmetrical components of the phase a voltage drops across the open-circuit points p and p'.

the resulting network yields the symmetrical components of current and voltage *at the fault*. The connections of the sequence networks to simulate various types of short-circuit faults, including a symmetrical three-phase fault, are shown in Fig. 12.31. The sequence networks are indicated schematically by rectangles enclosing a heavy line to represent the reference of the network and a point marked bus ⓚ to represent the point in the network where the fault occurs. The positive-sequence network contains emfs that represent the internal voltages of the machines.

Regardless of the prefault voltage profile or the particular type of short-circuit fault occurring, the only current which causes positive-sequence voltage *changes* at the buses of the system is the symmetrical component $I_{fa}^{(1)}$ of the current I_{fa} coming out of phase a of the system at the fault bus ⓚ. These positive-sequence voltage changes can be calculated simply by multiplying column k of the positive-sequence bus impedance matrix $\mathbf{Z}_{\text{bus}}^{(1)}$ by the *injected* current $-I_{fa}^{(1)}$. Similarly, the negative- and zero-sequence components of the voltage changes due to the short-circuit fault on the system are caused, respectively, by the symmetrical components $I_{fa}^{(2)}$ and $I_{fa}^{(0)}$ of the fault current I_{fa} *out* of bus ⓚ. These sequence voltage changes are also calculated by multiplying columns ⓚ of $\mathbf{Z}_{\text{bus}}^{(2)}$ and $\mathbf{Z}_{\text{bus}}^{(0)}$ by the respective current injections $-I_{fa}^{(2)}$ and $-I_{fa}^{(0)}$.

In a very real sense, therefore, there is only one procedure for calculating the symmetrical components of the voltage changes at the buses of the system when a short-circuit fault occurs at bus ⓚ—that is, to find $I_{fa}^{(0)}$, $I_{fa}^{(1)}$, and $I_{fa}^{(2)}$ and to multiply columns ⓚ of the corresponding bus impedance matrices by the negative values of these currents. For the common types of short-circuit faults the only differences in the calculations concern the method of simulating the fault at bus ⓚ and of formulating the equations for $I_{fa}^{(0)}$, $I_{fa}^{(1)}$, and $I_{fa}^{(2)}$. The connections of the Thévenin equivalents of the sequence networks, which provide a ready means of deriving the equations for $I_{fa}^{(0)}$, $I_{fa}^{(1)}$, and $I_{fa}^{(2)}$, are summarized in Fig. 12.31 and the equations themselves are set forth in Table 12.1.

Faults due to open conductors involve *two* injections into each sequence network at the buses nearest the opening in the conductor. Otherwise, the procedure for calculating sequence voltage changes in the system is the same as that for short-circuit faults. Equations for the sequence voltages and currents at the fault are also summarized in Table 12.1.

The reader is reminded of the need to adjust the phase angles of the symmetrical components of the currents and the voltages in those parts of the system which are separated from the fault bus by Δ-Y transformers.

PROBLEMS

12.1. A 60-Hz turbogenerator is rated 500 MVA, 22 kV. It is Y-connected and solidly grounded and is operating at rated voltage at no load. It is disconnected from the rest of the system. Its reactances are $X_d'' = X_1 = X_2 = 0.15$ and $X_0 = 0.05$ per unit. Find the ratio of the subtransient line current for a single line-to-ground fault to the subtransient line current for a symmetrical three-phase fault.

12.2. Find the ratio of the subtransient line current for a line-to-line fault to the subtransient current for a symmetrical three-phase fault on the generator of Prob. 12.1.

12.3. Determine the inductive reactance in ohms to be inserted in the neutral connection of the generator of Prob. 12.1 to limit the subtransient line current for a single line-to-ground fault to that for a three-phase fault.

12.4. With the inductive reactance found in Prob. 12.3 inserted in the neutral of the generator of Prob. 12.1, find the ratios of the subtransient line currents for the following faults to the subtransient line current for a three-phase fault: (*a*) single line-to-ground fault, (*b*) line-to-line fault, and (*c*) double line-to-ground fault.

12.5. How many ohms of resistance in the neutral connection of the generator of Prob. 12.1 would limit the subtransient line current for a single line-to-ground fault to that for a three-phase fault?

12.6. A generator rated 100 MVA, 20 kV has $X_d'' = X_1 = X_2 = 20\%$ and $X_0 = 5\%$. Its neutral is grounded through a reactor of 0.32 Ω. The generator is operating at rated voltage without load and is disconnected from the system when a single line-to-ground fault occurs at its terminals. Find the subtransient current in the faulted phase.

12.7. A 100-MVA 18-kV turbogenerator having $X_d'' = X_1 = X_2 = 20\%$ and $X_0 = 5\%$ is about to be connected to a power system. The generator has a current-limiting reactor of 0.162 Ω in the neutral. Before the generator is connected to the system, its voltage is adjusted to 16 kV when a double line-to-ground fault develops at terminals *b* and *c*. Find the initial symmetrical root-mean-square (rms) current in the ground and in line *b*.

12.8. The reactances of a generator rated 100 MVA, 20 kV are $X_d'' = X_1 = X_2 = 20\%$ and $X_0 = 5\%$. The generator is connected to a Δ-Y transformer rated 100 MVA, 20Δ-230Y kV, with a reactance of 10%. The neutral of the transformer is solidly grounded. The terminal voltage of the generator is 20 kV when a single line-to-ground fault occurs on the open-circuited, high-voltage side of the transformer. Find the initial symmetrical rms current in all phases of the generator.

12.9. A generator supplies a motor through a Y-Δ transformer. The generator is connected to the Y side of the transformer. A fault occurs between the motor terminals and the transformer. The symmetrical components of the subtransient current in the motor flowing toward the fault are

$$I_a^{(1)} = -0.8 - j2.6 \text{ per unit}$$

$$I_a^{(2)} = -j2.0 \text{ per unit}$$

$$I_a^{(0)} = -j3.0 \text{ per unit}$$

From the transformer toward the fault

$$I_a^{(1)} = 0.8 - j0.4 \text{ per unit}$$

$$I_a^{(2)} = -j1.0 \text{ per unit}$$

$$I_a^{(0)} = 0 \text{ per unit}$$

Assume $X_d'' = X_1 = X_2$ for both the motor and the generator. Describe the type of fault. Find (*a*) the prefault current, if any, in line *a*; (*b*) the subtransient fault current in per unit; and (*c*) the subtransient current in each phase of the generator in per unit.

12.10. Using Fig. 12.18, calculate the bus impedance matrices $\mathbf{Z}_{\text{bus}}^{(1)}$, $\mathbf{Z}_{\text{bus}}^{(2)}$, and $\mathbf{Z}_{\text{bus}}^{(0)}$ for the network of Example 12.6.

12.11. Solve for the subtransient current in a single line-to-ground fault first on bus ① and then on bus ② of the network of Example 12.6. Use the bus impedance matrices of Prob. 12.10. Also, find the voltages to neutral at bus ② with the fault at bus ①.

12.12. Calculate the subtransient currents in all parts of the system of Example 12.6 with prefault current neglected if the fault on the low-voltage side of the transformer is a line-to-line fault. Use $\mathbf{Z}_{\text{bus}}^{(1)}$, $\mathbf{Z}_{\text{bus}}^{(2)}$, and $\mathbf{Z}_{\text{bus}}^{(0)}$ of Prob. 12.10.

12.13. Repeat Prob. 12.12 for a double line-to-ground fault.

12.14. Each of the machines connected to the two high-voltage buses shown in the single-line diagram of Fig. 12.32 is rated 100 MVA, 20 kV with reactances of $X_d'' = X_1 = X_2 = 20\%$ and $X_0 = 4\%$. Each three-phase transformer is rated 100 MVA, 345Y/20Δ kV, with leakage reactance of 8%. On a base of 100 MVA, 345 kV the reactances of the transmission line are $X_1 = X_2 = 15\%$ and $X_0 = 50\%$. Find the 2 × 2 bus impedance matrix for each of the three sequence networks. If no prefault current is flowing in the network, find the subtransient current to ground for a double line-to-ground fault on lines *B* and *C* at bus ①. Repeat for a fault at bus ②. When the fault is at bus ②, determine the current in phase *b* of machine 2 if the lines are named so that $V_A^{(1)}$ leads $V_a^{(1)}$ by 30°. If the phases are named so that $I_a^{(1)}$ leads $I_A^{(1)}$ by 30°, what letter (*a*, *b*, or *c*) would identify the phase of machine 2 which would carry the current found for phase *b* above?

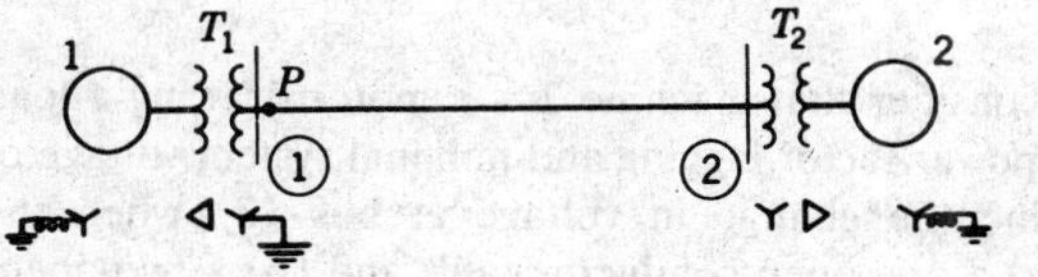

FIGURE 12.32
Single-line diagram for Prob. 12.14.

12.15. Two generators G_1 and G_2 are connected, respectively, through transformers T_1 and T_2 to a high-voltage bus which supplies a transmission line. The line is open at the far end at which point *F* a fault occurs. The prefault voltage at point *F* is

515 kV. Apparatus ratings and reactances are

G_1	1000 MVA, 20 kV, $X_s = 100\%$	$X_d'' = X_1 = X_2 = 10\%$	$X_0 = 5\%$
G_2	800 MVA, 22 kV, $X_s = 120\%$	$X_d'' = X_1 = X_2 = 15\%$	$X_0 = 8\%$
T_1	1000 MVA, 500Y/20Δ kV, $X = 17.5\%$		
T_2	800 MVA, 500Y/22Y kV, $X = 16.0\%$		
Line	$X_1 = 15\%$, $X_0 = 40\%$ on a base of 1500 MVA, 500 kV		

The neutral of G_1 is grounded through a reactance of 0.04 Ω. The neutral of G_2 is not grounded. Neutrals of all transformers are solidly grounded. Work on a base of 1000 MVA, 500 kV in the transmission line. Neglect prefault current and find subtransient current (*a*) in phase *c* of G_1 for a three-phase fault at *F*, (*b*) in phase *B* at *F* for a line-to-line fault on lines *B* and *C*, (*c*) in phase *A* at *F* for a line-to-ground fault on line *A*, and (*d*) in phase *c* of G_2 for a line-to-ground fault on line *A*. Assume $V_A^{(1)}$ leads $V_a^{(1)}$ by 30° in T_1.

12.16. In the network shown in Fig. 10.17 Y-Y-connected transformers, each with grounded neutrals, are at the ends of each transmission line that is not terminating at bus ③. The transformers connecting the lines to bus ③ are Y-Δ, with the neutral of the Y solidly grounded and the Δ sides connected to bus ③. All line reactances shown in Fig. 10.17 between buses include the reactances of the transformers. Zero-sequence values for these lines including transformers are 2.0 times those shown in Fig. 10.17.

Both generators are Y-connected. Zero-sequence reactances of the generators connected to buses ① and ③ are 0.04 and 0.08 per unit, respectively. The neutral of the generator at bus ① is connected to ground through a reactor of 0.02 per unit; the generator at bus ③ has a solidly grounded neutral.

Find the bus impedance matrices $\mathbf{Z}_{\text{bus}}^{(1)}$, $\mathbf{Z}_{\text{bus}}^{(2)}$, and $\mathbf{Z}_{\text{bus}}^{(0)}$ for the given network and then compute the subtransient current in per unit (*a*) in a single line-to-ground fault on bus ② and (*b*) in the faulted phase of line ①–②. Assume no prefault current is flowing and all prefault voltages at all the buses is $1.0\angle 0°$ per unit.

12.17. The network of Fig. 9.2 has the line data specified in Table 9.2. Each of the two generators connected to buses ① and ④ has $X_d'' = X_1 = X_2 = 0.25$ per unit. Making the usual simplifying assumptions of Sec. 10.6, determine the sequence matrices $\mathbf{Z}_{\text{bus}}^{(1)} = \mathbf{Z}_{\text{bus}}^{(2)}$ and use them to calculate

(*a*) The subtransient current in per unit in a line-to-line fault on bus ② of the network and

(*b*) The fault current contributions from line ①–② and line ③–②.

Assume that lines ①–② and ③–② are connected to bus ② directly (not through transformers) and that all positive- and negative-sequence reactances are identical.

12.18. In the system of Fig. 12.9(*a*), consider that machine 2 is a motor drawing a load equivalent to 80 MVA at 0.85 power-factor lagging and nominal system voltage of 345 kV at bus ③. Determine the change in voltage at bus ③ when the transmission line undergoes (*a*) a one-open-conductor fault and (*b*) a two-open-conductor fault along its span between buses ② and ③. Choose a base of 100 MVA, 345 kV in the transmission line. Consult Examples 12.1 and 12.2 for $\mathbf{Z}_{\text{bus}}^{(0)}$, $\mathbf{Z}_{\text{bus}}^{(1)}$, and $\mathbf{Z}_{\text{bus}}^{(2)}$.

CHAPTER 13

ECONOMIC OPERATION OF POWER SYSTEMS

Economic operation is very important for a power system to return a profit on the capital invested. Rates fixed by regulatory bodies and the importance of conservation of fuel place pressure on power companies to achieve maximum possible efficiency. Maximum efficiency minimizes the cost of a kilowatthour to the consumer and the cost to the company of delivering that kilowatthour in the face of constantly rising prices for fuel, labor, supplies, and maintenance.

Operational economics involving power generation and delivery can be subdivided into two parts—one dealing with minimum cost of power production called *economic dispatch* and the other dealing with *minimum-loss* delivery of the generated power to the loads. For any specified load condition economic dispatch determines the power output of each plant (and each *generating unit* within the plant) which will minimize the overall cost of fuel needed to serve the system load. Thus, economic dispatch focuses upon coordinating the production costs at all power plants operating on the system and is the major emphasis of this chapter. The minimum-loss problem can assume many forms depending on how control of the power flow in the system is exercised. The economic dispatch problem and also the minimum-loss problem can be solved by means of the *optimal power-flow* (OPF) program. The OPF calculation can be viewed as a sequence of conventional Newton-Raphson power-flow calculations in which certain controllable parameters are automatically adjusted to satisfy the network constraints while minimizing a specified objective function. In this chapter we consider the *classical* approach to economic dispatch and refer the interested reader to the citation below for details of the OPF approach.[1]

We study first the most economic distribution of the output of a plant between the generators, or units, within that plant. The method that we develop also applies to economic scheduling of plant outputs for a given loading of the system without consideration of transmission losses. Next, we express transmission loss as a function of the outputs of the various plants. Then, we determine how the output of each of the plants of a system is scheduled to achieve the minimum cost of power delivered to the load.

Because the total load of the power system varies throughout the day, coordinated control of the power plant outputs is necessary to ensure generation-to-load balance so that the system frequency will remain as close as possible to the nominal operating value, usually 50 or 60 Hz. Accordingly, the problem of *automatic generation control* (AGC) is developed from the steady-state viewpoint. Also, because of the daily load variation, the utility has to decide on the basis of economics which generators to start up, which to shut down, and in what order. The computational procedure for making such decisions, called *unit commitment*, is also developed at an introductory level in this chapter.

13.1 DISTRIBUTION OF LOAD BETWEEN UNITS WITHIN A PLANT

An early attempt at economic dispatch called for supplying power from only the most efficient plant at light loads. As load increased, power would be supplied by the most efficient plant until the point of maximum efficiency of that plant was reached. Then, for further increase in load the next most efficient plant would start to feed power to the system and a third plant would not be called upon until the point of maximum efficiency of the second plant was reached. Even with transmission losses neglected, this method fails to minimize cost.

To determine the economic distribution of load between the various generating units (consisting of a turbine, generator, and steam supply), the variable operating costs of the unit must be expressed in terms of the power output. Fuel cost is the principal factor in fossil-fuel plants, and cost of nuclear fuel can also be expressed as a function of output. We base our discussion on the economics of fuel cost with the realization that other costs which are a function of power output can be included in the expression for fuel cost. A typical input-output curve which is a plot of fuel input for a fossil-fuel plant in British thermal units (Btu) per hour versus power output of the unit in megawatts is shown in Fig. 13.1. The ordinates of the curve are converted to

[1] H. W. Dommel and W. F. Tinney, "Optimal Power-Flow Solutions," *IEEE Transactions on Power Apparatus and Systems*, vol. PAS-87, October 1968, pp. 1866–1876.

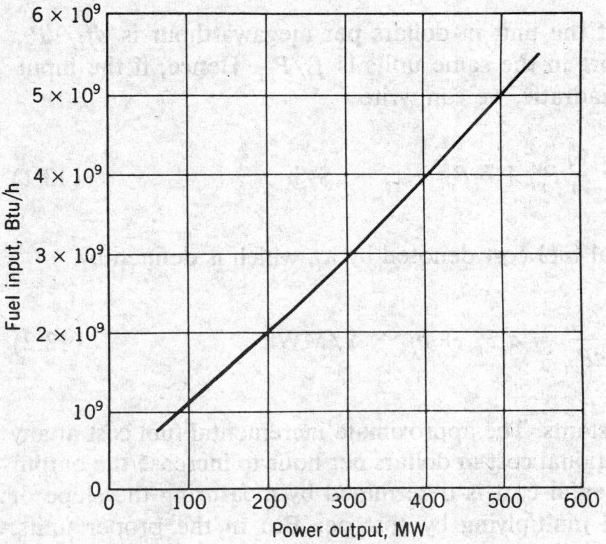

FIGURE 3.1
Input-output curve for a generating unit showing fuel input versus power output.

dollars per hour by multiplying the fuel input by the cost of fuel in dollars per million Btu.

If a line is drawn through the origin to any point on the input-output curve, the slope can be expressed in millions of Btu per hour divided by the output in megawatts, or the ratio of fuel input in Btu to energy output in kilowatthours. This ratio is called the *heat rate* and its reciprocal is the *fuel efficiency*. Hence, lower heat rates imply higher fuel efficiency. Maximum fuel efficiency occurs at that point where the slope of the line from the origin to a point on the curve is a minimum, that is, at the point where the line is tangent to the curve. For the unit whose input-output curve is shown in Fig. 13.1 the maximum efficiency is at an output of approximately 280 MW, which requires an input of 2.8×10^9 Btu/h. The heat rate is 10,000 Btu/kWh, and since 1 kWh = 3412 Btu, the fuel efficiency is approximately 34%. By comparison, when the output of the unit is 100 MW, the heat rate is 11,000 Btu/kWh and the fuel efficiency is 31%.

Of course, the fuel requirement for a given output is easily converted into dollars per megawatthour. As we shall see, the criterion for distribution of the load between any two units is based on whether increasing the load on one unit as the load is decreased on the other unit by the same amount results in an increase or decrease in total cost. Thus, we are concerned with *incremental fuel cost*, which is determined by the slopes of the input-output curves of the two units. If we express the ordinates of the input-output curve in dollars per hour and let

$$f_i = \text{input to Unit } i\text{, dollars per hour (\$/h)}$$

$$P_{gi} = \text{output of Unit } i\text{, megawatts (MW)}$$

the incremental fuel cost of the unit in dollars per megawatthour is df_i/dP_{gi}, whereas the *average fuel cost* in the same units is f_i/P_{gi}. Hence, if the input-output curve of Unit i is quadratic, we can write

$$f_i = \frac{a_i}{2}P_{gi}^2 + b_iP_{gi} + c_i \qquad \$/\text{h} \tag{13.1}$$

and the unit has incremental fuel cost denoted by λ_i, which is defined by

$$\lambda_i = \frac{df_i}{dP_{gi}} = a_iP_{gi} + b_i \qquad \$/\text{MWh} \tag{13.2}$$

where a_i, b_i, and c_i are constants. The approximate incremental fuel cost at any particular output is the additional cost in dollars per hour to increase the output by 1 MW. Actually, incremental cost is determined by measuring the slope of the input-output curve and multiplying by cost per Btu in the proper units. Since mills (tenths of a cent) per kilowatthour are equal to dollars per megawatthour, and since a kilowatt is a very small amount of power in comparison with the usual output of a unit of a steam plant, incremental fuel cost may be considered as the cost of fuel in *mills per hour* to supply an additional kilowatt output.

A typical plot of incremental fuel cost versus power output is shown in Fig. 13.2. This figure is obtained by measuring the slope of the input-output curve of Fig. 13.1 for various outputs and by applying a fuel cost of \$1.30 per million Btu. However, the cost of fuel in terms of Btu is not very predictable, and the reader should not assume that cost figures throughout this chapter are applicable at any particular time. Figure 13.2 shows that incremental fuel cost is quite linear with respect to power output over an appreciable range. In analytical work the curve is usually approximated by one or two straight lines. The dashed line in the figure is a good representation of the curve. The equation of the line is

$$\lambda_i = \frac{df_i}{dP_{gi}} = 0.0126P_{gi} + 8.9$$

so that when the power output is 300 MW, the incremental cost determined by the linear approximation is \$12.68/MWh. This value of λ_i is the approximate *additional cost* per hour of increasing the output P_{gi} by 1 MW and the *saving* in cost per hour of reducing the output P_{gi} by 1 MW. The actual incremental cost at 300 MW is \$12.50/MWh, but this power output is near the point of maximum deviation between the actual value and the linear approximation of incremental cost. For greater accuracy two straight lines may be drawn to represent this curve in its upper and lower range.

We now have the background to understand the principle of economic dispatch which guides distribution of load among the units within one or more

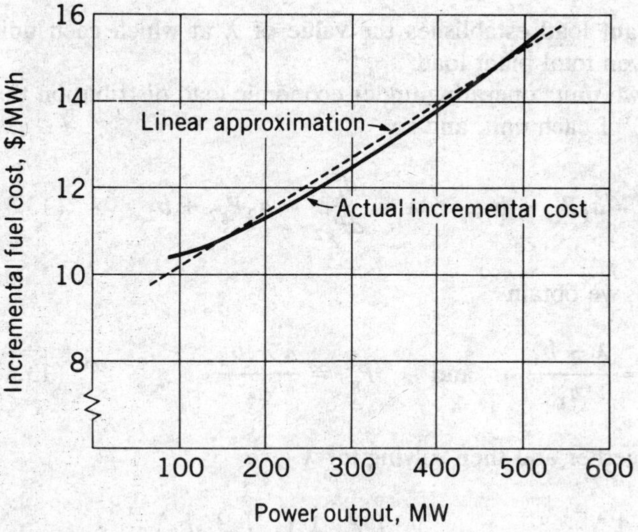

FIGURE 13.2
Incremental fuel cost versus power output for the unit whose input-output curve is shown in Fig. 13.1.

plants of the system. For instance, suppose that the total output of a particular plant is supplied by two units and that the division of load between these units is such that the incremental fuel cost of one is higher than that of the other. Now suppose that some of the load is transferred from the unit with the higher incremental cost to the unit with the lower incremental cost. Reducing the load on the unit with the higher incremental cost will result in a greater reduction of cost than the increase in cost for adding the same amount of load to the unit with the lower incremental cost. The transfer of load from one to the other can be continued with a reduction in total fuel cost until the incremental fuel costs of the two units are equal. The same reasoning can be extended to a plant with more than two units. Thus, for economical division of load between units *within* a plant, the criterion is that *all units must operate at the same incremental fuel cost*.

When the incremental fuel cost of each of the units in a plant is nearly linear with respect to power output over a range of operation under consideration, equations that represent incremental fuel costs as linear functions of power output will simplify the computations. An economic dispatch schedule for assigning loads to each unit in a plant can be prepared by:

1. Assuming various values of total plant output,
2. Calculating the corresponding incremental fuel cost λ of the plant, and
3. Substituting the value of λ for λ_i in the equation for the incremental fuel cost of each unit to calculate its output.

A curve of λ versus plant load establishes the value of λ at which each unit should operate for a given total plant load.

For a plant with two units operating under economic load distribution the λ of the plant equals λ_i of each unit, and so

$$\lambda = \frac{df_1}{dP_{g1}} = a_1 P_{g1} + b_1; \qquad \lambda = \frac{df_2}{dP_{g2}} = a_2 P_{g2} + b_2 \tag{13.3}$$

Solving for P_{g1} and P_{g2}, we obtain

$$P_{g1} = \frac{\lambda - b_1}{a_1} \qquad \text{and} \qquad P_{g2} = \frac{\lambda - b_2}{a_2} \tag{13.4}$$

Adding these results together and then solving for λ give

$$\lambda = \left(\sum_{i=1}^{2} \frac{1}{a_i} \right)^{-1} (P_{g1} + P_{g2}) + \left(\sum_{i=1}^{2} \frac{1}{a_i} \right)^{-1} \left(\sum_{i=1}^{2} \frac{b_i}{a_i} \right) \tag{13.5}$$

or

$$\lambda = a_T P_{gT} + b_T \tag{13.6}$$

where

$$a_T = \left(\sum_{i=1}^{2} \frac{1}{a_i} \right)^{-1}$$

$$b_T = a_T \left(\sum_{i=1}^{2} \frac{b_i}{a_i} \right)$$

and

$$P_{gT} = (P_{g1} + P_{g2})$$

is the total plant output. Equation (13.6) is a closed-form solution for λ which applies to a plant with more than two units on economic dispatch when the appropriate number of terms is added to the summations of Eq. (13.5). For instance, if the plant has K units operating on economic dispatch, then the coefficients of Eq. (13.6) are given by

$$a_T = \left(\sum_{i=1}^{K} \frac{1}{a_i} \right)^{-1} = \left(\frac{1}{a_1} + \frac{1}{a_2} + \cdots + \frac{1}{a_K} \right)^{-1} \tag{13.7}$$

$$b_T = a_T \sum_{i=1}^{K} \frac{b_i}{a_i} = a_T \left(\frac{b_1}{a_1} + \frac{b_2}{a_2} + \cdots + \frac{b_K}{a_K} \right) \tag{13.8}$$

and the total plant output P_{gT} is $(P_{g1} + P_{g2} + \cdots + P_{gK})$. The individual output of each of the K units is then calculated from the common value of λ

given by Eq. (13.6). If maximum and minimum loads are specified for each unit, some units will be unable to operate at the same incremental fuel cost as the other units and still remain within the limits specified for light and heavy loads. Suppose that this occurs for $K = 4$ and that the calculated value of P_{g4} violates a specified limit of Unit 4. We then discard the calculated outputs of all four units and set the operating value of P_{g4} equal to the violated limit of Unit 4. Returning to Eq. (13.6), we recalculate the coefficients a_T and b_T for the other three units and set the *effective* economic dispatch value of P_{gT} equal to the total plant load minus the limit value of P_{g4}. The resulting value of λ then governs the economic dispatch of Units 1, 2, and 3 when the actual plant output is to be increased or decreased, so long as Unit 4 remains as the only unit at a limit.

The equal-incremental-fuel-cost criterion which we have developed intuitively, and now illustrate numerically, will be established mathematically in Sec. 13.2.

Example 13.1. Incremental fuel costs in dollars per megawatthour for a plant consisting of two units are given by

$$\lambda_1 = \frac{df_1}{dP_{g1}} = 0.0080P_{g1} + 8.0 \qquad \lambda_2 = \frac{df_2}{dP_{g2}} = 0.0096P_{g2} + 6.4$$

Assume that both units are operating at all times, that total load varies from 250 to 1250 MW, and that maximum and minimum loads on each unit are to be 625 and 100 MW, respectively. Find the incremental fuel cost of the *plant* and the allocation of load between units for the minimum cost of various total loads.

Solution. At light loads Unit 1 will have the higher incremental fuel cost and operate at its lower limit of 100 MW for which df_1/dP_{g1} is \$8.8/MWh. When the output of Unit 2 is also 100 MW, df_2/dP_{g2} is \$7.36/MWh. Therefore, as plant output increases, the additional load should come from Unit 2 until df_2/dP_{g2} equals \$8.8/MWh. Until that point is reached, the incremental fuel cost λ of the *plant* is determined by Unit 2 alone. When the plant load is 250 MW, Unit 2 will supply 150 MW with df_2/dP_{g2} equal to \$7.84/MWh. When df_2/dP_{g2} equals \$8.8/MWh,

$$0.0096P_{g2} + 6.4 = 8.8$$

$$P_{g2} = \frac{2.4}{0.0096} = 250 \text{ MW}$$

and the total plant output P_{gT} is 350 MW. From this point on the required output of each unit for economic load distribution is found by assuming various values of P_{gT}, calculating the corresponding plant λ from Eq. (13.6), and substituting the value of λ in Eqs. (13.4) to compute each unit's output. Results are shown in Table 13.1. When P_{gT} is in the range from 350 to 1175 MW, the plant λ is determined by Eq. (13.6). At $\lambda = 12.4$ Unit 2 is operating at its upper limit and additional load

TABLE 13.1
The plant λ and outputs of each unit for various values of total output P_{gT} for Example 13.1

Plant		Unit 1	Unit 2
P_{gT}, MW	λ, \$/MWh	P_{g1}, MW	P_{g2}, MW
250	7.84	100†	150
350	8.80	100†	250
500	9.45	182	318
700	10.33	291	409
900	11.20	400	500
1100	12.07	509	591
1175	12.40	550	625†
1250	13.00	625	625†

†Indicates the output of the unit at its minimum (or maximum) limit and the plant λ is then equal to the incremental fuel cost of the unit not at a limit.

must come from Unit 1, which then determines the plant λ. Figure 13.3 shows plant λ plotted versus plant output.

If we wish to know the distribution of load between the units for a plant output of 500 MW, we could plot the output of each individual unit versus plant output, as shown in Fig. 13.4, from which each unit's output can be read for any plant output. The correct output of each of many units can be easily computed from Eq. (13.6) by requiring all unit incremental costs to be equal for any total plant output. For the two units of the example, given a total output of 500 MW,

$$P_{gT} = P_{g1} + P_{g2} = 500 \text{ MW}$$

$$a_T = \left(\frac{1}{a_1} + \frac{1}{a_2}\right)^{-1} = \left(\frac{1}{0.008} + \frac{1}{0.0096}\right)^{-1} = 4.363636 \times 10^{-3}$$

$$b_T = a_T\left(\frac{b_1}{a_1} + \frac{b_2}{a_2}\right) = a_T\left(\frac{8.0}{0.008} + \frac{6.4}{0.0096}\right) = 7.272727$$

and then for each unit

$$\lambda = a_T P_{gT} + b_T = 9.454545 \text{ \$/MWh}$$

which yields

$$P_{g1} = \frac{\lambda - b_1}{a_1} = \frac{9.454545 - 8.0}{0.008} = 181.8182 \text{ MW}$$

$$P_{g2} = \frac{\lambda - b_2}{a_2} = \frac{9.454545 - 6.4}{0.0096} = 318.1818 \text{ MW}$$

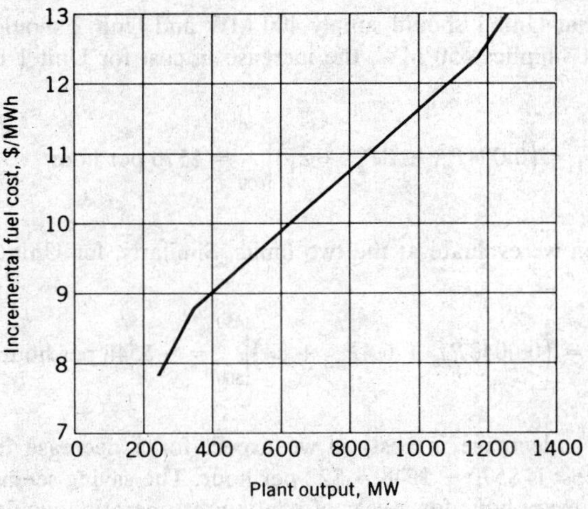

FIGURE 13.3
Incremental fuel cost versus plant output with total plant load economically distributed between units, as found in Example 13.1.

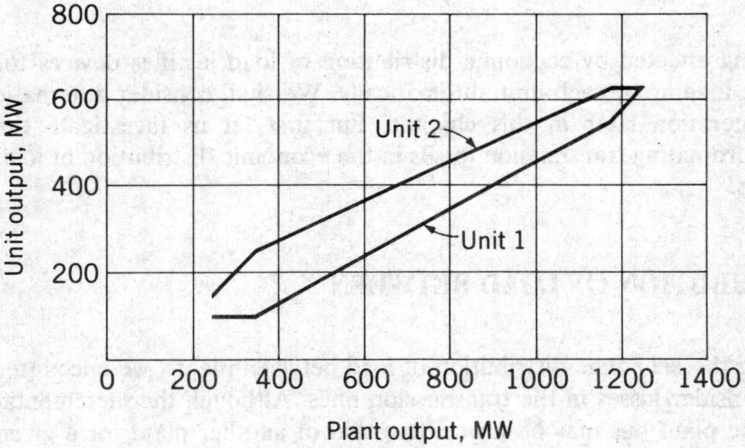

FIGURE 13.4
Output of each unit versus plant output for economical operation of the plant of Example 13.1.

Such accuracy, however, is not necessary because of the uncertainty in determining exact costs and the use of an approximate equation in this example to express the incremental costs.

The savings effected by economic distribution of load rather than some arbitrary distribution can be found by integrating the expression for incremental fuel cost and by comparing increases and decreases of cost for the units as load is shifted from the most economical allocation.

Example 13.2. Determine the saving in fuel cost in dollars per hour for the economic distribution of a total load of 900 MW between the two units of the plant described in Example 13.1 compared with equal distribution of the same total load.

Solution. Table 13.1 shows that Unit 1 should supply 400 MW and Unit 2 should supply 500 MW. If each unit supplies 450 MW, the increase in cost for Unit 1 is

$$\int_{400}^{450}(0.008P_{g1}+8)\,dP_{g1}=\left(0.004P_{g1}^2+8P_{g1}+c_1\right)\Big|_{400}^{450}=\$570\text{ per hour}$$

The constant c_1 cancels when we evaluate at the two limits. Similarly, for Unit 2

$$\int_{500}^{450}(0.0096P_{g2}+6.4)\,dP_{g2}=\left(0.0048P_{g2}^2+6.4P_{g2}+c_2\right)\Big|_{500}^{450}=-\$548\text{ per hour}$$

The negative sign indicates a decrease in cost, as we expect for a decrease in output. The net increase in cost is \$570 − \$548 = \$22 per hour. The saving seems small, but this amount saved every hour for a year of continuous operation would reduce fuel cost by \$192,720 for the year.

The saving effected by economic distribution of load justifies devices for controlling the loading of each unit automatically. We shall consider automatic control of generation later in this chapter. But first let us investigate the problem of coordinating transmission losses in the economic distribution of load between plants.

13.2 DISTRIBUTION OF LOAD BETWEEN PLANTS

In determining the economic distribution of load between plants, we encounter the need to consider losses in the transmission lines. Although the incremental fuel cost at one plant bus may be lower than that of another plant for a given distribution of load between the plants, the plant with the lower incremental cost at its bus may be much farther from the load center. The losses in transmission from the plant having the lower incremental cost may be so great that economy may dictate lowering the load at the plant with the lower incremental cost and increasing it at the plant with the higher incremental cost. Thus, we need to coordinate transmission loss into the scheduling of the output of each plant for maximum economy at a given level of system load.

For a system with K generating units let

$$f=f_1+f_2+\cdots+f_K=\sum_{i=1}^{K}f_i \tag{13.9}$$

where f is the *cost function* giving the total cost of all the fuel for the entire system and is the sum of the fuel costs of the individual units $f_1, f_2, \ldots, f_K$. The total megawatt power input to the network from all the units is the sum

$$P_{g1} + P_{g2} + \cdots + P_{gK} = \sum_{i=1}^{K} P_{gi} \tag{13.10}$$

where $P_{g1}, P_{g2}, \ldots, P_{gK}$ are the individual outputs of the units injected into the network. The total fuel cost f of the system is a function of all the power plant outputs. The constraining equation on the minimum value of f is given by the power balance of Eq. (9.10), which for the present purpose we rewrite in the form

$$P_L + P_D - \sum_{i=1}^{K} P_{gi} = 0 \tag{13.11}$$

where $P_D = \sum_{i=1}^{N} P_{di}$ is the total power received by the loads and P_L is the transmission loss of the system. Our objective is to obtain a minimum f for a fixed system load P_D subject to the power-balance constraint of Eq. (13.11). We now present the procedure for solving such minimization problems called the *method of Lagrange multipliers*.

The new cost function F is formed by combining the total fuel cost and the equality constraint of Eq. (13.11) in the following manner:

$$F = (f_1 + f_2 + \cdots + f_K) + \lambda\left(P_L + P_D - \sum_{i=1}^{K} P_{gi}\right) \tag{13.12}$$

The augmented cost function F is often called the *lagrangian*, and we shall see that the parameter λ, which we now call the *Lagrange multiplier*, is the effective incremental fuel cost of the *system* when transmission-line losses are taken into account. When f_i is given in dollars per hour and P is in megawatts, F and λ are expressed in dollars per hour and dollars per megawatthour, respectively. The original problem of minimizing f subject to Eq. (13.11) is transformed by means of Eq. (13.12) into an unconstrained problem in which it is required to minimize F with respect to λ and the generator outputs. Therefore, for minimum cost we require the derivative of F with respect to each P_{gi} to equal zero, and so

$$\frac{\partial F}{\partial P_{gi}} = \frac{\partial}{\partial P_{gi}}\left[(f_1 + f_2 + \cdots + f_K) + \lambda\left(P_L + P_D - \sum_{i=1}^{K} P_{gi}\right)\right] = 0 \tag{13.13}$$

Since P_D is fixed and the fuel cost of any one unit varies only if the power output of that unit is varied, Eq. (13.13) yields

$$\frac{\partial F}{\partial P_{gi}} = \frac{\partial f_i}{\partial P_{gi}} + \lambda\left(\frac{\partial P_L}{\partial P_{gi}} - 1\right) = 0 \tag{13.14}$$

for each of the generating-unit outputs $P_{g1}, P_{g2}, \ldots, P_{gK}$. Because f_i depends on only P_{gi}, the partial derivative of f_i can be replaced by the full derivative, and Eq. (13.14) then gives

$$\lambda = \left(\frac{1}{1 - \dfrac{\partial P_L}{\partial P_{gi}}}\right)\frac{df_i}{dP_{gi}} \tag{13.15}$$

for every value of i. This equation is often written in the form

$$\lambda = L_i \frac{df_i}{dP_{gi}} \tag{13.16}$$

where L_i is called the *penalty factor* of plant i and is given by

$$L_i = \frac{1}{1 - \dfrac{\partial P_L}{\partial P_{gi}}} \tag{13.17}$$

The result of Eq. (13.16) means that minimum fuel cost is obtained when the incremental fuel cost of each unit multiplied by its penalty factor is the same for all generating units in the system. The products $L_i(df_i/dP_{gi})$ are each equal to λ, called the *system* λ, which is approximately the cost in dollars per hour to increase the total delivered load by 1 MW. For a system of three units, not necessarily in the same power plant, Eq. (13.16) yields

$$\lambda = L_1 \frac{df_1}{dP_{g1}} = L_2 \frac{df_2}{dP_{g2}} = L_3 \frac{df_3}{dP_{g3}} \tag{13.18}$$

The penalty factor L_i depends on $\partial P_L/\partial P_{gi}$, which is a measure of the sensitivity of the transmission-system losses to changes in P_{gi} alone. Generating units connected to the *same* bus within a particular power plant have equal access to the transmission system, and so the change in system losses must be the same for a small change in the output of any one of those units. This means that the penalty factors are the same for such units located in the same power plant. Therefore, for a plant having, say, three generating units with outputs P_{g1}, P_{g2}, and P_{g3}, the penalty factors L_1, L_2, and L_3 are equal, and Eq. (13.18) then shows that

$$\frac{df_1}{dP_{g1}} = \frac{df_2}{dP_{g2}} = \frac{df_3}{dP_{g3}} \tag{13.19}$$

Thus, for units connected to a common bus within the same physical power plant we have developed mathematically the same criterion which we reached intuitively in Sec. 13.1.

Equation (13.16) governs the coordination of transmission loss into the problem of economic loading of units in plants which are geographically dispersed throughout the system. Accordingly, the penalty factors of the different plants need to be determined which requires that we first express the total transmission loss of the system as a function of plant loadings. This formulation is undertaken in Sec. 13.3.

13.3 THE TRANSMISSION-LOSS EQUATION

To derive the transmission-loss equation in terms of power output of the plants, we consider a simple system consisting of two generating plants and two loads with the transmission network represented by its bus impedance matrix. The derivation is undertaken in two stages. In the first stage we apply a power-invariant transformation to $\mathbf{Z}_{\text{bus}}$ of the system in order to express the system loss in terms of only generator currents. Thus, the reader may find it useful to review Sec. 8.6. In the second stage we transform the generator currents into the power outputs of the plants which leads to the desired form of the loss equation for a system with any number K of sources.

We begin the formulation using, as example, the four-bus system of Fig. 13.5(a) in which nodes ① and ② are generator buses, nodes ③ and ④ are load buses, and node ⓝ is the system neutral. The case where both generation and load are at the same bus is shown in Fig. 13.5(c), which is explained at the end of this section. The current injections I_3 and I_4 at the load buses of Fig. 13.5(a) are combined together to form the composite system load I_D given by

$$I_3 + I_4 = I_D \tag{13.20}$$

Assuming that each load is a constant fraction of the total load, we set

$$I_3 = d_3 I_D \qquad \text{and} \qquad I_4 = d_4 I_D \tag{13.21}$$

from which it follows that

$$d_3 + d_4 = 1 \tag{13.22}$$

By adding more terms, Eqs. (13.20) through (13.22) can be generalized to systems with more than two load buses.

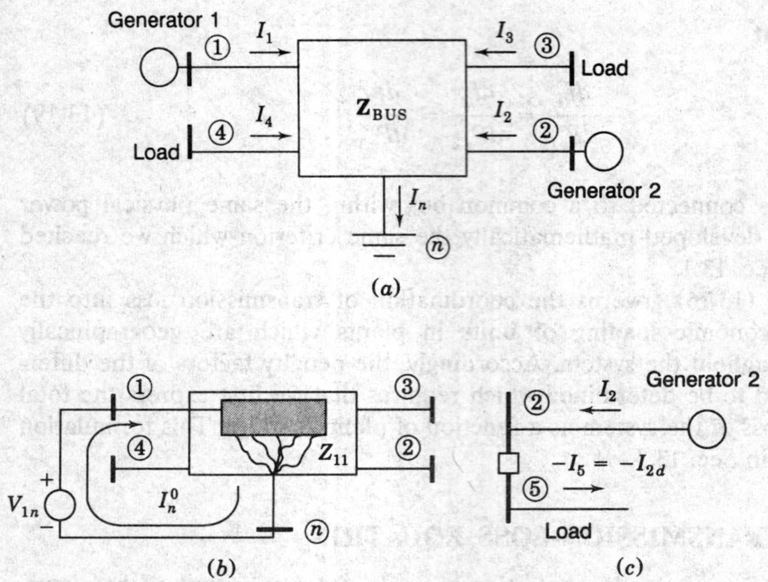

FIGURE 13.5
(*a*) The example four-bus system of Sec. 13.3; (*b*) the interpretation of no-load current I_n^0 of Eq. (13.26); (*c*) the treatment of load current $-I_{2d}$ at generator bus ②.

We now choose node ⓝ of Fig. 13.5(*a*) as reference for the nodal equations

$$\begin{array}{c} \\ \\ \end{array}\begin{bmatrix} V_{1n} \\ V_{2n} \\ V_{3n} \\ V_{4n} \end{bmatrix} = \begin{array}{c} \\ ① \\ ② \\ ③ \\ ④ \end{array}\begin{array}{c} \begin{array}{cccc} ① & ② & ③ & ④ \end{array} \\ \begin{bmatrix} Z_{11} & Z_{12} & Z_{13} & Z_{14} \\ Z_{21} & Z_{22} & Z_{23} & Z_{24} \\ Z_{31} & Z_{32} & Z_{33} & Z_{34} \\ Z_{41} & Z_{42} & Z_{43} & Z_{44} \end{bmatrix} \end{array} \begin{bmatrix} I_1 \\ I_2 \\ I_3 \\ I_4 \end{bmatrix} \tag{13.23}$$

Double-subscript notation emphasizes the fact that the bus voltages are measured with respect to reference node ⓝ. Expanding the first row of Eq. (13.23) gives

$$V_{1n} = Z_{11}I_1 + Z_{12}I_2 + Z_{13}I_3 + Z_{14}I_4 \tag{13.24}$$

Substituting in this equation for $I_3 = d_3 I_D$ and $I_4 = d_4 I_D$, then solving the resultant equation for I_D yield

$$I_D = \frac{-Z_{11}}{d_3 Z_{13} + d_4 Z_{14}} I_1 + \frac{-Z_{12}}{d_3 Z_{13} + d_4 Z_{14}} I_2 + \frac{-Z_{11}}{d_3 Z_{13} + d_4 Z_{14}} I_n^0 \tag{13.25}$$

in which the current I_n^0, called the *no-load current*, is simply

$$I_n^0 = -\frac{V_{1n}}{Z_{11}} \tag{13.26}$$

We shall soon see the physical meaning of I_n^0, which is a constant current injected into node ⓝ of the system whenever V_{1n} is constant. By denoting

$$t_1 = \frac{Z_{11}}{d_3 Z_{13} + d_4 Z_{14}} \quad \text{and} \quad t_2 = \frac{Z_{12}}{d_3 Z_{13} + d_4 Z_{14}} \tag{13.27}$$

we may simplify the coefficients of Eq. (13.25), which then becomes

$$I_D = -t_1 I_1 - t_2 I_2 - t_1 I_n^0 \tag{13.28}$$

Substituting in Eqs. (13.21) for I_D from Eq. (13.28) yields

$$I_3 = -d_3 t_1 I_1 - d_3 t_2 I_2 - d_3 t_1 I_n^0 \tag{13.29}$$

$$I_4 = -d_4 t_1 I_1 - d_4 t_2 I_2 - d_4 t_1 I_n^0 \tag{13.30}$$

We may regard Eqs. (13.29) and (13.30) as defining the transformation **C** of "old" currents I_1, I_2, I_3, and I_4 to the set of "new" currents I_1, I_2, and I_n^0, just as in Eq. (8.55); that is,

$$\begin{bmatrix} I_1 \\ I_2 \\ I_3 \\ I_4 \end{bmatrix} = \begin{array}{c} \\ ① \\ ② \\ ③ \\ ④ \end{array} \overset{\qquad ① \qquad\quad ② \qquad\quad ⓝ}{\begin{bmatrix} 1 & \cdot & \cdot \\ \cdot & 1 & \cdot \\ -d_3 t_1 & -d_3 t_2 & -d_3 t_1 \\ -d_4 t_1 & -d_4 t_2 & -d_4 t_1 \end{bmatrix}} \begin{bmatrix} I_1 \\ I_2 \\ I_n^0 \end{bmatrix} = \mathbf{C} \begin{bmatrix} I_1 \\ I_2 \\ I_n^0 \end{bmatrix} \tag{13.31}$$

We shall explain the row and column labels when we provide a physical interpretation of transformation **C** in Sec. 13.4.

As a result of Eq. (13.31), the expression for the real power loss of the network takes the form of Eq. (8.68), which we now write as

$$P_L = \begin{bmatrix} I_1 & I_2 & I_n^0 \end{bmatrix} \begin{bmatrix} \mathbf{C}^T \mathbf{R}_{\text{bus}} \mathbf{C}^* \end{bmatrix} \begin{bmatrix} I_1 \\ I_2 \\ I_n^0 \end{bmatrix}^* \tag{13.32}$$

where $\mathbf{R}_{\text{bus}}$ is the symmetrical real part of $\mathbf{Z}_{\text{bus}}$ of Eq. (13.23). Because of the power-invariant nature of transformation **C**, Eq. (13.32) fully represents the real power loss of the system in terms of the generator currents I_1 and I_2 and the no-load current I_n^0. By fixing upon bus ① as the slack bus in power-flow studies of the system, the current $I_n^0 = -V_{1n}/Z_{11}$ becomes a constant complex number, which leaves I_1 and I_2 as the only variables in the loss expression of Eq. (13.32).

Figure 13.5(*b*) helps to explain why I_n^0 is called the no-load current. If all load and generation were removed from the system and the voltage V_{1n} were applied at bus ①, only the current I_n^0 would flow through the shunt connections to node ⓝ. This current is normally small and relatively constant since it is determined by Thévenin impedance Z_{11}, which includes the high impedances of paths associated with line-charging and transformer magnetizing currents but not load.

At each generator bus we now assume that the reactive power Q_{gi} is a constant fraction s_i of the real power P_{gi} over the time period of interest. This is equivalent to assuming that each generator operates at a constant power factor over the same period, and so we write

$$P_{g1} + jQ_{g1} = (1 + js_1)P_{g1}; \qquad P_{g2} + jQ_{g2} = (1 + js_2)P_{g2} \tag{13.33}$$

where $s_1 = Q_{g1}/P_{g1}$ and $s_2 = Q_{g2}/P_{g2}$ are real numbers. The output currents from the generators are then given by

$$I_1 = \frac{(1 - js_1)}{V_1^*}P_{g1} = \alpha_1 P_{g1}; \qquad I_2 = \frac{(1 - js_2)}{V_2^*}P_{g2} = \alpha_2 P_{g2} \tag{13.34}$$

in which α_1 and α_2 have obvious definitions. From Eqs. (13.34) the currents I_1, I_2, and I_n^0 can be expressed in matrix form by

$$\begin{bmatrix} I_1 \\ I_2 \\ I_n^0 \end{bmatrix} = \begin{bmatrix} \alpha_1 & \cdot & \cdot \\ \cdot & \alpha_2 & \cdot \\ \cdot & \cdot & I_n^0 \end{bmatrix} \begin{bmatrix} P_{g1} \\ P_{g2} \\ 1 \end{bmatrix} \tag{13.35}$$

and substituting from this equation into Eq. (13.32), we obtain

$$P_L = \begin{bmatrix} P_{g1} \\ P_{g2} \\ 1 \end{bmatrix}^T \underbrace{\begin{bmatrix} \alpha_1 & \cdot & \cdot \\ \cdot & \alpha_2 & \cdot \\ \cdot & \cdot & I_n^0 \end{bmatrix} \mathbf{C}^T \mathbf{R}_{\text{bus}} \mathbf{C}^* \begin{bmatrix} \alpha_1 & \cdot & \cdot \\ \cdot & \alpha_2 & \cdot \\ \cdot & \cdot & I_n \end{bmatrix}^*}_{\mathbf{T}_\alpha} \begin{bmatrix} P_{g1} \\ P_{g2} \\ 1 \end{bmatrix}^* \tag{13.36}$$

We recall that the transpose of a product of matrices equals the *reverse-order* product of their transposes. For instance, if there are three matrices **A**, **B**, and **C**, we have $(\mathbf{ABC})^T = \mathbf{C}^T\mathbf{B}^T\mathbf{A}^T$, and taking the complex conjugate of each side gives $(\mathbf{ABC})^{T*} = \mathbf{C}^{T*}\mathbf{B}^{T*}\mathbf{A}^{T*}$. Thus, we can show that the matrix $\mathbf{T}_\alpha$ of Eq. (13.36) has the convenient property of being equal to the complex conjugate of its own transpose. A matrix with this characteristic is called *hermitian*.[2] Each off-diagonal element m_{ij} of a hermitian matrix is equal to the complex conjugate of the corresponding element m_{ji} and all the diagonal elements are real numbers. Consequently, adding $\mathbf{T}_\alpha$ to $\mathbf{T}_\alpha^*$ cancels out the imaginary parts of the off-diagonal elements, and we obtain twice the symmetrical real part of $\mathbf{T}_\alpha$, which we denote by

$$\left[\begin{array}{cc|c} B_{11} & B_{12} & B_{10}/2 \\ B_{21} & B_{22} & B_{20}/2 \\ \hline B_{10}/2 & B_{20}/2 & B_{00} \end{array}\right] = \frac{\mathbf{T}_\alpha + \mathbf{T}_\alpha^*}{2} \tag{13.37}$$

To conform to industry practices, we use the symbols $B_{10}/2$, $B_{20}/2$, and B_{00} here. Adding Eq. (13.36) to its complex conjugate and applying Eq. (13.37) to the result give

$$P_L = \left[\begin{array}{cc|c} P_{g1} & P_{g2} & 1 \end{array}\right] \left[\begin{array}{cc|c} B_{11} & B_{12} & B_{10}/2 \\ B_{21} & B_{22} & B_{20}/2 \\ \hline B_{10}/2 & B_{20}/2 & B_{00} \end{array}\right] \left[\begin{array}{c} P_{g1} \\ P_{g2} \\ \hline 1 \end{array}\right] \tag{13.38}$$

in which B_{12} equals B_{21}. Expanding Eq. (13.38) by row-column multiplication gives

$$\begin{aligned} P_L &= B_{11}P_{g1}^2 + 2B_{12}P_{g1}P_{g2} + B_{22}P_{g2}^2 + B_{10}P_{g1} + B_{20}P_{g2} + B_{00} \\ &= \sum_{i=1}^{2}\sum_{j=1}^{2} P_{gi}B_{ij}P_{gj} + \sum_{i=1}^{2} B_{i0}P_{gi} + B_{00} \end{aligned} \tag{13.39}$$

which we can rearrange into the equivalent form

$$P_L = \begin{bmatrix} P_{g1} & P_{g2} \end{bmatrix} \begin{bmatrix} B_{11} & B_{12} \\ B_{21} & B_{22} \end{bmatrix} \begin{bmatrix} P_{g1} \\ P_{g2} \end{bmatrix} + \begin{bmatrix} P_{g1} & P_{g2} \end{bmatrix} \begin{bmatrix} B_{10} \\ B_{20} \end{bmatrix} + B_{00} \tag{13.40}$$

or the more general vector-matrix formulation

$$P_L = \mathbf{P}_\mathbf{G}^T\mathbf{B}\mathbf{P}_\mathbf{G} + \mathbf{P}_\mathbf{G}^T\mathbf{B}_0 + B_{00} \tag{13.41}$$

---

[2]An example of a hermitian matrix is $\begin{bmatrix} 1 & 1+j1 \\ 1-j1 & 1 \end{bmatrix}$

When the system has K sources rather than just two as in our example, the vectors and matrices of Eq. (13.41) have K rows and/or K columns and the summations of Eq. (13.39) range from 1 to K to yield the general form of the transmission-loss equation.

$$P_L = \sum_{i=1}^{K} \sum_{j=1}^{K} P_{gi} B_{ij} P_{gj} + \sum_{i=1}^{K} B_{i0} P_{gi} + B_{00} \tag{13.42}$$

The B terms are called *loss coefficients* or *B-coefficients* and the $K \times K$ square matrix $\mathbf{B}$, which is always symmetrical, is known simply as the **B**-*matrix*. The units of the loss coefficients are reciprocal megawatts when the three-phase power P_{g1} to P_{gK} are expressed in megawatts, in which case P_L will be in megawatts also. The units of B_{00} match those of P_L while B_{i0} is dimensionless. Of course, per-unit coefficients are used in normalized computations.

For the system for which they are derived B-coefficients yield the exact loss only for the particular load and operating conditions used in the derivation. The B-coefficients of Eq. (13.40) are constant as P_{g1} and P_{g2} vary, only insofar as bus voltages at the loads and plants maintain constant magnitude and plant power factors remain constant. Fortunately, the use of constant values for the loss coefficients yields reasonably accurate results when the coefficients are calculated for some average operating conditions and if extremely wide shifts of load between plants or in total load do not occur. In practice, large systems are economically loaded using different sets of loss coefficients calculated for various load conditions.

Example 13.3. The four-bus system of Fig. 13.5 has line and bus data given in Table 13.2. The base-case power-flow solution is shown in Table 13.3. Calculate the B-coefficients of the system and show that the transmission loss computed by the loss formula checks with the power-flow results.

TABLE 13.2
Line data and bus data for Example 13.3†

Bus to Bus	Line data: Series Z, R	Line data: Series Z, X	Line data: Shunt Y, B	Bus data: Bus	Generation: P	Generation: $\lvert V\rvert\angle\delta°$	Load: P	Load: Q
Line ①–④	.00744	.0372	.0775	①		$1.0\angle 0°$	·	·
Line ①–③	.01008	.0504	.1025	②	3.18	1.0	·	·
Line ②–③	.00744	.0372	.0775	③	·	·	2.20	1.3634
Line ②–④	.01272	.0636	.1275	④	·	·	2.80	1.7352

†All values in per unit on 230-kV, 100-MVA base.

TABLE 13.3
Power-flow solution for Example 13.3†

	Base case			
	Generation		Voltage	
Bus	P	Q	Magnitude (per unit)	Angle (deg)
①	1.913152	1.872240	1.0	0.0
②	3.18	1.325439	1.0	2.43995
③	·	·	0.96051	−1.07932
④	·	·	0.94304	−2.62658
Total	5.093152	3.197679		

†All values in per unit on 230-kV, 100-MVA base.

Solution. Each transmission line is represented by its equivalent-π circuit with half of the line-charging susceptance to neutral at both ends. Choosing the neutral node ⓝ as reference, we construct the bus impedance matrix $\mathbf{Z}_{\text{bus}} = \mathbf{R}_{\text{bus}} + j\mathbf{X}_{\text{bus}}$, where

$$\mathbf{R}_{\text{bus}} = \begin{array}{c} \\ ① \\ ② \\ ③ \\ ④ \end{array}\begin{array}{cccc} ① & ② & ③ & ④ \\ \left[\begin{array}{cccc} +2.911963 & -1.786620 & -0.795044 & -0.072159 \\ -1.786620 & +2.932995 & -0.072159 & -1.300878 \\ -0.795044 & -0.072159 & +2.911963 & -1.786620 \\ -0.072159 & -1.300878 & -1.786620 & +2.932995 \end{array}\right] \end{array} \times 10^{-3}$$

$$\mathbf{X}_{\text{bus}} = \begin{array}{c} \\ ① \\ ② \\ ③ \\ ④ \end{array}\begin{array}{cccc} ① & ② & ③ & ④ \\ \left[\begin{array}{cccc} -2.582884 & -2.606321 & -2.601379 & -2.597783 \\ -2.606321 & -2.582784 & -2.597783 & -2.603899 \\ -2.601379 & -2.597783 & -2.582884 & -2.606321 \\ -2.597783 & -2.603899 & -2.606321 & -2.582784 \end{array}\right] \end{array}$$

From the power-flow results of Table 13.3 the load currents are calculated,

$$I_3 = \frac{P_3 - jQ_3}{V_3^*} = \frac{-2.2 + j1.36340}{0.96051\angle 1.07932^\circ} = 2.694641\angle 147.1331^\circ$$

$$I_4 = \frac{P_4 - jQ_4}{V_4^*} = \frac{-2.8 + j1.73520}{0.94304\angle 2.62658^\circ} = 3.493043\angle 145.5863^\circ$$

and we then find that

$$d_3 = \frac{I_3}{I_3 + I_4} = 0.435473 + j0.006637$$

$$d_4 = \frac{I_4}{I_3 + I_4} = 0.564527 - j0.006637$$

The quantities t_1 and t_2 of Eq. (13.27) are calculated from d_3, d_4, and the row-one elements of $\mathbf{Z}_{\text{bus}}$ as follows:

$$t_1 = \frac{Z_{11}}{d_3 Z_{13} + d_4 Z_{14}} = 0.993664 + j0.001259$$

$$t_2 = \frac{Z_{12}}{d_3 Z_{13} + d_4 Z_{14}} = 1.002681 - j0.000547$$

Based on the above results, we compute the $-d_i t_j$ terms of Eq. (13.31) to obtain the current transformation $\mathbf{C}$

$$\mathbf{C} = \begin{array}{c} \textcircled{1} \\ \textcircled{2} \\ \textcircled{3} \\ \textcircled{4} \end{array} \overset{\begin{array}{ccc} \quad\textcircled{1}\quad & \quad\textcircled{2}\quad & \quad\textcircled{n}\quad \end{array}}{\begin{bmatrix} 1 & \cdot & \cdot \\ \cdot & 1 & \cdot \\ -0.432705 - j0.007143 & -0.436644 - j0.006416 & -0.432705 - j0.007143 \\ -0.560958 + j0.005884 & -0.566037 + j0.006964 & -0.560958 + j0.005884 \end{bmatrix}}$$

and we then find that

$$\mathbf{C}^T \mathbf{R}_{\text{bus}} \mathbf{C}^* = \begin{array}{c} \textcircled{1} \\ \textcircled{2} \\ \textcircled{n} \end{array} \overset{\begin{array}{ccc} \quad\textcircled{1}\quad & \quad\textcircled{2}\quad & \quad\textcircled{n}\quad \end{array}}{\begin{bmatrix} 4.282185 + j0 & -0.030982 - j0.010638 & 0.985724 - j0.005255 \\ --0.030982 + j0.010638 & 5.080886 + j0 & 1.367642 + j0.006039 \\ 0.\ 0.985724 + j0.005255 & 1.367642 - j0.006039 & 0.601225 + j0 \end{bmatrix}} \times 10^{-3}$$

The row and column numbers of the last two equations are explained in Sec. 13.4. The per-unit no-load current is calculated to be

$$I_n^0 = \frac{-V_1}{Z_{11}} = -\frac{1.0 + j0.0}{0.002912 - j2.582884} = -0.000436 - j0.387164$$

and using the base-case power-flow results, we compute from Eq. (13.34)

$$\alpha_1 = \frac{1 - js_1}{V_1^*} = \frac{1 - j\left(\dfrac{1.872240}{1.913152}\right)}{1.0\angle 0^\circ} = 1.0 - j0.978615$$

$$\alpha_2 = \frac{1 - js_2}{V_2^*} = \frac{1 - j\left(\dfrac{1.325439}{3.180000}\right)}{1.0\angle -2.43995^\circ} = 1.016838 - j0.373855$$

The hermitian matrix $\mathbf{T}_\alpha$ of Eq. (13.36) is given by

$$\mathbf{T}_\alpha = \begin{bmatrix} \alpha_1 & \cdot & \cdot \\ \cdot & \alpha_2 & \cdot \\ \cdot & \cdot & I_n^0 \end{bmatrix} \mathbf{C}^T \mathbf{R}_{\text{bus}} \mathbf{C}^* \begin{bmatrix} \alpha_1 & \cdot & \cdot \\ \cdot & \alpha_2 & \cdot \\ \cdot & \cdot & I_n^0 \end{bmatrix}^*$$

$$\mathbf{T}_\alpha = \begin{bmatrix} 8.383183 + j0.0 & -0.049448 + j0.004538 & 0.375082 + j0.380069 \\ -0.049448 - j0.004538 & 5.963568 + j0.0 & 0.194971 + j0.539511 \\ 0.375082 - j0.380069 & 0.194971 - j0.539511 & 0.090121 + j0.0 \end{bmatrix} \times 10^{-3}$$

By extracting the real parts of the respective elements of $\mathbf{T}_\alpha$, we obtain the desired **B**-matrix of per-unit loss coefficients

$$\left[\begin{array}{cc|c} B_{11} & B_{12} & B_{10}/2 \\ B_{21} & B_{22} & B_{20}/2 \\ \hline B_{10}/2 & B_{20}/2 & B_{00} \end{array}\right] = \left[\begin{array}{cc|c} 8.383183 & -0.049448 & 0.375082 \\ -0.049448 & 5.963568 & 0.194971 \\ \hline 0.375082 & 0.194971 & 0.090121 \end{array}\right] \times 10^{-3}$$

from which we calculate the power loss

$$P_L = \left[1.913152 \quad 3.18 \mid 1\right] \left[\begin{array}{cc|c} B_{11} & B_{12} & B_{10}/2 \\ B_{21} & B_{22} & B_{20}/2 \\ \hline B_{10}/2 & B_{20}/2 & B_{00} \end{array}\right] \left[\begin{array}{c} 1.913152 \\ 3.18 \\ \hline 1 \end{array}\right]$$

$$= 0.093153 \text{ per unit}$$

This checks with the power-flow result of Table 13.3.

In Example 13.3 exact agreement between methods of loss calculation is expected since the loss coefficients were determined for the power-flow conditions for which loss was calculated. The amount of error introduced by using the loss coefficients of Example 13.3 for two other operating conditions may be seen by examining the results of two converged power-flow solutions shown in Table 13.4. The load levels correspond to 90 and 80% of the base-case load of Example 13.3 and P_{g2} is assigned by economic dispatch, as described in Sec.

TABLE 13.4
Comparison of transmission loss calculated by the *B*-coefficients of Example 13.3 and by power-flow solutions for different operating conditions of the example system

	Load level		P_{g1}	P_{g2}	P_L	
	P_{d3}	P_{d4}	(slack bus)	(econ. disp.)	Power flow	*B*-coeffs.
base:	2.2	2.8	1.913152	3.18	0.093152	0.093152
90%	1.98	2.52	1.628151	2.947650	0.075801	0.076024
80%	1.76	2.24	1.354751	2.705671	0.060422	0.060842

13.5. In practice, the loss coefficients are recalculated and updated on a periodic basis using data acquired from the physical power system.

So far we have not considered generator buses having local loads. Suppose that bus ② of Fig. 13.5(*a*) has a load component $-I_{2d}$ in addition to the network injection I_2. Since all currents are considered as injections, we can regard the load current $-I_{2d}$ as a current I_{2d} entering the network at a dummy bus, say, number ⑤, as shown in Fig. 13.5(*c*). $\mathbf{R}_{\text{bus}}$ is then expanded to include a row and column for bus ⑤ with off-diagonal elements identical to those of row 2 and column 2, and $Z_{55} = Z_{22}$. We now proceed to develop the transformation **C** exactly as before, mechanically treating bus ⑤ as a load bus with injected current $I_5 = I_{2d} = d_5 I_D$, where $I_D = I_3 + I_4 + I_5$. This strategy can be followed in solving some of the problems at the end of this chapter.

13.4 AN INTERPRETATION OF TRANSFORMATION C

Transformation **C** of Eq. (13.31) has a useful interpretation which provides insight and physical meaning to its use. Let us consider that **C** is a combination of *two* sequential transformations from the "old" currents of Fig. 13.6(*a*) to the "new" currents, as similarly discussed in Sec. 8.6. The first transformation $\mathbf{C}_1$ follows from Eqs. (13.21) and is written

$$\begin{bmatrix} I_1 \\ I_2 \\ I_3 \\ I_4 \end{bmatrix} = \begin{matrix} ① \\ ② \\ ③ \\ ④ \end{matrix} \underbrace{\overset{\begin{matrix} ① & ② & Ⓓ \end{matrix}}{\begin{bmatrix} 1 & \cdot & \cdot \\ \cdot & 1 & \cdot \\ \cdot & \cdot & d_3 \\ \cdot & \cdot & d_4 \end{bmatrix}}}_{\mathbf{C}_1} \begin{bmatrix} I_1 \\ I_2 \\ I_D \end{bmatrix} \tag{13.43}$$

By relating individual load currents to the total load current I_D, $\mathbf{C}_1$ introduces the mathematical concept of *system load center* with current I_D injected into a hypothetical (that is, nonphysical) node Ⓓ of the network, as shown in

Fig. 13.6(b). The current I_D is the algebraic sum of *all* the load currents injected into the network at the buses. The transformer turns ratios $d_i : 1$ shown in Fig. 13.6(b) are *current* ratios which are the complex-conjugate reciprocals of their corresponding voltage ratios, as explained in connection with Eq. (2.37). The second transformation $\mathbf{C}_2$ defined by

$$\begin{bmatrix} I_1 \\ I_2 \\ I_D \end{bmatrix} = \begin{matrix} \textcircled{1} \\ \textcircled{2} \\ \textcircled{D} \end{matrix} \underbrace{\overset{\begin{matrix} \textcircled{1} & \quad\textcircled{2} & \quad\textcircled{n} \end{matrix}}{\begin{bmatrix} 1 & \cdot & \cdot \\ \cdot & 1 & \cdot \\ -t_1 & -t_2 & -t_1 \end{bmatrix}}}_{\mathbf{C}_2} \begin{bmatrix} I_1 \\ I_2 \\ I_n^0 \end{bmatrix} \tag{13.44}$$

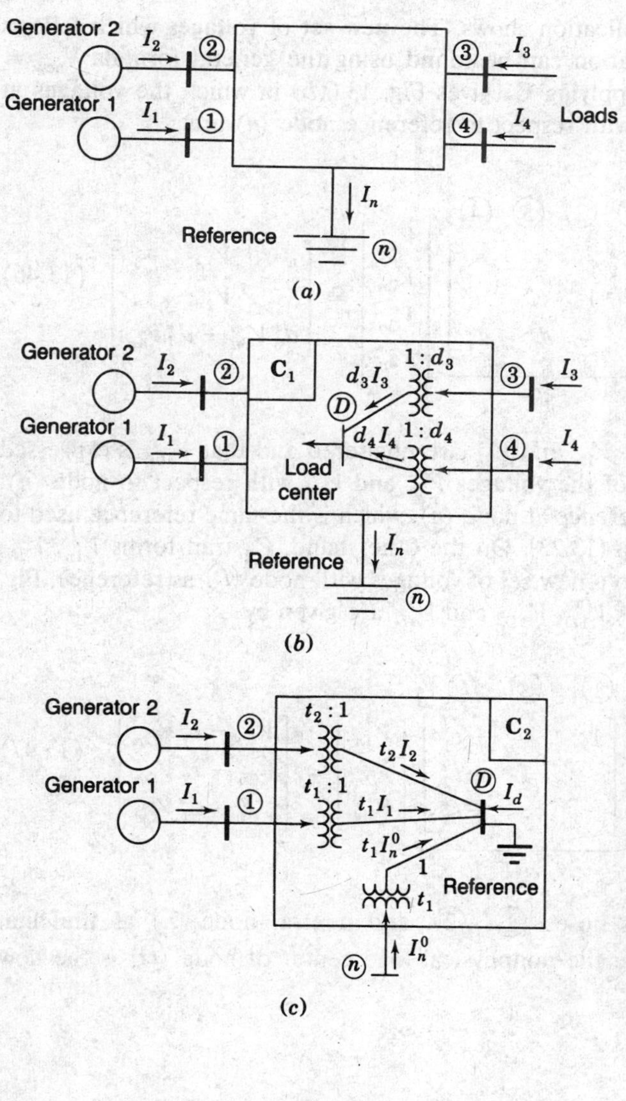

FIGURE 13.6
Physical interpretation of transformation $\mathbf{C} = \mathbf{C}_1\mathbf{C}_2$: ($a$) the original system of two generators and two load buses; (b) $\mathbf{C}_1$ eliminates load currents and forms system load center D; (c) $\mathbf{C}_2$ switches reference from node ⓝ to system load center D. All tap-changing transformers are ideal and *current* ratios are shown.

removes I_D as an independent current and replaces it by the no-load current I_n^0. This is achieved by switching the system reference from node ⓝ to the load center at node Ⓓ, as depicted in Fig. 13.6(c). Here also the transformer turns ratios $t_i : 1$ are current transformation ratios. The combination of $\mathbf{C}_1$ and $\mathbf{C}_2$ in sequence is equivalent to the single transformation $\mathbf{C}$ shown in Eq. (13.31); that is,

$$\mathbf{C} = \mathbf{C}_1\mathbf{C}_2 = \begin{array}{c} ① \\ ② \\ ③ \\ ④ \end{array} \overset{\begin{array}{ccc} ① & ② & ⓝ \end{array}}{\begin{bmatrix} 1 & \cdot & \cdot \\ \cdot & 1 & \cdot \\ -d_3t_1 & -d_3t_2 & -d_3t_1 \\ -d_4t_1 & -d_4t_2 & -d_4t_1 \end{bmatrix}} \tag{13.45}$$

as straightforward multiplication shows. The new set of voltages which follows each separate transformation can be found using the general formula $\mathbf{V}_{\text{new}} = \mathbf{C}^{*T}\mathbf{V}_{\text{old}}$ of Eq. (8.63). Applying $\mathbf{C}_1$ gives Fig. 13.6(b) in which the voltages at buses ①, ②, and Ⓓ with respect to reference node ⓝ are

$$\begin{bmatrix} V_{1n} \\ V_{2n} \\ V_{Dn} \end{bmatrix} = \begin{array}{c} ① \\ ② \\ Ⓓ \end{array} \underbrace{\overset{\begin{array}{cccc} ① & ② & ③ & ④ \end{array}}{\begin{bmatrix} 1 & \cdot & \cdot & \cdot \\ \cdot & 1 & \cdot & \cdot \\ \cdot & \cdot & d_3^* & d_4^* \end{bmatrix}}}_{\mathbf{C}_1^{*T}} \begin{bmatrix} V_{1n} \\ V_{2n} \\ V_{3n} \\ V_{4n} \end{bmatrix} = \begin{bmatrix} V_{1n} \\ V_{2n} \\ d_3^*V_{3n} + d_4^*V_{4n} \end{bmatrix} \tag{13.46}$$

This equation shows that V_{1n} and V_{2n} are unaltered and that V_{Dn} is expressed as a linear combination of the voltages V_{3n} and V_{4n} with respect to node ⓝ. That is, $\mathbf{C}_1$ keeps the reference at node ⓝ, which is the same reference used to construct the $\mathbf{Z}_{\text{bus}}$ of Eq. (13.23). On the other hand, $\mathbf{C}_2$ transforms V_{1n}, V_{2n}, and V_{Dn} of Fig. 13.6(b) to a new set of voltages with node Ⓓ as reference, Fig. 13.6(c). The new voltages V_{1D}, V_{2D}, and V_{nD} are given by

$$\begin{bmatrix} V_{1D} \\ V_{2D} \\ V_{nD} \end{bmatrix} = \begin{array}{c} ① \\ ② \\ ⓝ \end{array} \underbrace{\overset{\begin{array}{ccc} ① & ② & Ⓓ \end{array}}{\begin{bmatrix} 1 & \cdot & -t_1^* \\ \cdot & 1 & -t_2^* \\ \cdot & \cdot & -t_1^* \end{bmatrix}}}_{\mathbf{C}_2^{*T}} \begin{bmatrix} V_{1n} \\ V_{2n} \\ V_{Dn} \end{bmatrix} = \begin{bmatrix} V_{1n} - t_1^*V_{Dn} \\ V_{2n} - t_2^*V_{Dn} \\ 0 - t_1^*V_{Dn} \end{bmatrix} \tag{13.47}$$

Equation (13.47) reveals buses ①, ②, and neutral node ⓝ as the final independent nodes, while the nonphysical load center of node Ⓓ is the new reference. Accordingly, the row-column labels of **C** are those shown in Eq. (13.45), while the rows and columns of the square matrix $[\mathbf{C}^T\mathbf{R}_{\text{bus}}\mathbf{C}^*]$ of Eq. (13.32) can be likewise labeled sequentially with the node numbers ①, ②, and ⓝ since load center D is now the reference node.

13.5 CLASSICAL ECONOMIC DISPATCH WITH LOSSES

When system loss is neglected, the transmission network is equivalent to a single node to which all generation and load is connected. The penalty factor for each plant then becomes unity and the system λ is given in the same literal form as Eq. (13.6). When transmission loss is included, however, the economic dispatch strategy has to be iteratively determined by solving the nonlinear coordination equations represented by Eq. (13.15), which can be written in the form

$$\frac{df_i}{dP_{gi}} - \lambda + \lambda\frac{\partial P_L}{\partial P_{gi}} = 0 \qquad (13.48)$$

We consider each generating unit of the system to have a second-order fuel-cost characteristic in the form of Eq. (13.1) and a linear incremental fuel cost given by Eq. (13.2). The partial derivative term of Eq. (13.48), called the *incremental loss*, is a measure of the sensitivity of the *system losses* to an incremental change in the output of plant i when all other plant outputs are kept fixed. For example, in a system of two plants the incremental loss of Unit 1 is found from the loss expression of Eq. (13.39), which gives

$$\frac{\partial P_L}{\partial P_{g1}} = 2B_{11}P_{g1} + 2B_{12}P_{g2} + B_{10} \qquad (13.49)$$

Setting i equal to 1 in Eq. (13.48), and then substituting for df_1/dP_{g1} from Eq. (13.2) and for $\partial P_L/\partial P_{g1}$ from Eq. (13.49), we obtain

$$(a_1P_{g1} + b_1) - \lambda + \lambda(2B_{11}P_{g1} + 2B_{12}P_{g2} + B_{10}) = 0 \qquad (13.50)$$

Collecting terms involving P_{g1} and dividing the resultant equation by λ give

$$\left(\frac{a_1}{\lambda} + 2B_{11}\right)P_{g1} + 2B_{12}P_{g2} = (1 - B_{10}) - \frac{b_1}{\lambda} \qquad (13.51)$$

Following exactly the same procedure for $\partial P_L/\partial P_{g2}$, we obtain the analogous equation for Unit 2

$$2B_{21}P_{g1} + \left(\frac{a_2}{\lambda} + 2B_{22}\right)P_{g2} = (1 - B_{20}) - \frac{b_2}{\lambda} \qquad (13.52)$$

Equations (13.51) and (13.52) can be rearranged into vector-matrix form

$$\begin{bmatrix} \left(\frac{a_1}{\lambda} + 2B_{11}\right) & 2B_{12} \\ 2B_{21} & \left(\frac{a_2}{\lambda} + 2B_{22}\right) \end{bmatrix} \begin{bmatrix} P_{g1} \\ P_{g2} \end{bmatrix} = \begin{bmatrix} (1 - B_{10}) - \frac{b_1}{\lambda} \\ (1 - B_{20}) - \frac{b_2}{\lambda} \end{bmatrix} \tag{13.53}$$

and this is the set of equations for Units 1 and 2. When the system has K sources as in Eq. (13.42), the partial derivative of P_L with respect to P_{gi} gives the general formula for Unit i

$$\left(\frac{a_i}{\lambda} + 2B_{ii}\right)P_{gi} + \sum_{\substack{j=1 \\ j \neq i}}^{K} 2B_{ij}P_{gj} = (1 - B_{i0}) - \frac{b_i}{\lambda} \tag{13.54}$$

Letting i range from 1 to K, we obtain a system of linear equations for all K sources, which takes on the form of Eq. (13.53); that is,

$$\begin{bmatrix} \left(\frac{a_1}{\lambda} + 2B_{11}\right) & 2B_{12} & \cdots & 2B_{1K} \\ 2B_{21} & \left(\frac{a_2}{\lambda} + 2B_{22}\right) & \cdots & 2B_{2K} \\ \vdots & \vdots & \ddots & \vdots \\ 2B_{K1} & 2B_{K2} & \cdots & \left(\frac{a_K}{\lambda} + 2B_{KK}\right) \end{bmatrix} \begin{bmatrix} P_{g1} \\ P_{g2} \\ \vdots \\ P_{gK} \end{bmatrix} = \begin{bmatrix} (1 - B_{10}) - \frac{b_1}{\lambda} \\ (1 - B_{20}) - \frac{b_2}{\lambda} \\ \vdots \\ (1 - B_{K0}) - \frac{b_K}{\lambda} \end{bmatrix} \tag{13.55}$$

Substituting in Eq. (13.11) for P_L from Eq. (13.42), we obtain

$$\left(\sum_{i=1}^{K}\sum_{j=1}^{K} P_{gi}B_{ij}P_{gj} + \sum_{i=1}^{K} B_{i0}P_{gi} + B_{00}\right) + P_D - \sum_{i=1}^{K} P_{gi} = 0 \tag{13.56}$$

which is the power-balance requirement for the system as a whole in terms of B-coefficients, the plant loadings, and the total load. The *economic dispatch strategy* consists of solving the K equations represented by Eq. (13.55) for those values of power output which also satisfy the power loss and load requirements of Eq. (13.56). There are many different ways to solve Eqs. (13.55) and (13.56) for the unknowns $P_{g1}, P_{g2}, \ldots, P_{gK}$ and λ. When an initial value of λ is chosen in Eq. (13.55), the resulting set of equations becomes linear. The values of $P_{g1}, P_{g2}, \ldots, P_{gK}$ can then be found by any number of solution techniques, such as inverting the coefficient matrix, within the following iterative procedure:

Step 1

Specify the system load level $P_D = \Sigma_{j=1}^{N} P_{dj}$.

Step 2

For the first iteration, choose initial values for the system λ. [One way to make this initializing choice is to assume that losses are zero and calculate initial values of λ from Eq. (13.6).]

Step 3

Substitute the value of λ into Eq. (13.55) and solve the resultant system of linear simultaneous equations for the values of P_{gi} by some efficient means.

Step 4

Compute the transmission loss of Eq. (13.42) using values of P_{gi} from Step 3.

Step 5

Compare the quantity $(\Sigma_{i=1}^{K} P_{gi} - P_L)$ with P_D to check the power balance of Eq. (13.56). If power balance is not achieved within a specified tolerance, update the system λ by setting

$$\lambda^{(k+1)} = \lambda^{(k)} + \Delta\lambda^{(k)} \tag{13.57}$$

One possible formula for the increment $\Delta\lambda^{(k)}$ is

$$\Delta\lambda^{(k)} = \frac{\lambda^{(k)} - \lambda^{(k-1)}}{\sum_{i=1}^{K} P_{gi}^{(k)} - \sum_{i=1}^{K} P_{gi}^{(k-1)}} \left[P_D + P_L^{(k)} - \sum_{i=1}^{K} P_{gi}^{(k)} \right] \tag{13.58}$$

In Eqs. (13.57) and (13.58) superscript $(k+1)$ indicates the next iteration being started, superscript (k) indicates the iteration just completed, and $(k-1)$ denotes the immediately preceding iteration.

Step 6

Return to Step 3 and continue the calculations of Steps 3, 4, and 5 until final convergence is achieved.

The final results from the above procedure determine both the system λ and the economic dispatch output of each generating unit for the specified level of system load. It is interesting to note that during *each* iteration of the overall solution Step 3 provides an economic dispatch answer, which is correct at one load level even though it may not be the *specified* load level of the system.

Example 13.4. The generating units at buses ① and ② of the system of Example 13.3 have incremental fuel costs given in Example 13.1. Calculate the economic loading of each unit to meet a total customer load of 500 MW. What is the system λ and what is the transmission loss of the system? Determine the penalty factor for each unit and the incremental fuel cost at each generating bus.

Solution. The generating units at the two different power plants have incremental fuel costs in dollars per megawatthour given by

$$\frac{df_1}{dP_{g1}} = 0.0080P_{g1} + 8.0; \qquad \frac{df_2}{dP_{g2}} = 0.0096P_{g2} + 6.4$$

where P_{g1} and P_{g2} are expressed in megawatts. At the specified load level of 500 MW the loss coefficients in per unit on a 100-MVA base are given in Example 13.3 as

$$\begin{bmatrix} B_{11} & B_{12} & B_{10}/2 \\ B_{21} & B_{22} & B_{20}/2 \\ B_{10}/2 & B_{20}/2 & B_{00} \end{bmatrix} = \begin{bmatrix} 8.383183 & -0.049448 & 0.375082 \\ -0.049448 & 5.963568 & 0.194971 \\ 0.375082 & 0.194971 & 0.090121 \end{bmatrix} \times 10^{-3}$$

To begin the solution, we must estimate initial values for λ for the first iteration. The results of Example 13.1 at the 500-MW load level can be used for this purpose:

Step 1

We are given $P_D = 5.00$ per unit on a 100-MVA base.

Step 2

From Example 13.1 we choose $\lambda^{(1)} = 9.454545$.

Step 3

Based on the estimated value of $\lambda^{(1)}$, we calculate the outputs P_{g1} and P_{g2} from the equation

$$\begin{bmatrix} \dfrac{0.8}{\lambda^{(1)}} + 2 \times 8.383183 \times 10^{-3} & -2 \times 0.049448 \times 10^{-3} \\ -2 \times 0.049448 \times 10^{-3} & \dfrac{0.96}{\lambda^{(1)}} + 2 \times 5.963568 \times 10^{-3} \end{bmatrix} \begin{bmatrix} P_{g1} \\ P_{g2} \end{bmatrix}$$

$$= \begin{bmatrix} (1 - 0.750164 \times 10^{-3}) - \dfrac{8.0}{\lambda^{(1)}} \\ (1 - 0.389942 \times 10^{-3}) - \dfrac{6.4}{\lambda^{(1)}} \end{bmatrix}$$

Note that per unit a_1 and a_2 are used in this calculation because all other quantities are in per unit. Owing to the simplicity of the example, this equation for P_{g1} and P_{g2} can be solved directly to yield results for the first iteration

$$P_{g1}^{(1)} = 1.512870 \text{ per unit}; \qquad P_{g2}^{(1)} = 2.845238 \text{ per unit}$$

Step 4

From the results of Step 3 and the given values of the B-coefficients the system power loss is calculated as follows:

$$\begin{aligned} P_L &= B_{11}P_{g1}^2 + 2B_{12}P_{g1}P_{g2} + B_{22}P_{g2}^2 + B_{10}P_{g1} + B_{20}P_{g2} + B_{00} \\ &= B_{11}(1.512870)^2 + 2B_{12}(1.512870)(2.845238) \\ &\quad + B_{22}(2.845238)^2 + B_{10}(1.512870) + B_{20}(2.845238) + B_{00} \\ &= 0.069373 \text{ per unit} \end{aligned}$$

Step 5

Checking the power balance for $P_D = 5.00$ per unit, we find

$$P_D + P_L^{(1)} - \left(P_{g1}^{(1)} + P_{g2}^{(1)}\right) = 5.069373 - 4.358108 = 0.711265$$

which exceeds $\varepsilon = 10^{-6}$, and so a new value of λ must be supplied. The incremental change in λ is calculated from Eq. (13.58) as follows:

$$\Delta\lambda^{(1)} = \left(\lambda^{(1)} - \lambda^{(0)}\right)\left[\frac{P_D + P_L^{(1)} - \left(P_{g1}^{(1)} + P_{g2}^{(1)}\right)}{\left(\sum_{i=1}^{2} P_{gi}^{(1)}\right) - \left(\sum_{i=1}^{2} P_{gi}^{(0)}\right)}\right]$$

Since this is the first iteration, $\lambda^{(0)}$ and $\Sigma_{i=1}^{2} P_{gi}^{(0)}$ are both set equal to zero, which gives

$$\Delta\lambda^{(1)} = (9.454545 - 0)\left[\frac{0.711265}{4.358108 - 0}\right] = 1.543035$$

and the updated λ then becomes

$$\lambda^{(2)} = \lambda^{(1)} + \Delta\lambda^{(1)} = 9.454545 + 1.543035 = 10.99758$$

Step 6

We now return to Step 3 to repeat the above calculations using $\lambda^{(2)}$ during the second iteration, and so on.

The final converged solution for both system λ and the economic loading of the two generating units is found to be

$$\lambda = 9.839863 \text{ \$/MWh}$$

$$P_{g1} = 190.2204 \text{ MW} \qquad P_{g2} = 319.1015 \text{ MW}$$

For illustrative purposes a convergence criterion of $\varepsilon = 10^{-6}$ was used in this example, but such accuracy is not warranted in practice.

The transmission loss computed from the solved values of P_{g1} and P_{g2} is 9.321914 MW at Step 4 of the final iteration, and so the total generation by the two plants amounts to 509.32 MW for load and losses. The incremental losses of the two plants are

$$\frac{\partial P_L}{\partial P_{g1}} = 2(B_{11}P_{g1} + B_{12}P_{g2} + B_{10}/2)$$

$$= 2(8.383183 \times 1.902204 - 0.049448 \times 3.191015 + 0.375083) \times 10^{-3}$$

$$= 0.032328$$

$$\frac{\partial P_L}{\partial P_{g2}} = 2(B_{22}P_{g2} + B_{21}P_{g1} + B_{20}/2)$$

$$= 2(5.963568 \times 3.191015 - 0.049448 \times 1.902204 + 0.194971) \times 10^{-3}$$

$$= 0.038261$$

and so the penalty factors are given by

$$L_1 = \frac{1}{1 - 0.032328} = 1.03341; \qquad L_2 = \frac{1}{1 - 0.038261} = 1.03978$$

The incremental fuel costs at the two plant buses are calculated to be

$$\frac{df_1}{dP_{g1}} = a_1 P_{g1} + b_1 = 0.80(1.902204) + 8.0 = 9.521763 \text{ \$/MWh}$$

$$\frac{df_2}{dP_{g2}} = a_2 P_{g2} + b_2 = 0.96(3.191015) + 6.4 = 9.463374 \text{ \$/MWh}$$

In this example plant 2 has the lower incremental fuel cost at its bus and carries the bigger share of the 500-MW load. The reader can confirm that the *effective* incremental cost of supplying the system load (often called the *incremental cost of power delivered*) checks with the calculations $L_1(df_1/dP_{g1}) = L_2(df_2/dP_{g2}) =$ 9.839863 \$/MWh.

We noted earlier that Step 3 of each iteration of the above procedure provides valid answers for economic loading of the units. Those answers are correct at the particular load level which gives power balance within that iteration. For instance, at Step 3 of the first iteration in Example 13.4 the system λ is 9.454545 \$/MWh and the generator outputs are computed to be $P_{g1}^{(1)} =$ 151.287 MW and $P_{g2}^{(1)} = 284.5238$ MW. At Step 4 of the same iteration the corresponding value of $P_L^{(1)}$ is found to be 6.9373 MW. Therefore, a system load P_D given by

$$\left(P_{g1}^{(1)} + P_{g2}^{(1)}\right) - P_L^{(1)} = (435.8108 - 6.9373) = 428.8735 \text{ MW}$$

would satisfy the power balance of the system. We make use of this observation in the following example.

Example 13.5. Calculate the decrease in production costs of the two plants of Example 13.4 when the system load is reduced from 500 to 429 MW.

Solution. The economic loading of the two plants for a system load level of 500 MW is found in Example 13.4 to be $P_{g1} = 190.2$ MW and $P_{g2} = 319.1$ MW. In the first iteration of the same example we also find that the plant outputs $P_{g1} = 151.3$ MW and $P_{g2} = 284.5$ MW ensure economic dispatch of the units when the load level is essentially 429 MW. These results are sufficiently accurate for calculating the production-cost reduction between the two load levels as follows:

$$\Delta f_1 = \int_{190.2}^{151.3} (0.0080 P_{g1} + 8.0)\, dP_{g1}$$

$$= \left(0.0040 P_{g1}^2 + 8.0 P_{g1} + c_1\right)\Big|_{190.2}^{151.3} = -364.34 \text{ \$/h}$$

$$\Delta f_2 = \int_{319.1}^{284.5} (0.0096 P_{g2} + 6.4)\, dP_{g2}$$

$$= \left(0.0048 P_{g2}^2 + 6.4 P_{g2} + c_2\right)\Big|_{319.1}^{284.5} = -321.69 \text{ \$/h}$$

Thus, the total reduction in system fuel cost amounts to \$686 per hour.

Procedures have now been developed for coordinating transmission loss of the system into the economic dispatch of those units which are already on line. In Sec. 13.6 we consider automatic generation control before investigating the unit commitment problem which determines the units to be connected on line in the first place.

13.6 AUTOMATIC GENERATION CONTROL

Almost all generating companies have tie-line interconnections to neighboring utilities. Tie lines allow the sharing of generation resources in emergencies and economies of power production under normal conditions of operation. For purposes of control the entire interconnected system is subdivided into *control areas* which usually conform to the boundaries of one or more companies. The *net interchange* of power over the tie lines of an area is the algebraic difference between area generation and area load (plus losses). A schedule is prearranged with neighboring areas for such tie-line flows, and as long as an area maintains the interchange power on schedule, it is evidently fulfilling its primary responsibility to absorb its own load changes. But since each area shares in the benefits of interconnected operation, it is expected also to share in the responsibility to maintain system frequency.

Frequency changes occur because system load varies randomly throughout the day so that an exact forecast of real power demand cannot be assured. The imbalance between real power generation and load demand (plus losses) throughout the daily load cycle causes kinetic energy of rotation to be either added to or taken from the on-line generating units, and frequency throughout the interconnected system varies as a result. Each control area has a central facility called the *energy control center*, which monitors the system frequency and the actual power flows on its tie lines to neighboring areas. The deviation between desired and actual system frequency is then combined with the deviation from the scheduled net interchange to form a composite measure called the *area control error*, or simply ACE. To remove area control error, the energy control center sends command signals to the generating units at the power plants within its area to control the generator outputs so as to restore the net interchange power to scheduled values and to assist in restoring the system frequency to its desired value. The monitoring, telemetering, processing, and control functions are coordinated within the individual area by the computer-based automatic generation control (AGC) system at the energy control center.

To help understand the control actions at the power plants, let us first consider the boiler-turbine-generator combination of a thermal generating unit. Most steam turbogenerators (and also hydroturbines) now in service are equipped with turbine *speed governors*. The function of the speed governor is to monitor continuously the turbine-generator speed and to control the throttle valves which adjust steam flow into the turbine (or the gate position in hydroturbines) in response to changes in "system speed" or frequency. We use *speed* and *frequency* interchangeably since they describe proportional quantities. To permit parallel operation of generating units, the speed-versus-power output governing characteristic of each unit has droop, which means that a decrease in speed should accompany an increase in load, as depicted by the straight line of Fig. 13.7(*a*). The *per-unit droop* or *speed regulation* R_u of the generating unit is defined as the magnitude of the change in steady-state speed, expressed in per unit of rated speed, when the output of the unit is gradually reduced from 1.00 per-unit rated power to zero. Thus, per-unit regulation is

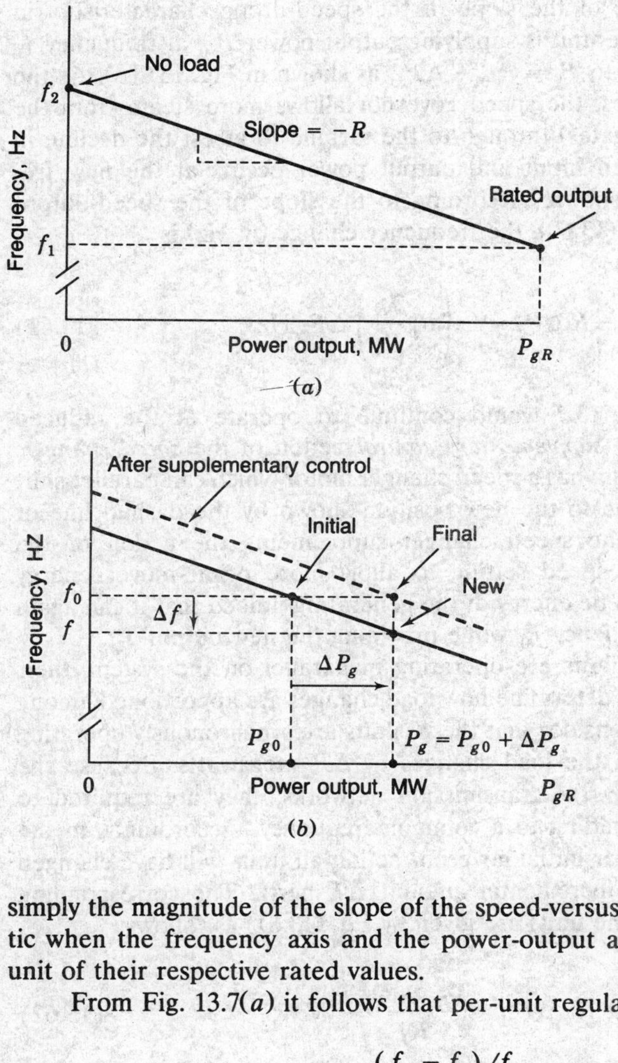

FIGURE 13.7
(*a*) Speed-governing characteristic of a generating unit; (*b*) before/after load increase ΔP_g and supplementary control.

simply the magnitude of the slope of the speed-versus-power output characteristic when the frequency axis and the power-output axis are each scaled in per unit of their respective rated values.

From Fig. 13.7(*a*) it follows that per-unit regulation is given by

$$R_u = \frac{(f_2 - f_1)/f_R}{P_{gR}/S_R} \text{ per unit} \tag{13.59}$$

where f_2 = frequency (in Hz) at no load
f_1 = frequency (in Hz) at rated megawatt output P_{gR}
f_R = rated frequency (in Hz) of the unit
S_R = megawatt base

Multiplying each side of Eq. (13.59) by f_R/S_R gives

$$R = R_u \frac{f_R}{S_R} = \frac{f_2 - f_1}{P_{gR}} \text{ Hz/MW} \tag{13.60}$$

where R is the magnitude of the slope of the speed-droop characteristic (in Hz/MW). Suppose that the unit is supplying output power P_{g0} at frequency f_0 when the load is increased to $P_g = P_{g0} + \Delta P_g$, as shown in Fig. 13.7($b$). As the speed of the unit decreases, the speed governor allows more steam from the boiler (or water from the gates) through to the turbine to arrest the decline in speed. Equilibrium between input and output power occurs at the new frequency $f = (f_0 + \Delta f)$ as shown. According to the slope of the speed-output characteristic given by Eq. (13.60), the frequency change (in Hz) is

$$\Delta f = -R\,\Delta P_g = -\left(R_u \frac{f_R}{S_R}\right) \Delta P_g \text{ Hz} \tag{13.61}$$

The isolated unit of Fig. 13.7 would continue to operate at the reduced frequency f except for the *supplementary control* action of the *speed changer*. The speed control mechanism has a speed changer motor which can parallel-shift the regulation characteristic to the new position shown by the dashed line of Fig. 13.7(b). Effectively, the speed changer supplements the action of the governor by changing the speed setting to allow more prime-mover energy through to increase the kinetic energy of the generating unit so that it can again operate at the desired frequency f_0 while providing the new output P_g.

When K generating units are operating in parallel on the system, their speed-droop characteristics determine how load changes are apportioned among them in the steady state. Consider that the K units are synchronously operating at a given frequency when the load changes by ΔP megawatts. Because the units are interconnected by the transmission networks, they are required to operate at speeds corresponding to a common frequency. Accordingly, in the steady-state equilibrium after initial governor action all units will have changed in frequency by the same incremental amount Δf hertz. The corresponding changes in the outputs of the units are given by Eq. (13.61) as follows:

$$\text{Unit 1:}\quad \Delta P_{g1} = -\frac{S_{R1}}{R_{1u}} \frac{\Delta f}{f_R} \text{ MW} \tag{13.62}$$

$$\vdots$$

$$\text{Unit } i\text{:}\quad \Delta P_{gi} = -\frac{S_{Ri}}{R_{iu}} \frac{\Delta f}{f_R} \text{ MW} \tag{13.63}$$

$$\vdots$$

$$\text{Unit } K\text{:}\quad \Delta P_{gK} = -\frac{S_{RK}}{R_{Ku}} \frac{\Delta f}{f_R} \text{ MW} \tag{13.64}$$

Adding these equations together gives the total change in output

$$\Delta P = -\left(\frac{S_{R1}}{R_{1u}} + \cdots + \frac{S_{Ri}}{R_{iu}} + \cdots + \frac{S_{RK}}{R_{Ku}}\right)\frac{\Delta f}{f_R} \tag{13.65}$$

from which the system frequency change is

$$\frac{\Delta f}{f_R} = -\frac{\Delta P}{\left(\dfrac{S_{R1}}{R_{1u}} + \cdots + \dfrac{S_{Ri}}{R_{iu}} + \cdots + \dfrac{S_{RK}}{R_{Ku}}\right)} \text{ per unit} \tag{13.66}$$

Substituting from Eq. (13.66) into Eq. (13.63), we find the *additional* output ΔP_{gi} of Unit i

$$\Delta P_{gi} = \frac{S_{Ri}/R_{iu}}{\left(\dfrac{S_{R1}}{R_{1u}} + \cdots + \dfrac{S_{Ri}}{R_{iu}} + \cdots + \dfrac{S_{RK}}{R_{Ku}}\right)}\Delta P \quad \text{MW} \tag{13.67}$$

which combines with the additional outputs of the other units to satisfy the load change ΔP of the system. The units would continue to operate in synchronism at the new system frequency except for the supplementary control exercised by the AGC system at the energy control center of the area in which the load change occurs. *Raise* or *lower* signals are sent to some or all the speed changers at the power plants of the particular area. Through such coordinated control of the set points of the speed governors it is possible to bring all the units of the system back to the desired frequency f_0 and to achieve any desired load division within the capabilities of the generating units.

Therefore, the governors on units of the interconnected system tend to maintain load-generation balance rather than a specific speed and the supplementary control of the AGC system within the individual control area functions so as to:

- Cause the area to absorb its own load changes,
- Provide the prearranged net interchange with neighbors,
- Ensure the desired economic dispatch output of each area plant, and
- Allow the area to do its share to maintain the desired system frequency.

The ACE is continuously recorded within the energy control center to show how well the individual area is accomplishing these tasks.

The block diagram of Fig. 13.8 indicates the flow of information in a computer controlling a particular area. The numbers enclosed by circles adjacent to the diagram identify positions on the diagram to simplify our discussion of the control operation. The larger circles on the diagram enclosing the

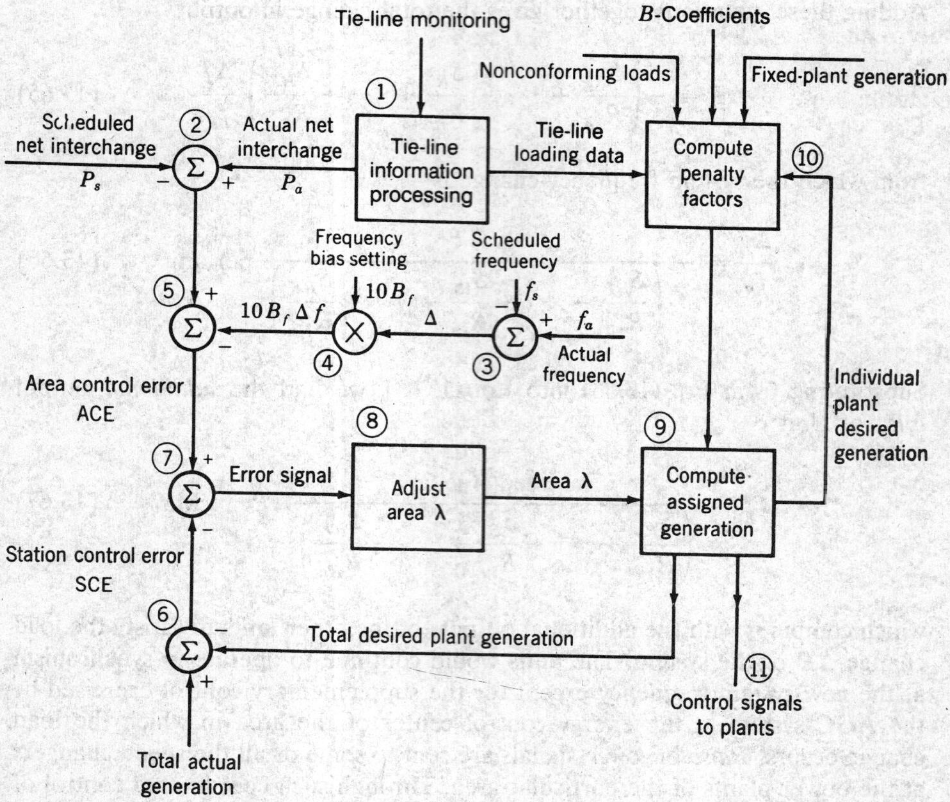

FIGURE 13.8
Block diagram illustrating the computer-controlled operation of a particular area.

symbols × or Σ indicate points of multiplication or algebraic summation of incoming signals.

At position 1 processing of information about power flow on tie lines to other control areas is indicated. The *actual* net interchange P_a is positive when net power is out of the area. The *scheduled* net interchange is P_s. At position 2 the scheduled net interchange is subtracted from the actual net interchange.[3] We shall discuss the condition where both actual and scheduled net interchange are out of the system and therefore positive.

Position 3 on the diagram indicates the subtraction of the scheduled frequency f_s (for instance, 60 Hz) from the actual frequency f_a to obtain Δf,

[3]Subtraction of standard or reference value from actual value to obtain the error is the accepted convention of power system engineers and is the negative of the definition of control error found in the literature of control theory.

the frequency deviation. Position 4 on the diagram indicates that the frequency bias setting B_f, a factor with a *negative sign* and the units MW/0.1 Hz, is multiplied by $10\ \Delta f$ to obtain a value of megawatts called the *frequency bias* $(10\ B_f \Delta f)$.

The frequency bias, which is positive when the actual frequency is less than the scheduled frequency, is subtracted from $(P_a - P_s)$ at position 5 to obtain the ACE, which may be positive or negative. As an equation

$$\text{ACE} = (P_a - P_s) - 10B_f(f_a - f_s) \text{ MW} \tag{13.68}$$

A negative ACE means that the area is not generating enough power to send the desired amount out of the area. There is a deficiency in net power output. Without frequency bias, the indicated deficiency would be less because there would be no positive offset $(10\ B_f \Delta f)$ added to P_s (subtracted from P_a) when actual frequency is less than scheduled frequency and the ACE would be less. The area would produce sufficient generation to supply its own load and the prearranged interchange but would not provide the additional output to assist neighboring interconnected areas to raise the frequency.

Station control error (SCE) is the amount of actual generation of all the area plants minus the desired generation, as indicated at position 6 of the diagram. This SCE is negative when desired generation is greater than existing generation.

The key to the whole control operation is the comparison of ACE and SCE. Their difference is an error signal, as indicated at position 7 of the diagram. If both ACE and SCE are negative and equal, the deficiency in the output from the area equals the excess of the desired generation over the actual generation and no error signal is produced. However, this excess of desired generation will cause a signal, indicated at position 11, to go to the plants to increase their generation and to reduce the magnitude of the SCE; the resulting increase in output from the area will reduce the magnitude of the ACE at the same time.

If ACE is more negative than SCE, there will be an error signal to increase the λ of the area, and this increase will in turn cause the desired plant generation to increase (position 9). Each plant will receive a signal to increase its output as determined by the principles of economic dispatch.

This discussion has considered specifically only the case of scheduled net interchange out of the area (positive scheduled net interchange) that is greater than actual net interchange with ACE equal to or more negative than SCE. The reader should be able to extend the discussion to the other possibilities by referring to Fig. 13.8.

Position 10 on the diagram indicates the computation of penalty factors for each plant. Here the *B*-coefficients are stored to calculate $\partial P_L/\partial P_{gi}$ and the penalty factors. The penalty factors are transmitted to the section (position 9), which establishes the individual plant outputs for economic dispatch and the total desired plant generation.

One other point of importance (not indicated on Fig. 13.8) is the offset in scheduled net interchange of power that varies in proportion to the *time error*, which is the integral of the per-unit frequency error over time in seconds. The offset is in the direction to help in reducing the integrated difference to zero and thereby to keep electric clocks accurate.

Example 13.6. Two thermal generating units are operating in parallel at 60 Hz to supply a total load of 700 MW. Unit 1, with a rated output of 600 MW and 4% speed-droop characteristic, supplies 400 MW, and Unit 2, which has a rated output of 500 MW and 5% speed droop, supplies the remaining 300 MW of load. If the total load increases to 800 MW, determine the new loading of each unit and the common frequency change before any supplementary control action occurs. Neglect losses.

Solution. The initial point a of operation on the speed regulation characteristic of each unit is shown in Fig. 13.9. For the load increase of 100 MW, Eq. (13.66) gives the per-unit frequency deviation

$$\frac{\Delta f}{f_R} = \frac{-100}{\dfrac{600}{0.04} + \dfrac{500}{0.05}} = -0.004 \text{ per unit}$$

Since f_R equals 60 Hz, the frequency change is 0.24 Hz and the new frequency of operation is 59.76 Hz. The load change allocated to each unit is given by Eq. (13.67)

$$\Delta P_{g1} = \frac{600/0.04}{\dfrac{600}{0.04} + \dfrac{500}{0.05}} 100 = 60 \text{ MW}$$

$$\Delta P_{g2} = \frac{500/0.05}{\dfrac{600}{0.04} + \dfrac{500}{0.05}} 100 = 40 \text{ MW}$$

and so Unit 1 supplies 460 MW, whereas Unit 2 supplies 340 MW at the new operating points b shown in Fig. 13.9. If supplementary control were applied to Unit 1 alone, the entire 100-MW load increase could be absorbed by that unit by shifting its characteristic to the final 60-Hz position at point c of Fig. 13.9. Unit 2 would then automatically return to its original operating point to supply 300 MW at 60 Hz.

The large number of generators and governors within a control area combine to yield an aggregate governing speed-power characteristic for the area as a whole. For relatively small load changes this area characteristic is often assumed linear and then treated like that of a single unit of capacity equal to that of the prevailing on-line generation in the area. On this basis, the following

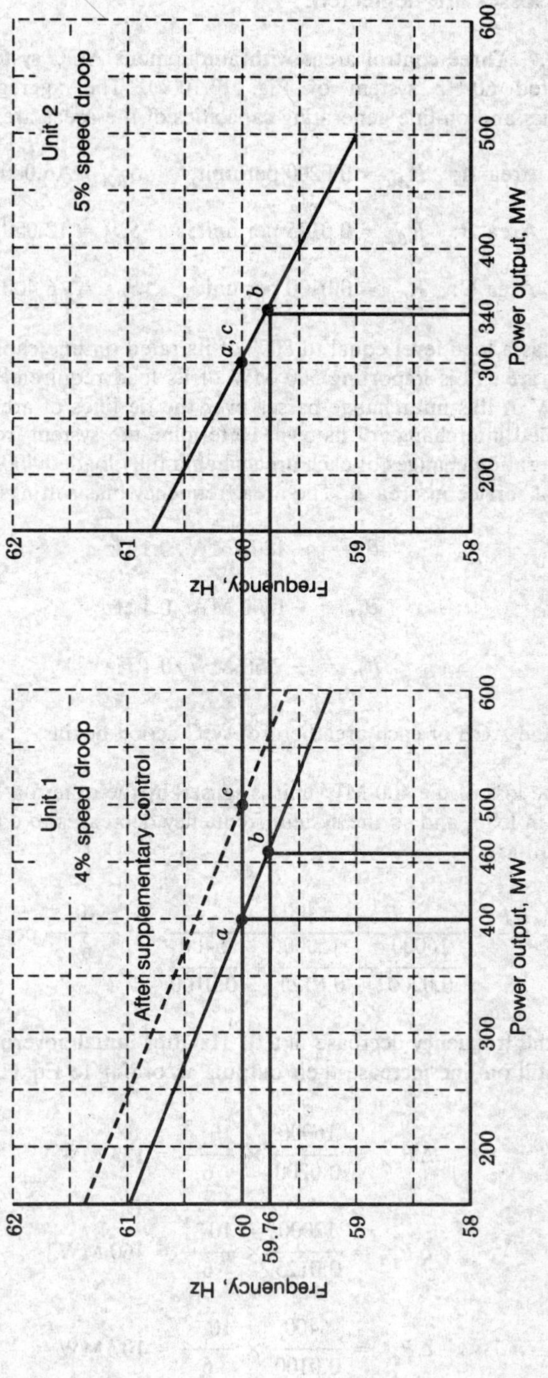

FIGURE 13.9
Load division between two isolated units of different speed-droop characteristics. Points *a* show initial distribution of 700-MW load; points *b* show distribution of 800-MW load at 59.76 Hz, and points *c* show final operating points of the units after supplementary control of Unit 1.

example demonstrates the steady-state operation of AGC for a three-area system in which losses are neglected.

Example 13.7. Three control areas with autonomous AGC systems comprise the interconnected 60-Hz system of Fig. 13.10(a). The aggregate speed-droop characteristics and on-line generating capacities of the areas are

$$\text{Area } A\text{:} \quad R_{Au} = 0.0200 \text{ per unit;} \qquad S_{RA} = 16{,}000 \text{ MW}$$

$$\text{Area } B\text{:} \quad R_{Bu} = 0.0125 \text{ per unit;} \qquad S_{RB} = 12{,}000 \text{ MW}$$

$$\text{Area } C\text{:} \quad R_{Cu} = 0.0100 \text{ per unit;} \qquad S_{RC} = 6{,}400 \text{ MW}$$

Each area has a load level equal to 80% of its rated on-line capacity. For reasons of economy area C is importing 500 MW of its load requirements from area B, and 100 MW of this interchange passes over the tie lines of area A, which has a zero scheduled interchange of its own. Determine the system frequency deviation and the generation changes of each area when a fully loaded 400-MW generator is forced out of service in area B. The area frequency bias settings are

$$B_{fA} = -1200 \text{ MW}/0.1 \text{ Hz}$$

$$B_{fB} = -1500 \text{ MW}/0.1 \text{ Hz}$$

$$B_{fC} = -\ \ 950 \text{ MW}/0.1 \text{ Hz}$$

Determine the ACE of each area before AGC action begins.

Solution. The loss of the 400-MW unit is sensed by the other on-line generators as an increase in load, and so the system frequency decreases to a value determined by Eq. (13.66) as

$$\frac{\Delta f}{f_R} = \frac{-400}{\dfrac{16000}{0.0200} + \dfrac{12000}{0.0125} + \dfrac{6400}{0.0100}} = \frac{-10^{-3}}{6} \text{ per unit}$$

Therefore, the frequency decrease in 0.01 Hz after initial governor action and the generators still on-line increase their outputs according to Eq. (13.63); that is,

$$\Delta P_{gA} = \frac{16000}{0.0200} \times \frac{10^{-3}}{6} = 133 \text{ MW}$$

$$\Delta P_{gB} = \frac{12000}{0.0125} \times \frac{10^{-3}}{6} = 160 \text{ MW}$$

$$\Delta P_{gC} = \frac{6400}{0.0100} \times \frac{10^{-3}}{6} = 107 \text{ MW}$$

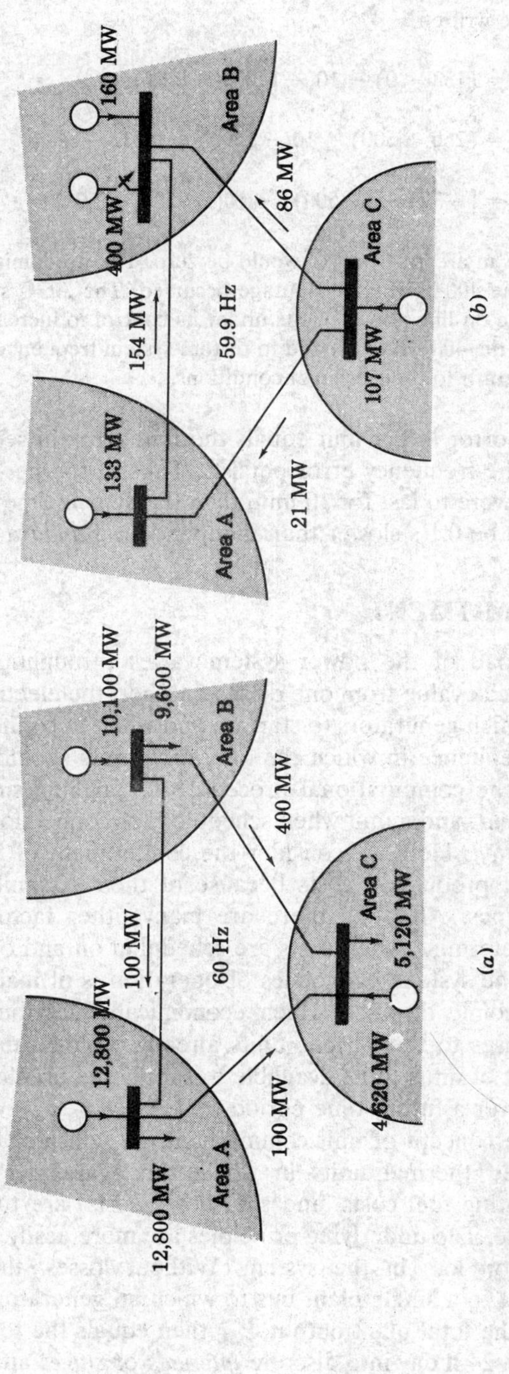

FIGURE 13.10
(*a*) Normal 60-Hz operation of three-area system of Example 13.9; (*b*) the incremental generation and tie-line flows resulting from the loss of 400-MW unit in area *B* before AGC action.

Let us suppose that these incremental changes are distributed to the interarea tie lines, as shown in Fig. 13.10(*b*). Then, by inspection, the area control error for each area may be written

$$(\text{ACE})_A = (133 - 0) - 10(-1200)(-0.01) \qquad = \quad 13\ \text{MW}$$

$$(\text{ACE})_B = (260 - 500) - 10(-1500)(-0.01) \qquad = -390\ \text{MW}$$

$$(\text{ACE})_C = [-393 - (-500)] - 10(-950)(-0.01) = \quad 12\ \text{MW}$$

Ideally, the ACE in areas A and C would be zero. The predominating ACE is in area B where the 400-MW forced outage occurred. The AGC system of area B will command the on-line power plants under its control to increase generation to offset the loss of the 400-MW unit and to restore system frequency of 60 Hz. Areas A and C then return to their original conditions.

The frequency error in per unit equals the *time error* in seconds for every second over which the frequency error persists. Thus, if the per-unit frequency error of $(-10^{-3}/6)$ were to last for 10 min, then the *system* time (as given by an electric clock) would be 0.1 s slower than independent *standard* time.

13.7 UNIT COMMITMENT

Because the total load of the power system varies throughout the day and reaches a different peak value from one day to another, the electric utility has to decide in advance which generators to start up and when to connect them to the network—and the sequence in which the operating units should be shut down and for how long. The computational procedure for making such decisions is called *unit commitment*, and a unit when scheduled for connection to the system is said to be *committed*. Here we consider the commitment of fossil-fuel units which have different production costs because of their dissimilar efficiencies, designs, and fuel types. Although there are many other factors of practical significance which determine when units are scheduled on and off to satisfy the operating needs of the system, economics of operation is of major importance. Unlike on-line economic dispatch which economically distributes the *actual* system load as it arises to the various units already on-line, unit commitment plans for the best set of units to be available to supply the predicted or *forecast* load of the system over a future time period.

To develop the concept of unit commitment, we consider the problem of scheduling fossil-fired thermal units in which the aggregate costs (such as start-up costs, operating fuel costs, and shut-down costs) are to be minimized over a daily load cycle. The underlying principles are more easily explained if we disregard transmission loss in the system. Without losses, the transmission network is equivalent to a single plant bus to which all generators and all loads are connected, and the total plant output P_{gT} then equals the total system load P_D. We subdivide the 24-h day into discrete *intervals* or *stages* and the predicted

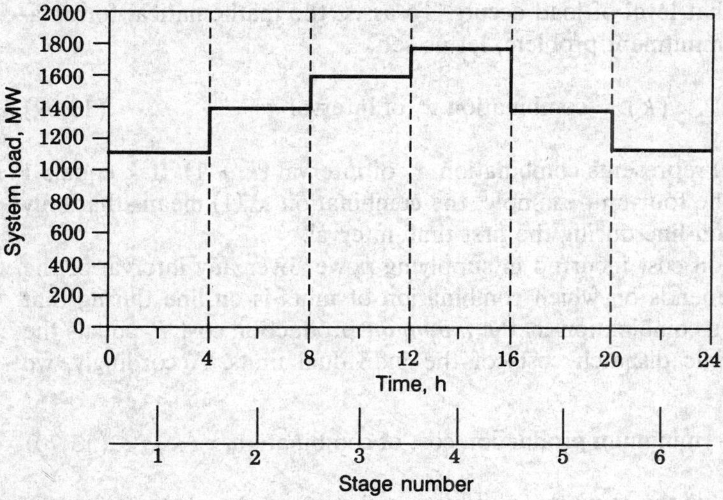

FIGURE 13.11
Discrete levels of system load for an example daily load cycle.

load of the system will be considered constant over each interval, as exemplified in Fig. 13.11. The unit commitment procedure then searches for the most economic *feasible combination* of generating units to serve the forecast load of the system at each stage of the load cycle.

The power system with K generating units (no two identical) must have at least one unit on-line to supply the system load which is never zero over the daily load cycle. If each unit can be considered either *on* (denoted by 1) or *off* (denoted by 0), there are $2^K - 1$ candidate combinations to be examined in each stage of the study period. For example, if $K = 4$, the 15 theoretically possible combinations for each interval are

Combinations

Unit	x_1	x_2	x_3	x_4	x_5	x_6	x_7	x_8	x_9	x_{10}	x_{11}	x_{12}	x_{13}	x_{14}	x_{15}
1	1	1	1	1	0	0	1	0	1	1	0	1	0	0	0
2	1	1	1	0	1	1	0	0	1	0	1	0	1	0	0
3	1	1	0	1	1	0	0	1	0	1	1	0	0	1	0
4	1	0	1	1	1	1	1	1	0	0	0	0	0	0	1

where x_i denotes combination i of the four units. Of course, not all combinations are feasible because of the constraints imposed by the load level and other practical operating requirements of the system. For example, a combination of units of total capability less than 1400 MW cannot serve a load of 1400 MW or greater; any such combination is *infeasible* and can be disregarded over any time interval in which that level of load occurs. To assist the mathematical formulation of the unit commitment problem, let us set

$$x_i(k) = \text{combination } x_i \text{ of interval } k \tag{13.69}$$

and then $x_j(k+1)$ represents combination x_j of interval $(k+1)$. If k equals 1 and i equals 9 in the four-unit example, the combination $x_9(1)$ means that only Units 1 and 2 are on-line during the first time interval.

The production cost incurred in supplying power over any interval of the daily load cycle depends on which combination of units is on-line during that interval. For a given combination x_i the *minimum* production cost P_i equals the sum of the economic dispatch costs of the individual units. Accordingly, we designate

$$P_i(k) = \text{minimum production cost of combination } x_i(k) \tag{13.70}$$

and then $P_j(k+1)$ is the minimum production cost of combination $x_j(k+1)$. Besides production cost, in the unit commitment problem we must also consider *transition cost*, which is the cost associated with changing from one combination of power-producing units to another combination. Usually, a fixed cost is assigned to shutting down a unit which has been operating on the system since *shut-down* cost is generally independent of the length of time the unit has been running. However, in practical situations the *start-up* cost of a unit depends on how long the unit has been shut down from previous operation. This is to be expected since the boiler temperature of the unit and the fuel required to restore operating temperature depend on the duration of cooling. To explain more easily the unit commitment concept, we assume a fixed start-up cost for each unit and refer the reader to other practical considerations treated elsewhere.[4] Thus, transition cost associated with changing from one combination of operating units to another will have fixed start-up and shut-down components denoted by

$$T_{ij}(k) = \text{cost of transition from combination } x_i(k) \text{ to combination } x_j(k+1) \text{ between intervals } k \text{ and } k+1 \tag{13.71}$$

If each unit could be started up or shut down without incurring any transition cost, then from the economic viewpoint the problem of scheduling units to operate in any one hour would become disjoint from and totally unrelated to the scheduling problem in any other hour of the load cycle. On the other hand, suppose that it costs \$1500 to shut down a unit and \$3000 for each unit start-up. Then, to change from combination $x_2(k)$ to combination $x_3(k+1)$ in the above four-unit example, the transition cost $T_{2,3}(k)$ becomes (\$1500 + \$3000) = \$4500

[4]See, for example, C. K. Pang and H. C. Chen, "Optimal Short-Term Thermal Unit Commitment," *IEEE Transactions on Power Apparatus and Systems*, vol. PAS-95, no. 4, 1976, pp. 1336–1346.

because Unit 3 is to shut down and Unit 4 is to start up at the beginning of interval $(k + 1)$. Furthermore, the status of units in interval $(k + 1)$ affects the cost of transition to interval $(k + 2)$, and so on. Therefore, transition cost links the scheduling decision of any one interval to the scheduling decisions of all the other intervals of the load cycle. Accordingly, the problem of minimizing costs at one stage is tied to the combinations of units chosen for all the other stages, and we say that unit commitment is a *multistage* or *dynamic* cost-minimization problem.

The dynamic nature of the unit commitment problem complicates its solution. Suppose that 10 units are available for scheduling within any one-hour interval, which is not unlikely in practice. Then, theoretically a total of $2^{10} - 1 = 1023$ combinations can be listed. If it were possible to link each prospective combination of any one hour to each prospective combination of the next hour of the day, the total number of candidate combinations becomes $(1023)^{24} = 1.726 \times 10^{72}$, which is enormously large and unrealistic to handle. Fortunately, however, the multistage decision process of the unit commitment problem can be dimensionally reduced by practical constraints of system operations and by a search procedure based on the following observations:

- The daily schedule has N discrete time intervals or stages, the durations of which are not necessarily equal. Stage 1 precedes stage 2, and so on to the final stage N.
- A decision must be made for each stage k regarding which particular combination of units to operate during that stage. This is the stage k subproblem.
- To solve for the N decisions, N subproblems are solved sequentially in such a way (called the *principle of optimality* to be explained later) that the combined best decisions for the N subproblems yield the best overall solution for the original problem.

This strategy greatly reduces the amount of computation to solve the original unit commitment problem, as we shall see.

The cost $f_{ij}(k)$ associated with any stage k has two components given by

$$f_{ij}(k) = P_i(k) + T_{ij}(k) \tag{13.72}$$

which is the combined transition and production cost incurred by combination x_i during interval k plus the transition cost to combination x_j of the next interval. For ease of explanation it is assumed that the system load levels at the beginning and end of the day are the same. Consequently, it is reasonable to expect that the state of the system is the same at the beginning and at the end of the day. Because of this, when $k = N$, the transition cost $T_{ij}(N)$ becomes zero. We note that the cost $f_{ij}(k)$ is tied by $T_{ij}(k)$ to the decision of the next stage $(k + 1)$. Suppose, for the moment, that we happen to know the *best policy* or set of decisions (in the sense of *minimum cost*) over the first $(N-1)$ stages of the daily load cycle; this is equivalent to assuming that we know how to choose the best combination of units for each of the first $(N-1)$ intervals. If we agree that combination x_{i^*} is the *best* combination for stage $(N-1)$, then by searching among all the feasible combinations x_j of the final stage N, we can find

$$F_{i^*}(N-1) = \min_{\{x_j(N)\}} \left\{P_{i^*}(N-1) + T_{i^*j}(N-1) + F_j(N)\right\} \quad (13.73)$$

where $F_{i^*}(N-1)$ is the *minimum cumulative cost* of the *final two* stages starting with combination $x_{i^*}(N-1)$ and ending with combination $x_j(N)$; the cumulative cost $F_j(N)$ of stage N equals the production cost $P_j(N)$ since there is no further transition cost involved. The notation of Eq. (13.73) means that the search for the minimum-cost decision is made over all feasible combinations x_j at stage N. Of course, we do not yet know which combination is $x_{i^*}(N-1)$. But for *each* possible starting combination $x_i(N-1)$ it is straightforward to solve for the best stage-N decision and to store the corresponding minimum-cost results in a table for later retrieval when the best combination $x_{i^*}(N-1)$ has been identified.

Similarly, starting with the combination $x_{i^*}(N-2)$ at interval $(N-2)$, the minimum cumulative cost of the *final three* stages of the study period is given by

$$F_{i^*}(N-2) = \min_{\{x_j(N-1)\}} \left\{P_{i^*}(N-2) + T_{i^*j}(N-2) + F_i(N-1)\right\} \quad (13.74)$$

where the search is now made among the feasible combinations x_j of stage $(N-1)$. Continuing the above logic, we find the *recursive formula*

$$F_{i^*}(k) = \min_{\{x_j(k+1)\}} \left\{P_{i^*}(k) + T_{i^*j}(k) + F_j(k+1)\right\} \quad (13.75)$$

for the minimum cumulative cost at stage k, where k ranges from 1 to N. For the reasons stated earlier regarding the state being the same at the beginning and at the end of the day when k equals N in Eq. (13.75), $T_{i^*j}(N)$ and $F_j(N+1)$ are set to zero. When k equals 1, the combination $x_{i^*}(1)$ is the known initial-condition input to the unit commitment problem. The combinations corresponding to subscripts i^* and j of Eq. (13.75) change roles from one stage to the next; the combination $x_{i^*}(k)$, which initiates one search among the feasible combinations $x_j(k+1)$, becomes one of the feasible combinations $x_j(k)$ which enter into all searches of stage $(k-1)$. This is graphically illustrated in Fig. 13.12, which shows a grid constructed for three typical consecutive stages $(k-1)$, k, and $(k+1)$. Suppose that each stage is associated with the load level shown and has a choice of one of the combinations x_1, x_2, x_3, and x_9 previously listed. Then, each node of the grid represents one of the combina-

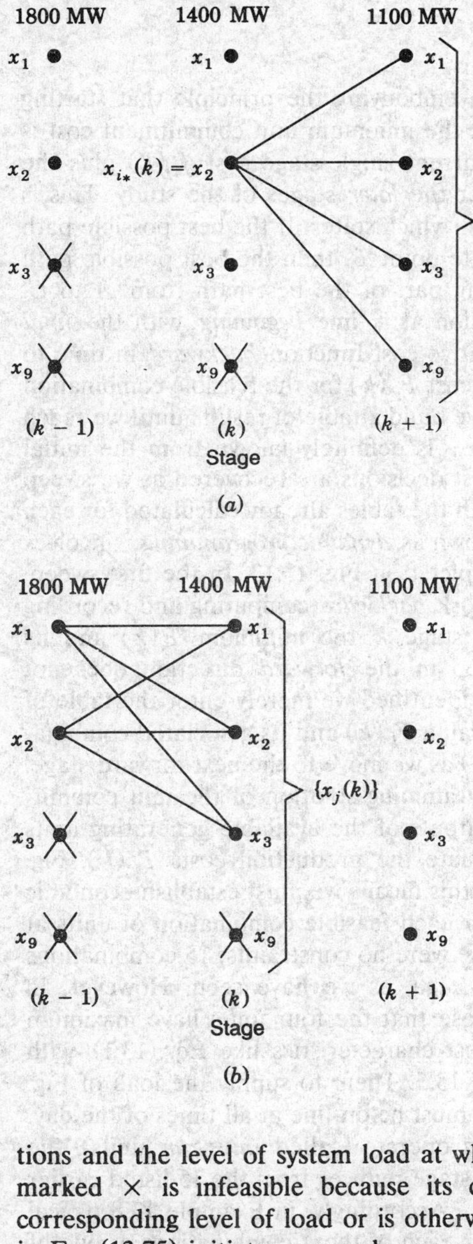

FIGURE 13.12
Illustration of the minimum cost search procedure associated with Eq. (13.75) for: (*a*) $x_{i^*}(k)$ set equal to x_2; (*b*) $x_{i^*}(k-1)$ set equal to x_1 and then to x_2.

tions and the level of system load at which its units are to operate. Any node marked $\times$ is infeasible because its combination either cannot supply the corresponding level of load or is otherwise inadmissible. Choosing $x_{i^*}(k) = x_2$ in Eq. (13.75) initiates a search among the feasible set $\{x_1, x_2, x_3, x_9\}$ for the minimum cumulative cost $F_{i^*}(k)$, as depicted in Fig. 13.12(*a*); a similar picture applies when each of the other two feasible combinations x_1 and x_3 of stage k is chosen as $x_{i^*}(k)$. Replacing k by $(k-1)$ in Eq. (13.75) steps us through to Fig. 13.12(*b*), which shows that the previous search-initiating combination has now the role of $x_j(k)$ and enters into every search for $F_{i^*}(k-1)$.

13.8 SOLVING THE UNIT COMMITMENT PROBLEM

Equation (13.75) is an iterative relation embodying the principle that starting with a given combination x_{i^*} at stage k, the minimum unit commitment cost is found by minimizing the *sum* of the current single-stage cost $f_{ij}(k)$ plus the minimum cumulative cost $F_j(k+1)$ over the *later* stages of the study. This is one example of the *principle of optimality*, which states: If the best possible path from A to C passes through intermediate point B, then the best possible path from B to C must be the corresponding part of the best path from A to C. Computationally, we evaluate one decision at a time *beginning* with the *final* stage N and carry the minimum cumulative cost function *backward* in time to stage k to find the minimum cumulative cost $F_{i^*}(k)$ for the feasible combination x_{i^*} of stage f, Fig. 13.13. At each stage we build a table of results until we reach stage 1, where the input combination x_{i^*} is definitely known from the initial conditions. The minimum cumulative cost decisions are recovered as we sweep from stage 1 to stage N searching through the tables already calculated for each stage. The computational procedure, known as *dynamic programming*,[5] involves *two* sweeps through each stage k, as depicted in Fig. 13.13. In the first sweep, which is computationally intensive, we work *backward* computing and recording for each candidate combination x_i of stage k the minimum $F_i(k)$ and its associated $x_j(k+1)$. The second sweep in the *forward* direction does not involve any processing since with $x_{i^*}(k)$ identified we merely enter the table of results already recorded to retrieve the value $F_{i^*}(k)$ and its associated combination $x_j(k+1)$, which becomes $x_{i^*}(k+1)$ as we move to the next forward stage.

At each stage in the dynamic programming solution of the unit commitment problem the economic dispatch outputs of the available generating units must be calculated before we can evaluate the production costs $P_{i^*}(k)$, Fig. 13.13. For a system with four generators this means we must establish economic dispatch tables, similar to Table 13.1, for each feasible combination of units at every load level of the daily cycle. If there were no constraints, 15 combinations would have to be considered at *each stage*, as we have seen. However, in practice, constraints always apply. Suppose that the four units have maximum and minimum loading limits and fuel-cost characteristics like Eq. (13.1), with coefficients a_i, b_i, and c_i given in Table 13.5. Then, to supply the load of Fig. 13.11, at least two of the four generators must be on-line at all times of the day. By specifying that Units 1 and 2 always operate (called *must-run* units), the number of feasible combinations at each stage reduces from the 15 listed earlier to the 4 combinations x_1, x_2, x_3, and x_9. Accordingly, in Example 13.8 typical economic dispatch results are derived for each of the 4 combinations at one of the load levels of Fig. 13.11. The more complete results of Table 13.6 are then

[5]See, for example, R. Bellman, *Dynamic Programming*, Princeton University Press, Princeton, NJ, 1957.

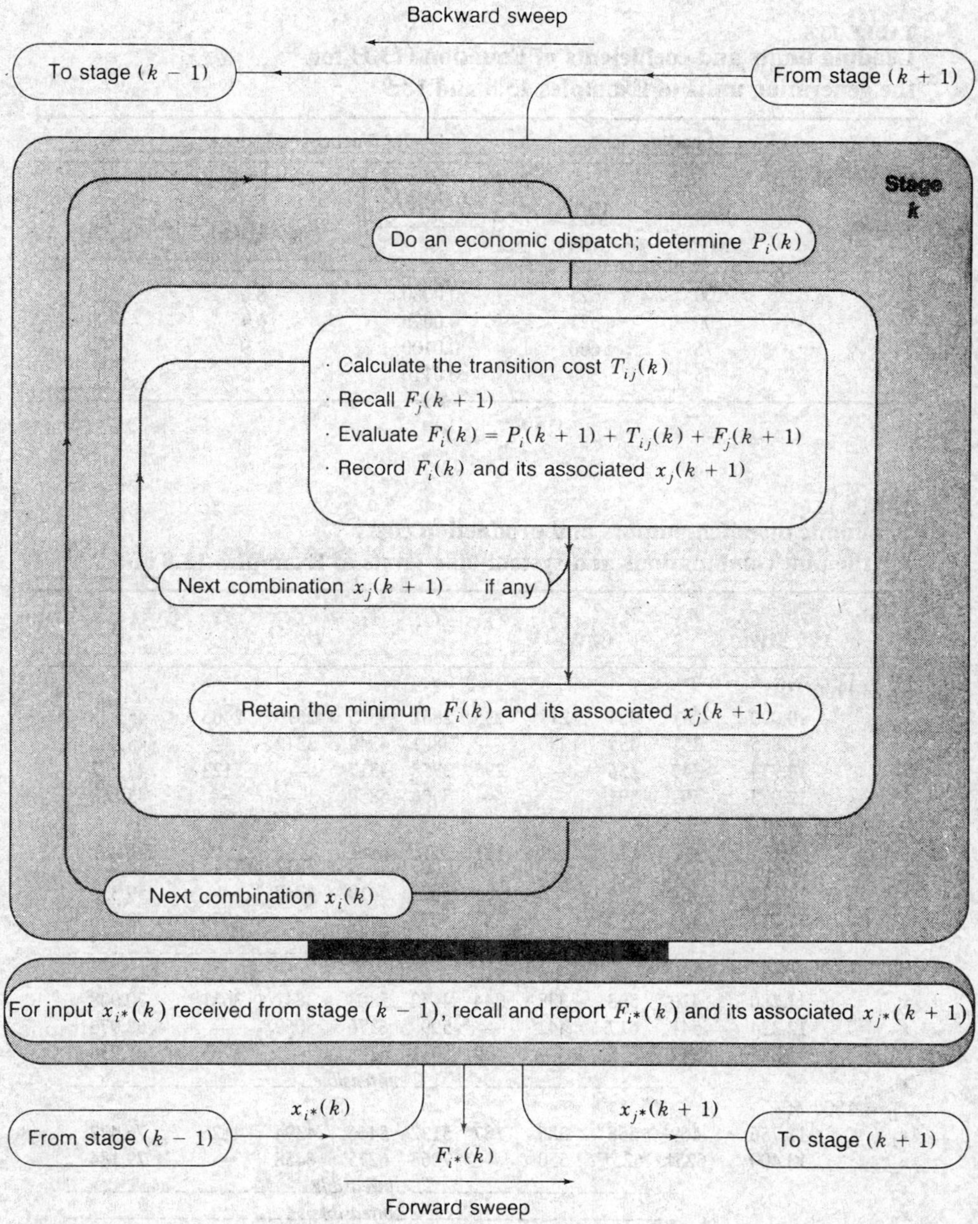

FIGURE 13.13
Summary of stage k calculations in the dynamic programming solution of the unit commitment problem of Eq. (13.75).

TABLE 13.5
Loading limits and coefficients of Equation (13.1) for the generating units of Examples 13.8 and 13.9

Generating unit number	Loading limits Min, (MW)	Loading limits Max, (MW)	Fuel cost parameters a_i, $\left(\frac{\$/(\text{MW})^2}{\text{h}}\right)$	Fuel cost parameters b_i, ($/MWh)	Fuel cost parameters c_i, ($/h)
1	100	625	0.0080	8.0	500
2	100	625	0.0096	6.4	400
3	75	600	0.0100	7.9	600
4	75	500	0.0110	7.5	400

TABLE 13.6
Economic dispatch outputs and production costs for the unit combinations and system load levels of Examples 13.8 and 13.9

Comb. x_i	System λ ($/MWh)	P_{g1} (MW)	P_{g2} (MW)	P_{g3} (MW)	P_{g4} (MW)	f_1 ($/h)	f_2 ($/h)	f_3 ($/h)	f_4 ($/h)	$P_i(k) = 4\Sigma f_i$ ($)
P_D = 1100 *MW*										
x_1	10.090	261	384	219	235	2861	3565	2570	2466	45,848
x_2	10.805	351	459	290	—	3801	4349	3312	—	45,848
x_3	10.774	347	456	—	298	3758	4317	—	3123	44,792
x_9	12.073	509	591	—	—	5608	5859	—	—	45,868
P_D = 1400 *MW*										
x_1	10.804	351	459	290	300	3801	4349	3312	3145	58,428
x_2	11.717	464	554	382	—	5073	5419	4347	—	59,356
x_3	11.711	464	553	—	383	5073	5407	—	4079	58,236
x_9		*infeasible*								
P_D = 1600 *MW*										
x_1	11.280	410	508	338	344	4452	5803	3841	3631	70,908
x_2	12.324	541	617	442	—	5999	6176	5069	—	68,976
x_3	12.336	542	618	—	440	6011	6188	—	4765	67,856
x_9		*infeasible*								
P_D = 1800 *MW*										
x_1	11.756	469	558	386	387	5132	5466	4394	4126	76,472
x_2	13.400	625†	625†	550	—	7063	6275	6458	—	79,184
x_3		*infeasible*								
x_9		*infeasible*								

†Daggers denote loading limits.

used in Example 13.9 to solve the unit commitment problem for the same four-unit system and load cycle. We neglect transmission loss in both of the following examples to simplify the calculations.

Example 13.8. The system load of Fig. 13.11 is to be supplied by combinations of the four generating units of Table 13.5. Treating Units 1 and 2 as must-run units, determine the power supplied by the generators of each combination and the corresponding production cost in economically loading the units when the system load level is 1100 MW.

Solution. Because Units 1 and 2 must operate at all times, the four combinations of units to be considered at each stage of the daily cycle are x_1, x_2, x_3, and x_9, as indicated above. Since losses are neglected, economic dispatch of each of these combinations is determined directly from Eq. (13.6), provided no unit is operating at a maximum or minimum loading limit. We consider each combination separately.

Combination x_9. Units 1 and 2 have incremental fuel costs identical to those of Example 13.1. Accordingly, with only Units 1 and 2 operating, Table 13.1 applies and the economic dispatch outputs at the 1100-MW load level are $P_{g1} = 509$ MW and $P_{g2} = 591$ MW. The corresponding hourly production costs of the two units are calculated from Eq. (13.1) using the given parameters as follows:

$$f_1 = 0.004P_{g1}^2 + 8.0P_{g1} + 500\Big|_{P_{g1}=509} = \$5608 \text{ per hour}$$

$$f_2 = 0.0048P_{g2}^2 + 6.4P_{g2} + 400\Big|_{P_{g2}=591} = \$5859 \text{ per hour}$$

Hence, at stage 1 of Fig. 13.11 the production cost of combination x_9 amounts to $P_9(1) = \$45{,}868$, which is the accumulation of (\$5608 + \$5859) for each hour of the 4-h interval.

Combination x_3. With Units 1, 2, and 4 operating, we follow Eq. (13.7) to calculate the coefficients

$$a_T = \left(a_1^{-1} + a_2^{-1} + a_4^{-1}\right)^{-1} = \left(0.008^{-1} + 0.0096^{-1} + 0.011^{-1}\right)^{-1} = 3.1243 \times 10^{-3}$$

$$b_T = a_T\left(\frac{b_1}{a_1} + \frac{b_2}{a_2} + \frac{b_4}{a_4}\right) = a_T\left(\frac{8}{0.008} + \frac{6.4}{0.0096} + \frac{7.5}{0.011}\right) = 7.3374$$

The incremental fuel cost for the three units at the load level of 1100 MW is then given by Eq. (13.6) as

$$\lambda = a_T P_{gT} + b_T = 3.1243 \times 10^{-3}(1100) + 7.3374 = 10.774 \text{ MWh}$$

and the corresponding economic dispatch outputs of Units 1, 2, and 4 have the (rounded-off) values

$$P_{g1} = \frac{\lambda - b_1}{a_1} = \frac{10.774 - 8.0}{0.008} = 347 \text{ MW}$$

$$P_{g2} = \frac{\lambda - b_2}{a_2} = \frac{10.774 - 6.4}{0.0096} = 456 \text{ MW}$$

$$P_{g4} = \frac{\lambda - b_4}{a_4} = \frac{10.774 - 7.5}{0.011} = 298 \text{ MW}$$

The hourly production costs of the three units are calculated to be

$$f_1 = 0.004P_{g1}^2 + 8.0P_{g1} + 500|_{P_{g1}=347} = \$3758 \text{ per hour}$$

$$f_2 = 0.0048P_{g2}^2 + 6.4P_{g2} + 400|_{P_{g2}=456} = \$4317 \text{ per hour}$$

$$f_4 = 0.0055P_{g4}^2 + 7.5P_{g4} + 400|_{P_{g4}=298} = \$3123 \text{ per hour}$$

Therefore, if Units 1, 2, and 4 of combination x_3 were to operate during interval 1 of Fig. 13.11, the production cost $P_3(1)$ would equal \$44,792, which is four times the sum of hourly costs.

Combination x_2. The economic dispatch outputs and the production costs of Units 1, 2, and 3 are similarly found by replacing a_4, b_4, and c_4 in the last set of calculations by the Unit 3 parameters a_3, b_3, and c_3 of Table 13.5. The results are

$$a_T = 3.038 \times 10^{-3} \qquad P_{g1} = 351 \text{ MW} \qquad f_1 = 3801 \text{ \$/h}$$

$$b_T = 7.4634 \qquad P_{g2} = 459 \text{ MW} \qquad f_2 = 4349 \text{ \$/h}$$

$$\lambda = 10.805 \text{ \$/MWh} \qquad P_{g3} = 290 \text{ MW} \qquad f_3 = 3312 \text{ \$/h}$$

Thus, if only Units 1, 2, and 3 are operating at the load level of 1100 MW, the production cost is $P_2(1) = \$45{,}848$.

Combination x_1. If all four generating units of combination x_1 were to serve the load of 1100 MW, the corresponding results are

$$a_T = 2.3805 \times 10^{-3} \qquad P_{g1} = 261 \text{ MW} \qquad f_1 = 2861 \text{ \$/h}$$

$$b_T = 7.4712 \qquad P_{g2} = 385 \text{ MW} \qquad f_2 = 3565 \text{ \$/h}$$

$$\lambda = 10.090 \text{ \$/MWh} \qquad P_{g3} = 219 \text{ MW} \qquad f_3 = 2570 \text{ \$/h}$$

$$P_{g4} = 235 \text{ MW} \qquad f_4 = 2466 \text{ \$/h}$$

$$P_1(1) = \$45{,}848$$

Since the preceding results apply whenever the system load is 1100 MW, they are also applicable to stages 1 and 6 of Fig. 13.11. Table 13.6 provides the complete results required in the next example.

Example 13.9. Assuming the start-up cost of each thermal generating unit is $3000 and the shut-down cost is $1500, determine the optimal unit commitment policy for the four thermal units of Example 13.8 to serve the system load of Fig. 13.11. Only the must-run Units 1 and 2 are to operate at the first *and* final stages of the load cycle.

Solution. The dynamic programming solution will be represented graphically, as shown in Fig. 13.14. A grid is first constructed for the six stages of the load cycle and the production cost is entered to the right and below each node from the economic dispatch results of Table 13.6. Infeasible nodes occur at the left and right boundaries of the grid because of specified input and terminating conditions of this example. Each diagonal line of the grid has an associated transition cost because of the change from one combination of units to another combination. A horizontal line has zero transition cost because it connects two nodes of the same combination, even though at different load levels. The transition costs of Fig. 13.14 assign $3000 and $1500, respectively, to each unit start-up and shut-down as specified.

The graphical solution of this example is the path of least cumulative cost linking the initial-condition node of stage 1 and the destination node of stage 6. We begin at stage 6, which has only Units 1 and 2 of combination x_9 operating. Setting N equal to 6 and i^* equal to 9 in Eq. (13.75) gives

$$F_9(6) = \min_{x_j(7)} \{ P_9(6) + 0 + 0 \} = \$45{,}868$$

which is the minimum production cost at stage 6 since the set $\{x_j(7)\}$ has no elements. The set of feasible combinations at stage 6 has only one element, namely, x_9. We now move to stage 5 to evaluate the minimum cumulative cost of the two final stages. Setting k equal to 5 and $\{x_j(6)\}$ equal to $\{x_9(6)\}$ in Eq. (13.75), we obtain

$$F_{i^*}(5) = \min_{x_9(6)} \{ P_{i^*}(5) + T_{i^*9}(5) + F_9(6) \} = \{ P_{i^*}(5) + T_{i^*9}(5) + \$45{,}868 \}$$

At stage 5 the node corresponding to combination x_9 is infeasible because Units 1 and 2 alone cannot supply 1400 MW of system load. Thus, there are three feasible nodes at stage 5 corresponding to the combinations x_1, x_2, and x_3, but we do not yet know which of them is on the path of least cumulative cost. Systematically, we set x_1, then x_2, and finally x_3 equal to x_{i^*} in the last equation, which gives

$$F_1(5) = \{ P_1(5) + T_{1,9}(5) + \$45{,}868 \} = [\$58{,}428 + \$3000 + \$45{,}868] = \$107{,}296$$

$$F_2(5) = \{ P_2(5) + T_{2,9}(5) + \$45{,}868 \} = [\$59{,}356 + \$1500 + \$45{,}868] = \$106{,}724$$

$$F_3(5) = \{ P_3(5) + T_{3,9}(5) + \$45{,}868 \} = [\$58{,}236 + \$1500 + \$45{,}868] = \$105{,}604$$

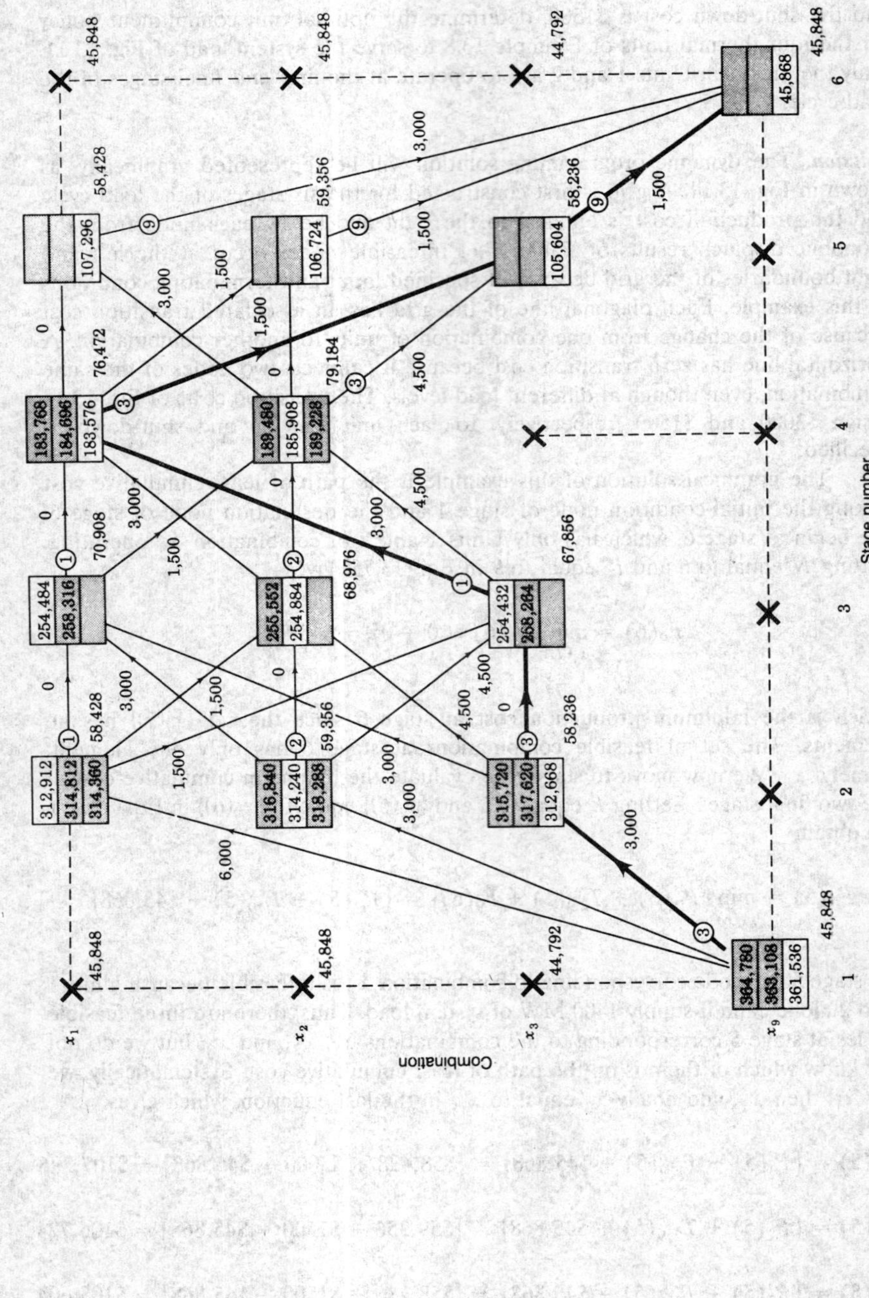

FIGURE 13.14
Dynamic programming solution of the unit commitment problem of Example 13.9.

The preceding calculations are straightforward because the search for the minimum cumulative cost involves only the single combination $x_9(6)$ of the final stage. We record these results at each of the stage 5 nodes, as shown in Fig. 13.14, and proceed to stage 4 where only combinations x_1 and x_2 have sufficient capacity to serve the load of 1800 MW. Hence, the best combination x_{i^*} at stage 4 must be either x_1 or x_2, but we do not yet know which one. The stage 4 version of Eq. (13.75) is

$$F_{i^*}(4) = \min_{x_j(5)} \{P_{i^*}(4) + T_{i^*,j}(4) + F_j(5)\}$$

where $x_j(5)$ ranges over the set of feasible combinations $\{x_1, x_2, x_3\}$ of stage 5. Setting x_{i^*} first equal to x_1, we obtain

$$F_1(4) = \min_{x_j(5)} \{P_1(4) + T_{1,j}(4) + F_j(5)\}$$

$$= \min\{[P_1(4) + T_{1,1}(4) + F_1(5)]; [P_1(4) + T_{1,2}(4) + F_2(5)];$$

$$[P_1(4) + T_{1,3}(4) + F_3(5)]\}$$

The numerical values for this equation are available from Fig. 13.14, which gives

$$F_1(4) = \min\{[76,472 + 0 + 107,296]; [76,472 + 1500 + 106,724];$$

$$[76,472 + 1500 + 105,604]\}$$

$$= \min\{[183,768]; [184,696]; [183,576]\} = \$183,576$$

This result must now be recorded, as shown in Fig. 13.14, and the two values \$183,768 and \$184,696, which could be discarded, are also shown for information only.

At this point we have determined the cumulative cost from stage 4 to stage 6 for the path beginning with combination x_1 at stage 4. To evaluate the path starting with the feasible combination x_2 of stage 4, we set x_{i^*}, equal to x_2 in the stage 4 version of Eq. (13.75) and perform calculations similar to those just demonstrated. The result is

$$F_2(4) = \min\{[189,480]; [185,908]; [189,288]\} = \$185,908$$

which is recorded at the appropriate node of Fig. 13.14. No conclusion can be drawn yet, but when we identify which stage 4 node is on the path of least overall cumulative cost, we select from the numerical results $F_1(4)$ and $F_2(4)$. We now proceed to stage 3 and on to stages 2 and 1, repeating the evaluation of Eq. (13.75) at each feasible node and recording the corresponding minimum cumulative cost. One of the transition paths leaving each node is marked with the (encircled) subscript of the combination from which the least cumulative cost of the node is derived.

At stage 1 it is determined that the least overall cumulative cost is \$361,536 from stage 6 to the initial-condition combination (and vice versa). This total cost is shown in Fig. 13.14 to derive from combination x_3 of stage 2, which in turn derives from combination x_3 of stage 3, and so on back to stage 6. The least cost path retraced in Fig. 13.14 shows that the optimal unit commitment schedule is:

Stage	Load level, MW	Combination	Units
1	1100	x_9	1, 2
2	1400	x_3	1, 2, 4
3	1600	x_3	1, 2, 4
4	1800	x_1	1, 2, 3, 4
5	1400	x_3	1, 2, 4
6	1100	x_9	1, 2

and the total cost to supply the daily forecast load of Fig. 13.11 amounts to \$361,536 in this example.

Example 13.9 demonstrates the great reduction in computational effort which is possible by the dynamic programming approach. As shown by the recorded values at the nodes of Fig. 13.14, the cumulative cost function $F_i(k)$ was evaluated only 27 times, which is the *sum* $(3 + 9 + 6 + 6 + 3)$ of the interstage transitions from stage 1 to stage 6. A brute-force enumeration approach to the same problem would have involved a total of 2916 interstage transitions, which is the *product* $(3 \times 9 \times 6 \times 6 \times 3)$ of line segments shown in Fig. 13.14. Indeed, given the starting and ending states as in Example 13.9, and if there were no infeasible nodes at any of the intermediate stages, the number of interstage transitions between the 15 combinations x_1 to x_{15} would increase enormously to $15 \times 15^2 \times 15^2 \times 15^2 \times 15 = (225)^4 = 2.563 \times 10^9$. Thus, dynamic programming offers a viable approach to the solution of the unit commitment problem when practical constraints of system operations are taken into account.

13.9 SUMMARY

The classical solution of the economic dispatch problem is provided by the method of Lagrange multipliers. The solution states that the minimum fuel cost is obtained when the incremental fuel cost df_i/dP_{gi} of each unit multiplied by its penalty factor L_i is the same for all operating units of the system. Each of the products $L_i(df_i/dP_{gi})$ is equal to the system λ, which is approximately the cost in dollars per hour to increase the total delivered load by 1 MW; that is,

$$\lambda = L_1 \frac{df_1}{dP_{g1}} = L_2 \frac{df_2}{dP_{g2}} = L_3 \frac{df_3}{dP_{g3}} \tag{13.76}$$

The penalty factor for plant i is defined by

$$L_i = \frac{1}{1 - \partial P_L / \partial P_{gi}} \tag{13.77}$$

where P_L is the total real power transmission loss. The incremental loss $\partial P_L / \partial P_{gi}$ is a measure of the sensitivity of the system losses to an incremental change in the output of plant i when all other plant outputs are kept fixed.

Transmission loss P_L can be expressed in terms of B-coefficients and the power outputs P_{gi} by

$$P_L = \sum_{i=1}^{K} \sum_{j=1}^{K} P_{gi} B_{ij} P_{gj} + \sum_{i=1}^{K} B_{i0} P_{gi} + B_{00} \tag{13.78}$$

The B-coefficients, which must have units consistent with those of P_{gi}, can be determined from the results of a converged power-flow solution by means of a power-invarient transformation based on the real part ($\mathbf{R}_{\text{bus}}$) of the system $\mathbf{Z}_{\text{bus}}$. An algorithm for solving the classical economic dispatch problem is provided in Sec. 13.5.

The fundamentals of automatic generation control are explained in Sec. 13.6, where the definitions of area control error (ACE) and time error are introduced.

The rudiments of unit commitment are provided in Sec. 13.7, which explains the principle of optimality underlying the dynamic programming solution approach. Figure 13.13 summarizes the overall solution procedure.

PROBLEMS

13.1. For a generating unit the fuel input in millions of Btu/h is expressed as a function of output P_g in megawatts by $0.032P_g^2 + 5.8P_g + 120$. Determine

(*a*) The equation for incremental fuel cost in dollars per megawatthour as a function of P_g in megawatts based on a fuel cost of \$2 per million Btu.

(*b*) The average cost of fuel per megawatthour when $P_g = 200$ MW.

(*c*) The approximate additional fuel cost per hour to raise the output of the unit from 200 to 201 MW. Also, find this additional cost accurately and compare it with the approximate value.

13.2. The incremental fuel cost in \$/MWh for four units of a plant are

$$\lambda_1 = \frac{df_1}{dP_{g1}} = 0.012P_{g1} + 9.0 \qquad \lambda_2 = \frac{df_2}{dP_{g2}} = 0.0096P_{g2} + 6.0$$

$$\lambda_3 = \frac{df_3}{dP_{g3}} = 0.008P_{g3} + 8.0 \qquad \lambda_4 = \frac{df_4}{dP_{g4}} = 0.0068P_{g4} + 10.0$$

Assuming that all four units operate to meet the total plant load of 80 MW, find the incremental fuel cost λ of the plant and the required output of each unit for economic dispatch.

13.3. Assume that maximum load on each of the four units described in Prob. 13.2 is 200, 400, 250, and 300 MW, respectively, and that minimum load on each unit is 50, 100, 80, and 110 MW, respectively. With these maximum and minimum output limits, find the plant λ and MW output of each unit for economic dispatch.

13.4. Solve Prob. 13.3 when the minimum load on Unit 4 is 50 MW rather than 110 MW.

13.5. The incremental fuel costs for two units of a plant are

$$\lambda_1 = \frac{df_1}{dP_{g1}} = 0.012P_{g1} + 8.0 \qquad \lambda_2 = \frac{df_2}{dP_{g2}} = 0.008P_{g2} + 9.6$$

where f is in dollars per hour (\$/h) and P_g is in megawatts (MW). If both units operate at all times and maximum and minimum loads on each unit are 550 and 100 MW, respectively, plot λ of the plant in \$/MWh versus plant output in MW for economic dispatch as total load varies from 200 to 1100 MW.

13.6. Find the savings in \$/h for economic dispatch of load between the units of Prob. 13.5 compared with their sharing the output equally when the total plant output is 600 MW.

13.7. A power system is supplied by three plants, all of which are operating on economic dispatch. At the bus of plant 1 the incremental cost is \$10.0 per MWh, at plant 2 it is \$9.0 per MWh, and at plant 3 it is \$11.0 per MWh. Which plant has the highest penalty factor and which one has the lowest penalty factor? Find the penalty factor of plant 1 if the cost per hour to increase the total delivered load by 1 MW is \$12.0.

13.8. A power system has two generating plants and B-coefficients corresponding to Eq. (13.37), which are given in per unit on a 100-MVA base by

$$\left[\begin{array}{cc|c} 5.0 & -0.03 & 0.15 \\ -0.03 & 8.0 & 0.20 \\ \hline 0.15 & 0.20 & 0.06 \end{array}\right] \times 10^{-3}$$

The incremental fuel cost in \$/MWh of the generating units at the two plants are

$$\lambda_1 = \frac{df_1}{dP_{g1}} = 0.012P_{g1} + 6.6 \qquad \lambda_2 = \frac{df_2}{dP_{g2}} = 0.0096P_{g2} + 6.0$$

If plant 1 presently supplies 200 MW and plant 2 supplies 300 MW, find the penalty factors of each plant. Is the present dispatch most economical? If not, which plant output should be increased and which one should be decreased? Explain why.

13.9. Using \$10.0/MWh as the starting value of system λ in Example 13.4, perform the necessary calculations during the first iteration to obtain an updated λ.

13.10. Suppose that bus ② of a four-bus system is a generator bus and at the same time a load bus. By defining both a generation current and a load current at bus ②, as shown in Fig. 13.5(*c*), find the transformation matrix **C** for this case in the form shown in Eq. (13.31).

13.11. The four-bus system depicted in Fig. 13.5 has bus and line data given in Table 13.2. Suppose that the bus data are slightly modified such that at bus ②, *P*-generation is 4.68 per unit and *P*-load and *Q*-load are 1.5 per unit and 0.9296 per unit, respectively. Using the results of Table 13.3, find the power-flow solution corresponding to this modified bus data. Using the solution to Prob. 13.10, also find the *B*-coefficients of this modified problem in which there is load as well as generation at bus ②.

13.12. Three generating units operating in parallel at 60 Hz have ratings of 300, 500, and 600 MW and have speed-droop characteristics of 5, 4, and 3%, respectively. Due to a change in load, an increase in system frequency of 0.3 Hz is experienced before any supplementary control action occurs. Determine the amount of the change in system load and also the amount of the change in generation of each unit to absorb the load change.

13.13. A 60-Hz system consisting of the three generating units described in Prob. 13.12 is connected to a neighboring system via a tie line. Suppose that a generator in the neighboring system is forced out of service, and that the tie-line flow is observed to increase from the scheduled value of 400 to 631 MW. Determine the amount of the increase in generation of each of the three units and find the ACE of this system whose frequency bias setting is -58 MW/0.1 Hz.

13.14. Suppose that it takes 5 min for the AGC of the power system of Prob. 13.13 to command the three units to increase their generation to restore system frequency to 60 Hz. What is the time error in seconds incurred during this 5-min period? Assume that the initial frequency deviation remains the same throughout this restoration period.

13.15. Solve Example 13.8 when the system load level is 1300 MW.

13.16. If the start-up costs of the four units of Example 13.9 are changed to \$2500, \$3000, \$3400, and \$2600, and the shut-down costs are changed to \$1500, \$1200, \$1000, and \$1400, respectively, find the optimal unit commitment policy. Assume that all other conditions remain unchanged.

13.17. Due to a 400-MW short-term purchase request from the neighboring utility, the demand during the second interval of the day is expected to increase from 1400 to 1800 MW for the system described in Example 13.9. Assuming that other conditions remain unchanged, find the optimal unit commitment policy and the associated total operating cost for the day.

13.18. Suppose Unit 4 of Example 13.9 will have to be taken off line for 8 h, beginning at the fifth interval of the day to undergo minor repair work. Determine the optimal unit commitment policy to serve the system load of Fig. 13.11 and the increase in the operating cost for the day.

13.19. A diagram similar to Fig. 13.14 is shown in Fig. 13.15 in which directed branches represent transitions from one state, represented by a node, to another state. Associated with each directed branch (i, j) is the cost $f_{ij}(k)$, as defined in Eq. (13.72). The values of $f_{ij}(k)$ are given in Table 13.7. Note that index k of $f_{ij}(k)$ does not play any role here, and consequently will now be omitted. If the

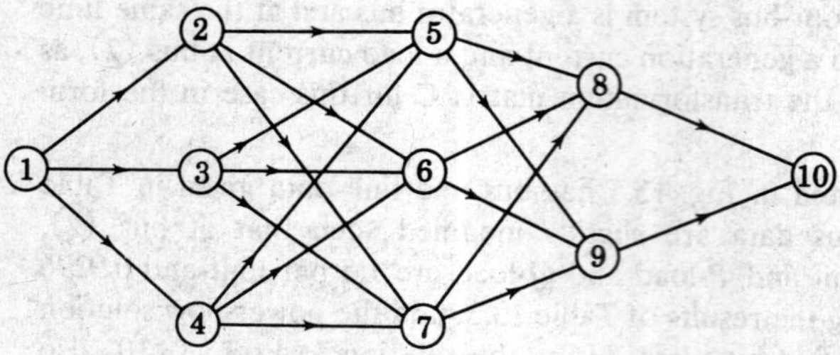

FIGURE 13.15
State-transition diagram for Prob. 13.19.

TABLE 13.7
Matrix of costs (or distances) f_{ij} between states (or nodes) ⓘ and ⓙ of Fig. 13.15

		ⓙ							
	②	③	④	⑤	⑥	⑦	⑧	⑨	⑩
①	20	15	17						
②				35	31	38			
③				39	42	33			
④				36	40	34			
ⓘ ⑤							26	22	
⑥							29	25	
⑦							41	44	
⑧									15
⑨									18

value of f_{ij} is interpreted to be the distance between states i and j, then the unit commitment problem becomes that of finding the *shortest path* from the origin, represented by node ①, to the destination, represented by node ⑩. The problem of this nature is called the *stagecoach problem*. Write the backward recurrence equation similar to Eq. (13.75) and solve the problem by commencing calculations at the destination and then moving toward the origin.

In forward recurrence the process starts with the origin and moves toward the destination. Write the forward recurrence equation, solve the problem, and check the result with that of the backward dynamic programming procedure.

CHAPTER
14

$\mathbf{Z}_{\text{bus}}$ METHODS IN CONTINGENCY ANALYSIS

When a line is switched onto or off the system through the action of circuit breakers, line currents are redistributed throughout the network and bus voltages change. The new steady-state bus voltages and line currents can be predicted by what is called the *contingency analysis* program. The large-scale network models used for contingency evaluation, like those used for fault calculations, do not have to be exact because the system planners and operators, who must undertake hundreds of studies in a short time period, are more concerned with knowing if *overload levels* of current and *out-of-limit* voltages exist than with finding the exact values of those quantities. On this account, approximations are made. Often resistance is considered negligible and the network model becomes purely reactive. Line charging and off-nominal tap changing of transformers are also frequently omitted. In many cases linear models are considered satisfactory and the principle of superposition is then employed. Methods of contingency analysis which use the system $\mathbf{Z}_{\text{bus}}$, and also $\mathbf{Y}_{\text{bus}}$, become attractive from the computational viewpoint—especially if loads can be treated as constant current injections into the various buses of the system.

Removing a line from service can be simulated in the system model by adding the *negative* of the series impedance of the line between its two end buses. Thus, methods to examine the steady-state effects of adding lines to an existing system are developed in Sec. 14.1. The concept of *compensating currents* is introduced which allows use of the existing system $\mathbf{Z}_{\text{bus}}$ without having to modify it. A particular application of this approach to interconnected power systems is shown in Sec. 14.2.

In order to quickly evaluate the effects of line outages and generation shifts aimed at relieving overloads, we develop the concept of *distribution factors* in Secs. 14.3 and 14.4. It is shown how the distribution factors can be formulated from the existing $\mathbf{Z}_{\text{bus}}$ of the system and how they can be used to study multiple contingencies. The similarity of the distribution-factor approach to the dc power-flow method of Sec. 9.7 is explained in Sec. 14.5.

When undertaking the contingency analysis of one part of an interconnected power system, we need to represent the remaining portion of the overall system by an equivalent network. Section 14.6 discusses the basics of equivalencing and compares the network equivalents obtainable from both the system $\mathbf{Z}_{\text{bus}}$ and the system $\mathbf{Y}_{\text{bus}}$.

14.1 ADDING AND REMOVING MULTIPLE LINES

When considering line additions to or removals from an existing system, it is not always necessary to build a new $\mathbf{Z}_{\text{bus}}$ or to calculate new triangular factors of $\mathbf{Y}_{\text{bus}}$—especially if the only interest is to establish the impact of the changes on the existing bus voltages and line flows. An alternative procedure is to consider the injection of *compensating currents* into the existing system to account for the effect of the line changes. To illustrate basic concepts, let us consider *adding* two lines of impedances Z_a and Z_b to an existing system with known $\mathbf{Z}_{\text{bus}}$. Later we shall consider three or more lines.

Suppose that impedances Z_a and Z_b are to be added between buses Ⓜ–Ⓝ and Ⓟ–Ⓠ, respectively, of Fig. 14.1. We assume that the bus voltages $V_1, V_2, \ldots, V_N$ produced in the original system (without Z_a and Z_b) by the current injections $I_1, I_2, \ldots, I_n$ are known, and that these injections are fixed in value and therefore are unaffected by the addition of Z_a and Z_b. On a per-phase basis, the bus impedance equations for the original system are then given by

$$\mathbf{V} = \begin{bmatrix} V_1 \\ \vdots \\ V_m \\ V_n \\ V_p \\ V_q \\ \vdots \\ V_N \end{bmatrix} = \begin{array}{c} \textcircled{1} \\ \vdots \\ \textcircled{m} \\ \textcircled{n} \\ \textcircled{p} \\ \textcircled{q} \\ \vdots \\ \textcircled{N} \end{array} \overset{\begin{array}{cccccccc} \textcircled{1} & & \textcircled{m} & \textcircled{n} & \textcircled{p} & \textcircled{q} & & \textcircled{N} \end{array}}{\begin{bmatrix} Z_{11} & \cdots & Z_{1m} & Z_{1n} & Z_{1p} & Z_{1q} & \cdots & Z_{1N} \\ \vdots & & \vdots & \vdots & \vdots & \vdots & & \vdots \\ Z_{m1} & \cdots & Z_{mm} & Z_{mn} & Z_{mp} & Z_{mq} & \cdots & Z_{mN} \\ Z_{n1} & \cdots & Z_{nm} & Z_{nn} & Z_{np} & Z_{nq} & \cdots & Z_{nN} \\ Z_{p1} & \cdots & Z_{pm} & Z_{pn} & Z_{pp} & Z_{pq} & \cdots & Z_{pN} \\ Z_{q1} & \cdots & Z_{qm} & Z_{qn} & Z_{qp} & Z_{qq} & \cdots & Z_{qN} \\ \vdots & & \vdots & \vdots & \vdots & \vdots & & \vdots \\ Z_{N1} & \cdots & Z_{Nm} & Z_{Nn} & Z_{Np} & Z_{Nq} & \cdots & Z_{NN} \end{bmatrix}} \begin{bmatrix} I_1 \\ \vdots \\ I_m \\ I_n \\ I_p \\ I_q \\ \vdots \\ I_N \end{bmatrix} \tag{14.1}$$

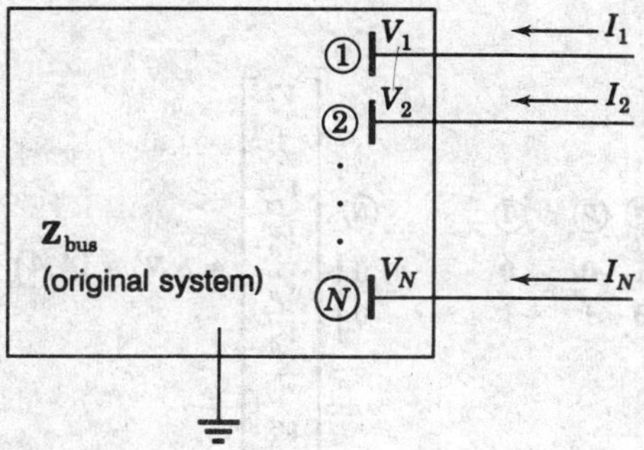

FIGURE 14.1
Original system with voltages $V_1, V_2, \ldots, V_N$ due to the current injections $I_1, I_2, \ldots, I_N$.

and we wish to determine the changes in the bus voltages due to adding the two new line impedances. Let $\mathbf{V}' = [V_1', V_2', \ldots, V_N']^T$ denote the vector of bus voltages which apply *after* Z_a and Z_b have been added. The voltage change at a typical bus ⓚ is then given by

$$\Delta V_k = V_k' - V_k \tag{14.2}$$

The currents I_a and I_b in the added branch impedances Z_a and Z_b are related to the new bus voltages by the equations

$$Z_a I_a = V_m' - V_n' \qquad\qquad Z_b I_b = V_p' - V_q' \tag{14.3}$$

Figure 14.2(a) shows the new branch currents flowing from bus ⓜ to bus ⓝ and from bus ⓟ to bus ⓠ. We can rewrite Eqs. (14.3) in the vector-matrix form

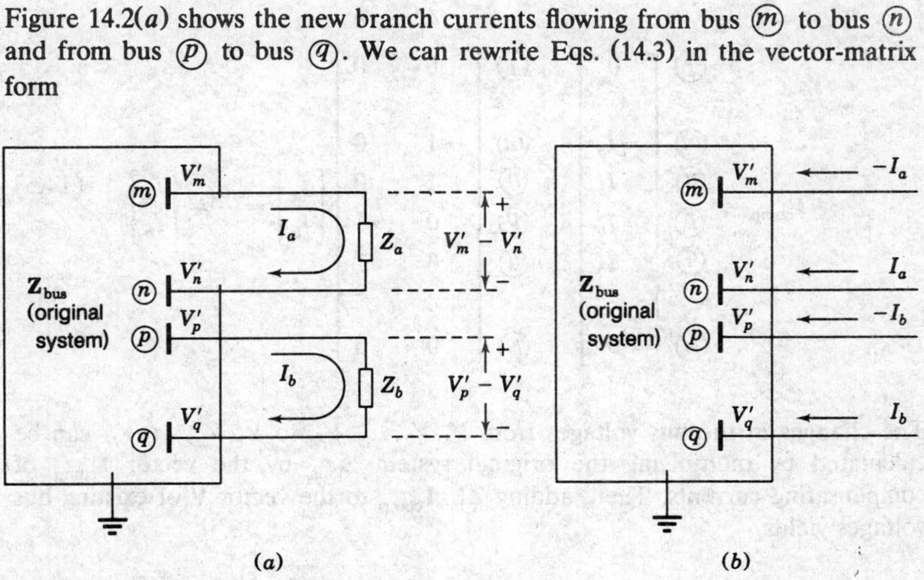

FIGURE 14.2
Original system of Fig. 14.1 with bus voltages V_i changed to V_i' by adding (a) impedances Z_a and Z_b or by (b) equivalent compensating current injections.

form

$$\begin{bmatrix} Z_a & 0 \\ 0 & Z_b \end{bmatrix}\begin{bmatrix} I_a \\ I_b \end{bmatrix} = \begin{matrix} \\ a \\ b \end{matrix}\overset{\begin{matrix} ① & & ⓜ & ⓝ & ⓟ & ⓠ & & Ⓝ \end{matrix}}{\begin{bmatrix} 0 & \cdots & 1 & -1 & 0 & 0 & \cdots & 0 \\ 0 & \cdots & 0 & 0 & 1 & -1 & \cdots & 0 \end{bmatrix}}\begin{bmatrix} V_1' \\ \vdots \\ V_m' \\ V_n' \\ V_p' \\ V_q' \\ \vdots \\ V_N' \end{bmatrix} = \mathbf{A}_c\mathbf{V}' \tag{14.4}$$

where $\mathbf{A}_c$ is the branch-to-node incidence matrix, which shows the incidence of the two new branches to the nodes of the system. The new branch currents I_a and I_b have the same effect on the voltages of the original system as two sets of injected currents—$\{-I_a$ at bus ⓜ, I_a at bus ⓝ$\}$ and $\{-I_b$ at bus ⓟ, I_b at bus ⓠ$\}$, as shown in Fig. 14.2(b). These equivalent current injections combine with actual current injections into the original system to produce the bus voltages $V_1', V_2', \ldots, V_N'$—the same as if the branch impedances Z_a and Z_b had been actually added to the network. In other words, currents I_a and I_b compensate for *not* modifying $\mathbf{Z}_{\text{bus}}$ of the original system to include Z_a and Z_b. On this account, they are called *compensating currents*.

We can express the compensating currents in vector-matrix form as follows:

$$\mathbf{I}_{\text{comp}} = \begin{matrix} ① \\ \\ ⓜ \\ ⓝ \\ ⓟ \\ ⓠ \\ \\ Ⓝ \end{matrix}\begin{bmatrix} 0 \\ \vdots \\ -I_a \\ I_a \\ -I_b \\ I_b \\ \vdots \\ 0 \end{bmatrix} = \begin{matrix} ① \\ \\ ⓜ \\ ⓝ \\ ⓟ \\ ⓠ \\ \\ Ⓝ \end{matrix}\overset{\begin{matrix} a & \; & b \end{matrix}}{\begin{bmatrix} 0 & 0 \\ \vdots & \vdots \\ -1 & 0 \\ 1 & 0 \\ 0 & -1 \\ 0 & 1 \\ \vdots & \vdots \\ 0 & 0 \end{bmatrix}}\begin{bmatrix} I_a \\ I_b \end{bmatrix} = -\mathbf{A}_c^T\begin{bmatrix} I_a \\ I_b \end{bmatrix} \tag{14.5}$$

The changes in the bus voltages from $V_1, V_2, \ldots, V_N$ to $V_1', V_2', \ldots, V_N'$ can be calculated by multiplying the original system $\mathbf{Z}_{\text{bus}}$ by the vector $\mathbf{I}_{\text{comp}}$ of compensating currents. Then, adding $\mathbf{Z}_{\text{bus}}\mathbf{I}_{\text{comp}}$ to the vector $\mathbf{V}$ of existing bus voltages yields

$$\mathbf{V}' = \mathbf{V} + \mathbf{Z}_{\text{bus}}\mathbf{I}_{\text{comp}} = \mathbf{V} - \mathbf{Z}_{\text{bus}}\mathbf{A}_c^T\begin{bmatrix} I_a \\ I_b \end{bmatrix} \tag{14.6}$$

This equation shows that the voltage changes at the buses of the original system due to the addition of the branch impedances Z_a and Z_b between the buses ⓜ–ⓝ and ⓟ–ⓠ, respectively, are given by

$$\Delta\mathbf{V} = \mathbf{V}' - \mathbf{V} = -\mathbf{Z}_{\text{bus}}\mathbf{A}_c^T\begin{bmatrix} I_a \\ I_b \end{bmatrix} \tag{14.7}$$

where I_a and I_b are the compensating currents. The reader will find it useful to check the dimensions of each term in Eq. (14.7) from which the voltage changes $\Delta\mathbf{V} = \mathbf{V}' - \mathbf{V}$ can be calculated directly once values for the currents I_a and I_b are determined. We now show how this determination can be made.

Premultiplying Eq. (14.6) by $\mathbf{A}_c$ and then substituting for $\mathbf{A}_c\mathbf{V}'$ from Eq. (14.4), we obtain

$$\begin{bmatrix} Z_a & 0 \\ 0 & Z_b \end{bmatrix}\begin{bmatrix} I_a \\ I_b \end{bmatrix} = \mathbf{A}_c\mathbf{V} - \mathbf{A}_c\mathbf{Z}_{\text{bus}}\mathbf{A}_c^T\begin{bmatrix} I_a \\ I_b \end{bmatrix} \tag{14.8}$$

Collecting terms which involve I_a and I_b gives

$$\underbrace{\left(\begin{bmatrix} Z_a & 0 \\ 0 & Z_b \end{bmatrix} + \mathbf{A}_c\mathbf{Z}_{\text{bus}}\mathbf{A}_c^T\right)}_{\mathbf{Z}}\begin{bmatrix} I_a \\ I_b \end{bmatrix} = \mathbf{A}_c\mathbf{V} = \begin{bmatrix} V_m - V_n \\ V_p - V_q \end{bmatrix} \tag{14.9}$$

where $\mathbf{Z}$ is a *loop* impedance matrix which can be formed directly from the original bus impedance matrix of the system, as we shall soon see. Solving Eq. (14.9) for I_a and I_b, we find that

$$\begin{bmatrix} I_a \\ I_b \end{bmatrix} = \mathbf{Z}^{-1}\mathbf{A}_c\mathbf{V} = \mathbf{Z}^{-1}\begin{bmatrix} V_m - V_n \\ V_p - V_q \end{bmatrix} \tag{14.10}$$

Note that $V_m - V_n$ and $V_p - V_q$ are *open-circuit* voltage drops between buses ⓜ–ⓝ and ⓟ–ⓠ in the original network, that is, with branch impedances Z_a and Z_b open in Fig. 14.2(a). These open-circuit voltages are either known or can be easily calculated from Eq. (14.1). The definition of matrix $\mathbf{Z}$ in Eq. (14.9) includes the term $\mathbf{A}_c\mathbf{Z}_{\text{bus}}\mathbf{A}_c^T$, which can be determined as follows:

$$\mathbf{A}_c\mathbf{Z}_{\text{bus}}\mathbf{A}_c^T = \begin{array}{c} \\ a \\ b \end{array}\overset{\begin{matrix} ⓜ & ⓝ & ⓟ & ⓠ \end{matrix}}{\begin{bmatrix} 1 & -1 & 0 & 0 \\ 0 & 0 & 1 & -1 \end{bmatrix}} \begin{array}{c} ⓜ \\ ⓝ \\ ⓟ \\ ⓠ \end{array}\overset{\begin{matrix} ⓜ & ⓝ & ⓟ & ⓠ \end{matrix}}{\begin{bmatrix} Z_{mm} & Z_{mn} & Z_{mp} & Z_{mq} \\ Z_{nm} & Z_{nn} & Z_{np} & Z_{nq} \\ Z_{pm} & Z_{pn} & Z_{pp} & Z_{pq} \\ Z_{qm} & Z_{qn} & Z_{qp} & Z_{qq} \end{bmatrix}} \begin{array}{c} ⓜ \\ ⓝ \\ ⓟ \\ ⓠ \end{array}\overset{\begin{matrix} a & b \end{matrix}}{\begin{bmatrix} 1 & 0 \\ -1 & 0 \\ 0 & 1 \\ 0 & -1 \end{bmatrix}} \tag{14.11}$$

In this equation we have shown only those elements of $\mathbf{A}_c$ and $\mathbf{Z}_{\text{bus}}$ which contribute to the calculation. Since all the other elements of $\mathbf{A}_c$ are zeros, it is not necessary to display the full $\mathbf{Z}_{\text{bus}}$. The indicated multiplications yield

$$\mathbf{A}_c \mathbf{Z}_{\text{bus}} \mathbf{A}_c^T = \begin{array}{c} a \\ b \end{array} \left[\begin{array}{c|c} (Z_{mm} - Z_{mn}) - (Z_{nm} - Z_{nn}) & (Z_{mp} - Z_{mq}) - (Z_{np} - Z_{nq}) \\ \hline (Z_{pm} - Z_{pn}) - (Z_{qm} - Z_{qn}) & (Z_{pp} - Z_{pq}) - (Z_{qp} - Z_{qq}) \end{array} \right] \tag{14.12}$$

(columns labelled a and b)

The diagonal elements in this equation can be recognized from Fig. 14.3 as the Thévenin impedances $Z_{\text{th},\,mn}$ and $Z_{\text{th},\,pq}$ seen when looking into the original system between buses ⓜ–ⓝ and ⓟ–ⓠ, respectively. That is,

$$\begin{aligned} Z_{\text{th},\,mn} &= Z_{mm} + Z_{nn} - Z_{mn} - Z_{nm} \\ Z_{\text{th},\,pq} &= Z_{pp} + Z_{qq} - Z_{pq} - Z_{qp} \end{aligned} \tag{14.13}$$

Substituting from Eq. (14.12) into Eq. (14.9), we obtain

$$\underbrace{\begin{array}{c} a \\ b \end{array} \left[\begin{array}{c|c} (Z_{mm} - Z_{mn}) - (Z_{nm} - Z_{nn}) + Z_a & (Z_{mp} - Z_{mq}) - (Z_{np} - Z_{nq}) \\ \hline (Z_{pm} - Z_{pn}) - (Z_{qm} - Z_{qn}) & (Z_{pp} - Z_{pq}) - (Z_{qp} - Z_{qq}) + Z_b \end{array} \right]}_{\mathbf{Z}} \begin{bmatrix} I_a \\ I_b \end{bmatrix} = \begin{bmatrix} V_m - V_n \\ V_p - V_q \end{bmatrix} \tag{14.14}$$

(columns labelled a and b)

which shows that the compensating currents I_a and I_b can be calculated by using the known bus voltages V_m, V_n, V_p, and V_q of the original network and the elements of its $\mathbf{Z}_{\text{bus}}$ shown in Eq. (14.14).

Thus, Eqs. (14.14) and (14.7), in that order, constitute a two-step procedure for the closed-form solution of the voltage changes at the buses of the original system due to the simultaneous addition of branch impedances Z_a and Z_b. Under the assumption of constant externally injected currents into the original system, we first calculate the compensating currents using Eq. (14.14) and then substitute for these currents in Eq. (14.7) to find the new bus voltages which result from adding the new branches. The *removal* of branch impedances Z_a and Z_b from the original system can be analyzed in a similar manner simply by treating the removals as additions of the negative impedances $-Z_a$ and $-Z_b$, as explained in Chap. 8.

The elements in the 2×2 matrix of Eq. (14.12) can be calculated by using the appropriate elements of columns m, n, p, and q of $\mathbf{Z}_{\text{bus}}$, or they can be generated from the triangular factors $\mathbf{L}$ and $\mathbf{U}$ of $\mathbf{Y}_{\text{bus}}$, as described in Sec. 8.5.

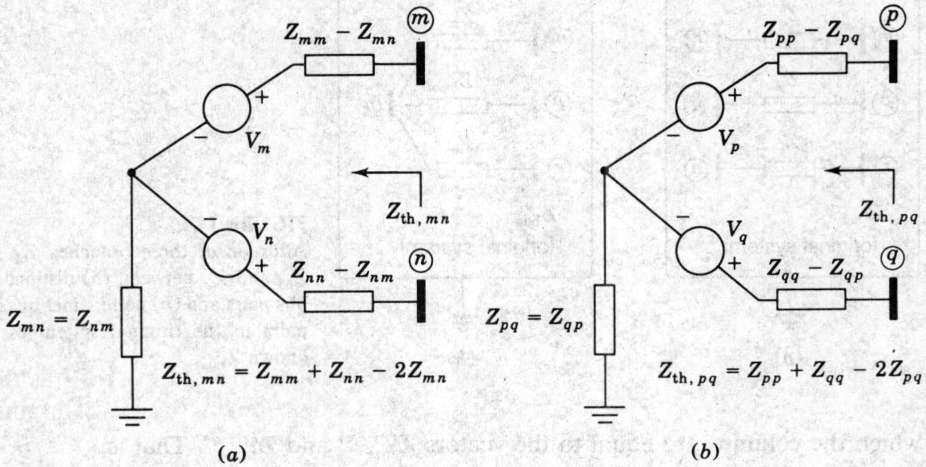

FIGURE 14.3
Thévenin equivalent impedances looking into the system between (*a*) buses ⓜ and ⓝ and (*b*) buses ⓟ and ⓠ.

Since the vectors $\mathbf{Z}_{\text{bus}}^{(m-n)}$ and $\mathbf{Z}_{\text{bus}}^{(p-q)}$ are the respective solutions of the equations

$$\mathbf{LUZ}_{\text{bus}}^{(m-n)} = \begin{bmatrix} 0 \\ \vdots \\ 1_m \\ \vdots \\ -1_n \\ \vdots \\ 0 \end{bmatrix} \qquad \mathbf{LUZ}_{\text{bus}}^{(p-q)} = \begin{bmatrix} 0 \\ \vdots \\ 1_p \\ \vdots \\ -1_q \\ \vdots \\ 0 \end{bmatrix} \tag{14.15}$$

we can write

$$\begin{aligned} (Z_{mm} - Z_{mn}) - (Z_{nm} - Z_{nn}) &= (\text{row } m - \text{row } n) \text{ of } \mathbf{Z}_{\text{bus}}^{(m-n)} \\ (Z_{pm} - Z_{pn}) - (Z_{qm} - Z_{qn}) &= (\text{row } p - \text{row } q) \text{ of } \mathbf{Z}_{\text{bus}}^{(m-n)} \\ (Z_{mp} - Z_{mq}) - (Z_{np} - Z_{nq}) &= (\text{row } m - \text{row } n) \text{ of } \mathbf{Z}_{\text{bus}}^{(p-q)} \\ (Z_{pp} - Z_{pq}) - (Z_{qp} - Z_{qq}) &= (\text{row } p - \text{row } q) \text{ of } \mathbf{Z}_{\text{bus}}^{(p-q)} \end{aligned} \tag{14.16}$$

It also follows from Eqs. (14.15) that $\mathbf{Z}_{\text{bus}}\mathbf{A}_c^T$ in Eq. (14.7) is an $N \times 2$ matrix in

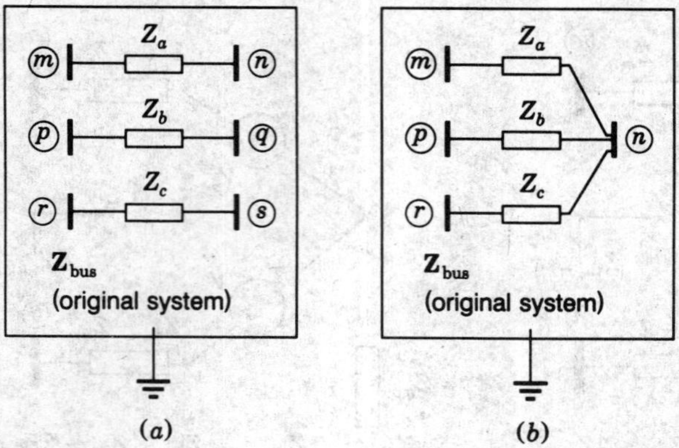

FIGURE 14.4
Addition of three branches Z_a, Z_b, and Z_c between (*a*) distinct bus pairs and (*b*) nondistinct bus pairs in the original system of known $\mathbf{Z}_{bus}$.

which the columns are equal to the vectors $\mathbf{Z}_{\text{bus}}^{(m-n)}$ and $\mathbf{Z}_{\text{bus}}^{(p-q)}$. That is,

$$\mathbf{Z}_{\text{bus}}\mathbf{A}_c^T = \left[\mathbf{Z}_{\text{bus}}^{(m-n)} \middle| \mathbf{Z}_{\text{bus}}^{(p-q)}\right] \tag{14.17}$$

When more than two lines are to be added to the original system, the matrix $\mathbf{Z}$ is formed in the manner described above. For instance, to add three line impedances Z_a, Z_b, and Z_c between the buses ⓜ–ⓝ, ⓟ–ⓠ, and ⓡ–ⓢ, respectively, of Fig. 14.4(a), we find that

$$\mathbf{Z} = \begin{bmatrix} Z_a & 0 & 0 \\ 0 & Z_b & 0 \\ 0 & 0 & Z_c \end{bmatrix} + \begin{matrix} a \\ b \\ c \end{matrix} \overset{\begin{matrix} \qquad\qquad a \qquad\qquad\qquad\qquad & \qquad b \qquad\qquad\qquad & \qquad c \end{matrix}}{\begin{bmatrix} Z_{\text{th},mn} & Z_{mp} - Z_{mq} - Z_{np} + Z_{nq} & Z_{mr} - Z_{ms} - Z_{nr} + Z_{ns} \\ Z_{pm} - Z_{pn} - Z_{qm} + Z_{qn} & Z_{\text{th},pq} & Z_{pr} - Z_{ps} - Z_{qr} + Z_{qs} \\ Z_{rm} - Z_{rn} - Z_{sm} + Z_{sn} & Z_{rp} - Z_{rq} - Z_{sp} + Z_{sq} & Z_{\text{th},rs} \end{bmatrix}} \tag{14.18}$$

It is instructive for the reader to determine the elements of $\mathbf{Z}$ directly from $\mathbf{Z}_{\text{bus}}$ of the system. Other line additions, exemplified in Fig. 14.4(b), can be handled with the same basic procedures simply by treating each line addition as described here.

There are many applications for the procedures presented in this section, such as in developing distribution factors for *contingency evaluation* later in this chapter and in *piecewise methods*, which are discussed in the section immediately following.

Example 14.1. Current injections into the five-bus system of Fig. 8.9 cause voltages at selected buses given by $V_2 = 0.98\angle 0^\circ$, $V_4 = 0.985\angle 0^\circ$, and $V_5 = 1.0\angle 0^\circ$, all in per unit. Predict the changes in the bus voltages when line ②–⑤ with 0.04 per-unit reactance and line ④–⑤ with 0.08 per-unit reactance are simultaneously added to the system. Use the triangular factors of $\mathbf{Y}_{\text{bus}}$ specified in Example 8.6.

Solution. To account for the addition of line ②–⑤ and line ④–⑤, we need to calculate the corresponding vectors $\mathbf{Z}_{\text{bus}}^{(2-5)}$ and $\mathbf{Z}_{\text{bus}}^{(4-5)}$ using forward and back substitution, as explained in Sec. 8.5. Continuing the calculations of Example 8.6, the reader can readily show that

$$\mathbf{Z}_{\text{bus}}^{(4-5)} = \begin{bmatrix} Z_{14} - Z_{15} \\ Z_{24} - Z_{25} \\ Z_{34} - Z_{35} \\ Z_{44} - Z_{45} \\ Z_{54} - Z_{55} \end{bmatrix} = \begin{bmatrix} j0.1 \\ 0 \\ -j0.0625 \\ j0.15 \\ -j0.1125 \end{bmatrix} \quad \text{and}$$

$$\mathbf{Z}_{\text{bus}}^{(2-5)} = \begin{bmatrix} Z_{12} - Z_{15} \\ Z_{22} - Z_{25} \\ Z_{32} - Z_{35} \\ Z_{42} - Z_{45} \\ Z_{52} - Z_{55} \end{bmatrix} = \begin{bmatrix} 0 \\ 0 \\ -j0.0625 \\ 0 \\ -j0.1125 \end{bmatrix}$$

From Eq. (14.14) with $m = 2$, $n = 5$, $p = 4$, and $q = 5$ the elements of the $\mathbf{Z}$ matrix are determined to be

$$\mathbf{Z} = \begin{array}{c} \\ a \\ b \end{array}\!\!\begin{array}{c} \begin{array}{cc} \qquad\qquad a \qquad\qquad & \qquad\qquad b \qquad\qquad \end{array} \\ \left[\begin{array}{c|c} (Z_{22} - Z_{25}) - (Z_{52} - Z_{55}) + Z_a & (Z_{24} - Z_{25}) - (Z_{54} - Z_{55}) \\ \hline (Z_{42} - Z_{45}) - (Z_{52} - Z_{55}) & (Z_{44} - Z_{45}) - (Z_{54} - Z_{55}) + Z_b \end{array}\right] \end{array}$$

$$= \begin{array}{c} \\ a \\ b \end{array}\!\!\begin{array}{c} \begin{array}{cc} \qquad a \qquad & \qquad b \qquad \end{array} \\ \left[\begin{array}{c|c} 0 - (-j0.1125) + j0.04 & 0 - (-j0.1125) \\ \hline 0 - (-j0.1125) & j0.15 - (-j0.1125) + j0.08 \end{array}\right] \end{array}$$

$$= \begin{array}{c} \\ a \\ b \end{array}\!\!\begin{array}{c} \begin{array}{cc} a & \quad b \end{array} \\ \left[\begin{array}{c|c} j0.1525 & j0.1125 \\ \hline j0.1125 & j0.3425 \end{array}\right] \end{array}$$

Using Eq. (14.14), we can calculate I_a and I_b as the solution of

$$\left[\begin{array}{c|c} j0.1525 & j0.1125 \\ \hline j0.1125 & j0.3425 \end{array}\right] \begin{bmatrix} I_a \\ I_b \end{bmatrix} = \begin{bmatrix} V_2 - V_5 \\ V_4 - V_5 \end{bmatrix} = \begin{bmatrix} 1.0 - 0.98 \\ 0.985 - 0.98 \end{bmatrix} = \begin{bmatrix} 0.02 \\ 0.005 \end{bmatrix}$$

from which we find $I_a = -j0.158876$ and $I_b = j0.037587$. According to Eqs. (14.7) and (14.17), the changes in the voltages at all buses are

$$\Delta\mathbf{V} = -I_a\mathbf{Z}_{\text{bus}}^{(2-5)} - I_b\mathbf{Z}_{\text{bus}}^{(4-5)}$$

$$= j0.158876\begin{bmatrix} 0 \\ 0 \\ -j0.0625 \\ 0 \\ -j0.1125 \end{bmatrix} - j0.037587\begin{bmatrix} j0.1 \\ 0 \\ -j0.0625 \\ j0.15 \\ -j0.1125 \end{bmatrix} = \begin{bmatrix} 0.003759\angle 0^\circ \\ 0 \\ 0.007581\angle 0^\circ \\ 0.005638\angle 0^\circ \\ 0.013645\angle 0^\circ \end{bmatrix}$$

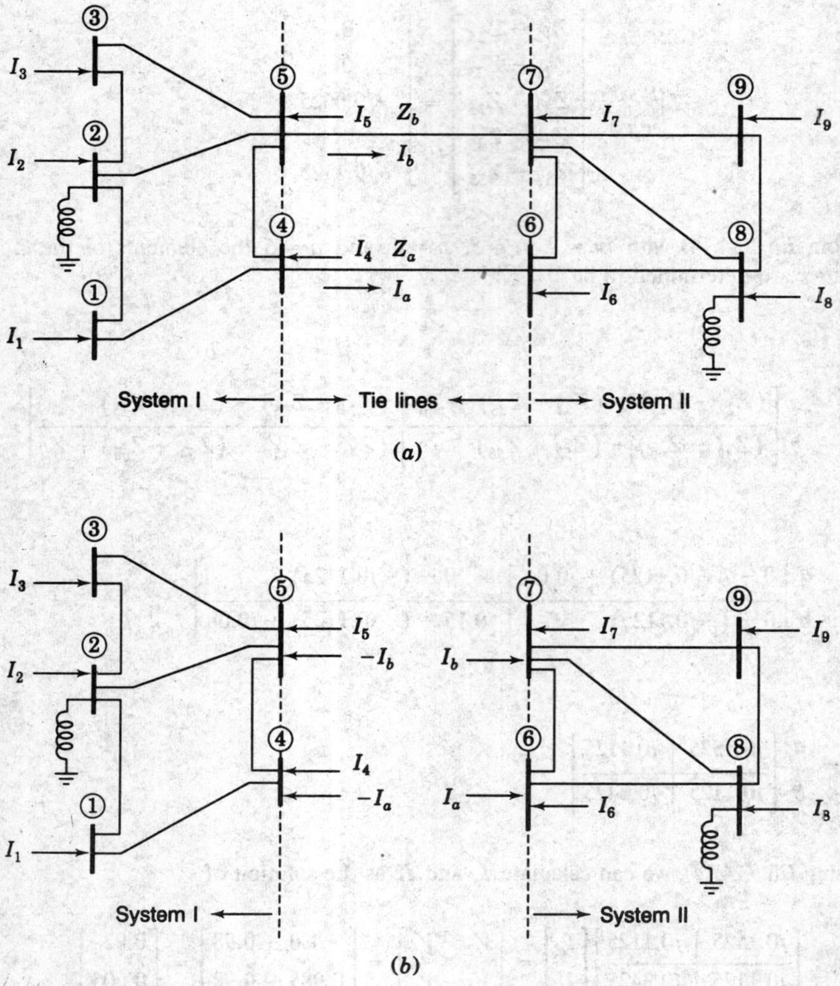

FIGURE 14.5
Two systems with: (*a*) interconnecting tie-line impedances Z_a and Z_b; (*b*) equivalent compensating current injections.

14.2 PIECEWISE SOLUTION OF INTERCONNECTED SYSTEMS

The procedures developed in Sec. 14.1 have a particular application in the *piecewise* solution of a large network. Piecewise methods can be of benefit when individual power companies, joined together by tie lines to form a large interconnected system, are using linear system models for problem solution such as in fault analysis. We recall that a system model is linear if all currents are directly proportional to the corresponding voltages. As a first step, each utility could form its own $\mathbf{Z}_{\text{bus}}$ and solve its own network as if it were isolated. In a subsequent step results can be exchanged between utilities so that each can modify its solutions to account for the effects of the interconnections.

Figure 14.5(*a*) shows two systems with a *common* reference node. The systems are independent except for the branches *a* and *b*, which are tie lines interconnecting the two systems. System I contains buses ① to ⑤ whereas System II contains buses ⑥ to ⑨. Current injections I_1 to I_9 are assumed known at the buses of the two systems, and it is required to calculate the bus voltages $V_1', V_2', \ldots, V_9'$ for the overall interconnected system, *including* the tie-line impedances Z_a and Z_b, using the known voltages available for each separate system.

Let Z_{I} denote the 5×5 bus impedance matrix for the stand-alone network of System I containing buses ①–⑤. Likewise, let Z_{II} denote the 4×4 bus impedance matrix constructed separately for the stand-alone network of System II containing buses ⑥–⑨. Suppose that the current injections I_1 to I_5 cause voltages $V_1, V_2, \ldots, V_5$ at the buses of System I when the tie lines are open, and that injections I_6 and I_9 cause voltages $V_6, V_7, \ldots, V_9$ in System II under the same condition of open ties. The bus impedance equations for the overall system *without* interconnecting tie lines are then given by

$$\mathbf{V} = \begin{bmatrix} V_1 \\ V_2 \\ \vdots \\ V_5 \\ \hline V_6 \\ V_7 \\ \vdots \\ V_9 \end{bmatrix} = \underbrace{\left[\begin{array}{c|c} \mathbf{Z}_{\text{I}} & \mathbf{0} \\ \hline \mathbf{0} & \mathbf{Z}_{\text{II}} \end{array}\right]}_{\mathbf{Z}_{\text{bus}}} \begin{bmatrix} I_1 \\ I_2 \\ \vdots \\ I_5 \\ \hline I_6 \\ I_7 \\ \vdots \\ I_9 \end{bmatrix} \tag{14.19}$$

When tie-line impedances Z_a and Z_b are added between the two systems, tie-line currents I_a and I_b will flow from buses ④ and ⑤ of System I to buses ⑥ and ⑦ of System II, as shown in Fig. 14.5(a). We can calculate the ensuing bus-voltage changes from $V_1, V_2, \ldots, V_9$ to $V_1', V_2', \ldots, V_9'$ in the interconnected system by using the compensating currents of Fig. 14.5(b) in the manner described in Sec. 14.1. Since I_a and I_b are both assumed to flow from System I to System II, we set $m = 4$, $n = 6$, $p = 5$, and $q = 7$ in forming matrix **Z** of Eq. (14.14) for the present illustrative example, which yields

$$\mathbf{Z} = \begin{array}{c} \\ a \\ b \end{array}\left[\begin{array}{c|c} \multicolumn{1}{c}{a} & \multicolumn{1}{c}{b} \\ Z_{44}^{(1)} + Z_{66}^{(2)} + Z_a & Z_{45}^{(1)} + Z_{67}^{(2)} \\ \hline Z_{54}^{(1)} + Z_{76}^{(2)} & Z_{55}^{(1)} + Z_{77}^{(2)} + Z_b \end{array}\right] \tag{14.20}$$

We have added superscripts (1) and (2) to identify more readily the respective elements of $\mathbf{Z}_{\text{I}}$ and $\mathbf{Z}_{\text{II}}$. Since voltages are known for the separate systems, the tie-line currents can be calculated as in Eq. (14.14) as follows:

$$\begin{bmatrix} I_a \\ I_b \end{bmatrix} = \begin{array}{c} a \\ b \end{array}\left[\begin{array}{c|c} Z_{44}^{(1)} + Z_{66}^{(2)} + Z_a & Z_{45}^{(1)} + Z_{67}^{(2)} \\ \hline Z_{54}^{(1)} + Z_{76}^{(2)} & Z_{55}^{(1)} + Z_{77}^{(2)} + Z_b \end{array}\right]^{-1} \begin{bmatrix} V_4 - V_6 \\ V_5 - V_7 \end{bmatrix} \tag{14.21}$$

(columns labelled a, b)

The unknown voltages at the buses of the overall interconnected system, including the tie lines, can now be calculated from Eq. (14.6), which here assumes the form

$$\begin{bmatrix} V_1' \\ V_2' \\ \vdots \\ V_5' \\ \hline V_6' \\ V_7' \\ \vdots \\ V_9' \end{bmatrix} = \begin{bmatrix} V_1 \\ V_2 \\ \vdots \\ V_5 \\ \hline V_6 \\ V_7 \\ \vdots \\ V_9 \end{bmatrix} - \left[\begin{array}{c|c} \mathbf{Z}_{\text{I}} & \mathbf{0} \\ \hline \mathbf{0} & \mathbf{Z}_{\text{II}} \end{array}\right] \mathbf{A}_{\text{c}}^T \begin{bmatrix} I_a \\ I_b \end{bmatrix} \tag{14.22}$$

where in this case

$$\mathbf{A}_{\text{c}} = [\mathbf{A}_{\text{cI}} \,|\, \mathbf{A}_{\text{cII}}] = \begin{array}{c} \\ a \\ b \end{array}\left[\begin{array}{ccccc|cccc} ① & ② & ③ & ④ & ⑤ & ⑥ & ⑦ & ⑧ & ⑨ \\ 0 & 0 & 0 & 1 & 0 & -1 & 0 & 0 & 0 \\ 0 & 0 & 0 & 0 & 1 & 0 & -1 & 0 & 0 \end{array}\right]$$

$$\underbrace{\qquad\qquad}_{\mathbf{A}_{\text{cI}}} \qquad \underbrace{\qquad\qquad}_{\mathbf{A}_{\text{cII}}} \tag{14.23}$$

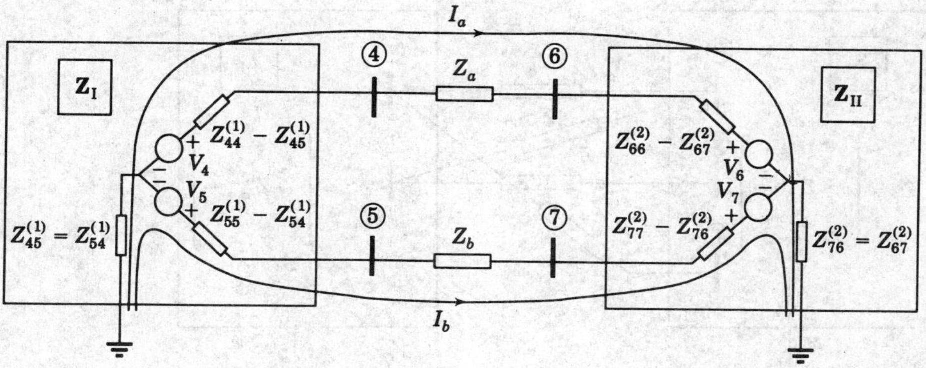

FIGURE 14.6
The Thévenin equivalent circuits of System I and System II of Fig. 14.5 and the paths traced by the line currents I_a and I_b. The elements $Z_{ij}^{(1)}$ belong to the bus impedance matrix $\mathbf{Z}_I$ of System I and the elements $Z_{ij}^{(2)}$ belong to $\mathbf{Z}_{II}$ of System II.

The amount of computation involved in finding the overall solution by piecewise methods can be considerably reduced, especially since the only matrix inversion is $\mathbf{Z}^{-1}$, which is of small dimension depending on the number of tie lines.

It is worthwhile to rearrange matrix $\mathbf{Z}$ of Eq. (14.21) in the following form:

$$\mathbf{Z} = \underbrace{\begin{matrix} & ④ & ⑤ \\ ④ & Z_{44}^{(1)} & Z_{45}^{(1)} \\ ⑤ & Z_{54}^{(1)} & Z_{55}^{(1)} \end{matrix}}_{\text{from } \mathbf{Z}_I} + \underbrace{\begin{bmatrix} Z_a & 0 \\ 0 & Z_b \end{bmatrix}}_{\text{tie lines}} + \underbrace{\begin{matrix} & ⑥ & ⑦ \\ ⑥ & Z_{66}^{(2)} & Z_{67}^{(2)} \\ ⑦ & Z_{76}^{(2)} & Z_{77}^{(2)} \end{matrix}}_{\text{from } \mathbf{Z}_{II}} \tag{14.24}$$

since then it shows how $\mathbf{Z}$ is formed by inspection of the bus impedance matrices $\mathbf{Z}_I$ and $\mathbf{Z}_{II}$ of the separate systems. Evidently, we need only extract from $\mathbf{Z}_I$ the submatrix of elements with subscripts corresponding to the *boundary buses* to which the tie lines are incident, and likewise for $\mathbf{Z}_{II}$. Then, we add together the two extracted submatrices and the impedances of the tie lines so that the diagonal entries combine in $\mathbf{Z}$ to give the impedance of the total *path to reference* through each tie line. The Thévenin equivalent circuits plus the tie lines of Fig. 14.6 graphically explain the interpretation of Eq. (14.21) for our example and also show that the off-diagonal elements of Eq. (14.20) represent the total impedances *common* to the paths through the reference node of the circulating tie-line currents.

In the particular system of Fig. 14.5 the tie lines have distinct pairs of boundary buses— ④-⑥ for line a and ⑤-⑦ for line b. Figure 14.7 shows a more general situation with the systems now interconnected by four tie lines a, b, c, and d between the boundary buses ⓟ, ⓠ, and ⓡ of System I, and

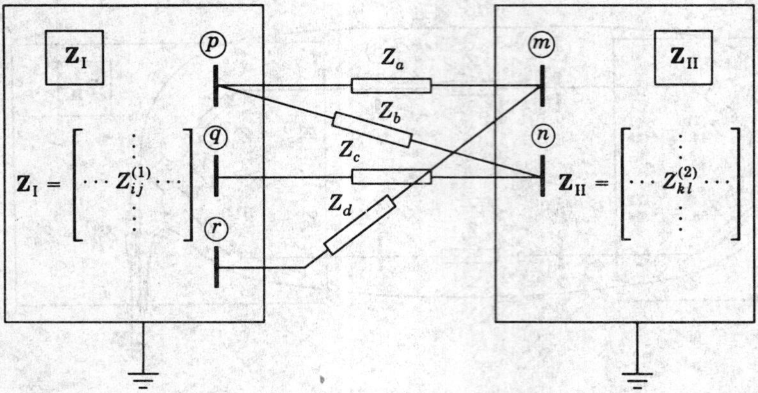

FIGURE 14.7
Two systems interconnected by four tie lines having impedances Z_a, Z_b, Z_c, and Z_d. The typical element of bus impedance matrix $\mathbf{Z}_{\text{I}}$ is $Z_{ij}^{(1)}$ and of $\mathbf{Z}_{\text{II}}$ is $Z_{kl}^{(2)}$.

boundary buses ⓜ and ⓝ of System II. The two submatrices to be formed separately from the elements of $\mathbf{Z}_{\text{I}}$ and $\mathbf{Z}_{\text{II}}$ are always square and symmetrical with row-column dimensions equal to the number of tie lines. Thus, for Fig. 14.7 we have the 4×4 submatrices

$$
\underbrace{\begin{array}{c} \\ a \\ b \\ c \\ d \end{array}\!\!\begin{array}{c} \begin{matrix} a & b & c & d \end{matrix} \\ \begin{bmatrix} Z_{pp}^{(1)} & Z_{pp}^{(1)} & Z_{pq}^{(1)} & Z_{pr}^{(1)} \\ Z_{pp}^{(1)} & Z_{pp}^{(1)} & Z_{pq}^{(1)} & Z_{pr}^{(1)} \\ Z_{qp}^{(1)} & Z_{qp}^{(1)} & Z_{qq}^{(1)} & Z_{qr}^{(1)} \\ Z_{rp}^{(1)} & Z_{rp}^{(1)} & Z_{rq}^{(1)} & Z_{rr}^{(1)} \end{bmatrix} \end{array}}_{\text{From } \mathbf{Z}_{\text{I}}}
\qquad
\underbrace{\begin{array}{c} \\ a \\ b \\ c \\ d \end{array}\!\!\begin{array}{c} \begin{matrix} a & b & c & d \end{matrix} \\ \begin{bmatrix} Z_{mm}^{(2)} & Z_{mn}^{(2)} & Z_{mn}^{(2)} & Z_{mm}^{(2)} \\ Z_{nm}^{(2)} & Z_{nn}^{(2)} & Z_{nn}^{(2)} & Z_{nm}^{(2)} \\ Z_{nm}^{(2)} & Z_{nn}^{(2)} & Z_{nn}^{(2)} & Z_{nm}^{(2)} \\ Z_{mm}^{(2)} & Z_{mn}^{(2)} & Z_{mn}^{(2)} & Z_{mm}^{(2)} \end{bmatrix} \end{array}}_{\text{From } \mathbf{Z}_{\text{II}}}
$$

The rows and columns of both submatrices are labeled a, b, c, and d corresponding to the four tie lines, and the entries in each submatrix have subscripts relating to the boundary buses of its corresponding system. Note that the diagonal elements in each submatrix are diagonal elements of the bus impedance matrix for the corresponding system. Each submatrix can be formulated from the corresponding bus impedance matrix by using an equation similar to Eq. (14.11) or by applying the following rules sequentially:

1. For each individual system of Fig. 14.7, arrange a submatrix with one row and a corresponding column for each tie line.

2. In each submatrix, set the *diagonal* element for a given tie line equal to the driving-point impedance with subscripts the same as the boundary node to which the tie line is incident within the particular system. For example, in Fig. 14.7 tie line c is incident to boundary bus ⓠ of System I and boundary bus ⓝ of System II. Hence, in the submatrix for System I we set the diagonal element for tie line c equal to the driving-point impedance $Z_{qq}^{(1)}$ of $\mathbf{Z}_{\mathrm{I}}$, whereas in the submatrix for System II we set the diagonal element equal to $Z_{nn}^{(2)}$ of $\mathbf{Z}_{\mathrm{II}}$.
3. In each submatrix, set the off-diagonal elements equal to the transfer impedances with row-column subscripts aligned and consistent with those of the diagonal elements. For instance, in the submatrix for System I all row c elements have the first subscript q corresponding to the diagonal element $Z_{qq}^{(1)}$, whereas column b entries all have the second subscript p corresponding to the diagonal element $Z_{pp}^{(1)}$.

The submatrices so formed are added together and the result is then added to the diagonal matrix of tie-line impedances. For the tie-line connections of Fig. 14.7 we obtain

$$\mathbf{Z} = \begin{array}{c} \\ a \\ b \\ c \\ d \end{array} \begin{array}{c} \begin{array}{cccc} a & b & c & d \end{array} \\ \left[\begin{array}{c|c|c|c} Z_a + Z_{pp}^{(1)} + Z_{mm}^{(2)} & Z_{pp}^{(1)} + Z_{mn}^{(2)} & Z_{pq}^{(1)} + Z_{mn}^{(2)} & Z_{pr}^{(1)} + Z_{mm}^{(2)} \\ Z_{pp}^{(1)} + Z_{nm}^{(2)} & Z_b + Z_{pp}^{(1)} + Z_{nn}^{(2)} & Z_{pq}^{(1)} + Z_{nn}^{(2)} & Z_{pr}^{(1)} + Z_{nm}^{(2)} \\ Z_{qp}^{(1)} + Z_{nm}^{(2)} & Z_{qp}^{(1)} + Z_{nn}^{(2)} & Z_c + Z_{qq}^{(1)} + Z_{nn}^{(2)} & Z_{qr}^{(1)} + Z_{nm}^{(2)} \\ Z_{rp}^{(1)} + Z_{mm}^{(2)} & Z_{rp}^{(1)} + Z_{mn}^{(2)} & Z_{rq}^{(1)} + Z_{mn}^{(2)} & Z_d + Z_{rr}^{(1)} + Z_{mm}^{(2)} \end{array}\right] \end{array} \tag{14.25}$$

and the tie-line currents are given by

$$\begin{bmatrix} I_a \\ I_b \\ I_c \\ I_d \end{bmatrix} = \mathbf{Z}^{-1} \underbrace{\begin{bmatrix} V_p - V_m \\ V_p - V_n \\ V_q - V_n \\ V_r - V_m \end{bmatrix}}_{\substack{\text{open-circuit} \\ \text{voltage drops}}} \tag{14.26}$$

Each diagonal element of $\mathbf{Z}$ is a *path* or *loop* impedance associated with the corresponding tie-line current which circulates from the reference bus through System I, then through the tie line, and back to the reference bus through System II. The off-diagonal elements of $\mathbf{Z}$ are the impedances common to the paths of those circulating currents which are the cause of the bus-voltage changes when the system is interconnected. Thus, **Z** can be interpreted as a symmetrical *loop impedance* matrix formed from the *bus impedance* submatrices

$$\underbrace{\begin{array}{c} \\ (p) \\ (q) \\ (r) \end{array}\begin{array}{c}\begin{array}{ccc} (p) & (q) & (r) \end{array}\\ \begin{bmatrix} Z_{pp} & Z_{pq} & Z_{pr} \\ Z_{qp} & Z_{qq} & Z_{qr} \\ Z_{rp} & Z_{rq} & Z_{rr} \end{bmatrix}\end{array}}_{\text{Extracted from } \mathbf{Z}_{\mathrm{I}}} \qquad \underbrace{\begin{array}{c} \\ (m) \\ (n) \end{array}\begin{array}{c}\begin{array}{cc} (m) & (n) \end{array}\\ \begin{bmatrix} Z_{mm} & Z_{mn} \\ Z_{nm} & Z_{nn} \end{bmatrix}\end{array}}_{\text{Extracted from } \mathbf{Z}_{\mathrm{II}}}$$

By inverting these submatrices, we obtain bus admittance submatrices which represent System I and System II at their boundary buses on which the tie lines terminate. Equivalent mesh circuits, called *Ward equivalents*, can be drawn corresponding to the bus admittance submatrices. Ward equivalents are further discussed in Sec. 14.6.

Example 14.2. The reactance diagram of Fig. 14.8 shows two systems, System I and System II, interconnected by three tie lines a, b, and c. The per-unit reactances of the elements are as marked. Note that each system is connected through a reactance to ground, which serves as the common reference for the bus impedance matrices $\mathbf{Z}_{\mathrm{I}}$ and $\mathbf{Z}_{\mathrm{II}}$ of the separate systems without tie lines. Table 14.1 gives the numerical values of $\mathbf{Z}_{\mathrm{I}}$ and $\mathbf{Z}_{\mathrm{II}}$, along with known bus voltages for each individual system without tie lines. Predict the steady-state voltage changes at the buses of System I when the three tie lines are simultaneously closed between the systems. Use the piecewise method based on $\mathbf{Z}_{bus}$.

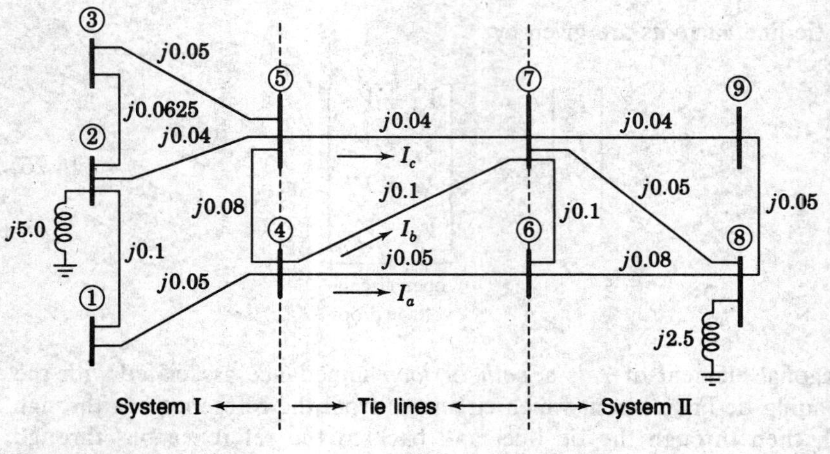

FIGURE 14.8
Reactance diagram for Example 14.2 showing per-unit values of branch impedances. Boundary buses ④ and ⑤ of System I are connected by three tie lines to boundary buses ⑥ and ⑦ of System II.

TABLE 14.1
Bus impedance matrices and voltages for Example 14.2

$$\mathbf{Z}_{\mathrm{I}} = \begin{matrix} & ① & ② & ③ & ④ & ⑤ \\ ① & j5.061466 & j5.000000 & j5.006317 & j5.042198 & j5.011371 \\ ② & j5.000000 & j5.000000 & j5.000000 & j5.000000 & j5.000000 \\ ③ & j5.006317 & j5.000000 & j5.035850 & j5.009476 & j5.014529 \\ ④ & j5.042198 & j5.000000 & j5.009476 & j5.063298 & j5.017056 \\ ⑤ & j5.011371 & j5.000000 & j5.014529 & j5.017056 & j5.026153 \end{matrix}$$

$$\mathbf{Z}_{\mathrm{II}} = \begin{matrix} & ⑥ & ⑦ & ⑧ & ⑨ \\ ⑥ & j2.549832 & j2.512121 & j2.500000 & j2.506734 \\ ⑦ & j2.512121 & j2.527273 & j2.500000 & j2.515152 \\ ⑧ & j2.500000 & j2.500000 & j2.500000 & j2.500000 \\ ⑨ & j2.506734 & j2.515152 & j2.500000 & j2.530640 \end{matrix}$$

$$\mathbf{V}_{\mathrm{I}} = \begin{matrix} ① \\ ② \\ ③ \\ ④ \\ ⑤ \end{matrix} \begin{bmatrix} 1.000000 + j0.000000 \\ 0.986301 - j0.083834 \\ 0.984789 - j0.095108 \\ 0.993653 - j0.045583 \\ 0.998498 - j0.054795 \end{bmatrix} \qquad \mathbf{V}_{\mathrm{II}} = \begin{matrix} ⑥ \\ ⑦ \\ ⑧ \\ ⑨ \end{matrix} \begin{bmatrix} 0.991467 - j0.039757 \\ 0.999943 + j0.010640 \\ 1.000000 + j0.000000 \\ 0.993820 - j0.020792 \end{bmatrix}$$

Solution. The main steps in the solution involve forming the matrix **Z**, finding the tie-line currents, and using these currents as injections into System I to determine the bus-voltage changes.

Buses ④ and ⑤ are the boundary buses of System I since tie lines a and b connect to bus ④ and tie line c connects to bus ⑤. Therefore, the submatrix for System I according to our rules is

$$\begin{matrix} & a & b & c \\ a & Z_{44}^{(1)} & Z_{44}^{(1)} & Z_{45}^{(1)} \\ b & Z_{44}^{(1)} & Z_{44}^{(1)} & Z_{45}^{(1)} \\ c & Z_{54}^{(1)} & Z_{54}^{(1)} & Z_{55}^{(1)} \end{matrix} = \begin{matrix} & a & b & c \\ a & j5.063298 & j5.063298 & j5.017056 \\ b & j5.063298 & j5.063298 & j5.017056 \\ c & j5.017056 & j5.017056 & j5.026153 \end{matrix}$$

In System II buses ⑥ and ⑦ are the boundary buses since the tie line a

connects to bus ⑥ whereas tie lines b and c connect to bus ⑦. Hence, we have

$$\begin{array}{c} \\ a \\ b \\ c \end{array}\begin{array}{c} \begin{array}{ccc} a & b & c \end{array} \\ \begin{bmatrix} Z_{66}^{(2)} & Z_{67}^{(2)} & Z_{67}^{(2)} \\ Z_{76}^{(2)} & Z_{77}^{(2)} & Z_{77}^{(2)} \\ Z_{76}^{(2)} & Z_{77}^{(2)} & Z_{77}^{(1)} \end{bmatrix} \end{array} = \begin{array}{c} \\ a \\ b \\ c \end{array}\begin{array}{c} \begin{array}{ccc} a & b & c \end{array} \\ \begin{bmatrix} j2.549832 & j2.512121 & j2.512121 \\ j2.512121 & j2.527273 & j2.527273 \\ j2.512121 & j2.527273 & j2.527273 \end{bmatrix} \end{array}$$

The diagonal matrix of tie line impedances is

$$\begin{array}{c} \\ a \\ b \\ c \end{array}\begin{array}{c} \begin{array}{ccc} a & b & c \end{array} \\ \begin{bmatrix} j0.05 & \cdot & \cdot \\ \cdot & j0.10 & \cdot \\ \cdot & \cdot & j0.04 \end{bmatrix} \end{array}$$

The $\mathbf{Z}$ matrix is therefore calculated as follows:

$$\mathbf{Z} = \begin{array}{c} \\ a \\ b \\ c \end{array}\begin{array}{c} \begin{array}{ccc} a & b & c \end{array} \\ \begin{bmatrix} Z_{44}^{(1)} & Z_{44}^{(1)} & Z_{45}^{(1)} \\ Z_{44}^{(1)} & Z_{44}^{(1)} & Z_{45}^{(1)} \\ Z_{54}^{(1)} & Z_{54}^{(1)} & Z_{55}^{(1)} \end{bmatrix} \end{array} + \begin{bmatrix} Z_a & \cdot & \cdot \\ \cdot & Z_b & \cdot \\ \cdot & \cdot & Z_c \end{bmatrix} + \begin{array}{c} \\ a \\ b \\ c \end{array}\begin{array}{c} \begin{array}{ccc} a & b & c \end{array} \\ \begin{bmatrix} Z_{66}^{(2)} & Z_{67}^{(2)} & Z_{67}^{(2)} \\ Z_{76}^{(2)} & Z_{77}^{(2)} & Z_{77}^{(2)} \\ Z_{76}^{(2)} & Z_{77}^{(2)} & Z_{77}^{(2)} \end{bmatrix} \end{array}$$

$$= \begin{array}{c} \\ a \\ b \\ c \end{array}\begin{array}{c} \begin{array}{ccc} a & b & c \end{array} \\ \begin{bmatrix} j7.663130 & j7.575419 & j7.529177 \\ j7.575419 & j7.690571 & j7.544329 \\ j7.529177 & j7.544329 & j7.593426 \end{bmatrix} \end{array}$$

Since the tie-line currents I_a, I_b, and I_c are chosen with positive direction from System I to System II, we have, as in Eq. (14.26)

$$\begin{bmatrix} I_a \\ I_b \\ I_c \end{bmatrix} = \mathbf{Z}^{-1}\begin{bmatrix} V_4 - V_6 \\ V_4 - V_7 \\ V_5 - V_7 \end{bmatrix} = \mathbf{Z}^{-1}\begin{bmatrix} (0.993653 - j0.045583) - (0.991467 - j0.039757) \\ (0.993653 - j0.045583) - (0.999943 + j0.010640) \\ (0.998498 - j0.054795) - (0.999943 + j0.010640) \end{bmatrix}$$

After substituting numerical values for $\mathbf{Z}$ and inverting, we then compute

$$\begin{bmatrix} I_a \\ I_b \\ I_c \end{bmatrix} = \begin{bmatrix} 0.361673 - j0.039494 \\ -0.130972 + j0.044114 \\ -0.237105 - j0.004479 \end{bmatrix} \text{ per unit}$$

The branch-to-node incidence matrix for the tie lines has a distinct part for each system,

$$\mathbf{A}_c = \left[\mathbf{A}_{cI} \mid \mathbf{A}_{cII}\right] = \begin{matrix} & ① & ② & ③ & ④ & ⑤ & ⑥ & ⑦ & ⑧ & ⑨ \\ a & 0 & 0 & 0 & 1 & 0 & -1 & 0 & 0 & 0 \\ b & 0 & 0 & 0 & 1 & 0 & 0 & -1 & 0 & 0 \\ c & 0 & 0 & 0 & 0 & 1 & 0 & -1 & 0 & 0 \end{matrix}$$

and from Eqs. (14.22) and (14.23) the bus-voltage changes $\Delta\mathbf{V}_\mathrm{I}$ in System I are given by

$$\Delta\mathbf{V}_\mathrm{I} = \mathbf{V}_\mathrm{I}' - \mathbf{V}_\mathrm{I} = -\mathbf{Z}_\mathrm{I}\mathbf{A}_{c\mathrm{I}}^T \begin{bmatrix} I_a \\ I_b \\ I_c \end{bmatrix} = -\underbrace{\begin{bmatrix} \text{col. 4} & \text{col. 4} & \text{col. 5} \end{bmatrix}}_{\text{From } \mathbf{Z}_\mathrm{I}} \begin{bmatrix} I_a \\ I_b \\ I_c \end{bmatrix}$$

(columns labelled a, b, c)

$$= -\begin{matrix} ① \\ ② \\ ③ \\ ④ \\ ⑤ \end{matrix} \overset{\qquad a \qquad\qquad b \qquad\qquad c}{\begin{bmatrix} j5.042198 & j5.042198 & j5.011371 \\ j5.000000 & j5.000000 & j5.000000 \\ j5.009476 & j5.009476 & j5.014529 \\ j5.063298 & j5.063298 & j5.017056 \\ j5.017056 & j5.017056 & j5.026153 \end{bmatrix}} \begin{bmatrix} 0.361673 - j0.039494 \\ -0.130972 + j0.044114 \\ -0.237105 - j0.004479 \end{bmatrix}$$

$$= \begin{matrix} ① \\ ② \\ ③ \\ ④ \\ ⑤ \end{matrix} \begin{bmatrix} 0.000849 + j0.024981 \\ 0.000705 + j0.032020 \\ 0.000684 + j0.033279 \\ -0.000921 + j0.021461 \\ -0.000667 + j0.034286 \end{bmatrix} \text{ per unit}$$

Example 14.3. Draw the Thévenin equivalent circuits seen at the boundary buses of Systems I and II of Example 14.2 and show their interconnection through the tie lines *a*, *b*, and *c*. Determine the Ward equivalent of System II at boundary buses ⑥ and ⑦.

Solution. The Thévenin equivalents are shown in Fig. 14.9 with numerical values for the impedances. The open-circuit voltages V_4, V_5, V_6, and V_7 have the numerical values given in Table 14.1. Replacing the impedances in the System II equivalent of Fig. 14.10(*a*) by their reciprocal values gives the admittance circuit shown in Fig. 14.10(*b*). Using the Y-Δ transformation formulas of Table 1.2, we

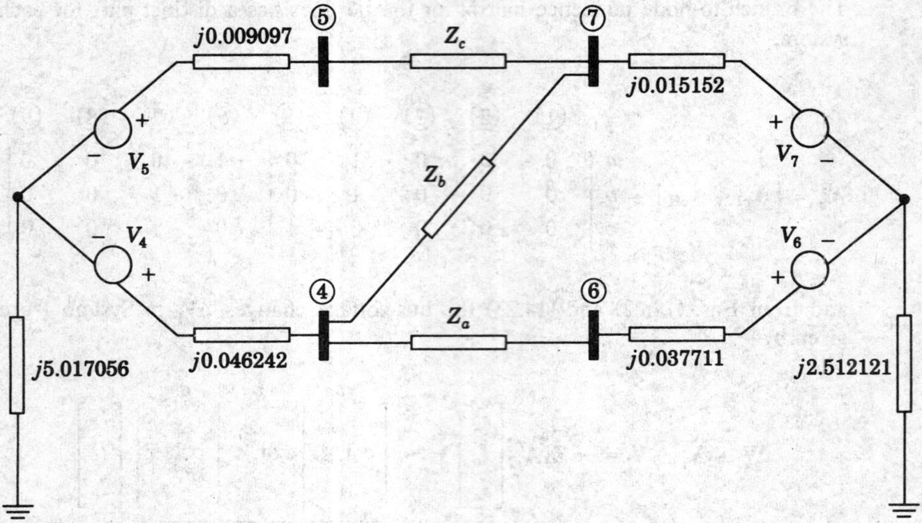

FIGURE 14.9
Thévenin equivalent circuits for Example 14.3.

now convert to the Ward equivalent mesh circuit shown in Fig. 14.10(*c*). The reader can easily check that the bus admittance matrix for the Ward equivalent agrees (within round-off error) with the inverse of the bus impedance submatrix extracted from $\mathbf{Z}_{II}$ at the boundary buses of System II. That is,

$$\begin{matrix} & ⑥ & ⑦ \\ ⑥ & \\ ⑦ & \end{matrix}\begin{bmatrix} Z_{66} & Z_{67} \\ Z_{76} & Z_{77} \end{bmatrix}^{-1} = \begin{bmatrix} j2.549832 & j2.512121 \\ j2.512121 & j2.527273 \end{bmatrix}^{-1} = \begin{matrix} ⑥ \\ ⑦ \end{matrix}\begin{bmatrix} -j18.949386 & j18.835777 \\ j18.835777 & -j19.118533 \end{bmatrix}$$

Piecewise methods give exact results for the overall system if currents and voltages are directly proportional. For then the system is linear and the principle of superposition applies, which is the basis of the piecewise approach. The assumption of linearity has to be reevaluated in power-flow studies, for example, where the active power P and reactive power Q are modeled at the buses rather than the current injections. The load current is therefore a nonlinear function of the applied voltages and the operating point of the overall system may not be consistent with the individual operating points of the separate systems. In such a case the methods developed in Chap. 9 for solving the power flows in large-scale systems should be used.

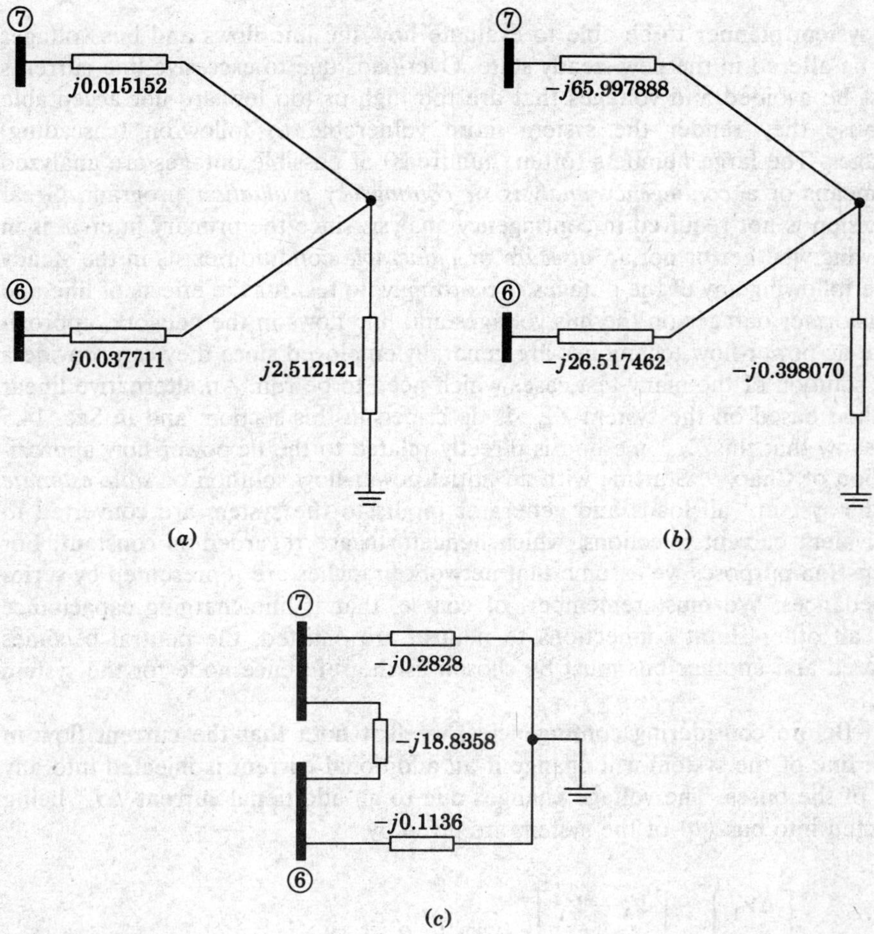

FIGURE 14.10
Equivalent circuits for System II of Example 14.3 showing: (*a*) Thévenin impedances; (*b*) reciprocals of Thévenin impedances; (*c*) admittance circuit of Ward equivalent.

14.3 ANALYSIS OF SINGLE CONTINGENCIES

Whenever a transmission line or transformer is removed from service, we say that an *outage* has occurred. Outages may be *planned* for purposes of scheduled maintenance or they may be *forced* by weather conditions, faults or other contingencies. A line or transformer is deenergized and isolated from the network by tripping the appropriate circuit breakers. The ensuing current and voltage transients in the network quickly die away and new steady-state operating conditions are established. It is important for both the system operator and the system planner to be able to evaluate how the line flows and bus voltages will be altered in the new steady state. Overloads due to excessive line currents must be avoided and voltages that are too high or too low are not acceptable because they render the system more vulnerable to follow-on (cascading) outages. The large numbers (often, hundreds) of possible outages are analyzed by means of a *contingency analysis* or *contingency evaluation* program. Great precision is not required in contingency analysis since the primary interest is in knowing whether or not an *insecure* or *vulnerable* condition exists in the steady state following any of the outages. Accordingly, to test for the effects of line and transformer outages on the bus voltages and line flows in the network, approximate ac power-flow techniques are generally employed since they can provide a fast solution of the many test cases which need to be run. An alternative linear method based on the system $\mathbf{Z}_{bus}$ is described in this section, and in Sec. 14.5 we show that this $\mathbf{Z}_{bus}$ method is directly related to the dc power-flow approximation of Chap. 9. Starting with an initial power-flow solution or *state estimate* of the system,[1] all loads and generator inputs to the system are converted to equivalent current injections, which henceforth are regarded as constant. For discussion purposes we assume that network branches are represented by series impedances. We must remember, of course, that if line-charging capacitance and all other shunt connections to neutral are omitted, the neutral becomes isolated and another bus must be chosen as the reference node for the system $\mathbf{Z}_{bus}$.

Before considering contingencies, we first note that the current flow in each line of the system will change if an additional current is injected into any one of the buses. The voltage changes due to an additional current ΔI_m being injected into bus ⓜ of the system are given by

$$\begin{bmatrix} \Delta V_1 \\ \vdots \\ \Delta V_i \\ \Delta V_j \\ \vdots \\ \Delta V_N \end{bmatrix} = \begin{bmatrix} V_1' - V_1 \\ \vdots \\ V_i' - V_i \\ V_j' - V_j \\ \vdots \\ V_N' - V_N \end{bmatrix} = \mathbf{Z}_{bus} \begin{bmatrix} 0 \\ \vdots \\ \Delta I_m \\ \vdots \\ 0 \end{bmatrix} = \begin{bmatrix} \text{column } m \\ \text{of} \\ \mathbf{Z}_{bus} \end{bmatrix} \Delta I_m \tag{14.27}$$

where $\mathbf{Z}_{bus}$ is for the normal configuration of the system. The changes in the voltages of buses ⓘ and ⓙ can be written

$$\Delta V_i = Z_{im}\,\Delta I_m \qquad \Delta V_j = Z_{jm}\,\Delta I_m \tag{14.28}$$

[1]See Chap. 15.

If the line connecting buses ⓘ and ⓙ has impedance Z_c, then the change in the line current from bus ⓘ to bus ⓙ is

$$\Delta I_{ij} = \frac{\Delta V_i - \Delta V_j}{Z_c} = \frac{Z_{im} - Z_{jm}}{Z_c} \Delta I_m \tag{14.29}$$

This equation leads us to define the *current-injection distribution factor* $K_{ij,m}$,

$$K_{ij,m} \triangleq \frac{\Delta I_{ij}}{\Delta I_m} = \frac{Z_{im} - Z_{jm}}{Z_c} \tag{14.30}$$

Thus, whenever the current injected into bus ⓜ changes by ΔI_m, the current flow in line ⓘ-ⓙ changes by the amount $\Delta I_{ij} = K_{ij,m} \Delta I_m$. We are now in a position to consider contingencies.

In the normal configuration of the system if a particular line current is excessive, it may be possible to remove the overload by reducing the current injected into the system at one bus and correspondingly increasing the current injection at another bus. Suppose that the current injected at bus ⓟ is changed by ΔI_p whereas the current injected at bus ⓠ is changed by ΔI_q. By the principle of superposition the shift in the current injection from bus ⓟ to bus ⓠ causes the current in line ⓘ-ⓙ to change by the amount

$$\Delta I_{ij} = K_{ij,p} \Delta I_p + K_{ij,q} \Delta I_q \tag{14.31}$$

Due to this application, the current-injection distribution factor is also called the *current-shift distribution factor*. In Sec. 14.5 we shall see that shifting current injection from one bus to another is equivalent to shifting real power inputs between the same two buses in the dc power-flow model. On this account, current-shift distribution factors are often called *generation-shift distribution factors*. Equation (14.31) shows that precomputed tables of such factors can be used to evaluate how a proposed shift in current injection from bus ⓟ to bus ⓠ changes the level of current in the line ⓘ-ⓙ.

Let us now consider removing from the network a line or transformer of series impedance Z_a between buses ⓜ and ⓝ. We know from Sec. 8.3 that the outage can be simulated by *adding* impedance $-Z_a$ between the buses in the preoutage Thévenin equivalent circuit shown in Fig. 14.11. Closing switch S connects the impedance $-Z_a$ between the buses, and the loop current I_a then flows as shown. With $Z_{mn} = Z_{nm}$, we see from the figure that

$$I_a = \frac{V_m - V_n}{(Z_{mm} + Z_{nn} - 2Z_{mn}) - Z_a} = \frac{V_m - V_n}{Z_{\text{th},mn} - Z_a} \tag{14.32}$$

Here V_m and V_n are preoutage bus voltages and $Z_{\text{th},mn} = Z_{mm} + Z_{nn} - 2Z_{mn}$ is the preoutage Thévenin impedance between buses ⓜ and ⓝ. The loop

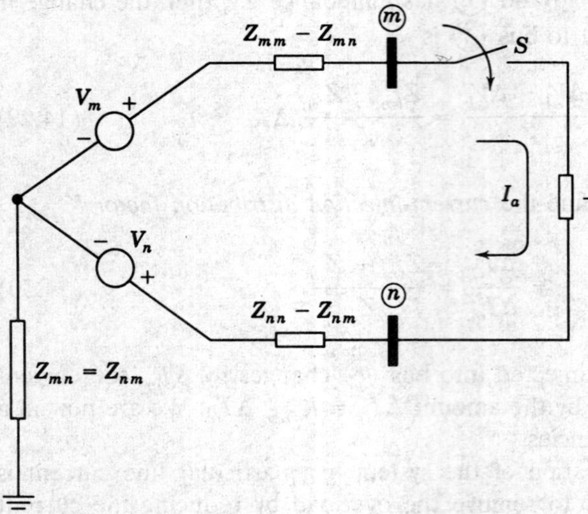

FIGURE 14.11
Preoutage Thévenin equivalent circuit for simulating outage of line ⓜ–ⓝ.

current I_a has the same effect on the voltages of the preoutage network as two compensating currents, $\Delta I_m = -I_a$ into bus ⓜ and $\Delta I_n = I_a$ into bus ⓝ. The expression for the resulting current change ΔI_{ij} from bus ⓘ to bus ⓙ is given by

$$\Delta I_{ij} = K_{ij,m}\,\Delta I_m + K_{ij,n}\,\Delta I_n = \frac{\left[(Z_{in} - Z_{im}) - (Z_{jn} - Z_{jm})\right]}{Z_c} I_a \quad (14.33)$$

Substituting for I_a from Eq. (14.32) gives

$$\Delta I_{ij} = \frac{\left[(Z_{in} - Z_{im}) - (Z_{jn} - Z_{jm})\right]}{Z_c} \frac{V_m - V_n}{(Z_{\text{th},mn} - Z_a)} \quad (14.34)$$

Before the outage of line ⓜ–ⓝ the voltage drop $V_m - V_n$ is due to the current

$$I_{mn} = \frac{V_m - V_n}{Z_a} \quad (14.35)$$

From Eqs. (14.34) and (14.35) we find that the current change in the line ⓘ–ⓙ due to the outage of the line ⓜ–ⓝ is

$$\Delta I_{ij} = \frac{Z_a}{Z_c}\left[\frac{(Z_{in} - Z_{im}) - (Z_{jn} - Z_{jm})}{Z_{\text{th},mn} - Z_a}\right] I_{mn}$$

Expressing this change as a fraction of the preoutage current I_{mn} gives

$$\frac{\Delta I_{ij}}{I_{mn}} = -\frac{Z_a}{Z_c}\left[\frac{(Z_{im} - Z_{in}) - (Z_{jm} - Z_{jn})}{Z_{\text{th},mn} - Z_a}\right] \triangleq L_{ij,mn} \qquad (14.36)$$

where $L_{ij,mn}$ is called the *line-outage distribution factor*. A similar distribution factor can be defined for the change in current in any other line of the network. For instance, the line-outage distribution factor expressing the change in current in line ⓟ-ⓠ of impedance Z_b due to the outage of line ⓜ-ⓝ of impedance Z_a is

$$L_{pq,mn} = \frac{\Delta I_{pq}}{I_{mn}} = -\frac{Z_a}{Z_b}\left[\frac{(Z_{pm} - Z_{pn}) - (Z_{qm} - Z_{qn})}{Z_{\text{th},mn} - Z_a}\right] \qquad (14.37)$$

The distribution factors $L_{ij,mn}$ and $L_{pq,mn}$ contain numerator and denominator terms which can be easily generated from the triangle factors of $\mathbf{Y}_{\text{bus}}$, as demonstrated in Secs. 8.5 and 14.1. It is also evident from Eqs. (14.36) and (14.37) that tables of distribution factors can be precomputed from the line impedances and the elements of $\mathbf{Z}_{\text{bus}}$ for the normal system configuration. These tables can be used to evaluate the impact of any single-line outage on the current flows in the remaining lines. Following the outage of line ⓜ-ⓝ, the new currents in the lines are given by equations of the form

$$\begin{aligned} I'_{ij} &= I_{ij} + \Delta I_{ij} = I_{ij} + L_{ij,mn} I_{mn} \\ I'_{pq} &= I_{pq} + \Delta I_{pq} = I_{pq} + L_{pq,mn} I_{mn} \end{aligned} \qquad (14.38)$$

Thus, if the actual preoutage current I_{mn} in line ⓜ-ⓝ is measured or estimated, line overloads due to excessive current changes caused by the loss of that line can be identified from Eq. (14.38).

The numerical examples which follow illustrate the close agreement often found between the values of the line flows calculated by distribution factors and by ac power-flow methods. The same level of agreement is generally not found in corresponding calculations of bus voltages.

Example 14.4. Figure 14.12 shows a small power system of five buses and six lines, which is the same as System I of Fig. 14.8. When the lines are represented by their series reactances, the system $\mathbf{Z}_{\text{bus}}$ is the same as that given for $\mathbf{Z}_{\text{I}}$ in Table 14.1. The *P-Q* loads and generation shown in Fig. 14.12 correspond to a *base-case* power-flow solution of the system, which yields the bus voltages tabulated under Case *a* in Table 14.2. Using distribution factors, predict the current in line ⑤-③ when line ⑤-② is outaged under the given operating conditions. Compare the prediction with the exact value obtained from an ac power-flow solution of the same *change case*.

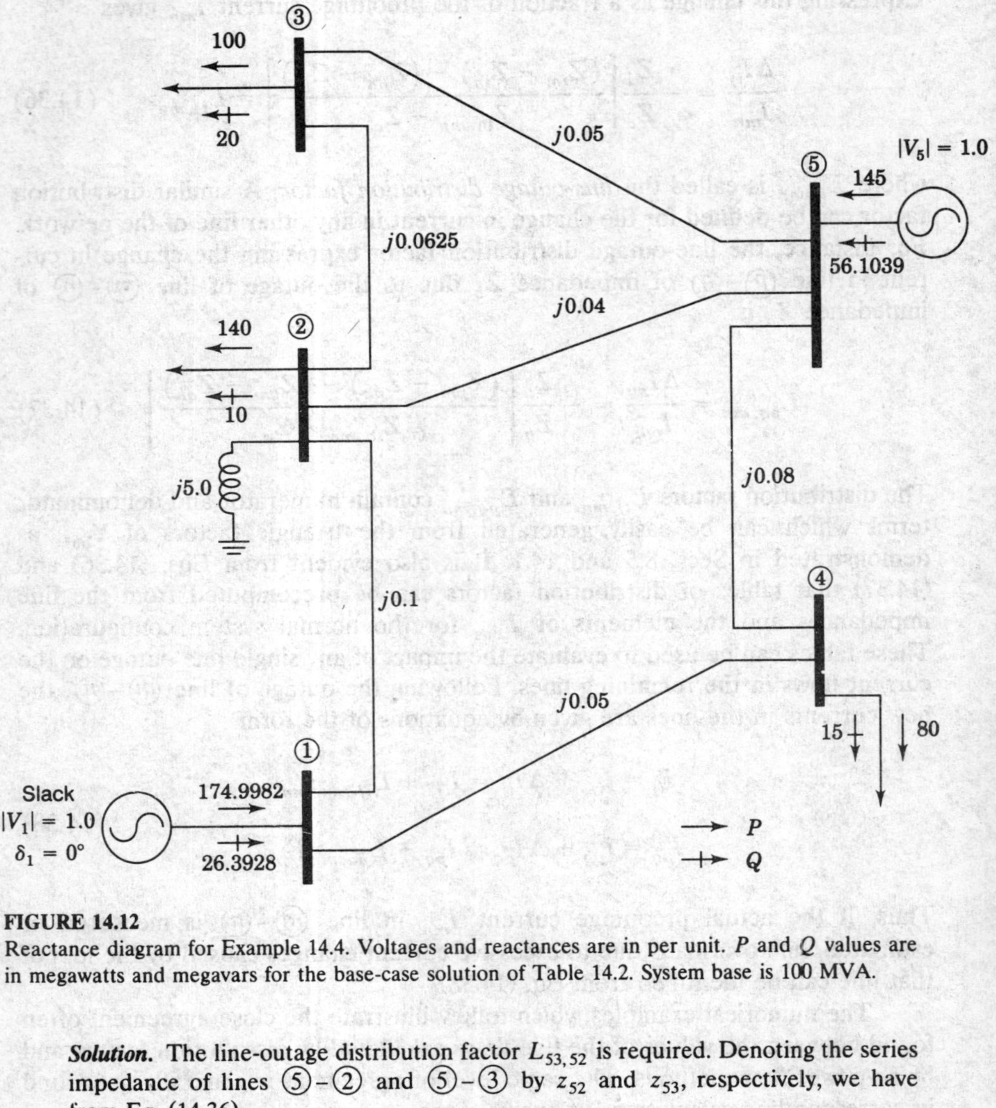

FIGURE 14.12
Reactance diagram for Example 14.4. Voltages and reactances are in per unit. P and Q values are in megawatts and megavars for the base-case solution of Table 14.2. System base is 100 MVA.

Solution. The line-outage distribution factor $L_{53,52}$ is required. Denoting the series impedance of lines ⑤–② and ⑤–③ by z_{52} and z_{53}, respectively, we have from Eq. (14.36)

$$L_{53,52} = \frac{\Delta I_{53}}{I_{52}} = -\frac{z_{52}}{z_{53}}\left[\frac{(Z_{55} - Z_{52}) - (Z_{35} - Z_{32})}{(Z_{55} + Z_{22} - 2Z_{52}) - z_{52}}\right]$$

Substituting numerical values, we obtain

$$L_{53,52} = -\frac{j0.04}{j0.05}\left[\frac{(j5.026153 - j5.000000) - (j5.014529 - j5.000000)}{\underbrace{(j5.026153 + j5.000000 - 2 \times j5.000000)}_{j0.026153} - j0.04}\right]$$

$$= 0.671533$$

TABLE 14.2
Bus voltages from power-flow solutions of the system of Fig. 14.12

Bus number	Base case *a*	Change cases *b*	*c*	*d*	*e*
①	1.000000 +j0.000000	1.000000 +j0.000000	1.000000 +j0.000000	1.000000 +j0.000000	1.000000 +j0.000000
②	0.986301 −j0.083834	0.968853 −j0.108108	0.983733 −j0.106285	0.966186 −j0.124878	0.970492 −j0.095000
③	0.984789 −j0.095108	0.977822 −j0.088536	0.981780 −j0.121225	0.975005 −j0.116235	0.979301 −j0.066882
④	0.993653 −j0.045583	0.994430 −j0.033446	0.992582 −j0.056858	0.993489 −j0.047561	0.990816 −j0.040000
⑤	0.998498 −j0.054795	0.999734 −j0.023079	0.996444 −j0.084253	0.998201 −j0.059964	0.999984 +j0.005738

Change from base case:
Case *b*: line ⑤–② out of service.
Case *c*: 45-MW shift from bus ⑤ to bus ①
Case *d*: line ⑤–② out of service plus Case *c* change
Case *e*: lines ⑤–② and ⑤–④ out of service

Other line-outage distribution factors calculated in a similar manner are shown in Table 14.3.

From the Case *a* power-flow results in Table 14.2 (now indicated by superscript *a*) the preoutage currents in lines ⑤–② and ⑤–③ are calculated as follows:

$$I_{52}^a = \frac{V_5^a - V_2^a}{z_{52}} = \frac{(0.998498 - j0.054795) - (0.986301 - j0.083834)}{j0.04}$$

$$= 0.725975 - j0.304925 \text{ per unit}$$

$$I_{53}^a = \frac{V_5^a - V_3^a}{z_{53}} = \frac{(0.998498 - j0.054795) - (0.984789 - j0.095108)}{j0.05}$$

$$= 0.806260 - j0.274180 \text{ per unit}$$

The predicted current change in line ⑤–③ therefore is

$$\Delta I_{53} = L_{53,52} I_{52}^a = 0.671533(0.725975 - j0.304925)$$

$$= 0.487516 - j0.204767 \text{ per unit}$$

TABLE 14.3
Line-outage and generation-shift distribution factors for the system of Figure 14.12

Line-outage distribution factors—$L_{ij,mn}$
Outage of line ⓜ–ⓝ

Current change in line ⓘ–ⓙ	②–①	④–①	③–②	⑤–②	⑤–③	⑤–④
②–①		1.000000	− 0.148148	− 0.328467	− 0.148148	1.000000
④–①	1.000000		0.148148	0.328467	0.148148	− 1.000000
③–②	− 0.262295	0.262295		0.671533	− 1.000000	0.262295
⑤–②	− 0.737705	0.737705	0.851852		0.851852	0.737705
⑤–③	− 0.262295	0.262295	− 1.000000	0.671533		0.262295
⑤–④	1.000000	− 1.000000	0.148148	0.328467	0.148148	

Generation-shift distribution factors—$K_{ij,p}$
Current change in line ⓘ–ⓙ

Injected into bus ⓟ	②–①	④–①	③–②	⑤–②	⑤–③	⑤–④
①	− 0.614656	− 0.385344	0.101074	0.284270	0.101074	− 0.385344
⑤	− 0.113708	0.113708	0.232470	0.653822	0.232470	0.113708

and the predicted postoutage current from bus ⑤ to bus ③ is

$$I'_{53} = I^a_{53} + \Delta I_{53} = (0.80626 - j0.274180) + (0.487516 - j0.204767)$$

$$= 1.293776 - j0.478947 = 1.380\angle{-20.31^\circ} \text{ per unit}$$

With line ⑤–② out of service and all operating conditions otherwise the same as in the base case, a new ac power-flow solution gives the change-case results listed under Case *b* in Table 14.2. From these results we can calculate the corresponding current in line ⑤–③ as follows:

$$I^b_{53} = \frac{V^b_5 - V^b_3}{z_{53}} = \frac{(0.999734 - j0.023079) - (0.977822 - j0.088536)}{j0.05}$$

$$= 1.30914 - j0.438240 = 1.381\angle{-18.51^\circ} \text{ per unit}$$

Thus, we see that the predicted magnitude of the postoutage current compares very favorably with that calculated by the ac power-flow method.

Example 14.5. The only change from the input data of the base-case power-flow solution of Example 14.4 is a shift of 45 MW from the generator at bus ⑤ to the generator at bus ① in the system of Fig. 14.12. The bus impedance matrix $\mathbf{Z}_1$ shown in Table 14.1 therefore applies. Using distribution factors, predict the magnitude of the resulting current in line ⑤-④ and compare the predicted value with the exact value calculated from an ac power-flow solution of the same change case.

Solution. To facilitate a quick comparison of results, let us assume that the changes in the megawatt inputs at buses ⑤ and ① are exactly equal in per unit to the corresponding changes in the current injections at those two buses. That is, $\Delta I_1 = 0.45 + j0.0$ per unit and $\Delta I_5 = -0.45 + j0.0$ per unit. According to Eq. (14.31), the current change in line ⑤-④ is given by

$$\Delta I_{54} = K_{54,1}\,\Delta I_1 + K_{54,5}\,\Delta I_5$$

where $K_{54,1} = \dfrac{Z_{51} - Z_{41}}{z_{54}}$ and $K_{54,5} = \dfrac{Z_{55} - Z_{45}}{z_{54}}$

Substituting the impedance values from $\mathbf{Z}_1$ of Table 14.1, we obtain

$$K_{54,1} = \left(\frac{j5.011371 - j5.042198}{j0.08}\right) = -0.38534 \qquad K_{54,5} = \left(\frac{j5.026153 - j5.017056}{j0.08}\right) = 0.11371$$

These and other current-shift distribution factors are already tabulated in Table 14.3. The current change now is

$$\Delta I_{54} = -0.38534 \times (0.45 + j0.0) + 0.11371(-0.45 + j0.0)$$

$$= -0.225 + j0.0 \text{ per unit}$$

The predicted value of I_{54} is found by adding this change to the preoutage current I_{54}^a calculated from the base-case solution of Table 14.2. That is,

$$I_{54}^a = \frac{V_5^a - V_4^a}{z_{54}} = \frac{(0.998498 - j0.054795) - (0.993653 - j0.045583)}{j0.08}$$

$$= -0.115150 - j0.060563 \text{ per unit}$$

$$I'_{54} = I_{54}^a + \Delta I_{54}^a = (-0.115150 - j0.060563) + (-0.225 + j0.0)$$

$$= -0.340150 - j0.060563 = 0.345\angle{-170^\circ} \text{ per unit}$$

With bus ① retained as the slack bus and the 45-MW reduction at bus ⑤ as the only change from the base case, a change-case power-flow calculation gives the results listed under Case c in Table 14.2, from which we calculate the exact value of I_{54},

$$I_{54}^c = \frac{V_5^c - V_4^c}{z_{54}} = \frac{(0.996444 - j0.084253) - (0.992582 - j0.056858)}{j0.08}$$

$$= -0.342438 - j0.048275 = 0.346\angle -172^\circ \text{ per unit}$$

Thus, we see that the current-shift distribution factors give a predicted value of I_{54}, which is acceptably close to the value obtained from an ac power-flow solution.

14.4 ANALYSIS OF MULTIPLE CONTINGENCIES

Contingencies can arise in which two or more lines are tripped simultaneously, or where a line outage has already occurred and a shift in generation is being considered to determine if a line overload caused by the outage can be relieved. If tables of distribution factors for *first* contingencies are available, it is not necessary to recompute the tables in order to study the effect of two simultaneous contingencies. Although the existing distribution factors assume a normal system configuration before the first contingency occurs, they can be combined into formulas for evaluating *double* contingencies. Examples follow which illustrate how this can be achieved.

Suppose that line Ⓜ–ⓝ is carrying current I_{mn} before it trips out of service. If an overload occurs on another line ⓘ–ⓙ due to the outage, it may be possible to reduce the overload by decreasing the current injected into the system at some bus ⓟ and correspondingly increasing the current injected at another bus ⓠ. Such current-shifting occurs when generation is shifted from a power plant at bus ⓟ to a power plant at bus ⓠ. The distribution factors of the preceding section can be employed to study this possibility of overload relief. Since the $\mathbf{Z}_{\text{bus}}$ model is linear, the line outage and the proposed shift in current will have the same combined effect on the overloaded line regardless of the sequence in which the two events are considered to occur. For our purposes it is more convenient to consider first the proposed current shift and then the outage of the line Ⓜ–ⓝ.

According to Eq. (14.31), a current shift from bus ⓟ to bus ⓠ changes the line currents I_{mn} and I_{ij} by the amounts

$$\begin{aligned} \Delta I_{mn} &= K_{mn,p}\,\Delta I_p + K_{mn,q}\,\Delta I_q \\ \Delta I_{ij} &= K_{ij,p}\,\Delta I_p + K_{ij,q}\,\Delta I_q \end{aligned} \tag{14.39}$$

The distribution factors in this equation are those already precomputed for the normal system configuration and ΔI_p and ΔI_q are the proposed changes in the injected currents at buses ⓟ and ⓠ. As a result of shifting the current, the new line currents become

$$I'_{mn} = I_{mn} + \Delta I_{mn}$$
$$I'_{ij} = I_{ij} + \Delta I_{ij} \tag{14.40}$$

Then, if line ⓜ–ⓝ is opened, the *additional* change in the current of line ⓘ–ⓙ can be calculated using the single line-outage distribution factor $L_{ij,mn}$ as follows:

$$\Delta I'_{ij} = L_{ij,mn} I'_{mn} = L_{ij,mn} I_{mn} + L_{ij,mn}\,\Delta I_{mn} \tag{14.41}$$

This is the total change in the current I_{ij} due to the combination of the line-opening and the current-shifting. The final value of the steady-state current in line ⓘ–ⓙ becomes

$$\begin{aligned} I''_{ij} = I'_{ij} + \Delta I'_{ij} &= \left(I_{ij} + \Delta I_{ij}\right) + \left(L_{ij,mn} I_{mn} + L_{ij,mn}\,\Delta I_{mn}\right) \\ &= \left[I_{ij} + L_{ij,mn} I_{mn}\right] + \Delta I_{ij} + L_{ij,mn}\,\Delta I_{mn} \end{aligned} \tag{14.42}$$

According to Eq. (14.38), the quantity within the brackets represents the current which would flow in line ⓘ–ⓙ by opening line ⓜ–ⓝ *without* any generation shift also occurring. Therefore, the net change of current in the overloaded line due to shifting generation in the network, with line ⓜ–ⓝ already out of service, is given by Eq. (14.42) as

$$\Delta I^{sh}_{ij} = I''_{ij} - \left[I_{ij} + L_{ij,mn} I_{mn}\right] = \Delta I_{ij} + L_{ij,mn}\,\Delta I_{mn} \tag{14.43}$$

Substituting for ΔI_{ij} and ΔI_{mn} from Eqs. (14.39) and rearranging, we obtain

$$\Delta I^{sh}_{ij} = \underbrace{\left(K_{ij,p} + L_{ij,mn} K_{mn,p}\right)}_{K'_{ij,p}} \Delta I_p + \underbrace{\left(K_{ij,q} + L_{ij,mn} K_{mn,q}\right)}_{K'_{ij,q}} \Delta I_q \tag{14.44}$$

Thus, $K'_{ij,p}$ and $K'_{ij,q}$ are new generation-shift distribution factors which take into account the preexistng outage of line ⓜ–ⓝ. The new distribution factors can be quickly calculated from those already precomputed for the normal system configuration. Example 14.7 in Sec. 14.5 illustrates the simplicity of the calculations. By evaluating $\Delta I''_{ij}$, the system operators can determine the level of overload relief available from the proposed shift of injections.

As another application, let us consider what happens if a second line ⓟ–ⓠ is tripped out of service when the line ⓜ–ⓝ is already out of service in the network. If line ⓜ–ⓝ has series impedance Z_a and line ⓟ–ⓠ has series impedance Z_b, then the simultaneous outage of the two lines can be simulated by *adding* $-Z_a$ and $-Z_b$ between the respective end buses of the lines. We have a similar situation in Sec. 14.1 where equations are developed for adding new lines to an existing system. Those equations apply directly to the present situation. If we allow for the fact that the impedances to be added now have negative signs, Eq. (14.14) becomes

$$\left[\begin{array}{c|c} Z_{\text{th},mn} - Z_a & (Z_{mp} - Z_{mq}) - (Z_{np} - Z_{nq}) \\ \hline (Z_{pm} - Z_{pn}) - (Z_{qm} - Z_{qn}) & Z_{\text{th},pq} - Z_b \end{array}\right]\left[\begin{array}{c} I_a \\ \hline I_b \end{array}\right] = \left[\begin{array}{c} V_m - V_n \\ \hline V_p - V_q \end{array}\right] \tag{14.45}$$

The symbols in this equation have the same meaning as in Sec. 14.1. For example, V_m, V_n, V_p, and V_q represent bus voltages in the normal system configuration before any line outages have occurred. The voltage drops $(V_m - V_n)$ and $(V_p - V_q)$ are related to the preoutage currents, I_{mn} in line ⓜ–ⓝ and I_{pq} in line ⓟ–ⓠ, by

$$I_{mn} = \frac{V_m - V_n}{Z_a} \qquad I_{pq} = \frac{V_p - V_q}{Z_b} \tag{14.46}$$

New distribution factors for the two simultaneous outages can be developed by using Eqs. (14.45) and (14.46). It is helpful to present the development in a sequence of steps as follows:

Step 1

To find expressions for currents I_a and I_b in terms of the line-outage distribution factors:

- Divide each row of Eq. (14.45) by its diagonal element to obtain

$$\begin{bmatrix} 1 & \left[\dfrac{(Z_{mp} - Z_{mq}) - (Z_{np} - Z_{nq})}{Z_{\text{th},mn} - Z_a}\right] \\ \left[\dfrac{(Z_{pm} - Z_{pn}) - (Z_{qm} - Z_{qn})}{Z_{\text{th},pq} - Z_b}\right] & 1 \end{bmatrix}\begin{bmatrix} I_a \\ I_b \end{bmatrix} = \begin{bmatrix} \dfrac{V_m - V_n}{Z_{\text{th},mn} - Z_a} \\ \dfrac{V_p - V_q}{Z_{\text{th},pq} - Z_b} \end{bmatrix} \tag{14.47}$$

- Express the new off-diagonal elements in terms of the line-outage distribution factors

$$\begin{aligned} L_{pq,mn} &= -\frac{Z_a}{Z_b}\left[\frac{(Z_{pm} - Z_{pn}) - (Z_{qm} - Z_{qn})}{Z_{\text{th},mn} - Z_a}\right] \\ L_{mn,pq} &= -\frac{Z_b}{Z_a}\left[\frac{(Z_{mp} - Z_{mq}) - (Z_{np} - Z_{nq})}{Z_{\text{th},pq} - Z_b}\right] \end{aligned} \tag{14.48}$$

The expression obtained is

$$\begin{bmatrix} 1 & -\dfrac{Z_b}{Z_a}L_{pq,mn} \\ -\dfrac{Z_a}{Z_b}L_{mn,pq} & 1 \end{bmatrix}\begin{bmatrix} I_a \\ I_b \end{bmatrix} = \begin{bmatrix} \dfrac{V_m - V_n}{Z_{\text{th},mn} - Z_a} \\ \dfrac{V_p - V_q}{Z_{\text{th},pq} - Z_b} \end{bmatrix} \tag{14.49}$$

- Solve Eq. (14.49) for I_a and I_b

$$\begin{bmatrix} I_a \\ I_b \end{bmatrix} = \frac{1}{1 - L_{pq,mn}L_{mn,pq}}\begin{bmatrix} 1 & \dfrac{Z_b}{Z_a}L_{pq,mn} \\ \dfrac{Z_a}{Z_b}L_{mn,pq} & 1 \end{bmatrix}\begin{bmatrix} \dfrac{V_m - V_n}{Z_{\text{th},mn} - Z_a} \\ \dfrac{V_p - V_q}{Z_{\text{th},pq} - Z_b} \end{bmatrix} \tag{14.50}$$

- Substitute for $V_m - V_n$ and $V_p - V_q$ from Eqs. (14.46)

$$\begin{bmatrix} I_a \\ I_b \end{bmatrix} = \frac{1}{1 - L_{pq,mn}L_{mn,pq}}\begin{bmatrix} \dfrac{Z_a}{Z_{\text{th},mn} - Z_a} & \dfrac{Z_b^2}{Z_a}\dfrac{L_{pq,mn}}{Z_{\text{th},pq} - Z_b} \\ \dfrac{Z_a^2}{Z_b}\dfrac{L_{mn,pq}}{Z_{\text{th},mn} - Z_a} & \dfrac{Z_b}{Z_{\text{th},pq} - Z_a} \end{bmatrix}\begin{bmatrix} I_{mn} \\ I_{pq} \end{bmatrix} \tag{14.51}$$

In this equation I_{mn} and I_{pq} represent the actual currents in lines ⓜ–ⓝ and ⓟ–ⓠ prior to their simultaneous outage. The currents I_a and I_b are the (nonphysical) loop currents discussed in Sec. 14.1. They represent the outage of the two lines and can be considered as compensating current injections into the *normal* configuration of the system.

Step 2

To find the current change ΔI_{ij} due to I_a and I_b:

- Determine the bus-voltage changes due to the current injections ($-I_a$ into bus ⓜ, I_a into bus ⓝ) and ($-I_b$ into bus ⓟ, I_b into bus ⓠ). At buses ⓘ and ⓙ we obtain

$$\begin{aligned} \Delta V_i &= (Z_{in} - Z_{im})I_a + (Z_{iq} - Z_{ip})I_b \\ \Delta V_j &= (Z_{jn} - Z_{jm})I_a + (Z_{jq} - Z_{jp})I_b \end{aligned} \tag{14.52}$$

- Determine the current change $\Delta I_{ij} = (\Delta V_i - \Delta V_j)/Z_c$ in line ⓘ-ⓙ of impedance Z_c,

$$\Delta I_{ij} = \left[\frac{(Z_{in} - Z_{im}) - (Z_{jn} - Z_{jm})}{Z_c} \quad \frac{(Z_{iq} - Z_{ip}) - (Z_{jp} - Z_{jq})}{Z_c}\right]\begin{bmatrix} I_a \\ I_b \end{bmatrix} \tag{14.53}$$

- Express Eq. (14.53) in terms of the line-outage distribution factors

$$L_{ij,mn} = -\frac{Z_a}{Z_c}\left[\frac{(Z_{im} - Z_{in}) - (Z_{jm} - Z_{jn})}{Z_{\text{th},mn} - Z_a}\right]$$

$$L_{ij,pq} = -\frac{Z_b}{Z_c}\left[\frac{(Z_{ip} - Z_{iq}) - (Z_{jp} - Z_{jq})}{Z_{\text{th},pq} - Z_b}\right] \tag{14.54}$$

The resultant equation is

$$\Delta I_{ij} = \left[\frac{Z_{\text{th},mn} - Z_a}{Z_a} L_{ij,mn} \quad \frac{Z_{\text{th},pq} - Z_b}{Z_b} L_{ij,pq}\right]\begin{bmatrix} I_a \\ I_b \end{bmatrix} \tag{14.55}$$

- Substitute for I_a and I_b from Eq. (14.51) to obtain

$$\Delta I_{ij} = \left[\underbrace{\left(\frac{L_{ij,mn} + L_{ij,pq}L_{pq,mn}}{1 - L_{pq,mn}L_{mn,pq}}\right)}_{L'_{ij,mn}} \quad \underbrace{\left(\frac{L_{ij,pq} + L_{ij,mn}L_{mn,pq}}{1 - L_{mn,pq}L_{pq,mn}}\right)}_{L'_{ij,pq}}\right]\begin{bmatrix} I_{mn} \\ I_{pq} \end{bmatrix} \tag{14.56}$$

$L'_{ij,mn}$ is the effective line-outage distribution factor which expresses the change in the steady-state current of line ⓘ-ⓙ due to the outage of line ⓜ-ⓝ *when line* ⓟ-ⓠ *is already removed from the network*. A similar statement applies to $L'_{ij,pq}$.

Thus, we see that the distribution factors already precomputed for single-line outages can be combined to produce distribution factors for *two* simultaneous line outages. The above development can be generalized to three or more simultaneous outages, provided the assumption of constant current injections remains acceptable—which may not always be the case in practice.

Example 14.6. Using distribution factors, predict the change in the current of line ⑤-③ when lines ⑤-② and ⑤-④ are simultaneously outaged in the system of Fig. 14.12. Use the base-case bus data of Example 14.4 and compare results with those obtained by an ac power-flow calculation.

Solution. By setting $i = 5$, $j = 3$; $m = 5$, $n = 2$; $p = 5$, $q = 4$ in Eq. (14.56), we find that

$$\Delta I_{53} = L'_{53,52} I^a_{52} + L'_{53,54} I^a_{54}$$

where I^a_{52} and I^a_{54} are line currents of the base case and

$$L'_{53,52} = \frac{L_{53,52} + L_{53,54} L_{54,52}}{1 - L_{54,52} L_{52,54}}$$

$$L'_{53,54} = \frac{L_{53,54} + L_{53,52} L_{52,54}}{1 - L_{52,54} L_{54,52}}$$

The line-outage distribution factors for single contingencies are now substituted into these formulas from Table 14.3 to obtain

$$L'_{53,52} = \frac{0.671533 + 0.262295 \times 0.328467}{1 - 0.328467 \times 0.737705} = 1$$

$$L'_{53,54} = \frac{0.262295 + 0.671533 \times 0.737705}{1 - 0.737705 \times 0.328467} = 1$$

To confirm understanding of these particular numerical results, the reader should refer to the system diagram of Fig. 14.12 and consider line ⑤-② being taken out of service when line ⑤-④ is already outaged, and conversely.

From Examples 14.4 and 14.5 we have

$$I^a_{52} = 0.725975 - j0.304925 \text{ per unit}$$

$$I^a_{53} = 0.806260 - j0.274180 \text{ per unit}$$

$$I^a_{54} = -0.115150 - j0.060563 \text{ per unit}$$

Therefore,

$$\Delta I_{53} = (1.0)(0.725975 - j0.304925) + (1.0)(-0.115150 - j0.060563)$$

$$= 0.610825 - j0.365488 \text{ per unit}$$

Adding this current change to the preoutage current I^a_{53} of the base case, we obtain

$$I''_{53} = I^a_{53} + \Delta I_{53} = (0.806260 - j0.274180) + (0.610825 - j0.365488)$$

$$= 1.417085 - j0.639668 = 1.555\angle -24.3^\circ \text{ per unit}$$

The ac power-flow results for the system with both lines ⑤-② and ⑤-④ out of service are shown in Table 14.2 under Case *e*, which yields

$$I_{53}^e = \frac{V_5^e - V_3^e}{z_{53}} = \frac{(0.999984 + j0.005738) - (0.979301 - j0.066882)}{j0.05}$$

$$= 1.452400 - j0.413660 = 1.510\angle -15.9^\circ \text{ per unit}$$

Thus, the distribution factors predict a result for $|I_{53}|$, which is comparable to that of the ac power-flow calculation.

14.5 CONTINGENCY ANALYSIS BY DC MODEL

We have already discussed the fact that approximate ac power-flow methods are also used to analyze contingencies. The simplest method is based on the so-called dc power-flow model described in Sec. 9.7. The dc model assumes that:

- The system is lossless and each line is represented by its series reactance.
- Each bus has rated system voltage of 1.0 per unit.
- Angular differences between voltages at adjacent buses Ⓜ and Ⓝ are so small that $\cos \delta_m = \cos \delta_n$ and $\sin \delta_m - \sin \delta_n = \delta_m - \delta_n$ rad.

On the basis of these assumptions, the per-unit current flow from bus Ⓜ to bus Ⓝ in a line of series reactance jX_a per unit is a real quantity given by

$$I_{mn} = \frac{V_m - V_n}{jX_a} = \frac{|V_m|(\cos \delta_m + j \sin \delta_m) - |V_n|(\cos \delta_n + j \sin \delta_n)}{jX_a}$$

$$\cong \frac{(\cos \delta_m - \cos \delta_n) + j(\sin \delta_m - \sin \delta_n)}{jX_a} \cong \frac{\delta_m - \delta_n}{X_a} \text{ per unit} \quad (14.57)$$

and the per-unit power flow in the same line is

$$P_{mn} = \frac{|V_m V_n|}{X_a} \sin(\delta_m - \delta_n) \cong \frac{\delta_m - \delta_n}{X_a} \text{ per unit} \quad (14.58)$$

It is evident, therefore, that per-unit power and per-unit current are synonymous in the dc power-flow model where nominal system voltages are assumed. For convenience, let us agree that equality signs in the equations which follow replace the more accurate approximation signs. Then, the current in the line of per-unit impedance jX_c between buses ⓘ and ⓙ can be written

$$I_{ij} = \frac{\delta_i - \delta_j}{X_c} \text{ per unit} \tag{14.59}$$

and if the angles of the bus voltages change, the line current changes by the amount

$$\Delta I_{ij} = \frac{\Delta(\delta_i - \delta_j)}{X_c} = \frac{\Delta\delta_i - \Delta\delta_j}{X_c} \text{ per unit} \tag{14.60}$$

If all impedances in the definition of the line-outage distribution factor $L_{ij,mn}$ of Eq. (14.36) are replaced by their corresponding reactances, we obtain

$$L_{ij,mn} = \frac{\Delta I_{ij}}{I_{mn}} = -\frac{X_a}{X_c}\left[\frac{(X_{im} - X_{in}) - (X_{jm} - X_{jn})}{X_{\text{th},mn} - X_a}\right] \tag{14.61}$$

The correspondence between per-unit power and per-unit current in the lines then enables us to write

$$L_{ij,mn} = \frac{\Delta P_{ij}}{P_{mn}} = -\frac{X_a}{X_c}\left[\frac{(X_{im} - X_{in}) - (X_{jm} - X_{jn})}{X_{\text{th},mn} - X_a}\right] \tag{14.62}$$

It is apparent that the line-outage distribution factors developed in the preceding sections by $\mathbf{Z}_{\text{bus}}$ methods are essentially the same as those for the dc power-flow method, provided series line reactances are used to represent the lines. With the lines so represented, the current-shift distribution factors of the $\mathbf{Z}_{\text{bus}}$ method become numerically equal to the *generation-shift distribution factors* of the dc power-flow model. If measurements are made of the actual *power* flows on the physical lines of a system, tables of precomputed distribution factors can be used, as described in the preceding sections. In fact, in the preceding numerical examples the magnitudes of the predicted changes in the current flows are very nearly equal in per unit to the magnitudes of the corresponding megawatt changes obtained by means of distribution factors based on the dc power-flow model. The following numerical example illustrates this fact.

Example 14.7. Use distribution factors based on the dc power-flow model to predict the change in the power flow P_{53} from bus ⑤ to bus ③ in the system of Fig. 14.12 due to a generation shift of 45 MW from bus ⑤ to bus ① when line ⑤-② has already been taken out of service. All other conditions of load and generation are the same as in Case *b* of Example 14.4.

Solution. Because the lines in the system of Fig. 14.12 are represented by series reactances, the distribution factors of Table 14.3 apply directly here. With per-unit power replacing per-unit current in Eq. (14.44), and setting $i = 5$, $j = 3$; $m = 5$, $n = 2$; $p = 5$, $q = 1$ in that equation, we obtain

$$\Delta P_{53} = K'_{53,5}\,\Delta P_5 + K'_{53,1}\,\Delta P_1$$

$$= (K_{53,5} + L_{53,52}K_{52,5})\,\Delta P_5 + (K_{53,1} + L_{53,52}K_{52,1})\,\Delta P_1$$

From Table 14.3 we can calculate

$$K'_{53,5} = K_{53,5} + L_{53,52}K_{52,5} = 0.232470 + 0.671533(0.653822)$$

$$= 0.671533$$

$$K'_{53,1} = K_{53,1} + L_{53,52}K_{52,1} = 0.101074 + 0.671533(0.284270)$$

$$= 0.291971$$

and so

$$\Delta P_{53} = 0.671533 \times (-0.45) + 0.291971 \times 0.45 = -0.170803 \text{ per unit}$$

Hence, with line ⑤–② already out of service, the power flow in the line from bus ⑤ to bus ③ is predicted to decrease by 17.1 MW due to the shift in generation.

The power-flow results for Case b show that $P^b_{53} = 131.89$ MW before the generation shift. Therefore, the predicted value of P_{53} after the 45-MW shift from bus ⑤ to bus ① is

$$P_{53} = P^b_{53} + \Delta P_{53} = 131.89 + (-17.1) = 114.8 \text{ MW}$$

Using the full ac power-flow solution for the combined line outage and generation shift (Case d of Table 14.2), the reader can show that $P^d_{53} = 115.12$ MW, and so the predicted value of P_{53} approximates the more accurate ac calculation very closely.

Previous examples involving distribution factors yield current flows with magnitudes very nearly the same in per unit as the real power flows obtainable by the dc power-flow method.

14.6 SYSTEM REDUCTION FOR CONTINGENCY AND FAULT STUDIES

For economic and security reasons individual power companies are connected together by tie lines to form a large interconnected system. If one of the companies wishes to study faults and contingencies on its own *internal* system, then it must represent the adjoining *external* system in some manner. Since the individual utility is mostly interested in the line flows and voltages of its own system over which it has some local control, it seeks to represent the external system by a reduced equivalent network.

Consider the system of Fig. 14.13(a), which shows buses ① to ③ in the internal System I and the remaining buses ④ to ⑦ in the external System II. The systems are connected by two tie lines. The starting point in the contingency analysis of System I is an ac power-flow solution for the entire interconnected system including tie lines and the external system. If the power-flow results are used to convert the loads and generation inputs of *all* the buses into equivalent current injections, then bus admittance equations can be written as follows:

$$\begin{array}{c} \\ ⑦ \\ ⑥ \\ ⑤ \\ ④ \\ ③ \\ ② \\ ① \end{array} \underbrace{\left[\begin{array}{cccc|ccc} ⑦ & ⑥ & ⑤ & ④ & ③ & ② & ① \\ Y_{77} & Y_{76} & Y_{75} & Y_{74} & 0 & 0 & 0 \\ Y_{67} & Y_{66} & Y_{65} & Y_{64} & 0 & 0 & 0 \\ Y_{57} & Y_{56} & Y_{55} & Y_{54} & Y_{53} & 0 & 0 \\ Y_{47} & Y_{46} & Y_{45} & Y_{44} & 0 & Y_{42} & 0 \\ \hline 0 & 0 & Y_{35} & 0 & Y_{33} & Y_{32} & Y_{31} \\ 0 & 0 & 0 & Y_{24} & Y_{23} & Y_{22} & Y_{21} \\ 0 & 0 & 0 & 0 & Y_{13} & Y_{12} & Y_{11} \end{array}\right]}_{\mathbf{Y}_{\text{bus}}} \left[\begin{array}{c} V_7' \\ V_6' \\ V_5' \\ V_4' \\ \hline V_3' \\ V_2' \\ V_1' \end{array}\right] = \left[\begin{array}{c} I_7 \\ I_6 \\ I_5 \\ I_4 \\ \hline I_3 \\ I_2 \\ I_1 \end{array}\right] \tag{14.63}$$

The voltages are shown with primed superscripts to emphasize that they are numerical results in the power-flow solution of the *entire* interconnected system. The currents I_1 to I_7 are the corresponding equivalent current injections calculated for each bus ⓘ from the equation $I_i = (P_i - jQ_i)/V_i'^*$. The zeros in $\mathbf{Y}_{\text{bus}}$ show that the tie lines are the only connections between System I and System II, and elements Y_{42} and Y_{53} equal the negatives of the corresponding tie-line admittances. In order to reduce the size of the external system representation, suppose we eliminate buses ⑦ and ⑥ using gaussian-elimination formulas like those in Eqs. (7.59) and (7.60), namely,

$$Y_{ij}' = Y_{ij} - \frac{Y_{ik}Y_{kj}}{Y_{kk}} \qquad I_i' = I_i - \frac{Y_{ik}}{Y_{kk}}I_k \tag{14.64}$$

The reduced system of equations is then given by

$$\begin{array}{c} \\ ⑤ \\ ④ \\ ③ \\ ② \\ ① \end{array} \left[\begin{array}{cc|ccc} ⑤ & ④ & ③ & ② & ① \\ Y_{55}' & Y_{54}' & Y_{53} & 0 & 0 \\ Y_{45}' & Y_{44}' & 0 & Y_{42} & 0 \\ \hline Y_{35} & 0 & Y_{33} & Y_{32} & Y_{31} \\ 0 & Y_{24} & Y_{23} & Y_{22} & Y_{21} \\ 0 & 0 & Y_{13} & Y_{12} & Y_{11} \end{array}\right] \left[\begin{array}{c} V_5' \\ V_4' \\ \hline V_3' \\ V_2' \\ V_1' \end{array}\right] = \left[\begin{array}{c} I_5' \\ I_4' \\ \hline I_3 \\ I_2 \\ I_1 \end{array}\right] \tag{14.65}$$

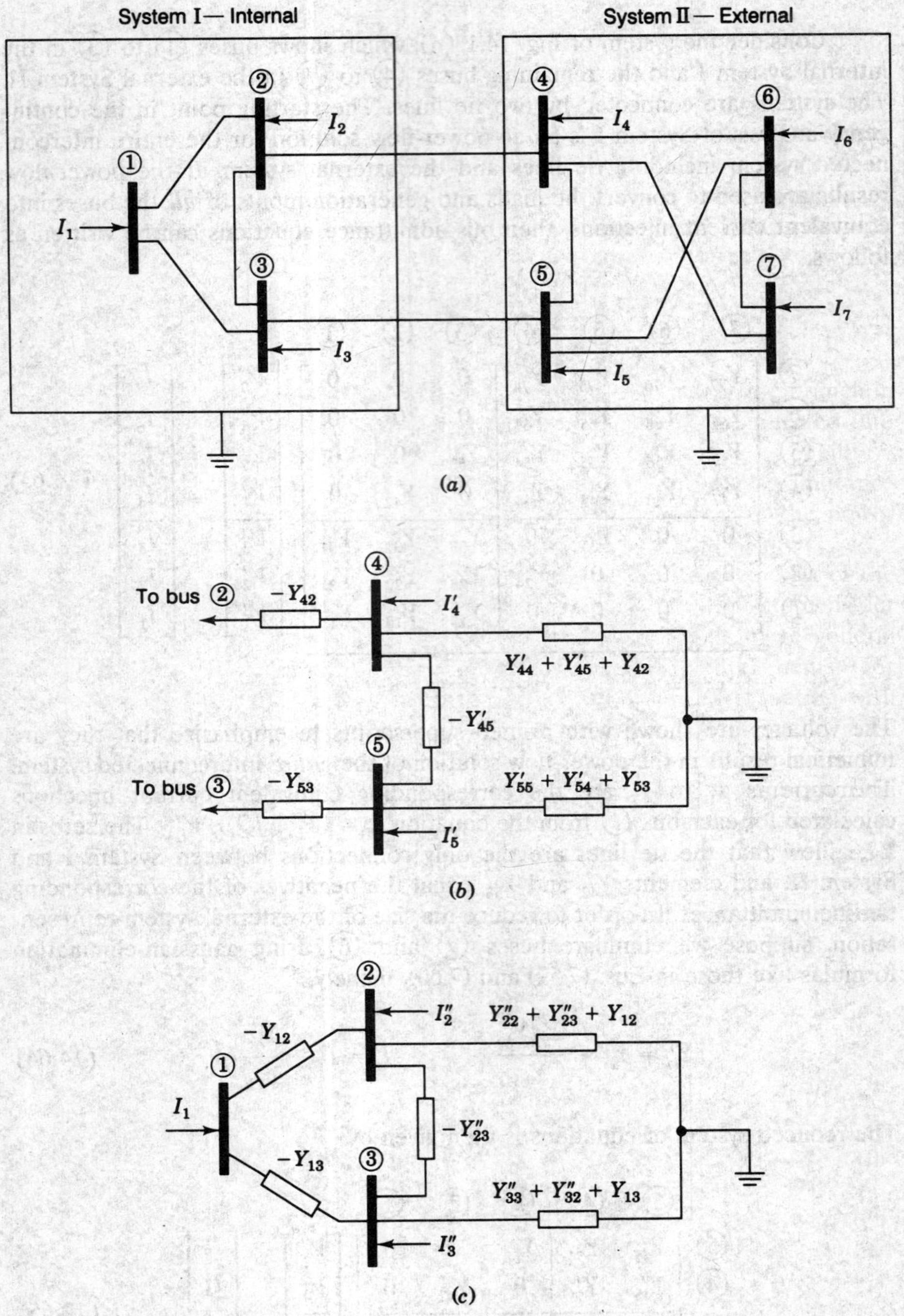

FIGURE 14.13
(*a*) One-line diagram showing System I and System II interconnected by two tie lines; (*b*) Ward equivalent circuit of System II at buses ④ and ⑤; (*c*) Ward equivalent circuit of System II plus tie lines at buses ② and ③.

Note that the only new elements are Y'_{44}, Y'_{45}, Y'_{54}, and Y'_{55} associated with buses ④ and ⑤ at which the tie lines terminate. The new current injections into those buses are I'_4 and I'_5, which take into account the current injections of the eliminated buses ⑥ and ⑦ in accordance with Eqs. (14.64). Rows ⑤ and ④ of Eq. (14.65) enable us to draw the mesh equivalent circuit at buses ⑤ and ④ of Fig. 14.13(*b*). This mesh circuit is usually called a *Ward equivalent*.[2] If buses ⑤ and ④ are not of particular interest to System I, then the node elimination process can be continued until only buses ③, ②, and ① remain. The new bus admittance equations are given by

$$\begin{array}{c} \\ ③ \\ ② \\ ① \end{array}\begin{array}{c} \begin{array}{ccc} ③ & ② & ① \end{array} \\ \begin{bmatrix} Y''_{33} & Y''_{32} & Y_{31} \\ Y''_{23} & Y''_{22} & Y_{21} \\ Y_{13} & Y_{12} & Y_{11} \end{bmatrix} \end{array} \begin{bmatrix} V'_3 \\ V'_2 \\ V'_1 \end{bmatrix} = \begin{bmatrix} I''_3 \\ I''_2 \\ I_1 \end{bmatrix} \tag{14.66}$$

The corresponding reduced equivalent circuit then includes a new Ward equivalent, as shown in Fig. 14.13(*c*). The currents I''_2 and I''_3 at the boundary buses of System I are equivalent injections which take into account the actual injections of *all* the eliminated buses of System II, as explained in Sec. 7.6. Multiplying Eq. (14.66) by the inverse of its coefficient matrix, we obtain bus impedance equations in the form

$$\begin{array}{c} \\ ③ \\ ② \\ ① \end{array}\begin{array}{c} \begin{array}{ccc} ③ & ② & ① \end{array} \\ \underbrace{\begin{bmatrix} Z''_{33} & Z''_{32} & Z''_{31} \\ Z''_{23} & Z''_{22} & Z''_{21} \\ Z''_{13} & Z''_{12} & Z''_{11} \end{bmatrix}}_{\mathbf{Z}''_{\text{bus}}} \end{array} \begin{bmatrix} I''_3 \\ I''_2 \\ I_1 \end{bmatrix} = \begin{bmatrix} V'_3 \\ V'_2 \\ V'_1 \end{bmatrix} \tag{14.67}$$

The impedance elements have double-primed superscripts to show that they account for the impedances of the entire interconnected system including those of System I. Distribution factors for analyzing contingencies in System I can be computed from $\mathbf{Z}''_{\text{bus}}$, as demonstrated in the preceding sections.

The above admittance equations can be expressed in more general terms. By sequentially numbering the internal (I) buses, the boundary (B) buses, and the external (E) buses, we can write the $\mathbf{Y}_{\text{bus}}$ equations for the entire intercon-

[2] J. B. Ward, "Equivalent Circuits for Power-Flow Studies," *Transactions of AIEE*, vol. 98, 1949, pp. 373–382.

nected system in the more general form

$$\begin{bmatrix} \mathbf{Y}_{EE} & \mathbf{Y}_{EB} & 0 \\ \mathbf{Y}_{BE} & \mathbf{Y}_{BB} & \mathbf{Y}_{BI} \\ 0 & \mathbf{Y}_{IB} & \mathbf{Y}_{II} \end{bmatrix} \begin{bmatrix} \mathbf{V}'_E \\ \mathbf{V}'_B \\ \mathbf{V}'_I \end{bmatrix} = \begin{bmatrix} \mathbf{I}_E \\ \mathbf{I}_B \\ \mathbf{I}_I \end{bmatrix} \tag{14.68}$$

If the current injections are considered constants, we can eliminate the external nodes as previously described to obtain

$$\begin{bmatrix} \overbrace{\mathbf{Y}_{BB} - \mathbf{Y}_{BE}\mathbf{Y}_{EE}^{-1}\mathbf{Y}_{EB}}^{\mathbf{Y}'_{BB}} & \mathbf{Y}_{BI} \\ \mathbf{Y}_{IB} & \mathbf{Y}_{II} \end{bmatrix} \begin{bmatrix} \mathbf{V}'_B \\ \mathbf{V}'_I \end{bmatrix} = \begin{bmatrix} \overbrace{\mathbf{I}_B - \mathbf{Y}_{BE}\mathbf{Y}_{EE}^{-1}\mathbf{I}_E}^{\mathbf{I}'_B} \\ \mathbf{I}_I \end{bmatrix} \tag{14.69}$$

Example 14.8. The nine-bus interconnected system of Fig. 14.8 has the bus admittance matrix shown in Table 14.4. A Newton-Raphson power-flow solution of the entire system is given in Table 14.5. Use this solution to convert the *P-Q* loads and generation into current injections at the buses. By gaussian elimination reduce the representation of System II down to its own boundary buses ⑥ and ⑦, and then down to the boundary buses ④ and ⑤ of System I. At each of the two stages of reduction draw the Ward equivalent corresponding to the reduced representation of System II.

TABLE 14.4
Elements of bus admittance matrix for interconnected system of Fig. 14.8†

	⑨	⑧	⑦	⑥	⑤	④	③	②	①
⑨	$-j45.0$	$j20.0$	$j25.0$	0	0	0	0	0	0
⑧	$j20.0$	$-j52.9$	$j20.0$	$j12.5$	0	0	0	0	0
⑦	$j25.0$	$j20.0$	$-j90.0$	$j10.0$	$j25.0$	$j10.0$	0	0	0
⑥	0	$j12.5$	$j10.0$	$-j42.5$	0	$j20.0$	0	0	0
⑤	0	0	$j25.0$	0	$-j82.5$	$j12.5$	$j20.0$	$j25.0$	0
④	0	0	$j10.0$	$j20.0$	$j12.5$	$-j62.5$	0	0	$j20.0$
③	0	0	0	0	$j20.0$	0	$-j36.0$	$j16.0$	0
②	0	0	0	0	$j25.0$	0	$j16.0$	$-j51.2$	$j10.0$
①	0	0	0	0	0	$j20.0$	0	$j10.0$	$-j30.0$

†Admittances are in per unit.

TABLE 14.5
Bus data and power-flow solution for the system of Example 14.8†

Bus number	Generation P	Generation Q	Load P	Load Q	Voltage	Current injection
①	1.75	0.2511			1.000000 +j0.000000	1.749950 −j0.251110
②			1.40	0.10	0.986973 −j0.076735	−1.402114 +j0.210339
③			1.00	0.20	0.985577 −j0.086848	−0.989048 +j0.290096
④			0.80	0.15	0.993958 −j0.049130	−0.795430 +j0.190238
⑤	1.45	0.5553			0.998965 −j0.045486	1.423215 −j0.620698
⑥			1.00	0.15	0.991606 −j0.067204	−0.993630 +j0.218623
⑦	1.50	0.2759			0.999343 −j0.036254	1.488995 −j0.330145
⑧	0.70	0.2058			0.999143 −j0.041392	0.674313 −j0.634250
⑨			1.20	0.25	0.991924 −j0.064941	−1.188155 +j0.329855

†All data are in per unit.

Solution. From the given data we now calculate the equivalent current injections at buses ⑨ and ⑧ as follows:

$$I_9 = \frac{P_9 - jQ_9}{V_9'^*} = \frac{-1.200000 + j0.250000}{0.991924 + j0.064941} = -1.188155 + j0.329855$$

$$I_8 = \frac{P_8 - jQ_8}{V_8'^*} = \frac{0.700000 - j0.605790}{0.999143 + j0.041392} = 0.674313 - j0.634250$$

A complete set of current injections is shown in Table 14.5.

The first stage of reduction does not involve buses of System I, and so we show only the System II (external) portion of the $\mathbf{Y}_{\text{bus}}$ equations as follows:

$$\begin{array}{c} \textcircled{9} \\ \textcircled{8} \\ \textcircled{7} \\ \textcircled{6} \\ \\ \end{array} \overset{\begin{array}{cccccc} \textcircled{9} & \quad\textcircled{8} & \quad\textcircled{7} & \quad\textcircled{6} & & \end{array}}{\begin{bmatrix} -j45.0 & j20.0 & j25.0 & 0 & \cdots \\ j20.0 & -j52.9 & j20.0 & j12.5 & \cdots \\ j25.0 & j20.0 & -j90.0 & j10.0 & \cdots \\ 0 & j12.5 & j10.0 & -j42.5 & \cdots \\ \vdots & \vdots & \vdots & \vdots & \ddots \end{bmatrix}} \begin{bmatrix} V_9' \\ V_8' \\ V_7' \\ V_6' \\ \vdots \end{bmatrix} = \begin{bmatrix} -1.188155 + j0.329855 \\ 0.674313 - j0.634250 \\ 1.488995 - j0.330145 \\ -0.993630 + j0.218623 \\ \vdots \end{bmatrix}$$

From Eqs. (14.64) the new current injection at bus ⑧ upon elimination of bus ⑨ becomes

$$I_8' = I_8 - \frac{Y_{89}}{Y_{99}} I_9 = (0.674313 - j0.634250) - \frac{j20}{-j45}(-1.188155 + j0.329855)$$

$$= 0.146244 - j0.487647 \text{ per unit}$$

The new per-unit current injection at bus ⑦ is similarly calculated to be (0.828909 − j0.146892), and so gaussian elimination of bus ⑨ gives

$$\begin{array}{c} \\ ⑧ \\ ⑦ \\ ⑥ \\ ⑤ \\ \\ \end{array}
\begin{array}{c} \begin{array}{ccc|cc} ⑧ & ⑦ & ⑥ & ⑤ & \end{array} \\
\left[\begin{array}{ccc|cc}
-j44.011111 & j31.111111 & j12.5 & 0 & \cdots \\
j31.111111 & -j76.111111 & j10.0 & j25.0 & \cdots \\
j12.5 & j10.0 & -j42.5 & 0 & \cdots \\ \hline
0 & j10.0 & 0 & -j82.5 & \cdots \\
\vdots & \vdots & \vdots & \vdots & \ddots
\end{array}\right] \end{array}
\left[\begin{array}{c} V_8' \\ V_7' \\ V_6' \\ \hline V_5' \\ \vdots \end{array}\right]
=
\left[\begin{array}{c} 0.146244 - j0.487647 \\ 0.828909 - j0.146892 \\ -0.993630 + j0.218623 \\ \hline 1.423215 - j0.620698 \\ \vdots \end{array}\right]$$

Further elimination of bus ⑧ gives the system of equations

$$\begin{array}{c} \\ ⑦ \\ ⑥ \\ ⑤ \\ ④ \\ \\ \end{array}
\begin{array}{c} \begin{array}{cc|cc|c} ⑦ & ⑥ & ⑤ & ④ & \end{array} \\
\left[\begin{array}{cc|cc|c}
-j54.118909 & j18.836152 & j25.0 & j10.0 & \cdots \\
j18.836152 & -j38.949760 & 0 & j20.0 & \cdots \\ \hline
j25.0 & 0 & -j82.5 & j12.5 & \cdots \\
j10.0 & j20.0 & j12.5 & -j62.5 & \cdots \\ \hline
\vdots & \vdots & \vdots & \vdots & \ddots
\end{array}\right] \end{array}
\left[\begin{array}{c} V_7' \\ V_6' \\ \hline V_5' \\ V_4' \\ \hline \vdots \end{array}\right]
=
\left[\begin{array}{c} 0.932288 - j0.491606 \\ -0.952094 + j0.080121 \\ \hline 1.423215 - j0.620698 \\ -0.795430 + j0.190238 \\ \hline \vdots \end{array}\right]$$

As expected, the only new current injections are at buses ⑥ and ⑦. Continuing on with the second stage of external system elimination, we obtain

$$\begin{array}{c} \\ ⑤ \\ ④ \\ ③ \\ ② \\ ① \end{array}
\begin{array}{c} \begin{array}{cc|ccc} ⑤ & ④ & ③ & ② & ① \end{array} \\
\left[\begin{array}{cc|ccc}
-j68.614111 & j23.426542 & j20.0 & j25.0 & 0 \\
j23.426542 & -j43.632472 & 0 & 0 & j20.0 \\ \hline
j20.0 & 0 & -j36.0 & j16.0 & 0 \\
j25.0 & 0 & j16.0 & -j51.2 & j10.0 \\
0 & j20.0 & 0 & j10.0 & -j30.0
\end{array}\right] \end{array}
\left[\begin{array}{c} V_5' \\ V_4' \\ \hline V_3' \\ V_2' \\ V_1' \end{array}\right]
=
\left[\begin{array}{c} 1.685300 - j0.872232 \\ -1.078084 + j0.033451 \\ \hline -0.989048 + j0.290096 \\ -1.402114 + j0.210339 \\ 1.749950 - j0.251110 \end{array}\right]$$

When buses ⑧ and ⑨ are eliminated, the Ward equivalent of System II between buses ⑥ and ⑦ is that shown in Fig. 14.14(*a*). The reader may wish to

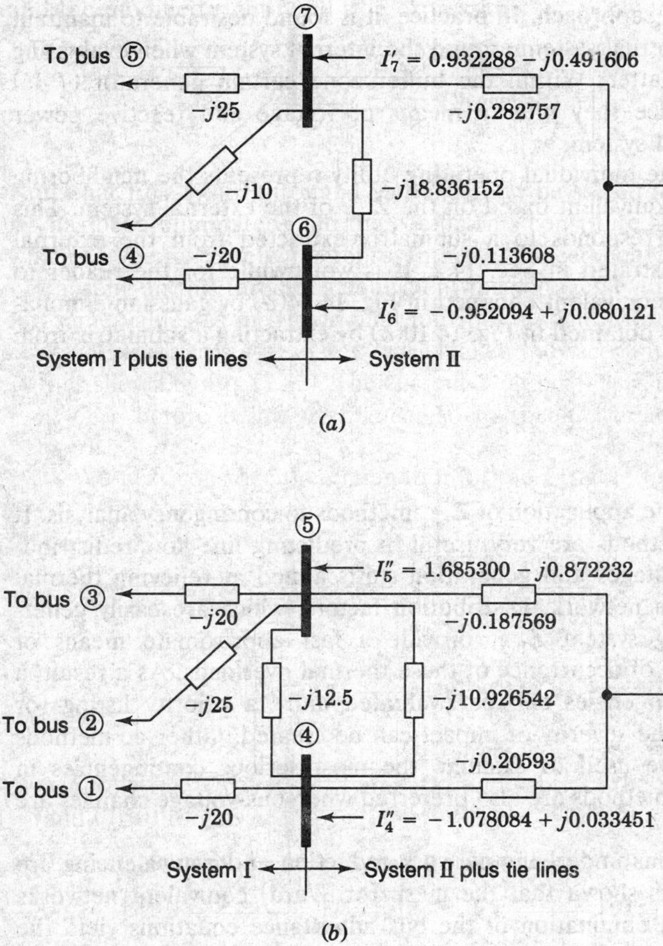

FIGURE 14.14
(*a*) Ward equivalent buses ⑥ and ⑦ of System II; (*b*) Ward equivalent of System II plus tie lines.

confirm that the numerical values of the elements in the figure are consistent with the numerical entries in the corresponding $\mathbf{Y}_{bus}$ equations given above. Similarly, Fig. 14.14(*b*) shows the Ward equivalent of the entire System II plus tie lines as seen at the boundary buses ④ and ⑤ of System I. Note that the internal line ④–⑤ of System I is shown separately.

If the current injections can be regarded as constant, then contingency analysis studies of System I can proceed with the reduced system of equations. The bus admittance representation and triangular factors of $\mathbf{Y}_{bus}$ can be employed or else a $\mathbf{Z}_{bus}$ approach. In practice, it is found desirable to maintain a *buffer zone* of the external system around the internal system when evaluating contingencies on the latter. Within the buffer zone certain generator (P-V) buses are retained since they offer a means of voltage and reactive power support for the internal system.

In fault studies the individual operating utility represents the neighboring systems by a reduced equivalent based on the $\mathbf{Z}_{bus}$ of the external system. This reduced equivalent corresponds to a submatrix extracted from the external system $\mathbf{Z}_{bus}$, as demonstrated in Sec. 14.2. It is worthwhile for the reader to observe that the Ward equivalent obtained in Fig. 14.14(a) by gaussian elimination coincides with that obtained in Fig. 14.10(c) by extracting a submatrix from the $\mathbf{Z}_{bus}$ of System II.

14.7 SUMMARY

This chapter explains the application of $\mathbf{Z}_{bus}$ methods to contingency analysis. It is shown that these methods are very useful in predicting line-flow redistributions caused by line outages and generation shifts aimed at relieving thermal overloads in the system network. Distribution factors, which are easily generated from the existing system $\mathbf{Z}_{bus}$, provide a fast approximate means of assessing the possibility of occurrence of these thermal overloads. As a result, a large number of contingencies can be evaluated and a priority listing (or ranking) according to the severity of impact can be created. Other ac methods of analysis can then be used to examine the most serious contingencies in greater detail. Such ac methods are also preferred when bus-voltage changes are to be predicted.

Basic considerations underlying network reduction and equivalencing are also discussed, and it is shown that the mesh (or Ward) equivalent networks resulting from gaussian elimination of the bus admittance equations yield the same results as the matrix-based equivalents extracted from the system $\mathbf{Z}_{bus}$. The discussion provides a basis for understanding the many other approaches to equivalencing which are found in industry literature.

PROBLEMS

14.1. A four-bus system with $\mathbf{Z}_{bus}$ given in per unit by

$$\begin{array}{c} \\ ① \\ ② \\ ③ \\ ④ \end{array} \begin{array}{c} \begin{array}{cccc} ① & ② & ③ & ④ \end{array} \\ \begin{bmatrix} j0.041 & j0.031 & j0.027 & j0.018 \\ j0.031 & j0.256 & j0.035 & j0.038 \\ j0.027 & j0.035 & j0.158 & j0.045 \\ j0.018 & j0.038 & j0.045 & j0.063 \end{bmatrix} \end{array}$$

has bus voltages $V_1 = 1.0\angle 0°$, $V_2 = 0.98\angle 0°$, $V_3 = 0.96\angle 0°$, and $V_4 = 1.04\angle 0°$. Using the compensation current method, determine the change in voltage at bus ② due to the outage of line ①-③ with series impedance $j0.3$ per unit.

14.2. Solve Prob. 14.1 when the outage involves (*a*) only line ①-④ of series impedance $j0.2$ per unit and (*b*) both line ①-④ and line ①-③ of series impedance $j0.3$ per unit.

14.3. Suppose that line ③-④ of the system of Prob. 14.1 is actually a double-circuit line of combined impedance $j0.2$ per unit and that one of the circuits of series impedance $j0.4$ per unit is to be removed. Using the compensation current method, determine the change in voltage at bus ② due to this outage.

14.4. Consider a portion of a large power system, whose $\mathbf{Z}_{\text{bus}}$ elements corresponding to the selected buses ① to ⑤ are given in per unit by

	①	②	③	④	⑤
①	$j0.038$	$j0.034$	$j0.036$	$j0.018$	$j0.014$
②	$j0.034$	$j0.057$	$j0.044$	$j0.019$	$j0.013$
③	$j0.036$	$j0.044$	$j0.062$	$j0.018$	$j0.014$
④	$j0.018$	$j0.019$	$j0.018$	$j0.028$	$j0.010$
⑤	$j0.014$	$j0.013$	$j0.014$	$j0.010$	$j0.018$

The base-case bus voltages at those selected buses are $V_1 = 1.0\angle 0°$, $V_2 = 1.1\angle 0°$, $V_3 = 0.98\angle 0°$, $V_4 = 1.0\angle 0°$, and $V_5 = 0.99\angle 0°$, all in per unit. Using the compensating current method, determine the change in voltage at bus ① when line ②-③ of series impedance $j0.05$ per unit and line ④-⑤ of series impedance $j0.08$ per unit are both outaged.

14.5. Redo Prob. 14.4 when line ②-③ and line ①-④ of series impedance $j0.10$ per unit are outaged.

14.6. Two systems are connected by three branches *a*, *b*, and *c* in the network of Fig. 14.15. Using the sequential rules of Sec. 14.2, write the elements of the loop impedance matrix **Z** in symbolic form. Assume that System I contains buses ① through ④ and System II contains buses ⑤ through ⑦. Use superscripts $^{(1)}$ and $^{(2)}$ to denote elements of $\mathbf{Z}_{\text{I}}$ and $\mathbf{Z}_{\text{II}}$, respectively.

14.7. Form the branch-to-node incidence matrix $\mathbf{A}_c$ for branches *a*, *b*, and *c* of the network of Fig. 14.15. Then, for the network, determine the matrix **Z** defined in Eq. (14.9). Compare your answer with that of Prob. 14.6.

14.8. Use the piecewise method to determine the bus voltages for the reactance network of Fig. 14.16 where System I and System II are defined. Each of current injections I_1 to I_4 is equal to $1.0\angle -90°$ per unit and each of current injections I_5 and I_6 is equal to $4.0\angle -90°$ per unit.

14.9. Consider the four-bus system of Fig. 14.17 in which line impedances are shown in per unit. Using a generation-shift distribution factor based on the dc power-flow method, find the change in power flow in line ①-② when the power genera-

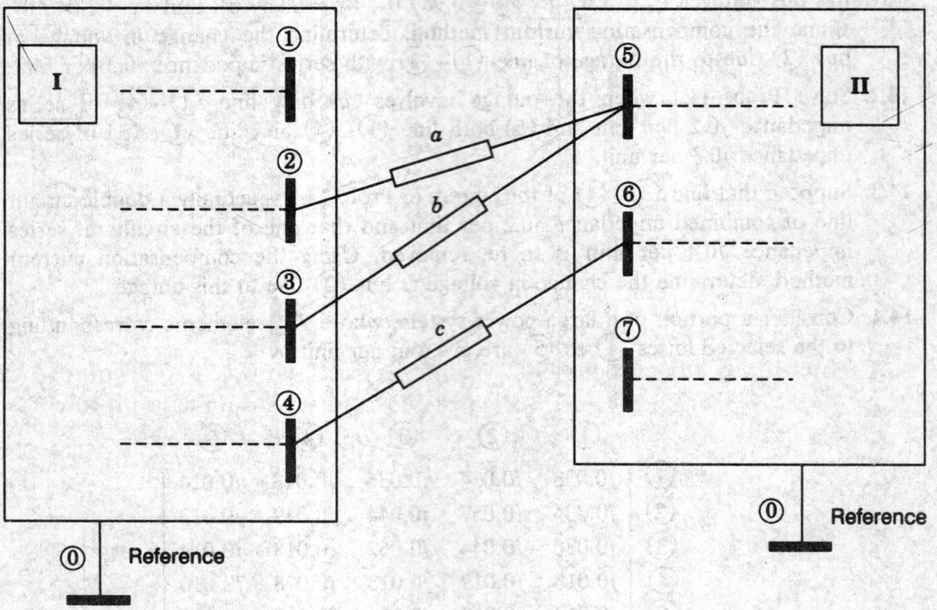

FIGURE 14.15
Branches *a*, *b*, and *c* interconnect two subsystems in Probs. 14.6 and 14.7.

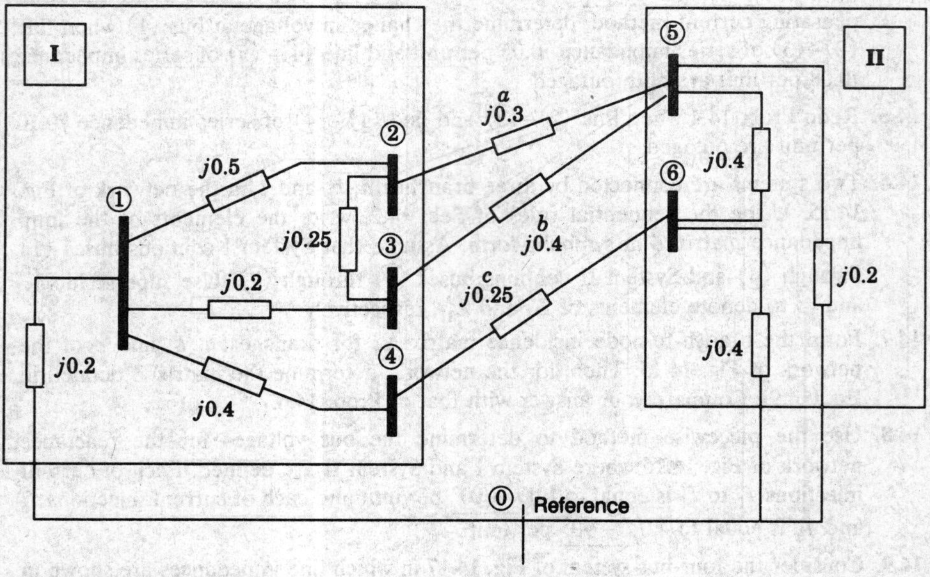

FIGURE 14.16
Network for Prob. 14.8. Parameters are in per unit.

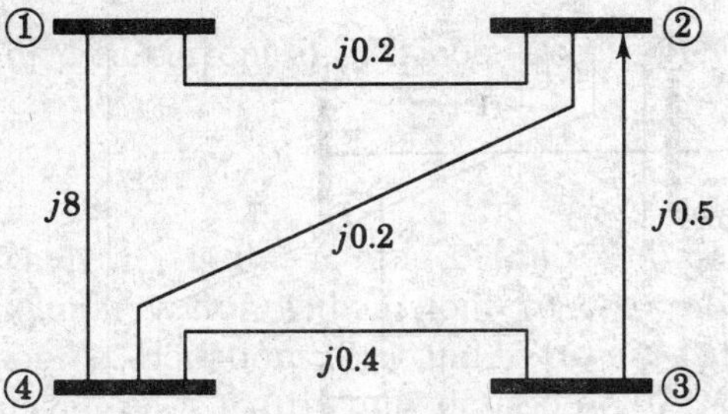

FIGURE 14.17
A four-bus system for Prob. 14.9.

tion at bus ④ is incrementally increased by 0.1 per unit. *Note*: An appropriate bus should be selected as reference to find $\mathbf{Z}_{\text{bus}}$ of the system.

14.10. Consider the power system of Fig. 14.12 discussed in Example 14.4. Using distribution factors, predict the current in line ①-④ when line ④-⑤ is outaged. Assume that the operating condition specified remains the same as in the example.

14.11. The base-case bus voltages at buses ⓜ and ⓝ in a large power system are $1.02\angle 0°$ and $0.99\angle 0°$ per unit, respectively. Suppose that line ⓜ-ⓝ has been removed from service and that the following selected elements are extracted from the $\mathbf{Z}_{\text{bus}}$ of the system with line ⓜ-ⓝ excluded.

	ⓘ	ⓙ	ⓜ	ⓝ
ⓘ	$j0.019$	$j0.015$	$j0.017$	$j0.014$
ⓙ	$j0.015$	$j0.044$	$j0.025$	$j0.030$
ⓜ	$j0.017$	$j0.025$	$j0.075$	$j0.052$
ⓝ	$j0.014$	$j0.030$	$j0.052$	$j0.064$

The impedances of lines ⓘ-ⓙ and ⓜ-ⓝ are $j0.05$ and $j0.1$ per unit, respectively. Using the principle of superposition, determine the change in the line current of line ⓘ-ⓙ due to an outage of line ⓜ-ⓝ.

14.12. Solve Example 14.6 for the case where only lines ①-② and ②-⑤ are simultaneously outaged.

14.13. In the five-bus system of Fig. 14.12, consider that line ④-⑤ has been taken out of service when an additional incremental load of $1 + j0$ at bus ③ has to be met by an additional 1-MW generation at bus ⑤. Use Eq. (14.44) to predict the change in current flow from bus ② to bus ③.

14.14. A five-bus system consisting of buses ① through ⑤ is connected to a partially depicted large power system through three tie lines, as shown in Fig. 14.18 in which line admittances are indicated in per unit. The equivalent current injections at those five buses are $I_1 = 1.5 + j0$, $I_2 = -0.8 + j0$, $I_3 = 0.5 + j0$, $I_4 = -1.2 + j0$, and $I_5 = 0.4 + j0$, all in per unit. By gaussian elimination, reduce the $\mathbf{Y}_{\text{bus}}$ representation of the five-bus system down to the boundary buses ④ and ⑤ and draw the corresponding Ward equivalent network.

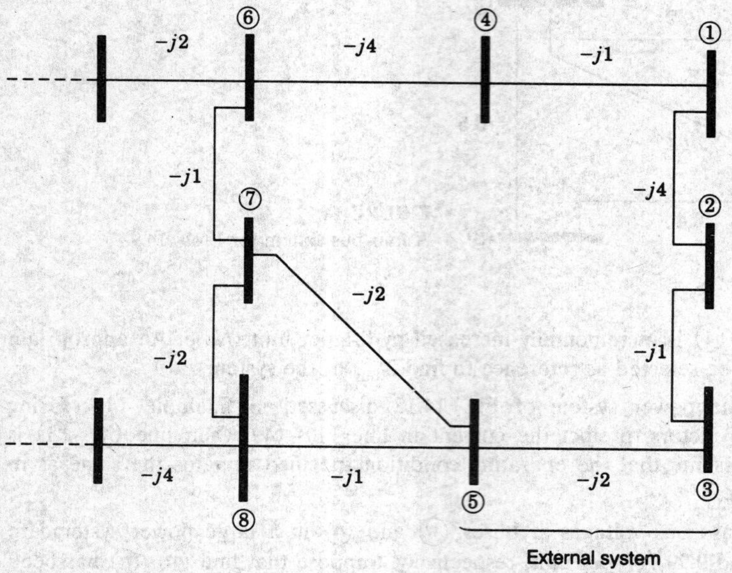

FIGURE 14.18
An interconnected system for Prob. 14.14.

14.15. Redo Prob. 14.14 using matrix inversion as in Eq. (14.69). Compare the results with the solution to Prob. 14.14. Now, using gaussian elimination, continue the elimination of all five buses in the external system and then find the resulting Ward equivalent at buses ⑥, ⑦, and ⑧. Draw an equivalent admittance circuit and mark the values of all quantities.

CHAPTER 15

STATE ESTIMATION OF POWER SYSTEMS

Selective monitoring of the generation and transmission system has been providing the data needed for economic dispatch and load frequency control. More recently, however, interconnected power networks have become more complex and the task of securely operating the system has become more difficult. To help avoid major system failures and regional power blackouts, electric utilities have installed more extensive supervisory control and data acquisition (SCADA) throughout the network to support computer-based systems at the energy control center. The data bank created is intended for a number of application programs—some to ensure economic system operation and others to assess how secure the system would be if equipment failures and transmission-line outages were to occur.

Before any security assessment can be made or control actions taken, a reliable estimate of the *existing* state of the system must be determined. For this purpose the number of physical measurements cannot be restricted to only those quantities required to support conventional power-flow calculations. The inputs to the conventional power-flow program are confined to the P, Q injections at load buses and $P, |V|$ values at voltage-controlled buses. If even one of these inputs is unavailable, the conventional power-flow solution cannot be obtained. Moreover, gross errors in one or more of the input quantities can cause the power-flow results to become useless. In practice, other conveniently measured quantities such as P, Q *line flows* are available, but they cannot be used in conventional power-flow calculations. These limitations can be removed by *state estimation* based on weighted least-squares calculations.

The techniques developed in this chapter provide an estimate of the system state and a quantitative measure of how good the estimate is *before* it is used for real-time power-flow calculations or on-line system security assessment. Besides the inputs required for conventional power-flow analysis, additional measurements are made which usually include megawatt and megavar flow in the transmission lines of the system. The unavoidable errors of the measurements are assigned statistical properties and the estimates of the states are subjected to statistical testing before being accepted as satisfactory. Thus, gross errors detected in the course of state estimation are automatically filtered out.

15.1 THE METHOD OF LEAST SQUARES

The electric power transmission system uses wattmeters, varmeters, voltmeters, and current meters to measure real power, reactive power, voltages, and currents, respectively. These continuous or *analog* quantities are monitored by current and potential transformers (or other equivalent devices) installed on the lines and on transformers and buses of the power plants and substations of the system. The analog quantities pass through transducers and analog-to-digital converters, and the digital outputs are then telemetered to the energy control center over various communication links. The data received at the energy control center is processed by computer to inform the system operators of the present state of the system. The acquired data always contains inaccuracies which are unavoidable since physical measurements (as opposed to numerical calculations) cannot be entirely free of random errors or *noise*. These errors can be quantified in a statistical sense and the estimated values of the quantities being measured are then either accepted as reasonable or rejected if certain measures of accuracy are exceeded.

Because of noise, the true values of physical quantities are never known and we have to consider how to calculate the best possible *estimates* of the unknown quantities. The *method of least squares* is often used to "best fit" measured data relating two or more quantities. Here we apply the method to a simple set of dc measurements which contain errors, and Sec. 15.4 extends the estimation procedures to the ac power system. The best estimates are chosen as those which minimize the weighted sum of the squares of the measurement errors.

Consider the simple dc circuit of Fig. 15.1 with five resistances of 1 Ω each and two voltage sources V_1 and V_2 of unknown values which are to be estimated. The measurement set consists of ammeter readings z_1 and z_2 and voltmeter readings z_3 and z_4. The symbol z is normally used for measurements regardless of the physical quantity being measured, and likewise, the symbol x applies to quantities being estimated. The *system model* based on elementary circuit analysis expresses the true values of the measured quantities in terms of the network parameters and the *true* (but *unknown*) source voltages $x_1 = V_1$ and $x_2 = V_2$. Then, *measurement equations* characterizing the meter readings are found by adding error terms to the system model. For Fig. 15.1 we obtain

$$z_1 = \tfrac{5}{8}x_1 - \tfrac{1}{8}x_2 + e_1 \tag{15.1}$$

$$z_2 = -\tfrac{1}{8}x_1 + \tfrac{5}{8}x_2 + e_2 \tag{15.2}$$

$$z_3 = \tfrac{3}{8}x_1 + \tfrac{1}{8}x_2 + e_3 \tag{15.3}$$

$$\underbrace{z_4}_{\text{Measurements}} = \underbrace{\tfrac{1}{8}x_1 + \tfrac{3}{8}x_2}_{\text{True values from system model}} + \underbrace{e_4}_{\text{Errors}} \tag{15.4}$$

in which the numerical coefficients are determined by the circuit resistances and the terms e_1, e_2, e_3, and e_4 represent *errors* in measuring the two currents z_1 and z_2 and the two voltages z_3 and z_4. Some authors use the term *residuals* instead of errors, so we use both terms interchangeably.

If e_1, e_2, e_3, and e_4 were zero (the ideal case), then any two of the meter readings would give exact and consistent readings from which the true values x_1 and x_2 of V_1 and V_2 could be determined. But in any measurement scheme there are unknown errors which generally follow a statistical pattern, as we shall discuss in Sec. 15.2. Labeling the coefficients of Eqs. (15.1) through (15.4) in an obvious way, we obtain

$$z_1 = h_{11}x_1 + h_{12}x_2 + e_1 = z_{1,\,\text{true}} + e_1 \tag{15.5}$$

$$z_2 = h_{21}x_1 + h_{22}x_2 + e_2 = z_{2,\,\text{true}} + e_2 \tag{15.6}$$

$$z_3 = h_{31}x_1 + h_{32}x_2 + e_3 = z_{3,\,\text{true}} + e_3 \tag{15.7}$$

$$z_4 = h_{41}x_1 + h_{42}x_2 + e_4 = z_{4,\,\text{true}} + e_4 \tag{15.8}$$

where $z_{j,\text{true}}$ denotes the *true* value of the measured quantity z_j. We now rearrange Eqs. (15.5) through (15.8) into the vector-matrix form

$$\begin{bmatrix} e_1 \\ e_2 \\ e_3 \\ e_4 \end{bmatrix} = \begin{bmatrix} z_1 \\ z_2 \\ z_3 \\ z_4 \end{bmatrix} - \begin{bmatrix} z_{1,\,\text{true}} \\ z_{2,\,\text{true}} \\ z_{3,\,\text{true}} \\ z_{4,\,\text{true}} \end{bmatrix} = \begin{bmatrix} z_1 \\ z_2 \\ z_3 \\ z_4 \end{bmatrix} - \begin{bmatrix} h_{11} & h_{12} \\ h_{21} & h_{22} \\ h_{31} & h_{32} \\ h_{41} & h_{42} \end{bmatrix} \begin{bmatrix} x_1 \\ x_2 \end{bmatrix} \tag{15.9}$$

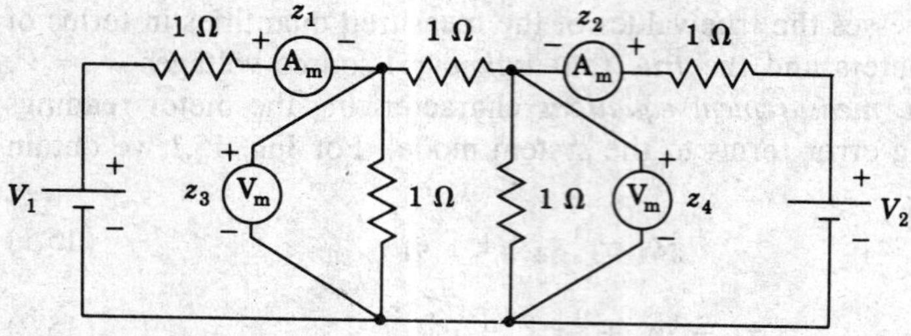

FIGURE 15.1
Simple dc circuit with two ammeters (A_m) measuring z_1 and z_2 and two voltmeters (V_m) measuring z_3 and z_4.

In more compact notation Eq. (15.9) can be written as

$$\mathbf{e} = \mathbf{z} - \mathbf{z}_{\text{true}} = \mathbf{z} - \mathbf{H}\mathbf{x} \tag{15.10}$$

which represents the errors between the actual measurements $\mathbf{z}$ and the true (but unknown) values $\mathbf{z}_{\text{true}} \triangleq \mathbf{H}\mathbf{x}$ of the measured quantities. The true values of x_1 and x_2 cannot be determined, but we can calculate *estimates* $\hat{x}_1$ and $\hat{x}_2$, as we shall soon see. Substituting these estimates in Eq. (15.9) gives estimated values of the errors in the form

$$\underbrace{\begin{bmatrix} \hat{e}_1 \\ \hat{e}_2 \\ \hat{e}_3 \\ \hat{e}_4 \end{bmatrix}}_{\text{Estimated errors}} = \underbrace{\begin{bmatrix} z_1 \\ z_2 \\ z_3 \\ z_4 \end{bmatrix}}_{\text{Measurements}} - \underbrace{\begin{bmatrix} h_{11} & h_{12} \\ h_{21} & h_{22} \\ h_{31} & h_{32} \\ h_{41} & h_{42} \end{bmatrix} \begin{bmatrix} \hat{x}_1 \\ \hat{x}_2 \end{bmatrix}}_{\text{Estimates of } z_j} \tag{15.11}$$

Quantities with hats,[1] such as $\hat{e}_j$ and $\hat{x}_i$, are estimates of the corresponding quantities without hats. In Eq. (15.11) the left-hand vector is $\hat{\mathbf{e}}$, which represents the differences between the actual measurements $\mathbf{z}$ and their *estimated* values $\hat{\mathbf{z}} \triangleq \mathbf{H}\hat{\mathbf{x}}$ so that we can write

$$\hat{\mathbf{e}} = \mathbf{z} - \hat{\mathbf{z}} = \mathbf{z} - \mathbf{H}\hat{\mathbf{x}} = \mathbf{e} - \mathbf{H}(\hat{\mathbf{x}} - \mathbf{x}) \tag{15.12}$$

[1]The symbol is more properly called a *circumflex*.

We must now decide upon a criterion for calculating the estimates $\hat{x}_1$ and $\hat{x}_2$ from which $\hat{\mathbf{e}} = [\hat{e}_1\ \hat{e}_2\ \hat{e}_3\ \hat{e}_4]^T$ and $\hat{\mathbf{z}} = [\hat{z}_1\ \hat{z}_2\ \hat{z}_3\ \hat{z}_4]^T$ are to be computed. It is not desirable to choose the algebraic sum of the errors to be minimized since positive and negative errors could then offset one another and the estimates would not necessarily be acceptable. It is preferable to minimize the direct sum of the *squares* of the errors. However, to ensure that measurements from meters of known greater accuracy are treated more favorably than less accurate measurements, each term in the sum of squares is multiplied by an appropriate *weighting factor* w to give the objective function

$$f = \sum_{j=1}^{4} w_j e_j^2 = w_1 e_1^2 + w_2 e_2^2 + w_3 e_3^2 + w_4 e_4^2 \tag{15.13}$$

We select the *best estimates* of the state variables as those values $\hat{x}_1$ and $\hat{x}_2$ which cause the objective function f to take on its minimum value. According to the usual necessary conditions for minimizing f, the estimates $\hat{x}_1$ and $\hat{x}_2$ are those values of x_1 and x_2 which satisfy the equations

$$\left.\frac{\partial f}{\partial x_1}\right|_{\hat{\mathbf{x}}} = 2\left[w_1 e_1 \frac{\partial e_1}{\partial x_1} + w_2 e_2 \frac{\partial e_2}{\partial x_1} + w_3 e_3 \frac{\partial e_3}{\partial x_1} + w_4 e_4 \frac{\partial e_4}{\partial x_1}\right]\Bigg|_{\hat{\mathbf{x}}} = 0 \tag{15.14}$$

$$\left.\frac{\partial f}{\partial x_2}\right|_{\hat{\mathbf{x}}} = 2\left[w_1 e_1 \frac{\partial e_1}{\partial x_2} + w_2 e_2 \frac{\partial e_2}{\partial x_2} + w_3 e_3 \frac{\partial e_3}{\partial x_2} + w_4 e_4 \frac{\partial e_4}{\partial x_2}\right]\Bigg|_{\hat{\mathbf{x}}} = 0 \tag{15.15}$$

The notation $|\hat{\mathbf{x}}$ indicates that the equations have to be evaluated from the state estimates $\hat{\mathbf{x}} = [\hat{x}_1\ \hat{x}_2]^T$ since the true values of the states are not known. The unknown actual errors e_j are then replaced by estimated errors $\hat{e}_j$, which can be calculated once the state estimates $\hat{x}_i$ are known. Equations (15.14) and (15.15) in vector-matrix form become

$$\left.\begin{bmatrix} \dfrac{\partial e_1}{\partial x_1} & \dfrac{\partial e_2}{\partial x_1} & \dfrac{\partial e_3}{\partial x_1} & \dfrac{\partial e_4}{\partial x_1} \\[2ex] \dfrac{\partial e_1}{\partial x_2} & \dfrac{\partial e_2}{\partial x_2} & \dfrac{\partial e_3}{\partial x_2} & \dfrac{\partial e_4}{\partial x_2} \end{bmatrix}\right|_{\hat{\mathbf{x}}} \underbrace{\begin{bmatrix} w_1 & \cdot & \cdot & \cdot \\ \cdot & w_2 & \cdot & \cdot \\ \cdot & \cdot & w_3 & \cdot \\ \cdot & \cdot & \cdot & w_4 \end{bmatrix}}_{\mathbf{W}} \begin{bmatrix} \hat{e}_1 \\ \hat{e}_2 \\ \hat{e}_3 \\ \hat{e}_4 \end{bmatrix} = \begin{bmatrix} 0 \\ 0 \end{bmatrix} \tag{15.16}$$

where $\mathbf{W}$ is the diagonal matrix of weighting factors which have special significance, as shown in Sec. 15.2. The partial derivatives for substitution in Eq. (15.16) are found from Eqs. (15.5) through (15.8) to be constants given by the elements of **H**, and so we obtain

$$\begin{bmatrix} h_{11} & h_{21} & h_{31} & h_{41} \\ h_{12} & h_{22} & h_{32} & h_{42} \end{bmatrix} \begin{bmatrix} w_1 & \cdot & \cdot & \cdot \\ \cdot & w_2 & \cdot & \cdot \\ \cdot & \cdot & w_3 & \cdot \\ \cdot & \cdot & \cdot & w_4 \end{bmatrix} \begin{bmatrix} \hat{e}_1 \\ \hat{e}_2 \\ \hat{e}_3 \\ \hat{e}_4 \end{bmatrix} = \begin{bmatrix} 0 \\ 0 \end{bmatrix} \tag{15.17}$$

Using the compact notation of Eq. (15.12) in Eq. (15.17) yields

$$\mathbf{H}^T\mathbf{W}\hat{\mathbf{e}} = \mathbf{H}^T\mathbf{W}(\mathbf{z} - \mathbf{H}\hat{\mathbf{x}}) = \mathbf{0} \tag{15.18}$$

Multiplying through in this equation and solving for $\hat{\mathbf{x}} = [\hat{x}_1\, \hat{x}_2]^T$ give

$$\hat{\mathbf{x}} = \begin{bmatrix} \hat{x}_1 \\ \hat{x}_2 \end{bmatrix} = \big(\underbrace{\mathbf{H}^T\mathbf{W}\mathbf{H}}_{\mathbf{G}}\big)^{-1}\mathbf{H}^T\mathbf{W}\mathbf{z} = \mathbf{G}^{-1}\mathbf{H}^T\mathbf{W}\mathbf{z} \tag{15.19}$$

where $\hat{x}_1$ and $\hat{x}_2$ are the *weighted least-squares estimates* of the state variables. Because **H** is rectangular, the symmetrical matrix $\mathbf{H}^T\mathbf{W}\mathbf{H}$ (often called the *gain matrix* **G**) must be inverted as a single entity to yield $\mathbf{G}^{-1} = (\mathbf{H}^T\mathbf{W}\mathbf{H})^{-1}$, which is also symmetrical. Later in this chapter we discuss the case where **G** is *not* invertible due to the lack of sufficient measurements.

We expect the weighted least-squares procedure to yield estimates $\hat{x}_i$, which are close to the true values x_i of the state variables. An expression for the differences $(\hat{x}_i - x_i)$ is found by substituting for $\mathbf{z} = \mathbf{H}\mathbf{x} + \mathbf{e}$ in Eq. (15.19) to obtain

$$\hat{\mathbf{x}} = \mathbf{G}^{-1}\mathbf{H}^T\mathbf{W}(\mathbf{H}\mathbf{x} + \mathbf{e}) = \mathbf{G}^{-1}\underbrace{(\mathbf{H}^T\mathbf{W}\mathbf{H})}_{\mathbf{G}}\mathbf{x} + \mathbf{G}^{-1}\mathbf{H}^T\mathbf{W}\mathbf{e} \tag{15.20}$$

Canceling **G** with $\mathbf{G}^{-1}$ and rearranging the result lead to the equation

$$\hat{\mathbf{x}} - \mathbf{x} = \begin{bmatrix} \hat{x}_1 - x_1 \\ \hat{x}_2 - x_2 \end{bmatrix} = \mathbf{G}^{-1}\mathbf{H}^T\mathbf{W}\mathbf{e} \tag{15.21}$$

It is useful to check the dimensions of each term in the matrix product of Eq. (15.21), which is important for developing properties of the weighted least-squares estimation. For the example circuit of Fig. 15.1 the matrix $\mathbf{G}^{-1}\mathbf{H}^T\mathbf{W}$ has the overall row × column dimensions 2 × 4, which means that any one or more of the four errors e_1, e_2, e_3, and e_4 can influence the difference between each state estimate and its true value. In other words, the weighted least-squares calculation spreads the effect of the error in any one measurement to some or all the estimates; this characteristic is the basis of the test for detecting bad data, which is explained in Sec. 15.3.

In a quite similar manner, we can compare the calculated values $\hat{\mathbf{z}} = \mathbf{H}\hat{\mathbf{x}}$ of the measured quantities with their actual measurements $\mathbf{z}$ by substituting for $\hat{\mathbf{x}} - \mathbf{x}$ from Eq. (15.21) into Eq. (15.12) to obtain

$$\hat{\mathbf{e}} = \mathbf{z} - \hat{\mathbf{z}} = \mathbf{e} - \mathbf{H}\mathbf{G}^{-1}\mathbf{H}^T\mathbf{W}\mathbf{e} = [\mathbf{I} - \mathbf{H}\mathbf{G}^{-1}\mathbf{H}^T\mathbf{W}]\mathbf{e} \qquad (15.22)$$

where $\mathbf{I}$ is the *unit* or *identity matrix*. Immediate use of Eqs. (15.21) and (15.22) is not possible without knowing the actual errors e_i, which is never the practical case. But we can use those equations for analytical purposes after we have introduced the statistical properties of the errors in Sec. 15.2. For now let us numerically exercise some of the other formulas developed above in the following example.

Example 15.1. In the dc circuit of Fig. 15.1 the meter readings are $z_1 = 9.01$ A, $z_2 = 3.02$ A, $z_3 = 6.98$ V, and $z_4 = 5.01$ V. Assuming that the ammeters are more accurate than the voltmeters, let us assign the measurement weights $w_1 = 100$, $w_2 = 100$, $w_3 = 50$, and $w_4 = 50$, respectively. Determine the weighted least-squares estimates of the voltage sources V_1 and V_2.

Solution. The measured currents and voltages can be expressed in terms of the two voltage sources by elementary circuit analysis using superposition. The results are shown by the system model of Eqs. (15.1) through (15.4), which have the coefficient matrix $\mathbf{H}$ given by

$$\mathbf{H} = \begin{bmatrix} 0.625 & -0.125 \\ -0.125 & 0.625 \\ 0.375 & 0.125 \\ 0.125 & 0.375 \end{bmatrix}$$

Calculating the matrix $\mathbf{H}^T\mathbf{W}$, we obtain

$$\mathbf{H}^T\mathbf{W} = \begin{bmatrix} 0.625 & -0.125 & 0.375 & 0.125 \\ -0.125 & 0.625 & 0.125 & 0.375 \end{bmatrix} \begin{bmatrix} 100 & \cdot & \cdot & \cdot \\ \cdot & 100 & \cdot & \cdot \\ \cdot & \cdot & 50 & \cdot \\ \cdot & \cdot & \cdot & 50 \end{bmatrix}$$

$$= \begin{bmatrix} 62.50 & -12.50 & 18.75 & 6.25 \\ -12.50 & 62.50 & 6.25 & 18.75 \end{bmatrix}$$

and using this result to evaluate the symmetrical gain matrix **G** gives

$$\mathbf{G}=\mathbf{H}^T\mathbf{W}\mathbf{H}=\begin{bmatrix} 62.50 & -12.50 & 18.75 & 6.25 \\ -12.50 & 62.50 & 6.25 & 18.75 \end{bmatrix}\begin{bmatrix} 0.625 & -0.125 \\ -0.125 & 0.625 \\ 0.375 & 0.125 \\ 0.125 & 0.375 \end{bmatrix}$$

$$=\begin{bmatrix} 48.4375 & -10.9375 \\ -10.9375 & 48.4375 \end{bmatrix}$$

Substituting numerical values[2] in Eq. (15.19) yields

$$\begin{bmatrix} \hat{x}_1 \\ \hat{x}_2 \end{bmatrix} = \mathbf{G}^{-1}\mathbf{H}^T\mathbf{W}\mathbf{z}$$

$$=\begin{bmatrix} 48.4375 & -10.9375 \\ -10.9375 & 48.4375 \end{bmatrix}^{-1}\begin{bmatrix} 62.50 & -12.50 & 18.75 & 6.25 \\ -12.50 & 62.50 & 6.25 & 18.75 \end{bmatrix}\mathbf{z}$$

$$=\begin{bmatrix} 0.0218 & 0.0049 \\ 0.0049 & 0.0218 \end{bmatrix}\begin{bmatrix} 62.50 & -12.50 & 18.75 & 6.25 \\ -12.50 & 62.50 & 6.25 & 18.75 \end{bmatrix}\begin{bmatrix} z_1 \\ z_2 \\ z_3 \\ z_4 \end{bmatrix}$$

$$=\begin{bmatrix} 1.2982 & 0.0351 & 0.4386 & 0.2281 \\ 0.0351 & 1.2982 & 0.2281 & 0.4386 \end{bmatrix}\begin{bmatrix} 9.01\text{ A} \\ 3.02\text{ A} \\ 6.98\text{ V} \\ 5.01\text{ V} \end{bmatrix}=\begin{bmatrix} 16.0072\text{ V} \\ 8.0261\text{ V} \end{bmatrix}$$

which are the estimates of the voltage sources V_1 and V_2. The estimated measurements are calculated from $\hat{\mathbf{z}} = \mathbf{H}\hat{\mathbf{x}}$ as follows:

$$\begin{bmatrix} \hat{z}_1 \\ \hat{z}_2 \\ \hat{z}_3 \\ \hat{z}_4 \end{bmatrix}=\begin{bmatrix} 0.625 & -0.125 \\ -0.125 & 0.625 \\ 0.375 & 0.125 \\ 0.125 & 0.375 \end{bmatrix}\begin{bmatrix} 16.0072\text{ V} \\ 8.0261\text{ V} \end{bmatrix}=\begin{bmatrix} 9.00123\text{ A} \\ 3.01544\text{ A} \\ 7.00596\text{ V} \\ 5.01070\text{ V} \end{bmatrix}$$

[2]Numerical results in this chapter were generated by computer. Answers based on the rounded-off values displayed here may differ slightly.

and the estimated errors in the measurements are then given by Eq. (15.11) as

$$\begin{bmatrix} \hat{e}_1 \\ \hat{e}_2 \\ \hat{e}_3 \\ \hat{e}_4 \end{bmatrix} = \begin{bmatrix} 9.01 \\ 3.02 \\ 6.98 \\ 5.01 \end{bmatrix} - \begin{bmatrix} 9.00123 \\ 3.01544 \\ 7.00596 \\ 5.01070 \end{bmatrix} = \begin{bmatrix} 0.00877\text{ A} \\ 0.00456\text{ A} \\ -0.02596\text{ V} \\ -0.00070\text{ V} \end{bmatrix}$$

The preceding example calculates the state of the system by the method of weighted least squares. But knowing the estimated state is of little value if we cannot measure by some means how good the estimate is. Consider, for instance, that the voltmeter reading z_4 of Fig. 15.1 is 4.40 V rather than the 5.01 V used in Example 15.1. If the other three meter readings are unchanged, we can calculate the estimates of the state variables from Eq. (15.19) as follows:

$$\begin{bmatrix} \hat{x}_1 \\ \hat{x}_2 \end{bmatrix} = \begin{bmatrix} 1.2982 & 0.0351 & 0.4386 & 0.2281 \\ 0.0351 & 1.2982 & 0.2281 & 0.4386 \end{bmatrix} \begin{bmatrix} 9.01\text{ A} \\ 3.02\text{ A} \\ 6.98\text{ V} \\ 4.40\text{ V} \end{bmatrix}$$

$$= \begin{bmatrix} 15.86807\text{ V} \\ 7.75860\text{ V} \end{bmatrix}$$

and the estimated measurement errors are again given by Eq. (15.11) as

$$\begin{bmatrix} \hat{e}_1 \\ \hat{e}_2 \\ \hat{e}_3 \\ \hat{e}_4 \end{bmatrix} = \begin{bmatrix} 9.01 \\ 3.02 \\ 6.98 \\ 4.40 \end{bmatrix} - \begin{bmatrix} 0.625 & -0.125 \\ -0.125 & 0.625 \\ 0.375 & 0.125 \\ 0.125 & 0.375 \end{bmatrix} \begin{bmatrix} 15.86807\text{ V} \\ 7.75860\text{ V} \end{bmatrix} = \begin{bmatrix} 0.06228\text{ A} \\ 0.15439\text{ A} \\ 0.05965\text{ V} \\ -0.49298\text{ V} \end{bmatrix}$$

From the practical viewpoint it is not meaningful to compare this second set of numerical answers with those of Example 15.1 in order to decide which set is acceptable. In actual practice, only one set of measurements is available at any one time, and so there is no opportunity to make comparisons. Given a set of measurements, how then can we decide if the corresponding results should be accepted as good estimates of the true values? What criterion for such acceptance is reasonable and if a grossly erroneous meter reading is present, can we *detect* that fact and *identify* the bad measurement? These questions can be answered within a quantifiable *level of confidence* by attaching statistical meaning to the measurement errors in the least-squares calculations.

15.2 STATISTICS, ERRORS, AND ESTIMATES

Computer simulations such as power-flow studies give exact answers, but in reality we never know the absolutely true state of a physically operating system. Even when great care is taken to ensure accuracy, unavoidable random noise enters into the measurement process to distort more or less the physical results. However, repeated measurements of the same quantity under carefully controlled conditions reveal certain statistical properties from which the true value can be estimated.

If the measured values are plotted as a function of their relative frequency of occurrence, a histogram is obtained to which a continuous curve can be fitted as the number of measurements increases (theoretically, to an infinite number). The continuous curve most commonly encountered is the bell-shaped function $p(z)$ shown in Fig. 15.2. The function $p(z)$, called the *gaussian* or *normal probability density function*, has the formula

$$p(z) = \frac{1}{\sigma\sqrt{2\pi}}\varepsilon^{-\frac{1}{2}\left(\frac{z-\mu}{\sigma}\right)^2} \tag{15.23}$$

and the variable representing the values of the measured quantity along the horizontal axis is known as the *gaussian* or *normal random variable* z. Areas

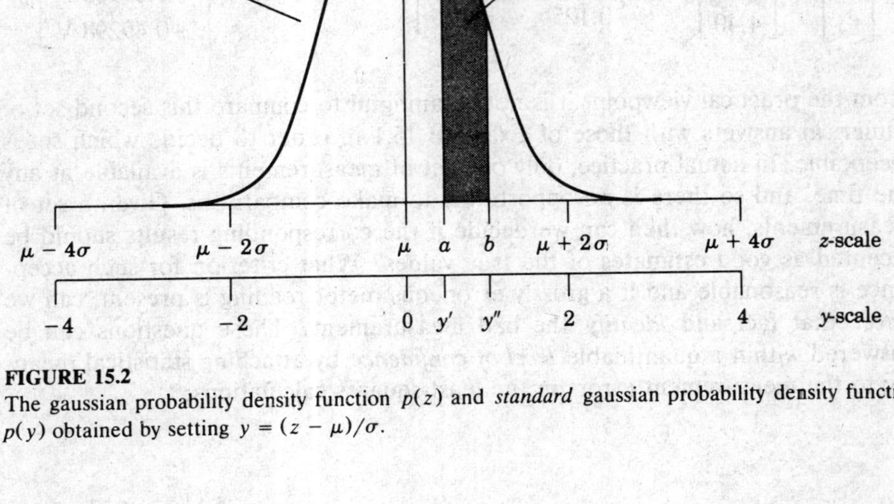

FIGURE 15.2
The gaussian probability density function $p(z)$ and *standard* gaussian probability density function $p(y)$ obtained by setting $y = (z - \mu)/\sigma$.

under the curve give the probabilities associated with the corresponding intervals of the horizontal axis. For example, the probability that z takes on values between the points a and b of Fig. 15.2 is the shaded area given by

$$Pr(a < z < b) = \int_a^b p(z)\,dz = \frac{1}{\sigma\sqrt{2\pi}} \int_a^b \varepsilon^{-\frac{1}{2}\left(\frac{z-\mu}{\sigma}\right)^2} dz \tag{15.24}$$

where the symbol $Pr(\cdot)$ means *the probability of* $(\cdot)$ *occurring*. The total area under the curve $p(z)$ between $-\infty$ and $+\infty$ equals 1 because the value of z is certain (with probability equal to 1 or 100%) to lie between arbitrarily large extreme values.

The gaussian probability density function of Eq. (15.23) is completely determined once the parameters μ and σ are known. Figure 15.2 shows that $p(z)$ has its maximum value when z equals μ, which is the *expected value* of z denoted by $E[z]$ and defined by

$$\mu = E[z] = \int_{-\infty}^{+\infty} z p(z)\,dz \tag{15.25}$$

The expected value μ is often called the *mean* since the values of z are symmetrically clustered about μ. The degree to which the curve of $p(z)$ spreads out about μ (that is, the width) depends on the *variance* σ^2 of z defined by

$$\sigma^2 = E\left[(z-\mu)^2\right] = \int_{-\infty}^{+\infty} (z-\mu)^2 p(z)\,dz \tag{15.26}$$

Thus, the variance of z is the expected value of the square of the deviations of z from its mean; smaller values of σ^2 provide narrower and higher curves with each having unity area centered about μ. The positive square root of the variance is the *standard deviation* σ of z. The mean and variance are important parameters of many commonly encountered probability density functions. Each such function has a curve with underlying areas which quantify the probabilities of the associated random variable.

The gaussian distribution plays a very important role in measurement statistics. Because the density $p(z)$ cannot be directly integrated, the areas under the curve, such as that shown in Fig. 15.2, have been tabulated for the *standard gaussian density function* $p(y)$, which has the parameters $\mu_y = 0$ and $\sigma_y = 1$. Separate tables of the gaussian distribution for other values of μ and σ are unnecessary since we can rescale the horizontal axis of Fig. 15.2, and thereby transform from z to y, using the change-of-variable formula

$$y = \frac{z-\mu}{\sigma} \tag{15.27}$$

TABLE 15.1
The standard gaussian distribution†

a	Pr(a)	a	Pr(a)	a	Pr(a)	a	Pr(a)
.05	0.01994	.80	0.28814	1.55	0.43943	2.30	0.48928
.10	0.03983	.85	0.30234	1.60	0.44520	2.35	0.49061
.15	0.05962	.90	0.31594	1.65	0.45053	2.40	0.49180
.20	0.07926	.95	0.32894	1.70	0.45543	2.45	0.49286
.25	0.09871	1.00	0.34134	1.75	0.45994	2.50	0.49379
.30	0.11791	1.05	0.35314	1.80	0.46407	2.55	0.49461
.35	0.13683	1.10	0.36433	1.85	0.46784	2.60	0.49534
.40	0.15542	1.15	0.37493	1.90	0.47128	2.65	0.49597
.45	0.17364	1.20	0.38493	1.95	0.47441	2.70	0.49653
.50	0.19146	1.25	0.39435	2.00	0.47726	2.75	0.49702
.55	0.20884	1.30	0.40320	2.05	0.47982	2.80	0.49744
.60	0.22575	1.35	0.41149	2.10	0.48214	2.85	0.49781
.65	0.24215	1.40	0.41924	2.15	0.48422	2.90	0.49813
.70	0.25804	1.45	0.42647	2.20	0.48610	2.95	0.49841
.75	0.27337	1.50	0.43319	2.25	0.48778	3.00	0.49865

†$Pr(a)$ is the value of $\dfrac{1}{\sqrt{2\pi}}\displaystyle\int_0^a \varepsilon^{-\frac{1}{2}y^2} dy$.

It is evident from Eq. (15.27) that $dz = \sigma dy$, and if z takes on values between a and b, then y must lie between the corresponding numbers $y' = (a - \mu)/\sigma$ and $y'' = (b - \mu)/\sigma$. Hence, substituting for y in Eq. (15.24), we obtain

$$Pr(a < z < b) = \frac{1}{\sigma\sqrt{2\pi}} \int_{y'}^{y''} \varepsilon^{-\frac{1}{2}y^2} \sigma\, dy$$

$$= \frac{1}{\sqrt{2\pi}} \int_{y'}^{y''} \varepsilon^{-\frac{1}{2}y^2} dy = Pr(y' < y < y'') \qquad (15.28)$$

which shows that y (with $\mu_y = 0$ and $\sigma_y = 1$) is the random variable with standard gaussian probability density. Thus, changing scale according to Eq. (15.27) allows us to use the tabulated values of y, given in Table 15.1, for any random variable z with normal probability density function. In the new scale y tells us how many standard deviations from the mean the corresponding value of z lies, Fig. 15.2. The area under the gaussian density curve is distributed as follows: 68% within the z-limits $\mu \pm \sigma$, 95% within the z-limits $\mu \pm 2\sigma$, and 99% within the z-limits $\mu \pm 3\sigma$. Accordingly, the probability is 99% that the value of the gaussian random variable z lies within 3σ of its mean μ.

Henceforth, we assume that the noise terms e_1, e_2, e_3, and e_4 are *independent* gaussian random variables with zero means and the respective variances σ_1^2, σ_2^2, σ_3^2, and σ_4^2. Two random variables e_i and e_j are independent when $E(e_i e_j) = 0$ for $i \neq j$. The zero-mean assumption implies that the error in each measurement has equal probability of taking on a positive or negative value of a given magnitude.

In Sec. 15.3 we encounter the product of the vector $\mathbf{e}$ and its transpose $\mathbf{e}^T = [e_1\, e_2\, e_3\, e_4]$, which is given by

$$\mathbf{e}\mathbf{e}^T = \begin{bmatrix} e_1 \\ e_2 \\ e_3 \\ e_4 \end{bmatrix} [e_1 \quad e_2 \quad e_3 \quad e_4] = \begin{bmatrix} e_1^2 & e_1e_2 & e_1e_3 & e_1e_4 \\ e_2e_1 & e_2^2 & e_2e_3 & e_2e_4 \\ e_3e_1 & e_3e_2 & e_3^2 & e_3e_4 \\ e_4e_1 & e_4e_2 & e_4e_3 & e_4^2 \end{bmatrix} \tag{15.29}$$

The expected value of $\mathbf{e}\mathbf{e}^T$ is found by calculating the expected value of each entry in the matrix of Eq. (15.29). The expected values of all the off-diagonal elements are zero because the errors are assumed to be independent; the expected values of the diagonal elements are nonzero and correspond to the variance $E[e_i^2] = \sigma_i^2$ for i from 1 to 4. The resultant diagonal matrix is usually assigned the symbol $\mathbf{R}$, and so the expected value of Eq. (15.29) becomes

$$E[\mathbf{e}\mathbf{e}^T] = \mathbf{R} = \begin{bmatrix} E[e_1^2] & \cdot & \cdot & \cdot \\ \cdot & E[e_2^2] & \cdot & \cdot \\ \cdot & \cdot & E[e_3^2] & \cdot \\ \cdot & \cdot & \cdot & E[e_4^2] \end{bmatrix} = \begin{bmatrix} \sigma_1^2 & \cdot & \cdot & \cdot \\ \cdot & \sigma_2^2 & \cdot & \cdot \\ \cdot & \cdot & \sigma_3^2 & \cdot \\ \cdot & \cdot & \cdot & \sigma_4^2 \end{bmatrix} \tag{15.30}$$

With this background, we can now consider some statistical properties of the weighted least-squares estimation procedure.

In Eq. (15.5) the measurement z_1 is the sum of the gaussian random variable e_1 and the *constant* term $(h_{11}x_1 + h_{12}x_2)$, which represents the true value $z_{1,\,\text{true}}$ of z_1. The addition of the constant term to e_1 shifts the curve of e_1 to the right by the amount of the true value (or to the left by the same amount if the true value is negative). But the shift in no way alters the shape and spread of the probability density function of e_1. Hence, z_1 also has a gaussian probability density function with a mean value μ_1 equal to the true value $z_{1,\,\text{true}}$ and a variance σ_1^2 equal to that of e_1. Similar remarks apply to z_2, z_3, and z_4, which also have means equal to their true values, and variances σ_2^2, σ_3^2, and σ_4^2 corresponding to the variances of e_2, e_3, and e_4, respectively. Thus, meters with smaller error variances have narrower curves and provide more accurate measurements. In formulating the objective function f of Eq. (15.13), preferential weighting is given to the more accurate measurements by choosing the weight w_i as the *reciprocal* of the corresponding variance σ_i^2. Errors of smaller variance thus have greater weight. Henceforth, we specify the weighting matrix $\mathbf{W}$ of

Eqs. (15.21) and (15.22) as

$$\mathbf{W} = \mathbf{R}^{-1} = \begin{bmatrix} \frac{1}{\sigma_1^2} & \cdot & \cdot & \cdot \\ \cdot & \frac{1}{\sigma_2^2} & \cdot & \cdot \\ \cdot & \cdot & \frac{1}{\sigma_3^2} & \cdot \\ \cdot & \cdot & \cdot & \frac{1}{\sigma_4^2} \end{bmatrix} \tag{15.31}$$

and the gain matrix **G** then becomes

$$\mathbf{G} = \mathbf{H}^T\mathbf{R}^{-1}\mathbf{H} \tag{15.32}$$

We can regard the elements in the weighting matrix of Example 15.1 as the reciprocals of the error variances $\sigma_1^2 = \sigma_2^2 = 1/100$ for meters 1 and 2 and $\sigma_3^2 = \sigma_4^2 = 1/50$ for meters 3 and 4. The corresponding standard deviations of the meter errors are $\sigma_1 = \sigma_2 = 0.1$ A and $\sigma_3 = \sigma_4 = (\sqrt{2}/10)$ V. Thus, the probability is 99% that ammeters 1 and 2 of Fig. 15.1, *when functioning properly*, will give readings within 3σ or 0.3 A of the true values of their measured currents. Comparable statements apply to the two voltmeters of Fig. 15.1 if their errors are also gaussian random variables.

From Eq. (15.25), which defines the expected value of the random variable z, we can show that $E[az + b] = aE[z] + b$ if a and b are constants. Therefore, taking the expected value of each side of Eq. (15.21) gives

$$\begin{bmatrix} E[\hat{x}_1 - x_1] \\ E[\hat{x}_2 - x_2] \end{bmatrix} = \begin{bmatrix} E[\hat{x}_1] - x_1 \\ E[\hat{x}_2] - x_2 \end{bmatrix} = \mathbf{G}^{-1}\mathbf{H}^T\mathbf{R}^{-1} \begin{bmatrix} E[e_1] \\ E[e_2] \\ E[e_3] \\ E[e_4] \end{bmatrix} = \begin{bmatrix} 0 \\ 0 \end{bmatrix} \tag{15.33}$$

This equation uses the fact that x_1 and x_2 are definite *numbers* equal to the true (but unknown) values of the state variables while e_1, e_2, e_3, and e_4 are gaussian random *variables* with zero expected values. It follows from Eq. (15.33) that

$$E[\hat{x}_i] = x_i \tag{15.34}$$

which implies that the weighted least-squares estimate of each state variable has an expected value equal to the true value. In a similar manner, we can use

Eq. (15.22) to show that

$$E\begin{bmatrix}\hat{e}_1\\ \hat{e}_2\\ \hat{e}_3\\ \hat{e}_4\end{bmatrix} = E\begin{bmatrix}z_1-\hat{z}_1\\ z_2-\hat{z}_2\\ z_3-\hat{z}_3\\ z_4-\hat{z}_4\end{bmatrix} = [\mathbf{I}-\mathbf{H}\mathbf{G}^{-1}\mathbf{H}^T\mathbf{R}^{-1}]E\begin{bmatrix}e_1\\ e_2\\ e_3\\ e_4\end{bmatrix} = \begin{bmatrix}0\\0\\0\\0\end{bmatrix} \tag{15.35}$$

and since z_j equals $z_{j,\,\text{true}} + e_j$, it follows from Eq. (15.35) that

$$E[\hat{z}_j] = E[z_j] = z_{j,\,\text{true}} \tag{15.36}$$

Equations (15.34) and (15.36) state that the weighted least-squares estimates of the state variables and the measured quantities, on the average, are equal to their true values—an obviously desirable property, and the estimates are said to be *unbiased*. Accordingly, in Example 15.1 the solutions of 16.0072 V for V_1 and 8.0261 V for V_2 can be accepted as unbiased estimates of the true values, *provided* no bad measurement data are present. The presence of bad data can be detected as described in the next section.

15.3 TEST FOR BAD DATA

When the system model is correct and the measurements are accurate, there is good reason to accept the state estimates calculated by the weighted least-squares estimator. But if a measurement is grossly erroneous or *bad*, it should be detected and then identified so that it can be removed from the estimator calculations. The statistical properties of the measurement errors facilitate such detection and identification.[3]

Each estimated measurement error $\hat{e}_j = (z_j - \hat{z}_j)$ is a gaussian random variable with zero mean, as shown by Eq. (15.35). A formula for the variance of $\hat{e}_j$ can be determined from Eq. (15.22) in two steps. First, with $\mathbf{W}$ replaced by $\mathbf{R}^{-1}$, we postmultiply $\hat{\mathbf{e}}$ by its transpose $\hat{\mathbf{e}}^T$ to obtain

$$\hat{\mathbf{e}}\hat{\mathbf{e}}^T = (\mathbf{z}-\hat{\mathbf{z}})(\mathbf{z}-\hat{\mathbf{z}})^T = [\mathbf{I}-\mathbf{H}\mathbf{G}^{-1}\mathbf{H}^T\mathbf{R}^{-1}]\mathbf{e}\mathbf{e}^T[\mathbf{I}-\mathbf{R}^{-1}\mathbf{H}\mathbf{G}^{-1}\mathbf{H}^T] \tag{15.37}$$

The left-hand side of this equation represents a square matrix like that of Eq. (15.29); the two matrices in brackets on the right-hand side contain only constant elements and are transposes of one another. The term $\mathbf{e}\mathbf{e}^T$ is actually the same matrix given in Eq. (15.29) and its expected value $\mathbf{R}$ is the diagonal matrix of Eq. (15.30). In the second step we take the expected value of each side

[3]This section is based on J. F. Dopazo, O. A. Klitin, and A. M. Sasson, "State Estimation of Power Systems: Detection and Identification of Gross Measurement Errors," in *Proceedings of the IEEE PICA Conference*, 1973.

of Eq. (15.37) to find that

$$E[\hat{\mathbf{e}}\hat{\mathbf{e}}^T] = [\mathbf{I} - \mathbf{H}\mathbf{G}^{-1}\mathbf{H}^T\mathbf{R}^{-1}]\underbrace{E[\mathbf{e}\mathbf{e}^T]}_{\mathbf{R}}[\mathbf{I} - \mathbf{R}^{-1}\mathbf{H}\mathbf{G}^{-1}\mathbf{H}^T] \quad (15.38)$$

Multiplying $\mathbf{R}$ by the bracketed matrix $[\mathbf{I} - \mathbf{R}^{-1}\mathbf{H}\mathbf{G}^{-1}\mathbf{H}^T]$ and then factoring out $\mathbf{R}$ to the right from the result give

$$\begin{aligned} E[\hat{\mathbf{e}}\hat{\mathbf{e}}^T] &= [\mathbf{I} - \mathbf{H}\mathbf{G}^{-1}\mathbf{H}^T\mathbf{R}^{-1}][\mathbf{R} - \mathbf{H}\mathbf{G}^{-1}\mathbf{H}^T] \\ &= [\mathbf{I} - \mathbf{H}\mathbf{G}^{-1}\mathbf{H}^T\mathbf{R}^{-1}][\mathbf{I} - \mathbf{H}\mathbf{G}^{-1}\mathbf{H}^T\mathbf{R}^{-1}]\mathbf{R} \end{aligned} \quad (15.39)$$

The reader should show that the matrix $[\mathbf{I} - \mathbf{H}\mathbf{G}^{-1}\mathbf{H}^T\mathbf{R}^{-1}]$ multiplied by itself remains unaltered (it is called an *idempotent* matrix), and so Eq. (15.39) simplifies as follows:

$$E[\hat{\mathbf{e}}\hat{\mathbf{e}}^T] = [\mathbf{I} - \mathbf{H}\mathbf{G}^{-1}\mathbf{H}^T\mathbf{R}^{-1}]\mathbf{R} = \mathbf{R} - \mathbf{H}\mathbf{G}^{-1}\mathbf{H}^T \quad (15.40)$$

The square matrix $\hat{\mathbf{e}}\hat{\mathbf{e}}^T$ takes the form of Eq. (15.29) with typical entries given by $\hat{e}_i\hat{e}_j$. Therefore, substituting for $\hat{e}_j = (z_j - \hat{z}_j)$ in the diagonal entries, we obtain

$$E\left[\hat{e}_j^2\right] = E\left[(z_j - \hat{z}_j)^2\right] = R'_{jj} \quad (15.41)$$

where R'_{jj} is the symbol for the jth diagonal element of the matrix $\mathbf{R}' = \mathbf{R} - \mathbf{H}\mathbf{G}^{-1}\mathbf{H}^T$. Since $\hat{e}_j$ has zero mean value, Eq. (15.41) is actually the formula for the variance of $\hat{e}_j$, which makes $\sqrt{R'_{jj}}$ the standard deviation. Dividing each side of Eq. (15.41) by the number R'_{jj} gives

$$E\left[\left(\frac{\hat{e}_j}{\sqrt{R'_{jj}}}\right)^2\right] = E\left[\left(\frac{z_j - \hat{z}_j}{\sqrt{R'_{jj}}}\right)^2\right] = 1 \quad (15.42)$$

Thus, in the manner of Eq. (15.27) the deviation of the estimated error ($\hat{e}_j$) from its mean (0), divided by the standard deviation ($\sqrt{R'_{jj}}$), yields the *standard* gaussian random variable

$$\frac{\hat{e}_j - 0}{\sqrt{R'_{jj}}} = \frac{z_j - \hat{z}_j}{\sqrt{R'_{jj}}} \quad (15.43)$$

with zero mean and variance equal to 1, as shown by Eq. (15.42).

Matrices $\mathbf{R}$ of Eq. (15.30) and $\mathbf{R}' = (\mathbf{R} - \mathbf{H}\mathbf{G}^{-1}\mathbf{H}^T)$ of Eq. (15.40) are called *covariance matrices*. It can be also shown using Eq. (15.21) that the covariance matrix of $\mathbf{x} - \hat{\mathbf{x}}$ given by $E[(\mathbf{x} - \hat{\mathbf{x}})(\mathbf{x} - \hat{\mathbf{x}})^T]$ is $\mathbf{G}^{-1} = (\mathbf{H}^T\mathbf{R}^{-1}\mathbf{H})^{-1}$.

TABLE 15.2
Summary of residue terms and covariant matrices†

Term	Statement	Expression	Covariance matrix‡
$\mathbf{e} = \mathbf{z} - \mathbf{Hx}$	$(\text{measured} - \text{true})_{\mathbf{z}}$	$\mathbf{z} - \mathbf{Hx}$	$\mathbf{R}$
$\hat{\mathbf{e}}_{\mathbf{x}} = \mathbf{x} - \hat{\mathbf{x}}$	$(\text{true} - \text{estimated})_{\mathbf{x}}$	$-\mathbf{G}^{-1}\mathbf{H}^T\mathbf{R}^{-1}\mathbf{e}$	$\mathbf{G}^{-1} = (\mathbf{H}^T\mathbf{R}^{-1}\mathbf{H})^{-1}$
$\hat{\mathbf{e}} = \mathbf{z} - \hat{\mathbf{z}}$	$(\text{measured} - \text{estimated})_{\mathbf{z}}$	$(\mathbf{I} - \mathbf{HG}^{-1}\mathbf{H}^T\mathbf{R}^{-1})\mathbf{e}$	$\mathbf{R}' = (\mathbf{R} - \mathbf{HG}^{-1}\mathbf{H}^T)$

where $\hat{\mathbf{x}} = \mathbf{G}^{-1}\mathbf{H}^T\mathbf{R}^{-1}\mathbf{z}$ and $\hat{\mathbf{z}} = \mathbf{H}\hat{\mathbf{x}}$

†All residues have zero means.
‡Diagonal terms are the error variances.

A summary of some of the terms encountered above and their statistical properties is presented in Table 15.2.

We have already discussed the fact that the true measurement error e_j is never known in engineering applications. The best that can be done is to calculate the estimated error $\hat{e}_j$, which then replaces e_j in the objective function. Accordingly, we substitute from Eq. (15.43) into Eq. (15.13) to obtain

$$\hat{f} = \sum_{j=1}^{N_m} w_j \hat{e}_j^2 = \sum_{j=1}^{N_m} \frac{\hat{e}_j^2}{\sigma_j^2} = \sum_{j=1}^{N_m} \frac{\left(z_j - \hat{z}_j\right)^2}{\sigma_j^2} \tag{15.44}$$

where N_m is the number of measurements and the weighting factor w_j is set equal to $1/\sigma_j^2$. The weighted sum of squares $\hat{f}$ is itself a random variable which has a well-known probability distribution with values (areas) already tabulated in many textbooks on statistics. In order to use those tables, we need to know the mean value of $\hat{f}$, which is simple to determine, as we now demonstrate.

Multiplying the numerator and the denominator of the jth term in Eq. (15.44) by the calculated variance R'_{jj} of the measurement error and taking the expected value of the result, gives

$$E\left[\hat{f}\right] = E\left[\sum_{j=1}^{N_m} \frac{R'_{jj}}{\sigma_j^2} \frac{\hat{e}_j^2}{R'_{jj}}\right] = \sum_{j=1}^{N_m} \frac{R'_{jj}}{\sigma_j^2} E\left[\frac{\left(z_j - \hat{z}_j\right)^2}{R'_{jj}}\right] \tag{15.45}$$

The expected value in the right-hand summation of Eq. (15.45) equals 1 according to Eq. (15.42), and so we obtain for the mean of $\hat{f}$

$$E\left[\hat{f}\right] = E\left[\sum_{j=1}^{N_m} \frac{\hat{e}_j^2}{\sigma_j^2}\right] = \sum_{j=1}^{N_m} \frac{R'_{jj}}{\sigma_j^2} \tag{15.46}$$

It is shown in Example 15.2 that the right-hand side of this equation has the numerical value $2 = (4 - 2)$ for the *four* measurements and *two* state variables of the circuit of Fig. 15.1. More generally, it can be demonstrated (Prob. 15.4) that the expected value of $\hat{f}$ is always numerically equal to the *number of degrees of freedom* $(N_m - N_s)$; that is,

$$E\left[\hat{f}\right] = \sum_{j=1}^{N_m} E\left[\frac{\left(z_j - \hat{z}_j\right)^2}{\sigma_j^2}\right] = N_m - N_s \tag{15.47}$$

when there is a set of N_m measurements and N_s independent state variables to be estimated. Thus, *the mean value of $\hat{f}$ is an integer which is found simply by subtracting the number of state variables from the number of measurements*. The number $(N_m - N_s)$ is also called the *redundancy* of the measurement scheme, and it is obviously large when there are many more measurements than state variables.

Just as $\hat{e}_j$ has the standard gaussian distribution, statistical theory shows that the weighted sum of squares $\hat{f}$ of Eq. (15.44) has the *chi-square distribution* $\chi^2_{k,\alpha}$, where χ is the Greek letter *chi*, $k = (N_m - N_s)$ is the number of degrees of freedom, and α relates to the area under the $\chi^2_{k,\alpha}$ curve. The chi-square distribution very closely matches the standard gaussian distribution when k is large ($k > 30$), which is often the case in power system applications. Figure 15.3 shows the probability density function of $\chi^2_{k,\alpha}$ for a representative small value of k. As usual, the total area under the curve equals 1, but here it is *not*

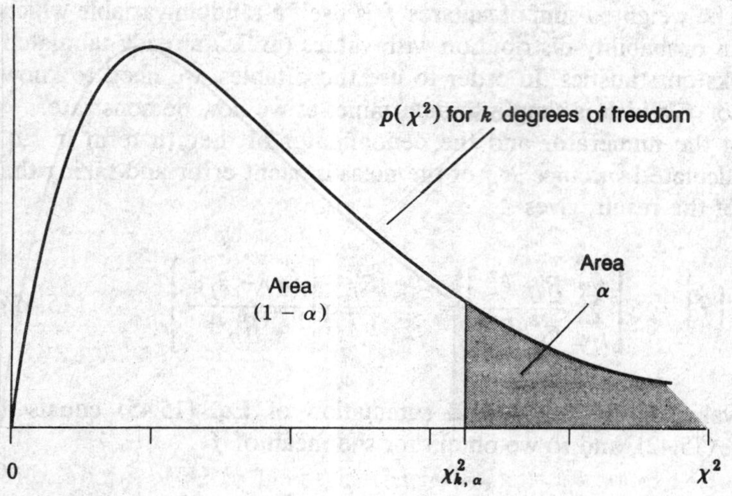

FIGURE 15.3
Probability density function $p(\chi^2)$ of the chi-square distribution $\chi^2_{k,\alpha}$ for a (small) value of k degrees of freedom.

TABLE 15.3
Values of area α to the right of $\chi^2 = \chi^2_{k,\alpha}$ in Fig. 15.3

	α					α			
k	0.05	0.025	0.01	0.005	k	0.05	0.025	0.01	0.005
1	3.84	5.02	6.64	7.88	11	19.68	21.92	24.73	26.76
2	5.99	7.38	9.21	10.60	12	21.03	23.34	26.22	28.30
3	7.82	9.35	11.35	12.84	13	22.36	24.74	27.69	29.82
4	9.49	11.14	13.28	14.86	14	23.69	26.12	29.14	31.32
5	11.07	12.83	15.09	16.75	15	25.00	27.49	30.58	32.80
6	12.59	14.45	16.81	18.55	16	26.30	28.85	32.00	34.27
7	14.07	16.01	18.48	20.28	17	27.59	30.19	33.41	35.72
8	15.51	17.54	20.09	21.96	18	28.87	31.53	34.81	37.16
9	16.92	19.02	21.67	23.59	19	30.14	32.85	36.19	38.58
10	18.31	20.48	23.21	25.19	20	31.41	34.17	37.57	40.00

symmetrically distributed. The area under the curve to the *right* of $\chi^2_{k,\alpha}$ in Fig. 15.3 equals α, which is the probability that $\hat{f}$ exceeds $\chi^2_{k,\alpha}$. The remaining area under the curve is the probability $(1-\alpha)$ that the calculated value of the weighted sum of squares $\hat{f}$, with k degrees of freedom, will take on a value less than $\chi^2_{k,\alpha}$; that is,

$$Pr\left(\hat{f} < \chi^2_{k,\alpha}\right) = (1-\alpha) \tag{15.48}$$

Based on this equation, the critical value of the statistic $\hat{f}$ can be determined using the tabulated values of $\chi^2_{k,\alpha}$ given in Table 15.3. For example, choosing $\alpha = 0.01$ and $k = (N_m - N_s) = 2$, we can conclude that the calculated value of $\hat{f}$ is less than the critical value of 9.21 with probability $(1 - 0.01)$ or 99% confidence since $\chi^2_{2,0.01} = 9.21$ in Table 15.3. Thus, the chi-square distribution of $\hat{f}$ provides a test for detection of bad measurements. The procedure is as follows:

- Use the raw measurements z_j from the system to determine the weighted least-squares estimates $\hat{x}_i$ of the system states from Eq. (15.19).
- Substitute the estimates $\hat{x}_i$ in the equation $\hat{\mathbf{z}} = \mathbf{H}\hat{\mathbf{x}}$ to calculate the estimated values $\hat{z}_j$ of the measurements and hence the estimated errors $\hat{e}_j = z_j - \hat{z}_j$.
- Evaluate the weighted sum of squares $\hat{f} = \sum_{j=1}^{N_m} \hat{e}_j^2/\sigma_j^2$.
- For the appropriate number of degrees of freedom $k = (N_m - N_s)$ and a *specified* probability α, determine whether or not the value of $\hat{f}$ is less than the critical value corresponding to α. In practice, this means we check that the inequality

$$\hat{f} < \chi^2_{k,\alpha} \tag{15.49}$$

is satisfied. If it is, then the measured raw data and the state estimates are accepted as being accurate.

- When the requirement of inequality (15.49) is not met, there is reason to suspect the presence of at least one bad measurement. Upon such detection, omit the measurement corresponding to the largest *standardized* error, namely, $(z_j - \hat{z}_j)/\sqrt{R'_{jj}}$ of Eq. (15.43), and reevaluate the state estimates along with the sum of squares $\hat{f}$. If the new value of $\hat{f}$ satisfies the chi-square test of inequality (15.49), then the omitted measurement has been successfully identified as the bad data point.

The sum of squares of the estimated errors will be large when bad measurements are present. Thus, *detection* of bad data points is readily accomplished by the chi-square test. The *identification* of the particular bad data is not so easily undertaken. In practical power system applications the number of degrees of freedom is large, which allows discarding a *group* of measurements corresponding to the largest standardized residuals. Although there is still no guarantee that the largest standardized errors always indicate the bad measurements, the problem of identifying the gross errors is thereby eased.

In each of the following examples we assume that all the measurement errors are gaussian random variables. In Example 15.2 the results of Example 15.1 are used to determine $\hat{f}$ when the meter accuracies are known. Then, in Examples 15.3 and 15.4 we check for the presence of bad data in the measurement set.

Example 15.2. Suppose that the weighting factors w_1 to w_4 in Example 15.1 are the corresponding reciprocals of the error variances for the four meters of Fig. 15.1. Evaluate the expected value of the sum of squares $\hat{f}$ of the measurement residuals.

Solution. First, let us evaluate the diagonal elements of $\mathbf{R}' = \mathbf{R} - \mathbf{H}\mathbf{G}^{-1}\mathbf{H}^T$ before forming the sum of squares for Eq. (15.46). To take advantage of previous calculations in Example 15.1, we write the matrix $\mathbf{R}'$ in the form $\mathbf{R}' = (\mathbf{I} - \mathbf{H}\mathbf{G}^{-1}\mathbf{H}^T\mathbf{R}^{-1})\mathbf{R}$. Then, from Example 15.1 the matrix $\mathbf{H}\mathbf{G}^{-1}\mathbf{H}^T\mathbf{R}^{-1}$ takes on the value

$$\mathbf{H}\mathbf{G}^{-1}\mathbf{H}^T\mathbf{R}^{-1} = \begin{bmatrix} 0.625 & -0.125 \\ -0.125 & 0.625 \\ 0.375 & 0.125 \\ 0.125 & 0.375 \end{bmatrix} \begin{bmatrix} 1.2982 & 0.0351 & 0.4386 & 0.2281 \\ 0.0351 & 1.2982 & 0.2281 & 0.4386 \end{bmatrix}$$

Only the diagonal elements resulting from this equation are required for Eq. (15.46), and so multiplying the first row of $\mathbf{H}$ by the first column of $\mathbf{G}^{-1}\mathbf{H}^T\mathbf{R}^{-1}$, the second row by the second column, and so on, we obtain

$$\mathbf{H}\mathbf{G}^{-1}\mathbf{H}^T\mathbf{R}^{-1} = \begin{bmatrix} 0.8070 & \times & \times & \times \\ \times & 0.8070 & \times & \times \\ \times & \times & 0.1930 & \times \\ \times & \times & \times & 0.1930 \end{bmatrix}$$

where $\times$ denotes the nonzero elements which do not have to be evaluated. The diagonal elements of $\mathbf{R}'$ are then found to be

$$\begin{bmatrix} R'_{11} & \cdot & \cdot & \cdot \\ \cdot & R'_{22} & \cdot & \cdot \\ \cdot & \cdot & R'_{33} & \cdot \\ \cdot & \cdot & \cdot & R'_{44} \end{bmatrix}$$

$$= \left\{ \begin{bmatrix} 1 & \cdot & \cdot & \cdot \\ \cdot & 1 & \cdot & \cdot \\ \cdot & \cdot & 1 & \cdot \\ \cdot & \cdot & \cdot & 1 \end{bmatrix} - \begin{bmatrix} 0.807 & \cdot & \cdot & \cdot \\ \cdot & 0.807 & \cdot & \cdot \\ \cdot & \cdot & 0.193 & \cdot \\ \cdot & \cdot & \cdot & 0.193 \end{bmatrix} \right\} \begin{bmatrix} \sigma_1^2 & \cdot & \cdot & \cdot \\ \cdot & \sigma_2^2 & \cdot & \cdot \\ \cdot & \cdot & \sigma_3^2 & \cdot \\ \cdot & \cdot & \cdot & \sigma_4^2 \end{bmatrix}$$

where σ_j^2 is the error variance of the jth meter. The expected value of $\hat{f}$ corresponding to Eq. (15.46) is calculated for $N_m = 4$, as follows:

$$\sum_{j=1}^{4} \frac{R'_{jj}}{\sigma_j^2} = \frac{(1-0.807)\sigma_1^2}{\sigma_1^2} + \frac{(1-0.807)\sigma_2^2}{\sigma_2^2} + \frac{(1-0.193)\sigma_3^2}{\sigma_3^2} + \frac{(1-0.193)\sigma_4^2}{\sigma_4^2}$$

$$= (1 + 1 + 1 + 1) - (0.807 + 0.807 + 0.193 + 0.193)$$

$$= 4 - 2 = 2$$

which is the number of degrees of freedom when the system of Example 15.1 has four measurements from which to estimate the two state variables. It is apparent from the way in which the solution is set forth in this example that the sum of the diagonal elements of the matrix $\mathbf{H}\mathbf{G}^{-1}\mathbf{H}^T\mathbf{R}^{-1}$ numerically equals the number of state variables, N_s.

Example 15.3. Using the chi-square test of inequality (15.49), check for the presence of bad data in the raw measurements of Example 15.1. Choose $\alpha = 0.01$.

Solution. Since there are four measurements and two state variables in Example 15.1, we choose $k = 2$ and find from Table 15.3 that $\chi^2_{k,\alpha} = 9.21$. Based on the results for $\hat{e}_j$ in Example 15.1, the estimated sum of squares $\hat{f}$ is calculated as follows:

$$\hat{f} = \sum_{j=1}^{4} \frac{\hat{e}_j^2}{\sigma_j^2} = 100\hat{e}_1^2 + 100\hat{e}_2^2 + 50\hat{e}_3^2 + 50\hat{e}_4^2$$

$$= 100(0.00877)^2 + 100(0.00456)^2 + 50(0.02596)^2 + 50(0.00070)^2$$

$$= 0.043507$$

which is obviously less than 9.21. Therefore, we conclude (with 99% confidence) that the raw measurement set of Example 15.1 has no bad measurements.

Example 15.4. Suppose that the raw measurement set for the system of Fig. 15.1 is given by

$$\begin{bmatrix} z_1 & z_2 & z_3 & z_4 \end{bmatrix}^T = \begin{bmatrix} 9.01\,A & 3.02\,A & 6.98\text{ V} & 4.40\text{ V} \end{bmatrix}^T$$

and the meters are the same as in Example 15.2. Check for the presence of bad data using the chi-square test for $\alpha = 0.01$. Eliminate any bad data detected and calculate the resultant state estimates from the *reduced* data set.

Solution. The raw measurement set is the same as at the end of Sec. 15.1 where we estimate the errors to be

$$\begin{bmatrix} \hat{e}_1 & \hat{e}_2 & \hat{e}_3 & \hat{e}_4 \end{bmatrix}^T = \begin{bmatrix} 0.06228 & 0.15439 & 0.05965 & -0.49298 \end{bmatrix}^T$$

Substituting these numerical values in Eq. (15.44), we find that

$$\hat{f} = \sum_{j=1}^{4} \frac{\hat{e}_j^2}{\sigma_j^2} = 100\hat{e}_1^2 + 100\hat{e}_2^2 + 50\hat{e}_3^2 + 50\hat{e}_4^2$$

$$= 100(0.06228^2 + 0.15439^2) + 50(0.05965^2 + 0.49298^2)$$

$$= 15.1009$$

This value of $\hat{f}$ exceeds $\chi^2_{2,0.01} = 9.21$, and so we conclude that there is at least one bad measurement. The standardized error estimates are next calculated using the diagonal elements R'_{jj} from Example 15.2 as follows:

$$\frac{\hat{e}_1}{\sqrt{R'_{11}}} = \frac{0.06228}{\sqrt{(1 - 0.807)\sigma_1^2}} = \frac{0.06228}{\sqrt{0.193/100}} = 1.4178$$

$$\frac{\hat{e}_2}{\sqrt{R'_{22}}} = \frac{0.15439}{\sqrt{(1 - 0.807)\sigma_2^2}} = \frac{0.15439}{\sqrt{0.193/100}} = 3.5144$$

$$\frac{\hat{e}_3}{\sqrt{R'_{33}}} = \frac{0.05965}{\sqrt{(1-0.193)\sigma_3^2}} = \frac{0.05965}{\sqrt{0.807/50}} = 0.4695$$

$$\frac{\hat{e}_4}{\sqrt{R'_{44}}} = \frac{-0.49298}{\sqrt{(1-0.193)\sigma_4^2}} = \frac{-0.49298}{\sqrt{0.807/50}} = -3.8804$$

The magnitude of the largest standardized error corresponds to measurement z_4 in this case. Therefore, we identify z_4 as the bad measurement and omit it from the state-estimation calculations. For the *three* remaining measurements the revised matrices **H** and $\mathbf{H}^T\mathbf{R}^{-1}$ are given by

$$\mathbf{H} = \begin{bmatrix} 0.625 & -0.125 \\ -0.125 & 0.625 \\ 0.375 & 0.125 \end{bmatrix} \qquad \mathbf{H}^T\mathbf{R}^{-1} = \begin{bmatrix} 62.50 & -12.50 & 18.75 \\ -12.50 & 62.50 & 6.25 \end{bmatrix}$$

and the gain matrix along with its inverse become

$$\mathbf{G} = \begin{bmatrix} 47.65625 & -13.28125 \\ -13.28125 & 41.40625 \end{bmatrix} \qquad \mathbf{G}^{-1} = \begin{bmatrix} 0.023043 & 0.007391 \\ 0.007391 & 0.026522 \end{bmatrix}$$

The state estimates based on the three retained measurements are

$$\begin{bmatrix} \hat{x}_1 \\ \hat{x}_2 \end{bmatrix} = \mathbf{G}^{-1}\mathbf{H}^T\mathbf{R}^{-1}\begin{bmatrix} z_1 \\ z_2 \\ z_3 \end{bmatrix} = \begin{bmatrix} 1.34783 & 0.17391 & 0.47826 \\ 0.13044 & 1.56522 & 0.30435 \end{bmatrix}\begin{bmatrix} 9.01 \\ 3.02 \\ 6.98 \end{bmatrix}$$

$$= \begin{bmatrix} 16.0074 \text{ V} \\ 8.0265 \text{ V} \end{bmatrix}$$

and the revised estimates of the same three measured quantities are

$$\begin{bmatrix} \hat{z}_1 \\ \hat{z}_2 \\ \hat{z}_3 \end{bmatrix} = \begin{bmatrix} 0.625 & -0.125 \\ -0.125 & 0.625 \\ 0.375 & 0.125 \end{bmatrix}\begin{bmatrix} 16.0074 \\ 8.0265 \end{bmatrix} = \begin{bmatrix} 9.0013 \text{ A} \\ 3.0157 \text{ A} \\ 7.0061 \text{ V} \end{bmatrix}$$

The estimated errors are calculated to be

$$\begin{bmatrix} \hat{e}_1 & \hat{e}_2 & \hat{e}_3 \end{bmatrix}^T = \begin{bmatrix} 0.0087 & 0.0043 & -0.0261 \end{bmatrix}^T$$

from which the sum of squares $\hat{f} = \Sigma_{j=1}^{3} \hat{e}_j^2/\sigma_j^2$ is found to equal 0.0435. Since there are now only three measurements from which to estimate the two state variables, we compare $\hat{f} = 0.0435$ with the chi-square value $\chi^2_{1,0.01} = 6.64$ for *one* degree of freedom and conclude that no more bad data exist. The reader is encouraged to check that the matrix $\mathbf{HG}^{-1}\mathbf{H}^T\mathbf{R}^{-1}$ of this example has diagonal elements which numerically add up to 2, the number of state variables (see Prob. 15.3). The new state estimates based on the three good measurements essentially match those of Example 15.1.

15.4 POWER SYSTEM STATE ESTIMATION

The state of the ac power system is expressed by the voltage magnitudes and phase angles at the buses. Although relative phase angles of bus voltages cannot be measured, they can be calculated using real-time data acquired from the system. These data are processed by the *state estimator*, a computer program which calculates voltage magnitudes and relative phase angles of the system buses. While the state estimator gives results similar to those available from a conventional power-flow program, the input data and calculation procedure are quite different.

The state estimator operates on real-time inputs which are of two kinds—*data* and *status* information. The on/off status of switching devices (such as circuit breakers, disconnect switches, and transformer taps) determines the network configuration, which changes whenever those devices operate. Remote terminal units located on the system to report such changes also monitor (1) analog data in the form of megawatt and megavar flows on all major lines, (2) P and Q loading of generators and transformers, and (3) voltage magnitudes at most of the buses of the system. Every few seconds the remote units are scanned and the complete set of measurements is telemetered to the energy control center. Thus, the state estimator draws on a larger data base than that previously described for the conventional power flow.

In practical state estimation the number of actual measurements is far greater than the number of data inputs required by the planning-type power flow of Table 9.1. As a result, there are many more equations to solve than there are unknown state variables. This *redundancy* is necessary, as we have seen, because measurements are sometimes grossly in error or unavailable owing to malfunctions in the data-gathering systems. Direct use of the raw measurements is not advisable, and some form of data *filtering* is needed before the raw data are used in computations. State estimation performs this filtering function by processing the set of measurements *as a whole* so as to obtain a mean or average-value estimate of all the system state variables. In this way the best fit of the entire set of input data is obtained rather than a separate processing of the individual measurements.

The estimated state of the simple dc circuit of Sec. 15.1 is given by the closed-form solution of Eq. (15.19) because the circuit equations are linear. For the ac power system the measurement equations are nonlinear and iterative solutions are required as in the Newton-Raphson power-flow procedure.

Suppose, for instance, that the measurement equations are of the form

$$e_1 = z_1 - h_1(x_1, x_2) \tag{15.50}$$

$$e_2 = z_2 - h_2(x_1, x_2) \tag{15.51}$$

$$e_3 = z_3 - h_3(x_1, x_2) \tag{15.52}$$

$$e_4 = z_4 - h_4(x_1, x_2) \tag{15.53}$$

where, in general, h_1, h_2, h_3, and h_4 are nonlinear functions which express the measured quantities in terms of the state variables, and e_1, e_2, e_3, and e_4 are gaussian random variable noise terms. The true values of x_1 and x_2 are not known and have to be estimated from the measurements z_1, z_2, z_3, and z_4.

We begin by forming the weighted sum of squares of the errors with weights w_j chosen equal to the reciprocal of the corresponding error variances σ_j^2, which gives

$$f = \sum_{j=1}^{4} \frac{e_j^2}{\sigma_j^2} = \frac{(z_1 - h_1(x_1, x_2))^2}{\sigma_1^2} + \frac{(z_2 - h_2(x_1, x_2))^2}{\sigma_2^2}$$
$$+ \frac{(z_3 - h_3(x_1, x_2))^2}{\sigma_3^2} + \frac{(z_4 - h_4(x_1, x_2))^2}{\sigma_4^2} \tag{15.54}$$

As before, the estimates $\hat{x}_1$ and $\hat{x}_2$ which minimize f must satisfy Eq. (15.16) and that equation now takes on the form

$$\left[\begin{array}{cccc} \dfrac{\partial h_1}{\partial x_1} & \dfrac{\partial h_2}{\partial x_1} & \dfrac{\partial h_3}{\partial x_1} & \dfrac{\partial h_4}{\partial x_1} \\ \dfrac{\partial h_1}{\partial x_2} & \dfrac{\partial h_2}{\partial x_2} & \dfrac{\partial h_3}{\partial x_2} & \dfrac{\partial h_4}{\partial x_2} \end{array}\right]_{\hat{\mathbf{x}}} \underbrace{\left[\begin{array}{cccc} \dfrac{1}{\sigma_1^2} & \cdot & \cdot & \cdot \\ \cdot & \dfrac{1}{\sigma_2^2} & \cdot & \cdot \\ \cdot & \cdot & \dfrac{1}{\sigma_3^2} & \cdot \\ \cdot & \cdot & \cdot & \dfrac{1}{\sigma_4^2} \end{array}\right]}_{\mathbf{R}^{-1}} \left[\begin{array}{c} z_1 - h_1(\hat{x}_1, \hat{x}_2) \\ z_2 - h_2(\hat{x}_1, \hat{x}_2) \\ z_3 - h_3(\hat{x}_1, \hat{x}_2) \\ z_4 - h_4(\hat{x}_1, \hat{x}_2) \end{array}\right] = \begin{bmatrix} 0 \\ 0 \end{bmatrix} \tag{15.55}$$

Noting that the partial derivative terms depend on x_1 and x_2, let us define

$$\mathbf{H_x} = \begin{bmatrix} \dfrac{\partial h_1}{\partial x_1} & \dfrac{\partial h_1}{\partial x_2} \\ \dfrac{\partial h_2}{\partial x_1} & \dfrac{\partial h_2}{\partial x_2} \\ \dfrac{\partial h_3}{\partial x_1} & \dfrac{\partial h_3}{\partial x_2} \\ \dfrac{\partial h_4}{\partial x_1} & \dfrac{\partial h_4}{\partial x_2} \end{bmatrix} \tag{15.56}$$

and then we can write Eq. (15.55) in the more compact form

$$\mathbf{H}_\mathbf{x}^T\mathbf{R}^{-1}\begin{bmatrix} z_1 - h_1(\hat{x}_1, \hat{x}_2) \\ z_2 - h_2(\hat{x}_1, \hat{x}_2) \\ z_3 - h_3(\hat{x}_1, \hat{x}_2) \\ z_4 - h_4(\hat{x}_1, \hat{x}_2) \end{bmatrix} = \begin{bmatrix} 0 \\ 0 \end{bmatrix} \tag{15.57}$$

To solve this equation for the state estimates $\hat{x}_1$ and $\hat{x}_2$, we follow the same procedure as for the Newton-Raphson power flow. For example, using Eq. (9.31) to linearize $h_1(x_1, x_2)$ about the initial point $(x_1^{(0)}, x_2^{(0)})$ gives

$$h_1(x_1, x_2) = h_1\left(x_1^{(0)}, x_2^{(0)}\right) + \Delta x_1^{(0)} \frac{\partial h_1}{\partial x_1}\bigg|^{(0)} + \Delta x_2^{(0)} \frac{\partial h_1}{\partial x_2}\bigg|^{(0)} \tag{15.58}$$

where, as before, $\Delta x_i^{(0)} = x_i^{(1)} - x_i^{(0)}$ represents the typical state-variable correction and $x_i^{(1)}$ is the first calculated value of x_i. Similarly, expanding $h_2(x_1, x_2)$, $h_3(x_1, x_2)$, and $h_4(x_1, x_2)$; substituting the expansions in Eq. (15.57); and rearranging yield

$$\mathbf{H}_\mathbf{x}^{(0)T}\mathbf{R}^{-1}\begin{bmatrix} z_1 - h_1\left(x_1^{(0)}, x_2^{(0)}\right) \\ z_2 - h_2\left(x_1^{(0)}, x_2^{(0)}\right) \\ z_3 - h_3\left(x_1^{(0)}, x_2^{(0)}\right) \\ z_4 - h_4\left(x_1^{(0)}, x_2^{(0)}\right) \end{bmatrix} = \mathbf{H}_\mathbf{x}^{(0)T}\mathbf{R}^{-1}\mathbf{H}_\mathbf{x}^{(0)}\begin{bmatrix} \Delta x_1^{(0)} \\ \Delta x_2^{(0)} \end{bmatrix} \tag{15.59}$$

in which all quantities with the superscript (0) are computed at the initial values $x_1^{(0)}$ and $x_2^{(0)}$. We require the corrections Δx_1 and Δx_2 to be (approximately) zero in order to satisfy Eq. (15.57), and so the calculations of Eq. (15.59) are continued by using $\Delta x_i^{(k)} = x_i^{(k+1)} - x_i^{(k)}$ to form the more general iterative equation

$$\begin{bmatrix} x_1^{(k+1)} \\ x_2^{(k+1)} \end{bmatrix} - \begin{bmatrix} x_1^{(k)} \\ x_2^{(k)} \end{bmatrix} = \left(\mathbf{H}_\mathbf{x}^T \mathbf{R}^{-1} \mathbf{H}_\mathbf{x}\right)^{-1} \mathbf{H}_\mathbf{x}^T \mathbf{R}^{-1} \begin{bmatrix} z_1 - h_1\left(x_1^{(k)}, x_2^{(k)}\right) \\ z_2 - h_2\left(x_1^{(k)}, x_2^{(k)}\right) \\ z_3 - h_3\left(x_1^{(k)}, x_2^{(k)}\right) \\ z_4 - h_4\left(x_1^{(k)}, x_2^{(k)}\right) \end{bmatrix} \tag{15.60}$$

For convenience the iteration counter k has been omitted from $\mathbf{H}_\mathbf{x}$ in Eq. (15.60). However, at each iteration the elements of the jacobian $\mathbf{H}_\mathbf{x}$ and the quantities $z_j - h_j(x_1^{(k)}, x_2^{(k)})$ are evaluated from the latest available values of the state variables until two successive solutions have converged to within a specified precision index ϵ; that is, until $|x_i^{(k+1)} - x_i^{(k)}| < \epsilon$ for every i.

When there are N_s state variables and a larger number N_m of measurements, the rectangular form of $\mathbf{H}_\mathbf{x}$ for iteration k is given by

$$\mathbf{H}_\mathbf{x} = \begin{bmatrix} \left.\dfrac{\partial h_1}{\partial x_1}\right|^{(k)} & \left.\dfrac{\partial h_1}{\partial x_2}\right|^{(k)} & \cdots & \left.\dfrac{\partial h_1}{\partial x_{N_s}}\right|^{(k)} \\ \left.\dfrac{\partial h_2}{\partial x_1}\right|^{(k)} & \left.\dfrac{\partial h_2}{\partial x_2}\right|^{(k)} & \cdots & \left.\dfrac{\partial h_2}{\partial x_{N_s}}\right|^{(k)} \\ \vdots & \vdots & \ddots & \vdots \\ \left.\dfrac{\partial h_{N_m}}{\partial x_1}\right|^{(k)} & \left.\dfrac{\partial h_{N_m}}{\partial x_2}\right|^{(k)} & \cdots & \left.\dfrac{\partial h_{N_m}}{\partial x_{N_s}}\right|^{(k)} \end{bmatrix} \tag{15.61}$$

and Eq. (15.60) assumes the general iterative form

$$\mathbf{x}^{(k+1)} - \mathbf{x}^{(k)} = \left(\mathbf{H}_\mathbf{x}^T \mathbf{R}^{-1} \mathbf{H}_\mathbf{x}\right)^{-1} \mathbf{H}_\mathbf{x}^T \mathbf{R}^{-1} \begin{bmatrix} z_1 - h_1\left(x_1^{(k)}, x_2^{(k)}, \cdots, x_{N_s}^{(k)}\right) \\ z_2 - h_2\left(x_1^{(k)}, x_2^{(k)}, \cdots, x_{N_s}^{(k)}\right) \\ \vdots \\ z_{N_m} - h_{N_m}\left(x_1^{(k)}, x_2^{(k)}, \cdots, x_{N_s}^{(k)}\right) \end{bmatrix} \tag{15.62}$$

At convergence the solution $\mathbf{x}^{(k+1)}$ corresponds to the weighted least-squares estimates of the state variables, which we denote by

$$\mathbf{x}^{(k+1)} = \hat{\mathbf{x}} = \begin{bmatrix} \hat{x}_1 & \hat{x}_2 & \cdots & \hat{x}_{N_s} \end{bmatrix}^T \tag{15.63}$$

To check the estimates using the same statistical tests as before, we

- Compute the corresponding measurement errors using

$$\hat{e}_j = z_j - h_j\left(\hat{x}_1, \hat{x}_2, \cdots, \hat{x}_{N_s}\right) \tag{15.64}$$

- Evaluate the sum of squares $\hat{f}$ according to Eq. (15.44), and
- Apply the chi-square test of Sec. 15.3 to check for the presence of bad measurements.

To illustrate the procedures for linearizing and iteratively solving the ac power system equations to obtain estimates of the states, let us consider Examples 15.5 and 15.6, which now follow.

Example 15.5. Two voltmeters, two varmeters, and one wattmeter are installed on the system of Fig. 9.3 to measure the following five quantities:

bus ② voltage magnitude: $z_1 = |V_2|$

bus ① voltage magnitude: $z_2 = |V_1|$

bus ① reactive-power injection: $z_3 = Q_1$

line P-flow from bus ① to bus ②: $z_4 = P_{12}$

line Q-flow from bus ② to bus ①: $z_5 = Q_{21}$

as shown in Fig. 15.4 on page 670. Formulate the linearized equations for calculating the weighted least-squares estimates of the system states.

Solution. In this two-bus example we choose $\delta_1 = 0°$ as the reference angle and the three unknown state variables as $x_1 = \delta_2$, $x_2 = |V_2|$, and $x_3 = |V_1|$. The first two state variables are the same as for the power-flow study of Example 9.3. The third state variable $x_3 = |V_1|$ is necessary here (unlike in the power-flow problems) because the magnitude of the voltage at bus ① is uncertain and has to be estimated.

Expressions for the line flows in terms of the state variables are given by Eqs. (9.36) and (9.37) as follows:

$$P_{12} = -P_2 = -4|V_1||V_2|\sin(\delta_2 - \delta_1) = -4x_3x_2 \sin x_1$$

$$Q_{21} = Q_2 = 4\left[|V_2|^2 - |V_1||V_2|\cos(\delta_2 - \delta_1)\right] = 4\left(x_2^2 - x_3x_2 \cos x_1\right)$$

For the reactive power Q_1 injected into the network at bus ① of Fig. 15.4,

Eq. (9.39) leads to

$$Q_1 = |V_1|^2\left(\tfrac{1}{6} + 4\right) - 4|V_1||V_2|\cos(\delta_2 - \delta_1) = \tfrac{25}{6}x_3^2 - 4x_3x_2\cos x_1$$

Note that Q_1 includes the reactive power in the reactance of $j6.0$ per unit from bus ① to neutral. The above equations allow us to write the following expressions for the measurement errors in the kth iteration as

$$e_1^{(k)} = z_1 - h_1\left(x_1^{(k)}, x_2^{(k)}, x_3^{(k)}\right) = z_1 - x_2^{(k)}$$

$$e_2^{(k)} = z_2 - h_2\left(x_1^{(k)}, x_2^{(k)}, x_3^{(k)}\right) = z_2 - x_3^{(k)}$$

$$e_3^{(k)} = z_3 - h_3\left(x_1^{(k)}, x_2^{(k)}, x_3^{(k)}\right) = z_3 - \left(\tfrac{25}{6}x_3^{(k)2} - 4x_3^{(k)}x_2^{(k)}\cos x_1^{(k)}\right)$$

$$e_4^{(k)} = z_4 - h_4\left(x_1^{(k)}, x_2^{(k)}, x_3^{(k)}\right) = z_4 - \left(-4x_3^{(k)}x_2^{(k)}\sin x_1^{(k)}\right)$$

$$e_5^{(k)} = z_5 - h_5\left(x_1^{(k)}, x_2^{(k)}, x_3^{(k)}\right) = z_5 - \left(4x_2^{(k)2} - 4x_3^{(k)}x_2^{(k)}\cos x_1^{(k)}\right)$$

In the same iteration the partial derivatives of h_1, h_2, and h_3 are evaluated according to Eq. (15.61), which now becomes

$$\mathbf{H}_{\mathbf{x}}^{(k)} = \begin{bmatrix} \dfrac{\partial h_1}{\partial x_1}\bigg|^{(k)} & \dfrac{\partial h_1}{\partial x_2}\bigg|^{(k)} & \dfrac{\partial h_1}{\partial x_3}\bigg|^{(k)} \\ \dfrac{\partial h_2}{\partial x_1}\bigg|^{(k)} & \dfrac{\partial h_2}{\partial x_2}\bigg|^{(k)} & \dfrac{\partial h_2}{\partial x_3}\bigg|^{(k)} \\ \dfrac{\partial h_3}{\partial x_1}\bigg|^{(k)} & \dfrac{\partial h_3}{\partial x_2}\bigg|^{(k)} & \dfrac{\partial h_3}{\partial x_3}\bigg|^{(k)} \\ \dfrac{\partial h_4}{\partial x_1}\bigg|^{(k)} & \dfrac{\partial h_4}{\partial x_2}\bigg|^{(k)} & \dfrac{\partial h_4}{\partial x_3}\bigg|^{(k)} \\ \dfrac{\partial h_5}{\partial x_1}\bigg|^{(k)} & \dfrac{\partial h_5}{\partial x_2}\bigg|^{(k)} & \dfrac{\partial h_5}{\partial x_3}\bigg|^{(k)} \end{bmatrix}$$

$$= \begin{bmatrix} 0 & 1 & 0 \\ 0 & 0 & 1 \\ 4x_3^{(k)}x_2^{(k)}\sin x_1^{(k)} & -4x_3^{(k)}\cos x_1^{(k)} & \tfrac{25}{3}x_3^{(k)} - 4x_2^{(k)}\cos x_1^{(k)} \\ -4x_3^{(k)}x_2^{(k)}\cos x_1^{(k)} & -4x_3^{(k)}\sin x_1^{(k)} & -4x_2^{(k)}\sin x_1^{(k)} \\ 4x_3^{(k)}x_2^{(k)}\sin x_1^{(k)} & 8x_2 - 4x_3^{(k)}\cos x_1^{(k)} & -4x_2^{(k)}\cos x_1^{(k)} \end{bmatrix}$$

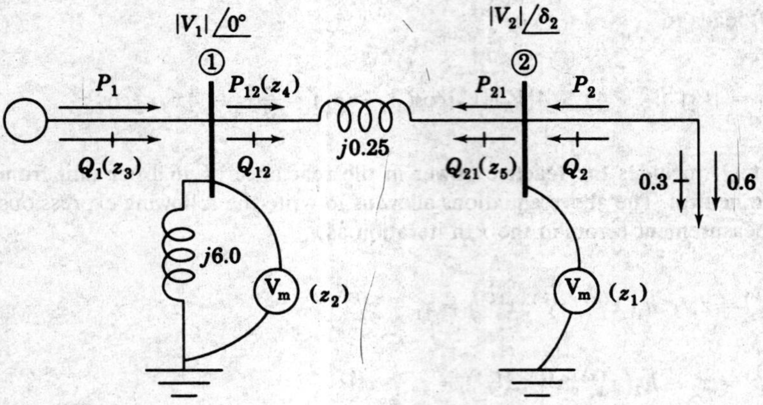

FIGURE 15.4
The one-line diagram of Example 15.5 showing the quantities $|V_2|$, $|V_1|$, Q_1, P_{12}, and Q_{21} constituting the measurement set $\{z_1, z_2, z_3, z_4, z_5\}$.

Based on $\mathbf{H}_{\mathbf{x}}^{(k)}$ and $e_j^{(k)}$, the counterpart of Eq. (15.62) for calculating the state estimates is of the form

$$\begin{bmatrix} x_1^{(k+1)} \\ x_2^{(k+1)} \\ x_3^{(k+1)} \end{bmatrix} - \begin{bmatrix} x_1^{(k)} \\ x_2^{(k)} \\ x_3^{(k)} \end{bmatrix} = \left(\mathbf{H}_{\mathbf{x}}^{(k)T}\mathbf{R}^{-1}\mathbf{H}_{\mathbf{x}}^{(k)}\right)^{-1}\mathbf{H}_{\mathbf{x}}^{(k)T}\mathbf{R}^{-1}\begin{bmatrix} e_1^{(k)} \\ e_2^{(k)} \\ e_3^{(k)} \\ e_4^{(k)} \\ e_5^{(k)} \end{bmatrix}$$

with the weighting matrix of reciprocal variances given by

$$\mathbf{R}^{-1} = \begin{bmatrix} \frac{1}{\sigma_1^2} & \cdot & \cdot & \cdot & \cdot \\ \cdot & \frac{1}{\sigma_2^2} & \cdot & \cdot & \cdot \\ \cdot & \cdot & \frac{1}{\sigma_3^2} & \cdot & \cdot \\ \cdot & \cdot & \cdot & \frac{1}{\sigma_4^2} & \cdot \\ \cdot & \cdot & \cdot & \cdot & \frac{1}{\sigma_5^2} \end{bmatrix}$$

The equations developed in the preceding example are important illustrations of the general procedure and form the basis of the numerical calculations in Examples 15.6 and 15.7 for estimating the system state.

Example 15.6. Telemetered measurements on the physical system corresponding to Fig. 15.4 show that the per-unit values of the five measured quantities are

$$z_1 = |V_2| = 0.92 \qquad z_2 = |V_1| = 1.02$$

$$z_3 = Q_1 = 0.605 \qquad z_4 = P_{12} = 0.598$$

$$z_5 = Q_{21} = 0.305$$

The variances of the measurement errors are specified in per unit as

$$\sigma_1^2 = \sigma_2^2 = (0.01)^2 \qquad \sigma_3^2 = \sigma_5^2 = (0.02)^2 \qquad \sigma_4^2 = (0.015)^2$$

Using the equations of Example 15.5, compute the weighted least-squares estimates of the state variables $x_1 = \delta_2$, $x_2 = |V_2|$, and $x_3 = |V_1|$ of Fig. 15.4. Check the results using the statistical tests of Sec. 15.3.

Solution. We begin the iterative calculations by using flat-start values $x_1^{(0)} = 0°$ and $x_2^{(0)} = x_3^{(0)} = 1.0$ per unit for the three state variables which yield $\sin x_1^{(0)} = 0$ and $\cos x_1^{(0)} = 1$. Substituting these numerical values in the expressions of Example 15.5 gives the measurement errors of the first iteration:

$$e_1^{(0)} = z_1 - x_2^{(0)} = 0.92 - 1.00 = -0.08$$

$$e_2^{(0)} = z_2 - x_3^{(0)} = 1.02 - 1.00 = 0.02$$

$$e_3^{(0)} = z_3 - \left(\tfrac{25}{6} x_3^{(0)2} - 4x_3^{(0)} x_2^{(0)} \cos x_1^{(0)}\right) = 0.605 - \left(\tfrac{1}{6}\right) = 0.4383$$

$$e_4^{(0)} = z_4 - \left(-4x_3^{(0)} x_2^{(0)} \sin x_1^{(0)}\right) = 0.598 - (0.0) = 0.5980$$

$$e_5^{(0)} = z_5 - \left(4x_2^{(0)2} - 4x_3^{(0)} x_2^{(0)} \cos x_1^{(0)}\right) = 0.305 - (0.0) = 0.3050$$

Similarly, the numerical form of the jacobian becomes

$$\mathbf{H}_{\mathbf{x}}^{(0)} = \begin{bmatrix} 0 & 1 & 0 \\ 0 & 0 & 1 \\ 0 & -4 & \frac{13}{3} \\ -4 & 0 & 0 \\ 0 & 4 & -4 \end{bmatrix}$$

The gain matrix $\mathbf{G}_{\mathbf{x}}^{(0)} = \mathbf{H}_{\mathbf{x}}^{(0)T}\mathbf{R}^{-1}\mathbf{H}_{\mathbf{x}}^{(0)}$ is now evaluated as follows:

$$\mathbf{G}_{\mathbf{x}}^{(0)} = \begin{bmatrix} 0 & 0 & 0 & -4 & 0 \\ 1 & 0 & -4 & 0 & 4 \\ 0 & 1 & \frac{13}{3} & 0 & -4 \end{bmatrix}$$

$$\times \begin{bmatrix} \frac{1}{(0.01)^2} & \cdot & \cdot & \cdot & \cdot \\ \cdot & \frac{1}{(0.01)^2} & \cdot & \cdot & \cdot \\ \cdot & \cdot & \frac{1}{(0.02)^2} & \cdot & \cdot \\ \cdot & \cdot & \cdot & \frac{1}{(0.015)^2} & \cdot \\ \cdot & \cdot & \cdot & \cdot & \frac{1}{(0.02)^2} \end{bmatrix} \mathbf{H}_{\mathbf{x}}^{(0)}$$

$$= 10^4 \times \begin{bmatrix} 0 & 0 & 0 & -1.7778 & 0 \\ 1 & 0 & -1 & 0 & 1 \\ 0 & 1 & 1.0833 & 0 & -1 \end{bmatrix} \begin{bmatrix} 0 & 1 & 0 \\ 0 & 0 & 1 \\ 0 & -4 & \frac{13}{3} \\ -4 & 0 & 0 \\ 0 & 4 & -4 \end{bmatrix}$$

$$= \begin{bmatrix} 7.1111 & 0 & 0 \\ 0 & 9.0000 & -8.3333 \\ 0 & -8.3333 & 9.6944 \end{bmatrix} \times 10^4$$

The numerical form of Eq. (15.62) for the first iteration becomes

$$\begin{bmatrix} x_1^{(1)} \\ x_2^{(1)} \\ x_3^{(1)} \end{bmatrix} = \begin{bmatrix} x_1^{(0)} \\ x_2^{(0)} \\ x_3^{(0)} \end{bmatrix} + \mathbf{G}_{\mathbf{x}}^{(0)^{-1}}\mathbf{H}_{\mathbf{x}}^{(0)T}\mathbf{R}^{-1} \begin{bmatrix} e_1^{(0)} \\ e_2^{(0)} \\ e_3^{(0)} \\ e_4^{(0)} \\ e_5^{(0)} \end{bmatrix}$$

$$= \begin{bmatrix} 0 \\ 1 \\ 1 \end{bmatrix} + \mathbf{G}_{\mathbf{x}}^{(0)^{-1}} \times 10^4 \begin{bmatrix} 0 & 0 & 0 & -1.7778 & 0 \\ 1 & 0 & -1 & 0 & 1 \\ 0 & 1 & 1.0833 & 0 & -1 \end{bmatrix} \begin{bmatrix} -0.0800 \\ 0.0200 \\ 0.4383 \\ 0.5980 \\ 0.3050 \end{bmatrix}$$

$$= \begin{bmatrix} 0 \\ 1 \\ 1 \end{bmatrix} + \begin{bmatrix} 7.1111 & 0 & 0 \\ 0 & & -8.3333 \\ 0 & -8.3333 & 9.6944 \end{bmatrix}^{-1} \begin{bmatrix} -1.0631 \\ -0.2133 \\ 0.1898 \end{bmatrix}$$

$$= \begin{bmatrix} -0.1495 \\ 0.9727 \\ 0.9961 \end{bmatrix}$$

These values of the state variables act as input data to the second iteration and computations continue until convergence is reached. The converged values of the x_i's are chosen as the state-variable estimates $\hat{x}_i$ with

$$\hat{x}_1 = \delta_2 = -0.1762 \text{ rad} = -10.0955°$$

$$\hat{x}_2 = |V_2| = 0.9578 \text{ per unit}$$

$$\hat{x}_3 = |V_1| = 0.9843 \text{ per unit}$$

The corresponding estimates of the measurement errors in per unit are

$$\hat{e}_1 = z_1 - \hat{x}_2 = 0.92 - 0.9578 = -0.0378$$

$$\hat{e}_2 = z_2 - \hat{x}_3 = 1.02 - 0.9843 = 0.0357$$

$$\hat{e}_3 = z_3 - \left(\tfrac{25}{6}\hat{x}_3^2 - 4\hat{x}_3\hat{x}_2 \cos \hat{x}_1\right) = 0.605 - 0.3240 = 0.2810$$

$$\hat{e}_4 = z_4 - \left(-4\hat{x}_3\hat{x}_2 \sin \hat{x}_1\right) = 0.598 - (0.6610) = -0.0630$$

$$\hat{e}_5 = z_5 - \left(4\hat{x}_2^2 - 4\hat{x}_3\hat{x}_2 \cos \hat{x}_1\right) = 0.305 - (-0.0430) = 0.3480$$

and the weighted sum of squares of these errors becomes

$$\hat{f} = \sum_{j=1}^{5} \left(\frac{\hat{e}_j}{\sigma_j}\right)^2 = \frac{(-0.0378)^2}{(0.01)^2} + \frac{(0.0357)^2}{(0.01)^2} + \frac{(0.2810)^2}{(0.02)^2}$$

$$+ \frac{(-0.0630)^2}{(0.015)^2} + \frac{(0.3480)^2}{(0.02)^2} = 545$$

If $\hat{f}$ is chi-square distributed, the number of degrees of freedom k equals 2 since there are five measurements and three state variables to be estimated. For a 99% confidence interval ($\alpha = 0.01$) the requirement $\hat{f} \leq \chi^2_{k,\alpha}$ must be satisfied. In Table 15.3 for $k = 2$ and $\alpha = 0.01$, we find that $\chi^2_{2,0.01} = 9.21$, and it is now clear that the calculated value $\hat{f} = 545$ is excessive, indicating the presence of bad data. Consequently, we cannot accept the calculated values of the state variables as being accurate. If there is only one bad measurement, we can follow the procedure of Sec. 15.3 in an attempt to identify it. But if there are two or more gross errors in the measurement set, our procedures break down since $N_m - N_s = 5 - 3 = 2$ and redundancy is lost when two or more measurements are discarded. The calculations underlying the identification procedure are provided in Example 15.7.

Example 15.7. Identify the bad data detected in the measurement set of Example 15.6. Discard the erroneous data, and if sufficient good/measurements remain, calculate the new system state.

Solution. In an effort to identify the bad measurements let us compute the standardized residuals in the given measurement set.

To do so, we first determine the diagonal elements of the covariance matrix $\mathbf{R}' = \mathbf{R} - \mathbf{H}_x\mathbf{G}_x^{-1}\mathbf{H}_x^T = (\mathbf{I} - \mathbf{H}_x\mathbf{G}_x^{-1}\mathbf{H}_x^T\mathbf{R}^{-1})\mathbf{R}$ using the estimates $\hat{x}_i$ of Example 15.6 in the following sequence of calculations:

$$\mathbf{H}_x = \begin{bmatrix} 0 & 1 & 0 \\ 0 & 0 & 1 \\ 4\hat{x}_3\hat{x}_2 \sin \hat{x}_1 & -4\hat{x}_3 \cos \hat{x}_1 & \frac{25}{3}\hat{x}_3 - 4\hat{x}_2 \cos \hat{x}_1 \\ -4\hat{x}_3\hat{x}_2 \cos \hat{x}_1 & -4\hat{x}_3 \sin \hat{x}_1 & -4\hat{x}_2 \sin \hat{x}_1 \\ 4\hat{x}_3\hat{x}_2 \sin \hat{x}_1 & 8\hat{x}_2 - 4\hat{x}_3 \cos \hat{x}_1 & -4\hat{x}_2 \cos \hat{x}_1 \end{bmatrix}$$

$$= \begin{bmatrix} 0 & 1 & 0 \\ 0 & 0 & 1 \\ -0.6610 & -3.8761 & 4.4303 \\ -3.7125 & 0.6901 & 0.6716 \\ -0.6610 & 3.7863 & -3.7719 \end{bmatrix}$$

Based on $\mathbf{R}^{-1}$ of Example 15.6, the corresponding numerical elements of the gain matrix are

$$\mathbf{G}_x = \mathbf{H}_x^T\mathbf{R}^{-1}\mathbf{H}_x = \begin{bmatrix} 6.3442 & -1.1239 & -1.2169 \\ -1.1239 & 8.5518 & -7.6575 \\ -1.2169 & -7.6575 & 9.6641 \end{bmatrix} \times 10^4$$

and the reader can now determine the diagonal elements of the matrix

$$\mathbf{H}_x\mathbf{G}_x^{-1}\mathbf{H}_x^T\mathbf{R}^{-1} = \begin{bmatrix} 0.5618 & \times & \times & \times & \times \\ \times & 0.4976 & \times & \times & \times \\ \times & \times & 0.5307 & \times & \times \\ \times & \times & \times & 0.9656 & \times \\ \times & \times & \times & \times & 0.4443 \end{bmatrix}$$

which sum to $N_m = 3$. Diagonal entries in the covariant matrix $\mathbf{R}'$ are readily computed as follows:

$$\mathbf{R}' = \left(\mathbf{I} - \mathbf{H_x G_x^{-1} H_x^T R^{-1}}\right)\mathbf{R}$$

$$= \begin{bmatrix} 0.4382 & \cdot & \cdot & \cdot & \cdot \\ \cdot & 0.5024 & \cdot & \cdot & \cdot \\ \cdot & \cdot & 0.4693 & \cdot & \cdot \\ \cdot & \cdot & \cdot & 0.0344 & \cdot \\ \cdot & \cdot & \cdot & \cdot & 0.5557 \end{bmatrix}$$

$$\times \begin{bmatrix} 0.01^2 & \cdot & \cdot & \cdot & \cdot \\ \cdot & 0.01^2 & \cdot & \cdot & \cdot \\ \cdot & \cdot & 0.02^2 & \cdot & \cdot \\ \cdot & \cdot & \cdot & 0.015^2 & \cdot \\ \cdot & \cdot & \cdot & \cdot & 0.02^2 \end{bmatrix}$$

$$= \begin{bmatrix} 0.4382 & \cdot & \cdot & \cdot & \cdot \\ \cdot & 0.5024 & \cdot & \cdot & \cdot \\ \cdot & \cdot & 1.8772 & \cdot & \cdot \\ \cdot & \cdot & \cdot & 0.0773 & \cdot \\ \cdot & \cdot & \cdot & \cdot & 2.2229 \end{bmatrix} \times 10^{-4}$$

The calculated $\hat{e}_j$ of Example 15.6 correspond to standardized errors

$$\frac{\hat{e}_1}{\sqrt{R'_{11}}} = \frac{-0.0378}{\sqrt{0.4382 \times 10^{-4}}} = -5.7106$$

$$\frac{\hat{e}_2}{\sqrt{R'_{22}}} = \frac{0.0357}{\sqrt{0.5024 \times 10^{-4}}} = 5.0419$$

$$\frac{\hat{e}_3}{\sqrt{R'_{33}}} = \frac{0.2810}{\sqrt{1.8772 \times 10^{-4}}} = 20.5079$$

$$\frac{\hat{e}_4}{\sqrt{R'_{44}}} = \frac{-0.0630}{\sqrt{0.0773 \times 10^{-4}}} = -22.6559$$

$$\frac{\hat{e}_5}{\sqrt{R'_{55}}} = \frac{0.3480}{\sqrt{2.2229 \times 10^{-4}}} = 23.3403$$

We note that large standardized errors are associated with measurements z_3, z_4, and z_5. In order to preserve redundancy, only one measurement can be discarded, which we choose to be z_5 corresponding to the largest standardized error. From

z_1, z_2, z_3, and z_4 in the original measurement set and the flat-start values $x_1^{(0)} = 0$, $x_2^{(0)} = x_3^{(0)} = 1.0$ per unit we repeat the state-estimation calculations, which yield

$$\hat{x}_1 = \delta_2 = -0.1600 \text{ rad} = -9.1673^\circ$$

$$\hat{x}_2 = |V_2| = 0.9223 \text{ per unit}$$

$$\hat{x}_3 = |V_1| = 1.0174 \text{ per unit}$$

The reader can verify the following numerical values of the corresponding jacobian $\mathbf{H}_\mathbf{x}$ and gain matrix $\mathbf{G}_\mathbf{x}$:

$$\mathbf{H}_\mathbf{x} = \begin{bmatrix} 0 & 1 & 0 \\ 0 & 0 & 1 \\ -0.5978 & -4.0175 & 4.8358 \\ -3.7053 & 0.6482 & 0.5876 \end{bmatrix}$$

$$\mathbf{G}_\mathbf{x} = \begin{bmatrix} 6.1913 & -0.4670 & -1.6904 \\ -0.4670 & 5.2217 & -4.6877 \\ -1.6904 & -4.6877 & 6.9998 \end{bmatrix} \times 10^{-4}$$

The new estimates of the measurement errors are found to be

$$\hat{e}_1 = z_1 - \hat{x}_2 = 0.92 - 0.9223 = -0.0023$$

$$\hat{e}_2 = z_2 - \hat{x}_3 = 1.02 - 1.0174 = 0.0026$$

$$\hat{e}_3 = z_3 - \left(\tfrac{25}{6}\hat{x}_3^2 - 4\hat{x}_3\hat{x}_2 \cos \hat{x}_1\right) = 0.605 - 0.6072 = -0.0022$$

$$\hat{e}_4 = z_4 - \left(-4\hat{x}_3\hat{x}_2 \sin \hat{x}_1\right) = 0.509 - 0.5978 = 0.0002$$

The weighted sum of squares (with one degree of freedom) amounts to

$$\hat{f} = \left(\frac{-.0023}{0.01}\right)^2 + \left(\frac{0.0026}{0.01}\right)^2 + \left(\frac{-.0022}{0.02}\right)^2 + \left(\frac{0.0002}{0.015}\right)^2 = 0.1355$$

which is certainly less than $\chi^2_{1,0.01} = 6.64$ of Table 15.3. Consequently, the estimates $\hat{x}_1$, $\hat{x}_2$, and $\hat{x}_3$ of this example can be accepted with 99% confidence as reasonable values of the state variables. We conclude, therefore, that z_5 is the only bad data point in the original measurement set.

Bad data can originate from many causes. One such cause arises when a meter is improperly connected with the leads reversed so that it reads upscale (positive) rather than downscale (negative). For instance, in the original data set of Example 15.6 z_5 is incorrectly reported to be equal to 0.305 per unit. If we change the sign of this measurement by setting $z_5 = -0.305$ per unit in the original data and repeat the state-estimation calculations using all five measurements, we obtain

$$\hat{x}_1 = \delta_2 = -0.1599 \text{ rad} = -9.1616^\circ$$

$$\hat{x}_2 = |V_2| = 0.9222 \text{ per unit}$$

$$\hat{x}_3 = |V_1| = 1.0175 \text{ per unit}$$

which are acceptable results with the sum of squares $\hat{f} = 0.1421$. Thus, state estimation can sometimes reveal when properly functioning meters have been installed backward on the physical system.

15.5 THE STRUCTURE AND FORMATION OF $\mathbf{H_x}$

The phase angles of the voltages at different substations of the system cannot be economically measured, but voltage magnitudes are routinely monitored. In order to estimate both voltage angles and magnitudes, we must choose the angle at one of the N buses of the system as reference for all other angles, which leaves $N - 1$ angles and N magnitudes to be calculated by Eq. (15.62). Hence, the state-estimation jacobian $\mathbf{H_x}$, unlike the square jacobian $\mathbf{J}$ of conventional Newton-Raphson power flow, always has $(2N - 1)$ columns and a larger number N_m of rows, as shown in Eq. (15.61). Each row of $\mathbf{H_x}$ corresponds uniquely to one of the measured quantities indicated in the transmission-line equivalent circuit of Fig. 15.5, namely,

- The voltage magnitude $|V_i|$ at a typical bus ⓘ,
- The active power P_i injected into the network at bus ⓘ,
- The reactive power Q_i injected into the network at bus ⓘ,
- The active power flow P_{ij} at bus ⓘ or P_{ji} at bus ⓙ in the line connecting buses ⓘ and ⓙ, and
- The reactive power flow Q_{ij} at bus ⓘ or Q_{ji} at bus ⓙ in the line connecting buses ⓘ and ⓙ.

Although line-charging capacitance is shown in Fig. 15.5, line-charging megavars

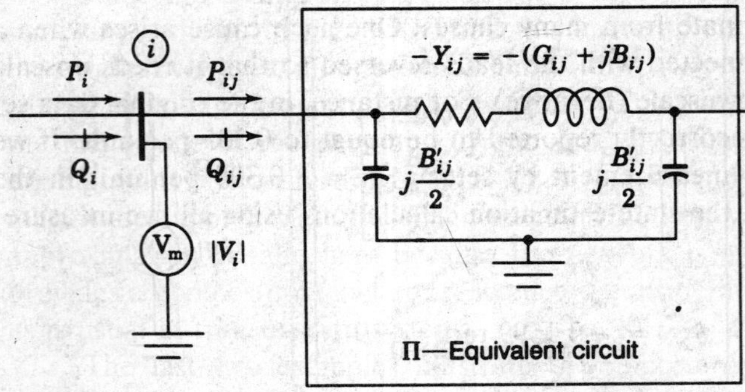

FIGURE 15.5
The per-phase representation of the transmission line from bus ⓘ to bus ⓙ showing the bus injections, voltage magnitudes, and line flows which can be measured. Line charging is represented by susceptance B'_{ij} and Y_{ij} is the element of $\mathbf{Y}_{\text{bus}}$ equal to the negative of the series admittance of the line.

cannot be separately measured since they are distributed along the length of the line and form part of the Q_{ij} and Q_{ji} measurements. For simplicity, other shunt-connected reactors or capacitors are not shown, but it should be understood that P and Q in any lumped external connections from the typical bus ⓘ to neutral can be measured and included as part of the net injections, P_i and Q_i. This is demonstrated in Example 15.5 where the injection Q_1 includes the reactive power flow in the reactance from bus ① to neutral.

If $|V_i|$, P_i, and Q_i are measured at every bus and the flows P_{ij}, P_{ji}, Q_{ij}, and Q_{ji} are measured in every line, the measurement set is *full* and $\mathbf{H}_\mathbf{x}$ has a total of $N_m = (3N + 4B)$ rows, where B is the number of network branches or lines. The ratio of rows to columns in $\mathbf{H}_\mathbf{x}$ (which is equivalent to the ratio of measurements to state variables) is called the *redundancy factor*, and it equals $(3N + 4B)/(2N - 1)$, or approximately 4.5 in a large fully monitored power system with an average of 1.5 branches per bus; that is, with $B/N = 1.5$. In practice, the redundancy factor can be somewhat smaller since the measurement set may be less than full while the number of state variables remains at $2N - 1$. When the number of measurements equals the number of state variables, there is no redundancy, $\mathbf{H}_\mathbf{x}$ is square, and the coefficient matrix $(\mathbf{H}_\mathbf{x}^T\mathbf{R}^{-1}\mathbf{H}_\mathbf{x})^{-1}\mathbf{H}_\mathbf{x}^T\mathbf{R}^{-1}$ of Eq. (15.62) reduces to $\mathbf{H}_\mathbf{x}^{-1}$ if $\mathbf{H}_\mathbf{x}$ is also invertible. In that case the weighting matrix $\mathbf{R}^{-1}$ has no effect; all the raw measurements must be used directly (even if erroneous); and if any one measurement is missing, the system cannot be solved. Thus, an adequate number of redundant measurements from strategic points of the system is absolutely essential to ensure a good state estimate, and this makes the jacobian rectangular in the state estimator.

When the full set of measurements is numbered so that those of the same type are grouped together, the measurement vector takes on the form

$$\mathbf{z} = \left[\begin{array}{c} z_1 \\ \vdots \\ z_N \\ \hline z_{N+1} \\ \vdots \\ z_{2N} \\ \hline z_{2N+1} \\ \vdots \\ z_{3N} \\ \hline z_{3N+1} \\ \vdots \\ z_{3N+B} \\ \hline z_{3N+B+1} \\ \vdots \\ z_{3N+2B} \\ \hline z_{3N+2B+1} \\ \vdots \\ z_{3N+3B} \\ \hline z_{3N+3B+1} \\ \vdots \\ z_{3N+4B} \end{array}\right] = \left[\begin{array}{c} \left.\begin{array}{c} |V_1| \\ \vdots \\ |V_N| \end{array}\right\} N \text{ voltage magnitudes } |V_i| \\ \hline \left.\begin{array}{c} P_1 \\ \vdots \\ P_N \end{array}\right\} N \text{ bus injections } P_i \\ \hline \left.\begin{array}{c} Q_1 \\ \vdots \\ Q_N \end{array}\right\} N \text{ bus injections } Q_i \\ \hline \left.\begin{array}{c} \vdots \\ P_{ij} \\ \vdots \end{array}\right\} B \text{ line flows } P_{ij} \\ \hline \left.\begin{array}{c} \vdots \\ P_{ji} \\ \vdots \end{array}\right\} B \text{ line flows } P_{ji} \\ \hline \left.\begin{array}{c} \vdots \\ Q_{ij} \\ \vdots \end{array}\right\} B \text{ line flows } Q_{ij} \\ \hline \left.\begin{array}{c} \vdots \\ Q_{ji} \\ \vdots \end{array}\right\} B \text{ line flows } Q_{ji} \end{array}\right] \tag{15.65}$$

and the corresponding jacobian $\mathbf{H}_x$ has the symbolic block form shown in Table 15.4 on page 681.

Note that $\mathbf{H}_x$ has a row for P_1 and for Q_1 and a column corresponding to $|V_1|$ but not for δ_1. This is because in state-estimation calculations a reference angle must be specified at one of the buses which we have chosen to be bus ①, as in Example 15.5. The voltage magnitude $|V_1|$ does *not* have to be specified, which makes bus ① different from the slack bus of the power-flow problem. The zero elements in block 1 of Table 15.4 result from the fact that the measurement $|V_i|$ does not explicitly depend on δ_j so that $\partial|V_i|/\partial\delta_j$ equals zero for all values of i and j. Also, for a full measurement set the diagonal elements of block 2 are unity because $\partial|V_i|/\partial|V_j| = 1$ when i equals j and otherwise is zero.

The real power injected into the system at bus ⓘ is given by Eq. (9.38), which is repeated here as

$$P_i = |V_i|^2 G_{ii} + \sum_{\substack{n=1\\ n\neq i}}^{N} |V_i V_n V_{in}|\cos(\theta_{in} + \delta_n - \delta_i) \tag{15.66}$$

and likewise, Eq. (9.39) expresses the reactive power injected into bus ⓘ

$$Q_i = -\left\{|V_i|^2 B_{ii} + \sum_{\substack{n=1\\ n\neq i}}^{N} |V_i V_n Y_{in}|\sin(\theta_{in} + \delta_n - \delta_i)\right\} \tag{15.67}$$

The injection P_i includes all the active power flows from bus ⓘ to other buses of the system so that for the particular line ⓘ-ⓙ the active power flow P_{ij} from bus ⓘ to bus ⓙ is given by

$$P_{ij} = -|V_i|^2 G_{ij} + |V_i V_j Y_{ij}|\cos(\theta_{ij} + \delta_j - \delta_i) \tag{15.68}$$

The first term $-|V_i|^2 G_{ij}$ is necessary because line ⓘ-ⓙ contributes the series conductance $-G_{ij}$ to the calculation of the self-conductance G_{ii} of Eq. (15.66). Likewise, the reactive power flow from bus ⓘ to bus ⓙ on line ⓘ-ⓙ can be written from Eq. (15.67) as

$$Q_{ij} = -\left\{|V_i|^2\left(\frac{B'_{ij}}{2} - B_{ij}\right) + |V_i V_j Y_{ij}|\sin(\theta_{ij} + \delta_j - \delta_i)\right\} \tag{15.69}$$

where $B'_{ij}/2$ is the line-charging susceptance and $-B_{ij}$ is the series susceptance contributed by line ⓘ-ⓙ to the self-susceptance B_{ii}. The expressions for P_{ji} and Q_{ji} are obtained from Eqs. (15.68) and (15.69) simply by interchanging

TABLE 15.4
Typical elements in the block form of state-estimation jacobian $\mathbf{H_x}$

$$
\begin{array}{c}
\\ \\ \\ \\ \\ \\ \\ \\ \\ \\ \\ \\ \\ \\ \mathbf{H_x} = \\ \\ \\ \\ \\ \\ \\ \\ \\ \\ \\ \\ \\
\end{array}
\left[
\begin{array}{cccc|cccc}
\textcircled{\delta_2} & \textcircled{\delta_3} & \cdots & \textcircled{\delta_N} & \textcircled{|V_1|} & \textcircled{|V_2|} & \cdots & \textcircled{|V_N|} \\
0 & & & & 1 & & & \\
 & 0 & & & & 1 & & \\
 & & \ddots & & & & \ddots & \\
 & & & 0 & & & & 1 \\
 & & & \boxed{1} & \boxed{2} & & & \\
\hline
\frac{\partial P_1}{\partial \delta_2} & \cdots & & \frac{\partial P_1}{\partial \delta_N} & \frac{\partial P_1}{\partial |V_1|} & \cdots & & \frac{\partial P_1}{\partial |V_N|} \\
\vdots & & & \vdots & \vdots & & & \vdots \\
\frac{\partial P_N}{\partial \delta_2} & \cdots & & \frac{\partial P_N}{\partial \delta_N} & \frac{\partial P_N}{\partial |V_1|} & \cdots & & \frac{\partial P_N}{\partial |V_N|} \\
 & & & \boxed{3} & \boxed{4} & & & \\
\hline
\frac{\partial Q_1}{\partial \delta_2} & \cdots & & \frac{\partial Q_1}{\partial \delta_N} & \frac{\partial Q_1}{\partial |V_1|} & \cdots & & \frac{\partial Q_1}{\partial |V_N|} \\
\vdots & & & \vdots & \vdots & & & \vdots \\
\frac{\partial Q_N}{\partial \delta_2} & \cdots & & \frac{\partial Q_N}{\partial \delta_N} & \frac{\partial Q_N}{\partial |V_1|} & \cdots & & \frac{\partial Q_N}{\partial |V_N|} \\
 & & & \boxed{5} & \boxed{6} & & & \\
\hline
\vdots & & \vdots & & \vdots & & \vdots & \\
\cdots\ \frac{\partial P_{ij}}{\partial \delta_i} & \cdots & \frac{\partial P_{ij}}{\partial \delta_j}\ \cdots & & \cdots\ \frac{\partial P_{ij}}{\partial |V_i|} & \cdots & \frac{\partial P_{ij}}{\partial |V_j|}\ \cdots & \\
\vdots & & \vdots & \boxed{7} & \boxed{8} & \vdots & \vdots & \\
\hline
\vdots & & \vdots & & \vdots & & \vdots & \\
\cdots\ \frac{\partial P_{ji}}{\partial \delta_i} & \cdots & \frac{\partial P_{ji}}{\partial \delta_j}\ \cdots & & \cdots\ \frac{\partial P_{ji}}{\partial |V_i|} & \cdots & \frac{\partial P_{ji}}{\partial |V_j|}\ \cdots & \\
\vdots & & \vdots & \boxed{9} & \boxed{10} & \vdots & \vdots & \\
\hline
\vdots & & \vdots & & \vdots & & \vdots & \\
\cdots\ \frac{\partial Q_{ij}}{\partial \delta_i} & \cdots & \frac{\partial Q_{ij}}{\partial \delta_j}\ \cdots & & \cdots\ \frac{\partial Q_{ij}}{\partial |V_i|} & \cdots & \frac{\partial Q_{ij}}{\partial |V_j|}\ \cdots & \\
\vdots & & \vdots & \boxed{11} & \boxed{12} & \vdots & \vdots & \\
\hline
\vdots & & \vdots & & \vdots & & \vdots & \\
\cdots\ \frac{\partial Q_{ji}}{\partial \delta_i} & \cdots & \frac{\partial Q_{ji}}{\partial \delta_j}\ \cdots & & \cdots\ \frac{\partial Q_{ji}}{\partial |V_i|} & \cdots & \frac{\partial Q_{ji}}{\partial |V_j|}\ \cdots & \\
\vdots & & \vdots & \boxed{13} & \boxed{14} & \vdots & \vdots & \\
\end{array}
\right]
\begin{array}{l}
\\ \\ \left.\right\} N \text{ voltage magnitudes } |V_i| \\ \\ \\ \\ \left.\right\} N \text{ bus injections } P_i \\ \\ \\ \\ \left.\right\} N \text{ bus injections } Q_i \\ \\ \\ \\ \left.\right\} B \text{ line flows } P_{ij} \\ \\ \\ \left.\right\} B \text{ line flows } P_{ji} \\ \\ \\ \left.\right\} B \text{ line flows } Q_{ij} \\ \\ \\ \left.\right\} B \text{ line flows } Q_{ji} \\ \\
\end{array}
$$

$$\underbrace{\qquad\qquad\qquad\qquad}_{N-1}\qquad\underbrace{\qquad\qquad\qquad\qquad}_{N}$$

TABLE 15.5
Formulas for elements of the state-estimation jacobian $\mathbf{H_x}$

$\frac{\partial \lvert V_i \rvert}{\partial \delta_j} = 0$ all i and j	$\frac{\partial \lvert V_i \rvert}{\partial \lvert V_j \rvert} = 0 (i \neq j)$; $\frac{\partial \lvert V_i \rvert}{\partial \lvert V_i \rvert} = 1$
1	2
$\frac{\partial P_i}{\partial \delta_j} = \lvert -V_iV_jY_{ij} \rvert \sin(\theta_{ij} + \delta_j - \delta_i)$	$\frac{\partial P_i}{\partial \lvert V_j \rvert} = \lvert V_iY_{ij} \rvert \cos(\theta_{ij} + \delta_j - \delta_i)$
$\frac{\partial P_i}{\partial \delta_i} = \sum_{\substack{n=1 \\ n \neq i}}^{N} \lvert V_iV_nY_{in} \rvert \sin(\theta_{in} + \delta_n - \delta_i)$	$\frac{\partial P_i}{\partial \lvert V_i \rvert} = 2\lvert V_i \rvert G_{ii} + \sum_{\substack{n=1 \\ n \neq i}}^{N} \lvert V_nY_{in} \rvert \cos(\theta_{in} + \delta_n - \delta_i)$
3	4
$\frac{\partial Q_i}{\partial \delta_j} = -\lvert V_iV_jY_{ij} \rvert \cos(\theta_{ij} + \delta_j - \delta_i)$	$\frac{\partial Q_i}{\partial \lvert V_j \rvert} = -\lvert V_iY_{ij} \rvert \sin(\theta_{ij} + \delta_j - \delta_i)$
$\frac{\partial Q_i}{\partial \delta_i} = \sum_{\substack{n=1 \\ n \neq i}}^{N} \lvert V_iV_nY_{in} \rvert \cos(\theta_{in} + \delta_n - \delta_i)$	$\frac{\partial}{\partial \lvert V_i \rvert} = -2\lvert V_i \rvert B_{ii} - \sum_{\substack{n=1 \\ n \neq i}}^{N} \lvert V_nY_{in} \rvert \sin(\theta_{in} + \delta_n - \delta_i)$
5	6
$\frac{\partial P_{ij}}{\partial \delta_j} = -\lvert V_iV_jY_{ij} \rvert \sin(\theta_{ij} + \delta_j - \delta_i)$	$\frac{\partial P_{ij}}{\partial \lvert V_j \rvert} = \lvert V_iY_{ij} \rvert \cos(\theta_{ij} + \delta_j - \delta_i)$
$\frac{\partial P_{ij}}{\partial \delta_i} = \lvert V_iV_jY_{ij} \rvert \sin(\theta_{ij} + \delta_j - \delta_i)$	$\frac{\partial P_{ij}}{\partial \lvert V_i \rvert} = 2\lvert V_i \rvert G_{ij} + \lvert V_jY_{ij} \rvert \cos(\theta_{ij} + \delta_j - \delta_i)$
7	8
$\frac{\partial P_{ji}}{\partial \delta_j} = \lvert V_iV_jY_{ij} \rvert \sin(\theta_{ij} + \delta_i - \delta_j)$	$\frac{\partial P_{ji}}{\partial \lvert V_j \rvert} = 2\lvert V_j \rvert G_{ij} + \lvert V_iY_{ij} \rvert \cos(\theta_{ij} + \delta_i - \delta_j)$
$\frac{\partial P_{ji}}{\partial \delta_i} = -\lvert V_iV_jY_{ij} \rvert \sin(\theta_{ij} + \delta_i - \delta_j)$	$\frac{\partial P_{ji}}{\partial \lvert V_i \rvert} = \lvert V_jY_{ij} \rvert \cos(\theta_{ij} + \delta_i - \delta_j)$
9	10
$\frac{\partial_j}{\partial \delta_j} = -\lvert V_iV_jY_{ij} \rvert \cos(\theta_{ij} + \delta_j - \delta_i)$	$\frac{\partial_j}{\partial \lvert V_j \rvert} = -\lvert V_iY_{ij} \rvert \sin(\theta_{ij} + \delta_j - \delta_i)$
$\frac{\partial_j}{\partial \delta_i} = \lvert V_iV_jY_{ij} \rvert \cos(\theta_{ij} + \delta_j - \delta_i)$	$\frac{\partial_j}{\partial \lvert V_i \rvert} = -2\lvert V_i \rvert \left(\frac{B'_{ij}}{2} + B_{ij}\right) - \lvert V_jY_{ij} \rvert \sin(\theta_{ij} + \delta_j - \delta_i)$
11	12
$\frac{\partial Q_{ji}}{\partial \delta_j} = \lvert V_iV_jY_{ij} \rvert \cos(\theta_{ij} + \delta_i - \delta_j)$	$\frac{\partial Q_{ji}}{\partial \lvert V_j \rvert} = -2\lvert V_j \rvert \left(\frac{B'_{ij}}{2} + B_{ij}\right) - \lvert V_iY_{ij} \rvert \sin(\theta_{ij} + \delta_i - \delta_j)$
$\frac{\partial Q_{ji}}{\partial \delta_i} = -\lvert V_iV_jY_{ij} \rvert \cos(\theta_{ij} + \delta_i - \delta_j)$	$\frac{\partial Q_{ji}}{\partial \lvert V_i \rvert} = -\lvert V_jY_{ij} \rvert \sin(\theta_{ij} + \delta_i - \delta_j)$
13	14

subscripts i and j on all voltage magnitudes and angles. Differentiating the equations for P_i, Q_i, P_{ij}, and Q_{ij} with respect to the state variables $\delta_2, \cdots, \delta_N$ and $|V_1|, \cdots, |V_N|$, we obtain expressions for the remaining elements of $\mathbf{H}_\mathbf{x}$, as shown in Table 15.5.

The equations of Table 15.5 determine the pattern of zero and nonzero elements in the jacobian $\mathbf{H}_\mathbf{x}$, which appears in Table 15.4. We have already remarked that in $\mathbf{H}_\mathbf{x}$ for a full measurement set:

- Each element of block 1 is zero.
- Each off-diagonal element of block 2 is zero and each diagonal element is 1.

Let us now examine the pattern of zero and nonzero elements in blocks 3, 4, 5, and 6, which involve derivatives of the real and reactive power injections into the buses. When the measurement set is full, blocks 4 and 6 are square ($N \times N$) and have off-diagonal elements with $|Y_{ij}|$ as a multiplying factor. Therefore, if all P_i and Q_i measurements are numbered according to their numbers, we find that:

- Blocks 4 and 6 follow exactly the same zero and nonzero pattern as the system $\mathbf{Y}_{\text{bus}}$. This is because the (row, column) locations (i, j) and (j, i) in those blocks have nonzero entries if, and only if, Y_{ij} is nonzero.

Allowing for the fact that there is no column for δ_1 in $\mathbf{H}_\mathbf{x}$, we note:

- Blocks 3 and 5 have the same zero and nonzero pattern as $\mathbf{Y}_{\text{bus}}$ with its first column omitted.

Again, assuming a full measurement set, let us examine blocks 7 through 14, which involve P and Q line flows. If the B measurements in each of the subsets $\{P_{ij}\}$, $\{P_{ji}\}$, $\{Q_{ij}\}$, and $\{Q_{ji}\}$ are numbered in exactly the same sequence as the network branches, then:

- Blocks 8, 10, 12, and 14 have exactly the same zero and nonzero structure as the network branch-to-bus incidence matrix **A**. This is because each line flow is incident to its two terminal buses.
- Blocks 7, 9, 11, and 13 have the same zero and nonzero structure as **A** with its first column omitted.

When the measurement set is not full, those rows corresponding to the missing measurements are simply deleted from the full jacobian $\mathbf{H}_\mathbf{x}$.

These observations yield the structure of the complete jacobian $\mathbf{H}_\mathbf{x}$ in a straightforward manner. Since $\mathbf{H}_\mathbf{x}$ has many zeros, the gain matrix $\mathbf{G}_\mathbf{x} =$

$\mathbf{H}_x^T \mathbf{R}^{-1} \mathbf{H}_x$ also has many zeros and further rules can be developed for identifying the locations of the nonzero elements of $\mathbf{G}_x$ in order to enhance computational efficiency.

In the example which follows we demonstrate the formation of $\mathbf{H}_x$ for the particular case where only line flows and one bus-voltage magnitude are measured. At least one voltage measurement is always required in order to establish the voltage *level* of the system.

Example 15.8. In the system of Fig. 15.6 (on page 686), P_{ij} and Q_{ij} measurements are made at the ends of each line and voltage magnitude $|V_3|$ is also measured. No bus injected powers and no other bus voltages are measured. Determine the structure of $\mathbf{H}_x$, write the partial derivative form of its nonzero elements, and express any four of them in literal form.

Solution. Figure 15.6 has $N = 4$ buses and $B = 4$ branches. Choosing $\delta_1 = 0°$ as the reference angle leaves δ_2, δ_3, δ_4, $|V_1|$, $|V_2|$, $|V_3|$, and $|V_4|$ as the seven ($N_s = 2N - 1$) state variables to be estimated from the 17 ($4B + 1$) measurements. Therefore, the redundancy factor is $(4B + 1)/N_s = 17/7 = 2.4$. Since $|V_3|$ is the only voltage measurement and neither P_i nor Q_i are being measured, $\mathbf{H}_x$ has only one row corresponding to $\partial|V_3|/\partial|V_3| = 1$ in blocks 1 through 6 of the general state-estimation jacobian. Let us mark the branches b_1 to b_4, as shown in Fig. 15.6, and order the line-flow measurements accordingly as follows:

Branch	Measurements			
b_1:	P_{14}	P_{41}	Q_{14}	Q_{41}
b_2:	P_{13}	P_{31}	Q_{13}	Q_{31}
b_3:	P_{23}	P_{32}	Q_{23}	Q_{32}
b_4:	P_{24}	P_{42}	Q_{24}	Q_{42}

Then, the branch-to-bus incidence matrix for the network is

$$\mathbf{A} = \begin{array}{c} \\ b_1 \\ b_2 \\ b_3 \\ b_4 \end{array} \begin{array}{c} \begin{array}{cccc} ① & ② & ③ & ④ \end{array} \\ \begin{bmatrix} 1 & 0 & 0 & -1 \\ 1 & 0 & -1 & 0 \\ 0 & 1 & -1 & 0 \\ 0 & 1 & 0 & -1 \end{bmatrix} \end{array}$$

which shows the zero and nonzero pattern of blocks 7 through 14 of $\mathbf{H}_x$ for this particular example. With a column for each state variable and a row for each measurement, $\mathbf{H}_x$ assumes the form

$$
\mathbf{H}_x = \begin{array}{c}
\begin{array}{ccccccc} \delta_2 & \delta_3 & \delta_4 & |V_1| & |V_2| & |V_3| & |V_4| \end{array} \\
\left[\begin{array}{ccc|cccc}
0 & 0 & 0 & 0 & 0 & 1 & 0 \\
0 & 0 & \dfrac{\partial P_{14}}{\partial \delta_4} & \dfrac{\partial P_{14}}{\partial |V_1|} & 0 & 0 & \dfrac{\partial P_{14}}{\partial |V_4|} \\
0 & \dfrac{\partial P_{13}}{\partial \delta_3} & 0 & \dfrac{\partial P_{13}}{\partial |V_1|} & 0 & \dfrac{\partial P_{13}}{\partial |V_3|} & 0 \\
\dfrac{\partial P_{23}}{\partial \delta_2} & \dfrac{\partial P_{23}}{\partial \delta_3} & 0 & 0 & \dfrac{\partial P_{23}}{\delta |V_2|} & \dfrac{\partial P_{23}}{\partial |V_3|} & 0 \\
\dfrac{\partial P_{24}}{\partial \delta_2} & 0 & \dfrac{\partial P_{24}}{\partial \delta_4} & 0 & \dfrac{\partial P_{24}}{\partial |V_2|} & 0 & \dfrac{\partial P_{24}}{\partial |V_4|} \\
\hline
0 & 0 & \dfrac{\partial P_{41}}{\partial \delta_4} & \dfrac{\partial P_{41}}{\partial |V_1|} & 0 & 0 & \dfrac{\partial P_{41}}{\partial |V_4|} \\
0 & \dfrac{\partial P_{31}}{\partial \delta_3} & 0 & \dfrac{\partial P_{31}}{\partial |V_1|} & 0 & \dfrac{\partial P_{31}}{\partial |V_3|} & 0 \\
\dfrac{\partial P_{32}}{\partial \delta_2} & \dfrac{\partial P_{32}}{\partial \delta_3} & 0 & 0 & \dfrac{\partial P_{32}}{\partial |V_2|} & \dfrac{\partial P_{32}}{\partial |V_3|} & 0 \\
\dfrac{\partial P_{42}}{\partial \delta_2} & 0 & \dfrac{\partial P_{42}}{\partial \delta_4} & 0 & \dfrac{\partial P_{42}}{\partial |V_2|} & 0 & \dfrac{\partial P_{42}}{\partial |V_4|} \\
\hline
0 & 0 & \dfrac{\partial Q_{14}}{\partial \delta_4} & \dfrac{\partial Q_{14}}{\partial |V_1|} & 0 & 0 & \dfrac{\partial Q_{14}}{\partial |V_4|} \\
0 & \dfrac{\partial Q_{13}}{\partial \delta_3} & 0 & \dfrac{\partial Q_{13}}{\partial |V_1|} & 0 & \dfrac{\partial Q_{13}}{\partial |V_3|} & 0 \\
\dfrac{\partial Q_{23}}{\partial \delta_2} & \dfrac{\partial Q_{23}}{\partial \delta_3} & 0 & 0 & \dfrac{\partial Q_{23}}{\partial |V_2|} & \dfrac{\partial Q_{23}}{\partial |V_3|} & 0 \\
\dfrac{\partial Q_{24}}{\partial \delta_2} & 0 & \dfrac{\partial Q_{24}}{\partial \delta_4} & 0 & \dfrac{\partial Q_{24}}{\partial |V_2|} & 0 & \dfrac{\partial Q_{24}}{\partial |V_4|} \\
\hline
0 & 0 & \dfrac{\partial Q_{41}}{\partial \delta_4} & \dfrac{\partial Q_{41}}{\partial |V_1|} & 0 & 0 & \dfrac{\partial Q_{41}}{\partial |V_4|} \\
0 & \dfrac{\partial Q_{31}}{\partial \delta_3} & 0 & \dfrac{\partial Q_{31}}{\partial |V_1|} & 0 & \dfrac{\partial Q_{31}}{\partial |V_3|} & 0 \\
\dfrac{\partial Q_{32}}{\partial \delta_2} & \dfrac{\partial Q_{32}}{\partial \delta_3} & 0 & 0 & \dfrac{\partial Q_{32}}{\partial |V_2|} & \dfrac{\partial Q_{32}}{\partial |V_3|} & 0 \\
\dfrac{\partial Q_{42}}{\partial \delta_2} & 0 & \dfrac{\partial Q_{42}}{\partial \delta_4} & 0 & \dfrac{\partial Q_{42}}{\partial |V_2|} & 0 & \dfrac{\partial Q_{42}}{\partial |V_4|}
\end{array}\right]
\end{array}
$$

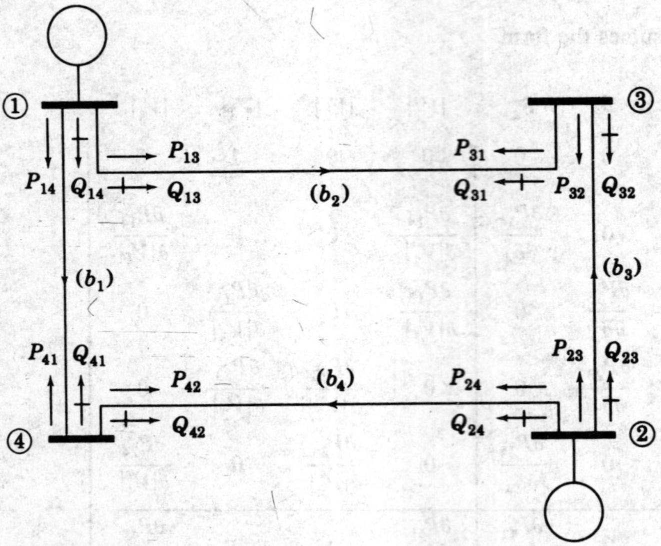

FIGURE 15.6
One-line diagram for Example 15.8 showing the P_{ij} and Q_{ij} measurements on lines marked b_1 to b_4.

The expressions for $\partial P_{23}/\partial\delta_3$, $\partial P_{23}/\partial|V_2|$, $\partial Q_{24}/\partial\delta_2$, and $\partial Q_{24}/\partial|V_4|$ are given by Table 15.5 as follows:

$$\frac{\partial P_{23}}{\partial \delta_3} = \frac{\partial}{\partial \delta_3}\Big[-|V_2|^2 G_{23} + |V_2V_3Y_{23}|\cos(\theta_{23} + \delta_3 - \delta_2)\Big]$$

$$= -|V_2V_3Y_{23}|\sin(\theta_{23} + \delta_3 - \delta_2)$$

$$\frac{\partial P_{23}}{\partial |V_2|} = -2|V_2|G_{23} + |V_3Y_{23}|\cos(\theta_{23} + \delta_3 - \delta_2)$$

$$\frac{\partial Q_{24}}{\partial \delta_2} = \frac{\partial}{\partial \delta_2}\left[-|V_2|^2\left(\frac{B'_{24}}{2} - B_{24}\right) - |V_2V_4Y_{24}|\sin(\theta_{24} + \delta_4 - \delta_2)\right]$$

$$= |V_2V_4Y_{24}|\cos(\theta_{24} + \delta_4 - \delta_2)$$

$$\frac{\partial Q_{24}}{\partial |V_4|} = -|V_2Y_{24}|\sin(\theta_{24} + \delta_4 - \delta_2)$$

The expressions for other elements can be similarly obtained.

A state-estimation algorithm based on measurements of only line flows and the required bus-voltage magnitude is described in the literature to which the reader is referred for further details.[4]

Because $\mathbf{H_x}$ has many zeros, the gain matrix $\mathbf{G_x} = \mathbf{H_x^T R^{-1} H_x}$ also has many zeros, and so $\mathbf{G_x}$ is never explicitly inverted as implied by Eq. (15.60). Instead **LDU** triangular factorization (described in Chap. 7), along with $P - \delta$ and $Q - |V|$ decoupling (described in Sec. 9.7), and optimal ordering (described in Sec. B.1 of the Appendix) can be applied for efficient computations and enhanced solution times of state estimation.

15.6 SUMMARY

State estimation determines the existing operating conditions of the system which are required for real-time control. The network model normally used covers the transmission portion of the operating area. Hence, the parameter values for all transmission lines, transformers, capacitor banks, and interconnections are required as in power-flow studies. In this chapter it is assumed that accurate parameters are known and on this basis state-estimation calculations employing the basic weighted least-squares approach are described. The state estimates are the bus-voltage angles and magnitudes which are computed by Eq. (15.62) from a set of redundant measurements. When bad measurements are detected, the state estimates are no longer reliable, which means that grossly erroneous data have to be identified and filtered out by statistical tests. The diagonal elements of the covariance matrix $\mathbf{R}' = \mathbf{R} - \mathbf{H_x G_x^{-1} H_x^T}$ are used to calculate the largest standardized residuals which help in identifying the bad measurements.

In industry studies most of the computation time is spent in evaluating the gain matrix $\mathbf{G_x} = \mathbf{H_x^T R^{-1} H_x}$, which is very large, is symmetric, and contains many zeros (very sparse). Because of its sparse nature, $\mathbf{G_x}$ is not explicitly inverted. Instead, optimal ordering and **LDU** factorization are employed. Convergence upon the solution of Eq. (15.62) can be influenced by where the meters are placed on the system and the types of measurements (whether injections, voltage magnitudes, or line flows). Redundancy is important but a full set of measurements is not necessary and may not be desirable from the computational viewpoint.

[4] J. F. Dopazo, O. A. Klitin, G. W. Stagg, and L. S. Van Slyck, "State Calculation of Power Systems from Line Flow Measurements," *IEEE Transactions on Power Apparatus and Systems*, vol. PAS-89. no. 7, September/October 1970.

PROBLEMS

15.1. The circuit of Fig. 15.1 is redrawn in Fig. 15.7, in which three loop current variables are identified as x_1, x_2, and x_3. Although not shown, ammeters and voltmeters with the same accuracy are assumed to be installed as in Fig. 15.1, and the meter readings are also assumed to be the same as those in Example 15.1. Determine the weighted least-squares estimates of the three loop currents. Using the estimated loop currents, determine the source voltages V_1 and V_2 and compare the results with those of Example 15.1.

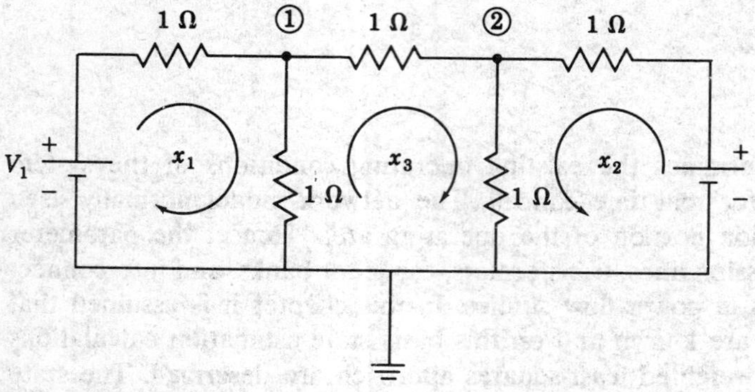

FIGURE 15.7
Circuit of Fig. 15.1 with three loop currents x_1, x_2, and x_3 and two nodes labeled ① and ②.

15.2. Show that $E[(\mathbf{x} - \hat{\mathbf{x}})(\mathbf{x} - \hat{\mathbf{x}})^T] = \mathbf{G}^{-1}$, where **G** is the gain matrix.

15.3. Show that the sum of the diagonal elements in the matrix $\mathbf{H}\mathbf{G}^{-1}\mathbf{H}^T\mathbf{R}^{-1}$ in Eq. (15.40) is numerically equal to the number of state variables.

15.4. Prove Eq. (15.47).

15.5. Consider the voltages at the two nodes labeled ① and ② in the circuit of Fig. 15.7 to be state variables. Using the ammeters and voltmeters connected as shown in Fig. 15.1 and their readings given in Example 15.1, determine the weighted least-squares estimates of these node voltages. Using the result, determine the source voltages V_1 and V_2 and compare the results with those of Example 15.1. Also, calculate the expected value of the sum of squares of the measurement residuals using Eq. (15.46) and check your answer using Eq. (15.47).

15.6. Five ammeters numbered A_1 to A_5 are used in the dc circuit of Fig. 15.8 to determine the two unknown source currents I_1 and I_2. The standard deviations of the meter errors are 0.2 A for meters A_2 and A_5 and 0.1 A for the other three meters. The readings of the five meters are 0.12, 1.18, 3.7, 0.81, and 7.1 A, respectively.

(*a*) Determine the weighted least-squares estimates of the source currents I_1 and I_2.

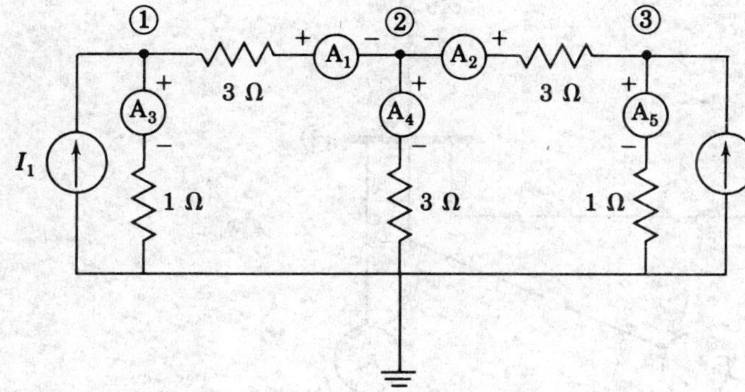

FIGURE 15.8
DC circuit with five ammeters.

(*b*) Using the chi-square test of Eq. (15.49) for $\alpha = 0.01$, check for the presence of bad data in the measurements.
(*c*) Eliminate any bad data detected in part (*b*) and find the weighted least-squares estimates of the source currents using the reduced data set.
(*d*) Apply the chi-square test for $\alpha = 0.01$ to the results of part (*c*) to check if the result is statistically acceptable.

15.7. Redo Prob. 15.6 when the unknowns to be determined are not the source currents but, rather, the voltages at the three nodes labeled ①, ②, and ③ in Fig. 15.8.

15.8. Consider the circuit of Fig. 15.8 for which accuracy of the ammeters and their readings are the same as those specified in Prob. 15.6. As in Prob. 15.7, the voltages at the three nodes labeled ①, ②, and ③ are to be estimated without first finding the source currents.
(*a*) Suppose that meters A_4 and A_5 are found to be out of order, and therefore only three measurements $z_1 = 0.12$, $z_2 = 1.18$, and $z_3 = 3.7$ are available. Determine the weighted least-squares estimates of the nodal voltages and the estimated errors $\hat{e}_1$, $\hat{e}_2$, and $\hat{e}_3$.
(*b*) This time suppose meters A_2 and A_5 are out of order and the remaining three meters are working. Using three measurements $z_1 = 0.12$, $z_3 = 3.7$, and $z_4 = 0.81$, can the nodal voltages be estimated without finding the source currents first? Explain why by examining the matrix **G**.

15.9. Suppose that the two voltage sources in Example 15.1 have been replaced with new ones, and the meter readings now show $z_1 = 2.9$ A, $z_2 = 10.2$ A, $z_3 = 5.1$ V, and $z_4 = 7.2$ V.
(*a*) Determine the weighted least-squares estimates of the new source voltages.
(*b*) Using the chi-square test for $\alpha = 0.005$, detect bad data.
(*c*) Eliminate the bad data and determine again the weighted least-squares estimates of the source voltages.
(*d*) Check your result in part (*c*) again using the chi-square test.

15.10. Five wattmeters are installed on the four-bus system of Fig. 15.9 to measure line real power flows, where per-unit reactances of the lines are $X_{12} = 0.05$, $X_{13} = 0.1$,

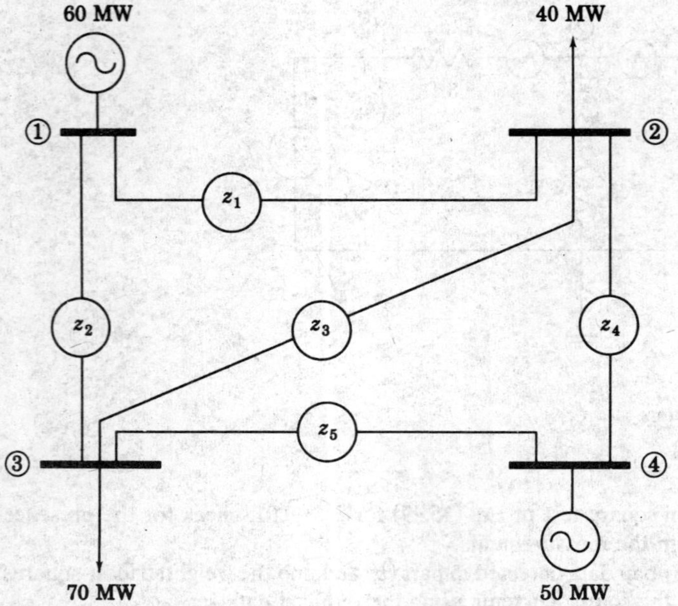

FIGURE 15.9
Four-bus network.

$X_{23} = 0.04$, $X_{24} = 0.0625$, and $X_{34} = 0.08$. Suppose that the meter readings show that

$$z_1 = P_{12} = \quad 0.34 \text{ per unit}$$

$$z_2 = P_{13} = \quad 0.26 \text{ per unit}$$

$$z_3 = P_{23} = \quad 0.17 \text{ per unit}$$

$$z_4 = P_{24} = -0.24 \text{ per unit}$$

$$z_5 = P_{34} = -0.22 \text{ per unit}$$

where the variances of the measurement errors in per unit are given by

$$\sigma_1^2 = \sigma_2^2 = \sigma_3^2 = \sigma_4^2 = \sigma_5^2 = (0.01)^2$$

(*a*) Apply the dc power-flow method of Sec. 9.7 to this network with bus ① as reference and determine the corresponding **H** matrix. Then, compute the weighted least-squares estimates of the phase angles of the bus voltages in radians.

(*b*) Using the chi-square test for $\alpha = 0.01$, identify two bad measurements. Between the two bad measurements one is not worse than the other as far as accuracy is concerned. Explain why. If both bad measurements are eliminated simultaneously, would it be possible to estimate the states of the system?

(*c*) For the two bad measurements identified in part (*b*), determine the relationship between the two error estimates in terms of the reactances of the corresponding two lines.

(*d*) Eliminate one of the bad measurements identified in part (*b*) and determine the weighted least-squares estimates of the phase angles of the bus voltages using the reduced data set. Do the same for the other bad measurement. By comparing the two results, identify the buses at which the estimated phase angles are equal in the two cases.

15.11. In the four-bus system of Prob. 15.10 suppose that the variance of the measurement error for z_5 is $(0.05)^2$ and that all the other data remain the same. Qualitatively describe how the newly estimated values $\hat{z}_4$ and $\hat{z}_5$ of the measurements will differ from those obtained in Prob. 15.10. Verify your answer by recalculating the weighted least-squares estimates of the phase angles (in radians) of the bus voltages and the corresponding $\hat{z}$.

15.12. Suppose that a line of impedance $j0.025$ per unit is added between buses ① and ④ in the network of Fig. 15.9, and that a wattmeter is installed on this line at bus ①. The variance of the measurement error for this added wattmeter is assumed to be the same as that of the others. The meter readings now show

$$z_1 = P_{12} = \quad 0.32 \text{ per unit}$$

$$z_2 = P_{13} = \quad 0.24 \text{ per unit}$$

$$z_3 = P_{23} = \quad 0.16 \text{ per unit}$$

$$z_4 = P_{24} = -0.29 \text{ per unit}$$

$$z_5 = P_{34} = -0.27 \text{ per unit}$$

$$z_6 = P_{14} = \quad 0.05 \text{ per unit}$$

(*a*) Find the **H** matrix that describes the dc power flow with bus ① as reference and compute the weighted least-squares estimates of the phase angles of the bus voltages in radians.

(*b*) Using the chi-square test for $\alpha = 0.01$, eliminate any bad data and recompute the weighted least-squares estimates of the phase angles of the bus voltages. Check your result again using the chi-square test for $\alpha = 0.01$.

15.13. In the four-bus system described in Prob. 15.12 suppose that the wattmeter on line ①–④ is out of order and that the readings of the remaining five wattmeters are the same as those specified in Prob. 15.12.

(*a*) Apply the dc power-flow analysis with bus ① as reference and determine the **H** matrix. Then, compute the weighted least-squares estimates of the phase angles of the bus voltages in radians.

(*b*) Using the chi-square test for $\alpha = 0.01$, identify two bad measurements. Eliminate one of them and compute the weighted least-squares estimates of the bus-voltage phase angles. Restore the eliminated bad measurement and remove the second one before recomputing the estimates of the bus-voltage phase angles. Compare the two sets of results and identify the buses at which the estimated angles are equal in the two cases. Does the presence of line ①-④ (but with no line measurement) affect the identification of those buses? Compare the identified buses with those identified in Prob. 15.10(*d*).

15.14. Three voltmeters and four wattmeters are installed on the three-bus system of Fig. 15.10, where per-unit reactances of the lines are $X_{12} = 0.1$, $X_{13} = 0.08$, and $X_{23} = 0.05$. The per-unit values of the three voltmeter measurements are $z_1 = |V_1| = 1.01$, $z_2 = |V_2| = 1.02$, and $z_3 = |V_3| = 0.98$. The readings of the two wattmeters measuring MW generation at buses ① and ② are $z_4 = 0.48$ per unit and $z_5 = 0.33$ per unit, respectively. The measurement of the wattmeter on line ①-③ at bus ① shows $z_6 = 0.41$ per unit and that of the wattmeter on line ②-③ at bus ② is $z_7 = 0.38$ per unit. The variances of the measurement errors are given in per unit as

$$\sigma_1^2 = \sigma_2^2 = \sigma_3^2 = (0.02)^2$$

$$\sigma_4^2 = \sigma_5^2 = \sigma_6^2 = \sigma_7^2 = (0.05)^2$$

(*a*) Use bus ① as reference to find expressions for the elements of the matrix

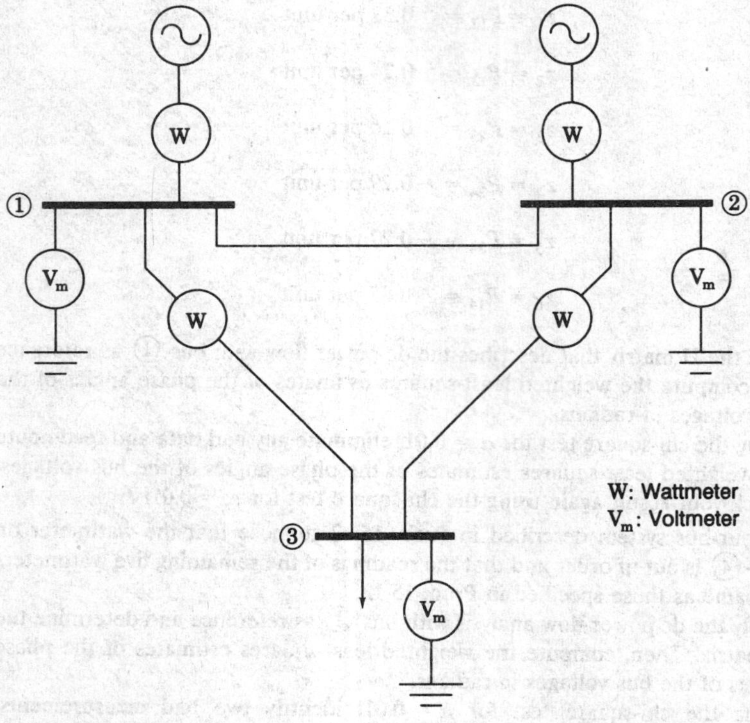

FIGURE 15.10
Three-bus network.

$\mathbf{H}_{\mathbf{x}}^{(k)}$ and those of the measurement errors $e_i^{(k)}$ in terms of state variables, as done in Example 15.5.

(*b*) Using the initial value of $1.0\underline{/0°}$ per unit for all bus voltages, find the values of the state variables that will be obtained at the end of the first iteration of the weighted least-squares state-estimation process.

15.15. Application of the weighted least-squares state estimation to the three-bus system with all the measurements described in Prob. 15.14 yields the following estimates of the states:

$$|V_1| = 1.0109 \text{ per unit}$$

$$|V_2| = 1.0187 \text{ per unit} \qquad \delta_2 = -0.0101 \text{ radians}$$

$$|V_3| = 0.9804 \text{ per unti} \qquad \delta_3 = -0.0308 \text{ radians}$$

The sequence of diagonal elements in the covariance matrix $\mathbf{R}'$ is 0.8637×10^{-6}, 0.1882×10^{-5}, 0.2189×10^{-6}, 0.7591×10^{-3}, 0.8786×10^{-3}, 0.1812×10^{-2}, and 0.1532×10^{-2}. Find the estimates of the measurement errors $\hat{e}_i$ and the corresponding standardized errors.

15.16. Solve Prob. 15.14 when the two wattmeters installed on lines ①-③ and ②-③ are replaced with two varmeters and their readings are 0.08 and 0.24 per unit, respectively.

15.17. Suppose that real and reactive power flows are measured at both ends of each of the five lines in the four-bus system of Fig. 15.9 using ten wattmeters and ten varmeters. The voltage magnitude is measured at bus ② only, and bus injected powers are not measured at all.

(*a*) Determine the structure of $\mathbf{H}_{\mathbf{x}}$ by writing the partial derivative form of its nonzero elements, as shown in Example 15.8. Assume that line-flow measurements are ordered in the following sequence: ①-②, ①-③, ②-③, ②-④, and ③-④ (and the same sequence also in reverse directions).

(*b*) Suppose that the elements of the $\mathbf{Y}_{\text{bus}}$ of the network are given by

$$Y_{ij} = G_{ij} + jB_{ij} = |Y_{ij}|\underline{/\theta_{ij}}$$

and that the total charging susceptance of line ⓘ-ⓙ is B'_{ij}. Write out nonlinear functions which express the measured quantities P_{21} and Q_{21} in terms of state variables.

(*c*) In terms of state variables write out the expressions, similar to those given in Example 15.8, for the nonzero elements in the rows of the matrix $\mathbf{H}_{\mathbf{x}}$ corresponding to measurements P_{21} and Q_{21}.

15.18. The method of Example 15.8 based on measurements of only line flows (plus a voltage measurement at one bus) is applied to the three-bus system of Fig. 15.10 using three wattmeters and three varmeters. The per-unit values of the measurements are

$$z_1 = |V_1| = 1.0 \qquad z_5 = Q_{12} = -0.101$$

$$z_2 = P_{12} = 0.097 \qquad z_6 = Q_{13} = 0.048$$

$$z_3 = P_{13} = 0.383 \qquad z_7 = Q_{23} = 0.276$$

$$z_4 = P_{23} = 0.427$$

where the variances of all the measurements are $(0.02)^2$. The per-unit reactances of the lines are as specified in Prob. 15.14. Using bus ① as reference and flat-start values, find the values of the state variables that will be obtained at the end of the first iteration of the weighted least-squares state estimation.

CHAPTER 16

POWER SYSTEM STABILITY

When ac generators were driven by reciprocating steam engines, one of the major problems in the operation of machinery was caused by sustained oscillations in speed or *hunting* due to the periodic variations in the torque applied to the generators. The resulting periodic variations in voltage and frequency were transmitted to the motors connected to the system. Oscillations of the motors caused by the variations in voltage and frequency sometimes caused the motors to lose synchronism entirely if their natural frequency of oscillation coincided with the frequency of oscillation caused by the engines driving the generators. Damper windings were first used to minimize hunting by the damping action of the losses resulting from the currents induced in the damper windings by any relative motion between the rotor and the rotating field set up by the armature current. The use of turbines has reduced the problem of hunting although it is still present where the prime mover is a diesel engine. Maintaining synchronism between the various parts of a power system becomes increasingly difficult, however, as the systems and interconnections between systems continue to grow.

16.1 THE STABILITY PROBLEM

Stability studies which evaluate the impact of disturbances on the electromechanical dynamic behavior of the power system are of two types—*transient* and *steady state*. Transient stability studies are very commonly undertaken by electric utility planning departments responsible for ensuring proper dynamic performance of the system. The system models used in such studies are extensive because present-day power systems are vast, heavily interconnected systems with hundreds of machines which can interact through the medium of their extra-high-voltage and ultra-high-voltage networks. These machines have associated excitation systems and turbine-governing control systems which in some but not all cases are modeled in order to reflect properly correct dynamic performance of the system. If the resultant nonlinear differential and algebraic equations of the overall system are to be solved, then either direct methods or iterative step-by-step procedures must be used. In this chapter we emphasize transient stability considerations and introduce basic iterative procedures used in transient stability studies. Before doing so, however, let us first discuss certain terms commonly encountered in stability analysis.[1]

A power system is in a *steady-state operating condition* if all the measured (or calculated) physical quantities describing the operating condition of the system can be considered *constant* for purposes of analysis. When operating in a steady-state condition if a sudden change or sequence of changes occurs in one or more of the parameters of the system, or in one or more of its operating quantities, we say that the system has undergone a *disturbance* from its steady-state operating condition. Disturbances can be large or small depending on their origin. A *large disturbance* is one for which the nonlinear equations describing the dynamics of the power system cannot be validly linearized for purposes of analysis. Transmission system faults, sudden load changes, loss of generating units, and line switching are examples of large disturbances. If the power system is operating in a steady-state condition and it undergoes change which can be properly analyzed by linearized versions of its dynamic and algebraic equations, we say that a *small disturbance* has occurred. A change in the gain of the automatic voltage regulator in the excitation system of a large generating unit could be an example of a small disturbance. The power system is *steady-state stable* for a particular steady-state operating condition if, following a small disturbance, it returns to essentially the same steady-state condition of operation. However, if following a large disturbance, a significantly different but acceptable steady-state operating condition is attained, we say that the system is *transiently stable*.

Steady-state stability studies are usually less extensive in scope than transient stability studies and often involve a single machine operating into an infinite bus or just a few machines undergoing one or more small disturbances. Thus, steady-state stability studies examine the stability of the system under small *incremental* variations in parameters or operating conditions about a steady-state equilibrium point. The nonlinear differential and algebraic equations of the system are replaced by a set of linear equations which are then

[1]For further discussion, see "Proposed Terms and Definitions for Power System Stability," A Task Force Report of the System Dynamic Performance Subcommittee, *IEEE Transactions on Power Apparatus and Systems*, vol. PAS 101, July 1982, pp. 1894–1898.

solved by methods of linear analysis to determine if the system is steady-state stable.

Since transient stability studies involve large disturbances, linearization of the system equations is not permitted. Transient stability is sometimes studied on a *first-swing* rather than a *multiswing* basis. First-swing transient stability studies use a reasonably simple generator model consisting of the transient internal voltage E_i' behind transient reactance X_d'; in such studies the excitation systems and turbine-governing control systems of the generating units are not represented. Usually, the time period under study is the first second following a system fault or other large disturbance. If the machines of the system are found to remain essentially in synchronism within the first second, the system is regarded as being transiently stable. Multiswing stability studies extend over a longer study period, and therefore the effects of the generating units' control systems must be considered since they can affect the dynamic performance of the units during the extended period. Machine models of greater sophistication are then needed to properly reflect the behavior of the system.

Thus, excitation systems and turbine-governing control systems may or may not be represented in steady-state and transient stability studies depending on the objectives. In all stability studies the objective is to determine whether or not the rotors of the machines being perturbed return to constant speed operation. Obviously, this means that the rotor speeds have departed at least temporarily from synchronous speed. To facilitate computation, three fundamental assumptions therefore are made in *all* stability studies:

1. Only synchronous frequency currents and voltages are considered in the stator windings and the power system. Consequently, dc offset currents and harmonic components are neglected.
2. Symmetrical components are used in the representation of unbalanced faults.
3. Generated voltage is considered unaffected by machine speed variations.

These assumptions permit the use of phasor algebra for the transmission network and solution by power-flow techniques using 60-Hz parameters. Also, negative- and zero-sequence networks can be incorporated into the positive-sequence network at the fault point. As we shall see, three-phase balanced faults are generally considered. However, in some special studies circuit-breaker clearing operation may be such that consideration of unbalanced conditions is unavoidable.[2]

---

[2]For information beyond the scope of this book, see P. M. Anderson and A. A. Fouad, *Power System Control and Stability*, The Iowa State University Press, Ames, IA, 1977.

16.2 ROTOR DYNAMICS AND THE SWING EQUATION

The equation governing rotor motion of a synchronous machine is based on the elementary principle in dynamics which states that accelerating torque is the product of the moment of inertia of the rotor times its angular acceleration. In the MKS (meter-kilogram-second) system of units this equation can be written for the synchronous generator in the form

$$J\frac{d^2\theta_m}{dt^2} = T_a = T_m - T_e \text{ N-m} \tag{16.1}$$

where the symbols have the following meanings:

- J the total moment of inertia of the rotor masses, in kg-m^2
- θ_m the angular displacement of the rotor with respect to a stationary axis, in mechanical radians (rad)
- t time, in seconds (s)
- T_m the mechanical or shaft torque supplied by the prime mover less retarding torque due to rotational losses, in N-m
- T_e the net electrical or electromagnetic torque, in N-m
- T_a the net accelerating torque, in N-m

The mechanical torque T_m and the electrical torque T_e are considered positive for the synchronous generator. This means that T_m is the resultant shaft torque which tends to accelerate the rotor in the positive θ_m direction of rotation, as shown in Fig. 16.1(*a*). Under steady-state operation of the generator T_m and T_e are equal and the accelerating torque T_a is zero. In this case there is no acceleration or deceleration of the rotor masses and the resultant constant speed is the *synchronous speed*. The rotating masses, which include the rotor of the generator and the prime mover, are said to be *in synchronism* with the other machines operating at synchronous speed in the power system. The prime

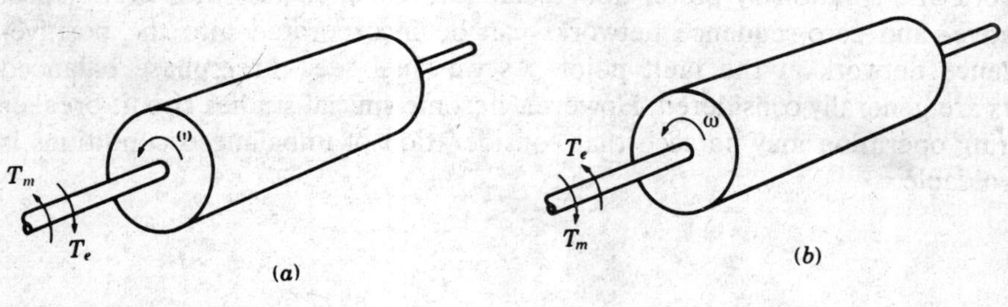

FIGURE 16.1
Representation of a machine rotor comparing direction of rotation and mechanical and electrical torques for: (*a*) a generator; (*b*) a motor.

mover may be a hydroturbine or a steam turbine, for which models of different levels of complexity exist to represent their effect on T_m. In this text T_m is considered constant at any given operating condition. This assumption is a fair one for generators even though input from the prime mover is controlled by governors. Governors do not act until after a change in speed is sensed, and so they are not considered effective during the time period in which rotor dynamics are of interest in our stability studies here. The electrical torque T_e corresponds to the net air-gap power in the machine, and thus accounts for the total output power of the generator plus $|I|^2R$ losses in the armature winding. In the synchronous motor the direction of power flow is opposite to that in the generator. Accordingly, for a motor both T_m and T_e in Eq. (16.1) are reversed in sign, as shown in Fig. 16.1(*b*). T_e then corresponds to the air-gap power supplied by the electrical system to drive the rotor, whereas T_m represents the counter torque of the load and rotational losses tending to retard the rotor.

Since θ_m is measured with respect to a stationary reference axis on the stator, it is an absolute measure of rotor angle. Consequently, it continuously increases with time even at constant synchronous speed. Since the rotor speed relative to synchronous speed is of interest, it is more convenient to measure the rotor angular position with respect to a reference axis which rotates at synchronous speed. Therefore, we define

$$\theta_m = \omega_{sm} t + \delta_m \tag{16.2}$$

where ω_{sm} is the synchronous speed of the machine in mechanical radians per second and δ_m is the angular displacement of the rotor, in mechanical radians, from the synchronously rotating reference axis. The derivatives of Eq. (16.2) with respect to time are

$$\frac{d\theta_m}{dt} = \omega_{sm} + \frac{d\delta_m}{dt} \tag{16.3}$$

and

$$\frac{d^2\theta_m}{dt^2} = \frac{d^2\delta_m}{dt^2} \tag{16.4}$$

Equation (16.3) shows that the rotor angular velocity $d\theta_m/dt$ is constant and equals the synchronous speed only when $d\delta_m/dt$ is zero. Therefore, $d\delta_m/dt$ represents the deviation of the rotor speed from synchronism and the units of measure are mechanical radians per second. Equation (16.4) represents the rotor acceleration measured in mechanical radians per second-squared.

Substituting Eq. (16.4) in Eq. (16.1), we obtain

$$J\frac{d^2\delta_m}{dt^2} = T_a = T_m - T_e \text{ N-m} \tag{16.5}$$

It is convenient for notational purposes to introduce

$$\omega_m = \frac{d\theta_m}{dt} \tag{16.6}$$

for the angular velocity of the rotor. We recall from elementary dynamics that power equals torque times angular velocity, and so multiplying Eq. (16.5) by ω_m, we obtain

$$J\omega_m \frac{d^2\delta_m}{dt^2} = P_a = P_m - P_e \text{ W} \tag{16.7}$$

where P_m = shaft power input to the machine less rotational losses
P_e = electrical power crossing its air gap
P_a = accelerating power which accounts for any unbalance between those two quantities

Usually, we neglect rotational losses and armature $|I|^2R$ losses and think of P_m as power supplied by the prime mover and P_e as the electrical power output.

The coefficient $J\omega_m$ is the angular momentum of the rotor; at synchronous speed ω_{sm} it is denoted by M and called the *inertia constant* of the machine. Obviously, the units in which M is expressed must correspond to those of J and ω_m. A careful check of the units in each term of Eq. (16.7) shows that M is expressed in joule-seconds per mechanical radian, and we write

$$M\frac{d^2\delta_m}{dt^2} = P_a = P_m - P_e \text{ W} \tag{16.8}$$

While we have used M in this equation, the coefficient is not a constant in the strictest sense because ω_m does not equal synchronous speed under all conditions of operation. However, in practice, ω_m does not differ significantly from synchronous speed when the machine is stable, and since power is more convenient in calculations than torque, Eq. (16.8) is preferred. In machine data supplied for stability studies another constant related to inertia is often encountered. This is the so-called *H constant*, which is defined by

$$H = \frac{\text{stored kinetic energy in megajoules at synchronous speed}}{\text{machine rating in MVA}}$$

and

$$H = \frac{\frac{1}{2}J\omega_{sm}^2}{S_{\text{mach}}} = \frac{\frac{1}{2}M\omega_{sm}}{S_{\text{mach}}} \text{ MJ/MVA} \tag{16.9}$$

where S_{mach} is the three-phase rating of the machine in megavoltamperes. Solving for M in Eq. (16.9), we obtain

$$M = \frac{2H}{\omega_{sm}} S_{\text{mach}} \text{ MJ/mech rad} \tag{16.10}$$

and substituting for M in Eq. (16.8), we find

$$\frac{2H}{\omega_{sm}} \frac{d^2\delta_m}{dt^2} = \frac{P_a}{S_{\text{mach}}} = \frac{P_m - P_e}{S_{\text{mach}}} \tag{16.11}$$

This equation leads to a very simple result.

Note that δ_m is expressed in mechanical radians in the numerator of Eq. (16.11), whereas ω_{sm} is expressed in mechanical radians per second in the denominator. Therefore, we can write the equation in the form

$$\frac{2H}{\omega_s} \frac{d^2\delta}{dt^2} = P_a = P_m - P_e \text{ per unit} \tag{16.12}$$

provided both δ and ω_s have consistent units, which may be mechanical or electrical degrees or radians. H and t have consistent units since megajoules per megavoltampere is in units of time in seconds and P_a, P_m, and P_e must be in per unit on the same base as H. When the subscript m is associated with ω, ω_s, and δ, it means *mechanical* units are being used; otherwise, *electrical* units are implied. Accordingly, ω_s is the synchronous speed in electrical units. For a system with an electrical frequency of f hertz Eq. (16.12) becomes

$$\frac{H}{\pi f} \frac{d^2\delta}{dt^2} = P_a = P_m - P_e \text{ per unit} \tag{16.13}$$

when δ is in electrical radians, while

$$\frac{H}{180f} \frac{d^2\delta}{dt^2} = P_a = P_m - P_e \text{ per unit} \tag{16.14}$$

applies when δ is in electrical degrees.

Equation (16.12), called the *swing equation* of the machine, is the fundamental equation which governs the rotational dynamics of the synchronous machine in stability studies. We note that it is a second-order differential equation, which can be written as the two first-order differential equations

$$\frac{2H}{\omega_s}\frac{d\omega}{dt} = P_m - P_e \text{ per unit} \tag{16.15}$$

$$\frac{d\delta}{dt} = \omega - \omega_s \tag{16.16}$$

in which ω, ω_s, and δ involve electrical radians or electrical degrees.

We use the various equivalent forms of the swing equation throughout this chapter to determine the stability of a machine within a power system. When the swing equation is solved, we obtain the expression for δ as a function of time. A graph of the solution is called the *swing curve* of the machine and inspection of the swing curves of all the machines of the system will show whether the machines remain in synchronism after a disturbance.

16.3 FURTHER CONSIDERATIONS OF THE SWING EQUATION

The megavoltampere (MVA) base used in Eq. (16.11) is the machine rating S_{mach} which is introduced by the definition of H. In a stability study of a power system with many synchronous machines only one MVA base common to all parts of the system can be chosen. Since the right-hand side of the swing equation for each machine must be expressed in per unit on this common system base, it is clear that H on the left-hand side of each swing equation must also be consistent with the system base. This is accomplished by converting H for each machine based on its own individual rating to a value determined by the system base, S_{system}. Equation (16.11), multiplied on each side by the ratio $(S_{\text{mach}}/S_{\text{system}})$, leads to the conversion formula

$$H_{\text{system}} = H_{\text{mach}}\frac{S_{\text{mach}}}{S_{\text{system}}} \tag{16.17}$$

in which the subscript for each term indicates the corresponding base being used. In industry studies the system base that is usually chosen is 100 MVA.

The inertia constant M is rarely used in practice and the forms of the swing equation involving H are more often encountered. This is because the value of M varies widely with the size and type of the machine, whereas H assumes a much narrower range of values, as shown in Table 16.1. Machine manufacturers also use the symbol WR^2 to specify for the rotating parts of a generating unit (including the prime mover) the weight in pounds multiplied by the square of the radius of gyration in feet. Hence, $WR^2/32.2$ is the moment of inertia of the machine in slug-feet squared.

TABLE 16.1
Typical inertia constants of synchronous machines†

Type of machine	Inertia constant, H‡ MJ / MVA
Turbine generator:	
Condensing, 1800 r/min	9–6
3600 r/min	7–4
Noncondensing, 3600 r/min	4–3
Waterwheel generator:	
Slow-speed, < 200 r/min	2–3
High-speed, > 200 r/min	2–4
Synchronous condenser;§	
Large	1.25
Small	1.00
Synchronous motor with load varies from 1.0 to 5.0 and higher for heavy flywheels	2.00

†Reprinted by permission of the ABB Power T & D Company, Inc. from *Electrical Transmission and Distribution Reference Book*.

‡Where range is given, the first figure applies to machines of smaller megavoltampere rating.

§Hydrogen-cooled, 25% less.

Example 16.1. Develop a formula to calculate the H constant from WR^2 and then evaluate H for a nuclear generating unit rated at 1333 MVA, 1800 r/min with $WR^2 = 5{,}820{,}000$ lb-ft^2.

Solution. The kinetic energy (KE) of rotation in foot-pounds at synchronous speed

$$\text{KE} = \frac{1}{2}\frac{WR^2}{32.2}\left[\frac{2\pi(\text{r/min})}{60}\right]^2 \text{ ft-lb}$$

Since 550 ft-lb/s equals 746 W, it follows that 1 ft-lb equals 746/550 J. Hence, converting foot-pounds to megajoules and dividing by the machine rating in megavoltamperes, we obtain

$$H = \frac{\left(\dfrac{746}{550} \times 10^{-6}\right)\dfrac{1}{2}\dfrac{WR^2}{32.2}\left[\dfrac{2\pi(\text{r/min})}{60}\right]^2}{S_{\text{mach}}}$$

which yields upon simplification

$$H = \frac{2.31 \times 10^{-10} WR^2(\text{r/min})^2}{S_{\text{mach}}}$$

Inserting the given machine data in this equation, we obtain

$$H = \frac{2.31 \times 10^{-10}(5.82 \times 10^6)(1800)^2}{1333} = 3.27 \text{ MJ/MVA}$$

Converting H to a 100-MVA system base, we obtain

$$H = 3.27 \times \frac{1333}{100} = 43.56 \text{ MJ/MVA}$$

In a stability study for a large system with many machines geographically dispersed over a wide area it is desirable to minimize the number of swing equations to be solved. This can be done if the transmission-line fault, or other disturbance on the system, affects the machines within a power plant so that their rotors swing together. In such cases the machines within the plant can be combined into a single equivalent machine just as if their rotors were mechanically coupled and only one swing equation must be written for them. Consider a power plant with two generators connected to the same bus which is electrically remote from the network disturbances. The swing equations on the common system base are

$$\frac{2H_1}{\omega_s}\frac{d^2\delta_1}{dt^2} = P_{m1} - P_{e1} \text{ per unit} \tag{16.18}$$

$$\frac{2H_2}{\omega_s}\frac{d^2\delta_2}{dt^2} = P_{m2} - P_{e2} \text{ per unit} \tag{16.19}$$

Adding the equations together and denoting δ_1 and δ_2 by δ since the rotors swing together, we obtain

$$\frac{2H}{\omega_s}\frac{d^2\delta}{dt^2} = P_m - P_e \text{ per unit} \tag{16.20}$$

where $H = (H_1 + H_2)$, $P_m = (P_{m1} + P_{m2})$, and $P_e = (P_{e1} + P_{e2})$. This single equation, which is in the form of Eq. (16.12), can be solved to represent the plant dynamics.

Example 16.2. Two 60-Hz generating units operate in parallel within the same power plant and have the following ratings:

Unit 1: 500 MVA, 0.85 power factor, 20 kV, 3600 r/min

$$H_1 = 4.8 \text{ MJ/MVA}$$

Unit 2: 1333 MVA, 0.9 power factor, 22 kV, 1800 r/min

$$H_2 = 3.27 \text{ MJ/MVA}$$

Calculate the equivalent H constant for the two units on a 100-MVA base.

Solution. The total kinetic energy (KE) of rotation of the two machines is

$$\text{KE} = (4.8 \times 500) + (3.27 \times 1333) = 6759 \text{ MJ}$$

Therefore, the H constant for the equivalent machine on 100-MVA base is

$$H = 67.59 \text{ MJ/MVA}$$

and this value can be used in a single swing equation, provided the machines swing together so that their rotor angles are in step at each instant of time.

Machines which swing together are called *coherent* machines. It is noted that when both ω_s and δ are expressed in electrical degrees or radians, the swing equations for coherent machines can be combined together even though, as in the example, the rated speeds are different. This fact is often used in stability studies involving many machines in order to reduce the number of swing equations which need to be solved.

For *any pair* of noncoherent machines in a system swing equations similar to Eqs. (16.18) and (16.19) can be written. Dividing each equation by its left-hand-side coefficient and subtracting the resultant equations, we obtain

$$\frac{d^2\delta_1}{dt^2} - \frac{d^2\delta_2}{dt^2} = \frac{\omega_s}{2}\left(\frac{P_{m1} - P_{e1}}{H_1} - \frac{P_{m2} - P_{e2}}{H_2}\right) \tag{16.21}$$

Multiplying each side by $H_1H_2/(H_1 + H_2)$ and rearranging, we find that

$$\frac{2}{\omega_s}\left(\frac{H_1H_2}{H_1 + H_2}\right)\frac{d^2(\delta_1 - \delta_2)}{dt^2} = \frac{P_{m1}H_2 - P_{m2}H_1}{H_1 + H_2} - \frac{P_{e1}H_2 - P_{e2}H_1}{H_1 + H_2} \tag{16.22}$$

which also may be written more simply in the form of the basic swing equation, Eq. (16.12), as follows:

$$\frac{2}{\omega_s}H_{12}\frac{d^2\delta_{12}}{dt^2} = P_{m12} - P_{e12} \tag{16.23}$$

Here the relative angle δ_{12} equals $\delta_1 - \delta_2$, and an equivalent inertia and weighted input and output powers are defined by

$$H_{12} = \frac{H_1 H_2}{H_1 + H_2} \tag{16.24}$$

$$P_{m12} = \frac{P_{m1} H_2 - P_{m2} H_1}{H_1 + H_2} \tag{16.25}$$

$$P_{e12} = \frac{P_{e1} H_2 - P_{e2} H_1}{H_1 + H_2} \tag{16.26}$$

A noteworthy application of these equations concerns a two-machine system having only one generator (machine one) and a synchronous motor (machine two) connected by a network of pure reactances. Whatever change occurs in the generator output is thus absorbed by the motor, and we can write

$$\begin{aligned} P_{m1} &= -P_{m2} = P_m \\ P_{e1} &= -P_{e2} = P_e \end{aligned} \tag{16.27}$$

Under these conditions, $P_{m12} = P_m$, $P_{e12} = P_e$, and Eq. (16.22) reduces to

$$\frac{2H_{12}}{\omega_s} \frac{d^2\delta_{12}}{dt^2} = P_m - P_e$$

which is also the format of Eq. (16.12) for a single machine.

Equation (16.22) demonstrates that stability of a machine within a system is a relative property associated with its dynamic behavior with respect to the other machines of the system. The rotor angle of one machine, say, δ_1, can be chosen for comparison with the rotor angle of each other machine, symbolized by δ_2. In order to be stable the angular differences between all machines must decrease after the final switching operation—such as the opening of a circuit breaker to clear a fault. Although we may choose to plot the angle between a machine's rotor and a synchronously rotating reference axis, it is the relative angles between machines which are important. Our discussion above emphasizes the relative nature of the system stability property and shows that the essential features of stability study are revealed by consideration of two-machine problems. Such problems are of two types: those having one machine of finite inertia swinging with respect to an infinite bus and those having two finite inertia machines swinging with respect to each other. An infinite bus may be considered for stability purposes as a bus at which there is located a machine of constant internal voltage, having zero impedance and infinite inertia. The point of connection of a generator to a large power system may be regarded as such a bus. In all cases the swing equation assumes the form of Eq. (16.12), each term of which must be explicitly described before it can be solved. The equation for P_e is essential to this description, and we now proceed to its characterization for a general two-machine system.

16.4 THE POWER-ANGLE EQUATION

In the swing equation for the generator the input mechanical power from the prime mover, P_m, will be considered constant. As we have mentioned previously, this is a reasonable assumption because conditions in the electrical network can be expected to change before the control governor can cause the turbine to react. Since P_m in Eq. (16.12) is constant, the electrical power output P_e will determine whether the rotor accelerates, decelerates, or remains at synchronous speed. When P_e equals P_m, the machine operates at steady-state synchronous speed; when P_e changes from this value, the rotor deviates from synchronous speed. Changes in P_e are determined by conditions on the transmission and distribution networks and the loads on the system to which the generator supplies power. Electrical network disturbances resulting from severe load changes, network faults, or circuit-breaker operations may cause the generator output P_e to change rapidly, in which case electromechanical transients exist. Our fundamental assumption is that the effect of machine speed variations upon the generated voltage is negligible so that the manner in which P_e changes is determined by the power-flow equations applicable to the state of the electrical network and by the model chosen to represent the electrical behavior of the machine. Each synchronous machine is represented for transient stability studies by its transient internal voltage E_i' in series with the transient reactance X_d', as shown in Fig. 16.2(a) in which V_t is the terminal voltage. This corresponds to the steady-state representation in which syn-

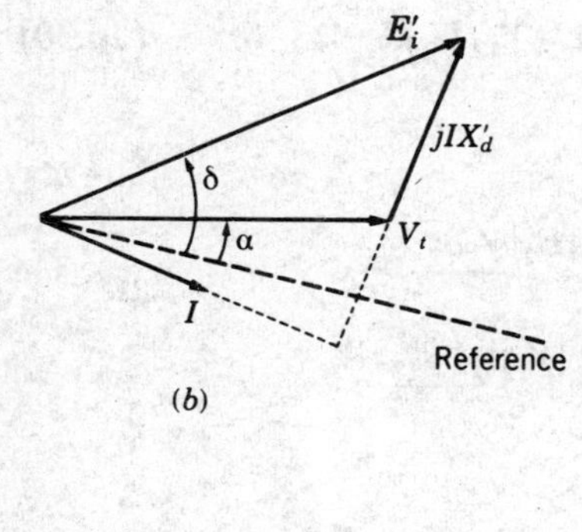

FIGURE 16.2
Phasor diagram of a synchronous machine for transient stability studies.

FIGURE 16.3
Schematic diagram for stability studies. Transient reactances associated with E_1' and E_2' are included in the transmission network.

chronous reactance X_d is in series with the synchronous internal or no-load voltage E_i. Armature resistance is negligible in most cases so that the phasor diagram of Fig. 16.2(*b*) applies. Since each machine must be considered relative to the system of which it is part, the phasor angles of the machine quantities are measured with respect to the common system reference.

Figure 16.3 schematically represents a generator supplying power through a transmission system to a receiving-end system at bus ①. The rectangle represents the transmission system of linear passive components, such as transformers, transmission lines, and capacitors, and includes the transient reactance of the generator. Therefore, the voltage E_1' represents the transient internal voltage of the generator at bus ①. The voltage E_2' at the receiving end is regarded here as that of an infinite bus or as the transient internal voltage of a synchronous motor whose transient reactance is included in the network. Later we shall consider the case of two generators supplying constant-impedance loads within the network. The bus admittance matrix for the network reduced to two nodes in addition to the reference node is

$$\mathbf{Y}_{bus} = \begin{matrix} & ① & ② \\ ① & Y_{11} & Y_{12} \\ ② & Y_{21} & Y_{22} \end{matrix} \tag{16.28}$$

From Eq. (9.4) we have

$$P_k + jQ_k = V_k \sum_{n=1}^{N} (Y_{kn}V_n)^* \tag{16.29}$$

Letting k and N equal 1 and 2, respectively, and substituting E' for V, we obtain

$$P_1 + jQ_1 = E_1'(Y_{11}E_1')^* + E_1'(Y_{12}E_2')^* \tag{16.30}$$

If we define

$$E_1' = |E_1'|\underline{/\delta_1} \qquad E_2' = |E_2'|\underline{/\delta_2}$$

$$Y_{11} = G_{11} + jB_{11} \qquad Y_{12} = |Y_{12}|\underline{/\theta_{12}}$$

Eq. (16.30) yields

$$P_1 = |E_1'|^2 G_{11} + |E_1'||E_2'||Y_{12}|\cos(\delta_1 - \delta_2 - \theta_{12}) \tag{16.31}$$

$$Q_1 = -|E_1'|^2 B_{11} + |E_1'||E_2'||Y_{12}|\sin(\delta_1 - \delta_2 - \theta_{12}) \tag{16.32}$$

Similar equations apply at bus ① by substituting subscripts 1 for 2 and 2 for 1 in the two preceding equations.

If we let

$$\delta = \delta_1 - \delta_2$$

and define a new angle γ such that

$$\gamma = \theta_{12} - \frac{\pi}{2}$$

we obtain from Eqs. (16.31) and (16.32)

$$P_1 = |E_1'|^2 G_{11} + |E_1'||E_2'||Y_{12}|\sin(\delta - \gamma) \tag{16.33}$$

$$Q_1 = -|E_1'|^2 B_{11} - |E_1'||E_2'||Y_{12}|\cos(\delta - \gamma) \tag{16.34}$$

Equation (16.33) may be written more simply as

$$P_e = P_c + P_{\max} \sin(\delta - \gamma) \tag{16.35}$$

where $$P_c = |E_1'|^2 G_{11} \qquad P_{\max} = |E_1'||E_2'||Y_{12}| \tag{16.36}$$

Since P_1 represents the electric power output of the generator (armature loss neglected), we have replaced it by P_e in Eq. (16.35), which is often called the *power-angle equation*; its graph as a function of δ is called the *power-angle curve*. The parameters P_c, $P_{\max}$, and γ are constants for a given network configuration and constant voltage magnitudes $|E_1'|$ and $|E_2'|$. When the network is considered without resistance, all the elements of $\mathbf{Y}_{\text{bus}}$ are susceptances, and then G_{11} and γ are both zero. The power-angle equation which then applies for the pure reactance network is simply the familiar equation

$$P_e = P_{\max} \sin \delta \tag{16.37}$$

where $P_{\max} = |E_1'||E_2'|/X$ and X is the transfer reactance between E_1' and E_2'.

Example 16.3. The single-line diagram of Fig. 16.4 shows a generator connected through parallel transmission lines to a large metropolitan system considered as an infinite bus. The machine is delivering 1.0 per-unit power and both the terminal voltage and the infinite-bus voltage are 1.0 per unit. Numbers on the diagram

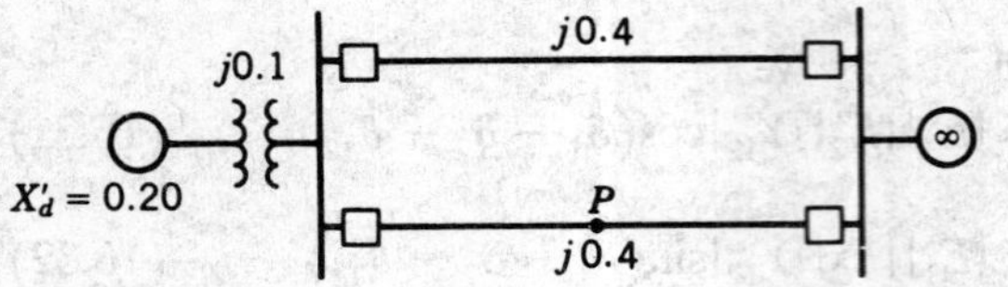

FIGURE 16.4
One-line diagram for Examples 16.3 and 16.4. Point P is at the center of the line.

indicate the values of the reactances on a common system base. The transient reactance of the generator is 0.20 per unit as indicated. Determine the power-angle equation for the given system operating conditions.

Solution. The reactance diagram for the system is shown in Fig. 16.5(a). The series reactance between the terminal voltage and the infinite bus is

$$X = 0.10 + \frac{0.4}{2} = 0.3 \text{ per unit}$$

and therefore the 1.0 per-unit power output of the generator is determined by

$$\frac{|V_t||V|}{X} \sin \alpha = \frac{(1.0)(1.0)}{0.3} \sin \alpha = 1.0$$

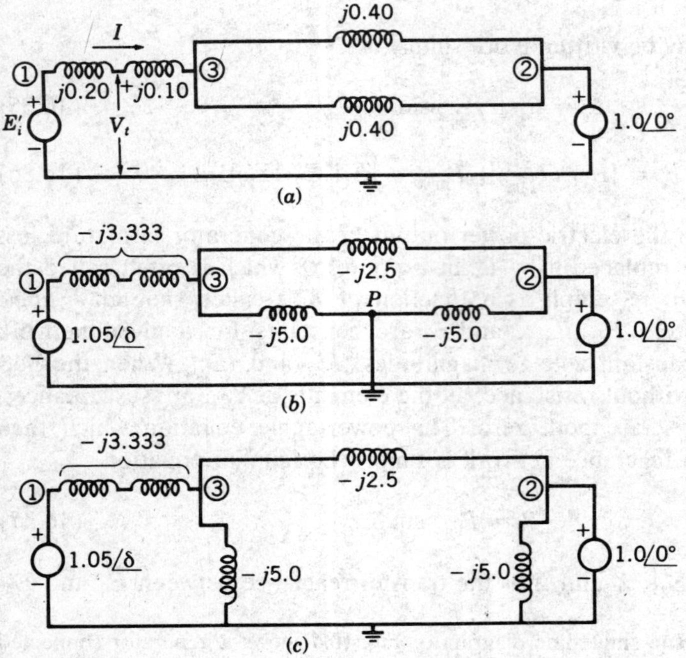

FIGURE 16.5
Reactance diagram for: (a) the prefault network for Example 16.3 with impedances in per unit; (b) and (c); the faulted network for Example 16.4 with the same impedances converted to admittances and marked in per unit.

where V is the voltage of the infinite bus and α is the angle of the terminal voltage relative to the infinite bus. Solving for α, we obtain

$$\alpha = \sin^{-1} 0.3 = 17.458°$$

so that the terminal voltage is

$$V_t = 1.0\angle 17.458° = 0.954 + j0.300 \text{ per unit}$$

The output current from the generator is now calculated as

$$I = \frac{1.0\angle \delta_2 - 1.0\angle 0°}{j0.3}$$

$$= 1.0 + j0.1535 = 1.012\angle 8.729° \text{ per unit}$$

and the transient internal voltage is then found to be

$$E_1' = (0.954 + j0.30) + j(0.2)(1.0 + j0.1535)$$

$$= 0.923 - j0.5 = 1.050\angle 28.44° \text{ per unit}$$

The power-angle equation relating the transient internal voltage E_i' and the infinite-bus voltage V is determined by the *total* series reactance

$$X = 0.2 + 0.1 + \frac{0.4}{2} = 0.5 \text{ per unit}$$

Hence, the desired equation is

$$P_e = \frac{(1.050)(1.0)}{0.5} \sin \delta = 2.10 \sin \delta \text{ per unit}$$

where δ is the machine rotor angle with respect to the infinite bus.

The power-angle equation of the preceding example is plotted in Fig. 16.6. Note that the mechanical input power P_m is constant and intersects the

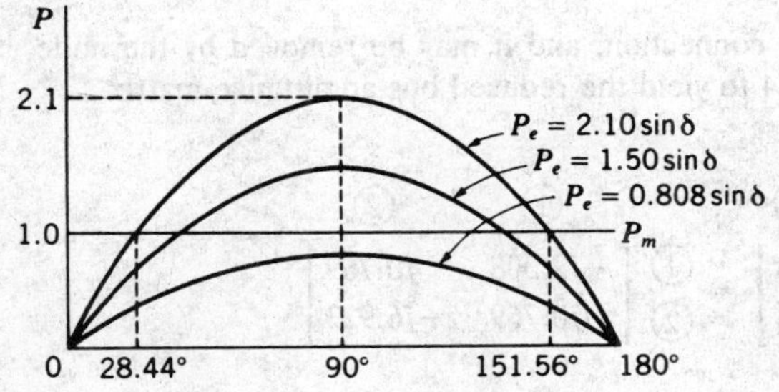

FIGURE 16.6
Plot of power-angle curves found in Examples 16.3 to 16.5.

sinusoidal power-angle curve at the operating angle $\delta_0 = 24.44°$. This is the initial angular position of the generator rotor corresponding to the given operating conditions. The swing equation for the machine may be written

$$\frac{H}{180f}\frac{d^2\delta}{dt^2} = 1.0 - 2.10\sin\delta \text{ per unit} \tag{16.38}$$

where H is in megajoules per megavoltampere, f is the electrical frequency of the system, and δ is in electrical degrees. We can easily check the results of Example 16.3 since under the given operating conditions, $P_e = 2.10\sin 28.44° = 1.0$ per unit, which corresponds exactly to the mechanical power input P_m and the acceleration is zero.

In the next example we determine the power-angle equation for the same system with a three-phase fault at P, the midpoint of one of the transmission lines. Positive acceleration is shown to exist due to the fault.

Example 16.4. The system of Example 16.3 is operating under the indicated conditions when a three-phase fault occurs at point P of Fig. 16.4. Determine the power-angle equation for the system with the fault on and the corresponding swing equation. Take $H = 5$ MJ/MVA.

Solution. The reactance diagram is shown in Fig. 16.5(b) with the fault on the system at point P. Values shown are admittances in per unit. The effect of the short circuit caused by the fault is clearly shown by redrawing the reactance diagram, as in Fig. 16.5(c). As calculated in Example 16.3, transient internal voltage of the generator remains at $E_i' = 1.05\angle 28.44°$ based on the assumption of constant flux linkage in the machine. The net transfer admittance connecting the voltage sources remains to be determined. The buses are numbered as shown and the $\mathbf{Y}_{\text{bus}}$ is formed by inspection of Fig. 16.5(c) as follows:

$$\mathbf{Y}_{\text{bus}} = \begin{matrix} & ① & ② & ③ \\ ① & -j3.333 & 0.00 & j3.333 \\ ② & 0.000 & -j7.50 & j2.500 \\ ③ & j3.333 & j2.50 & -j10.833 \end{matrix}$$

Bus ③ has no external source connection, and it may be removed by the node elimination procedure of Sec. 7.4 to yield the reduced bus admittance matrix

$$\begin{matrix} & ① & ② \\ ① & Y_{11} & Y_{12} \\ ② & Y_{21} & Y_{22} \end{matrix} = \begin{matrix} & ① & ② \\ ① & -j2.308 & j0.769 \\ ② & j0.769 & -j6.923 \end{matrix}$$

The magnitude of the transfer admittance is 0.769, and thus

$$P_{max} = |E_1'||E_2'||Y_{12}| = (1.05)(1.0)(0.769) = 0.808 \text{ per unit}$$

The power-angle equation with the fault on the system is therefore

$$P_e = 0.808 \sin \delta \text{ per unit}$$

and the corresponding swing equation is

$$\frac{5}{180f}\frac{d^2\delta}{dt^2} = 1.0 - 0.808 \sin \delta \text{ per unit} \tag{16.39}$$

Due to its inertia, the rotor cannot change position instantly upon occurrence of the fault. Therefore, the rotor angle δ is initially 28.44°, the same as in Example 16.3, and the electrical power output is $P_e = 0.808 \sin 28.44° = 0.385$. The initial accelerating power is

$$P_a = 1.0 - 0.385 = 0.615 \text{ per unit}$$

and the initial acceleration is positive with the value given by

$$\frac{d^2\delta}{dt^2} = \frac{180f}{5}(0.615) = 22.14f \text{ elec deg/s}^2$$

where f is the system frequency.

Relaying schemes sensing the fault on the line will act to clear the fault by simultaneous opening of the line-end breakers. When this occurs, another power-angle equation applies because of the network change.

Example 16.5. The fault on the system of Example 16.4 is cleared by simultaneous opening of the circuit breakers at each end of the affected line. Determine the power-angle equation and the swing equation for the postfault period.

Solution. Inspection of Fig. 16.5(a) shows that upon removal of the faulted line, the net transfer admittance across the system is

$$\frac{1}{j(0.2 + 0.1 + 0.4)} = -j1.429 \text{ per unit}$$

so that in the bus admittance matrix

$$Y_{12} = j1.429$$

Therefore, the postfault power-angle equation is

$$P_e = (1.05)(1.0)(1.429)\sin \delta = 1.500 \sin \delta$$

and the corresponding swing equation is

$$\frac{5}{180f}\frac{d^2\delta}{dt^2} = 1.0 - 1.500 \sin \delta$$

The acceleration at the instant of clearing the fault depends on the angular position of the rotor at that time. The power-angle curves for Examples 16.3 through 16.5 are compared in Fig. 16.6.

16.5 SYNCHRONIZING POWER COEFFICIENTS

In Example 16.3 the operating point on the sinusoidal P_e curve of Fig. 16.6 is found to be at $\delta_0 = 28.44°$, where the mechanical power input P_m equals the electrical power output P_e. In the same figure it is also seen that P_e equals P_m at $\delta = 151.56°$, and this might appear to be an equally acceptable operating point. However, this is not the case as now shown.

A commonsense requirement for an acceptable operating point is that the generator should not lose synchronism when small temporary changes occur in the electrical power output from the machine. To examine this requirement for fixed mechanical input power P_m, consider small incremental changes in the operating point parameters; that is, consider

$$\delta = \delta_0 + \delta_\Delta \qquad P_e = P_{e0} + P_{e\Delta} \tag{16.40}$$

where the subscript zero denotes the steady-state operating point values and the subscript delta (Δ) identifies the incremental variations from those values. Substituting Eqs. (16.40) in Eq. (16.37), we obtain the power-angle equation for the general two-machine system in the form

$$\begin{aligned} P_{e0} + P_{e\Delta} &= P_{\max} \sin(\delta_0 + \delta_\Delta) \\ &= P_{\max}(\sin \delta_0 \cos \delta_\Delta + \cos \delta_0 \sin \delta_\Delta) \end{aligned}$$

Since δ_Δ is a small incremental displacement from δ_0,

$$\sin \delta_\Delta \cong \delta_\Delta \qquad \text{and} \qquad \cos \delta_\Delta \cong 1 \tag{16.41}$$

and when strict equality is assumed, the previous equation becomes

$$P_{e0} + P_{e\Delta} = P_{\max} \sin \delta_0 + (P_{\max} \cos \delta_0)\, \delta_\Delta \tag{16.42}$$

At the initial operating point δ_0

$$P_m = P_{e0} = P_{\max} \sin \delta_0 \tag{16.43}$$

and it then follows from Eq. (16.42) that

$$P_m - (P_{e0} + P_{e\Delta}) = -(P_{\max} \cos \delta_0)\, \delta_\Delta \tag{16.44}$$

Substituting the incremental variables of Eq. (16.40) in the basic swing equation, Eq. (16.12), we obtain

$$\frac{2H}{\omega_s} \frac{d^2(\delta_0 + \delta_\Delta)}{dt^2} = P_m - (P_{e0} + P_{e\Delta}) \tag{16.45}$$

Replacing the right-hand side of this equation by Eq. (16.44) and transposing terms, we obtain

$$\frac{2H}{\omega_s} \frac{d^2\delta_\Delta}{dt^2} + (P_{\max} \cos \delta_0)\, \delta_\Delta = 0 \tag{16.46}$$

since δ_0 is a constant value. Noting that $P_{\max} \cos \delta_0$ is the slope of the power-angle curve at the angle δ_0, we denote this slope as S_p and define it as

$$S_p = \left.\frac{dP_e}{d\delta}\right|_{\delta=\delta_0} = P_{\max} \cos \delta_0 \tag{16.47}$$

where S_p is called the *synchronizing power coefficient*. When S_p is used in Eq. (16.46), the swing equation governing the incremental rotor-angle variations may be rewritten in the form

$$\frac{d^2\delta_\Delta}{dt^2} + \frac{\omega_s S_p}{2H} \delta_\Delta = 0 \tag{16.48}$$

This is a linear, second-order differential equation, the solution to which depends on the algebraic sign of S_p. When S_p is positive, the solution $\delta_\Delta(t)$ corresponds to that of simple harmonic motion; such motion is represented by the oscillations of an undamped swinging pendulum.[3] When S_p is negative, the solution $\delta_\Delta(t)$ increases exponentially without limit. Therefore, in Fig. 16.6 the operating point $\delta_0 = 28.44°$ is a point of stable equilibrium, in the sense that the rotor-angle swing is bounded following a small perturbation. In the physical situation damping will restore the rotor angle to δ_0 following the temporary electrical perturbation. On the other hand, the point $\delta = 151.56°$ is a point of

[3]The equation of simple harmonic motion is $d^2x/dt^2 + \omega_n^2 x = 0$, which has the general solution $A \cos \omega_n t + B \sin \omega_n t$ with constants A and B determined by the initial conditions. The solution when plotted is an undamped sinusoid of angular frequency ω_n.

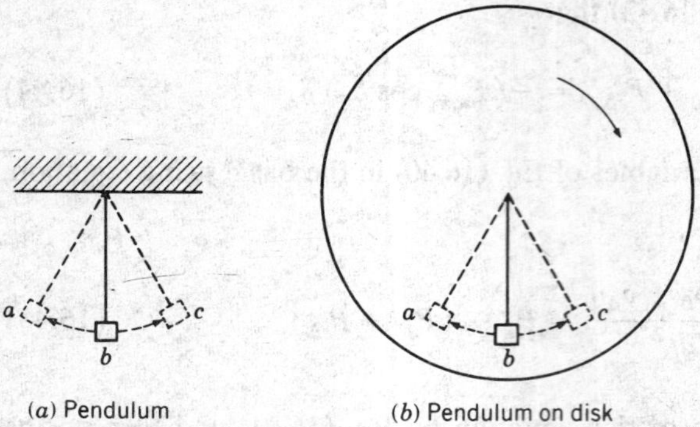

FIGURE 16.7
Pendulum and rotating disk to illustrate a rotor swinging with respect to an infinite bus.

unstable equilibrium since S_p is negative there. So, this point is not a valid operating point.

The changing position of the generator rotor swinging with respect to the infinite bus may be visualized by an analogy. Consider a pendulum swinging from a pivot on a stationary frame, as shown in Fig. 16.7(*a*). Points *a* and *c* are the maximum points of the oscillation of the pendulum about the equilibrium point *b*. Damping will eventually bring the pendulum to rest at *b*. Now imagine a disk rotating in a clockwise direction about the pivot of the pendulum, as shown in Fig. 16.7(*b*), and superimpose the motion of the pendulum on the motion of the disk. When the pendulum is moving from *a* to *c*, the combined angular velocity is slower than that of the disk. When the pendulum is moving from *c* to *a*, the combined angular velocity is faster than that of the disk. At points *a* and *c* the angular velocity of the pendulum alone is zero and the combined angular velocity equals that of the disk. If the angular velocity of the disk corresponds to the synchronous speed of the rotor, and if the motion of the pendulum alone represents the swinging of the rotor with respect to an infinite bus, the superimposed motion of the pendulum on that of the disk represents the actual angular motion of the rotor.

From the above discussion we conclude that the solution of Eq. (16.48) represents sinusoidal oscillations, provided the synchronizing power coefficient S_p is positive. The angular frequency of the undamped oscillations is given by

$$\omega_n = \sqrt{\frac{\omega_s S_p}{2H}} \text{ elec rad/s} \tag{16.49}$$

which corresponds to a frequency of oscillation given by

$$f_n = \frac{1}{2\pi}\sqrt{\frac{\omega_s S_p}{2H}} \text{ Hz} \tag{16.50}$$

Example 16.6. The machine of Example 16.3 is operating at $\delta = 28.44°$ when it is subjected to a slight temporary electrical system disturbance. Determine the frequency and period of oscillation of the machine rotor if the disturbance is removed before the prime mover responds. Take $H = 5$ MJ/MVA.

Solution. The applicable swing equation is Eq. (16.48) and the synchronizing power coefficient at the operating point is

$$S_p = 2.10 \cos 28.44° = 1.8466$$

The angular frequency of oscillation is therefore

$$\omega_n = \sqrt{\frac{\omega_s S_p}{2H}} = \sqrt{\frac{377 \times 1.8466}{2 \times 5}} = 8.343 \text{ elec rad/s}$$

The corresponding frequency of oscillation is

$$f_n = \frac{8.343}{2\pi} = 1.33 \text{ Hz}$$

and the period of oscillation is

$$T = \frac{1}{f_n} = 0.753 \text{ s}$$

The above example is an important one from the practical viewpoint since it indicates the order of magnitude of the frequencies which can be superimposed upon the nominal 60-Hz frequency in a large system having many interconnected machines. As load on the system changes randomly throughout the day, intermachine oscillations involving frequencies of the order of 1 Hz tend to arise, but these are quickly damped out by the various damping influences caused by the prime mover, the system loads, and the machine itself. It is worthwhile to note that even if the transmission system in our example contains resistance, nonetheless, the swinging of the rotor is harmonic and undamped. Problem 16.8 examines the effect of resistance on the synchronizing power coefficient and the frequency of oscillations. In a later section we again discuss the concept of synchronizing coefficients. In the next section we examine a method of determining stability under transient conditions caused by large disturbances.

16.6 EQUAL-AREA CRITERION OF STABILITY

In Sec. 16.4 we developed swing equations which are nonlinear in nature. Formal solutions of such equations cannot be explicitly found. Even in the case of a single machine swinging with respect to an infinite bus, it is very difficult to obtain literal-form solutions, and computer methods are therefore normally

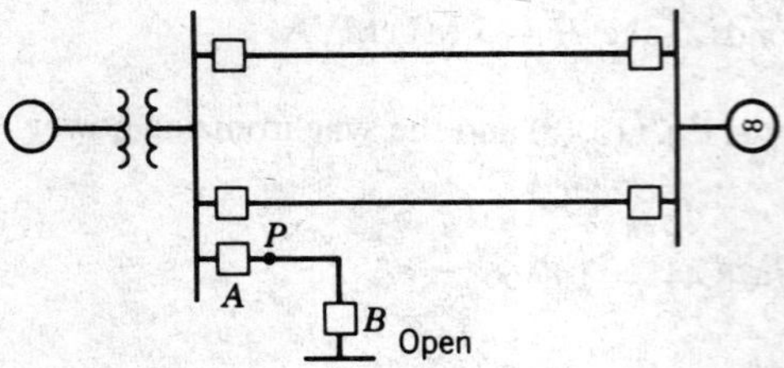

FIGURE 16.8
One-line diagram of the system of Fig. 16.4 with the addition of a short transmission line.

used. To examine the stability of a two-machine system without solving the swing equation, a direct approach is possible as now discussed.

The system shown in Fig. 16.8 is the same as that considered previously except for the addition of a short transmission line. Initially, circuit breaker A is considered to be closed while circuit breaker B at the opposite end of the short line is open. Therefore, the initial operating conditions of Example 16.3 may be considered unaltered. At point P close to the bus a three-phase fault occurs and is cleared by circuit breaker A after a short period of time. Thus, the effective transmission system is unaltered except while the fault is on. The short circuit caused by the fault is effectively at the bus, and so the electrical power output from the generator is zero until the fault is cleared. The physical conditions before, during, and after the fault can be understood by analyzing the power-angle curves of Fig. 16.9.

The generator is operating initially at synchronous speed with a rotor angle of δ_0, and the input mechanical power P_m equals the output electrical power P_e, as shown at point a in Fig. 16.9(a). When the fault occurs at $t = 0$, the electrical power output is suddenly zero while the input mechanical power is unaltered, as shown in Fig. 16.9(b). The difference in power must be accounted for by a rate of change of stored kinetic energy in the rotor masses. This can be accomplished only by an increase in speed which results from the constant accelerating power P_m. If we denote the time to clear the fault by t_c, then the acceleration is constant for time t less than t_c and is given by

$$\frac{d^2\delta}{dt^2} = \frac{\omega_s}{2H}P_m \tag{16.51}$$

While the fault is on, the velocity increase above synchronous speed is found by integrating this equation to obtain

$$\frac{d\delta}{dt} = \int_0^t \frac{\omega_s}{2H}P_m\,dt = \frac{\omega_s}{2H}P_m t \tag{16.52}$$

A further integration with respect to time yields for the rotor angle

$$\delta = \frac{\omega_s P_m}{4H}t^2 + \delta_0 \tag{16.53}$$

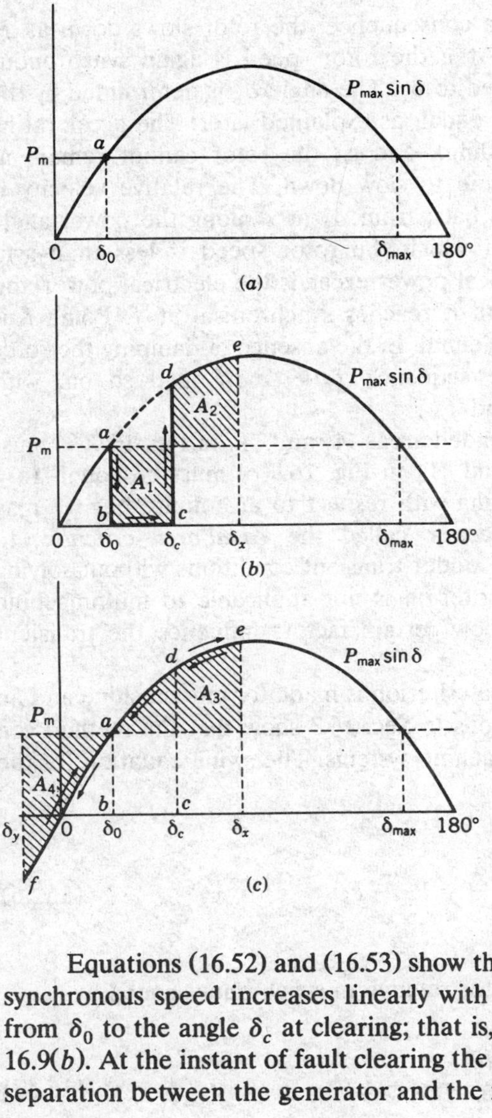

FIGURE 16.9
Power-angle curves for the generator shown in Fig. 16.8. Areas A_1 and A_2 are equal as are areas A_3 and A_4.

Equations (16.52) and (16.53) show that the velocity of the rotor relative to synchronous speed increases linearly with time while the rotor angle advances from δ_0 to the angle δ_c at clearing; that is, the angle δ goes from b to c in Fig. 16.9(b). At the instant of fault clearing the increase in rotor speed and the angle separation between the generator and the infinite bus are given, respectively, by

$$\left.\frac{d\delta}{dt}\right|_{t=t_c} = \frac{\omega_s P_m}{2H} t_c \tag{16.54}$$

and

$$\delta(t)\big|_{t=t_c} = \frac{\omega_s P_m}{4H} t_c^2 + \delta_0 \tag{16.55}$$

When the fault is cleared at the angle δ_c, the electrical power output abruptly increases to a value corresponding to point d on the power-angle curve. At d the electrical power output exceeds the mechanical power input, and thus the accelerating power is negative. As a consequence, the rotor slows down as P_e goes from d to e in Fig. 16.9(c). At e the rotor speed is again synchronous although the rotor angle has advanced to δ_x. The angle δ_x is determined by the fact that areas A_1 and A_2 must be equal, as explained later. The accelerating power at e is still negative (retarding), and so the rotor cannot remain at synchronous speed but must continue to slow down. The relative velocity is negative and the rotor angle moves back from δ_x at e along the power-angle curve of Fig. 16.9(c) to point a at which the rotor speed is less than synchronous. From a to f the mechanical power exceeds the electrical power and the rotor increases speed again until it reaches synchronism at f. Point f is located so that areas A_3 and A_4 are equal. In the absence of damping the rotor would continue to oscillate in the sequence f-a-e, e-a-f, and so on, with synchronous speed occurring at e and f.

We shall soon show that the shaded areas A_1 and A_2 in Fig. 16.9(b) must be equal, and similarly, areas A_3 and A_4 in Fig. 16.9(c) must be equal. In a system where one machine is swinging with respect to an infinite bus we may use this principle of equality of areas, called the *equal-area criterion*, to determine the stability of the system under transient conditions without solving the swing equation. Although the criterion is not applicable to multimachine systems, it helps in understanding how certain factors influence the transient stability of any system.

The derivation of the equal-area criterion is made for one machine and an infinite bus although the considerations in Sec. 16.3 show that the method can be readily adapted to general two-machine systems. The swing equation for the machine connected to the bus is

$$\frac{2H}{\omega_s}\frac{d^2\delta}{dt^2} = P_m - P_e \tag{16.56}$$

Define the angular velocity of the rotor relative to synchronous speed by

$$\omega_r = \frac{d\delta}{dt} = \omega - \omega_s \tag{16.57}$$

Differentiating Eq. (16.57) with respect to t and substituting in Eq. (16.56), we obtain

$$\frac{2H}{\omega_s}\frac{d\omega_r}{dt} = P_m - P_e \tag{16.58}$$

When the rotor speed is synchronous, it is clear that ω equals ω_s and ω_r is zero. Multiplying both sides of Eq. (16.58) by $\omega_r = d\delta/dt$, we have

$$\frac{H}{\omega_s} 2\omega_r \frac{d\omega_r}{dt} = (P_m - P_e)\frac{d\delta}{dt} \tag{16.59}$$

The left-hand side of this equation can be rewritten to give

$$\frac{H}{\omega_s} \frac{d(\omega_r^2)}{dt} = (P_m - P_e)\frac{d\delta}{dt} \tag{16.60}$$

Multiplying by dt and integrating, we obtain

$$\frac{H}{\omega_s}(\omega_{r2}^2 - \omega_{r1}^2) = \int_{\delta_1}^{\delta_2}(P_m - P_e)\, d\delta \tag{16.61}$$

The subscripts for the ω_r terms correspond to those for the δ limits. That is, the rotor speed ω_{r1} corresponds to that at the angle δ_1 and ω_{r2} corresponds to δ_2. Since ω_r represents the *departure* of the rotor speed from synchronous speed, we readily see that if the rotor speed is synchronous at δ_1 *and* δ_2, then, correspondingly, $\omega_{r1} = \omega_{r2} = 0$. Under this condition, Eq. (16.61) becomes

$$\int_{\delta_1}^{\delta_2}(P_m - P_e)\, d\delta = 0 \tag{16.62}$$

This equation applies to any two points δ_1 and δ_2 on the power-angle diagram, provided they are points at which the rotor speed is synchronous. In Fig. 16.9(b) two such points are a and e corresponding to δ_0 and δ_x, respectively. If we perform the integration of Eq. (16.62) in two steps, we can write

$$\int_{\delta_0}^{\delta_c}(P_m - P_e)\, d\delta + \int_{\delta_c}^{\delta_x}(P_m - P_e)\, d\delta = 0 \tag{16.63}$$

or

$$\int_{\delta_0}^{\delta_c}(P_m - P_e)\, d\delta = \int_{\delta_c}^{\delta_x}(P_e - P_m)\, d\delta \tag{16.64}$$

The left integral applies to the fault period, whereas the right integral corresponds to the immediate postfault period up to the point of maximum swing δ_x. In Fig. 16.9(b) P_e is zero during the fault. The shaded area A_1 is given by the left-hand side of Eq. (16.64) and the shaded area A_2 is given by the right-hand side. So, the two areas A_1 and A_2 are equal.

Since the rotor speed is also synchronous at δ_x and at δ_y in Fig. 16.9(c), the same reasoning as above shows that A_3 equals A_4. The areas A_1 and A_4 are directly proportional to the increase in kinetic energy of the rotor while it is

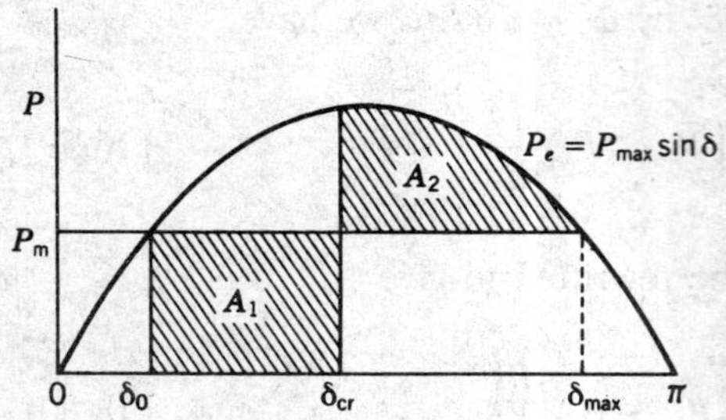

FIGURE 16.10
Power-angle curve showing the critical clearing angle δ_{cr}. Areas A_1 and A_2 are equal.

accelerating, whereas areas A_2 and A_3 are proportional to the decrease in kinetic energy of the rotor while it is decelerating. This can be seen by inspection of both sides of Eq. (16.61). Therefore, the equal-area criterion states that the kinetic energy added to the rotor following a fault must be removed after the fault in order to restore the rotor to synchronous speed.

The shaded area A_1 is dependent on the time taken to clear the fault. If there is delay in clearing, the angle δ_c is increased; likewise, the area A_1 increases and the equal-area criterion requires that area A_2 also increase to restore the rotor to synchronous speed at a larger angle of maximum swing δ_x. If the delay in clearing is prolonged so that the rotor angle δ swings beyond the angle δ_{max} in Fig. 16.9, then the rotor speed at that point on the power-angle curve is above synchronous speed when positive accelerating power is again encountered. Under the influence of this positive accelerating power, the angle δ will increase without limit and instability results. Therefore, there is a critical angle for clearing the fault in order to satisfy the requirements of the equal-area criterion for stability. This angle, called the *critical clearing angle* δ_{cr}, is shown in Fig. 16.10. The corresponding critical time for removing the fault is called the *critical clearing time* t_{cr}. Thus, the critical clearing time is the maximum elapsed time from the initiation of the fault until its isolation such that the power system is transiently stable.

In the particular case of Fig. 16.10 *both* the critical clearing angle *and* the critical clearing time can be calculated as follows. The rectangular area A_1 is

$$A_1 = \int_{\delta_0}^{\delta_{cr}} P_m \, d\delta = P_m(\delta_{cr} - \delta_0) \tag{16.65}$$

while the area A_2 is

$$\begin{aligned} A_2 &= \int_{\delta_{cr}}^{\delta_{max}} (P_{max} \sin \delta - P_m) \, d\delta \\ &= P_{max}(\cos \delta_{cr} - \cos \delta_{max}) - P_m(\delta_{max} - \delta_{cr}) \end{aligned} \tag{16.66}$$

Equating the expressions for A_1 and A_2 and transposing terms, yield

$$\cos\delta_{cr} = (P_m/P_{max})(\delta_{max} - \delta_0) + \cos\delta_{max} \tag{16.67}$$

We see from the sinusoidal power-angle curve that

$$\delta_{max} = \pi - \delta_0 \text{ elec rad} \tag{16.68}$$

and

$$P_m = P_{max}\sin\delta_0 \tag{16.69}$$

Substituting for δ_{max} and P_m in Eq. (16.67), simplifying the result, and solving for the critical clearing angle δ_{cr}, we obtain

$$\delta_{cr} = \cos^{-1}[(\pi - 2\delta_0)\sin\delta_0 - \cos\delta_0] \tag{16.70}$$

Substituting this value of δ_{cr} in the left-hand side of Eq. (16.55) yields

$$\delta_{cr} = \frac{\omega_s P_m}{4H}t_{cr}^2 + \delta_0 \tag{16.71}$$

from which we find the critical clearing time

$$t_{cr} = \sqrt{\frac{4H(\delta_{cr} - \delta_0)}{\omega_s P_m}} \tag{16.72}$$

Example 16.7. Calculate the critical clearing angle and the critical clearing time for the system of Fig. 16.8 when the system is subjected to a three-phase fault at point P on the short transmission line. The initial conditions are the same as those in Example 16.3, and $H = 5$ MJ/MVA.

Solution. In Example 16.3 the power-angle equation and initial rotor angle are

$$P_e = P_{max}\sin\delta = 2.10\sin\delta$$

$$\delta_0 = 28.44° = 0.496 \text{ elec rad}$$

Mechanical input power P_m is 1.0 per unit, and Eq. (16.70) then gives

$$\delta_{cr} = \cos^{-1}[(\pi - 2\times 0.496)\sin 28.44° - \cos 28.44°]$$

$$= 81.697° = 1.426 \text{ elec rad}$$

Using this value with the other known quantities in Eq. (16.72) yields

$$t_{cr} = \sqrt{\frac{4 \times 5(1.426 - 0.496)}{377 \times 1}} = 0.222 \text{ s}$$

which is equivalent to a critical clearing time of 13.3 cycles on a frequency base of 60 Hz.

This example serves to establish the concept of critical clearing time which is essential to the design of proper relaying schemes for fault clearing. In more general cases the critical clearing time cannot be explicitly found without solving the swing equations by computer simulation.

16.7 FURTHER APPLICATIONS OF THE EQUAL-AREA CRITERION

The equal-area criterion is a very useful means for analyzing stability of a system of two machines, or of a single machine supplied from an infinite bus. However, the computer is the only practical way to determine the stability of a large system. Because the equal-area criterion is so helpful in understanding transient stability, we continue to examine it briefly before discussing the determination of swing curves by the computer approach.

When a generator is supplying power to an infinite bus over parallel transmission lines, opening one of the lines may cause the generator to lose synchronism even though the load could be supplied over the remaining line under steady-state conditions. If a three-phase short circuit occurs on the bus to which two parallel lines are connected, no power can be transmitted over either line. This is essentially the case in Example 16.7. However, if the fault is at the end of one of the lines, opening breakers at both ends of the line will isolate the fault from the system and allow power to flow through the other parallel line. When a three-phase fault occurs at some point on a double-circuit line other than on the paralleling buses or at the extreme ends of the line, there is some impedance between the paralleling buses and the fault. Therefore, some power is transmitted while the fault is still on the system. The power-angle equation in Example 16.4 demonstrates this fact.

When power is transmitted during a fault, the equal-area criterion is applied, as shown in Fig. 16.11, which is similar to the power-angle diagram of Fig. 16.6. Before the fault $P_{max} \sin \delta$ is the power which can be transmitted; during the fault $r_1 P_{max} \sin \delta$ is the power which can be transmitted; and $r_2 P_{max} \sin \delta$ is the power which can be transmitted after the fault is cleared by switching at the instant when $\delta = \delta_{cr}$. Examination of Fig. 16.11 shows that δ_{cr} is the critical clearing angle in this case. By evaluating the areas A_1 and A_2

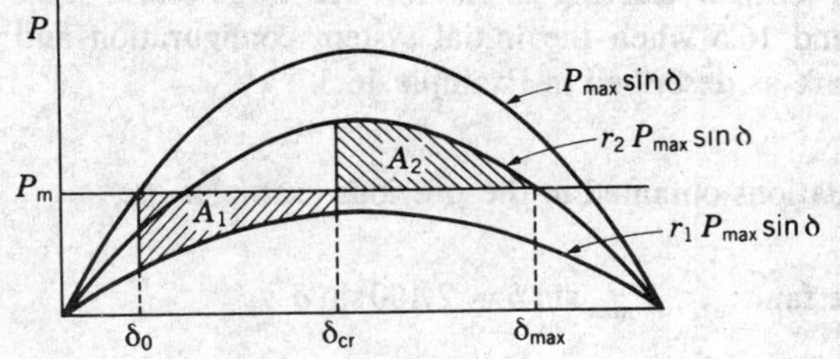

FIGURE 16.11
Equal-area criterion applied to fault clearing when power is transmitted during the fault. Areas A_1 and A_2 are equal.

using the procedural steps of the previous section, we find that

$$\cos \delta_{cr} = \frac{(P_m/P_{max})(\delta_{max} - \delta_0) + r_2 \cos \delta_{max} - r_1 \cos \delta_0}{r_2 - r_1} \tag{16.73}$$

A literal-form solution for the critical clearing time t_{cr} is not possible in this case. For the particular system and fault location shown in Fig. 16.8 the applicable values are $r_1 = 0$, $r_2 = 1$, and Eq. (16.73) then reduces to Eq. (16.67).

Short-circuit faults which do not involve all three phases allow some power to be transmitted over the unaffected phases. Such faults are represented by connecting an impedance (rather than a short circuit) between the fault point and the reference node in the positive-sequence impedance diagram. The larger the impedance shunted across the positive-sequence network to represent the fault, the larger the power transmitted during the fault. The amount of power transmitted during the fault affects the value of A_1 for any given clearing angle. Thus, smaller values of r_1 result in greater disturbances to the system, as smaller amounts of power are transmitted during the fault. Consequently, the area A_1 of acceleration is larger. In order of increasing severity (that is, decreasing $r_1 P_{max}$) the various faults are:

1. A single line-to-ground fault
2. A line-to-line fault
3. A double line-to-ground fault
4. A three-phase fault

The single line-to-ground fault occurs most frequently, and the three-phase fault is the least frequent. For complete reliability a system should be designed for transient stability for three-phase faults at the worst locations, and this is virtually the universal practice.

Example 16.8. Determine the critical clearing angle for the three-phase fault described in Examples 16.4 and 16.5 when the initial system configuration and prefault operating conditions are as described in Example 16.3.

Solution. The power-angle equations obtained in the previous examples are

$$\text{Before the fault:} \qquad P_{max} \sin \delta = 2.100 \sin \delta$$

$$\text{During the fault:} \quad r_1 P_{max} \sin \delta = 0.808 \sin \delta$$

$$\text{After the fault:} \qquad r_2 P_{max} \sin \delta = 1.500 \sin \delta$$

Hence,

$$r_1 = \frac{0.808}{2.100} = 0.385 \qquad r_2 = \frac{1.500}{2.100} = 0.714$$

From Example 16.3 we have

$$\delta_0 = 28.44° = 0.496 \text{ rad}$$

and from Fig. 16.11 we calculate

$$\delta_{max} = 180° - \sin^{-1}\left[\frac{1.000}{1.500}\right] = 138.190° = 2.412 \text{ rad}$$

Therefore, inserting numerical values in Eq. (16.73), we obtain

$$\cos \delta_{cr} = \frac{\left(\dfrac{1.0}{2.10}\right)(2.412 - 0.496) + 0.714 \cos(138.19°) - 0.385 \cos(28.44°)}{0.714 - 0.385}$$

$$= 0.127$$

Hence,

$$\delta_{cr} = 82.726°$$

To determine the critical clearing time for this example, we must obtain the swing curve of δ versus t. In Sec. 16.9 we discuss one method of computing such swing curves.

16.8 MULTIMACHINE STABILITY STUDIES: CLASSICAL REPRESENTATION

The equal-area criterion cannot be used directly in systems where three or more machines are represented. Although the physical phenomena observed in the two-machine problems are basically the same as in the multimachine case, nonetheless, the complexity of the numerical computations increases with the number of machines considered in a transient stability study. When a multimachine system operates under electromechanical transient conditions, intermachine oscillations occur through the medium of the transmission system connecting the machines. If any one machine could be considered to act alone as the single oscillating source, it would send into the interconnected system an electromechanical oscillation determined by its inertia and synchronizing power. A typical frequency of such an oscillation is of the order of 1–2 Hz, and this is superimposed upon the nominal 60-Hz frequency of the system. When many machine rotors are simultaneously undergoing transient oscillation, the swing curves reflect the combined presence of many such oscillations. Therefore, the transmission system frequency is not unduly perturbed from nominal frequency, and the assumption is made that the 60-Hz network parameters are still applicable. To ease the complexity of system modeling, and thereby the computational burden, the following additional assumptions are commonly made in transient stability studies:

1. The mechanical power input to each machine remains constant during the entire period of the swing curve computation.
2. Damping power is negligible.
3. Each machine may be represented by a constant transient reactance in series with a constant transient internal voltage.
4. The mechanical rotor angle of each machine coincides with δ, the electrical phase angle of the transient internal voltage.
5. All loads may be considered as shunt impedances to ground with values determined by conditions prevailing immediately prior to the transient conditions.

The system stability model based on these assumptions is called the *classical stability model*, and studies which use this model are called *classical stability studies*. These assumptions, which we shall adopt, are in addition to the fundamental assumptions set forth in Sec. 16.1 for *all* stability studies. Of course, detailed computer programs with more sophisticated machine and load models are available to modify one or more of assumptions 1 to 5. Throughout this chapter, however, the classical model is used to study system disturbances originating from three-phase faults.

The system conditions before the fault occurs, and the network configuration both during and after its occurrence, must be known in any transient stability study, as we have seen. Consequently, in the multimachine case two preliminary steps are required:

1. The steady-state prefault conditions for the system are calculated using a production-type power-flow program.
2. The prefault network representation is determined and then modified to account for the fault and for the postfault conditions.

From the first preliminary step we know the values of power, reactive power, and voltage at each generator terminal and load bus, with all angles measured with respect to the slack bus. The transient internal voltage of each generator is then calculated using the equation

$$E = V_t + jX'_d I \tag{16.74}$$

where V_t is the corresponding terminal voltage and I is the output current. Each load is converted into a constant admittance to ground at its bus using the equation

$$Y_L = \frac{P_L - jQ_L}{|V_L|^2} \tag{16.75}$$

where $P_L + jQ_L$ is the load and $|V_L|$ is the magnitude of the corresponding bus voltage. The bus admittance matrix which is used for the prefault power-flow calculation is now augmented to include the transient reactance of each generator and the shunt admittance of each load, as suggested in Fig. 16.12. Note that the injected current is zero at all buses except the internal buses of the generators.

In the second preliminary step the bus admittance matrix is modified to correspond to the faulted and postfault conditions. Since only the generator internal buses have injections, all other buses can be eliminated by Kron

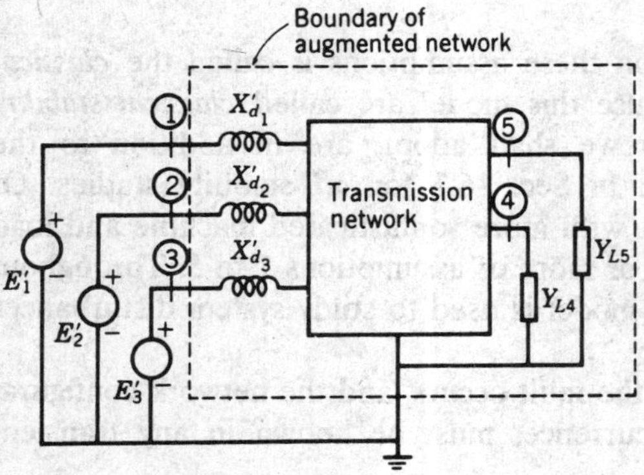

FIGURE 16.12
Augmented network of a power system.

reduction. The dimensions of the modified matrices then correspond to the number of generators. During and after the fault the power flow into the network from each generator is calculated by the corresponding power-angle equation. For example, in Fig. 16.12 the power out of generator 1 is given by

$$P_{e1} = |E_1'|^2 G_{11} + |E_1'||E_2'||Y_{12}|\cos(\delta_{12} - \theta_{12}) + |E_1'||E_3'||Y_{13}|\cos(\delta_{13} - \theta_{13}) \tag{16.76}$$

where δ_{12} equals $\delta_1 - \delta_2$. Similar equations are written for P_{e2} and P_{e3} using the Y_{ij} elements of the 3×3 bus admittance matrices appropriate to the fault or postfault condition. The P_{ei} expressions form part of the swing equations

$$\frac{2H_i}{\omega_s}\frac{d^2\delta_i}{dt^2} = P_{mi} - P_{ei} \qquad i = 1, 2, 3 \tag{16.77}$$

which represent the motion of each rotor during the fault and postfault periods. The solutions depend on the location and duration of the fault, and the $\mathbf{Y}_{\text{bus}}$ resulting when the faulted line is removed. The basic procedures used in computer programs for classical stability studies are revealed in the following examples.

Example 16.9. A 60-Hz, 230-kV transmission system shown in Fig. 16.13 has two generators of finite inertia and an infinite bus. The transformer and line data are given in Table 16.2. A three-phase fault occurs on line ④–⑤ near bus ④. Using the prefault power-flow solution given in Table 16.3, determine the swing equation for each machine during the fault period. The generators have reactances

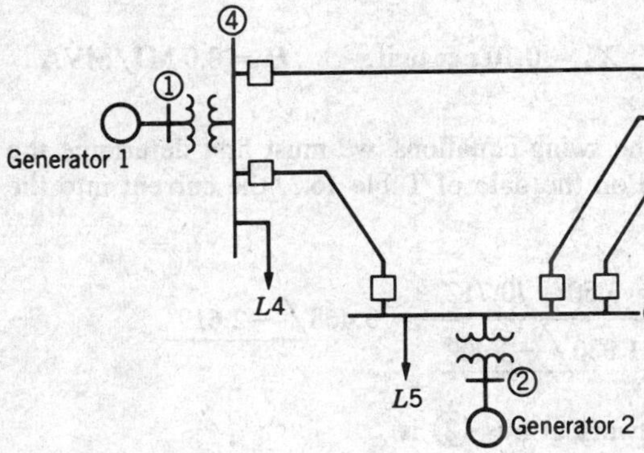

FIGURE 16.13
One-line diagram for Example 16.9.

TABLE 16.2
Line and transformer data for Example 16.9†

Bus to bus	Series Z R	Series Z X	Shunt Y B
Transformer ①-④	—	0.022	
Transformer ②-⑤	—	0.040	
Line ③-④	0.007	0.040	0.082
Line ③-⑤(1)	0.008	0.047	0.098
Line ③-⑤(2)	0.008	0.047	0.098
Line ④-⑤	0.018	0.110	0.226

†All values in per unit on 230-kV, 100-MVA base.

TABLE 16.3
Bus data and prefault load-flow values†

Bus	Voltage	Generation P	Generation Q	Load P	Load Q
①	$1.030\angle 8.88^\circ$	3.500	0.712		
②	$1.020\angle 6.38^\circ$	1.850	0.298		
③	$1.000\angle 0^\circ$	—	—		
④	$1.018\angle 4.68^\circ$	—	—	1.00	0.44
⑤	$1.011\angle 2.27^\circ$	—	—	0.50	0.16

†Values are in per unit on 230-kV, 100-MVA base.

and H values expressed on a 100-MVA base as follows:

Generator 1: 400 MVA, 20 kV, $X_d' = 0.067$ per unit, $H = 11.2$ MJ/MVA

Generator 2: 250 MVA, 18 kV, $X_d' = 0.10$ per unit, $H = 8.0$ MJ/MVA

Solution. In order to formulate the swing equations, we must first determine the transient internal voltages. Based on the data of Table 16.3, the current into the network at bus ① is

$$I_1 = \frac{(P_1 + jQ_1)^*}{V_1^*} = \frac{3.50 - j0.712}{1.030\angle -8.88^\circ} = 3.468\angle -2.619^\circ$$

Similarly, the current into the network at bus ② is

$$I_2 = \frac{(P_2 + jQ_2)^*}{V_2^*} = \frac{1.850 - j0.298}{1.020\angle -6.38^\circ} = 1.837\angle -2.771^\circ$$

From Eq. (16.74) we then calculate

$$E_1' = 1.030\angle 8.88^\circ + j0.067 \times 3.468\angle -2.619^\circ = 1.100\angle 20.82^\circ$$

$$E_2' = 1.020\angle 6.38^\circ + j0.10 \times 1.837\angle -2.771^\circ = 1.065\angle 16.19^\circ$$

At the infinite bus we have

$$E_3' = E_3 = 1.000\angle 0.0^\circ$$

and so $\delta_{13} = \delta_1 \qquad \delta_{23} = \delta_2$

The P-Q loads at buses ④ and ⑤ are converted into equivalent shunt admittances using Eq. (16.75), which yields

$$Y_{L4} = \frac{1.00 - j0.44}{(1.018)^2} = 0.9649 - j0.4246 \text{ per unit}$$

$$Y_{L5} = \frac{0.50 - j0.16}{(1.011)^2} = 0.4892 - j0.1565 \text{ per unit}$$

The prefault bus admittance matrix is now modified to include the load admittances and the transient reactances of the machines. Buses ① and ② designate the fictitious internal nodes behind the transient reactances of the machines. So, in the prefault bus admittance matrix, for example,

$$Y_{11} = \frac{1}{j0.067 + j0.022} = -j11.236 \text{ per unit}$$

$$Y_{34} = -\frac{1}{0.007 + j0.040} = -4.2450 + j24.2571 \text{ per unit}$$

The sum of the admittances connected to buses ③, ④, and ⑤ must include the shunt capacitances of the transmission lines. So, at bus ④ we have

$$Y_{44} = -j11.236 + \frac{j0.082}{2} + \frac{j0.226}{2} + 4.2450 - j24.2571$$

$$+ \frac{1}{0.018 + j0.110} + 0.9649 - j0.4246$$

$$= 6.6587 - j44.6175 \text{ per unit}$$

The new prefault bus admittance matrix is displayed as Table 16.4 on page 733.

Bus ④ must be short-circuited to the reference to represent the fault. Row 4 and column 4 of Table 16.4 thereby disappear because node ④ is now merged with the reference node. Next, the row and column representing bus ⑤ are eliminated by Kron reduction, and we obtain the bus admittance matrix shown in the upper half of Table 16.5. The faulted-system $\mathbf{Y}_{\text{bus}}$ shows that bus ① decouples from the other buses during the fault and that bus ② is connected directly to bus ③. This reflects the physical fact that the short circuit at bus ④ reduces to zero the power injected into the system from generator 1 and causes generator 2 to deliver its power radially to bus ③. Under fault conditions, the power-angle equations based on values from Table 16.5 are

$$P_{e1} = 0$$

$$\begin{aligned} P_{e2} &= |E_2'|^2 G_{22} + |E_2'||E_3||Y_{23}|\cos(\delta_{23} - \theta_{23}) \\ &= (1.065)^2(0.1362) + (1.065)(1.0)(5.1665)\cos(\delta_2 - 90.755°) \\ &= 0.1545 + 5.5023\sin(\delta_2 - 0.755°) \text{ per unit} \end{aligned}$$

Therefore, while the fault is on the system, the desired swing equations (values of P_{m1} and P_{m2} from Table 16.3) are

$$\begin{aligned} \frac{d^2\delta_1}{dt^2} &= \frac{180f}{H_1}(P_{m1} - P_{e1}) = \frac{180f}{H_1}P_{a1} \\ &= \frac{180f}{11.2}(3.5) \text{ elec deg/s}^2 \end{aligned}$$

$$\begin{aligned} \frac{d^2\delta_2}{dt^2} &= \frac{180f}{H_2}(P_{m2} - P_{e2}) = \frac{180f}{H_2}P_{a2} \\ &= \frac{180f}{8.0}\left\{\overbrace{1.85}^{P_m} - \left[\overbrace{0.1545}^{P_c} + \overbrace{5.5023}^{P_{\max}}\sin\left(\delta_2 - \overbrace{0.755°}^{\gamma}\right)\right]\right\} \\ &= \frac{180f}{8.0}\left[\underbrace{1.6955}_{P_m - P_c} - \underbrace{5.5023}_{P_{\max}}\sin\left(\delta_2 - \underbrace{0.755°}_{\gamma}\right)\right] \text{ elec deg/s}^2 \end{aligned}$$

Example 16.10. The three-phase fault in Example 16.9 is cleared by simultaneously opening the circuit breakers at the ends of the faulted line. Determine the swing equations for the postfault period.

Solution. Since the fault is cleared by removing line ④–⑤, the prefault $\mathbf{Y}_{\text{bus}}$ of Table 16.4 must be modified again. This is accomplished by substituting zero for Y_{45} and Y_{54} and by subtracting the series admittance of line ④–⑤ and the capacitive susceptance of one-half the line from elements Y_{44} and Y_{55} of Table 16.4. The reduced bus admittance matrix applicable to the postfault network is shown in the lower half of Table 16.5. The zero elements in the first and second rows reflect the fact that generators 1 and 2 are not interconnected when line

TABLE 16.4

Elements of prefault bus admittance matrix for Example 16.9†

Bus	①	②	③	④	⑤
①	$-j11.2360$	0.0	0.0	$j11.2360$	0.0
②	0.0	$-j7.1429$	0.0	0.0	$j7.1429$
③	0.0	0.0	11.2841 $-j65.4731$	-4.2450 $+j24.2571$	-7.0392 $+j41.3550$
④	$j11.2360$	0.0	-4.2450 $+j24.2571$	6.6588 $-j44.6175$	-1.4488 $+j8.8538$
⑤	0.0	$j7.1429$	-7.0392 $+j41.3550$	-1.4488 $+j8.8538$	8.9772 $-j57.2972$

†Admittances in per unit.

TABLE 16.5

Elements of faulted and postfault bus admittance matrices for Example 16.9†

Bus	①	②	③
		Faulted network	
①	$0.0000 - j11.2360$ $(11.2360\angle -90^\circ)$	$0.0 + j0.0$	$0.0 + j0.0$
②	$0.0 + j0.0$	$0.1362 - j6.2737$ $(6.2752\angle -88.7563^\circ)$	$-0.0681 + j5.1661$ $(5.1665\angle 90.7552^\circ)$
③	$0.0 + j0.0$	$-0.681 + j5.1661$ $(5.1665\angle 90.7552^\circ)$	$5.7986 - j35.6299$ $(36.0987\angle -80.7564^\circ)$
		Postfault network	
①	$0.5005 - j7.7897$ $(7.8058\angle -86.3237^\circ)$	$0.0 + j0.0$	$-0.2216 + j7.6291$ $(7.6323\angle 91.6638^\circ)$
②	$0.0 + j0.0$	$0.1591 - j6.1168$ $(6.1189\angle -88.5101^\circ)$	$-0.0901 + j6.0975$ $(6.0982\angle 90.8466^\circ)$
③	$-0.2216 + j7.6291$ $(7.6323\angle 91.6638^\circ)$	$-0.0901 + j6.0975$ $(6.0982\angle 90.8466^\circ)$	$1.3927 - j13.8728$ $(13.9426\angle -84.2672^\circ)$

†Admittances in per unit.

④–⑤ is removed. Accordingly, each generator is connected radially to the infinite bus, and we can write power-angle equations for the postfault conditions as follows:

$$P_{e1} = |E_1'|^2 G_{11} + |E_1'||E_3||Y_{13}|\cos(\delta_{13} - \theta_{13})$$

$$= (1.100)^2(0.5005) + (1.100)(1.0)(7.6323)\cos(\delta_1 - 91.664^\circ)$$

$$= 0.6056 + 8.3955\sin(\delta_1 - 1.664^\circ) \text{ per unit}$$

and $$P_{e2} = |E_2'|^2 G_{22} + |E_2'||E_3||Y_{23}|\cos(\delta_{23} - \theta_{23})$$

$$= (1.065)^2(0.1591) + (1.065)(1.0)(6.0982)\cos(\delta_2 - 90.847°)$$

$$= 0.1804 + 6.4934\sin(\delta_2 - 0.847°) \text{ per unit}$$

For the postfault period the applicable swing equations are given by

$$\frac{d^2\delta_1}{dt^2} = \frac{180f}{11.2}\{3.5 - [0.6056 + 8.3955\sin(\delta_1 - 1.664°)]\}$$

$$= \frac{180f}{11.2}[2.8944 - 8.3955\sin(\delta_1 - 1.664°)] \text{ elec deg/s}^2$$

and $$\frac{d^2\delta_2}{dt^2} = \frac{180f}{8.0}\{1.85 - [0.1804 + 6.4934\sin(\delta_2 - 0.847°)]\}$$

$$= \frac{180f}{8.0}[1.6696 - 6.4934\sin(\delta_2 - 0.847°)] \text{ elec deg/s}^2$$

The power-angle equations obtained in Examples 16.9 and 16.10 are of the form of Eq. (16.35), and the corresponding swing equations assume the form

$$\frac{d^2\delta}{dt^2} = \frac{180f}{H}[P_m - P_c - P_{\max}\sin(\delta - \gamma)] \tag{16.78}$$

The bracketed right-hand term represents the accelerating power of the rotor. Accordingly, we may write Eq. (16.78) as

$$\frac{d^2\delta}{dt^2} = \frac{180f}{H}P_a \text{ elec deg/s}^2 \tag{16.79}$$

where $$P_a = P_m - P_c - P_{\max}\sin(\delta - \gamma) \tag{16.80}$$

In the next section we discuss how to solve equations of the form of Eq. (16.79) in order to obtain δ as a function of time for specified clearing times.

16.9 STEP-BY-STEP SOLUTION OF THE SWING CURVE

For large systems we depend on the computer, which determines δ versus t for all machines in which we are interested; and δ may be plotted versus t for a machine to obtain the swing curve of that machine. The angle δ is calculated as a function of time over a period long enough to determine whether δ will increase without limit or reach a maximum and start to decrease. Although the latter result usually indicates stability, on an actual system where a number of variables are taken into account it may be necessary to plot δ versus t over a long enough interval to be sure δ will not increase again without returning to a low value.

By determining swing curves for various clearing times, we can find the length of time permitted before clearing a fault. Standard interrupting times for circuit breakers and their associated relays are commonly 8, 5, 3, or 2 cycles after a fault occurs, and thus breaker speeds may be specified. Calculations should be made for a fault in the position which will allow the least transfer of power from the machine and for the most severe type of fault for which protection against loss of stability is justified.

A number of different methods is available for the numerical evaluation of second-order differential equations in step-by-step computations for small increments of the independent variable. The more elaborate methods are practical only when the computations are performed on a computer. The step-by-step method used for hand calculations is necessarily simpler than some of the methods recommended for computers. In the methods for hand calculation the change in the angular position of the rotor during a short interval of time is computed by making the following assumptions:

1. The accelerating power P_a computed at the beginning of an interval is constant from the middle of the preceding interval to the middle of the interval considered.
2. Throughout any interval the angular velocity is constant at the value computed for the middle of the interval.

Of course, neither of the above assumptions is accurate since δ is changing continuously and both P_a and ω are functions of δ. As the time interval is decreased, the computed swing curve becomes more accurate. Figure 16.14(a) will help in visualizing the assumptions. The accelerating power is computed for the points enclosed in circles at the ends of the $n-2$, $n-1$, and n intervals, which are the beginnings of the $n-1$, n, and $n+1$ intervals. The step curve of P_a in Fig. 16.14(a) results from the assumptions that P_a is constant between the midpoints of the intervals. Similarly, ω_r, the *excess* of the angular velocity ω over the synchronous angular velocity ω_s, is shown in Fig. 16.14(b) as a step curve that is constant throughout the interval at the value computed for the midpoint. Between the ordinates $n-\frac{3}{2}$ and $n-\frac{1}{2}$ there is a change of speed caused by the constant accelerating power. The change in speed is the product of the acceleration and the time interval, and so

$$\omega_{r,n-1/2} - \omega_{r,n-3/2} = \frac{d^2\delta}{dt^2}\Delta t = \frac{180f}{H}P_{a,n-1}\Delta t \qquad (16.81)$$

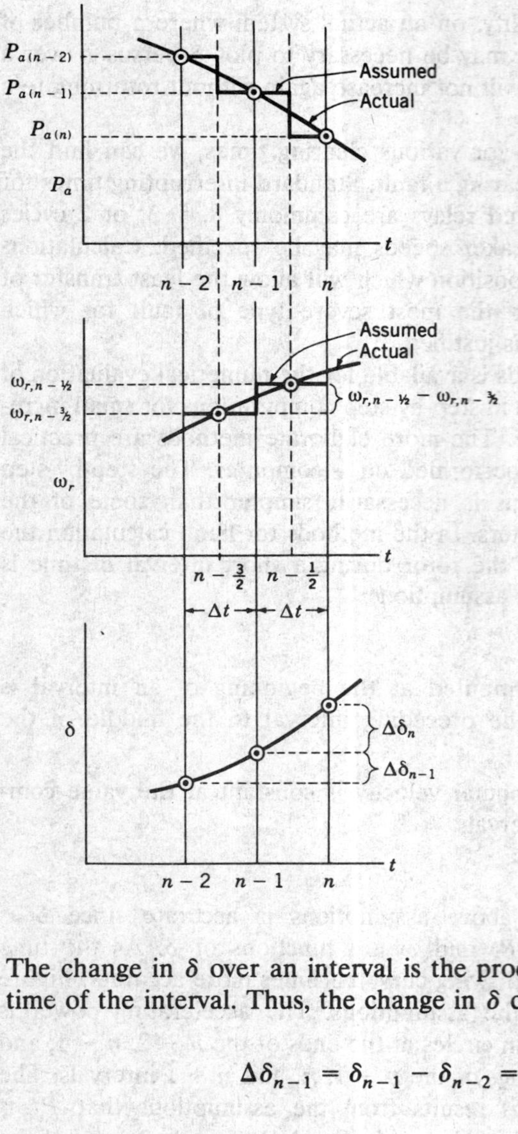

FIGURE 16.14
Actual and assumed values of P_a, ω_r, and δ as functions of time.

The change in δ over an interval is the product of ω_r for the interval and the time of the interval. Thus, the change in δ during the $n-1$ interval is

$$\Delta\delta_{n-1} = \delta_{n-1} - \delta_{n-2} = \Delta t \times \omega_{r,n-3/2} \tag{16.82}$$

and during the nth interval

$$\Delta\delta_n = \delta_n - \delta_{n-1} = \Delta t \times \omega_{r,n-1/2} \tag{16.83}$$

Subtracting Eq. (16.82) from Eq. (16.83) and substituting Eq. (16.81) in the resulting equation eliminate all values of ω_r, and we find that

$$\Delta\delta_n = \delta_{n-1} + kP_{a,n-1} \tag{16.84}$$

where
$$k = \frac{180f}{H}(\Delta t)^2 \tag{16.85}$$

Equation (16.84) is important for the step-by-step solution of the swing equation, for it shows how to calculate the change in δ during an interval based on the accelerating power for that interval and the change in δ for the preceding interval. The accelerating power is calculated at the beginning of each new interval and the solution progresses until enough points are obtained for plotting the swing curve. Greater accuracy is obtained when Δt is small. A value of $\Delta t = 0.05$ s is usually satisfactory.

The occurrence of a fault causes a discontinuity in the accelerating power P_a, which has a zero value before the fault and a nonzero value immediately following the fault. The discontinuity occurs at the beginning of the interval when $t = 0$. Reference to Fig. 16.14 shows that our method of calculation assumes that the accelerating power computed at the beginning of an interval is constant from the middle of the preceding interval to the middle of the interval being considered. When the fault occurs, we have two values of P_a at the beginning of an interval, and we must take the average of these two values as the constant accelerating power. The procedure is illustrated in the following example.

Example 16.11. Prepare a table showing the steps taken to plot the swing curve of machine 2 for the fault on the 60-Hz system of Examples 16.9 and 16.10. The fault is cleared at 0.225 s by simultaneously opening the circuit breakers at the ends of the faulted line.

Solution. Without loss of generality, we consider the detailed computations for machine 2. Computations to plot the swing curve for machine 1 are left to the student. Accordingly, we drop the subscript 2 as the indication of the machine number from all symbols in what follows. All calculations are made in per unit on a 100-MVA base. For the time interval $\Delta t = 0.05$ s the parameter k applicable to machine 2 is

$$k = \frac{180f}{H}(\Delta t)^2 = \frac{180 \times 60}{8.0} \times (0.05)^2 = 3.375 \text{ elec deg}$$

When the fault occurs at $t = 0$, the rotor angle of machine 2 cannot change instantly. Hence, from Example 16.9

$$\delta_0 = 16.19°$$

and during the fault

$$P_e = 0.1545 + 5.5023 \sin(\delta - 0.755°) \text{ per unit}$$

Therefore, as already seen in Example 16.9,

$$P_a = P_m - P_e = 1.6955 - 5.5023\sin(\delta - 0.755°) \text{ per unit}$$

At the beginning of the first interval there is a discontinuity in the accelerating power of each machine. Immediately before the fault occurs $P_a = 0$. Immediately after the fault occurs

$$P_a = 1.6955 - 5.5023\sin(16.19° - 0.755°) = 0.231 \text{ per unit}$$

The average value of P_a at $t = 0$ is $\frac{1}{2} \times 0.2310 = 0.1155$ per unit. We then find that

$$kP_a = 3.375 \times 0.1155 = 0.3898°$$

Identifying the time intervals by numerical subscripts, we find that the change in the rotor angle of machine 2 as time advances over the *first* interval from 0 to Δt is given by

$$\Delta\delta_1 = 0 + 0.3898 = 0.3898°$$

At the end of the first time interval we then have

$$\delta_1 = \delta_0 + \Delta\delta_1 = 16.19° + 0.3898° = 16.5798°$$

and $$\delta_1 - \gamma = 16.5798° - 0.755° = 15.8248°$$

At $t = \Delta t = 0.05$ s we find that

$$kP_{a,1} = 3.375[(P_m - P_c) - P_{\max}\sin(\delta_1 - \gamma)]$$

$$= 3.375[1.6955 - 5.5023\sin(15.8248°)] = 0.6583°$$

and it follows that the increase in the rotor angle over the *second* time interval is

$$\Delta\delta_2 = \Delta\delta_1 + kP_{a,1} = 0.3898° + 0.6583° = 1.0481°$$

At the end of the second time interval

$$\delta_2 = \delta_1 + \Delta\delta_2 = 16.5798° + 1.0481° = 17.6279°$$

The subsequent steps in the computations are shown in Table 16.6. Note that the postfault equation found in Example 16.10 is needed.

In Table 16.6 the terms $P_{\max}\sin(\delta - \gamma)$, P_a, and δ_n have values computed at the time t shown in the first column but $\Delta\delta_n$ is the *change* in the rotor angle *during* the interval that begins at the time indicated. For example, in the row for $t = 0.10$ s the angle 17.6279° is the first value calculated and is found by adding the change in angle during the preceding time interval (0.05 to 0.10 s) to the angle at $t = 0.05$ s. Next, $P_{\max}\sin(\delta - \gamma)$ is calculated for $\delta = 17.6279°$. Then, $P_a = (P_m - P_c) - P_{\max}\sin(\delta - \gamma)$ and kP_a are calculated. The value of kP_a is 0.3323°, which is added to the angular change of 1.0481° during the preceding interval to find the

TABLE 16.6

Computation of swing curve for machine 2 of Example 16.11 for clearing at 0.225 s

$k = (180f/H)(\Delta t)^2 = 3.375$ elec deg. Before clearing $P_m - P_c = 1.6955$ p.u., $P_{max} = 5.5023$ p.u., and $\gamma = 0.755°$. After clearing these values become 1.6696, 6.4934, and 0.847, respectively.

t, s	$(\delta_n - \gamma)$, elec deg	$P_{max} \sin(\delta_n - \gamma)$, per unit	P_a, per unit	$kP_{a,n-1}$, elec deg	$\Delta\delta_n$, elec deg	δ_n, elec deg
0 −	—	—	0.00	—		16.19
0 +	15.435	1.4644	0.2310	—		16.19
0 av	—	—	0.1155	0.3898	—	16.19
					0.3898	
0.05	15.8248	1.5005	0.1950	0.6583	—	16.5798
					1.0481	
0.10	16.8729	1.5970	0.0985	0.3323	—	17.6279
					1.3804	
0.15	18.2533	1.7234	−0.0279	−0.0942	—	19.0083
					1.2862	
0.20	19.5395	1.8403	−0.1448	−0.4886	—	20.2945
					0.7976	
0.25	20.2451	2.2470	−0.5774	−1.9487	—	21.0921
					−1.1511	
0.30	19.0940	2.1241	−0.4545	−1.534	—	19.9410
					−2.6852	
0.35	16.4088	1.8343	−0.1647	−0.5559	—	17.2558
					−3.2410	
0.40	13.1678	1.4792	0.1904	0.6425	—	14.0148
					−2.5985	
0.45	10.5693	1.1911	0.4785	1.6151	—	11.4163
					−0.9833	
0.50	9.5860	1.0813	0.5883	1.9854	—	10.4330
					1.0020	
0.55	10.5880	1.1931	0.4765	1.6081	—	11.4350
					2.6101	
0.60	13.1981	1.4826	0.1870	0.6312	—	14.0451
					3.2414	
0.65	16.4395	1.8376	−0.1680	−0.5672	—	17.2865
					2.6742	
0.70	19.1137	2.1262	−0.4566	−1.5411	—	19.9607
					1.1331	
0.75	20.2468	2.2471	−0.5775	−1.9492	—	21.0938
					−0.8161	
0.80	19.4307	2.1601	−0.4905	−1.6556	—	20.2777
					−2.4716	
0.85	—	—	—	—	—	17.8061

change of 1.3804° during the interval beginning at $t = 0.10$ s. This value added to 17.6279° gives the value $\delta = 19.0083°$ at $t = 0.15$ s. Note that the value of $P_m - P_c$ changes at 0.25 s because the fault is cleared at 0.225 s. The angle γ hasalso changed from 0.755° to 0.847°.

Whenever a fault is cleared, a discontinuity occurs in the accelerating power P_a. When clearing is at 0.225 s, as in Table 16.6, no special approach is required since our procedure assumes a discontinuity at the middle of an interval. At the beginning of the interval following clearing the assumed constant value of P_a is that determined for δ at the beginning of the interval following clearing.

When clearing is at the *beginning* of an interval such as at 3 cycles (0.05 s), two values of accelerating power result from the two expressions for the power output of the generator. One applies during and one after clearing the fault. For the system of Example 16.11 if the discontinuity occurs at 0.05 s, the average of the two values is assumed as the constant value of P_a from 0.025 to 0.075 s. The procedure is the same as that followed upon occurrence of the fault at $t = 0$, as demonstrated in Table 16.6.

Following the same procedures as in Table 16.6, we can determine δ versus t for machine 1 for clearing at 0.025 s and for both machines for clearing at 0.05 s. In the next section we see computer printouts of δ versus t for both machines calculated for clearing at 0.05 and 0.225 s.

Swing curves plotted for the two machines in Fig. 16.15 show that machine 1 is unstable for clearing at 0.225 s. For clearing at 0.20 s, however, it can be shown that the system is stable. The equal-area criterion confirms that the actual critical clearing time is between 0.20 and 0.225 s (see Prob. 16.16).

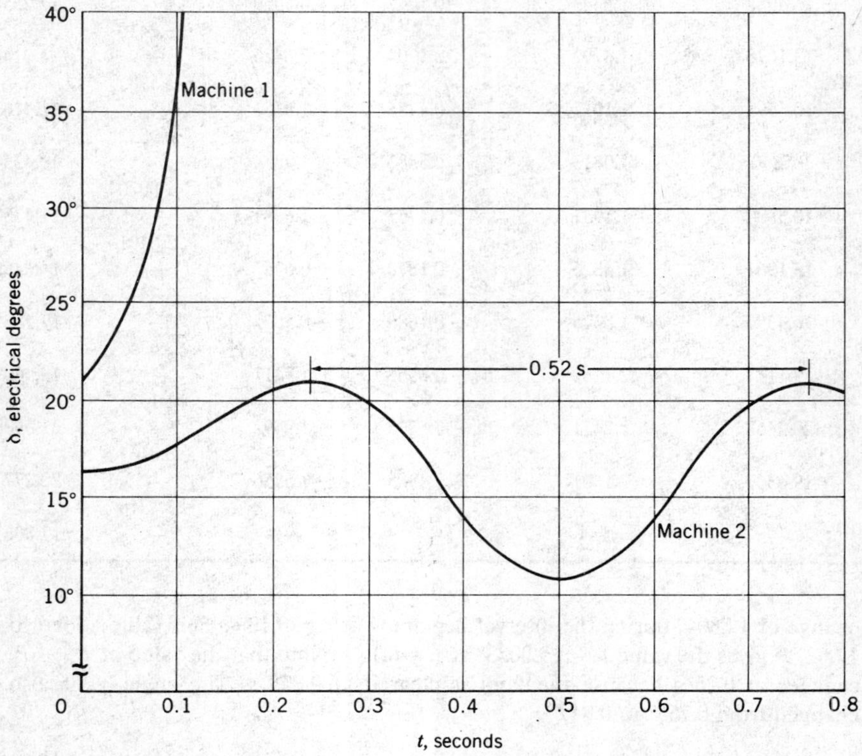

FIGURE 16.15
Swing curves for machines 1 and 2 of Examples 16.9 to 16.11 for clearing at 0.225 s.

Figure 16.15 shows that the change in the rotor angle of machine 2 is quite small, even though the fault is not cleared until 13.5 cycles after it occurs. It is interesting, therefore, to calculate the approximate frequency of oscillation of the rotor using the linearization procedure of Sec. 16.5. The synchronizing power coefficient calculated from the postfault power-angle equation for machine 2 is given by

$$S_p = \frac{dP_e}{d\delta} = \frac{d}{d\delta}[0.1804 + 6.4934\sin(\delta - 0.847°)]$$

$$= 6.4934\cos(\delta - 0.847°)$$

We note from Table 16.6 that the angle of machine 2 varies between 10.43° and 21.09°. Using either angle makes little difference in the value found for S_p. If we use the average value of 15.76°, we find that

$$S_p = 6.274 \text{ per-unit power/elec rad}$$

and by Eq. (16.50) the frequency of oscillation is

$$f_n = \frac{1}{2\pi}\sqrt{\frac{377 \times 6.274}{2 \times 8}} = 1.935 \text{ Hz}$$

ırom which the period of oscillation is calculated to be

$$T = \frac{1}{f_n} = \frac{1}{1.935} = 0.517 \text{ s}$$

Figure 16.15 and Table 16.6 confirm this value of T for machine 2. When faults are of shorter duration than 0.225 s, even more accurate results can be expected since the swing of the rotor is correspondingly smaller.

In the above examples it is possible to calculate the swing curves for each machine separately because of the fault location considered. When other fault locations are chosen, intermachine oscillations occur because the two generators do not decouple. The swing-curve computations are then more unwieldy. For such cases manual calculations are time-consuming and should be avoided. Computer programs of great versatility are generally available and should be used.

16.10 COMPUTER PROGRAMS FOR TRANSIENT STABILITY STUDIES

Present-day computer programs for transient stability studies have evolved from two basic needs (1) the requirement to study very large interconnected systems with numerous machines and (2) the need to represent machines and their associated control systems by more detailed models. The classical machine representation is suitable for many studies. However, more elaborate models may be required to represent modern turboalternators with dynamic characteristics determined by the many technological advances in the design of machine and control systems.

The simplest possible synchronous machine model is that used in classical stability studies. The more complicated two-axis machine models of Chap. 3 provide for direct- and quadrature-axis flux conditions during the subtransient and transient periods following a system disturbance. By providing for varying flux linkages of the field winding in the direct axis, these models allow representation of the action of the continuously acting automatic voltage regulator and excitation system with which all modern machines are equipped. Turbine control systems, which automatically govern the mechanical power input to the generating unit, also have dynamic response characteristics which can influence rotor dynamics. If these control schemes are to be represented, the model of the generating unit must be further extended.

The more complex generator models give rise to a larger number of differential and algebraic equations for each machine. In large system studies many generators supply widely dispersed load centers through extensive transmission systems whose performance also must be represented by a very large number of algebraic equations. Therefore, two sets of equations need to be solved simultaneously for each interval of time following the occurrence of a system disturbance. One set consists of the *algebraic* equations for the *steady-state* behavior of the network and its loads, and the algebraic equations relating V_t and E' of the synchronous machines. The other set consists of the *differential* equations which describe the *dynamic* electromechanical performance of the machines and their associated control systems.

The Newton-Raphson power-flow procedure described in Chap. 9 is the most commonly used solution technique for the network equations. Any one of several well-known step-by-step procedures may be chosen for numerical integration of the differential equations. The fourth-order Runge-Kutta method is very often used in production-type transient stability programs. Other methods known as the Euler method, the modified Euler method, the trapezoidal method, and predictor-corrector methods similar to the step-by-step method developed in Sec. 16.9 are alternatives. Each of these methods has advantages and disadvantages associated with numerical stability, time-step size, computational effort per integration step, and accuracy of solutions obtained.[4]

Table 16.7 shows the computer printout for swing curves of machines 1 and 2 of Example 16.11 for clearing at 0.225 s and at 0.05 s. These results were obtained by use of a production-type stability program, which couples a Newton-Raphson power-flow program with a fourth-order Runge-Kutta proce-

---

[4]For further information, see G. W. Stagg and A. H. El-Abiad, *Computer Methods in Power System Analysis*, Chaps. 9 and 10, McGraw-Hill, Inc. New York, 1968.

TABLE 16.7
Computer printout of swing curves for machines 1 and 2 of Examples 16.9 to 16.11 for clearing at 0.225 and 0.05 s

Clearing at 0.225 s			Clearing at 0.05 s		
Time	Mach. 1 angle	Mach. 2 angle	Time	Mach. 1 angle	Mach. 2 angle
0.00	20.8	16.2	0.00	20.8	16.2
0.05	25.1	16.6	0.05	25.1	16.6
0.10	37.7	17.6	0.10	32.9	17.2
0.15	58.7	19.0	0.15	37.3	17.2
0.20	88.1	20.3	0.20	36.8	16.7
0.25	123.1	20.9	0.25	31.7	15.9
0.30	151.1	19.9	0.30	23.4	15.0
0.35	175.5	17.4	0.35	14.6	14.4
0.40	205.1	14.3	0.40	8.6	14.3
0.45	249.9	11.8	0.45	6.5	14.7
0.50	319.3	10.7	0.50	10.1	15.6
0.55	407.0	11.4	0.55	17.7	16.4
0.60	489.9	13.7	0.60	26.6	17.1
0.65	566.0	16.8	0.65	34.0	17.2
0.70	656.4	19.4	0.70	37.6	16.8
0.75	767.7	20.8	0.75	36.2	16.0

dure. It is interesting to compare the closeness of the hand-calculated values of Table 16.6 with those for machine 2 in Table 16.7 for the case where the fault is cleared at 0.225 s.

The assumption of constant admittances for the loads allows us to absorb these admittances into $\mathbf{Y}_{bus}$ and thereby to avoid power-flow calculations, which are required when more accurate solutions using Runge-Kutta calculations are desired. The latter, being of the fourth order, require four iterative power-flow computations per time step.

16.11 FACTORS AFFECTING TRANSIENT STABILITY

Two factors which indicate the relative stability of a generating unit are (1) the angular swing of the machine during and following fault conditions and (2) the critical clearing time. It is apparent from this chapter that both the H constant and the transient reactance X_d' of the generating unit have a direct effect on both of these factors.

Equations (16.84) and (16.85) show that the smaller the H constant, the larger the angular swing during any time interval. On the other hand, Eq. (16.36) shows that P_{max} decreases as the transient reactance of the machine increases. This is so because the transient reactance forms part of the overall series reactance which is the reciprocal of the transfer admittance of the system.

Examination of Fig. 16.11 shows that all three power curves are lowered when P_{max} is decreased. Accordingly, for a given shaft power P_m, the initial rotor angle δ_0 is increased, δ_{max} is decreased, and a smaller difference between δ_0 and δ_{cr} exists for a smaller P_{max}. The net result is that a decreased P_{max} constrains a machine to swing through a smaller angle from its original position before it reaches the critical clearing angle. Thus, any developments which lower the H constant and increase transient reactance X_d' of the machine cause the critical clearing time to decrease and lessen the probability of maintaining stability under transient conditions. As power systems continually increase in size, there may be a corresponding need for higher-rated generating units. These larger units have advanced cooling systems which allow higher-rated capacities without comparable increase in rotor size. As a result, H constants continue to decrease with potential adverse impact on generating unit stability. At the same time this uprating process tends to result in higher transient and synchronous reactances, which makes the task of designing a reliable and stable system even more challenging.

Fortunately, stability control techniques and transmission system designs have also been evolving to increase overall system stability. The control schemes include:

- Excitation systems
- Turbine valve control
- Single-pole operation of circuit breakers
- Faster fault clearing times

System design strategies aimed at lowering system reactance include:

- Minimum transformer reactance
- Series capacitor compensation of lines
- Additional transmission lines

When a fault occurs, the voltages at all buses of the system are reduced. At generator terminals the reduced voltages are sensed by the automatic voltage regulators which act within the excitation system to restore generator terminal voltages. The general effect of the excitation system is to reduce the initial rotor angle swing following the fault. This is accomplished by boosting the voltage applied to the field winding of the generator through action of the amplifiers in the forward path of the voltage regulators. The increased air-gap flux exerts a restraining torque on the rotor, which tends to slow down its motion. Modern excitation systems employing thyristor controls can respond rapidly to bus-voltage reduction and can effect from 0.5 to 1.5 cycles gain in critical clearing times for three-phase faults on the high-side bus of the generator step-up transformer.

Modern electrohydraulic turbine-governing systems have the ability to close turbine valves to reduce unit acceleration during severe system faults near the unit. Immediately upon detecting differences between mechanical input and electrical output, control action initiates the valve closing, which reduces the power input. A gain of 1 to 2 cycles in critical clearing time can be achieved.

Reducing the reactance of the system during fault conditions increases $r_1 P_{max}$ and decreases the acceleration area of Fig. 16.11. The possibility of maintaining stability is thereby enhanced. Since single-phase faults occur more often than three-phase faults, relaying schemes which allow independent or selective circuit-breaker pole operation can be used to clear the faulted phase while keeping the unfaulted phases intact. Separate relay systems, trip coils, and operating mechanisms can be provided for each pole so as to mitigate *stuck-breaker* contingencies following three-phase faults. Independent-pole operation of critical circuit breakers can extend the critical clearing time by 2 to 5 cycles depending on whether one or two poles fail to open under fault conditions. Such gain in critical clearing time can be important especially if backup clearing times are a problem for system stability.

Reducing the reactance of a transmission line is another way of raising P_{max}. Compensating for line reactance by series capacitors is often an economical means of increasing stability. Increasing the number of parallel lines between two points is a common means of reducing reactance. When parallel transmission lines are used instead of a single line, some power is transferred over the remaining line even during a three-phase fault on one of the lines—unless the fault occurs at a paralleling bus. For other types of faults on one line more power is transferred during the fault if there are two lines in parallel than is transferred over a single faulted line. For more than two lines in parallel the power transferred during the fault is even greater. Power transferred into the system is subtracted from power input to the generator to obtain accelerating power. Thus, the more power is transferred into the system during a fault, the lower the acceleration of the machine rotor and the greater the degree of stability.

16.12 SUMMARY

This chapter presents the basics of power system stability analysis. Starting with elementary principles of rotational motion, the swing equation governing the electromechanical dynamic behavior of each generating unit is developed. The swing equation is shown to be nonlinear because the electrical power output from the generating unit is a nonlinear function of the rotor angle. Because of this nonlinearity, iterative step-by-step methods of solution of the swing equation are generally required. In the special case of two finite machines (or one machine operating into an infinite bus) the equal-area criterion of stability can be used to calculate the critical clearing angle. It is shown, however, that finding the critical clearing time (which is the maximum elapsed time from the initiation of a fault until its isolation such that the system is transiently stable) generally requires a numerical solution of the swing equation.

Classical stability studies and their underlying assumptions are explained for the multimachine case and a simple step-by-step procedure for solving the swing equations of the system is illustrated numerically. A basis is thereby provided for further study of the more powerful numerical techniques employed in industry-based production-type computer programs.

Transient stability of the power system is affected by many other factors related to the design of the system network, its protection system, and the control schemes associated with each of the generating units. These factors are discussed in summary form.

PROBLEMS

16.1. A 60-Hz four-pole turbogenerator rated 500 MVA, 22 kV has an inertia constant of $H = 7.5$ MJ/MVA. Find (*a*) the kinetic energy stored in the rotor at synchronous speed and (*b*) the angular acceleration if the electrical power developed is 400 MW when the input less the rotational losses is 740,000 hp.

16.2. If the acceleration computed for the generator described in Prob. 16.1 is constant for a period of 15 cycles, find the change in δ in electrical degrees in that period and the speed in revolutions per minute at the end of 15 cycles. Assume that the generator is synchronized with a large system and has no accelerating torque before the 15-cycle period begins.

16.3. The generator of Prob. 16.1 is delivering rated megavoltamperes at 0.8 power-factor lag when a fault reduces the electric power output by 40%. Determine the accelerating torque in newton-meters at the time the fault occurs. Neglect losses and assume constant power input to the shaft.

16.4. Determine the WR^2 of the generator of Prob. 16.1.

16.5. A generator having $H = 6$ MJ/MVA is connected to a synchronous motor having $H = 4$ MJ/MVA through a network of reactances. The generator is delivering power of 1.0 per unit to the motor when a fault occurs which reduces the delivered power. At the time when the reduced power delivered is 0.6 per unit, determine the angular acceleration of the generator with respect to the motor.

16.6. A power system is identical to that of Example 16.3, except that the impedance of each of the parallel transmission lines is $j0.5$ and the delivered power is 0.8 per unit when both the terminal voltage of the machine and the voltage of the infinite bus are 1.0 per unit. Determine the power-angle equation for the system during the specified operating conditions.

16.7. If a three-phase fault occurs on the power system of Prob. 16.6 at a point on one of the transmission lines at a distance of 30% of the line length away from the sending-end terminal of the line, determine (*a*) the power-angle equation during the fault and (*b*) the swing equation. Assume that the system is operating under the conditions specified in Prob. 16.6 when the fault occurs. Let $H = 5.0$ MJ/MVA, as in Example 16.4.

16.8. Series resistance in the transmission network results in positive values for P_c and γ in Eq. (16.80). For a given electrical power output, show the effects of resistance on the synchronizing coefficient S_p, the frequency of rotor oscillations, and the damping of these oscillations.

16.9. A generator having $H = 6.0$ MJ/MVA is delivering power of 1.0 per unit to an infinite bus through a purely reactive network when the occurrence of a fault reduces the generator output power to zero. The maximum power that could be delivered is 2.5 per unit. When the fault is cleared, the original network conditions again exist. Determine the critical clearing angle and critical clearing time.

16.10. A 60-Hz generator is supplying 60% of P_{max} to an infinite bus through a reactive network. A fault occurs which increases the reactance of the network between the generator internal voltage and the infinite bus by 400%. When the fault is cleared, the maximum power that can be delivered is 80% of the original maximum value. Determine the critical clearing angle for the condition described.

16.11. If the generator of Prob. 16.10 has an inertia constant of $H = 6$ MJ/MVA and P_m (equal to 0.6 P_{max}) is 1.0 per-unit power, find the critical clearing time for the condition of Prob. 16.10. Use $\Delta t = 0.05$ to plot the necessary swing curve.

16.12. For the system and fault conditions described in Probs. 16.6 and 16.7, determine the power-angle equation if the fault is cleared by the simultaneous opening of breakers at both ends of the faulted line at 4.5 cycles after the fault occurs. Then, plot the swing curve of the generator through $t = 0.25$ s.

16.13. Extend Table 16.6 to find δ at $t = 1.00$ s.

16.14. Calculate the swing curve for machine 2 of Examples 16.9 through 16.11 for fault clearing at 0.05 s by the method described in Sec. 16.9. Compare the results with the values obtained by the production-type program and listed in Table 16.7.

16.15. If the three-phase fault on the system of Example 16.9 occurs on line ④–⑤ at bus ⑤ and is cleared by the simultaneous opening of breakers at both ends of the line at 4.5 cycles after the fault occurs, prepare a table like that of Table 16.6 to plot the swing curve of machine 2 through $t = 0.30$ s.

16.16. By applying the equal-area criterion to the swing curves obtained in Examples 16.9 and 16.10 for machine 1, (*a*) derive an equation for the critical clearing angle, (*b*) solve the equation by trial and error to evaluate δ_{cr} and (*c*) use Eq. (16.72) to find the critical clearing time.

APPENDIX A

TABLE A.1
Typical range of transformer reactances†

Power transformers 25,000 kVA and larger

Nominal system voltage, kV	Forced-air-cooled, %	Forced-oil-cooled, %
34.5	5–8	9–14
69	6–10	10–16
115	6–11	10–20
138	6–13	10–22
161	6–14	11–25
230	7–16	12–27
345	8–17	13–28
500	10–20	16–34
700	11–21	19–35

† Percent on rated kilovoltampere base. Typical transformers are now designed for the minimum reactance value shown. Distribution transformers have considerably lower reactance. Resistances of transformers are usually lower than 1%.

TABLE A.2

Typical reactances of three-phase synchronous machines†

Values are per unit. For each reactance a range of values is listed below the typical value‡

	Turbine-generators					
	2-pole		4-pole		Salient-pole generators	
	Conventional cooled	Conductor cooled	Conventional cooled	Conductor cooled	With dampers	Without dampers
X_d	1.76 1.7–1.82	1.95 1.72–2.17	1.38 1.21–1.55	1.87 1.6–2.13	1 0.6–1.5	1 0.6–1.5
X_q	1.66 1.63–1.69	1.93 1.71–2.14	1.35 1.17–1.52	1.82 1.56–2.07	0.6 0.4–0.8	0.6 0.4–0.8
X'_d	0.21 0.18–0.23	0.33 0.264–0.387	0.26 0.25–0.27	0.41 0.35–0.467	0.32 0.25–0.5	0.32 0.25–0.5
X''_d	0.13 0.11–0.14	0.28 0.23–0.323	0.19 0.184–0.197	0.29 0.269–0.32	0.2 0.13–0.32	0.30 0.2–0.5
X_2	$=X''_d$	$=X''_d$	$=X''_d$	$=X''_d$	0.2 0.13–0.32	0.40 0.30–0.45
X_0§						

†Data furnished by ABB Power T & D Company, Inc.

‡Reactances of older machines will generally be close to minimum values.

§X_0 varies so critically with armature winding pitch that an average value can hardly be given. Variation is from 0.1 to 0.7 of X''_d.

TABLE A.3
Electrical characteristics of bare aluminum conductors steel-reinforced (ACSR)†

Code word	Aluminum area, cmil	Stranding Al/St	Layers of aluminum	Outside diameter, in	Resistance: Dc, 20°C, Ω/1,000 ft	Resistance: Ac, 60 Hz, 20°C, Ω/mi	Resistance: Ac, 60 Hz, 50°C, Ω/mi	GMR D_s, ft	Reactance per conductor 1-ft spacing, 60 Hz: Inductive X_a, Ω/mi	Reactance per conductor 1-ft spacing, 60 Hz: Capacitive X'_a, MΩ·mi
Waxwing	266,800	18/1	2	0.609	0.0646	0.3488	0.3831	0.0198	0.476	0.1090
Partridge	266,800	26/7	2	0.642	0.0640	0.3452	0.3792	0.0217	0.465	0.1074
Ostrich	300,000	26/7	2	0.680	0.0569	0.3070	0.3372	0.0229	0.458	0.1057
Merlin	336,400	18/1	2	0.684	0.0512	0.2767	0.3037	0.0222	0.462	0.1055
Linnet	336,400	26/7	2	0.721	0.0507	0.2737	0.3006	0.0243	0.451	0.1040
Oriole	336,400	30/7	2	0.741	0.0504	0.2719	0.2987	0.0255	0.445	0.1032
Chickadee	397,500	18/1	2	0.743	0.0433	0.2342	0.2572	0.0241	0.452	0.1031
Ibis	397,500	26/7	2	0.783	0.0430	0.2323	0.2551	0.0264	0.441	0.1015
Pelican	477,000	18/1	2	0.814	0.0361	0.1957	0.2148	0.0264	0.441	0.1004
Flicker	477,000	24/7	2	0.846	0.0359	0.1943	0.2134	0.0284	0.432	0.0992
Hawk	477,000	26/7	2	0.858	0.0357	0.1931	0.2120	0.0289	0.430	0.0988
Hen	477,000	30/7	2	0.883	0.0355	0.1919	0.2107	0.0304	0.424	0.0980
Osprey	556,500	18/1	2	0.879	0.0309	0.1679	0.1843	0.0284	0.432	0.0981
Parakeet	556,500	24/7	2	0.914	0.0308	0.1669	0.1832	0.0306	0.423	0.0969
Dove	556,500	26/7	2	0.927	0.0307	0.1663	0.1826	0.0314	0.420	0.0965
Rook	636,000	24/7	2	0.977	0.0269	0.1461	0.1603	0.0327	0.415	0.0950
Grosbeak	636,000	26/7	2	0.990	0.0268	0.1454	0.1596	0.0335	0.412	0.0946
Drake	795,000	26/7	2	1.108	0.0215	0.1172	0.1284	0.0373	0.399	0.0912
Tern	795,000	45/7	3	1.063	0.0217	0.1188	0.1302	0.0352	0.406	0.0925
Rail	954,000	45/7	3	1.165	0.0181	0.0997	0.1092	0.0386	0.395	0.0897
Cardinal	954,000	54/7	3	1.196	0.0180	0.0988	0.1082	0.0402	0.390	0.0890
Ortolan	1,033,500	45/7	3	1.213	0.0167	0.0924	0.1011	0.0402	0.390	0.0885
Bluejay	1,113,000	45/7	3	1.259	0.0155	0.0861	0.0941	0.0415	0.386	0.0874
Finch	1,113,000	54/19	3	1.293	0.0155	0.0856	0.0937	0.0436	0.380	0.0866
Bittern	1,272,000	45/7	3	1.345	0.0136	0.0762	0.0832	0.0444	0.378	0.0855
Pheasant	1,272,000	54/19	3	1.382	0.0135	0.0751	0.0821	0.0466	0.372	0.0847
Bobolink	1,431,000	45/7	3	1.427	0.0121	0.0684	0.0746	0.0470	0.371	0.0837
Plover	1,431,000	54/19	3	1.465	0.0120	0.0673	0.0735	0.0494	0.365	0.0829
Lapwing	1,590,000	45/7	3	1.502	0.0109	0.0623	0.0678	0.0498	0.364	0.0822
Falcon	1,590,000	54/19	3	1.545	0.0108	0.0612	0.0667	0.0523	0.358	0.0814
Bluebird	2,156,000	84/19	4	1.762	0.0080	0.0476	0.0515	0.0586	0.344	0.0776

†Most used multilayer sizes.

‡Data, by permission, from Aluminum Association, *Aluminum Electrical Conductor Handbook*, 2nd ed., Washington, D.C., 1982.

TABLE A.4 Inductive reactance spacing factor X_d at 60 Hz† (ohms per mile per conductor)

	Separation											
	Inches											
Feet	0	1	2	3	4	5	6	7	8	9	10	11
0	......	−0.3015	−0.2174	−0.1682	−0.1333	−0.1062	−0.0841	−0.0654	−0.0492	−0.0349	−0.0221	−0.0106
1	0	0.0097	0.0187	0.0271	0.0349	0.0423	0.0492	0.0558	0.0620	0.0679	0.0735	0.0789
2	0.0841	0.0891	0.0938	0.0984	0.1028	0.1071	0.1112	0.1152	0.1190	0.1227	0.1264	0.1299
3	0.1333	0.1366	0.1399	0.1430	0.1461	0.1491	0.1520	0.1549	0.1577	0.1604	0.1631	0.1657
4	0.1682	0.1707	0.1732	0.1756	0.1779	0.1802	0.1825	0.1847	0.1869	0.1891	0.1912	0.1933
5	0.1953	0.1973	0.1993	0.2012	0.2031	0.2050	0.2069	0.2087	0.2105	0.2123	0.2140	0.2157
6	0.2174	0.2191	0.2207	0.2224	0.2240	0.2256	0.2271	0.2287	0.2302	0.2317	0.2332	0.2347
7	0.2361	0.2376	0.2390	0.2404	0.2418	0.2431	0.2445	0.2458	0.2472	0.2485	0.2498	0.2511
8	0.2523											
9	0.2666											
10	0.2794											
11	0.2910											
12	0.3015											
13	0.3112											
14	0.3202											
15	0.3286											
16	0.3364											
17	0.3438											
18	0.3507											
19	0.3573											
20	0.3635											
21	0.3694											
22	0.3751											
23	0.3805											
24	0.3856											
25	0.3906											
26	0.3953											
27	0.3999											
28	0.4043											
29	0.4086											
30	0.4127											
31	0.4167											
32	0.4205											
33	0.4243											
34	0.4279											
35	0.4314											
36	0.4348											
37	0.4382											
38	0.4414											
39	0.4445											
40	0.4476											
41	0.4506											
42	0.4535											
43	0.4564											
44	0.4592											
45	0.4619											
46	0.4646											
47	0.4672											
48	0.4697											
49	0.4722											

At 60 Hz, in Ω/mi per conductor
$X_d = 0.2794 \log d$
d = separation, ft
For three-phase lines
$d = D_{eq}$

†From *Electrical Transmission and Distribution Reference Book*, by permission of the ABB Power T & D Company, Inc.

TABLE A.5 Shunt capacitance-reactance spacing factor X_d at 10 Hz (megaohm-miles per conductor)

Feet	Separation (Inches) 0	1	2	3	4	5	6	7	8	9	10	11
0	......	−0.0737	−0.0532	−0.0411	−0.0326	−0.0260	−0.0206	−0.0160	−0.0120	−0.0085	−0.0054	−0.0026
1	0	0.0024	0.0046	0.0066	0.0085	0.0103	0.0120	0.0136	0.0152	0.0166	0.0180	0.0193
2	0.0206	0.0218	0.0229	0.0241	0.0251	0.0262	0.0272	0.0282	0.0291	0.0300	0.0309	0.0318
3	0.0326	0.0334	0.0342	0.0350	0.0357	0.0365	0.0372	0.0379	0.0385	0.0392	0.0399	0.0405
4	0.0411	0.0417	0.0423	0.0429	0.0435	0.0441	0.0446	0.0452	0.0457	0.0462	0.0467	0.0473
5	0.0478	0.0482	0.0487	0.0492	0.0497	0.0501	0.0506	0.0510	0.0515	0.0519	0.0523	0.0527
6	0.0532	0.0536	0.0540	0.0544	0.0548	0.0552	0.0555	0.0559	0.0563	0.0567	0.0570	0.0574
7	0.0577	0.0581	0.0584	0.0588	0.0591	0.0594	0.0598	0.0601	0.0604	0.0608	0.0611	0.0614
8	0.0617											
9	0.0652											
10	0.0683											
11	0.0711											
12	0.0737											
13	0.0761											
14	0.0783											
15	0.0803											
16	0.0823											
17	0.0841											
18	0.0858											
19	0.0874											
20	0.0889											
21	0.0903											
22	0.0917											
23	0.0930											
24	0.0943											
25	0.0955											
26	0.0967											
27	0.0978											
28	0.0989											
29	0.0999											
30	0.1009											
31	0.1019											
32	0.1028											
33	0.1037											
34	0.1046											
35	0.1055											
36	0.1063											
37	0.1071											
38	0.1079											
39	0.1087											
40	0.1094											
41	0.1102											
42	0.1109											
43	0.1116											
44	0.1123											
45	0.1129											
46	0.1136											
47	0.1142											
48	0.1149											
49	0.1155											

At 60 Hz, in MΩ·mi per conductor
$X_d' = 0.06831 \log d$
d = separation, ft
For three-phase lines
$d = D_{eq}$

From *Electrical Transmission and Distribution Reference Book*, by permission of the ABB Power T & D Company, Inc.

TABLE A.6
***ABCD* constants for various networks**

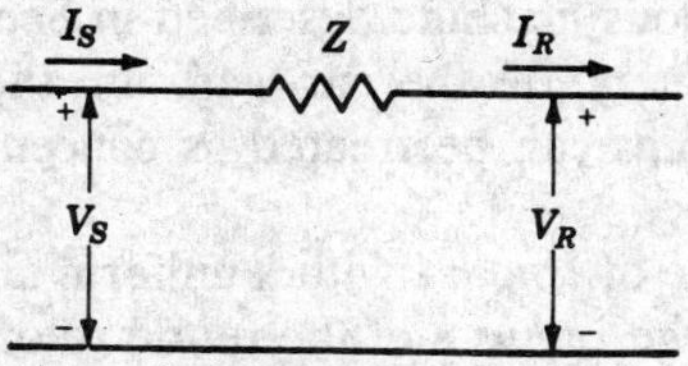

Series impedance

$$A = 1$$
$$B = Z$$
$$C = 0$$
$$D = 1$$

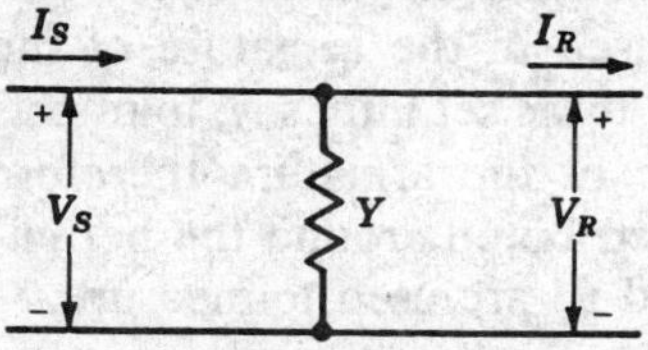

Shunt admittance

$$A = 1$$
$$B = 0$$
$$C = Y$$
$$D = 1$$

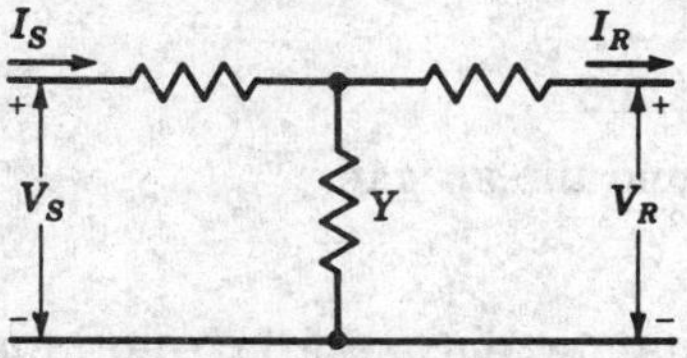

Unsymmetrical T

$$A = 1 + YZ_1$$
$$B = Z_1 + Z_2 + YZ_1Z_2$$
$$C = Y$$
$$D = 1 + YZ_2$$

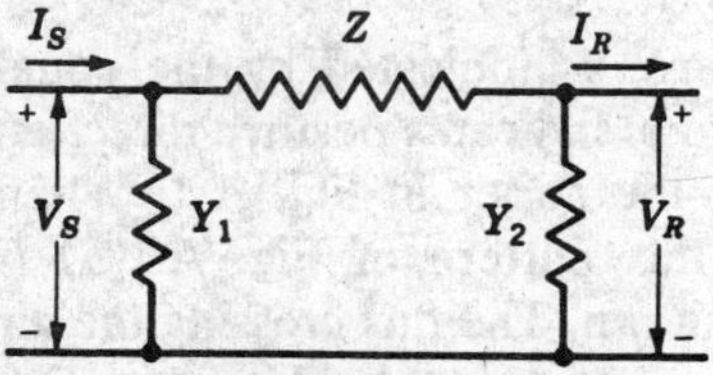

Unsymmetrical π

$$A = 1 + Y_2Z$$
$$B = Z$$
$$C = Y_1 + Y_2 + ZY_1Y_2$$
$$D = 1 + Y_1Z$$

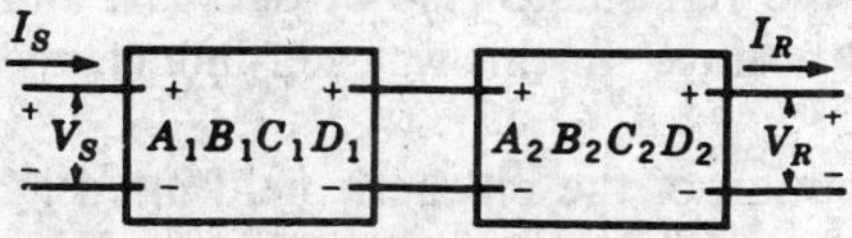

Networks in cascade

$$A = A_1A_2 + B_1C_2$$
$$B = A_1B_2 + B_1D_2$$
$$C = A_2C_1 + C_2D_1$$
$$D = B_2C_1 + D_1D_2$$

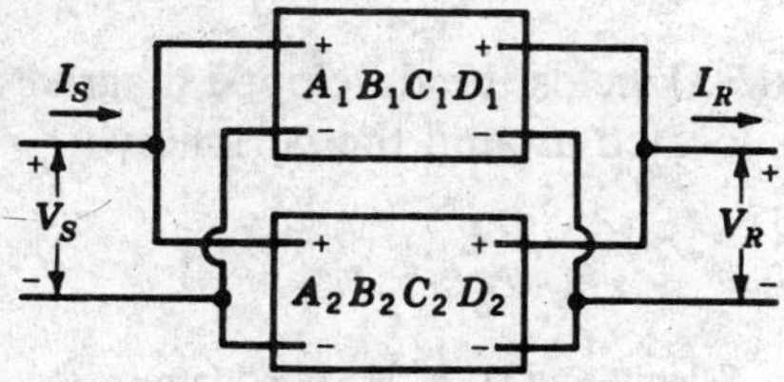

Networks in parallel

$$A = (A_1B_2 + A_2B_1)/(B_1 + B_2)$$
$$B = B_1B_2/(B_1 + B_2)$$
$$C = C_1 + C_2 + (A_1 - A_2)(D_2 - D_1)/(B_1 + B_2)$$
$$D = (B_2D_1 + B_1D_2)/(B_1 + B_2)$$

A.1 DISTRIBUTED WINDINGS OF THE SYNCHRONOUS MACHINE

The field and armature windings of the synchronous machine described in Sec. 3.1 are distributed in slots around the periphery of the air gap. In the round-rotor case we now show that these windings can be treated as concentrated coils in the idealized model of the machine.

Figure A.1(a) represents a two-pole machine of *length* l with a uniform air gap of *width* g, which is very much smaller than the *radius* r of the round rotor. Compared to the air gap, the stator and rotor iron parts have very high magnetic permeability and therefore have negligibly small magnetic field intensity. A distributed a phase winding of the type used in the armature of the three-phase synchronous machine has a total of N_a turns set into, say, four slots per pole as shown. All three phases a, b, and c of the armature therefore occupy a total of 24 slots that are symmetrically distributed around the periphery of the air gap. Cylindrical coordinates z, x, and θ_d are used to measure:

1. Distance z along the length l of the machine,
2. Radial distance x perpendicular to the length, and
3. Angular displacement θ_d *counterclockwise* around the air gap.

Because the air-gap width g is so small relative to the radius r of the rotor, for given values of z and θ_d we can indicate a position in the air gap by setting $x = r$.

The *net axial current* in each phase is zero, as indicated by the equal number of dots and crosses. The *dot* (and the cross) indicates positive direction of current flow *out of* (and into) the plane of the paper in Fig. A.1. When positive current i_a flows in the winding, the flux pattern of Fig. A.1(b) is produced with north (N) and south (S) poles as shown. The flux crossing the air gap passes radially through the elemental area $r \times d\theta_d \times dz$ shown in Fig. A.1(a). The flux lines are consistent with the right-hand rule of Sec. 2.1, and such flux linking the winding is considered positive. In this way self-inductance is positive.

Let $H_a(\theta_d)$ denote the air-gap distribution of the magnetic field intensity due to i_a. $H_a(\theta_d)$ may be determined in a straightforward manner by applying Ampere's and Gauss's laws as now explained:[1]

- Unrolling the air gap as depicted in Fig. A.2(a) yields the developed diagram of Fig. A.2(b), which shows the conductors located around the periphery.

---

[1]This section follows the development in Chap. 9 of N. L. Schmitz and D. W. Novotny, *Introductory Electromechanics*, The Ronald Press Co., New York, NY, 1965.

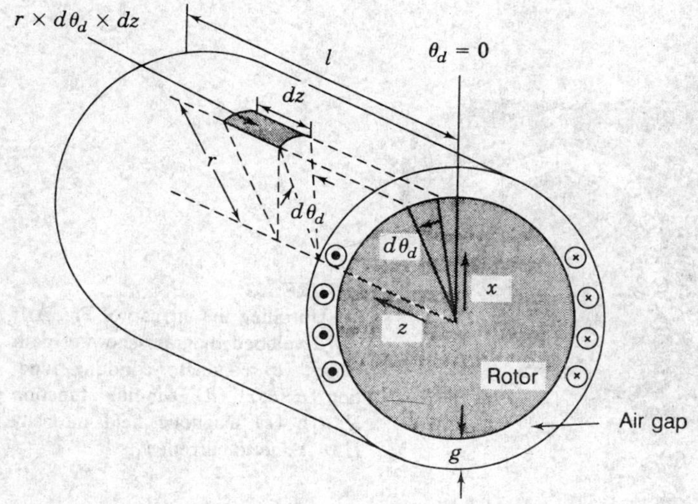

(a)

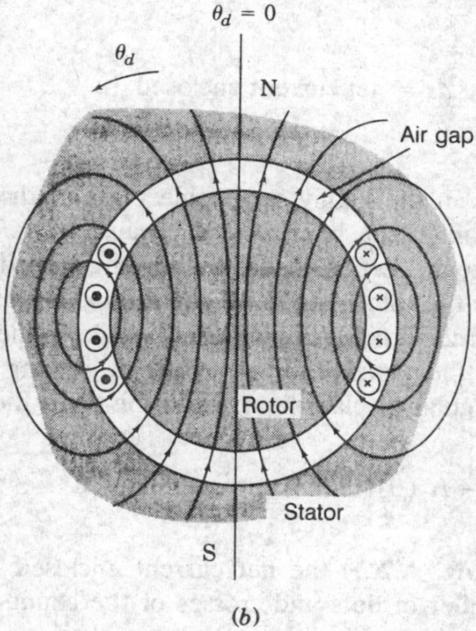

(b)

FIGURE A.1
Representation of a two-pole-nonsalient rotor machine showing: (*a*) elemental portion of an imaginary air-gap surface surrounding the rotor; (*b*) the pattern of flux due to positive current i_a into (cross) and out of (dot) plane of the paper.

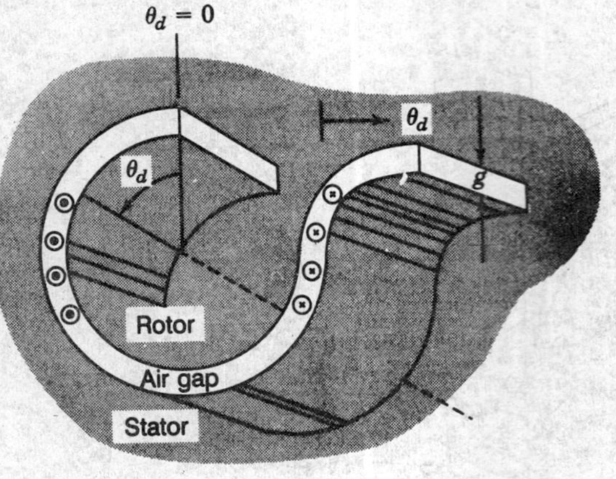

(a)

FIGURE A.2
(*a*) Unrolling the air gap of Fig. A.1; (*b*) developed diagram showing path *mnpq*; (*c*) conductor-counting function $n_a(\theta_d)$; (*d*) winding function $N_a(\theta_d)$; (*e*) magnetic field intensity $H_a(\theta_d)$ due to current i_a.

- Ampere's law applied along closed paths such as that marked *mnpq* in Fig. A.2(*b*) gives

$$\oint_{mnpq} H \cdot ds = \text{net current enclosed} \tag{A.1}$$

Those portions of the paths in the highly permeable parts of the rotor and stator make zero contribution to the line integral because H is considered negligible in iron compared to air. Let us choose the positive direction of H to be inward from the stator to the rotor. In the air gap at angular displacement θ_d the line integral of the *radial* field intensity $H_a(\theta_d)$ along the path m-n yields $H_a(\theta_d) \times g$. Likewise, the contribution is $-H_a(0) \times g$ along the air-gap portion of path p-q at 0° angular displacement. Equation (A.1) then gives

$$g\left[H_a(\theta_d) - H_a(0)\right] = \text{net current enclosed} \tag{A.2}$$

- For each choice of θ_d in Fig. A.2(*b*) the net current enclosed is obtained simply by counting the number of dots and crosses of the conductors inside the path *mnpq*. Counting conductors in this manner, we obtain

$$g\left[H_a(\theta_d) - H_a(0)\right] = n_a(\theta_d)i_a \tag{A.3}$$

where $n_a(\theta_d)$ denotes the number of conductors with dots minus the number with crosses inside the path of integration. Figure A.2(*c*) shows a sketch of $n_a(\theta_d)$ assuming that the current changes from zero to i_a (dot) or $-i_a$ (cross) at the center of the conductor.

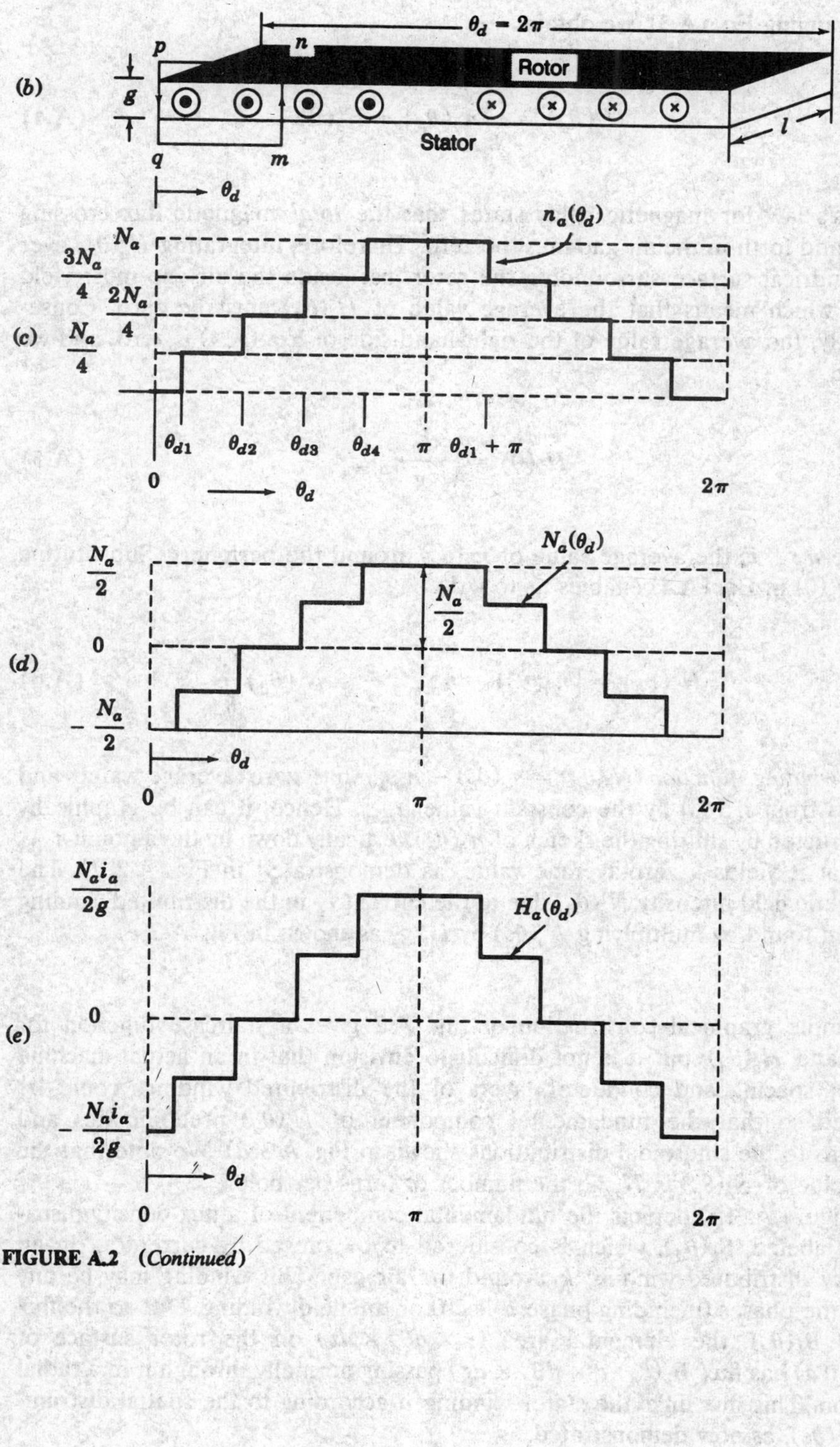

FIGURE A.2 (*Continued*)

- Rearranging Eq. (A.3), we obtain

$$H_a(\theta_d) = \frac{i_a}{g} n_a(\theta_d) + H_a(0) \tag{A.4}$$

Gauss's law for magnetic fields states that the *total* magnetic flux crossing back and forth in the air gap must be zero. Therefore, integrating $H_a(\theta_d)$ over a cylindrical surface surrounding the rotor just inside the air gap must yield zero, which means that the average value of $H_a(\theta_d)$ must be zero. Consequently, the average value of the right-hand side of Eq. (A.4) is zero, and we obtain

$$H_a(0) = -\frac{i_a}{g} n_{a,\,\mathrm{av}} \tag{A.5}$$

where $n_{a,\,\mathrm{av}}$ is the average value of $n_a(\theta_d)$ around the periphery. Substituting for $H_a(0)$ in Eq. (A.4) enables us to write

$$H_a(\theta_d) = [n_a(\theta_d) - n_{a,\,\mathrm{av}}]\frac{i_a}{g} = N_a(\theta_d)\frac{i_a}{g} \tag{A.6}$$

The *winding function* $N_a(\theta_d) = n_a(\theta_d) - n_{a,\,\mathrm{av}}$ has zero average value and differs from $n_a(\theta_d)$ by the constant value $n_{a,\mathrm{av}}$. Hence, it can be graphically constructed by shifting the sketch of $n_a(\theta_d)$ vertically down by the amount $n_{a,\,\mathrm{av}}$ so that it yields a zero-average value, as demonstrated in Fig. A.2(*d*). The magnetic field intensity $H_a(\theta_d)$ due to the current i_a in the distributed winding is then found by multiplying $N_a(\theta_d)$ by i_a/g, as shown in Fig. A.2(*e*).

The simple graphical construction of Fig. A.2 gives a staircase function for $N_a(\theta_d)$ and $H_a(\theta_d)$, but it is not difficult to envision that in an actual machine the slot spacing and conductor layers of the distributed windings could be arranged so that the fundamental component of $N_a(\theta_d)$ predominates and conforms to the sinusoidal distributions shown in Fig. A.3(*a*). We note that the peak value of $N_a(\theta_d)$ is $N_a/2$, the number of turns per pole.

Figure A.3(*b*) depicts the fundamental component of a flux-density distribution labeled $B_k(\theta_d)$, which is considered to be caused by current i_k in an arbitrary distributed winding k around the air gap. This winding may be any one of the phases (including phase a itself) or the field winding. Due to the flux density $B_k(\theta_d)$, the elemental area $(r \times d\theta_d \times dz)$ on the rotor surface of Fig. A.1(*a*) has flux $B_k(\theta_d)(r \times d\theta_d \times dz)$ passing normally through it in a radial direction. This flux links the stator winding a according to the spatial distribution $N_a(\theta_d)$, as now demonstrated.

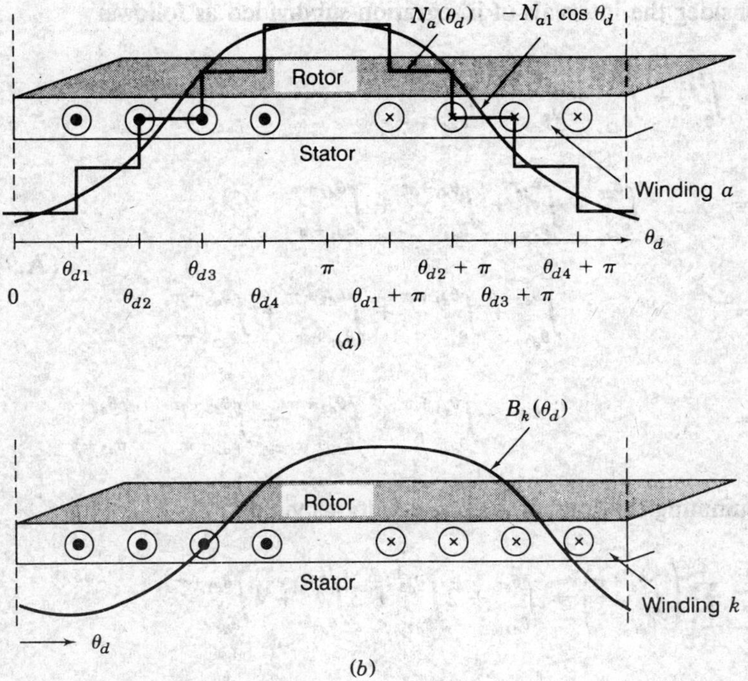

FIGURE A.3
Fundamental sinusoidal components of: (*a*) winding function $N_a(\theta_d)$; (*b*) flux density $B_k(\theta_d)$ due to current i_k in distributed winding k.

Integrating the flux density $B_k(\theta_d)$ over the entire length l of the rotor and around the circumference of the air gap, we obtain from Fig. A.3

$$\lambda_{ak} = \frac{N_a}{4} r \int_0^l \left[\int_{\theta_{d1}}^{\theta_{d1}+\pi} B_k(\theta_d)\, d\theta_d + \int_{\theta_{d2}}^{\theta_{d2}+\pi} B_k(\theta_d)\, d\theta_d + \int_{\theta_{d3}}^{\theta_{d3}+\pi} B_k(\theta_d)\, d\theta_d + \int_{\theta_{d4}}^{\theta_{d4}+\pi} B_k(\theta_d)\, d\theta_d \right] dz \tag{A.7}$$

where λ_{ak} represents the total flux linkage with the a winding due to the current i_k in winding k. Each cross section of the air gap along the axial length of the machine has the same winding distribution and the same flux density $B_k(\theta_d)$. Hence, the bracketed term in Eq. (A.7) is independent of z so that integration with respect to z yields

$$\lambda_{ak} = \frac{N_a}{4} rl \left[\int_{\theta_{d1}}^{\theta_{d1}+\pi} B_k(\theta_d)\, d\theta_d + \int_{\theta_{d2}}^{\theta_{d2}+\pi} B_k(\theta_d)\, d\theta_d + \int_{\theta_{d3}}^{\theta_{d3}+\pi} B_k(\theta_d)\, d\theta_d + \int_{\theta_{d4}}^{\theta_{d4}+\pi} B_k(\theta_d)\, d\theta_d \right] \tag{A.8}$$

Let us now consider the intervals of integration subdivided as follows:

$$\begin{aligned}
\int_{\theta_{d1}}^{\theta_{d1}+\pi} &= \int_{\theta_{d1}}^{\theta_{d2}} + \int_{\theta_{d2}}^{\theta_{d3}} + \int_{\theta_{d3}}^{\theta_{d4}} + \int_{\theta_{d4}}^{\theta_{d1}+\pi} \\
\int_{\theta_{d2}}^{\theta_{d2}+\pi} &= \int_{\theta_{d2}}^{\theta_{d3}} + \int_{\theta_{d3}}^{\theta_{d4}} + \int_{\theta_{d4}}^{\theta_{d1}+\pi} + \int_{\theta_{d1}+\pi}^{\theta_{d2}+\pi} \\
\int_{\theta_{d3}}^{\theta_{d3}+\pi} &= \int_{\theta_{d3}}^{\theta_{d4}} + \int_{\theta_{d4}}^{\theta_{d1}+\pi} + \int_{\theta_{d1}+\pi}^{\theta_{d2}+\pi} + \int_{\theta_{d2}+\pi}^{\theta_{d3}+\pi} \\
\int_{\theta_{d4}}^{\theta_{d4}+\pi} &= \int_{\theta_{d4}}^{\theta_{d1}+\pi} + \int_{\theta_{d1}+\pi}^{\theta_{d2}+\pi} + \int_{\theta_{d2}+\pi}^{\theta_{d3}+\pi} + \int_{\theta_{d3}+\pi}^{\theta_{d4}+\pi}
\end{aligned} \tag{A.9}$$

Symbolically summing the four preceding equations yields

$$\sum_{i=1}^{4}\left(\int_{\theta_{di}}^{\theta_{di}+\pi}\right) = \int_{\theta_{d1}}^{\theta_{d2}} + 2\int_{\theta_{d2}}^{\theta_{d3}} + 3\int_{\theta_{d3}}^{\theta_{d4}} + 4\int_{\theta_{d4}}^{\theta_{d1}+\pi} + 3\int_{\theta_{d1}}^{\theta_{d2}+\pi} + 2\int_{\theta_{d2}}^{\theta_{d3}+\pi} + \int_{\theta_{d3}}^{\theta_{d4}+\pi} \tag{A.10}$$

The right-hand side of Eq. (A.10) can be written more compactly in terms of the function $n_a(\theta_d)$ shown in Fig. A.2(*c*). Therefore, we have

$$\frac{N_a}{4}\sum_{i=1}^{4}\left(\int_{\theta_{di}}^{\theta_{di}+\pi} B_k(\theta_d)\,d\theta_d\right) = \int_0^{2\pi} n_a(\theta_d)B_k(\theta_d)\,d\theta_d \tag{A.11}$$

Substituting for $n_a(\theta_d) = N_a(\theta_d) + n_{a,\mathrm{av}}$, we obtain from Eqs. (A.8) and (A.11)

$$\lambda_{ak} = rl\int_0^{2\pi} N_a(\theta_d)B_k(\theta_d)\,d\theta_d + rln_{a,\mathrm{av}}\int_0^{2\pi} B_k(\theta_d)\,d\theta_d \tag{A.12}$$

The second integral term on the right-hand side of this equation equals zero since the average value of $B_k(\theta_d)$ around the periphery of the air gap is zero, and we obtain

$$\lambda_{ak} = rl\int_0^{2\pi} N_a(\theta_d)B_k(\theta_d)\,d\theta_d = \mu_0 rl\int_0^{2\pi} N_a(\theta_d)H_k(\theta_d)\,d\theta_d \tag{A.13}$$

where $B_k(\theta_d) = \mu_0 H_k(\theta_d)$. There are now three cases to be considered:

1. Suppose that the current i_k is actually in the a winding itself; that is, $k = a$. Then, substituting in Eq. (A.13) for $H_a(\theta_d)$ from Eq. (A.6), we obtain

$$\lambda_{aa} = \frac{\mu_0 r l i_a}{g} \int_0^{2\pi} (N_a(\theta_d))^2 \, d\theta_d \tag{A.14}$$

Considering only the fundamental component $-N_{a1} \cos \theta_d$ of $N_a(\theta_d)$, we find that

$$L_{aa} = \frac{\lambda_{aa}}{i_a} = \frac{\mu_0 r l}{g} \int_0^{2\pi} N_{a1}^2 \cos^2 \theta_d \, d\theta_d = \frac{\mu_0 \pi r l}{g} N_{a1}^2 \tag{A.15}$$

where L_{aa} is the self-inductance of a winding and N_{a1} equals the effective turns per pole of the fundamental component. Note that L_{aa} is a *positive constant* value.

2. Suppose that the current i_k is actually in a distributed b winding identical to, but shifted 120° from, the a winding. Setting $k = b$ in Eq. (A.13) and again considering only fundamental components, we obtain for the mutual inductance L_{ab} (or L_{ba}) between windings

$$L_{ab} = L_{ba} = \frac{\lambda_{ab}}{i_b} = \frac{\mu_0 r l}{g} \int_0^{2\pi} N_{a1}^2 \cos \theta_d \cos(\theta_d - 120°) \, d\theta_d \tag{A.16}$$

This equation can be simplified by using the identity

$$\cos \alpha \cos \beta = \tfrac{1}{2}[\cos(\alpha - \beta) + \cos(\alpha + \beta)] \tag{A.17}$$

to yield

$$L_{ab} = L_{ba} = \frac{\lambda_{ab}}{i_b} = -\frac{\mu_0 \pi r l}{2g} N_{a1}^2 \tag{A.18}$$

When $B_k(\theta_d)$ arises from current i_c in an identical c winding displaced 240° from the axis of the a phase, similar calculation shows that the mutual inductance L_{ac} (or L_{ca}) between the a and c windings is

$$L_{ac} = L_{ca} = \frac{\lambda_{ac}}{i_c} = -\frac{\mu_0 \pi r l}{2g} N_{a1}^2 \tag{A.19}$$

We note that L_{ab} and L_{ac} are equal *negative constants*.

3. If $B_k(\theta_d)$ is due to current i_f in the field winding at the instant when its axis is at an arbitrary counterclockwise angle θ_d with respect to the a winding, the mutual inductance L_{af} (or L_{fa}) due to the fundamental components of

the spatial distributions of the two windings is given by

$$L_{af} = L_{fa} = \frac{\lambda_{af}}{i_f} = \frac{\mu_0 rl}{g} \int_0^{2\pi} N_{a1} N_{f1} \cos \theta'_d \cos(\theta'_d - \theta_d) \, d\theta'_d \quad \text{(A.20)}$$

N_{f1} is the turns per pole of the fundamental component of the f winding and the dummy symbol θ'_d is used for the variable of integration because θ_d has an arbitrary but definite value in Eq. (A.20). Applying the trigonometric identity of Eq. (A.17) yields

$$L_{af} = L_{fa} = \frac{\lambda_{af}}{i_f} = \frac{\mu_0 rl N_{a1} N_{f1}}{2g} \int_0^{2\pi} [\cos \theta_d + \cos(2\theta'_d - \theta_d)] \, d\theta'_d \quad \text{(A.21)}$$

Integration over the 2π interval gives $2\pi \cos \theta_d$ for the first term of the integrand and zero for the second term so that

$$L_{af} = L_{fa} = \frac{\mu_0 \pi rl N_{a1} N_{f1}}{g} \cos \theta_d \quad \text{(A.22)}$$

We note that L_{af} varies cosinusoidally with the position θ_d of the rotor. When the field winding rotates counterclockwise, it causes flux to link the stator windings in the sequence *a-b-c*, and the linkage with phases *b* and *c* are exactly the same as with the *a* phase except that they occur later. This means that the field has exactly the same magnetizing effect on the *b* phase along the $\theta_d = 120°$ axis and on the *c* phase along the $\theta_d = 240°$ axis as it has on the *a* phase at $\theta_d = 0°$. Hence, the mutual inductance L_{bf} between the *b* phase and the field, and L_{cf} between the *c* phase and the field, must be of the same form as Eq. (A.22); that is,

$$L_{bf} = L_{fb} = \frac{\mu_0 \pi rl N_{a1} N_{f1}}{g} \cos(\theta_d - 120°) \quad \text{(A.23)}$$

$$L_{cf} = L_{fc} = \frac{\mu_0 \pi rl N_{a1} N_{f1}}{g} \cos(\theta_d - 240°) \quad \text{(A.24)}$$

Also, results similar to Eqs. (A.15) and (A.18) can be found for the *b* and *c* windings of the round-rotor machine. Summarizing, we have the following:

- The self-inductances of the distributed armature windings are positive constants, which we now call L_s so that

$$L_s = L_{aa} = L_{bb} = L_{cc} = \frac{\mu_0 \pi rl}{g} N_{a1}^2 \quad \text{(A.25)}$$

- The mutual inductances between the armature windings are negative constants, which we now call $-M_s$ so that

$$-M_s = L_{ab} = L_{bc} = L_{ca} = -\frac{\mu_0 \pi r l}{2g} N_{a1}^2 \tag{A.26}$$

- The mutual inductance between the field winding and each of the stator windings varies with the rotor position θ_d as a cosinusoidal function with maximum value $M_f = \mu_0 \pi r l N_{a1} N_{f1}/g$ so that

$$\begin{aligned} L_{af} &= M_f \cos\theta_d \\ L_{bf} &= M_f \cos(\theta_d - 120°) \\ L_{cf} &= M_f \cos(\theta_d - 240°) \end{aligned} \tag{A.27}$$

In the round-rotor machine (and, indeed, in the salient-pole machine also) the field winding has a constant self-inductance L_{ff}. This is because the field winding on the d-axis produces flux through a similar magnetic path in the stator for all positions of the rotor (neglecting the small effect of armature slots).

A.2 P-TRANSFORMATION OF STATOR QUANTITIES

To transform a-b-c stator flux linkages to d-q-0 quantities by means of matrix **P** of Eq. (3.42), rearrange the flux-linkage expressions of Eq. (3.41) as follows:

$$\begin{bmatrix} \lambda_a \\ \lambda_b \\ \lambda_c \end{bmatrix} = \begin{bmatrix} L_{aa} & L_{ab} & L_{ac} \\ L_{ba} & L_{bb} & L_{bc} \\ L_{ca} & L_{cb} & L_{cc} \end{bmatrix} \begin{bmatrix} i_a \\ i_b \\ i_c \end{bmatrix} + \begin{bmatrix} L_{af} \\ L_{bf} \\ L_{cf} \end{bmatrix} i_f \tag{A.28}$$

Now substitute for the a-b-c flux linkages and currents from Eqs. (3.43) to obtain

$$\mathbf{P}^{-1} \begin{bmatrix} \lambda_d \\ \lambda_q \\ \lambda_0 \end{bmatrix} = \begin{bmatrix} L_{aa} & L_{ab} & L_{ac} \\ L_{ba} & L_{bb} & L_{bc} \\ L_{ca} & L_{cb} & L_{cc} \end{bmatrix} \mathbf{P}^{-1} \begin{bmatrix} i_d \\ i_q \\ i_0 \end{bmatrix} + \begin{bmatrix} L_{af} \\ L_{bf} \\ L_{cf} \end{bmatrix} i_f \tag{A.29}$$

Multiplying across Eq. (A.29) by **P** gives

$$\begin{bmatrix} \lambda_d \\ \lambda_q \\ \lambda_0 \end{bmatrix} = \mathbf{P} \begin{bmatrix} L_{aa} & L_{ab} & L_{ac} \\ L_{ba} & L_{bb} & L_{bc} \\ L_{ca} & L_{cb} & L_{cc} \end{bmatrix} \mathbf{P}^{-1} \begin{bmatrix} i_d \\ i_q \\ i_0 \end{bmatrix} + \mathbf{P} \begin{bmatrix} L_{af} \\ L_{bf} \\ L_{cf} \end{bmatrix} i_f \tag{A.30}$$

According to Table 3.1, we have

$$\begin{bmatrix} L_{aa} & L_{ab} & L_{ac} \\ L_{ba} & L_{bb} & L_{bc} \\ L_{ca} & L_{cb} & L_{cc} \end{bmatrix} = (L_s + M_s) \begin{bmatrix} 1 & 0 & 0 \\ 0 & 1 & 0 \\ 0 & 0 & 1 \end{bmatrix} - M_s \begin{bmatrix} 1 & 1 & 1 \\ 1 & 1 & 1 \\ 1 & 1 & 1 \end{bmatrix}$$

$$- L_m \begin{bmatrix} -\cos 2\theta_d & \cos 2\left(\theta_d + \frac{\pi}{6}\right) & \cos 2\left(\theta_d + \frac{5\pi}{6}\right) \\ \cos 2\left(\theta_d + \frac{\pi}{6}\right) & -\cos 2\left(\theta_d + \frac{2\pi}{3}\right) & \cos 2\left(\theta_d + \frac{\pi}{2}\right) \\ \cos 2\left(\theta_d + \frac{5\pi}{6}\right) & \cos 2\left(\theta_d + \frac{\pi}{2}\right) & \cos 2\left(\theta_d + \frac{2\pi}{3}\right) \end{bmatrix} \tag{A.31}$$

and it can be shown that

$$\mathbf{P}^{-1} = \mathbf{P}^T = \sqrt{\frac{2}{3}} \begin{bmatrix} \cos\theta_d & \cos(\theta_d - 120°) & \cos(\theta_d - 240°) \\ \sin\theta_d & \sin(\theta_d - 120°) & \sin(\theta_d - 240°) \\ \frac{1}{\sqrt{2}} & \frac{1}{\sqrt{2}} & \frac{1}{\sqrt{2}} \end{bmatrix}^T \tag{A.32}$$

Substituting from Eqs. (A.31) and (A.32) into Eq. (A.30) and simplifying, we obtain Eq. (3.44).

To transform *a-b-c* stator voltages to *d-q-*0 quantities by means of matrix **P**, rearrange Eqs. (3.47) as follows:

$$\begin{bmatrix} v_a \\ v_b \\ v_c \end{bmatrix} = -R \begin{bmatrix} i_a \\ i_b \\ i_c \end{bmatrix} - \frac{d}{dt} \begin{bmatrix} \lambda_a \\ \lambda_b \\ \lambda_c \end{bmatrix} \tag{A.33}$$

Multiplying across Eq. (A.33) by **P** gives, according to Eqs. (3.43),

$$\begin{bmatrix} v_d \\ v_q \\ v_0 \end{bmatrix} = -R\begin{bmatrix} i_d \\ i_q \\ i_0 \end{bmatrix} - \mathbf{P}\frac{d}{dt}\left(\mathbf{P}^{-1}\begin{bmatrix} \lambda_d \\ \lambda_q \\ \lambda_0 \end{bmatrix}\right) \tag{A.34}$$

Preserving order while using the derivative-of-a-product rule in Eq. (A.34), we obtain

$$\begin{aligned} \begin{bmatrix} v_d \\ v_q \\ v_0 \end{bmatrix} &= -R\begin{bmatrix} i_d \\ i_q \\ i_0 \end{bmatrix} - \mathbf{P}\left\{\mathbf{P}^{-1} \times \frac{d}{dt}\begin{bmatrix} \lambda_d \\ \lambda_q \\ \lambda_0 \end{bmatrix} + \frac{d}{dt}(\mathbf{P}^{-1}) \times \begin{bmatrix} \lambda_d \\ \lambda_q \\ \lambda_0 \end{bmatrix}\right\} \\ &= -R\begin{bmatrix} i_d \\ i_q \\ i_0 \end{bmatrix} - \frac{d}{dt}\begin{bmatrix} \lambda_d \\ \lambda_q \\ \lambda_0 \end{bmatrix} - \mathbf{P}\frac{d}{dt}(\mathbf{P}^T) \times \begin{bmatrix} \lambda_d \\ \lambda_q \\ \lambda_0 \end{bmatrix} \end{aligned} \tag{A.35}$$

Substituting for **P** from Eq. (A.32) and simplifying yield Eqs. (3.48).

APPENDIX B

B.1 SPARSITY AND NEAR-OPTIMAL ORDERING

Section 7.6 shows that the equivalent network corresponding to the coefficient matrix at each step of the gaussian-elimination procedure has one node less than that for the previous step. However, the number of branches in the equivalent network at each step depends on the *order* in which the nodal equations are processed. To see the effect of ordering, consider the graph of the four-bus system shown in Fig. B.1(*a*). In this example the off-diagonal elements are nonzero in row 1 and column 1 of $\mathbf{Y}_{bus}$ because of the branches connected from bus ① to buses ②, ③, and ④. All the remaining off-diagonal elements of $\mathbf{Y}_{bus}$ are zero since buses ②, ③, and ④ are not directly connected to one another. If bus ① is processed first in gaussian elimination of the nodal equations, the effect on $\mathbf{Y}_{bus}$ is similar to the usual Y-Δ transformation symbolically indicated by

$$\underbrace{\begin{matrix} & ① & ② & ③ & ④ \\ ① & \times & \times & \times & \times \\ ② & \times & \times & \cdot & \cdot \\ ③ & \times & \cdot & \times & \cdot \\ ④ & \times & \cdot & \cdot & \times \end{matrix}}_{\text{Initial } \mathbf{Y}_{bus}} \Rightarrow \underbrace{\begin{matrix} & ② & ③ & ④ \\ ② & \times & \otimes & \otimes \\ ③ & \otimes & \times & \otimes \\ ④ & \otimes & \otimes & \times \end{matrix}}_{\mathbf{Y}_{bus} \text{ after Step 1}}$$

in which the symbol $\times$ indicates an element which is initially nonzero and $\otimes$ represents a fill-in for a previously zero element. The fill-ins are symmetrical

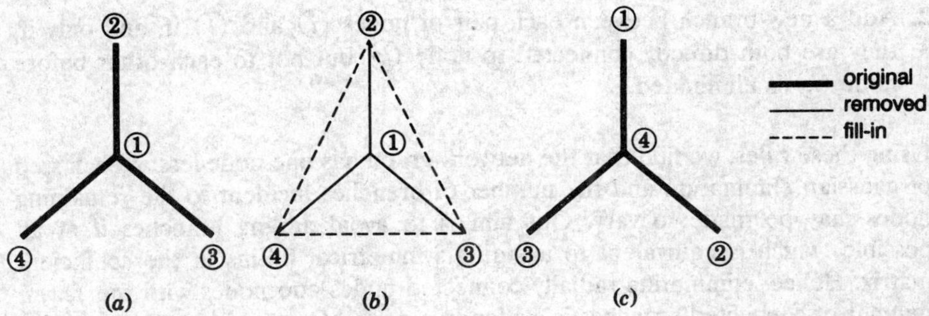

FIGURE B.1
The effects of ordering on node elimination: (*a*) the original graph and node numbers; (*b*) the three new branches after eliminating node ①; (*c*) the reordered nodes.

and result from the Step 1 calculations

$$Y_{ij(\text{new})} = Y_{ji(\text{new})} = Y_{ij} - \frac{Y_{i1}Y_{1j}}{Y_{11}} \qquad i \text{ and } j = 2, 3, 4 \qquad \text{(B.1)}$$

because the subtractive term on the right-hand side of this equation is nonzero. Since fill-ins and branch additions are equivalent, the new branches joining node-pairs ②–③, ②–④, and ③–④ of Fig. B.1(*b*) represent the fill-ins of the new 3×3 $\mathbf{Y}_{\text{bus}}$. If we now *renumber* the nodes as shown in Fig. B.1(*c*) and solve the equations in the sequence of the new node numbers, we find after Step 1 that

$$\underbrace{\begin{array}{c c} & \begin{array}{cccc} ① & ② & ③ & ④ \end{array} \\ \begin{array}{c} ① \\ ② \\ ③ \\ ④ \end{array} & \left[\begin{array}{cccc} \times & \cdot & \cdot & \times \\ \cdot & \times & \cdot & \times \\ \cdot & \cdot & \times & \times \\ \times & \times & \times & \times \end{array}\right] \end{array}}_{\text{Initial } \mathbf{Y}_{\text{bus}}} \Rightarrow \underbrace{\begin{array}{c c} & \begin{array}{ccc} ② & ③ & ④ \end{array} \\ \begin{array}{c} ② \\ ③ \\ ④ \end{array} & \left[\begin{array}{ccc} \times & \cdot & \times \\ \cdot & \times & \times \\ \times & \times & \times \end{array}\right] \end{array}}_{\mathbf{Y}_{\text{bus}} \text{ after Step 1}}$$

Therefore, by node renumbering fill-ins may be avoided and new branches need not be added to the network graph.

The above example demonstrates two general rules for altering the system graph when the voltage variable at node ⓚ is eliminated in the course of gaussian elimination:

1. Remove node ⓚ and all its incident branches from the system graph. In particular, remove branch *km* if node ⓚ is *radially* connected to the network, that is, if node ⓚ is connected to only one other node ⓜ.

2. Add a new branch between each pair of nodes ⓘ and ⓙ if, and only if, they are both directly connected to node ⓚ but not to each other before node ⓚ is eliminated.

Using these rules, we find that the network graph has one node less at each step of gaussian elimination and the number of branches incident to the remaining nodes may be made to vary. Our aim is to avoid adding branches if at all possible, which is equivalent to avoiding symmetrical fill-ins in the coefficient matrix. Hence, eliminating radially connected nodes and nodes with the *fewest* number of connected branches is evidently a desirable graphical strategy, which translates into the following scheme of ordering the system equations for gaussian elimination:

- At Step 1 of the forward elimination process, select the variable to be eliminated corresponding to the diagonal element of the row with the most zero entries. If two or more variables meet this condition, select one which causes the fewest fill-ins for the next step.
- At each subsequent step, select the variable to be eliminated by applying the same rule to the reduced coefficient matrix.

This ordering scheme is said to be *near-optimal* because in practical use it gives excellent reduction in the accumulation of fill-ins even though it does not guarantee the absolute minimum (optimum) number. To illustrate the ordering scheme graphically we:

- Draw a graph corresponding to $\mathbf{Y}_{\text{bus}}$ (thus omitting the reference node).
- At Step 1, select the first node for elimination from the graph which has the fewest incident branches and which creates the fewest new branches as determined by rules (1) and (2).
- At each subsequent step, update the count of branches at the remaining buses using rules (1) and (2) and then apply the Step 1 selection criterion to the updated graph.

More than one ordered sequence of nodes may be found since two or more nodes with exactly the same number of incident branches may qualify for elimination at some steps. In any event, limiting the branch count at a bus limits the fill-ins in the upper and lower (**LU**) factors of the corresponding coefficient matrix. These graphical features of gaussian elimination, which apply to any square matrix with symmetrical nonzero pattern, are illustrated in the following numerical examples.

Example B.1. The graph shown in Fig. B.2(a) is for a 5×5 $\mathbf{Y}_{\text{bus}}$ system (the reference node not explicitly represented). Graphically determine a sequence in which buses ⓐ, ⓑ, ⓒ, ⓓ, and ⓔ should be numbered so as to minimize the number of fill-ins in the **LU** factors of $\mathbf{Y}_{\text{bus}}$.

Solution. Node ⓒ of Fig. B.2(a) with only one incident branch is chosen for Step 1 elimination according to rule (1). With node ⓒ eliminated and branch ca effectively removed (indicated by thin line) as far as further calculations are concerned, we update the count of active branches (that is, heavy and fill-in lines) at each remaining node. We then have nodes ⓐ and ⓔ with three branches still active and nodes ⓓ and ⓑ with only two, Fig. B.2(b). The elimination of either node ⓓ or ⓑ in Step 2 is equally acceptable for two reasons. First, each has the fewest active branches, and second, the elimination of either one will not create a new branch because they connect to two nodes ⓐ and ⓔ, which are already connected to each other—rule (2). Let us agree to eliminate node ⓓ in Step 2, in which case nodes ⓐ, ⓑ, and ⓔ become equal candidates for Step 3 with two active branches remaining at each, Fig. B.2(c). In fact, nodes ⓐ, ⓑ, and ⓔ can be eliminated in any order without creating fill-ins, and so we label them nodes ③, ④, and ⑤, respectively, as shown in Figs. B.2(d) and B.2(e). When the nodal equations of the system are ordered according to the node numbers of Fig. B.2(e), the rows and columns of the corresponding $\mathbf{Y}_{\text{bus}}$ assume the nonzero pattern

$$
\begin{array}{c} \\ ① \\ ② \\ ③ \\ ④ \\ ⑤ \end{array}
\begin{array}{c}
\begin{array}{ccccc} ① & ② & ③ & ④ & ⑤ \end{array} \\
\begin{bmatrix}
\times & \cdot & \times & \cdot & \cdot \\
\cdot & \times & \times & \cdot & \times \\
\times & \times & \times & \times & \times \\
\cdot & \cdot & \times & \times & \times \\
\cdot & \times & \times & \times & \times
\end{bmatrix}
\end{array}
$$

It is readily shown that successive elimination of the node variables in the order V_c, V_d, V_a, V_b, and V_e creates no fill-ins in the triangular factors **L** and **U**. Thus, the original sparsity of $\mathbf{Y}_{\text{bus}}$ is preserved in its factors. On the other hand, eliminating nodes in the strict alphabetical order of Fig. B.2(a) yields after node ⓐ is eliminated the 4×4 $\mathbf{Y}_{\text{bus}}$ given by

$$
\begin{array}{c} \\ ⓐ \\ ⓑ \\ ⓒ \\ ⓓ \\ ⓔ \end{array}
\begin{array}{c}
\begin{array}{ccccc} ⓐ & ⓑ & ⓒ & ⓓ & ⓔ \end{array} \\
\begin{bmatrix}
\times & \times & \times & \times & \times \\
\times & \times & \cdot & \cdot & \times \\
\times & \cdot & \times & \cdot & \cdot \\
\times & \cdot & \cdot & \times & \times \\
\times & \times & \cdot & \times & \times
\end{bmatrix}
\end{array}
\Rightarrow
\begin{array}{c} \\ ⓑ \\ ⓒ \\ ⓓ \\ ⓔ \end{array}
\begin{array}{c}
\begin{array}{cccc} ⓑ & ⓒ & ⓓ & ⓔ \end{array} \\
\begin{bmatrix}
\times & \otimes & \otimes & \times \\
\otimes & \times & \otimes & \otimes \\
\otimes & \otimes & \times & \times \\
\times & \otimes & \times & \times
\end{bmatrix}
\end{array}
$$

corresponding to Fig. B.2(f) in which there are four fill-in branches. Thus, the alphabetical order of solution would not be desirable in this example since the original sparsity would then be completely lost and the triangular matrices **L** and **U** would require more storage and computer time for calculations.

Example B.2. Number the nodes of the graph of Fig. B.3(a) in an optimal order for the triangular factorization of the corresponding $\mathbf{Y}_{\text{bus}}$.

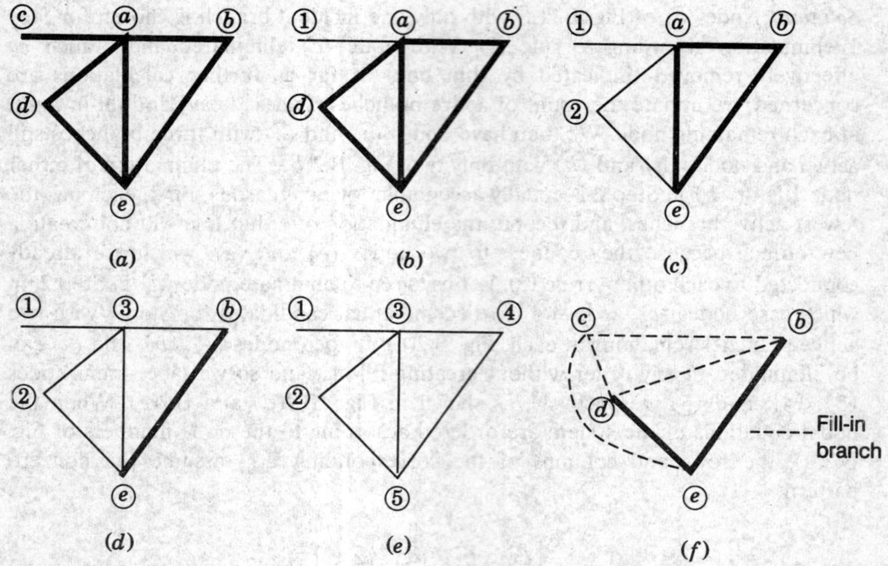

FIGURE B.2
A graphical illustration of optimal ordering in Example B.1; (*a*) the original graph; (*b*) after Step 1 elimination of node ⓒ; (*c*) after Step 2 elimination of node ⓓ; (*d*) after Step 3 elimination of node ⓐ; (*e*) optimal order of the node eliminations. Also, (*f*) the fill-in branches following Step 1 elimination of node ⓐ.

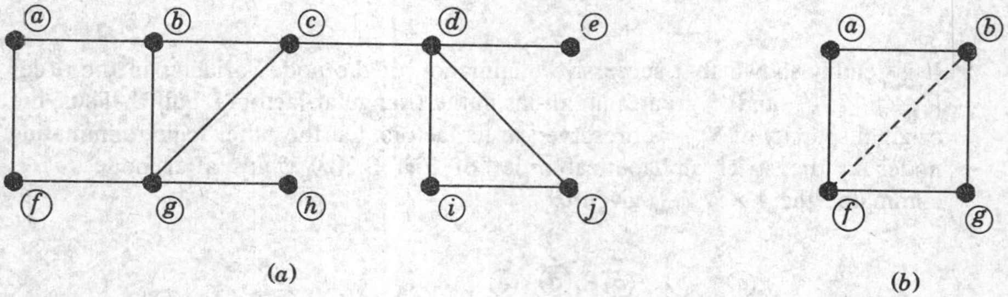

FIGURE B.3
The graph of Example B.2: (*a*) 10 nodes to be optimally ordered; (*b*) fill-in (dashed) branch after Step 6.

Solution. In this example more than one optimally ordered sequence can be expected since at some steps of elimination more than one node has the same number of incident active branches. For instance, nodes ⓔ and ⓗ of Fig. B.3(*a*) have one radial branch each, and so we may eliminate node ⓔ in Step 1 and node ⓗ in Step 2, or vice versa. After Step 2 branches *ed* and *hg* are made inactive and are effectively removed from the graph. Updating the count of active branches at the remaining nodes, we recognize nodes ⓐ, ⓕ, ⓘ, and ⓙ (each with two branches still active) as candidates for Step 3 elimination. To eliminate either node ⓐ or node ⓕ at this stage would create a fill-in branch in accordance with rule (2), and so we focus upon nodes ⓘ and ⓙ as the only acceptable choices in Step

3. Selecting node ⓙ removes branches *ji* and *jd* from the list of active branches and no new branch needs to be added since nodes ⓘ and ⓓ are already connected. In Step 4 we eliminate node ⓘ along with its single active branch *id*. Continuing this process of node elimination and active branch designation in Fig. B.3(*a*), we arrive at the solution

Step number	1	2	3	4	5	6	7	8	9	10
Node eliminated	ⓗ	ⓔ	ⓙ	ⓘ	ⓓ	ⓒ	ⓐ	ⓑ	ⓕ	ⓖ
No. of active branches	1	1	2	1	1	2	2	2	1	0
Resulting fill-ins	0	0	0	0	0	0	2	0	0	0

The above solution shows that two fill-ins are unavoidable. This is because in eliminating node ⓐ in Step 7 the new branch *bf* shown by the dashed line in Fig. B.3(*b*) is created between nodes ⓑ and ⓕ, which were not previously connected. Inspection of Fig. B.3(*b*) shows that one fill-in branch must occur regardless of which one of the nodes ⓐ, ⓑ, ⓕ, or ⓖ is chosen for elimination in Step 7. Besides, the nonzero pattern of the reduced coefficient matrix resulting from Step 6 is

$$\begin{array}{c} \\ ⓐ \\ ⓑ \\ ⓕ \\ ⓖ \end{array} \begin{array}{c} \begin{array}{cccc} ⓐ & ⓑ & ⓕ & ⓖ \end{array} \\ \begin{bmatrix} \times & \times & \times & \cdot \\ \times & \times & \cdot & \times \\ \times & \cdot & \times & \times \\ \cdot & \times & \times & \times \end{bmatrix} \end{array}$$

regardless of the order in which we choose the row and corresponding column of each of the remaining nodes ⓐ, ⓑ, ⓕ, or ⓖ. Thus, the reader can confirm by gaussian elimination that two symmetric fill-ins will occur even in the optimal ordering of the nodes.

When bus numbers are assigned to Fig. B.3(*a*) in accordance with the step numbers of the above solution, the rows and columns of $\mathbf{Y}_{\text{bus}}$ will be optimally ordered for gaussian elimination, and as a result, the triangular factors **L** and **U** will require minimum storage and computing time for solving the nodal equations.

B.2 SPARSITY OF THE JACOBIAN

The sparsity of bus admittance matrices of large-scale power systems is discussed in Secs. 7.9 and B.1. The pattern in which the zero elements occur in $\mathbf{Y}_{\text{bus}}$ is *almost* identical to that found in the four submatrices $\mathbf{J}_{11}$, $\mathbf{J}_{12}$, $\mathbf{J}_{21}$, and $\mathbf{J}_{22}$ of the jacobian. We can easily observe this sparsity by writing the mismatch equations for the four-bus system of Example 9.5 with symbols, rather than with numerical values, in the following form:

$$
\begin{array}{c}
\\ \textcircled{2} \\ \textcircled{3} \\ \textcircled{4} \\ \textcircled{2} \\ \textcircled{3}
\end{array}
\begin{array}{c}
\begin{array}{ccc|cc}
\textcircled{2} & \textcircled{3} & \textcircled{4} & \textcircled{2} & \textcircled{3}
\end{array} \\
\left[\begin{array}{ccc|cc}
M_{22} & 0 & M_{24} & N_{22} + 2|V_2|^2 G_{22} & 0 \\
0 & M_{33} & M_{34} & 0 & N_{33} + 2|V_3|^2 G_{33} \\
M_{42} & M_{43} & M_{44} & -N_{42} & -N_{43} \\
\hline
N_{22} & 0 & N_{24} & -M_{22} - 2|V_2|^2 B_{22} & 0 \\
0 & N_{33} & N_{34} & 0 & -M_{33} - 2|V_3|^2 B_{33}
\end{array}\right]
\end{array}
\left[\begin{array}{c}
\Delta\delta_2 \\ \Delta\delta_3 \\ \Delta\delta_4 \\ \hline \dfrac{\Delta|V_2|}{|V_2|} \\ \dfrac{\Delta|V_3|}{|V_3|}
\end{array}\right]
=
\left[\begin{array}{c}
\Delta P_2 \\ \Delta P_3 \\ \Delta P_4 \\ \hline \Delta Q_2 \\ \Delta Q_3
\end{array}\right]
\tag{B.2}
$$

There are zeros in the jacobian because the partial derivatives with Y_{23} or Y_{32} as a multiplier are zero since bus ② is not directly connected to bus ③ in the single-line diagram of Fig. 9.2. Of course, the slack bus has no rows and no columns in Eq. (B.2), and there is no row for Q_4 and no column for $\Delta|V_4|/|V_4|$ since bus ④ is voltage controlled. These missing rows and columns constitute the only sparsity differences between the four submatrices of the jacobian and the system $\mathbf{Y}_{\text{bus}}$ of this example.

If the five equations represented by Eq. (B.2) were solved by gaussian elimination in the same order in which they are shown, unnecessary fill-in elements would be generated because the equations are not yet ordered in accordance with the rules of Sec. B.1. By noting that the zero elements of Eq. (B.2) occur symmetrically in each row and column associated with the same bus, we are led to rearrange the equations into the form

$$
\begin{array}{c}
\\ \textcircled{2} \\ \\ \textcircled{3} \\ \\ \textcircled{4}
\end{array}
\begin{array}{c}
\begin{array}{cc|cc|c}
\multicolumn{2}{c|}{\textcircled{2}} & \multicolumn{2}{c|}{\textcircled{3}} & \textcircled{4}
\end{array} \\
\left[\begin{array}{cc|cc|c}
M_{22} & N_{22} + 2|V_2|^2 G_{22} & 0 & 0 & M_{24} \\
N_{22} & -M_{22} - 2|V_2|^2 B_{22} & 0 & 0 & N_{24} \\
\hline
0 & 0 & M_{33} & N_{33} + 2|V_3|^2 G_{33} & M_{34} \\
0 & 0 & N_{33} & -M_{33} - 2|V_3|^2 B_{33} & N_{34} \\
\hline
M_{42} & -N_{42} & M_{43} & -N_{43} & M_{44}
\end{array}\right]
\end{array}
\left[\begin{array}{c}
\Delta\delta_2 \\ \dfrac{\Delta|V_2|}{|V_2|} \\ \hline \Delta\delta_3 \\ \dfrac{\Delta|V_3|}{|V_3|} \\ \hline \Delta\delta_4
\end{array}\right]
=
\left[\begin{array}{c}
\Delta P_2 \\ \Delta Q_2 \\ \hline \Delta P_3 \\ \Delta Q_3 \\ \hline \Delta P_4
\end{array}\right]
\tag{B.3}
$$

In this equation we have rearranged the jacobian matrix so that the two state-variable corrections of each load bus and the associated power mismatches, ΔP and ΔQ, are grouped together in pairs. This pairwise grouping of the P and Q equations of the load buses clearly shows the correspondence between the sparsity of the jacobian and that of $\mathbf{Y}_{\text{bus}}$. Comparing the jacobian of Eq. (B.3) with the system $\mathbf{Y}_{\text{bus}}$ given in Table 9.4, we see that the sparsity of $\mathbf{Y}_{\text{bus}}$ is replicated in the jacobian on a 2×2 block basis for each of the buses—except for the slack bus and the voltage-controlled buses which have rows and columns excluded, as discussed previously. Therefore, the rules of Sec. B.1 for ordering the network nodes apply directly to the jacobian of the mismatch equations written in the form of Eq. (B.3). This sparsity correspondence between $\mathbf{Y}_{\text{bus}}$ and the jacobian is to be expected since those elements directly proportional to $|Y_{ij}|$ will be zero in both matrices if there is no branch admittance connecting buses ⓘ and ⓙ in the network.

The following numerical example illustrates a solution of the mismatch equations by optimal ordering and triangular factorization of the jacobian.

Example B.3. Order the rows of the numerical jacobian given in Sec. 9.4 following Example 9.5 so as to minimize fill-in elements when the lower- and upper-triangular factors are calculated. Numerically evaluate the triangular factors and from the mismatch equations calculate the voltage corrections in the first iteration of the Newton-Raphson procedure.

Solution. In this example the calculated results are displayed to three decimal figures although the calculations have been performed to much greater precision on the computer. Combining in pairs the P and Q mismatch equations for each of the load buses ② and ③ in Example 9.5 yields

$$
\underbrace{\begin{array}{c} \\ \textcircled{2} \\ \\ \textcircled{3} \\ \\ \textcircled{4} \end{array}
\left[\begin{array}{cc|cc|c}
\multicolumn{2}{c}{\textcircled{2}} & \multicolumn{2}{c}{\textcircled{3}} & \textcircled{4} \\
45.443 & 8.882 & 0 & 0 & -26.365 \\
-9.089 & 44.229 & 0 & 0 & 5.273 \\ \hline
0 & 0 & 41.269 & 8.133 & -15.421 \\
0 & 0 & -8.254 & 40.459 & 3.084 \\ \hline
-26.365 & -5.273 & -15.421 & -3.084 & 41.786
\end{array}\right]}_{\text{Jacobian}}
\underbrace{\left[\begin{array}{c}
\Delta\delta_2 \\ \dfrac{\Delta|V_2|}{|V_2|} \\ \hline \Delta\delta_3 \\ \dfrac{\Delta|V_3|}{|V_3|} \\ \hline \Delta\delta_4
\end{array}\right]}_{\text{Corrections}}
=
\underbrace{\left[\begin{array}{c}
-1.597 \\ -0.447 \\ \hline -1.940 \\ -0.835 \\ \hline 2.213
\end{array}\right]}_{\text{Mismatches}}
$$

The numerical values in the reordered jacobian are *not* symmetrically equal, but the zero elements do follow a symmetrical pattern. Consequently, the coefficient matrix has an *equivalent* linear graph to which the ordering rules of Sec. B.1 apply directly. We may thereby confirm that the rows and columns of the above equation are already optimally ordered for gaussian elimination. The elements in the first column of the jacobian constitute the first column of the lower-triangular factor **L**; and the elements of the first row, when divided by the initial pivot 45.443, constitute the first row of the upper-triangular factor **U**. Having recorded these elements, we follow the usual procedure of eliminating the first row and column from the jacobian and thereby obtain the second column of **L** and the second row of **U**, and so on. The calculations yield the matrix **L** given by

$$\underbrace{\begin{bmatrix} 45.443 & \cdot & \cdot & \cdot & \cdot \\ -9.089 & 46.005 & \cdot & \cdot & \cdot \\ 0 & 0 & 41.269 & \cdot & \cdot \\ 0 & 0 & -8.254 & 42.086 & \cdot \\ -26.365 & -0.120 & -15.421 & -0.045 & 20.727 \end{bmatrix}}_{\mathbf{L}} \underbrace{\begin{bmatrix} x_1 \\ x_2 \\ x_3 \\ x_4 \\ x_5 \end{bmatrix}}_{\mathbf{x}} = \begin{bmatrix} -1.597 \\ -0.447 \\ -1.940 \\ -0.835 \\ 2.213 \end{bmatrix}$$

Here the x's denote the intermediate results to be calculated by forward substitution. The solved values of the x's are shown along with the upper-triangular matrix **U** in the system of equations

$$\underbrace{\begin{bmatrix} 1 & 0.195 & 0 & 0 & -0.580 \\ \cdot & 1 & 0 & 0 & 0.000 \\ \cdot & \cdot & 1 & 0.197 & -0.374 \\ \cdot & \cdot & \cdot & 1 & 0.000 \\ \cdot & \cdot & \cdot & \cdot & 1 \end{bmatrix}}_{\mathbf{U}} \begin{bmatrix} \Delta\delta_2 \\ \dfrac{\Delta|V_2|}{|V_2|} \\ \Delta\delta_3 \\ \dfrac{\Delta|V_3|}{|V_3|} \\ \Delta\delta_4 \end{bmatrix} = \underbrace{\begin{bmatrix} -0.035 \\ -0.017 \\ -0.047 \\ -0.029 \\ 0.027 \end{bmatrix}}_{\mathbf{x}}$$

Back substitution yields the voltage corrections of the first iteration:

	②	③	④
$\Delta\delta$ (rad.)	−0.016	−0.031	0.027
$\dfrac{\Delta\|V\|}{\|V\|}$ (per unit)	−0.017	−0.029	—

In accordance with Eqs. (9.49) and (9.50), we add these corrections to the original (flat-start) values specified in Example 9.5 to obtain the updated voltages:

	②	③	④
δ (deg.)	−0.931	−1.788	1.544
$\|V\|$ (per unit)	0.983	0.971	—

from which the new jacobian elements and the power mismatches of the second iteration are calculated. The final converged solution is shown in Fig. 9.4.

Because of the order in which we have solved the mismatch equations of this example, no fill-ins have occurred and the zero elements of the original jacobian are repeated in the triangular matrices **L** and **U**. The elements 0.000 appearing in **U** are actually the result of truncating the numerical results to display only three significant decimal figures. It is straightforward to confirm that the numerical product of **L** and **U** equals the original jacobian matrix within the round-off accuracy displayed.

INDEX